Nonlinear PDEs

A Dynamical Systems Approach

Guido Schneider
Hannes Uecker

GRADUATE STUDIES IN MATHEMATICS 182

Nonlinear PDEs

A Dynamical Systems Approach

Guido Schneider
Hannes Uecker

American Mathematical Society
Providence, Rhode Island

2010 *Mathematics Subject Classification.* Primary 35-01, 35Bxx, 35Qxx, 37Kxx, 37Lxx.

For additional information and updates on this book, visit
www.ams.org/bookpages/gsm-182

Library of Congress Cataloging-in-Publication Data

Names: Schneider, Guido, 1964– | Uecker, Hannes, 1970–

Title: Nonlinear PDEs : a dynamical systems approach / Guido Schneider, Hannes Uecker.

Other titles: Nonlinear partial differential equations

Description: Providence, Rhode Island : American Mathematical Society, [2017] | Series: Graduate studies in mathematics ; volume 182 | Includes bibliographical references and index.

Identifiers: LCCN 2017012328 | ISBN 9781470436131 (alk. paper)

Subjects: LCSH: Differential equations, Partial. | Differential equations, Nonlinear. | AMS: Partial differential equations – Instructional exposition (textbooks, tutorial papers, etc.). msc | Partial differential equations – Qualitative properties of solutions – Qualitative properties of solutions. msc | Partial differential equations – Equations of mathematical physics and other areas of application – Equations of mathematical physics and other areas of application. msc | Dynamical systems and ergodic theory – Infinite-dimensional Hamiltonian systems – Infinite-dimensional Hamiltonian systems. msc | Dynamical systems and ergodic theory – Infinite-dimensional dissipative dynamical systems – Infinite-dimensional dissipative dynamical systems. msc

Classification: LCC QA377 .S35584 2017 | DDC 515/.353–dc23 LC record available at `https://lccn.loc.gov/2017012328`

∞ The paper used in this book is acid-free and falls within the guidelines
established to ensure permanence and durability.
Visit the AMS home page at `http://www.ams.org/`

10 9 8 7 6 5 4 3 2 1   22 21 20 19 18 17

To Daniela, Max and Jonas
and Anja, Franka and Henrike

Contents

Preface xi

Chapter 1. Introduction 1

§1.1. The three classical linear PDEs 1

§1.2. Nonlinear PDEs 4

§1.3. Our choice of equations and the idea of modulation equations 6

§1.4. Overview 11

Part I Nonlinear dynamics in $\mathbb{R}^d$

Chapter 2. Basic ODE dynamics 15

§2.1. Linear systems 17

§2.2. Local existence and uniqueness for nonlinear systems 34

§2.3. Special solutions 38

§2.4. ω-limit sets and attractors 49

§2.5. Chaotic dynamics 58

§2.6. Examples 64

Chapter 3. Dissipative dynamics 75

§3.1. Bifurcations 76

§3.2. Center manifold theory 85

§3.3. The Hopf bifurcation 91

§3.4. Routes to chaos 98

Chapter 4. Hamiltonian dynamics 109

§4.1. Basic properties 109
§4.2. Some celestial mechanics 116
§4.3. Completely integrable systems 121
§4.4. Perturbations of completely integrable systems 123
§4.5. Homoclinic chaos 128

Part II Nonlinear dynamics in countably many dimensions

Chapter 5. PDEs on an interval 133
§5.1. From finitely to infinitely many dimensions 134
§5.2. Basic function spaces and Fourier series 151
§5.3. The Chafee-Infante problem 167

Chapter 6. The Navier-Stokes equations 179
§6.1. Introduction 179
§6.2. The equations on a torus 186
§6.3. Other boundary conditions and more general domains 197

Part III PDEs on the infinite line

Chapter 7. Some dissipative PDE models 205
§7.1. The KPP equation 206
§7.2. The Allen-Cahn equation 222
§7.3. Intermezzo: Fourier transform 225
§7.4. The Burgers equation 237

Chapter 8. Three canonical modulation equations 249
§8.1. The NLS equation 250
§8.2. The KdV equation 259
§8.3. The GL equation 275

Chapter 9. Reaction-Diffusion systems 295
§9.1. Modeling, and existence and uniqueness 297
§9.2. Two classical examples 302
§9.3. The Turing instability 307

Part IV Modulation theory and applications

Chapter 10. Dynamics of pattern and the GL equation 315
§10.1. Introduction 316

§10.2. The Swift-Hohenberg equation 319
§10.3. The universality of the GL equation 332
§10.4. An abstract approximation result 337
§10.5. Reaction-Diffusion systems 347
§10.6. Convection problems 354
§10.7. The Couette-Taylor problem 370
§10.8. Attractors for pattern forming systems 378
§10.9. Further remarks 395

Chapter 11. Wave packets and the NLS equation 401
§11.1. Introduction 402
§11.2. Justification in case of cubic nonlinearities 404
§11.3. The universality of the NLS equation 411
§11.4. Quadratic nonlinearities 416
§11.5. Extension of the theory 421
§11.6. Pulse dynamics in photonic crystals 429
§11.7. Nonlinear optics 440

Chapter 12. Long waves and their modulation equations 451
§12.1. An approximation result 452
§12.2. The universality of the KdV equation 456
§12.3. Whitham, Boussinesq, BBM, etc. 465
§12.4. The long wave limit 468

Chapter 13. Center manifold reduction and spatial dynamics 473
§13.1. The center manifold theorem 473
§13.2. Local bifurcation theory on bounded domains 478
§13.3. Spatial dynamics for elliptic problems in a strip 482
§13.4. Applications 484

Chapter 14. Diffusive stability 497
§14.1. Linear and nonlinear diffusive behavior 498
§14.2. Diffusive stability of spatially periodic equilibria 507
§14.3. The critical case 523
§14.4. Phase diffusion equations 529
§14.5. Dispersive dynamics 535

Bibliography 541

List of symbols 567

Index 569

Preface

"If you want to build a ship, don't herd people together to collect wood and don't assign them tasks and work, but rather teach them to long for the endless immensity of the sea"

ANTOINE DE SAINT-EXUPERY

This is an introductory textbook about nonlinear dynamics of partial differential equations (PDEs), with a focus on problems over unbounded domains and modulation equations. We explain how dynamical systems methods can be used to analyze PDEs in order to get more insight into the real world phenomena behind the equations. Our presentation is example-oriented and the starting point is very often a real world problem. This means that new mathematical tools are developed step by step in order to analyze the equations. They are re-applied and improved in subsequent sections to handle more and more complicated systems. In the end the reader should have learned mathematical tools for the analysis of some important classes of nonlinear PDEs and gained insight into nonlinear dynamics phenomena which may occur in PDEs.

The book is divided into four parts. In order to keep the book as an introductory text and as self-contained as possible, Part I is an introduction into finite-dimensional dynamics, defined by ordinary differential equations (ODEs), including bifurcation theory, attractors, and the basics of Hamiltonian dynamics. In Part II we explain the major differences between finitely and infinitely many dimensions and that in principle a PDE on a bounded domain is isomorphic to a system of countably many ODEs. We give two main applications of this point of view. The first one is the characterization of the attractor for the Allen-Cahn equation on an interval, which is also known as the Chafee-Infante problem. The second one is a very basic introduction to the Navier-Stokes equations, with a focus on periodic boundary conditions.

Genuine PDE phenomena such as transport, diffusion, and dispersion can hardly be understood by the interpretation of PDEs as systems of infinitely many ODEs. In Part III we consider PDEs which are posed on the real line. We start with the linear heat equation, and then turn to nonlinear problems. For famous model equations such as the Kolmogorov-Petrovsky-Piskounov or Fisher equation, the Korteweg-de Vries (KdV) equation, the Nonlinear Schrödinger (NLS) equation, and the Ginzburg-Landau (GL) equation, we discuss the local existence and uniqueness of solutions, special solutions as fronts and pulses, their stability and instability, soliton dynamics, the construction of attractors, and some related results.

The equations from Part III all play an important role in mathematics and have entire monographs devoted to each. Moreover, they have many connections to physics and other fields of applications, where they are often used as simplest possible models for the description of some real world phenomena. In Part IV we explore these connections from a mathematical perspective. The scalar equations from Part III occur as asymptotic effective models, or more specifically as modulation equations, for the more complicated systems from physics considered in Part IV. Examples are pattern forming systems which can be described by the GL equation, light pulses in nonlinear optics which can be described by the NLS equation, or long waves in dispersive systems which can be described by the KdV equation. We discuss how the dynamics of the reduced model equations transfer to the more complicated systems. Thus, in Part IV we give a mathematically rigorous presentation of the formalism of modulation equations in the context of real world applications. While this last part is close to recent research, it is still in textbook style, and often we do not prove the sharpest or most general result possible, but instead refer to the literature for extensions.

All chapters are kept as self-contained as possible, such that the reader can start to read directly about his or her favorite equation. Having a good background in linear ODEs, cf. §2.1, a starting point for our goals and objectives are §2.2-§2.3 about basic nonlinear ODE dynamics combined with Part III. There are other possible combinations, for instance the sections about dissipative dynamics or the sections about conservative dynamics. Nevertheless the reader can also read the book from the beginning to the end. See the Grasshopper's Guide on page 12 for detailed proposals. All chapters contain exercises which we strongly recommend not to skip.

This book grew out of our manuscripts for the lectures and seminars we gave about ODEs and PDEs at the universities of Bayreuth, Karlsruhe, Oldenburg, and Stuttgart. We thank the students who attended our lectures and seminars and urged us to keep the presentation simple and accessible. Moreover, we thank all friends and colleagues with whom we cooperated over the years, mainly on topics from Part IV, in particular, Dirk Blömker, Tom Bridges, Kurt Busch, Martina Chirilus-Bruckner, Christopher Chong, Walter Craig, Markus Daub, Hannes de Witt, Arjen Doelman, Tomas Dohnal, Wolf-Patrick Düll, Wiktor Eckhaus, Jean-Pierre Eckmann, Bernold Fiedler, Thierry Gallay, Dieter Grass, Daniel Grieser, Mark Groves, Tobias Häcker, Mariana Haragus, Ronald Imbihl, Ralf Kaiser, Tasso Kaper, Klaus Kirchgässner, Markus Kunze, David Lannes, Vincent Lescarret, Karsten Matthies, Ian Melbourne, Andreas Melcher, Johannes Müller, Robert Pego, Dmitry Pelinovsky, Jens Rademacher, Björn Sandstede, Arnd Scheel, Zarif Sobirov, Aart van Harten, C. Eugene Wayne, Daniel Wetzel, Peter Wittwer, and Dominik Zimmermann. We thank Stefanie Siegert and the unknown referees for a number of additional proposals to improve the presentation. Especially we thank Alexander Mielke from whom we learned about nonlinear dynamics and PDEs.

Finally we would like to thank Ina Mette from the AMS for her never ending motivation to go on with this book project and many helpful comments to transform our lecture notes into a book.

Guido Schneider and Hannes Uecker
Stuttgart and Oldenburg, February 2017

Chapter 1

Introduction

Mathematicians want to classify things. However, with partial differential equations (PDEs) they had to stop on a rather unsatisfactory level. The reason for this is that almost all rules of theoretical physics and engineering, and many rules in life sciences and economics, are formulated as ordinary or partial differential equations (ODEs or PDEs). As different as the applications of differential equations are, as different is the behavior of their solutions. Therefore, a mathematical theory which wants to cover all differential equations can only cover the absolute basics. Hence, the books about PDEs necessarily differ strongly by the choice of examples and by the choice of mathematical theory which will be applied to the examples. There are entire books only covering one special important equation. Before we give the goals and objectives of this book we start with a short review of three important examples.

1.1. The three classical linear PDEs

In many courses about PDEs the following three examples, namely the Laplace equation, the heat equation, and the wave equation, play a major, sometimes exclusive, role.

Example 1.1.1. The Laplace equation is an equation for an unknown function $u : \Omega \to \mathbb{R}$ of two or more variables $x = (x_1, \ldots, x_d) \in \Omega \subset \mathbb{R}^d$ in terms of certain of its partial derivatives, namely

$$\Delta u = 0, \tag{1.1}$$

where $\Delta = \partial_{x_1}^2 + \ldots + \partial_{x_d}^2$. This PDE plays an important role in mathematics since the real and imaginary part of an analytic function in the complex plane satisfy the Laplace equation. It also plays a major role in applications. For instance the potential of an irrotational flow of an incompressible fluid such as water, or a stationary temperature field, or the potential of a stationary electric field in the absence of charges in Ω, satisfy this equation.

In order to solve this equation uniquely in a domain Ω, additional conditions are needed. To gain an intuition for the required boundary conditions we consider the factors which should determine a stationary temperature field u in a room $\Omega \subset \mathbb{R}^3$ as sketched from the side in Figure 1.1. The temperature will be determined by the temperature at the walls, the windows, the doors and the heating of the room, mathematically speaking by the conditions at the boundary $\partial\Omega$ of Ω.

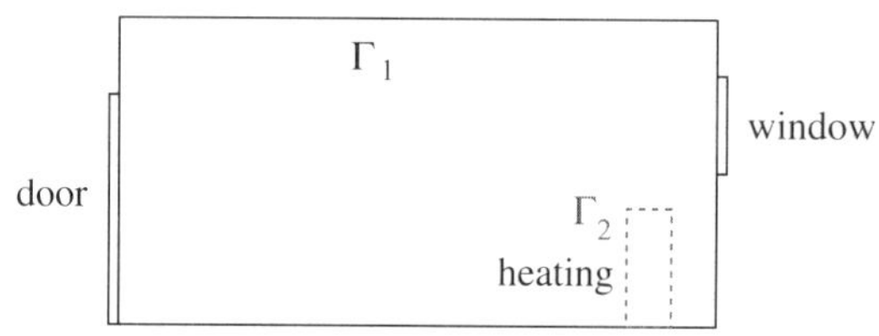

Figure 1.1. Different boundary conditions for the temperature field.

There are mainly two different kinds of boundary conditions. At the heating unit the temperature has a fixed value, while at a window or wall heat will go through the window or wall. Mathematically speaking the boundary $\partial\Omega = \Gamma_1 \cup \Gamma_2$ is split into two parts where in the first part we have so called Dirichlet conditions

$$u|_{\Gamma_1} = g_1,$$

and in the second part we have so called Neumann conditions

$$\partial_n u|_{\Gamma_2} = g_2,$$

with given functions $g_1 : \Gamma_1 \to \mathbb{R}$ and $g_2 : \Gamma_2 \to \mathbb{R}$ and $n : \partial\Omega \to \mathbb{R}^3$ the outer normals.

The Laplace equation is the paradigm of an **elliptic** PDE. It is of second order, i.e., the highest derivative is of order two. There is an extensive theory for elliptic systems, especially for second order elliptic systems. The equilibrium equation of linear elasticity

$$Lu := \mu\Delta u + (\lambda + \mu)\nabla(\nabla \cdot u) \stackrel{!}{=} 0,$$

for the displacement vector $u : \mathbb{R}^3 \supset \Omega \to \mathbb{R}^3$, with constants $\lambda, \mu \in \mathbb{R}$ depending on the material, is also a second order elliptic system. Like the negative Laplace operator $-\Delta$, the linear operator $-L$ defined in this equation is an example of a so called elliptic operator. Due to the important role of elasticity in the construction of cars, bridges, planes, etc., there are well developed numerical schemes such as the finite element method **(FEM)** or the boundary element method **(BEM)**, which are available for computing approximate solutions of such systems. Only in very special cases solutions can be found analytically. ⌋

Example 1.1.2. The heat equation or diffusion equation

$$\partial_t u = \Delta u, \tag{1.2}$$

with $u = u(x,t)$ and $u : \Omega \times \mathbb{R}^+ \to \mathbb{R}$, where t denotes time and x denotes space, describes the evolution of quantities such as heat, chemical concentrations, or the probability distribution of a particle obeying Brownian motion.

It can be derived as follows. Let $V \subset \Omega$ be an arbitrary subset with smooth boundary. The change of the total quantity within V equals the flux through ∂V, i.e.,

$$\frac{\mathrm{d}}{\mathrm{d}t}\int_V u\,\mathrm{d}x = -\int_{\partial V} \langle F, n\rangle\,\mathrm{d}S = -\int_V \nabla \cdot F\,\mathrm{d}x$$

with F the flux density, $\langle \cdot, \cdot \rangle$ the scalar product in $\mathbb{R}^d$, $n : \partial\Omega \to \mathbb{R}^d$ again the outer normal, and where we used the Gauss' integral theorem. Since this relation is true for all sets V, we find

$$\partial_t u = -\nabla \cdot F.$$

Very often the flux density F is proportional to the gradient ∇u of the concentration u, i.e., $F = -a\nabla u$ with a constant $a > 0$. By rescaling time we finally come to the diffusion equation (1.2).

In order to solve this equation uniquely in a domain $\Omega \times \mathbb{R}^+$ additional conditions are needed. As in Example 1.1.1 we need boundary conditions, but also the temperature field at time $t = 0$ has to be known, i.e., we need the initial condition $u|_{t=0} = u_0$ with $u_0 : \Omega \to \mathbb{R}$. Stationary solutions, i.e., time-independent solutions, satisfy the Laplace equation (1.1). The heat equation is the prototype **parabolic** PDE. There is an extensive theory for equations of the form $\partial_t u = Lu$ with an elliptic operator $-L$. ⌋

Example 1.1.3. The linear wave equation

$$\partial_t^2 u = \Delta u, \tag{1.3}$$

$u = u(x,t)$ with $u : \Omega \times \mathbb{R} \to \mathbb{R}$, is a simple model for, e.g., oscillations of a string ($\Omega \subset \mathbb{R}$) or of a membrane ($\Omega \subset \mathbb{R}^2$), or the propagation of light in vacuum. In order to solve this equation uniquely in a domain $\Omega \times (t_0, t_1)$, $t_0 < 0 < t_1$ we again need boundary and initial conditions. Like for scalar second order ODEs we need two initial conditions, namely $u|_{t=0} = u_0$ and $\partial_t u|_{t=0} = u_1$ with $u_0 : \Omega \to \mathbb{R}$ and $u_1 : \Omega \to \mathbb{R}$. The Dirichlet boundary condition $u|_{\partial\Omega}$ corresponds to a membrane which is fixed at the boundary. In this case, the boundary will reflect the waves.

For the wave equation the eigenmodes play a crucial role. An eigenmode is a solution $u(x,t) = e^{i\omega t} v(x)$. This yields the eigenvalue problem

$$-\Delta v = \omega^2 v.$$

Such problems play an important role in applications, especially in elasticity theory, where the evolution equations of linear elasticity

$$\partial_t^2 u = \mu \Delta u + (\lambda + \mu) \nabla (\nabla \cdot u)$$

yield to the eigenvalue problem

$$-\mu \Delta v + (\lambda + \mu) \nabla (\nabla \cdot v) = \omega^2 v.$$

If Ω is a bounded set then under suitable boundary conditions there are countably many real eigenvalues $\lambda_n = \omega_n^2$ with $0 \leq \lambda_1 \leq \lambda_2 \leq \ldots \to \infty$. In the construction of cars, bridges, planes, etc., one has to take care that these so called resonant modes are not periodically excited. Hence, there is a big industry using FEM and BEM in order to solve these elliptic eigenvalue problems. The wave equation is the prototype **hyperbolic** PDE. There is an extensive theory for equations of the form $\partial_t^2 u = Lu$ with an elliptic operator $-L$. ⌋

For reasons explained below we will focus on other examples than the three classical ones. The fundamental Examples 1.1.1-1.1.3 cannot be and will not be avoided. However, they will only occur as subproblems which will help to understand the nonlinear problems under consideration.

1.2. Nonlinear PDEs

We now start discussing our main objectives for this book, namely an introduction to nonlinear PDEs from a dynamical systems point of view, with a focus on reduction methods, in particular, the use of amplitude and modulation equations.

Many complications with ODEs or PDEs are due to the fact that **the world is nonlinear.** Ultimately, to solve a PDE means to look for solutions u of an abstract equation $F(u) = 0$. The problem is called linear if for all $\alpha, \beta \in \mathbb{R}$ we have

$$F(\alpha u + \beta v) = \alpha F(u) + \beta F(v).$$

As a consequence, for linear problems we have the superposition of solutions. With u, v being solutions, i.e., $F(u) = 0$ and $F(v) = 0$, also $\alpha u + \beta v$ is a solution, i.e., $F(\alpha u + \beta v) = 0$. Most "real life" problems are nonlinear, i.e., in general

$$F(\alpha u + \beta v) \neq \alpha F(u) + \beta F(v),$$

and therefore a sum of two solutions is no longer a solution of the ODE or PDE. A simple example of a nonlinear function is $F(u) = u^2$. As a consequence, the theory of linear algebra is not available, and the set of solutions in general is more complicated than that for linear problems. In science, for many decades linear problems played a dominating role. Examples 1.1.1 to 1.1.3 are linear. Next we present two famous examples of nonlinear PDEs.

Example 1.2.1. The Navier-Stokes equations

$$\begin{aligned}\partial_t u =& \frac{1}{R}\Delta u - \nabla p - (u\cdot\nabla)u,\\ 0 =& \nabla\cdot u,\end{aligned}$$

describe the evolution of the velocity field $u : \Omega \times \mathbb{R}^+ \to \mathbb{R}^3$ and the pressure field $p : \Omega \times \mathbb{R}^+ \to \mathbb{R}$ of an incompressible fluid, such as water, in a domain $\Omega \subset \mathbb{R}^3$. The Reynolds number R measures the ratio between inertial and viscous forces, and is in some sense proportional to the complexity of the flow. The global existence and uniqueness of smooth solutions of the three-dimensional (3D) Navier-Stokes equations is one of the seven one million dollar or Millennium problems in mathematics presented by the Clay-Foundation in the year 2000. There are a number of reasons for this choice. On the one hand, the Navier-Stokes equations describe the motion of fluids, and the answer to this question would allow us to understand fluids in a much better way. On the other hand, in mathematics the 3D Navier-Stokes equations are interesting PDEs, which so far have resisted all attempts to prove the global existence and uniqueness of solutions. This will be explained in more detail in Chapter 6. ⌋

Example 1.2.2. Maxwell's equations in a medium, for instance a glass fiber, are given by

$$\begin{aligned}\nabla\cdot B =& 0,\\ \nabla\times E + \partial_t B =& 0,\\ \nabla\cdot D =& \rho,\\ \nabla\times H - \partial_t D =& J.\end{aligned}$$

Here $E : \Omega \times \mathbb{R} \to \mathbb{R}^3$ is the electric field, $D = \varepsilon_0 E + P$ is the displacement field, with ε_0 the electric permeability of vacuum, $P : \Omega \times \mathbb{R} \to \mathbb{R}^3$ is the electric polarization of the material, $B : \Omega \times \mathbb{R} \to \mathbb{R}^3$ is the magnetic field, $H = B/\mu_0 - M$ is the magnetizing field, with μ_0 the magnetic permeability of vacuum and $M : \Omega \times \mathbb{R} \to \mathbb{R}^3$ the magnetic polarization of the material, ρ is the charge density, and $J : \Omega \to \mathbb{R}^3$ the charge flow density. Since the first and the third equation above are scalar, while the second and fourth equation are vector valued, so far we have eight equations for the twelve unknowns B_j, E_j, M_j and P_j for $j = 1, 2, 3$. Therefore, these equations have to be closed with constitutive laws $P = P(E, H)$ and $M = M(E, H)$ describing the reaction of the material to the electric and magnetic field. In general, these laws are nonlinear. Moreover, as an additional complication M and P may depend on the past, cf. §11.7. ⌋

The world is instationary, i.e., almost all systems evolve with time. Typical examples are a vibrating beam, the daily change of the weather, or the motion of the planets in the solar system. Hence, from the beginning we will consider **nonlinear time-dependent systems**.

A mathematical concept which is basic to the analytical understanding of all ODEs and PDEs is the concept of **Dynamical Systems**. Until the beginning of the 1960s, Laplace's principle that with the knowledge of all physical rules and the present state of the world, the past and future behavior of the world for all times can be computed, was widely accepted as a relevant philosophical foundation of science. Starting already with the work of H. Poincaré in the 1890s, cf. [**Poi57**], this principle was finally observed to be practically useless at the beginning of the 1960s, for instance by the work of the meteorologist E. Lorenz in 1963 [**Lor63**]. He observed with an analog computer for a three-dimensional model for the weather that the possible time for predictions goes logarithmically with the precision of the initial conditions, i.e., that long-time weather-forecasts are practically impossible. See Figure 1.2 for an illustration of the so called Lorenz attractor and of the sensitivity of solutions w.r.t. the initial conditions.

Certain ODEs and PDEs, or, more general, dynamical systems, can be classified as **chaotic**. The visualization of chaotic dynamical systems was in fashion in the 1980s. Famous examples are the Mandelbrot and the Julia sets. In this book, chaos will not play a central role, but one should keep in mind its existence already in low-dimensional dynamical systems.

1.3. Our choice of equations and the idea of modulation equations

PDEs play an important role in modern engineering. With the help of computer simulations, money can be saved, experiments can be replaced, and data can be gathered which are not available by classical experiments. However, **a numerical simulation of a PDE requires an analytic understanding of the PDE.** The reason for this is again the wide variety of different types of PDEs. Therefore, very often the numerical simulation of a PDE needs an adapted numerical scheme based on an analysis of the PDE. As the example of the crash of the Sleipner oil platform in 1991 shows, a misuse of numerical schemes can cost a lot of money. In the concrete example 700 million dollars [**JR94**].

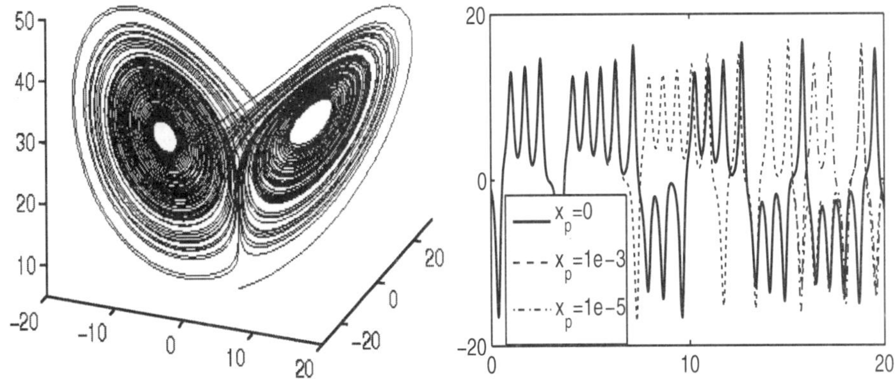

Figure 1.2. Left: Illustration of the attractor of the Lorenz system $\dot{x} = \sigma(y-x)$, $\dot{y} = \rho x - y - xz$, $\dot{z} = -\beta z + xy$ by one orbit in 3D phase space, $\sigma = 10$, $\beta = 8/3$, $\rho = 27$. Right: $x(t)$ for three nearby initial conditions, i.e., $x_1(0)$, $x_2(0) = x_1(0) + 10^{-3}$, and $x_3(0) = x_1(0) + 10^{-5}$, $y(0), z(0)$ always the same. The orbits behave completely different after a certain time, i.e., the orbits to $x_2(0)$ and $x_3(0)$ deviate from the unperturbed one after $t \approx 7$ and $t \approx 16$, respectively. It can be shown that the prediction time goes logarithmically with the precision of the initial conditions.

Moreover, **computers are fast, but never fast enough.** A three-dimensional body $[0,1]^3$ discretized with 100 points in each direction leads to 10^6 variables. A discretization in 1000 points in each direction yields 10^9 variables. Therefore, due to practical reasons one has to decide before what quantities shall be computed. Then the scheme can be adapted to the computation of these quantities.

We are especially interested in problems which cannot be directly studied numerically, i.e., where first **analysis is needed to reduce the dimensionality of the problem.** This is for instance the case in so called spatially extended domains, which means that the wave length of typical solutions is much smaller than the size of the underlying physical domain. In this case often the modeling over an unbounded domain is more reasonable. Then, via a multiple scaling perturbation ansatz simpler models can be derived to describe the phenomena under consideration. These models, called **modulation equations**, belong to the best studied nonlinear PDEs with a status in some scientific areas similar to the three classical linear PDEs from above. Besides the study of these basic nonlinear PDEs from a dynamical systems point of view, one of our main objectives will be the connections between these models and real world problems by going beyond the formal derivation of these modulation equations. This will be called the **justification** of the reduced models.

Example 1.3.1. The digital transport of information in glass fibers is done by sending 0s and 1s through the fiber. In most modern technologies the physical realization of a 1 is an electromagnetic pulsemodulating a carrier wave with a wave length of a few hundred nanometers. There are a number of relevant questions related to the transport of information. For instance:

- Which form is optimal for a pulse to travel a long distance?
- Which distance do two pulses initially need in order to stay separated during the complete journey through the fiber?
- How many kilometers can a pulse travel without an amplifier?
- How do pulses interact if the carrier waves have different frequencies?

There is dispersion in the fiber and thus in general the energy concentrated in a pulse will spread. Moreover, the fiber behaves nonlinearly. Hence, the answers to the above questions are not obvious at all. Numerical simulations, if possible, are much cheaper than experiments. However, suppose that the length of the fiber is 100km $= 10^5$m. Then, due to the wave length of light of approximately 10^{-7}m, a spatial discretization of Maxwell's equations in the fiber gives at least about 10^{12} points, still neglecting all three-dimensional effects. This number is too big for a direct numerical simulation.

A modulation equation helps. By perturbation analysis the **Nonlinear Schrödinger (NLS) equation**

$$\partial_\tau A = \mathrm{i}\nu_1 \partial_\xi^2 A + \mathrm{i}\nu_2 A|A|^2,$$

with $A(\xi, \tau) \in \mathbb{C}$, $\tau \in \mathbb{R}$, $\xi \in \mathbb{R}$ and coefficients $\nu_1, \nu_2 \in \mathbb{R}$, can be derived, describing the evolution of the envelope A of the pulse alone. On the relevant time scale the dynamics of the envelope of the pulse and the carrier wave which behaves linearly can approximately be separated. The properties of the original system, e.g., the refractive index of the material, and the underlying wave, condense to the coefficients $\nu_j \in \mathbb{R}$. The NLS equation is a **universal** modulation equation which describes slow modulations in time and space of the envelope of a spatially and temporarily oscillating underlying carrier wave in nonlinear dispersive equations.

The spatial discretization can thus be reduced from 10^{12} points to approximately 10^5 or less points, which is quite manageable for numerical schemes. Moreover, a number of problems can be solved analytically for the NLS equation, which is a so called **completely integrable system**. In particular, if $\nu_1\nu_2 > 0$ it has explicit so called **soliton solutions**. These solitons give the optimal form of pulses for the transport of information. These questions will be discussed in detail in Chapter 11. ⌋

Example 1.3.2. At the end of the 20th century a new generation of high speed ferries has caused serious problems, especially those that cross the Channel between England and France and those operating in the Marlborough sound in New Zealand. The waves created by these ferries can propagate without loss of energy over large distances, and thus retain the potential to create enormous havoc when they come ashore, and as a consequence of a fatal accident and other damage there are now speed limits for these ferries [**Ham99**].

Again a modulation equation gives an idea to understand these phenomena. The **Korteweg-de Vries (KdV) equation**

$$\partial_\tau A = \nu_1 \partial_\xi^3 A + \nu_2 A \partial_\xi A,$$

with $\tau \in \mathbb{R}$, $\xi \in \mathbb{R}$, $A(\xi, \tau) \in \mathbb{R}$ and coefficients $\nu_j \in \mathbb{R}$ can be derived with the help of a perturbation ansatz. The KdV equation is a **universal** modulation equation which describes long waves of small amplitude, where the original system condenses to the coefficients $\nu_j \in \mathbb{R}$.

Like the NLS equation, this famous nonlinear equation possesses soliton solutions, very robust solitary waves. These waves interact like particles, i.e., after some nonlinear interaction they reform and look exactly as before the interaction. This observation, made in the middle of the 1960s, that solutions of a PDE show simultaneously the behavior of a particle and a wave, had a big influence on nonlinear science due to the similarity with the particle-wave dualism in quantum mechanics.

For a long time the KdV equation has also been suggested as a model for the description of tsunamis, water waves of only a few meters height, but with a length of up to 100km, i.e., in the ocean they cannot be observed by eye. In the 5000m deep pacific ocean they move with a very high velocity of around 700km/h. If they approach land they become slower and steeper, and cause serious floodings. However, data which is now available from the tsunami at Christmas 2004 in the Indian Ocean seem to indicate that soliton dynamics had played at least for this tsunami no role on the open sea. The validity of the KdV equation will be discussed in Chapter 12. ⌋

Example 1.3.3. Since the 1960s, systems near the onset of a finite wave length instability have been analyzed in detail using modulation equations. These amplitude modulations describe slow changes in time and space of the envelope of the finite wave length pattern close to the first instability. The most famous and generic of such equations is the **Ginzburg-Landau (GL) equation**

$$\partial_\tau A = \nu_2 \partial_\xi^2 A + \nu_0 A + \nu_3 A|A|^2,$$

with $\tau \geq 0$, $\xi \in \mathbb{R}$, $A(\xi, \tau) \in \mathbb{C}$ and coefficients $\nu_j \in \mathbb{C}$. Famous **pattern forming systems** which can be described with the GL equation are reaction-diffusion systems such as the Brusselator, and hydrodynamical stability problems, such as the Couette-Taylor problem, Bénard's problem, or Poiseuille flow. A big part of Part IV, namely Chapter 10, is devoted to the justification of this so called GL approximation for various classes of systems. We explain and prove that the difference of true solutions of the pattern forming systems and the associated GL approximations remains small on the natural time scale of this approximation, and thus prove rigorously that the GL equation makes correct predictions about the dynamics of the original pattern forming systems.

Instead of "modulation equation", in particular the GL equation in the above context is also called the "amplitude equation". Although derived differently, the GL model also plays a crucial role in superconductivity.

A phenomenological model for pattern formation close to the first instability of a spatially homogeneous solution is the **Swift-Hohenberg** equation [**SH77**]

$$\partial_t u = -(1+\partial_x^2)^2 u + \alpha u - u^3, \tag{1.4}$$

with $u = u(x,t) \in \mathbb{R}$, $x \in \mathbb{R}$, $t \geq 0$ and control parameter $\alpha \in \mathbb{R}$. This fourth order scalar PDE is probably the simplest example to apply the "Ginzburg-Landau formalism". For small $\alpha =: \varepsilon^2 > 0$, plugging the ansatz

$$u(x,t) = \varepsilon A(\varepsilon x, \varepsilon^2 t)\mathrm{e}^{\mathrm{i}x} + \mathrm{c.c.} \tag{1.5}$$

into (1.4) and sorting w.r.t. powers of ε yields the GL equation

$$\partial_T A = 4\partial_X^2 A + A - 3|A|^2 A \tag{1.6}$$

at order ε^3. ⌋

As already said, the mathematical analysis of the approximation by these three 'generic' modulation equations, namely the KdV, the NLS, and the GL equation, will be one of the mathematical objectives of Part IV of this book. Beside these 'generic' equations there are many more.

Example 1.3.4. The Burgers equation

$$\partial_t u = \partial_x^2 u - \partial_x(u^2),$$

with $t \geq 0$, $x \in \mathbb{R}$, and $u(x,t) \in \mathbb{R}$ arises for instance as a modulation equation for small amplitude long waves on the surface of a viscous liquid running down an inclined plane. It describes this system in case when the trivial solution, the so called Nusselt solution, which possesses a parabolic flow profile and a flat top surface, is spectrally stable. This is the case when the inclination angle θ, which serves as a control parameter in this physical problem, is below a critical value θ_c. This model is used for instance for flood forecasts in rivers.

If the inclination angle is increased, the Nusselt solution becomes unstable via a so called sideband instability. Above the threshold of instability the Kuramoto-Shivashinsky-perturbed KdV equation serves as modulation equation. After some rescaling it has the form

$$\partial_t u = -\partial_x^3 u - \frac{1}{2}\partial_x(u^2) - \varepsilon(\partial_x^2 + \partial_x^4)u,$$

with $t \geq 0$, $x \in \mathbb{R}$, $u = u(x,t) \in \mathbb{R}$, and where $0 < \varepsilon \approx \sqrt{\theta - \theta_c} \ll 1$ is a small parameter. Therefore, complicated dynamics that are present in this equation occur directly at the first instability of the inclined plane problem. The dynamics is dominated by traveling pulse trains consisting of unstable pulses. Time series of the position of the pulses indicate the occurrence of chaotic dynamics. This situation is relevant for cooling units. Again the 3D Navier-Stokes equations for the water flowing down the unit is replaced by a simpler model still containing very complicated dynamics.

Another situation where the Burgers equation arises as a modulation equations are phase or wave number modulations of stable periodic pattern in a pattern forming system, while phase (or wave number) modulations of unstable pattern are generically described by Kuramoto-Shivashinsky type of equations. ⌋

In summary, modulation equations are simpler PDEs, which can be derived by perturbation analysis, and which serve as models for more complicated systems. Hence, modulation equations are a part of mathematical modeling. In Part IV of this book, the derivation and the approximation properties of the above equations will be explained. We will analyze the original system with the help of the modulation equations. We will explain to which extent conclusions based on the modulation equations can be proven to be correct. We will show how mathematics can decide which model of all possible proposed models is the right one. We will explain that modulation equations are universal models, i.e., exactly the same modulation equation describes the same phenomena in completely different physical systems. The much simpler modulation equations itself will be analyzed in Part III of this book.

1.4. Overview

In order to keep the book as an introductory text and as self-contained as possible, in Part I we explain basic dynamical systems concepts for ODEs, such as phase space, fixed points, periodic solutions, attractors, stability and instability, bifurcations and amplitude equations.

In Part II we start to transfer the dynamical systems concepts from finite to infinite dimensions. There are major differences due to the non-equivalence of norms in infinite-dimensional vector spaces and the loss of compactness of closed bounded sets. We explain that PDEs over bounded domains can be considered as dynamical systems with countably many degrees of freedom. As applications we discuss the Chafee-Infante problem and the Navier-Stokes equations.

We have already explained in the previous subsection our choice of equations for the Parts III and IV. In Part III we consider basic model PDEs posed on the real line, namely the Kolmogorov-Petrovsky-Piscounov (KPP) or Fisher equation, the Burgers equation, the Nonlinear Schrödinger (NLS) equation, the Korteweg-deVries (KdV) equation, and the Ginzburg-Landau (GL) equation. We explain fundamental PDE phenomena as diffusion, dispersion, and transport, discuss local and global existence and uniqueness, and construct stationary solutions, or traveling front and pulse solutions, using ODE techniques from Part I. We also give some first results for attractors on unbounded domains and a brief introduction to reaction-diffusion systems.

Part IV is devoted to the analysis of the more complicated systems with the help of the scalar model equations from Part III, which now reappear as modulation equations. Additionally we explain useful concepts such as diffusive stability and spatial dynamics.

At the end of each chapter we collect a number of **exercises**. We in general do not claim any originality for them, and many are taken from the literature, though in some cases we cannot trace back our source. As usual, the exercises are a crucial part of this book.

1.4.1. Grasshopper's Guide. To some extent the four parts of this book are intended to be independent. Moreover, the chapters are kept as self-contained as possible, such that the reader may start to read directly about his or her favorite equation. Therefore, we also give the following guide.

Part I can obviously be read independently of the rest of the book. It is an example-oriented basic course on finite-dimensional dynamical systems which together with Chapters 5 and 6 (and possibly Chapter 13) yields a two semester course about finite- and infinite-dimensional dynamical systems. Chapters 7 and 8 of Part III can subsequently serve as a basis for a seminar.

An alternative one or two semester course is given by §2.2-§2.3 about basic nonlinear ODE dynamics combined with (parts or all of) Part III and some parts of Part IV, for instance the beginning of Chapter 10. Other chapters of Part IV can then serve as a basis for a seminar.

There are other possibilities, for instance Chapter 3 about dissipative ODE dynamics combined with some dissipative PDE dynamics, chosen out of Chapters 5, 7, §8.3, Chapters 9-10, parts of Chapter 13, and Chapter 14. Similarly, Chapter 4 about conservative ODE dynamics could be combined with some conservative PDE dynamics, chosen for instance out of §8.1, §8.2, and Chapters 11 and 12. If the reader is familiar with the contents of Part I and Part II and is interested in an introduction to the mathematical theory of modulation equations, then we recommend to start reading in Part IV and going back to Part III where needed.

Nevertheless, the reader can also work through the book from the beginning to the end.

1.4.2. Recommended literature. Good classical books about PDEs are [**CH89, Joh91, Eva98, Sal08, Vas15**], while [**Str92, SVZZ13, Olv14, Log15a**] give more elementary introductions to PDEs. Books which look at PDEs from a dynamical systems point of view are [**Hen81, Tem97, RR04, Rob01**]. These books cover and extend material similar to that in the first two parts of our book, in particular Part II, while for instance [**SS99b, KP13**] discuss in more detail parts of what is treated in our Part III. For a general background on the functional analytic methods in our book we recommend [**Alt16, Wer00**], but the needed material can be found in most books on functional analysis. For more physically oriented introductions to PDEs see [**Fow97, BK00, TM05, Deb05**], for an overview of developments in the theory of PDEs in the 20th century see [**BB98**], and for an encyclopedic work on PDEs see [**Tay96**]. A "visual approach" to PDE with many motivating pictures is [**Mar07**]. For ODEs we refer for instance to [**Chi06, HSD04, Tes12, Log15b**]. Beginning in Part II, at the end of most Chapters we give an outlook and hints for further reading.

1.4.3. Software. There are many software packages for the numerical solution of ODEs and the graphical presentation of solutions. `Matlab`, `Maple`, and `Mathematica` have built in facilities, and there are various simple to use Java applets available. We strongly encourage the reader to do own experiments with any of these programs.

From the above remarks about the very different types of PDEs it readily follows that there cannot be a general tool for all types of PDEs. However, tools for specific types of PDEs, both commercial and free are widely available. We use some short self-written `matlab` scripts to illustrate some PDE dynamics, mostly for model problems. However, we do not discuss any numerical methods behind these programs and refer to [**Uec09**] and the references therein. For the computation of so called bifurcation diagrams we refer to `AUTO` [**Doe07, Dea16**] and `pde2path` [**UWR14**].

Exercises

1.1. Classify the following PDEs as linear or nonlinear.

a) $\partial_t u = \sum_{i,j=1}^{d} \partial_{x_i}\partial_{x_j}(a^{ij}u) + \sum_{i=1}^{d} \partial_{x_i}(b^i u)$, $a^{ij}, b^i : \mathbb{R}^d \to \mathbb{R}$ smooth functions.

b) $i\partial_t u = \Delta u$. c) $\partial_t V = rV - rS\partial_S V - \frac{1}{2}\sigma^2 S^2 \partial_S^2 V$, $(r, \sigma \in \mathbb{R})$.

d) $\partial_t^2 u = -\partial_x^4 u$. e) $\partial_t u = \Delta(u^\gamma)$, $(\gamma > 0)$.

f) $\partial_t u = \mathrm{div} F(u)$, $F : \mathbb{R} \to \mathbb{R}^d$ a smooth function. g) $\partial_t u = \partial_x^3 u + u\partial_x u$.

1.2. Constant coefficient second order linear partial differential equations in $\mathbb{R}^2$ can be written as

$$Lu = -\sum_{i,j=1,2} a^{ij}\partial_{x_i}\partial_{x_j} u + \sum_{i=1,2} b^i \partial_{x_i} u + c = 0.$$

The operator L is called elliptic if the eigenvalues of the symmetric matrix $A = (a^{ij})$ are strictly positive. It is called hyperbolic if they are nonzero, but have different signs. It is called parabolic if the associated quadratic form $(\partial_x \to \xi,\ \partial_y \to \eta)$ defines a parabola. Classify

a) $3\partial_x^2 u + 10\partial_x\partial_y u + 15\partial_y^2 u + 36\partial_x u + 12\partial_y u + 17 = 0$;

b) $3\partial_x^2 u + 4\partial_x u + \partial_y u + 2 = 0$.

1.3. Consider the PDE $\partial_t u = \partial_x u$ for $u = u(x, t)$.

a) Find the general solution for $x \in \mathbb{R}$.

b) Solve the PDE for $x \in (0, 1)$ with the initial condition $u(x, 0) = 1$ for $x \in (0, 1)$ under the boundary condition $u(1, t) = \cos t$.

c) Is it possible to solve the PDE for $x \in (0, 1)$ with the initial condition $u(x, 0) = 1$ for $x \in (0, 1)$ and the boundary condition $u(0, t) = \cos t$?

1.4. Consider a membrane $\Omega = (0, 1)^2$ which is fixed at the boundary $\partial\Omega$, i.e., $u|_{\partial\Omega} = 0$.

a) Make an ansatz $u(x, y, t) = v(t)\sin(m\pi x)\sin(n\pi y)$, $(n, m \in \mathbb{N})$ for the solutions of $\partial_t^2 u = \Delta u$. Which equation is satisfied by v?

b) Solve the equation for v with the initial conditions $v(0) = 0$ and $\dot{v}(0) = 1$.

c) Sketch for fixed $m, n \in \mathbb{N}$ the set of $(x, y) \in \Omega$, for which $u(x, y, t) = 0$ for all $t \in \mathbb{R}$.

Chapter 2

Basic ODE dynamics

The first part of this book is about nonlinear dynamics in $\mathbb{R}^d$. It consists of this chapter, Chapter 3 about dissipative dynamics, and Chapter 4 about Hamiltonian dynamics. In this part we provide some basic concepts of nonlinear dynamics. In order to avoid the various functional analytic difficulties associated with PDEs we restrict to the finite-dimensional situation, i.e., we consider ODEs

$$\dot{u}(t) = f(u(t), t),$$

with $u(t) \in \mathbb{R}^d$, $f : \mathbb{R}^d \times I \to \mathbb{R}^d$ a continuous vector field which is locally Lipschitz-continuous w.r.t. its first argument, where $I \subset \mathbb{R}$ is an open interval, usually $I = \mathbb{R}$, and where $\dot{u}(t)$ denotes the derivative of the function u w.r.t. time t. In general it is not possible to obtain explicit solutions, and so our main goal is to provide tools for the understanding of the qualitative behavior of solutions.

Some notation. The initial value problem consists in finding a solution of the ODE to an initial value u_0 given at an initial time $t_0 \in I$. A solution of the initial value problem is a function $u \in C^1(I_0, \mathbb{R}^d)$ which fulfills the ODE, where $I_0 \subset I$ is again an open interval, $t_0 \in I_0$, and $u(t_0) = u_0$. This solution is denoted by $u(t, t_0, u_0)$. If f on the right-hand side of the ODE does not depend explicitly on time, i.e., $f = f(u)$, then the ODE is called autonomous, and we may assume $t_0 = 0$ and write $u(t, u_0)$ for the solution of the initial value problem.

Absolutely fundamental for the understanding of nonlinear dynamics is the understanding of the dynamics of linear systems which we therefore consider first. Then we introduce basic concepts of nonlinear dynamics. These are the local and global existence and uniqueness of solutions, special solutions such as fixed points, periodic solutions, homoclinic and heteroclinic orbits, and further concepts such as stability and instability, invariant manifolds, ω-limit-sets, attractors, and chaotic dynamics. Many of these concepts will later be transferred to nonlinear PDEs. Moreover, the search for special solutions, such as front or pulse solutions for the PDEs, in later chapters very often lead to ODE problems as they are considered here.

The behavior and the analysis of an ODE or of a PDE strongly differ between dissipative and conservative systems. In Chapter 3 we provide the strategy and the tools to tackle dissipative systems. Such systems are typically characterized by the existence of compact absorbing sets, i.e., compact sets into which all solutions finally enter. In dissipative systems very often more complicated and eventually chaotic dynamics occur through bifurcations if some external parameter is varied. After introducing a number of elementary bifurcations for one- and two-dimensional systems we introduce with the Lyapunov-Schmidt reduction and the center manifold theorem two reduction methods which allow us to find these elementary bifurcations in higher dimensional systems, too. Chapter 3 is closed by presenting some routes of bifurcations to chaotic behavior in dissipative systems.

The systems considered in Chapter 3 change the volume in phase space, but many systems in nature preserve the volume in phase space, especially those of classical mechanics. Thus, Chapter 4 is devoted to Hamiltonian ODE dynamics. We provide some tools for their analysis and explain basic facts about their behavior, which shows fundamental differences compared to that of the systems of Chapter 3. For instance, a globally attracting fixed point cannot exist in conservative or volume-preserving systems. Therefore, stability and instability proofs or the route to chaotic behavior must be completely different. The starting point of the bifurcation analysis is not a globally attracting fixed point, but a so called completely integrable system. In Chapter 4 we also discuss KAM theory which allows to understand the behavior of systems which are small perturbations of completely integrable Hamiltonian systems.

The ideas presented in this first part will reappear in subsequent sections. For instance, Chapter 3 about dissipative ODE dynamics contains basic tools which will be used in Chapters 5-7, §8.3-Chapter 10, and Chapter 14 about dissipative PDE dynamics. Similarly, Chapter 4 about conservative ODE dynamics contains basic tools which will help to understand §8.1, §8.2, and Chapters 11-12 about conservative PDE dynamics.

We emphazise that the purpose of Part I is not to give a comprehensive overview about ODEs. Rather we present the basic ideas of nonlinear dynamics as needed in subsequent parts of this book in the analysis of PDEs.

There are a number of excellent textbooks on nonlinear ODE dynamics, many of them also reviewing basic linear ODE dynamics. An elementary and very readable account on ODE dynamics and bifurcations is [**HK91**], a very applied and example oriented approach is used in [**Str94**], and an excellent modern presentation is given in [**Tes12**]. Alternatives and complements to these textbooks are for instance [**Chi06, Ver96, Rob04a, Rob04b**]. More advanced texts include [**GH83, KH97, Wig03, HSD04**]. In [**SH96, Lyn04**] discrete dynamical systems and ODEs are treated from a numerical point of view, and [**Dev89**] focusses on discrete chaotical dynamical systems. For the bifurcation aspects of ODEs, and in particular center manifolds, we again refer to [**GH83, Wig03**], and to [**Kuz04, Erm02**] for invariant manifolds from a numerical point of view. Our favorite books on Hamiltonian systems and KAM theory are [**Arn78, Thi88, MH92**], see also [**Way96**].

2.1. Linear systems

Fundamental for the understanding of nonlinear dynamics is the understanding of linear dynamics. Linear ODEs occur for instance as linearizations around fixed points or periodic solutions. The solution of these linear ODEs and the variation of constant formula, which allow us to solve inhomogeneous linear problems, will be the basis for stability proofs for fixed points and periodic solutions of nonlinear ODEs in §2.3. Moreover, this technique will be generalized to semi-linear dissipative PDEs and a number of conservative PDEs in Parts II-IV for proving the local existence and uniqueness of solutions of PDEs.

A linear ODE is an equation

$$\dot{u}(t) = A(t)u(t) + g(t) \tag{2.1}$$

for an unknown function $u \in C^1(I, \mathbb{R}^d)$, where $I \subset \mathbb{R}$ is an interval, $A(t) \in \mathbb{R}^{d\times d}$ is a $d\times d$-matrix with entries $a_{ij}(t)$, and where $g(t) \in \mathbb{R}^d$ is an inhomogeneity. We generally think of t as time, and for simplicity we assume that A and g are at least continuous w.r.t. t. Together with an initial condition $u|_{t=t_0} = u_0 \in \mathbb{R}^d$ we have an initial value problem. Equation (2.1) is called homogeneous, if $g(t) \equiv 0$, i.e., if

$$\dot{u}(t) = A(t)u(t), \tag{2.2}$$

and (2.1), respectively (2.2), are called autonomous if A and g in (2.1), respectively A in (2.2), do not depend on t.

It is well-known that the initial value problems for (2.1) and (2.2) have unique solutions, and that the solutions of (2.1) form a d-dimensional affine space and the solutions of (2.2) a d-dimensional vector space. This will briefly be recalled in §2.1.2. We restrict ourselves to those parts of the theory which are needed in subsequent sections. That is, we restrict to A independent of time, and to A periodic in time, i.e., $A(t) = A(t+T)$ for a $T > 0$. Other important classes of linear systems are asymptotically constant systems, i.e., $A(t) \to A_\pm$ for $t \to \pm\infty$. They appear as linearization around so called homoclinic or heteroclinic orbits.

2.1.1. Notation. Let X be a real or complex vector space. A map $\|\cdot\| : X \to \mathbb{R}$ is called norm, if for all $u, v \in X$ and $\lambda \in \mathbb{R}$, respectively $\lambda \in \mathbb{C}$,

(i) $\|u\| \geq 0$ and $\|u\| = 0$ if and only if $u = 0$,
(ii) $\|\lambda u\| = |\lambda| \, \|u\|$,
(iii) $\|u + v\| \leq \|u\| + \|v\|$.

In $\mathbb{R}^d$, the major examples are the ℓ_1-norm $\|u\|_1 = \sum_{j=1}^d |u_j|$, the Euclidean or ℓ_2-norm $\|u\|_2 = (u^T u)^{1/2} = (\sum_{j=1}^d |u_j|^2)^{1/2}$, and the ℓ_∞- or maximum-norm $\|u\|_\infty = \max_{j=1,\dots,d} |u_j|$.

Concepts such as convergence in $\mathbb{R}^d$, or later on stability and instability for ODEs in $\mathbb{R}^d$, are independent of the chosen norm in $\mathbb{R}^d$. The reason for this is the equivalence of norms in finite-dimensional vector spaces.

Theorem 2.1.1. *All norms in $\mathbb{R}^d$ are equivalent, i.e., for two norms $\|\cdot\|$ and $\|\cdot\|_*$ there exist positive constants C_1, C_2 such that for all $u \in \mathbb{R}^d$ we have*

$$\|u\| \leq C_1 \|u\|_* \leq C_2 \|u\|.$$

Proof. Obviously, it is sufficient to establish the estimates between an arbitrary norm $\|\cdot\|$ and the $\|\cdot\|_\infty$-norm. By the triangle inequality we have

$$\|u\| = \left\| \sum_{j=1}^d u_j e_j \right\| \leq \sum_{j=1}^d |u_j| \|e_j\| \leq \left(\sum_{j=1}^d \|e_j\| \right) \|u\|_\infty.$$

For the second estimate let $M = \{u \in \mathbb{R}^d : \|u\|_\infty = 1\}$. Then $f : M \to \mathbb{R}$, $u \mapsto \|u\|^{-1}$ is a continuous map by definition. Suppose that f is unbounded on M, i.e., there exists a sequence $(u^n)_{n\in\mathbb{N}}$ with $\|u^n\| \to 0$ for $n \to \infty$. Since the finitely many coordinates u_j^n satisfy $|u_j^n| \leq 1$ there exists a convergent subsequence $u^{n_k} \to \xi \in M$ for $k \to \infty$. By the continuity of the norm we have $\|u^{n_k}\| \to \|\xi\| = 0$ for $k \to \infty$ which implies $\xi = 0$ contradicting $\xi \in M$. Hence, there exists a $C > 0$ such that $\sup_{u\in M} \|f(u)\| = C < \infty$ and so $\|u\| \geq \frac{1}{C}$ for all $u \in M$ which finally leads to $\|u\| \geq \frac{1}{C} \|u\|_\infty$. □

Remark 2.1.2. The reason why we gave a proof of this well known theorem is that in infinite dimensions there are infinitely many non-equivalent norms. This has a number of consequences for the subsequent analysis of PDEs. It is possible that uniqueness but no global existence of solutions is known in one space, and global existence but no uniqueness of solutions is known in another space, but there is no space known where both properties hold simultaneously. So far this is exactly the state of the art for one of the Millennium problems of the Clay Foundation, namely the global existence and uniqueness of smooth solutions of the 3D Navier-Stokes equations, which will be discussed in Chapter 6. ⌋

The $d \times d$-matrices form a normed vector space of dimension d^2. With the matrix multiplication AB they form an algebra. An important matrix norm is given by the operator norm

$$\|A\|^* = \sup\left\{\frac{\|Au\|}{\|u\|} : u \in \mathbb{R}^d \setminus \{0\}\right\}.$$

Obviously $\|Au\| \leq \|A\|^* \ \|u\|$, and so from

$$\|ABu\| \leq \|A\|^* \ \|Bu\| \leq \|A\|^* \ \|B\|^* \ \|u\|$$

it follows

$$\|AB\|^* \leq \|A\|^* \ \|B\|^*,$$

i.e., the matrices form a Banach algebra w.r.t. matrix multiplication and operator norm. For arbitrary norms the matrix norm and the vector norm are called compatible if

$$\|Au\| \leq \|A\| \ \|u\|.$$

Examples of such compatible norms are

$$\begin{aligned}
\|\cdot\| &= \|\cdot\|_1, & \|A\|^* &= \sup_{k=1,\dots,d} \sum_{j=1}^{d} |a_{jk}|, \\
\|\cdot\| &= \|\cdot\|_2, & \|A\|^* &= \left(\sum_{i,j=1}^{d} |a_{jk}|^2\right)^{1/2}, \\
\|\cdot\| &= \|\cdot\|_\infty, & \|A\|^* &= \sup_{j=1,\dots,d} \sum_{k=1}^{d} |a_{jk}|.
\end{aligned}$$

In the following we always use compatible norms, and $\mathbb{R}^d$ will be equipped with the Euclidean norm, if not indicated otherwise. Finally, we remark that for all norms

$$\left\|\int_{t_0}^{t} u(\tau)\,\mathrm{d}\tau\right\| \leq \int_{t_0}^{t} \|u(\tau)\|\,\mathrm{d}\tau.$$

2.1.2. Local existence and uniqueness. We briefly recall that the initial value problem for (2.2) has a unique solution, and that the solutions of (2.2) form a d-dimensional vector space. The following local existence and uniqueness result and many other results in this book are based on the contraction mapping principle which is absolutely fundamental in nonlinear analysis. We recall that a metric space M is called complete, if every Cauchy sequence in M possesses a limit in M.

Theorem 2.1.3. (Contraction mapping principle or fixed point theorem of Banach) *Let (M, d) be a complete metric space and $F : M \to M$ a contraction, i.e., there exists a $\kappa \in (0, 1)$ such that*

$$d(F(x), F(y)) \leq \kappa d(x, y)$$

for all $x, y \in M$. Then F has a unique fixed point $x^* \in M$, i.e., $x^* = F(x^*)$.

Proof. We first prove the uniqueness. Suppose that there exist two different fixed points x^* and y^*. Then

$$d(x^*, y^*) = d(F(x^*), F(y^*)) \leq \kappa d(x^*, y^*).$$

Since $\kappa \in (0, 1)$, it follows that $d(x^*, y^*) = 0$ and hence $x^* = y^*$, in contradiction to the assumption.

We define the sequence $x_{n+1} = F(x_n)$ with $x_0 \in M$ arbitrary, but fixed. Then for $m \geq n$

$$d(x_m, x_n) \leq \sum_{j=n}^{m-1} d(x_{j+1}, x_j) \leq \sum_{j=n}^{m-1} \kappa^j d(x_1, x_0) \leq \frac{\kappa^n}{1-\kappa} d(x_1, x_0).$$

Hence, for all $\varepsilon > 0$ there exists an $N > 0$, such that for all $n, m > N$:

$$d(x_m, x_n) \leq \frac{\kappa^N}{1-\kappa} d(x_1, x_0) \leq \varepsilon,$$

i.e., $(x_n)_{n \in \mathbb{N}}$ is a Cauchy-sequence. Since M is complete, there exists an $x^* \in M$, such that $x^* = \lim_{n \to \infty} x_n$.

The limit x^* is a fixed point due to the continuity of F, i.e.,

$$F(x^*) = F(\lim_{n \to \infty} x_n) = \lim_{n \to \infty} F(x_n) = \lim_{n \to \infty} x_{n+1} = x^*. \qquad \square$$

Corollary 2.1.4. *Let $(X, \|\cdot\|)$ be a Banach space, M be a closed subset of X, and $F : M \to M$ be a contraction. Then F has a unique fixed point $x^* \in M$.*

Our first version of the local existence and uniqueness of solutions for (2.2) is as follows.

Lemma 2.1.5. *Consider (2.2) with initial condition $u|_{t=t_0} = u_0$ and continuous $A = A(t)$. Then there exists a $\delta > 0$ independent of u_0 such that (2.2) has a unique solution $u \in C^1((t_0 - \delta, t_0 + \delta), \mathbb{R}^d)$ satisfying $u|_{t=t_0} = u_0$.*

Proof. The proof is based on the application of the contraction mapping theorem to the integrated ODE

$$u(t) = u_0 + \int_{t_0}^{t} A(s)u(s)\,\mathrm{d}s =: F(u)(t),$$

where $F : M \to M$ with $M = C^0([t_0 - \delta, t_0 + \delta], \mathbb{R}^d)$. Fix a $T_0 > 0$ and define $C_0 = \sup_{t\in[t_0-T_0,t_0+T_0]} \|A(t)\|$ which is finite due to the continuity of $t \mapsto A(t)$. Then we have $\|F(u) - F(v)\|_M \le \delta C_0 \|u - v\|_M$, where $\|u\|_M = \sup_{t\in[t_0-\delta,t_0+\delta]} \|u(t)\|_{\mathbb{R}^d}$, such that F is a contraction for instance for $\delta = \min\{1/(2C_0), T_0\}$. From $u \in M$ it follows that $F(u) \in C^1((t_0-\delta, t_0+\delta), \mathbb{R}^d)$. Since u is a fixed point we also have $u = F(u) \in C^1((t_0 - \delta, t_0 + \delta), \mathbb{R}^d)$. □

Remark 2.1.6. By the last argument, it is easy to see that the m-times differentiability of $t \mapsto A(t)$ implies inductively the $m+1$-times differentiability of $t \mapsto u(t)$. ⌋

Lemma 2.1.5 only asserts local existence and uniqueness. The way to show existence and uniqueness of solutions beyond $t_0 + \delta$ is to prove bounds on $u(t_0 + \delta)$. The key tool is Gronwall's inequality which will be used in many proofs below. We first restrict to a simple version [**Ver96**, Theorem 1.2].

Lemma 2.1.7. (Gronwall's inequality) *For $t \in (t_0, t_0 + a)$ with $a > 0$, and ϕ and ψ non-negative continuous functions assume that*

$$\phi(t) \le \int_{t_0}^{t} \psi(s)\phi(s)\,\mathrm{d}s + \delta. \tag{2.3}$$

Then for all $t \in (t_0, t_0 + a)$ we have

$$\phi(t) \le \delta \mathrm{e}^{\int_{t_0}^{t} \psi(s)\,\mathrm{d}s}.$$

Proof. Dividing (2.3) by its right-hand side and multiplication of both sides with $\psi(t)$ yields after integration that

$$\int_{t_0}^{t} \frac{\psi(\tau)\phi(\tau)}{\int_{t_0}^{\tau} \psi(s)\phi(s)\,\mathrm{d}s + \delta}\,\mathrm{d}\tau \le \int_{t_0}^{t} \psi(\tau)\,\mathrm{d}\tau$$

which implies that $\ln\left(\int_{t_0}^{t} \psi(s)\phi(s)\,\mathrm{d}s + \delta\right) - \ln\delta \le \int_{t_0}^{t} \psi(\tau)\,\mathrm{d}\tau$ and finally that

$$\int_{t_0}^{t} \psi(s)\phi(s)\,\mathrm{d}s + \delta \le \delta \exp\left(\int_{t_0}^{t} \psi(\tau)\,\mathrm{d}\tau\right).$$

By assumption $\phi(t)$ is smaller than the expression on the left-hand side. □

From the integrated ODE

$$u(t) = u_0 + \int_{t_0}^{t} A(s)u(s)\,\mathrm{d}s$$

we find the inequality

$$\|u(t)\|_{\mathbb{R}^d} \le \|u_0\|_{\mathbb{R}^d} + \left|\int_{t_0}^{t} \|A(s)\|\ \|u(s)\|_{\mathbb{R}^d}\,\mathrm{d}s\right|$$

and so by Gronwall's inequality

$$\|u(t)\|_{\mathbb{R}^d} \leq \|u_0\|_{\mathbb{R}^d} \exp\left(\left|\int_{t_0}^{t} \|A(s)\| \,\mathrm{d}s\right|\right). \tag{2.4}$$

Since continuous functions stay bounded on compact intervals, from (2.4) it follows that for continuous $t \mapsto A(t)$ the solutions $t \mapsto u(t)$ exist for all $t \in \mathbb{R}$.

Theorem 2.1.8. *Consider* (2.2) *with initial condition* $u|_{t=t_0}=u_0$ *and a continuous* $A=A(t)$ *for all* $t \in \mathbb{R}$. *Then there exists a unique solution* $u \in C^1(\mathbb{R}, \mathbb{R}^d)$ *satisfying* $u|_{t=t_0} = u_0$.

Proof. We choose an arbitrary, but fixed $T_0 > 0$. We apply Lemma 2.1.5 with the initial condition $u|_{t=t_0} = u_0$, which gives a unique solution on $[t_0, t_0 + \delta)$ and $u|_{t=t_0+\delta/2} = u_1$. Inductively we apply Lemma 2.1.5 now with the initial condition $u|_{t=t_0+n\delta/2} = u_n$, which gives a unique solution on $[t_0 + n\delta/2, t_0 + n\delta/2 + \delta)$ and $u|_{t=t_0+(n+1)\delta/2} = u_{n+1}$. Doing this until $(n+1)\delta/2 \geq T_0$ gives us the solution for all $t \in [t_0, t_0 + T_0)$. Proceeding similarly for negative $t - t_0$ gives the solution for all $t \in (t_0 - T_0, t_0 + T_0)$. Since $T_0 > 0$ was arbitrary we are done. $\square$

Lemma 2.1.9. *The solutions of* (2.2) *form a d-dimensional vector space.*

Proof. Since the solutions of (2.2) depend linearly and one-to-one on the initial conditions u_0 we have that the set of solutions of (2.2) is isomorphic via $u \mapsto u(t_0)$ to the space of initial conditions, i.e., $\mathbb{R}^d$. $\square$

Definition 2.1.10. *The matrix-valued function* $t \mapsto \phi(t)$ *is called fundamental matrix if* $\dot{\phi}(t) = A(t)\phi(t)$ *and if* $\phi(t_0)$ *has full rank for a* $t_0 \in \mathbb{R}$.

Remark 2.1.11. From the local existence and uniqueness Theorem 2.1.8, it immediately follows that $\phi(t)$ has full rank for all $t \in \mathbb{R}$. If $t \mapsto \phi(t)$ is a fundamental matrix then $u(t) = \phi(t)\phi(t_0)^{-1}u_0$ solves the initial value problem with $u|_{t=t_0} = u_0$. If $t \mapsto \psi(t)$ is another fundamental matrix then $\phi(t)\phi(t_0)^{-1} = \psi(t)\psi(t_0)^{-1}$ such that there exists an invertible matrix $C = \phi(t_0)^{-1}\psi(t_0)$ which is independent of t with $\psi(t) = \phi(t)C$. ⌋

2.1.3. The variation of constant formula. Associated to (2.2) we define the linear solution operator $S(t,s) : \mathbb{R}^d \mapsto \mathbb{R}^d$ through

$$S(t,s)u_0 = u(t,s,u_0),$$

where $u(t,s,u_0)$ is the solution of (2.2) with initial value $u|_{t=s} = u_0$. For fixed $s, t \in \mathbb{R}$ the linear map $S(t,s) : \mathbb{R}^d \to \mathbb{R}^d$ is one-to-one, i.e., an invertible matrix, with $S(t,s)^{-1} = S(s,t)$. The solution of the inhomogeneous problem (2.1), i.e.,

$$\dot{u}(t) = A(t)u(t) + g(t) \tag{2.5}$$

can be expressed in terms of the inhomogeneity $g = g(t)$ and the solution operator $S(t,s)$. Differentiation of $u(t) = S(t,s)y(t)$ w.r.t. t and using (2.5) shows that

$$\dot{u}(t) = \partial_t S(t,s)y(t) + S(t,s)\dot{y}(t) = A(t)S(t,s)y(t) + g(t).$$

Since $S(t,s)$ solves the homogeneous problem $\partial_t S(t,s) = A(t)S(t,s)$ we obtain

$$\dot{y}(t) = S(t,s)^{-1}g(t) = S(s,t)g(t).$$

Integration yields $y(t) - y(s) = \int_s^t \dot{y}(\tau)\,\mathrm{d}\tau = \int_s^t S(s,\tau)g(\tau)\,\mathrm{d}\tau$ and therefore

$$\begin{aligned} u(t) =& S(t,s)y(t) = S(t,s)y(s) + S(t,s)\int_s^t S(s,\tau)g(\tau)\,\mathrm{d}\tau \\ =& S(t,s)u(s) + \int_s^t S(t,\tau)g(\tau)\,\mathrm{d}\tau, \end{aligned}$$

since $S(s,s) = I$. This formula is called the variation of constant formula. It will be used in stability proofs and in proofs of the local existence and uniqueness of solutions of nonlinear problems. In case $S(t,s) = \mathrm{e}^{(t-s)A}$, see the subsequent §2.1.4, the variation of constant formula specializes to

$$u(t) = \mathrm{e}^{tA}u(0) + \int_0^t \mathrm{e}^{(t-\tau)A}g(\tau)\,\mathrm{d}\tau. \tag{2.6}$$

2.1.4. The exponential matrix. In general, (2.2) can only be solved explicitly in case $d = 1$. For $d \geq 2$, if a solution is known of the d-dimensional problem, then the dimension of the problem can be reduced by one, i.e., after the reduction a linear ODE in $d-1$ space dimensions has to be solved, cf. [**Cod61**, Page 118]. However, in case

$$\dot{u} = Au, \quad u|_{t=0} = u_0, \tag{2.7}$$

with $A \in \mathbb{R}^{d\times d}$ independent of t, all solutions can be computed explicitly. The ansatz $u(t) = \mathrm{e}^{\lambda t}\widehat{u}$ yields the eigenvalue problem $A\widehat{u} = \lambda\widehat{u}$, and in case that A has d linearly independent eigenvectors $\phi_1, \ldots, \phi_d \in \mathbb{R}^d$ with eigenvalues $\lambda_1, \ldots, \lambda_d$, the general solution reads $u(t) = \sum_{i=1}^d c_i \mathrm{e}^{\lambda_i t}\phi_i$ with $c_i, \ldots, c_d \in \mathbb{R}$. In case that there are complex eigenvalues or Jordan blocks this formula becomes slightly more complicated. Equation (2.7) appears as linearization around fixed points of nonlinear systems and hence plays a crucial role.

Remark 2.1.12. Linear scalar equations of nth order

$$y^{(n)}(t) + \sum_{j=0}^{n-1} a_j y^{(j)}(t) = 0,$$

with $a_j \in \mathbb{R}$, can always be written as a linear first order system. For the construction of the explicit solution this is not necessary. The r different zeroes λ_k with multiplicity k_r, i.e. $\sum_{k=1}^{r} k_r = n$, of the characteristic equation $\lambda^n + \sum_{j=0}^{n-1} a_j \lambda^j = 0$ allow to construct the general solution, namely

$$y(t) = \sum_{k=1}^{r} \sum_{j=0}^{k_r - 1} c_{k,j} e^{\lambda_k t} t^j$$

with n coefficients $c_{k,j}$. See [**Log15b**, Chapter 2]. ⌋

From a theoretical point of view, the following representation formula turns out to be useful. The solution of (2.7) is given by

$$u(t) = e^{tA} u_0 = \sum_{n=0}^{\infty} \frac{(tA)^n}{n!} u_0. \tag{2.8}$$

We have absolute and uniform convergence w.r.t. t on every compact interval due to

$$\sum_{n=0}^{\infty} \left\| \frac{(tA)^n}{n!} u_0 \right\|_{\mathbb{R}^d} \leq \sum_{n=0}^{\infty} \frac{\|A\|^n t^n}{n!} \|u_0\|_{\mathbb{R}^d} \leq e^{\|A\| t} \|u_0\|_{\mathbb{R}^d}.$$

Obviously this also holds for the series differentiated w.r.t. t. Similarly, we obtain the estimate

$$\|e^{tA} u_0\|_{\mathbb{R}^d} \leq e^{t\|A\|} \|u_0\|_{\mathbb{R}^d}.$$

Moreover, $e^{tA} u_0$ solves (2.7) due to

$$\frac{d}{dt} \left(\sum_{n=0}^{\infty} \frac{(tA)^n}{n!} u_0 \right) = \sum_{n=1}^{\infty} \frac{A(tA)^{n-1}}{(n-1)!} u_0 = A \sum_{n=0}^{\infty} \frac{(tA)^n}{n!} u_0,$$

where the time derivative and the infinite sums can be interchanged due to the uniform convergence w.r.t. t on compact intervals. Moreover, A and the infinite sum can be interchanged due to the boundedness and hence continuity of A.

The solution operator e^{tA} can be expressed in terms of the Jordan normal form J of A. The change of coordinates $u = Sy$ in $\dot{u} = Au$ yields $\dot{y} = S^{-1} A S y = Jy$. We have

$$e^{tA} u_0 = S e^{tJ} y_0 = S e^{tJ} S^{-1} u_0, \tag{2.9}$$

or equivalently

$$\begin{aligned} S^{-1}\mathrm{e}^{tA}S =& S^{-1}\left(1 + tA + \frac{t^2A^2}{2} + \dots\right)S \\ =& \left(1 + tS^{-1}AS + \frac{t^2S^{-1}ASS^{-1}AS}{2} + \dots\right) \\ =& \left(1 + tJ + \frac{t^2J^2}{2} + \dots\right) = \mathrm{e}^{tJ}. \end{aligned}$$

Hence, it is sufficient to consider e^{tJ} for J a matrix in Jordan normal form, i.e.,

$$J = \begin{pmatrix} J_1 & & & 0 \\ & J_2 & & \\ & & \ddots & \\ 0 & & & J_r \end{pmatrix} \text{ with } J_j = \begin{pmatrix} \lambda_j & 1 & & 0 \\ & \ddots & \ddots & \\ & & \ddots & 1 \\ 0 & & & \lambda_j \end{pmatrix}.$$

Since

$$\exp(tJ) = \exp\left(\begin{pmatrix} J_1 & & 0 \\ & \ddots & \\ 0 & & J_r \end{pmatrix} t\right) = \begin{pmatrix} \mathrm{e}^{tJ_1} & & 0 \\ & \ddots & \\ 0 & & \mathrm{e}^{tJ_r} \end{pmatrix}$$

it is sufficient to consider

$$\mathrm{e}^{tJ_j} = \mathrm{e}^{t(\lambda_j I + N_k)} = \mathrm{e}^{\lambda_j t I}\mathrm{e}^{tN_k}$$

due to $IN_k = N_k I$ with the $k \times k$-matrix

$$N_k = \begin{pmatrix} 0 & 1 & & 0 \\ & \ddots & \ddots & \\ & & \ddots & 1 \\ 0 & & & 0 \end{pmatrix}.$$

Hence, it remains to compute

$$\mathrm{e}^{tN_k} = \sum_{\nu=0}^{\infty} \frac{t^\nu}{\nu!} N_k^\nu.$$

We have

$$\begin{aligned} N_k^\mu &= (\delta_{i,j-\mu}), \quad \text{for } \mu = 0, ..., k-1, \\ N_k^\mu &= 0, \quad \text{for } \mu = k, k+1, ... \end{aligned}$$

and so finally

$$\mathrm{e}^{tN_k} = \begin{pmatrix} 1 & t & \dots & \dots & \frac{t^{k-1}}{(k-1)!} \\ 0 & 1 & \ddots & & \frac{t^{k-2}}{(k-2)!} \\ \vdots & \ddots & \ddots & \ddots & \vdots \\ & & \ddots & 1 & t \\ 0 & & \dots & 0 & 1 \end{pmatrix}.$$

This representation formula immediately yields a statement about the stability of the fixed point $u = 0$ of the ODE $\dot{u} = Au$.

Definition 2.1.13. *a) The fixed point $u = 0$ is called asymptotically stable for $\dot{u} = Au$ if for all $u_0 \in \mathbb{R}^d$ we have $u(t) = \mathrm{e}^{tA}u_0 \to 0$ for $t \to \infty$.*

b) The fixed point $u = 0$ is called stable for $\dot{u} = Au$ if for all $u_0 \in \mathbb{R}^d$ we have that $u(t) = \mathrm{e}^{tA}u_0$ stays bounded for all $t \geq 0$.

c) In all other cases the origin $u = 0$ is called unstable.

Theorem 2.1.14. *a) If all eigenvalues of A have strictly negative real parts, then $u = 0$ is asymptotically stable.*

b) If A possesses no eigenvalue with positive real part and if all eigenvalues with real part zero possess the same algebraic and geometric multiplicity, i.e., no non-trivial Jordan block, then the origin $u = 0$ is stable.

c) In all other cases, i.e., if A possesses at least one eigenvalue with strictly positive real part or at least one eigenvalue with real part zero with algebraic multiplicity bigger than the geometric multiplicity, then the origin $u = 0$ is unstable.

2.1.5. Linear planar systems. The behavior of the solutions of two-dimensional autonomous systems $\dot{u} = f(u)$ can be visualized with the help of so called phase portraits. As a first step we discuss and visualize the behavior of linear two-dimensional autonomous systems

$$\dot{u} = Au, \quad u(t) \in \mathbb{R}^2, \quad A \in \mathbb{R}^{2\times 2}. \tag{2.10}$$

In order to visualize the behavior of (2.10) we have a number of possibilities, which we shall later also apply to nonlinear systems. In doing so we also classify the different kinds of fixed points.

1) We plot the vector field $f : \mathbb{R}^2 \to \mathbb{R}^2$. This turns out to be not that helpful due to the in general strongly varying length of f.
2) Therefore, we plot in most cases the direction field $\alpha f/\|f\|_{\mathbb{R}^2} : \mathbb{R}^2 \to \mathbb{R}^2$ for a fixed $\alpha > 0$, i.e., in every point $u \in (h\mathbb{Z})^2$ of a grid with width h we plot a vector of $\mathbb{R}^2$ of fixed length α.

3) We plot the flow, i.e., a number of chosen orbits. These are curves which are defined by the solutions $t \mapsto u(t) \in \mathbb{R}^2$.

A combination of 2) and 3) is called a phase portrait. The choice of the size of h and of the orbits depends on the problem, and also is a matter of taste.

Remark 2.1.15. The vector field and the direction field are both tangent vectors of the solution $t \mapsto u(t)$. Hence, the differential equations $\dot{u} = f(u)$ and $\dot{u} = f(u)/\|f(u)\|_{\mathbb{R}^2}$ have the same orbits, i.e., solution curves in the phase plane, although their dynamics are very different. ⌋

One more option is to plot the nullclines. These are the sets

$$N_j := \{(u_1, u_2) \in \mathbb{R}^2 : f_j(u_1, u_2) = 0\},$$

for $j = 1, 2$, where the vector field is vertical, respectively horizontal. The intersection points of N_1 and N_2 give the fixed points u^* of the ODE, i.e., points with $f(u^*) = 0$. If the solution starts in a fixed point, the solution stays in that fixed point, i.e., $u(t) = u^*$ for all t. Often, nullclines at least partially coincide with coordinate axis, and, moreover, for (non-degenerate) linear systems we have $N_1 \cap N_2 = \{(0, 0)\}$ as the only fixed point.

Due to §2.1.4 it is sufficient to consider (2.2) with $A \in \mathbb{R}^{2\times 2}$ in Jordan normal form. There are the following cases.

a) The eigenvalues have the same algebraic and geometric multiplicity, i.e., $A = \begin{pmatrix} \lambda_1 & 0 \\ 0 & \lambda_2 \end{pmatrix}$ with **a1)** $0 \neq \lambda_j \in \mathbb{R}$ or **a2)** $0 \neq \lambda_1 = \overline{\lambda_2}$. In case a1) we distinguish three subcases i) $\lambda_1 = \lambda_2$, ii) $\lambda_1 > \lambda_2 > 0$ and iii) $\lambda_1 > 0 > \lambda_2$. All other cases are obtained from i)-iii) by a reversal of time $t \mapsto -t$.

b) The eigenvalue has geometric multiplicity one and algebraic multiplicity two, i.e., $A = \begin{pmatrix} \lambda & 1 \\ 0 & \lambda \end{pmatrix}$ is a Jordan block with $\lambda \in \mathbb{R}$.

c) The degenerate case of at least one eigenvalue $\lambda = 0$. Besides the trivial case $A = 0$ there are the cases **c1)** $0 = \lambda_1 < \lambda_2$ and **c2)** $\lambda = 0$ with geometric multiplicity one and algebraic multiplicity two.

In the following we consider a number of examples to visualize these cases, see Figure 2.1.5.

a1 i): Let $A = \begin{pmatrix} 1 & 0 \\ 0 & 1 \end{pmatrix}$, i.e., $\dot{u}_1 = u_1$, $\dot{u}_2 = u_2$. For the solutions we find $u_1(t) = e^t u_1(0)$, $u_2(t) = e^t u_2(0)$. The orbits are straight lines since $u_1(t)/u_2(t) = u_1(0)/u_2(0) = const.$. The nullclines are

$N_1 = \{(0,y) : y \in \mathbb{R}\}$ and $N_2 = \{(x,0) : x \in \mathbb{R}\}$. The fixed point $u^* = (0,0)$ is called a source or unstable node. If time is reversed, i.e., if $A = \begin{pmatrix} -1 & 0 \\ 0 & -1 \end{pmatrix}$, then $u^* = (0,0)$ is called a sink or stable node.

a1 ii): Let $A = \begin{pmatrix} 2 & 0 \\ 0 & 1 \end{pmatrix}$, i.e., $\dot{u}_1 = 2u_1$, $\dot{u}_2 = u_2$. For the solutions we find $u_1(t) = \mathrm{e}^{2t}u_1(0)$, $u_2(t) = \mathrm{e}^t u_2(0)$. The orbits are parabolas since $u_1(t)/(u_2(t))^2 = u_1(0)/(u_2(0))^2 = const.$. The phase portrait is robust w.r.t. small perturbations, i.e., $A = \mathrm{diag}(1.9, 1.1)$ has a similar phase portrait. Again the fixed point $u^* = (0,0)$ is called a source.

a1 iii): Let $A = \begin{pmatrix} 1 & 0 \\ 0 & -1 \end{pmatrix}$, i.e., $\dot{u}_1 = u_1$, $\dot{u}_2 = -u_2$. For the solutions we find $u_1(t) = \mathrm{e}^t u_1(0)$, $u_2(t) = \mathrm{e}^{-t}u_2(0)$. The orbits are hyperbolas since $u_1(t)u_2(t) = u_1(0)u_2(0) = const.$. The phase portrait is robust w.r.t. small perturbations, i.e., $A = \mathrm{diag}(1.1, -0.9)$ gives a similar phase portrait. The fixed point $u^* = (0,0)$ is called a saddle.

a2 i): Let $A = \begin{pmatrix} 0 & 1 \\ -1 & 0 \end{pmatrix}$, i.e., $\dot{u}_1 = u_2$, $\dot{u}_2 = -u_1$. Then $\widetilde{u} = u_1 + \mathrm{i}u_2$ solves $\dot{\widetilde{u}} = -\mathrm{i}\widetilde{u}$. The solution $\widetilde{u}(t) = \mathrm{e}^{-\mathrm{i}t}\widetilde{u}(0)$ leaves the circles $|\widetilde{u}(t)|^2 = u_1^2(t) + u_2^2(t) = |\widetilde{u}(0)|^2 = const.$ invariant. Hence, the orbits are circles, and $u^* = (0,0)$ is called a center. The phase portrait is not robust w.r.t. small perturbations. In general after the perturbation we obtain the phase portrait from a2 ii). However, in applications often additional effects such as a conserved quantity enforce the robustness of centers w.r.t. the class of possible perturbations. See for instance Chapter 4.

a2 ii): Let $A = \begin{pmatrix} 1 & 1 \\ -1 & 1 \end{pmatrix}$. For $\widetilde{u} = u_1 + \mathrm{i}u_2$ we obtain the equation $\dot{\widetilde{u}} = (1 - \mathrm{i})\widetilde{u}$ which is solved by $\widetilde{u}(t) = \mathrm{e}^t\mathrm{e}^{-\mathrm{i}t}\widetilde{u}(0)$. In polar coordinates $\widetilde{u}(t) = r(t)\mathrm{e}^{\mathrm{i}\phi(t)}$ with $r(t) \in \mathbb{R}$ and $\phi(t) \in S^1 = \mathbb{R}/(2\pi\mathbb{Z})$ we obtain $\dot{r} = r$ and $\dot{\phi} = -1$ with solution $r(t) = \mathrm{e}^t r(0)$ and $\phi(t) = (\phi(0) - t) \bmod 2\pi$. The orbits are spirals. The phase portrait is robust w.r.t. small perturbations. Here $u^* = (0,0)$ is called an unstable vortex or spiral.

b): Let $A = \begin{pmatrix} 1 & 1 \\ 0 & 1 \end{pmatrix}$, i.e., $\dot{u}_1 = u_1 + u_2$, $\dot{u}_2 = u_2$. For the second equation we obtain $u_2(t) = \mathrm{e}^t u_2(0)$. The variation of constant formula applied to the first equation yields

$$u_1(t) = \mathrm{e}^t u_1(0) + \int_0^t \mathrm{e}^{t-s}\mathrm{e}^s u_2(0)\,\mathrm{d}s = \mathrm{e}^t u_1(0) + \mathrm{e}^t t u_2(0).$$

Here $u^* = (0,0)$ is called a degenerate node. The phase portrait is not generic, since a Jordan block only occurs with probability zero under all matrices. However, it can make sense to keep the Jordan block as starting point of the analysis.

c1): Let $A = \begin{pmatrix} 1 & 0 \\ 1 & 0 \end{pmatrix}$. The general solution is $u_1(t) = \mathrm{e}^t u_1(0)$, $u_2(t) = u_2(0) + \mathrm{e}^t u_1(0)$, and there is the line of fixed points $u_1 = 0$.

c2): Let $A = \begin{pmatrix} 0 & 1 \\ 0 & 0 \end{pmatrix}$. The general solution is $u_1(t) = u_1(0) + u_2(0)t$, $u_2(t) = u_2(0)$, i.e., the flow is parallel to the line of fixed points $u_2 = 0$.

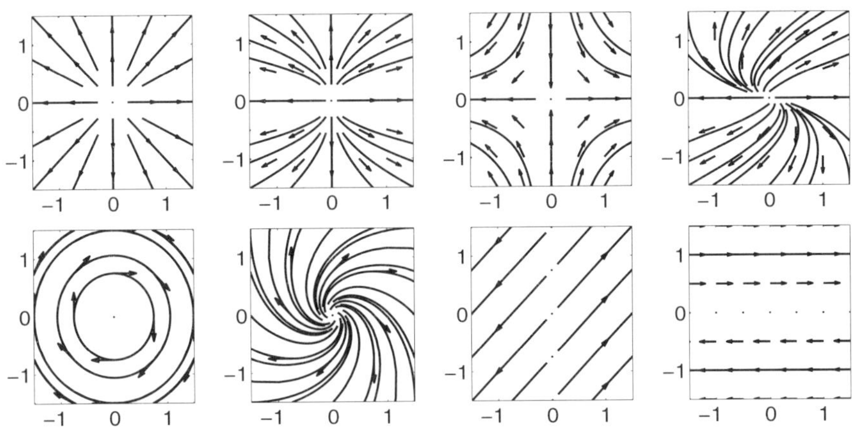

Figure 2.1. Phase portraits for a1i), a1ii), a1iii), and b) in the first row, for a2i), a2ii), c1) and c2) in the second row.

2.1.6. Linear systems with periodic coefficients.

Equations

$$\dot{u}(t) = A(t), u(t) \quad \text{with} \quad A(t) = A(t+T), \tag{2.11}$$

for a fixed $T > 0$, appear for instance as linearizations around time-periodic solutions of nonlinear systems. Hence, they will play an important role in the following. In contrast to the case of t-independent matrices A, where an arbitrary shift of time still gives the same system, in case of time-periodic $A = A(t)$ only integer multiples of the basic period T can be allowed.

Lemma 2.1.16. *With $u(t, t_0, u_0)$ also $u(t + nT, t_0 + nT, u_0)$ with $n \in \mathbb{N}$ solves* (2.11).

The following theorem is fundamental.

Theorem 2.1.17. ***(Floquet)****Each fundamental matrix $\phi(t)$ can be written as a product*

$$\phi(t) = P(t)\mathrm{e}^{tB}$$

of two $d \times d$-matrices, with $P(t) = P(t+T)$ and B a constant $d \times d$-matrix.

Proof. Since

$$\dot{\phi}(t+T) = A(t+T)\phi(t+T) = A(t)\phi(t+T)$$

with $\phi(t)$ also $\phi(t+T)$ is a fundamental matrix. Hence, there exists an invertible $d \times d$-matrix C, such that

$$\phi(t+T) = \phi(t)C.$$

Each invertible $d \times d$-matrix C can be written as $C = \mathrm{e}^{TB}$ with B a non-unique $d \times d$-matrix. As example consider $C = \mathrm{diag}(\lambda_1, \dots, \lambda_d)$ with $\lambda_j > 0$. Then a logarithm is given by $B = \mathrm{diag}(\ln \lambda_1, \dots, \ln \lambda_d)$. For the general case use $-1 = \mathrm{e}^{\mathrm{i}\pi}$ and the expansion of $\ln(1+\lambda)$ in case of Jordan blocks. See Exercise 2.5. For $P(t) = \phi(t)\mathrm{e}^{-tB}$ we obtain

$$P(t+T) = \phi(t+T)\mathrm{e}^{-(t+T)B} = \phi(t)C\mathrm{e}^{-TB}\mathrm{e}^{-tB} = \phi(t)\mathrm{e}^{-\mathrm{tB}} = \mathrm{P(t)}. \qquad \square$$

Definition 2.1.18. *The matrix $C = \mathrm{e}^{TB}$ is called monodromy matrix. The eigenvalues of C are called Floquet multipliers, and the eigenvalues of B are called Floquet exponents.*

Floquet exponents from a different matrix B differ only by adding integer multiples of $2\pi\mathrm{i}/T$. The Floquet multipliers are unique. Suppose that two fundamental matrices $\phi = \phi(t)$ and $\psi = \psi(t)$ are given. By Remark 2.1.11 then $\psi^{-1}(t)\phi(t) = S$ is independent of time and hence

$$C_\phi = \phi(t)^{-1}\phi(t+T) = S^{-1}\psi(t)^{-1}\psi(t+T)S = S^{-1}C_\psi S.$$

As a consequence the matrices C_ϕ and C_ψ have the same eigenvalues. The T-periodic transformation $u(t) = P(t)y(t)$ gives

$$P(t)\dot{y}(t) + \dot{P}(t)y(t) = \dot{u}(t) = A(t)u(t) = A(t)P(t)y(t)$$

and thus

$$\begin{aligned}
\dot{y}(t) =& P(t)^{-1}(A(t)P(t) - \dot{P}(t))y(t) \\
=& P(t)^{-1}(A(t)P(t) - \dot{\phi}(t)\mathrm{e}^{-tB} - \phi(t)(-B)\mathrm{e}^{-tB})y(t) \\
=& P(t)^{-1}(A(t)P(t) - A(t)\phi(t)\mathrm{e}^{-tB} + \phi(t)\mathrm{e}^{-tB}B)y(t) \\
=& P(t)^{-1}(A(t)P(t) - A(t)P(t) + P(t)B)y(t) = By(t).
\end{aligned}$$

For the stability of $u = 0$, it is therefore sufficient to consider B. If all eigenvalues λ of B satisfy $\mathrm{Re}\,\lambda < 0$, we have the asymptotic stability of $u = 0$, see Theorem 2.1.19.

Alternatively, by Lemma 2.1.16 for all $n \in \mathbb{N}$ and $\tau \in [0,T)$, we have

$$\begin{aligned}u(t,0,u_0) =&u(nT+\tau,0,u_0) = u(nT+\tau,nT,u(nT,0,u_0))\\ =&u(nT+\tau,nT,u(nT,(n-1)T,u((n-1)T,0,u_0)))\\ =&u(\tau,0,u(T,0,u(T,0,\ldots,u(T,0,u_0))\ldots)))\\ =&\Phi_\tau\circ\Phi_T\circ\ldots\circ\Phi_T(u_0),\end{aligned}$$

where $\Phi_t u_0 = u(t,0,u_0)$. Since $\tau \in [0,T)$ for the long-time dynamics only the iteration of the time T-map

$$\Phi_T = \phi(T)\phi(0)^{-1} = \left(P(T)e^{TB}\right)\left(P(0)e^{0B}\right)^{-1} = P(0)C_\phi P(0)^{-1},$$

is of interest, where we used $P(T) = P(0)$ and $e^{0B} = I$. The proof of the following theorem is an easy exercise.

Theorem 2.1.19. *In a discrete dynamical system $u_{n+1} = Cu_n$ we have:*

a) If all eigenvalues μ of C satisfy the condition $|\mu| < 1$, then $u = 0$ is asymptotically stable, i.e., $\lim_{n\to\infty} u(n,u_0) = 0$.

b) If C has an eigenvalue μ with $|\mu| > 1$, or a non-trivial Jordan block to an eigenvalue with $|\mu| = 1$, then $u = 0$ is unstable.

Figure 2.2. The eigenvalues of C in cases a) and b) of Theorem 2.1.19.

Example 2.1.20. Consider the iteration $x_{n+1} = Cx_n$, with $C = \begin{pmatrix} 2 & -1 \\ 0 & 1 \end{pmatrix}$. The solution can be computed explicitly by the transformation $x = Sy$, with $S = \begin{pmatrix} 1 & 1 \\ 0 & 1 \end{pmatrix}$. We find $x_{n+1} = SB^{n+1}S^{-1}x_0$, with $B = \begin{pmatrix} 2 & 0 \\ 0 & 1 \end{pmatrix}$. ⌋

Example 2.1.21. Consider the 1-periodic ODE $\dot{u}(t) = \cos^2(2\pi t)u(t)$ for $u(t) \in \mathbb{R}$. Using $\cos^2(2\pi t) = 1/2 + \cos(4\pi t)/2$ the solution with initial condition $u(0) = u_0$ is given by

$$u(t,0,u_0) = u_0 \exp\left(\frac{t}{2} + \frac{\sin 4\pi t}{8\pi}\right)$$

and therefore

$$P(t) = \exp\left(\frac{\sin 4\pi t}{8\pi}\right) \quad \text{and} \quad e^{tB} = \exp\left(\frac{t}{2}\right).$$

We find the Floquet multiplier $e^{1/2}$ and the Floquet exponent $1/2$. The time-one-map is given through $\Phi_1 u_0 = e^{1/2} u_0$. ⌋

The following example [**MY60**] shows that in the periodic case the eigenvalues of the matrix $A(t)$ have no significance for the stability of $u = 0$.

Example 2.1.22. Consider $\dot{u} = A(t)u$ with

$$A(t) = \begin{pmatrix} -1 + \frac{3}{2}\cos^2 t & 1 - \frac{3}{2}\sin t \cos t \\ -1 - \frac{3}{2}\sin t \cos t & -1 + \frac{3}{2}\sin^2 t \end{pmatrix}.$$

The characteristic polynomial is given by

$$\begin{aligned}
&\left(1 - \frac{3}{2}\cos^2 t + \lambda\right)\left(1 - \frac{3}{2}\sin^2 t + \lambda\right) - \left(1 - \frac{3}{2}\sin t \cos t\right)\left(1 + \frac{3}{2}\sin t \cos t\right) \\
&= \lambda^2 + 2\lambda - \frac{3}{2}\left(\cos^2 t + \sin^2 t\right)\lambda + 1 - \frac{3}{2}\left(\cos^2 t + \sin^2 t\right) \\
&\quad + \frac{9}{4}\cos^2 t \sin^2 t + 1 - \frac{9}{4}\cos^2 t \sin^2 t \\
&= \lambda^2 + \frac{1}{2}\lambda + \frac{1}{2},
\end{aligned}$$

i.e., the eigenvalues are independent of t and are given by $\lambda_{1,2} = (-1 \pm i\sqrt{7})/4$. Therefore, we expect $u = 0$ to be stable, but there is the solution

$$\begin{pmatrix} -\cos t \\ \sin t \end{pmatrix} e^{t/2},$$

which is unbounded for $t \to \infty$. With the help of this solution the 2×2-system of ODEs can be reduced to a scalar equation which can be solved with the method of separation of variables. Hence, the Floquet exponents can be computed explicitly. They are given by $\lambda_1 = \frac{1}{2}$ and $\lambda_2 = -1$. ⌋

2.1.7. An outlook on amplitude equations. We close this review of linear ODE theory with a first glimpse at what will be one of the main subjects of this book, namely reduction methods and amplitude (and modulation) equations. Consider the weakly damped linear oscillator

$$\ddot{u} + 2\varepsilon\dot{u} + u = 0, \qquad u(0) = a, \quad \dot{u}(0) = 0, \tag{2.12}$$

with $u(t) \in \mathbb{R}$ and $0 < \varepsilon \ll 1$. The explicit solution is

$$u(t) = e^{-\varepsilon t}\left(a\cos(\omega t) + \frac{\varepsilon a}{\omega}\sin(\omega t)\right), \text{ where } \omega = \sqrt{1 - \varepsilon^2},$$

cf. Remark 2.1.12. However, we might also try an expansion w.r.t. ε, i.e., $u(t) = u_0(t) + \varepsilon u_1(t) + \mathcal{O}(\varepsilon^2)$. Plugging this ansatz into (2.12) and sorting w.r.t. powers in ε yields

$$\begin{aligned}
\mathcal{O}(\varepsilon^0): &\quad \ddot{u}_0(0) + u_0 = 0, \quad u_0(0) = a, \ \dot{u}_0(0) = 0 \\
&\quad \Rightarrow u_0(t) = a\cos t, \\
\mathcal{O}(\varepsilon^1): &\quad \ddot{u}_1 + u_1 = 2a\sin t, \quad u_1(0) = 0, \ \dot{u}_1(0) = 0 \\
&\quad \Rightarrow u_1(t) = -at\cos t + a\sin t,
\end{aligned}$$

and hence $u_{\mathrm{app}_1}(t) = a\cos t - \varepsilon ta\cos t + \varepsilon a\sin t + \mathcal{O}(\varepsilon^2)$. Comparing with u shows that the expansion only makes sense for $t = \mathcal{O}(1)$, and becomes completely useless after that.

With some physical insight, we may however directly see from (2.12) that $2\varepsilon\partial_t u$ corresponds to a weak damping, and hence we suspect that there are two time scales involved in (2.12). Thus we may try a multi-scale ansatz of the form

$$u(t) = A(\varepsilon t)\mathrm{e}^{\mathrm{i}\omega_0 t} + \text{c.c.}, \tag{2.13}$$

with $\omega_0 \in \mathbb{R}$ an a priori unknown (fast) frequency, and where $A = A(\tau) \in \mathbb{C}$ is a slowly varying (complex valued) amplitude. Then, e.g, $\frac{\mathrm{d}}{\mathrm{d}t}u = (\mathrm{i}\omega_0 + \varepsilon\frac{\mathrm{d}}{\mathrm{d}\tau})A\mathrm{e}^{\mathrm{i}\omega_0 t} + \text{c.c.}$, and plugging into (2.12) we obtain

$$\begin{aligned}
\mathcal{O}(\varepsilon^0): &\quad -\omega_0^2 + 1 = 0, \quad A(0) = a/2 \qquad \Rightarrow \omega_0 = 1, \\
\mathcal{O}(\varepsilon^1): &\quad 0 = -2\mathrm{i}(\frac{\mathrm{d}}{\mathrm{d}\tau}A + A)\mathrm{e}^{\mathrm{i}t} + \text{c.c.}.
\end{aligned}$$

This yields $A(\tau) = \mathrm{e}^{-\tau}A(0)$, and thus

$$u_{\mathrm{app}_2}(t) = A(\tau)\mathrm{e}^{\mathrm{i}t} + \text{c.c.} + \mathcal{O}(\varepsilon) = a\mathrm{e}^{-\varepsilon t}\cos(t) + \mathcal{O}(\varepsilon),$$

which at least is a much better approximation of the true solution than u_{app_1}, see Fig. 2.3.

The equation $\frac{\mathrm{d}}{\mathrm{d}\tau}A = -A$ is called the amplitude equation for the ansatz (2.13) for the system (2.12), and here can be solved explicitly, like the original system. However, already in simple nonlinear ODEs in general neither the original equation nor the amplitude equation can be solved explicitly. Moreover, although the amplitude equation is usually a bit "simpler", this is not the essential characteristic. The main points are that the amplitude equation often falls into some universality class, and that it describes the system on long scales. Thus, if one has to use numerical methods, then the numerical costs are greatly reduced. For instance, in the present example we would reduce the numerical costs by a factor $1/\varepsilon$, e.g., by factor 10 if $\varepsilon = 0.1$. More drastic cost reductions may occur for PDEs, see Part IV of this book.

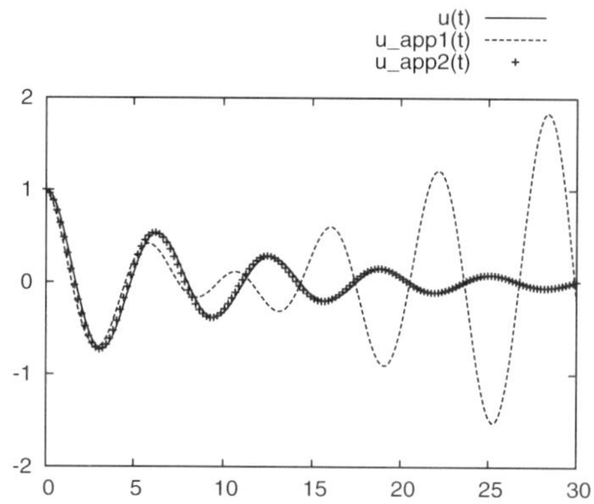

Figure 2.3. Exact solution and the two approximations for (2.12); $\varepsilon = 0.1$, $a = 1$.

2.2. Local existence and uniqueness for nonlinear systems

In this section we prove the local existence and uniqueness of solutions for nonlinear ODEs. We consider

$$\dot{u}(t) = f(u(t), t), \tag{2.14}$$

for an unknown function $u \in C^1(I, \mathbb{R}^d)$, where $I \subset \mathbb{R}$ is an interval and $f : \mathbb{R}^d \times \mathbb{R} \to \mathbb{R}^d$ is called the vector field, which is called autonomous if it does not depend explicitly on time t. An initial value problem consists in finding solutions of (2.14) to the initial condition $u_0 \in \mathbb{R}^d$ at some time t_0, i.e., $u|_{t=t_0} = u_0$. If $d > 1$, then (2.14) is sometimes called a system of ODEs.

Sometimes f is not defined for all $t \in \mathbb{R}$ or for all $u \in \mathbb{R}^d$; the latter is for instance always the case if we consider a planar ODE in polar coordinates $u{=}(r, \phi)$ with hence $r{\geq}0$. However, for a given initial condition u_0 at a time t_0 it is clear that for a local solution it is sufficient that f is only defined in a neighborhood of u_0 and t_0. The modifications needed in the theory below are obvious and thus for notational simplicity we generally assume that f is defined for all $u \in \mathbb{R}^d$ and all $t \in \mathbb{R}$.

For f locally Lipschitz-continuous w.r.t. u and continuous w.r.t. t we have the local existence and uniqueness of solutions. A function $f : \mathbb{R}^d \times I \to \mathbb{R}^d$ with $I \subset \mathbb{R}$ an open interval is called locally Lipschitz-continuous w.r.t. to the first variable if for all C_1 there exists a C_2 such that

$$\max\{\|u\|_{\mathbb{R}^d}, \|v\|_{\mathbb{R}^d}\} \leq C_1 \quad \Rightarrow \quad \sup_{t \in I} \|f(u,t) - f(v,t)\|_{\mathbb{R}^d} \leq C_2 \|u - v\|_{\mathbb{R}^d}.$$

Theorem 2.2.1. (Picard-Lindelöf) *Consider* (2.14) *with initial condition* $u|_{t=t_0} = u_0$ *and let* $f : \mathbb{R}^d \times I \to \mathbb{R}^d$ *be continuous w.r.t.* t *and locally Lipschitz-continuous w.r.t.* u*. For* $C_1 > 0$ *define* $M_0 = \{u \in \mathbb{R}^d : \|u -$

$u_0\|_{\mathbb{R}^d} \leq C_1\}$, $C_3 := \sup_{(u,t)\in M_0\times I} \|f(u,t)\|_{\mathbb{R}^d}$, *and denote the Lipschitz-constant in* M_0 *by* C_2. *Moreover, assume that there exists a* $\delta > 0$ *such that* $I \supset [t_0-\delta, t_0+\delta]$. *Then* (2.14) *has a unique solution* $u \in C^1([t_0-T_0, t_0+T_0], \mathbb{R}^d)$ *satisfying* $u|_{t=t_0} = u_0$, *where* $T_0 = \min(\delta, 1/(2C_2), C_1/C_3)$.

Proof. Similar to the proof of Theorem 2.1.5 we apply the contraction mapping Theorem 2.1.3 to the integrated ODE

$$u(t) = u_0 + \int_{t_0}^{t} f(u(s), s)\,\mathrm{d}s =: F(u)(t), \tag{2.15}$$

where $F : M \to M$ with

$$M = C^0([t_0 - T_0, t_0 + T_0], \{u \in \mathbb{R}^d \ : \ \|u - u_0\|_{\mathbb{R}^d} \leq C_1\})$$

which is equipped with the metric

$$d(u,v) = \sup_{t\in[t_0-T_0,t_0+T_0]} \|u(t) - v(t)\|_{\mathbb{R}^d} =: \|u - v\|_M.$$

We have

$$\begin{aligned}\|F(u) - u_0\|_M &\leq \sup_{t\in[t_0-T_0,t_0+T_0]} \|\int_{t_0}^{t} f(u(s), s)\,\mathrm{d}s\|_{\mathbb{R}^d} \\ &\leq \sup_{t\in[t_0-T_0,t_0+T_0]} |\int_{t_0}^{t} \|f(u(s), s)\|_{\mathbb{R}^d}\,\mathrm{d}s| \leq T_0 C_3 \leq C_1\end{aligned}$$

for $T_0 = \min(\delta, C_1/C_3)$ such that F maps M into M. Moreover,

$$\begin{aligned}\|F(u) - F(v)\|_M \leq & \sup_{t\in[t_0-T_0,t_0+T_0]} \|\int_{t_0}^{t} f(u(s), s) - f(v(s), s)\,\mathrm{d}s\|_{\mathbb{R}^d} \\ \leq & \sup_{t\in[t_0-T_0,t_0+T_0]} |\int_{t_0}^{t} \|f(u(s), s) - f(v(s), s)\|_{\mathbb{R}^d}\,\mathrm{d}s| \\ \leq & \sup_{t\in[t_0-T_0,t_0+T_0]} |\int_{t_0}^{t} C_2\|u(s) - v(s)\|_{\mathbb{R}^d}\,\mathrm{d}s| \\ \leq & T_0 C_2 \|u - v\|_M,\end{aligned}$$

such that F is a contraction for $T_0 = \min(\delta, 1/(2C_2))$. From $u \in M$ it follows that $F(u) \in C^1((t_0 - T_0, t_0 + T_0), \mathbb{R}^d)$. Since u is a fixed point we also have $u = F(u) \in C^1((t_0 - T_0, t_0 + T_0), \mathbb{R}^d)$. □

Remark 2.2.2. The last argument shows that $f \in C^m(\mathbb{R}^d \times I, \mathbb{R}^d)$ implies $u \in C^{m+1}(I, \mathbb{R}^d)$. ⌋

For $f : \mathbb{R}^d \times \mathbb{R} \to \mathbb{R}^d$ locally Lipschitz-continuous the solutions can only stop to exist if $\|u(t)\|_{\mathbb{R}^d}$ becomes infinitely large.

Theorem 2.2.3. *For locally Lipschitz-continuous f the solution u with $u|_{t=t_0} = u_0 \in \mathbb{R}^d$ exists for all $t \in (T_-, T_+)$, where*

$$T_- = \inf\{t \in \mathbb{R} : \|u(t)\|_{\mathbb{R}^d} < \infty\} \quad \text{and} \quad T_+ = \sup\{t \in \mathbb{R} : \|u(t)\|_{\mathbb{R}^d} < \infty\}.$$

Proof. If $\|u(t)\|_{\mathbb{R}^d}$ is finite, then the local existence and uniqueness Theorem 2.2.1 applies and also $\|u(t - T_0)\|_{\mathbb{R}^d}$ and $\|u(t + T_0)\|_{\mathbb{R}^d}$ are finite for some $T_0 > 0$. □

The following two examples show that f being only continuous is not sufficient for uniqueness, and that solutions in general do not exist globally.

Example 2.2.4. Consider the one-dimensional ODE $\dot{u} = \sqrt{|u|}$ with initial value $u|_{t=0} = 0$. The right-hand side is not Lipschitz-continuous at $u = 0$. This initial value problem has the solution $u = 0$, but also infinitely many other solutions, namely

$$u_\tau(t) = \begin{cases} 0, & \text{for} \quad 0 \le t \le \tau \\ (t-\tau)^2/4, & \text{for} \quad \tau \le t, \end{cases}$$

solves the ODE for each $\tau > 0$, i.e., there is no uniqueness of solutions. ⌋

Example 2.2.5. Consider the one-dimensional ODE $\dot{u} = 1 + u^2$ with the initial value $u|_{t=0} = 0$. This initial value problem has the solution $u(t) = \tan t$, i.e., the solution explodes, or blows up, in finite time, it becomes ∞ for $t = \pi/2$. Hence, the solution does not exist for all $t \in [0, \infty)$, i.e., there is no global existence of solutions. ⌋

Our next goal is to prove the continuity of the solutions w.r.t. the initial conditions. In order to do so we use Gronwall's inequality, cf. Lemma 2.1.7.

Lemma 2.2.6. *Let $f : \mathbb{R}^d \times \mathbb{R} \to \mathbb{R}^d$ be continuous and locally Lipschitz-continuous w.r.t. the first variable. Then each solution $u(t, t_0, u_0)$ is Lipschitz-continuous w.r.t. u_0 in the following sense: For every $T_0 > 0$ there exist $\delta > 0$ and $L > 0$ such that for all $u_1 \in \mathbb{R}^d$ with $\|u_0 - u_1\|_{\mathbb{R}^d} \le \delta$ we have, for all $t \in [t_0 - T_0, t_0 + T_0]$,*

$$\|u(t, t_0, u_0) - u(t, t_0, u_1)\|_{\mathbb{R}^d} \le L\|u_0 - u_1\|_{\mathbb{R}^d}. \tag{2.16}$$

Proof. We have

$$\begin{aligned} &\|u(t, t_0, u_0) - u(t, t_0, u_1)\|_{\mathbb{R}^d} \\ &\qquad \le \|u_0 - u_1\|_{\mathbb{R}^d} + \int_{t_0}^t \|f(u(\tau, t_0, u_0), \tau) - f(u(\tau, t_0, u_1), \tau)\|_{\mathbb{R}^d} \, \mathrm{d}\tau \\ &\qquad \le \|u_0 - u_1\|_{\mathbb{R}^d} + |C_2 \int_{t_0}^t \|u(\tau, t_0, u_0) - u(\tau, t_0, u_1)\|_{\mathbb{R}^d} \, \mathrm{d}\tau|. \end{aligned}$$

Hence, by Gronwall's inequality we find

$$\|u(t, t_0, u_0) - u(t, t_0, u_1)\|_{\mathbb{R}^d} \le \|u_0 - u_1\|_{\mathbb{R}^d} \mathrm{e}^{C_2|t-t_0|}.$$

□

Remark 2.2.7. For continuously differentiable f we have differentiability of $u(t, t_0, u_0)$ w.r.t. the data t_0, u_0, see, e.g., [**HSD04**, Page 402]. ⌋

Another application of Gronwall's inequality is the proof of bounds on the size of the solutions. We state it in a differential form and then present a simple, but fundamental, example.

Lemma 2.2.8. (Gronwall) *Let $I \subset \mathbb{R}$ be an interval, $\alpha, \beta \in \mathbb{R}$, and $\phi \in C^1(I, \mathbb{R})$ a non-negative function with*

$$\dot{\phi}(t) \leq \alpha + \beta\phi(t)$$

for all $t \in I$. Then, for all $t_0, t \in I$, $t \geq t_0$ we have

$$\phi(t) \leq \phi(t_0)\mathrm{e}^{\beta(t-t_0)} + \frac{\alpha}{\beta}\left(\mathrm{e}^{\beta(t-t_0)} - 1\right).$$

Proof. We introduce $\psi(t) = \phi(t)\mathrm{e}^{\beta t}$ which satisfies $\dot{\psi}(t) \leq \alpha\mathrm{e}^{-\beta t}$. Integration yields $\psi(t) \leq \psi(t_0) + \frac{\alpha}{\beta}(\mathrm{e}^{-\beta t_0} - \mathrm{e}^{-\beta t})$. Undoing the transformation gives the result. □

Example 2.2.9. Consider $\dot{u} = u - u^3$. For $\phi(t) = u^2(t)$ we obtain

$$\dot{\phi}(t) = 2u(t)\dot{u}(t) = 2u^2(t) - 2u^4(t) \leq 2 - 2u^2(t) = 2 - 2\phi(t).$$

Therefore, from Lemma 2.2.8

$$u^2(t) \leq u^2(0)\mathrm{e}^{-2t} + 2(1 - \mathrm{e}^{-2t})/2 \searrow 1 \text{ as } t \to \infty.$$

Thus, every solution exists globally (in forward time) since it stays bounded and enters for instance the interval $[-2, 2]$. ⌋

As long as they exist, solutions of ODEs (2.14) have the trivial, but fundamental property

$$u(t + s, t_0, u_0) = u(t, s, u(s, t_0, u_0)), \quad u(t_0, t_0, u_0) = u_0. \tag{2.17}$$

For autonomous systems we have

$$u(t, t_0, u_0) = u(t - t_0, 0, u_0) =: u(t - t_0, u_0),$$

i.e., w.l.o.g. we can always choose the initial time $t_0 = 0$. Then (2.17) transfers into

$$u(t + s, u_0) = u(t, u(s, u_0)), \quad u(0, u_0) = u_0. \tag{2.18}$$

Thus, it makes no difference whether we solve the ODE until the time $t+s$, or if we solve the ODE until the time s, start again, and solve until the time $t + s$. A similar structure occurs for iterations $u_{n+1} = f(u_n)$. In the following we focus on the autonomous case.

Definition 2.2.10. *A map $u : I \times M \to M$ which satisfies* (2.18), *where $I = \mathbb{R}_+$, $I = \mathbb{R}$, $I = \mathbb{N}$ or $I = \mathbb{Z}$, and where M is a set, is called a dynamical system or flow.*

If $I = \mathbb{R}_+$ or $I = \mathbb{R}$ then the dynamical system is called continuous.

If $I = \mathbb{N}$ or $I = \mathbb{Z}$ it is called discrete.

The set M is called the phase space.

The set $\gamma_+(u_0)=\{u(t, u_0) : t \geq 0\}$ is called the forward orbit through u_0, the set $\gamma_-(u_0) = \{u(t, u_0) : t \leq 0\}$ is called the backward orbit through u_0, and $\gamma(u_0) = \gamma_+(u_0) \cup \gamma_-(u_0)$ is called the orbit through u_0.

Since with this strict definition, only ODEs (2.14) with solutions existing globally forward in time in case $I = \mathbb{R}_+$, respectively, forward and backward in time in case $I = \mathbb{R}$ define continuous dynamical systems in the phase space $\mathbb{R}^d$, we shall not be that strict in the following and call any map u which fulfills (2.18) a dynamical system.

For ODEs (and PDEs) the dynamical systems property (2.18) expressed in terms of the family of (nonlinear) solution operators $(\mathcal{S}_t)_{t\in I}$ is given by

$$\mathcal{S}_{t+s} = \mathcal{S}_t\mathcal{S}_s, \quad \mathcal{S}_0 = \mathrm{I}, \tag{2.19}$$

where $\mathcal{S}_t$ is defined by $\mathcal{S}_t u_0 := u(t, u_0)$, and where I is here the identity on M. Due to (2.19), the family of solution operators $(\mathcal{S}_t)_{t\in I}$ is called a semigroup in case $I = \mathbb{R}^+$.

Remark 2.2.11. Except for the points which are mapped to infinity the map $u_0 \mapsto u(t, u_0)$ is bijective due to $u(t, u(-t, u_0))=u(t - t, u_0)=u_0$ in case $I = \mathbb{R}$ or $I = \mathbb{Z}$. Since additionally $u(t, u_0)$ depends continuously on u_0 dynamical systems can be interpreted as a flow of homeomorphisms, i.e., as flow of bijective bi-continuous maps from $\mathbb{R}^d$ into $\mathbb{R}^d$. If f is C^k then also $u(t, u_0)$ is C^k w.r.t. u_0, i.e., then the dynamical system can be interpreted as a flow of C^k-diffeomorphisms. ⌋

2.3. Special solutions

In this section we introduce special solutions such as fixed points, periodic solutions, and homoclinic and heteroclinic orbits and basic concepts such as stability and instability and invariant manifolds.

2.3.1. Fixed Points. Until further notice we consider the autonomous case. In order to explore the dynamics in phase space we start from the most simple dynamical objects, namely fixed points.

Definition 2.3.1. *A point $u^* \in \mathbb{R}^d$ is called fixed point for the ODE* (2.14) *if $f(u^*) = 0$.*

Example 2.3.2. We consider $\dot{u} = u - u^3$ with $u = u(t) \in \mathbb{R}$. From $f(u) = u - u^3 = 0$ we obtain the fixed points $u_1^* = 0$ and $u_{2,3}^* = \pm 1$. See Figure 2.4. ⌋

Figure 2.4. The phase portrait of $\dot{u} = u - u^3$ drawn in the 1D phase space. In one space dimension, i.e., $u = u(t) \in \mathbb{R}$, the complete qualitative behavior of the dynamics of an autonomous ODE is known with the knowledge of the fixed points due to topological reasons. In between two fixed points the "vector field" $f(u) \in \mathbb{R}$, which is a scalar function, cannot change sign. Therefore, the real line is divided by the fixed points which are connected by so called heteroclinic solutions. In case of an interval of fixed points the statement remains true with obvious modifications. Hence, the dynamics of autonomous one-dimensional ODEs is trivial.

In order to explore the dynamics near a fixed point u^* we write $u = u^* + v$ and make a Taylor expansion of the vector field f around u^*, i.e.,

$$\dot{v} = \frac{\mathrm{d}}{\mathrm{d}t}(u^* + v) = f(u^* + v) = 0 + \partial_u f(u^*)v + \mathcal{O}(\|v\|^2),$$

using $f(u^*) = 0$. The only approximate system which in general can be solved explicitly is the linearization at the fixed point u^*, namely

$$\dot{v} = \partial_u f(u^*)v.$$

In Example 2.3.2 we have $\partial_u f(u_j^*) = 1 - 3(u_j^*)^2$, and hence

$$\dot{v} = v \quad \text{for} \quad u_1^* = 0 \qquad \text{and} \qquad \dot{v} = -2v \quad \text{for} \quad u_{2,3}^* = \pm 1.$$

Therefore, from the linearization we expect that solutions which start close to $u_{2,3}^*$ converge towards $u_{2,3}^*$ for $t \to \infty$, while solutions which start close to u_1^* will leave any small neighborhood of u_1^*.

Definition 2.3.3. *A fixed point u^* is called stable for the ODE* (2.14) *if for any $\varepsilon > 0$ there is a $\delta > 0$ such that $\|u_0 - u^*\|_{\mathbb{R}^d} < \delta$ implies $\|u(t, u_0) - u^*\|_{\mathbb{R}^d} < \varepsilon$ for all $t \geq 0$. Otherwise, it is called unstable. A stable fixed point is called asymptotically stable if $\|u_0 - u^*\|_{\mathbb{R}^d} < \delta$ additionally implies $\lim_{t\to\infty} u(t, u_0) = u^*$.*

For linear systems the statements of Theorem 2.1.14 remain true with the more general Definition 2.3.3. The following theorem guarantees in many situations that stability or instability in the linearized system implies stability or instability for the full system (2.14).

Theorem 2.3.4. *Let u^* be a fixed point for* (2.14). *Let $A = \partial_u f(u^*) \in \mathbb{R}^{d\times d}$ be the linearization of f in u^*.*

a) If all eigenvalues λ_j of A satisfy $\operatorname{Re}\lambda_j < 0$, then u^* is asymptotically stable.

b) If A has an eigenvalue λ with $\operatorname{Re}\lambda > 0$, then u^* is unstable.

Proof. W.l.o.g. let $u^* = 0$. The proof is based in both situations on the fact that the nonlinear terms in a neighborhood of $u^* = 0$ are much smaller than the linear terms.

a) In order to use this fact in a) we use the variation of constant formula. Let e^{tA} be the solution operator of the linear system $\dot{u} = Au$. Since the eigenvalues of A have strictly negative real part there are positive constants μ_0 and C_0 (necessary due to possible Jordan blocks) with

$$\|\mathrm{e}^{tA}\|_{\mathbb{R}^d\to\mathbb{R}^d} \le \|S\mathrm{e}^{tJ}S^{-1}\| \le \|S\|\|S^{-1}\|\|\mathrm{e}^{tJ}\| \le C_0\mathrm{e}^{-\mu_0 t}$$

for all $t \ge 0$ where we used the notation of (2.9). The estimate follows by using $\|S\|\|S^{-1}\| < \infty$ and by estimating $\|\mathrm{e}^{tJ}\|$ in the $\|\cdot\|^*$ matrix norm associated to the $\|\cdot\|_2$ vector norm from page 19. We remark that $-\mu_0$ has to be larger than the largest real part of the eigenvalues of A in case of Jordan blocks. The closer $-\mu_0$ gets to the largest real part, the larger the constant C_0 becomes.

For the nonlinear terms $g(u) = f(u) - Au = \mathcal{O}(\|u\|^2_{\mathbb{R}^d})$ the following holds: For all $b > 0$ there exists a $\delta_0 > 0$ such that $\|u\|_{\mathbb{R}^d} \le \delta_0$ implies $\|g(u)\|_{\mathbb{R}^d} \le b\|u\|_{\mathbb{R}^d}$.

The variation of constant formula, cf. (2.6),

$$u(t) = \mathrm{e}^{tA}u_0 + \int_0^t \mathrm{e}^{(t-s)A}g(u(s))\,\mathrm{d}s$$

then implies

$$\begin{aligned}\|u(t)\|_{\mathbb{R}^d} &\le \|\mathrm{e}^{tA}\|_{\mathbb{R}^d\to\mathbb{R}^d}\|u_0\|_{\mathbb{R}^d} + \int_0^t \|\mathrm{e}^{(t-s)A}\|_{\mathbb{R}^d\to\mathbb{R}^d}\,\|g(u(s))\|_{\mathbb{R}^d}\,\mathrm{d}s\\ &\le C_0\mathrm{e}^{-\mu_0 t}\|u_0\|_{\mathbb{R}^d} + \int_0^t C_0\mathrm{e}^{-\mu_0(t-s)}b\|u(s)\|_{\mathbb{R}^d}\,\mathrm{d}s,\end{aligned}$$

and as a consequence

$$\mathrm{e}^{\mu_0 t}\|u(t)\|_{\mathbb{R}^d} \le C_0\|u_0\|_{\mathbb{R}^d} + \int_0^t C_0\mathrm{e}^{\mu_0 s}b\|u(s)\|_{\mathbb{R}^d}\,\mathrm{d}s.$$

Gronwall's inequality, cf. Lemma 2.2.8, applied to $\mathrm{e}^{\mu_0 t}\|u(t)\|_{\mathbb{R}^d}$ finally implies $\mathrm{e}^{\mu_0 t}\|u(t)\|_{\mathbb{R}^d} \le C_0\|u_0\|_{\mathbb{R}^d}\mathrm{e}^{C_0 bt}$, respectively

$$\|u(t)\|_{\mathbb{R}^d} \le C_0\|u_0\|_{\mathbb{R}^d}\mathrm{e}^{(C_0 b-\mu_0)t}.$$

Choosing $b = \mu_0/(2C_0)$ defines $\delta_0 = \delta_0(b)$. Therefore, with $\mu = \mu_0/2$, we find $\|u(t)\|_{\mathbb{R}^d} \leq C_0\|u_0\|_{\mathbb{R}^d}e^{-\mu t} \to 0$ for $t \to \infty$. Since this additionally implies that for any given $\varepsilon > 0$ we can choose $\delta = \frac{1}{2}\min\{C_0^{-1}\varepsilon, \delta_0\}$ such that $\|u_0\|_{\mathbb{R}^d} < \delta$ implies $\|u(t,u_0)\|_{\mathbb{R}^d} \leq C_0\delta < \varepsilon$, the asymptotic stability of $u^* = 0$ follows.

b) In order to prove b) we show that in the direction of the unstable subspaces there is a sector with radius ε which is entered by the solutions along the sides through the origin and which is left by the solutions along the side opposite to the origin. See Figure 2.5.

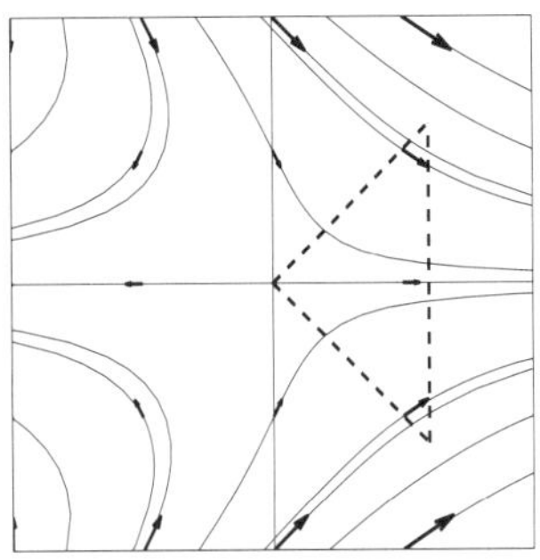

Figure 2.5. The phase portrait and the sector in the unstable case in a typical situation.

We start with a linear change of coordinates such that after the transform the linear part is of the form $A = \begin{pmatrix} A_1 & 0 \\ 0 & A_2 \end{pmatrix}$, where $A_1 \in \mathbb{R}^{k\times k}$ belongs to the part of the spectrum of A with positive real part and $A_2 \in \mathbb{R}^{(d-k)\times(d-k)}$ to the part of the spectrum of A with non-positive real part. Hence, there exists a $\sigma > 0$, such that for all eigenvalues λ_j of A_1 we have $\operatorname{Re}\lambda_j > \sigma$. Moreover, the change of coordinates is made in such a way that the norm of the off-diagonal elements of the transformed matrix A is less than γ. In order to do so, we assume $(\sum_{j=1}^k \sum_{m=1,m\neq j}^k |a_{jm}|^2)^{1/2} \leq \gamma$ for which we further assume $\gamma \leq \sigma/20$. From linear algebra it is known that this can always be achieved by using modified Jordan blocks. By changing the length of the vectors of the basis for instance the Jordan block $\begin{pmatrix} \lambda & 1 \\ 0 & \lambda \end{pmatrix}$ can be changed into $\begin{pmatrix} \lambda & r \\ 0 & \lambda \end{pmatrix}$ for every fixed $r > 0$. Like in a) we find for the nonlinear terms that for all $b > 0$ there exists a $\delta_0 > 0$ such that $\|u\|_{\mathbb{R}^d} \leq \delta_0$ implies $\|g(u)\|_{\mathbb{R}^d} \leq b\|u\|_{\mathbb{R}^d}$.

Next we define $R^2 = \sum_{j=1}^k |u_j|^2$ and $\rho^2 = \sum_{j=k+1}^d |u_j|^2$ and assume that $u^* = 0$ is stable. Then for all $\varepsilon > 0$ there exists a $\delta > 0$, such that $\rho(0) + R(0) < \delta$ implies $\rho(t) + R(t) < \varepsilon$ for all $t \geq 0$. For the transformed system we find

$$
\begin{aligned}
2R\frac{\mathrm{d}}{\mathrm{d}t}R =& \frac{\mathrm{d}}{\mathrm{d}t}(R^2) = \frac{\mathrm{d}}{\mathrm{d}t}\sum_{j=1}^{k}|u_j|^2 = 2\mathrm{Re}\sum_{j=1}^{k}\overline{u}_j\partial_t u_j \\
=& 2\mathrm{Re}\sum_{j=1}^{k}\overline{u}_j(\lambda_j u_j + \sum_{m\neq j} a_{jm}u_m + g_j).
\end{aligned}
$$

Using

$$2\mathrm{Re}\sum_{j=1}^{k}\overline{u}_j\lambda_j u_j \geq 2\sigma\sum_{j=1}^{k}\overline{u}_j u_j = 2\sigma R^2$$

and

$$|2\mathrm{Re}\sum_{j=1}^{k}\overline{u}_j\sum_{m\neq j} a_{jm}u_m| \leq 2(\sum_{j=1}^{k}\sum_{m=1,m\neq j}^{k}|a_{jm}|^2)^{1/2}R^2 \leq 2\gamma R^2,$$

together with

$$|2\mathrm{Re}\sum_{j=1}^{k}\overline{u}_j g_j| \leq 2Rb\sqrt{\rho^2+R^2} \leq 2Rb(\rho+R)$$

yields

$$2R\frac{\mathrm{d}}{\mathrm{d}t}R \geq 2\sigma R^2 - 2\gamma R^2 - 2bR(\rho+R).$$

Choosing $b = \sigma/10$ yields

$$\frac{\mathrm{d}}{\mathrm{d}t}R \geq \sigma R/2 - b\rho.$$

Similarly, we find

$$\frac{\mathrm{d}}{\mathrm{d}t}\rho \leq \sigma\rho/20 + b(\rho+R).$$

where we used $2\mathrm{Re}\sum_{j=k+1}^{d}\overline{u}_j\lambda_j u_j \leq 0$. Since

$$\sigma R/2 - b\rho - \sigma\rho/20 - b(\rho+R) \geq \sigma(R-\rho)/4$$

we finally obtain

$$\frac{\mathrm{d}}{\mathrm{d}t}(R-\rho) \geq \sigma(R-\rho)/4$$

and as consequence

$$R(t)-\rho(t) \geq (R(0)-\rho(0))\mathrm{e}^{\sigma t/4}.$$

For solutions with $R(0) = 2\rho(0)$ it follows that $R(t) \geq \rho(0)\mathrm{e}^{\sigma t/4}$. However, this contradicts the assumption of stability, since $R(t)+\rho(t) \leq \varepsilon$ for all $t \geq 0$ is not possible, independent of how small $\rho(0) > 0$ or $\delta > 0$ has been chosen. □

Example 2.3.5. We consider again the situation from Example 2.3.2. As a consequence of Theorem 2.3.4, the linear stability analysis is sufficient to determine the stability of the fixed points in the nonlinear system. Hence, the fixed point $u_1^* = 0$ is unstable, since the linearization $A = 1 \in \mathbb{R}^{1\times 1}$ has the eigenvalue 1, whereas the fixed points $u_{2,3}^* = \pm 1$ are asymptotically stable, since $A = -2 \in \mathbb{R}^{1\times 1}$ has the eigenvalue -2. ⌋

Example 2.3.6. An example of an ODE with a fixed point in the origin which is stable in the linearized system, but unstable in the full system is given by $\dot{u} = u^3$. ⌋

The following theorem states that near a hyperbolic fixed point the flow can be completely linearized by a change of coordinates h. For a proof we refer to [**Tes12**, Theorem 9.9].

Definition 2.3.7. *A fixed point u^* is called hyperbolic, if the linearization $A = \partial_u f(u^*)$ has no eigenvalues with* $\operatorname{Re}\lambda = 0$.

Theorem 2.3.8. (Hartman-Grobman) *Let u^* be a fixed point for* (2.14), *let $\mathcal{S}_t$ be the flow of* (2.14), *let $A = \partial_u f(u^*) \in \mathbb{R}^{d\times d}$ be the linearization of f in u^*, and assume that A has no eigenvalues with zero real part. Then there exists an homeomorphism h from a neighborhood U of u^* to a neighborhood V of u^* such that for all $u_0 \in U$ there exists an open interval $I_0 \subset \mathbb{R}$, $0 \in I_0$ such that for all $t \in I_0$ we have*

$$h \circ \mathcal{S}_t u_0 = \mathrm{e}^{tA} h(u_0).$$

Thus, h maps trajectories of (2.14) *near u^* to trajectories of the linearization $\dot{y} = Ay$.*

Remark 2.3.9. There is also a discrete version of the Hartman-Grobman theorem: Consider the nonlinear map $u_{n+1} = f(u_n)$ and assume that $f(u^*) = u^*$ and that all eigenvalues λ of linearization $A = \partial_u f(u^*)$ around u^* satisfy $|\lambda| \neq 1$. Then there exists a homeomorphism h in a neighborhood U of u^* such that $h(f(u)) = Ah(u)$ for all $u \in U$. ⌋

It is somewhat surprising that even for analytic f the map h is in general not differentiable; see Exercise 2.9.

2.3.2. Periodic solutions.

The first non-trivial dynamical object is a periodic solution.

Definition 2.3.10. *A solution $u = u(t)$ of the ODE* (2.14) *is called periodic if $u(t+T) = u(t)$ for a $T > 0$ and all $t \in \mathbb{R}$. If moreover, $u(t) \neq u(t+\tau)$ for all $0 < \tau < T$, then T is called the minimal period.*

Example 2.3.11. We consider the two-dimensional ODE

$$\begin{aligned}\dot{u}_1 &= -u_2 + u_1(1 - u_1^2 - u_2^2),\\ \dot{u}_2 &= u_1 + u_2(1 - u_1^2 - u_2^2).\end{aligned}$$

By introducing polar coordinates $u_1 = r\cos(\phi)$, $u_2 = r\sin(\phi)$ we obtain

$$\dot{r} = r - r^3 \quad \text{and} \quad \dot{\phi} = 1.$$

In order to understand the dynamics of this ODE we visualize its flow in the phase plane.

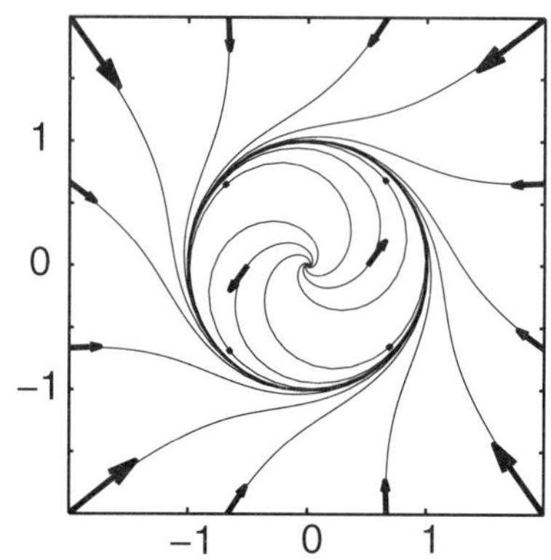

Figure 2.6. Flow for $\dot{u}_1 = -u_2 + u_1(1 - u_1^2 - u_2^2)$ and $\dot{u}_2 = u_1 + u_2(1 - u_1^2 - u_2^2)$.

From the phase portrait we find that all solutions converge towards the circle $r = 1$ which is a periodic solution with the minimal period $T = 2\pi$. Moreover, the origin $r = 0$ is an unstable fixed point in the r equation. As an exercise, we may consider the linearization around $(u_1, u_2) = 0$. We obtain

$$\begin{pmatrix}\dot{u}_1\\ \dot{u}_2\end{pmatrix} = A\begin{pmatrix}u_1\\ u_2\end{pmatrix} \quad \text{with} \quad A = \begin{pmatrix}1 & -1\\ 1 & 1\end{pmatrix},$$

with eigenvalues $\lambda_{1,2} = 1 \pm \mathrm{i}$. Since $\operatorname{Re}\lambda_{1,2} = 1 > 0$ we also have with Theorem 2.3.4 the instability of the origin. On the other hand, from the phase portrait the periodic solution $r = 1$ seems to be asymptotically stable. However, as we see in a moment we have to be more precise when we talk about stability of periodic solutions. ⌋

The stability or instability of non-trivial periodic solutions is a non-trivial task due to the fact that the derivative $\dot{u}_{per}$ of the periodic orbit u_{per} solves the linearization $\dot{v} = Df(u_{per})v$ around the periodic orbit u_{per}. Hence, the linearization possesses a Floquet exponent with real part zero, and so even a generalization of Theorem 2.3.4 to non-autonomous systems would not be applicable for proving stability in the nonlinear system. In order to study stability of periodic solutions we proceed as follows. We introduce a so called Poincaré section, a hyperplane which intersects the periodic orbit transversally. For our example we choose for instance

$$S = \{(x, y) \; : \; u_2 = 0, \; u_1 \in (1/2, 3/2)\}.$$

Transversality means that in the intersection point $u^* = (u_1, u_2) = (1, 0)$ the Poincaré section S and the vector field $f(1, 0) = (0, 1)$ span the complete phase space $\mathbb{R}^2$. Then we define the so called Poincaré map $\Pi : S \to S$ as follows: for $u_0 \in S$ we let $\Pi(u_0)$ be the first intersection point of $t \mapsto u(t, u_0)$ and S for $t > 0$, i.e., in the example $\Pi(u_0) = u(2\pi, u_0)$. As Figure 2.7 illustrates, Poincaré maps to different Poincaré sections are conjugated to each other in the following sense. Let Π_{S_1,S_2} be the map from section S_1 to S_2. Then we have $\Pi_{S_1,S_1} = \Pi_{S_1,S_2} \circ \Pi_{S_2,S_2} \circ \Pi_{S_2,S_1}$.

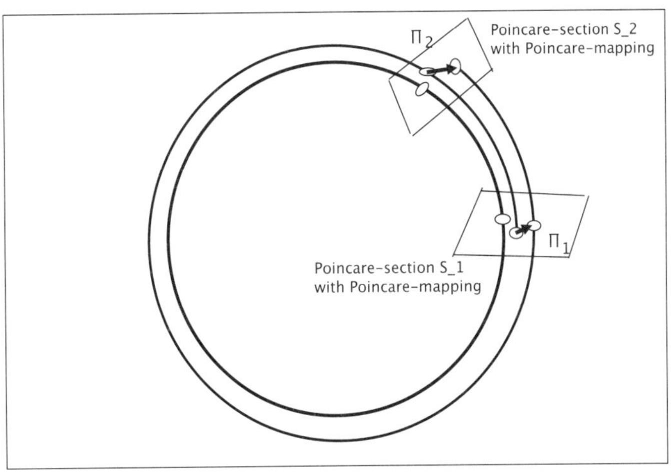

Figure 2.7. Two Poincaré maps to different Poincaré sections are conjugated to each other.

This fact and the fact that the intersection point u^* of the periodic solution is a fixed point of the Poincaré map Π, i.e., $\Pi(u^*) = u^*$, lead to the following definition.

Definition 2.3.12. *a) A fixed point u^* is called stable for the iteration $u_{n+1} = \Pi(u_n)$ with $\Pi : \mathbb{R}^d \to \mathbb{R}^d$ if for any $\varepsilon > 0$ there is a $\delta > 0$ such that $\|u_0 - u^*\|_{\mathbb{R}^d} < \delta$ implies $\|\Pi^n(u_0) - u^*\|_{\mathbb{R}^d} < \varepsilon$ for all $n \in \mathbb{N}$. Otherwise, it is called unstable. A stable fixed point is called asymptotically stable if additionally $\lim_{n\to\infty} \Pi^n(u_0) = u^*$ holds.*
b) A periodic solution for the ODE (2.14) is called stable, unstable, or asymptotically stable if the fixed point of the associated Poincaré map is stable, unstable, or asymptotically stable.

The eigenvalues of the linearization $D\Pi$ of the Poincaré map now play an analogous role as the eigenvalues of the linearization A around a fixed point, cf. Theorem 2.3.4.

Definition 2.3.13. *The eigenvalues of the linearization $D\Pi$ of the Poincaré map are called Floquet multipliers.*

Theorem 2.3.14. *Let u^* be a fixed point of the map $\Pi : \mathbb{R}^{d-1} \to \mathbb{R}^{d-1}$ and let $A = D\Pi \in \mathbb{R}^{d-1 \times d-1}$ be the linearization of Π in u^*.*

a) If all eigenvalues λ_j of $D\Pi$ satisfy $|\lambda_j| < 1$, then u^* is asymptotically stable.

b) If A has an eigenvalue λ with $|\lambda| > 1$, then u^* is unstable.

Proof. The proof goes along the lines of the proof of Theorem 2.3.4. □

Example 2.3.15. In order to prove the stability of the periodic solution $r = 1$ in Example 2.3.11 it remains to compute the Floquet multipliers. Since $\widetilde{r} = r - 1$ satisfies $\dot{\widetilde{r}} = -2\widetilde{r} + \mathcal{O}(\widetilde{r}^2)$ and since $\Pi(u_0) = u(2\pi, u_0)$ for $u_0 \in S$ due to ϕ solving $\dot{\phi} = 1$, we find

$$D\Pi((1,0)) = \mathrm{e}^{-2\cdot 2\pi} \in \mathbb{R}^{1\times 1}.$$

Thus we have one Floquet multiplier with $|\mathrm{e}^{-4\pi}| < 1$, which implies the stability of the periodic solution $r = 1$. ⌋

2.3.3. Homoclinic and heteroclinic solutions. Homoclinic and heteroclinic solutions connect fixed points with themselves or other fixed points. Pulse and front solutions in PDEs correspond to homoclinic and heteroclinic solutions in associated ODEs.

Definition 2.3.16. *A solution $u = u(t)$ of the ODE* (2.14) *is called heteroclinic, if $u_+ \neq u_-$, or homoclinic, if $u_+ = u_-$, connection between the fixed points u_- and u_+ if $\lim_{t\to-\infty} u(t) = u_-$ and $\lim_{t\to\infty} u(t) = u_+$.*

Homoclinic and heteroclinic solutions converge to the fixed points along special sets, namely to the fixed point u_- along the unstable manifold of u_- for $t \to -\infty$, and to the fixed point u_+ along the stable manifold of u_+ for $t \to \infty$.

Definition 2.3.17. *Let u^* be a fixed point of the ODE* (2.14). *The set*

$$W_s = \{u_s \in \mathbb{R}^d : \exists\, \beta > 0 \ : \ \lim_{t\to\infty} \|u(t, u_s) - u^*\|_{\mathbb{R}^d} \mathrm{e}^{\beta t} = 0\}$$

is called the stable manifold of u^*. The set

$$W_u = \{u_u \in \mathbb{R}^d : \exists\, \beta > 0 \ : \ \lim_{t\to-\infty} \|u(t, u_u) - u^*\|_{\mathbb{R}^d} \mathrm{e}^{\beta |t|} = 0\}$$

is called the unstable manifold of u^*.

Example 2.3.18. For $\dot{u}_1 = -u_1$, $\dot{u}_2 = u_2$ we have $W_s = \{(u_1, 0) : u_1 \in \mathbb{R}\}$ and $W_u = \{(0, u_2) : u_2 \in \mathbb{R}\}$. ⌋

The following theorem guarantees that the sets W_s and W_u from Definition 2.3.17 are smooth manifolds and that they are invariant under the flow of the ODE.

Theorem 2.3.19. (Invariant manifolds) *Let u^* be a fixed point of the ODE* (2.14), *$f \in C^k(\mathbb{R}^d, \mathbb{R}^d)$, and let $A = \partial_u f(u^*) \in \mathbb{R}^{d\times d}$ be the linearization of f in u^*. Let*

$$E_s = \operatorname{span}\{\varphi : \varphi \ \textit{ eigenvector of } A \textit{ to eigenvalues } \lambda \textit{ with } \operatorname{Re}\lambda < 0\}$$

be the so called stable subspace and let

$$E_u = \operatorname{span}\{\varphi : \varphi \ \textit{ eigenvector of } A \textit{ to eigenvalues } \lambda \textit{ with } \operatorname{Re}\lambda > 0\}$$

be the so called unstable subspace, where eigenvectors include here in all cases generalized eigenvectors. Then there exists a unique C^k-manifold $W_s = W_s(u^*)$ tangential to the stable subspace E_s, which coincides with the stable manifold from Definition 2.3.17*, and a unique C^k-manifold $W_u = W_u(u^*)$ tangential to the unstable subspace E_u, which coincides with the stable manifold from Definition* 2.3.17*. Moreover, there exists a (non-unique) C^{k-1}-center manifold tangential to the center subspace*

$$E_c = \operatorname{span}\{\varphi : \varphi \textit{ eigenvector of } A \textit{ to eigenvalues } \lambda \textit{ with } \operatorname{Re}\lambda = 0\}.$$

If $f \in C^\infty$, then $W_s, W_u \in C^\infty$. The center manifold W_c can be chosen to be in C^r for all $r < \infty$. All these manifolds are invariant under the flow of the ODE (2.14)*. A set M is called invariant if $u_0 \in M$ implies $u(t, u_0) \in M$ for all $t \in \mathbb{R}$.*

Proof. The lengthy proof of this theorem is well documented in [**Van89**]. A sketch of the proof of the existence of the center manifold can be found in §13.1. □

Example 2.3.20. Consider the equations for the mathematical pendulum without friction, namely

$$\begin{aligned} \dot{u}_1 &= u_2, \\ \dot{u}_2 &= -\sin(u_1), \end{aligned} \tag{2.20}$$

where u_1 is the angle between the pendulum and the vertical axis. We find the fixed points $u^* = (k\pi, 0)$ for $k \in \mathbb{Z}$. The linearization at $(k\pi, 0)$ is given by

$$\frac{\mathrm{d}}{\mathrm{d}t}\begin{pmatrix} v_1 \\ v_2 \end{pmatrix} = \begin{pmatrix} 0 & 1 \\ -\cos(k\pi) & 0 \end{pmatrix}\begin{pmatrix} v_1 \\ v_2 \end{pmatrix},$$

which yields the eigenvalues $\lambda \in \begin{cases} \{-1,1\} & \text{for } k \in 2\mathbb{Z}+1 \\ \{-\mathrm{i},\mathrm{i}\} & \text{for } k \in 2\mathbb{Z} \end{cases}$ implying saddles for odd k and centers for even k. The eigenvectors at the saddles are $\varphi_1 = \begin{pmatrix} 1 \\ 1 \end{pmatrix}$, $\varphi_2 = \begin{pmatrix} 1 \\ -1 \end{pmatrix}$, in the stable direction, and unstable direction, respectively. Physical intuition lets us suspect the existence of, for instance, heteroclinic orbits

$$\gamma_+ = W_s((\pi,0)) \cap W_u((-\pi,0)) \quad \text{and} \quad \gamma_- = W_u((\pi,0)) \cap W_s((-\pi,0)),$$

corresponding to one complete rotation of the pendulum from the unstable upper rest state to itself in infinite time. Since the points $(k\pi, 0)$ and $((k+2)\pi, 0)$ can be identified, γ_-, γ_+ can also be called homoclinic. These two orbits separate $\mathbb{R}^2$ into two domains, a bounded domain inside and an unbounded domain outside the two orbits. We further suspect the domain inside to be filled with periodic orbits corresponding to oscillations of the pendulum with amplitude $< \pi$. See Figure 2.8.

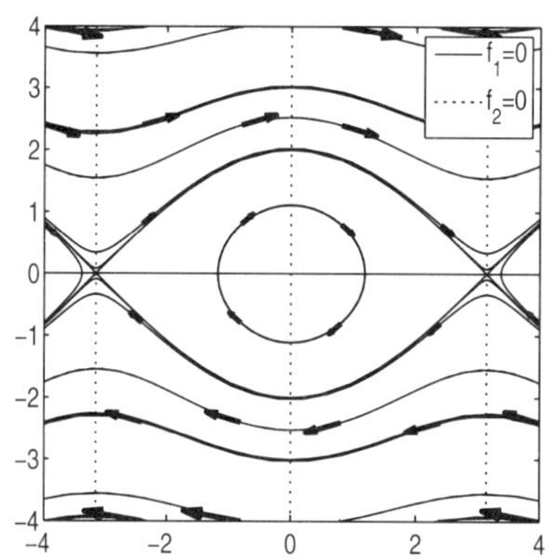

Figure 2.8. Phase portrait for the undamped pendulum.

The whole interior of the "eye" is filled with periodic solutions. Therefore, each of them is stable in the sense of Definition 2.3.12 with Floquet multiplier 1. ⌋

Remark 2.3.21. Instead of "physical intuition" we should rather use the fact that (2.20) is a Hamiltonian system, see also Chapter 4. Here, (2.20) is the first order system belonging to the second order equation

$$\ddot{u} = -\sin(u) = f(u) \tag{2.21}$$

corresponding to Newton's law, namely that the change of momentum equals the acting force. The qualitative behavior of equations such as (2.20) can be obtained by the following procedure independently of the concrete form of f. Multiplying (2.21) by $\dot{u}$ shows that

$$\frac{\mathrm{d}}{\mathrm{d}t}\left(\frac{1}{2}\dot{u}^2 - F(u)\right) = 0, \quad \text{where} \quad F' = f. \tag{2.22}$$

Here $\frac{1}{2}(\dot{u})^2$ and $-F(u)$ are called the kinetic and potential energy, respectively, and (2.22) shows that the total energy $E = \frac{1}{2}(\dot{u})^2 - F(u)$ is conserved. Hence, orbits of (2.20) lie on level sets of E. It turns out that all equations from classical mechanics without friction can be written as Hamiltonian systems in the form

$$\partial_t \begin{pmatrix} q \\ p \end{pmatrix} = J\nabla H(q,p), \quad \text{where} \quad J = \begin{pmatrix} 0 & \mathrm{I} \\ -\mathrm{I} & 0 \end{pmatrix} \tag{2.23}$$

is skew symmetric, where $q \in \mathbb{R}^d$ and $p \in \mathbb{R}^d$ are the position and the momentum coordinates, and where $H : \mathbb{R}^{2d} \to \mathbb{R}$ is called the Hamiltonian, see Chapter 4. Also PDEs can have a Hamiltonian structure, see for instance §8.1 and §8.2. ⌋

Stable, center, and unstable manifolds can also be generalized from fixed points to more complicated objects, for instance to periodic solutions. They exist for discrete dynamical systems, too.

Remark 2.3.22. Let u^* be a fixed point of the iteration $u_{n+1} = \Pi u_n$. The set

$$W_s = \{u_s \in \mathbb{R}^d : \exists\, \beta > 0 \ : \ \lim_{n\to\infty} \|\Pi^n(u_s) - u^*\|_{\mathbb{R}^d} \mathrm{e}^{\beta n} = 0\}$$

is called the stable manifold of u^*. The set

$$W_u = \{u_u \in \mathbb{R}^d : \exists\, \beta > 0 \ : \ \lim_{n\to -\infty} \|\Pi^n(u_u) - u^*\|_{\mathbb{R}^d} \mathrm{e}^{\beta |n|} = 0\}$$

is called the unstable manifold of u^*. They exist as smooth invariant manifolds with similar properties as the ones explained in Theorem 2.3.19. ⌋

2.4. ω-limit sets and attractors

In this section we are interested in objects which describe the dynamics for $t \to \infty$. These are so called ω-limit sets and attractors. We characterize them for two-dimensional autonomous systems and gradient systems. The concepts of this section will later on be applied in Part II of this book to PDEs on spatially bounded domains, which very often can be written as countably infinite-dimensional dynamical systems. Therefore, throughout this subsection, we consider a general dynamical system $X \ni u_0 \mapsto \mathcal{S}_t u_0$, where X is some possibly infinite-dimensional Banach space. The theory has to be modified in Part III and Part IV where PDEs on unbounded domains will be handled. For simplicity, the reader may think of $\mathcal{S}_t$ as being defined by the solutions of some ODE $\dot{u} = f(u)$ in the phase space $X = \mathbb{R}^d$. This section follows rather closely [**Rob01**, §10], including a number of examples.

2.4.1. ω-limit sets. Given some initial condition u_0 for a dynamical system $\mathcal{S}_t u_0 = u(t, u_0)$ on X, the behavior of the solution for $t \to \infty$ is described by the ω-limit set, defined by

$$\omega(u_0) = \{v \in X : \exists\ (t_n)_{n\in\mathbb{N}} \text{ with } t_n \to \infty \text{ and } \lim_{n\to\infty} u(t_n, u_0) = v\}. \tag{2.24}$$

Thus, the ω-limit set of u_0 consists of all limit points of the forward orbit through u_0. Hence, an equivalent characterization is

$$\omega(u_0) = \cap_{t\geq 0}\overline{\cup_{s\geq t} u(s, u_0)}. \tag{2.25}$$

If $\gamma_+ = \gamma_+(u_0)$ is the (forward) orbit through u_0 and $v \in \gamma_+$ then $\omega(v) = \omega(u_0)$ such that we also write $\omega(\gamma_+) := \omega(u_0)$.

Theorem 2.4.1. *The set $\omega(\gamma_+)$ is closed and invariant. If $X = \mathbb{R}^d$ and γ_+ is bounded, then $\omega(\gamma_+)$ is compact, connected and non-empty. In general, if $\overline{\cup_{t\geq t_0} \mathcal{S}_t u_0}$ is compact for some $t_0 \geq 0$, then $\omega(\gamma_+)$ is compact, connected and non-empty.*

Proof. We first consider the case $X = \mathbb{R}^d$.

a) $\omega(\gamma_+)$ as set of limit points is closed.

b) Next, we prove the invariance. Let $p \in \omega(\gamma_+)$. Then there exists a sequence $t_n \to \infty$ such that $\lim_{t_n\to\infty} u(t_n) = p$. We have to prove that $u(t, p) \in \omega(\gamma_+)$. Since $u(t + t_n, x_0) = u(t, u(t_n, x_0))$ it follows for $n \to \infty$ that

$$u(t + t_n, x_0) \to u(t, p),$$

which implies that $\gamma(p) \subset \omega(\gamma_+)$.

c) With γ_+ bounded, $\omega(\gamma_+)$ is bounded. Since $\omega(\gamma_+)$ is closed by a), compactness follows.

d) Suppose that γ_+ consists of more than one point, i.e., γ_+ is not a fixed point. Then γ_+ consists of infinitely many points. Hence, there exists at least one limit point p of the bounded set γ_+. Suppose that γ_+ consists only of a fixed point. Then $\gamma_+ = \omega(\gamma_+)$ is also non-empty.

e) Suppose that $\omega(\gamma_+)$ is not connected, i.e., there exist closed sets A_1 and A_2 satisfying $\omega(\gamma_+) = A_1 \cup A_2$ and $A_1 \cap A_2 = \emptyset$. Since γ_+ is bounded, there exists a $R > 0$, such that $\gamma_+ \subset B_R(0) = \{x \in \mathbb{R}^d : \|x\| \leq R\}$. Let $\delta > 0$ be the distance between A_1 and A_2. Define

$$A_3 = \{u \in B_R(0) : \delta/4 \leq \operatorname{dist}(u, \omega(\gamma_+))\}$$

where $\operatorname{dist}(u, A) = \inf_{a\in A} \|u - a\|$. Obviously the solution must pass A_3 infinitely many times, and hence there must be a limit point in A_3 which contradicts the assumption $\omega(\gamma) = A_1 \cup A_2$.

In the general case, i.e., with X some Banach space, the proof works the same way, with $\omega(\gamma_+)$ compact as a closed subset of the compact set $\overline{\cup_{t\geq t_0}\mathcal{S}_t u_0}$. Note that this compactness argument is used in d) and e). $\square$

2.4.2. Attractors. Attractors are compact sets describing the asymptotic dynamics of the system in the limit $t \to \infty$. They exist for so called dissipative systems.

Definition 2.4.2. *The flow $\mathcal{S}_t : X \to X$ is called dissipative if there exists a compact set B such that for any bounded set $M \subset X$ there exists a $t_0 = t_0(M)$ such that $\mathcal{S}_t M \subset B$ for all $t \geq t_0$. The set B is then called absorbing.*

The goal is to define the so called global attractor $\mathcal{A}$ which contains as much information as possible about the asymptotic behavior for $t \to \infty$. If the system is dissipative and thus has a compact absorbing set B, a first idea would be to take $\cup_{u\in B}\omega(u)$. However, this set in general does not contain homo- or heteroclinic connections, which we already know to be relevant for the asymptotic dynamics. Therefore, a better choice turns out to be

$$\begin{aligned} \mathcal{A} = \omega(B) \;=\; & \{v \in X : \exists\, t_n \to \infty,\ \exists\, (u_n)_{n\in\mathbb{N}} \subset B, \\ & \text{with } \lim_{n\to\infty} u(t_n, u_n) = v\}. \end{aligned} \tag{2.26}$$

Definition 2.4.3. *A set $\mathcal{A} \subset X$ is called an attractor of the dynamical system $\mathcal{S}_t : X \to X$ if*

(i) $\mathcal{A}$ is compact and invariant;

(ii) there is a neighborhood U of $\mathcal{A}$ such that $\mathcal{A}$ attracts U.

The basin of attraction of $\mathcal{A}$ is defined as

$$B(\mathcal{A}) = \{u \in X : \operatorname{dist}(\mathcal{S}_t u, \mathcal{A}) \to 0 \text{ as } t \to \infty\},$$

where $\operatorname{dist}(A, B) = \sup_{a\in A} \inf_{b\in B} \|a - b\|$. *A set $\mathcal{A} \subset X$ is called the global attractor for $\mathcal{S}_t$ if additionally*

(iii) $\mathcal{A}$ attracts all points in X, i.e., we have $B(\mathcal{A}) = X$.

Theorem 2.4.4. *If the flow $\mathcal{S}_t$ is dissipative and if B is a compact absorbing set, then $\mathcal{A} = \omega(B)$ is the unique global attractor. Moreover, $\mathcal{A}$ is connected, the maximal compact invariant set, and the minimal set that attracts all bounded sets.*

Proof. The fact that $\omega(B)$ is nonempty, compact, invariant and connected follows as in Theorem 2.4.1. To show that $\mathcal{A}$ is the maximal compact invariant set, let Y be a compact and invariant set. Then $\mathcal{S}_t Y = Y$ and, since B is absorbing, $\mathcal{S}_t Y \subset B$ for $t \geq t_0$, hence $Y \subset B$ and therefore $\omega(Y) = Y \subset \mathcal{A} = \omega(B)$. This shows that $\mathcal{A}$ is the maximal compact invariant set.

Next we show that $\mathcal{A}$ attracts all bounded sets. Suppose that this is not the case, then there is a bounded set Y, a $\delta > 0$, and a sequence $t_n \to \infty$ with

$$\operatorname{dist}(\mathcal{S}_{t_n} Y, \mathcal{A}) \geq \delta.$$

Thus, there are $u_n \in Y$ with $\operatorname{dist}(\mathcal{S}_{t_n} u_n, \mathcal{A}) \geq \delta/2$. Since Y is bounded and B absorbing we have $\mathcal{S}_{t_n} u_n \in B$ for n large enough, and as B is compact there is a subsequence with

$$\mathcal{S}_{t_{n_j}} u_{n_j} \to v \in B \quad \text{and} \quad \operatorname{dist}(v, \mathcal{A}) \geq \delta/2. \tag{2.27}$$

However, with $v_j = \mathcal{S}_{t_0} u_{n_j} \in B$ we have

$$v = \lim_{j\to\infty} \mathcal{S}_{t_{n_j}} u_{n_j} = \lim_{j\to\infty} \mathcal{S}_{t_{n_j}-t_0} \mathcal{S}_{t_0} u_{n_j}$$

and hence $v \in \mathcal{A}$, which contradicts (2.27). Obviously, $\mathcal{A}$ is also the minimal set that attracts all bounded sets since $\mathcal{S}_t \mathcal{A} = \mathcal{A}$. □

Example 2.4.5. Consider $\dot{u} = u - u^3$. The attractor is given by $\mathcal{A} = [-1, 1]$ and contains the heteroclinic connections between 0 and ± 1. Every set $[-1-\delta, 1+\delta]$ with $\delta > 0$ is an absorbing set. ⌋

2.4.3. Shadowing and upper-semicontinuity of attractors. Clearly, an important issue is the relation of the flow $\mathcal{S}_t$ in X to that on the attractor. The following theorem roughly says that given an initial condition in X, there exists a $\tau > 0$ such that after the time τ the flow can be approximated by a flow on the attractor for some finite time.

Theorem 2.4.6. *Let $\mathcal{A}$ be the global attractor for the flow $\mathcal{S}_t$ and let $u_0 \in X$. For all $\varepsilon > 0$ and $T > 0$ there exists a $\tau = \tau(\varepsilon, T) > 0$ and a point $v_0 \in \mathcal{A}$ such that*

$$\|\mathcal{S}_{\tau+t} u_0 - \mathcal{S}_t v_0\|_X \leq \varepsilon \quad \textit{for all} \quad 0 \leq t \leq T.$$

Proof. From the continuous dependence on the initial conditions for given $\varepsilon, T > 0$ there exists a $\delta = \delta(\varepsilon, T)$ such that

$$\|u_1 - v_0\| \leq \delta \quad \Rightarrow \quad \|\mathcal{S}_t u_1 - \mathcal{S}_t v_0\| \leq \varepsilon \quad \text{for } t \in [0, T].$$

Since $\mathcal{A}$ is the global attractor, for any $u_0 \in X$ and every $\delta > 0$ there exists a time τ and a $v_0 \in \mathcal{A}$ such that $\|u_1 - v_0\| \leq \delta$ where $u_1 = S(\tau) u_0$. □

An approximation of a solution $(\mathcal{S}_t u_0)_{t \geq \tau}$ by a single solution in the attractor cannot be expected in general. However, solutions can be approximated (or shadowed) by so called pseudo-orbits in the attractor. Moreover, due to the attractivity property of $\mathcal{A}$ the approximation becomes better and better for larger and larger times.

Corollary 2.4.7. (Shadowing) *For all $u_0 \in X$ there exists a sequence $(\varepsilon_n)_{n\in\mathbb{N}}$ of errors $\varepsilon_n > 0$ with $\varepsilon_n \to 0$, an increasing sequence $(t_n)_{n\in\mathbb{N}}$ of times with $t_{n+1} - t_n \to \infty$ for $n \to \infty$, and a sequence $(v_n)_{n\in\mathbb{N}}$ of points $v_n \in \mathcal{A}$ such that*

$$\|\mathcal{S}_t u_0 - \mathcal{S}_{t-t_n} v_n\|_X \le \varepsilon_n \quad \text{for all } t_n \le t \le t_{n+1},$$

and $\|v_{n+1} - \mathcal{S}_{t_{n+1}-t_n} v_n\|_X \to 0$ for $n \to \infty$.

However, in general a flow $\mathcal{S}_t u_0$ cannot be approximated by a single flow on the attractor as $t \to \infty$.

Example 2.4.8. For $(x, y, z) \in \mathbb{R}^3$ consider

$$\dot{x} = z(x+y) + x - xr^2, \quad \dot{y} = z(-x+y) + y - yr^2, \quad \dot{z} = -z|z|, \tag{2.28}$$

where $r = (x^2 + y^2)^{1/2}$ or equivalently in polar coordinates

$$\dot{r} = r - r^3, \quad \dot{\phi} = -z, \quad \dot{z} = -|z|z.$$

The global attractor is given by $\mathcal{A} = \{(x, y, z) \in \mathbb{R}^3 : z = 0, x^2 + y^2 \le 1\}$, and the dynamics on $\mathcal{A}$ is given by $\dot{r} = r - r^3$ and $\dot{\phi} = 0$. Hence, the attractor consists of the origin which is a fixed point, the circle of fixed points $\mathcal{S}^1 = \{x^2 + y^2 = 1\}$, and the radial heteroclinic connections between the origin and the points on $\mathcal{S}^1$.

However, given $z_0 \neq 0$ we obtain $z(t) = z_0/(1 + |z_0|t)$ and hence $\phi(t) = \phi_0 - \operatorname{sgn}(z_0)\ln(1 + |z_0|t)$. Thus, the solution converges (algebraically slow) to $\mathcal{A}$ but it does not converge to some particular solution on $\mathcal{A}$. It can only be approximated by a sequence of solutions (i.e., fixed points on $\mathcal{S}$), with smaller and smaller errors on longer and longer time intervals. ⌋

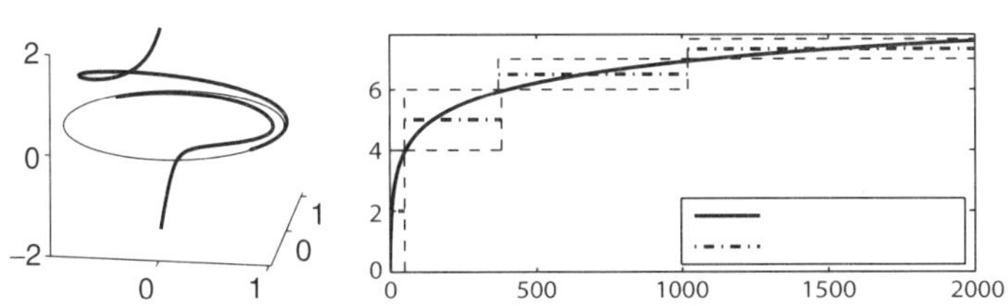

Figure 2.9. Left: Two orbits for (2.28) approaching the circle $S := \{(x, y, z) \in \mathbb{R}^3 : x^2 + y^2 = 1\}$ of fixed points from above and below, respectively. Right: illustration of the notion of pseudo orbits for (2.28), here consisting of fixed points on S.

Another important question is the robustness of attractors under perturbations of the dynamical system. As the following example shows, in general we can only expect upper semicontinuity.

Example 2.4.9. For $0 \le \varepsilon < 1$ consider $\dot{u} = f(u, \varepsilon)$ where

$$f(u,\varepsilon) = \begin{cases} -(u+1) & \text{for } u < -2, \\ 1-(1-\varepsilon)(u+2) & \text{for } -2 \le u < -1, \\ -\varepsilon u & \text{for } |u| \le 1, \\ -1-(1-\varepsilon)(u-2) & \text{for } 1 < u \le 2, \\ 1-u & \text{for } 2 < u, \end{cases}$$

cf. Figure 2.10. For all $\varepsilon > 0$ the global attractor is given by $\mathcal{A}_\varepsilon = \{0\}$. However, for $\varepsilon = 0$ we have $\mathcal{A}_0 = [-1, 1]$. ⌋

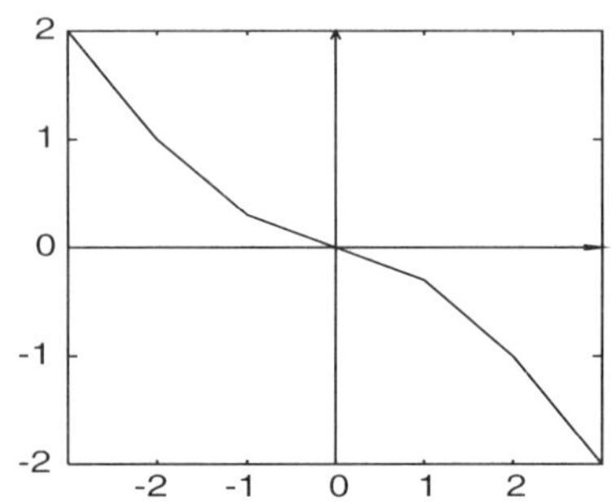

Figure 2.10. The "vector field" for Example 2.4.9 for $\varepsilon > 0$.

Theorem 2.4.10. (Attractor upper semicontinuity) *Assume that for $\mu \in [0, \mu_0)$ each of the flows $(\mathcal{S}_t^\mu)_{t \ge 0}$ has a global attractor $\mathcal{A}_\mu$ such that $\cup_{0 \le \mu < \mu_0} \mathcal{A}_\mu \subset Q$ for some bounded set Q, and that for each $t > 0$ the flows $\mathcal{S}_t^\mu$ converge to $\mathcal{S}_t^0$ uniformly on bounded subsets M, i.e.,*

$$\sup_{u_0 \in M} \|S_t^\mu u_0 - S_t^0 u_0\|_X \to 0 \quad \text{as} \quad \mu \to 0.$$

Then

$$\operatorname{dist}(\mathcal{A}_\mu, \mathcal{A}_0) \to 0 \quad \text{as} \quad \mu \to 0.$$

Proof. Let $\varepsilon > 0$. Since $\mathcal{A}_0$ attracts Q there exists a $t > 0$ such that $\mathcal{S}_t^0 Q$ is a subset of the $\varepsilon/2$-neighborhood $N(\mathcal{A}_0, \varepsilon/2)$ of $\mathcal{A}_0$, i.e., $\mathcal{S}_t^0 Q \subset N(\mathcal{A}_0, \varepsilon/2)$. Next, for $\mu > 0$ sufficiently small, we have

$$\sup_{u \in Q} \|\mathcal{S}_t^\mu u - S_t^0 u\| \le \varepsilon/2$$

for all $u \in Q$. Since $\mathcal{A}_\mu \subset Q$ we have $\mathcal{A}_\mu = \mathcal{S}_t^\mu \mathcal{A}_\mu \subset S_t^\mu Q \subset N(\mathcal{A}_0, \varepsilon)$. □

Only with a number of additional assumptions, cf. [**Rob01**, Theorem 10.17], lower semicontinuity and hence continuity can be obtained, too. In general, as the previous example has shown, lower semicontinuity is wrong.

2.4.4. Planar systems. For autonomous ODEs in two space dimensions the possible ω-limit sets and attractors are relatively easy.

Theorem 2.4.11. (Poincaré-Bendixson) *Consider the ODE* (2.14) *in* $\mathbb{R}^2$ *and assume that the positive semiorbit* $\gamma_+(u_0)$ *through* u_0 *is bounded. If* $\omega(u_0)$ *contains no fixed point, then* $\omega(u_0)$ *is a periodic solution. If* $\omega(u_0)$ *contains a fixed point, but only finitely many, then* $\omega(u_0)$ *is either a single fixed point or it consists of fixed points with the homoclinic and heteroclinic connections between the fixed points.*

Idea of the proof. We refrain from giving a complete proof of Theorem 2.4.11, cf. [**Tes12**, §7.3], since the ideas of the proof will not be used any more in the following. The proof is based on the Jordan curve theorem saying that a closed, non-self-intersecting curve separates $\mathbb{R}^2$ in an interior and an exterior part. Since this is only true in $\mathbb{R}^2$ this assumption is essential. As a consequence, Poincaré maps have to be monotonic. See Figure 2.11.

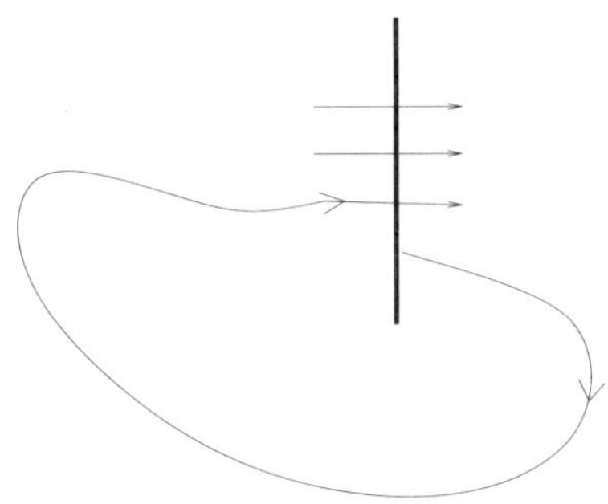

Figure 2.11. Monotonicity of the Poincaré maps.

This yields that the ω-limit set of an orbit γ intersects every Poincaré section in only one point. If there is no fixed point in $\omega(\gamma)$, then $\omega(\gamma)$ is a periodic orbit. □

Example 2.4.12. Using the Poincaré-Bendixon theorem allows us to prove the existence of a periodic solution for

$$\dot{x} = y \quad \text{and} \quad \dot{y} = -x + y(1 - x^2 - 2y^2). \tag{2.29}$$

A direct consequence of the Poincaré-Bendixon theorem is that a positively invariant bounded set for $\dot{x}=f(x)$, $x\in\mathbb{R}^2$, which does not contain a fixed point, must contain a periodic orbit. The set $A = \{(x, y) \in \mathbb{R}^2 : 1/4< x^2 + y^2 <1\}$ is positively invariant for (2.29). This follows by looking at the sign of

$$\frac{\mathrm{d}}{\mathrm{d}t}(x(t)^2 + y(t)^2) = 2y^2(1 - x^2 - 2y^2)$$

at the boundaries of A. We have $2y^2(1-x^2-2y^2)|_{x^2+y^2=1/4} \geq 0$ and $2y^2(1-x^2-2y^2)|_{x^2+y^2=1} \leq 0$. ⌋

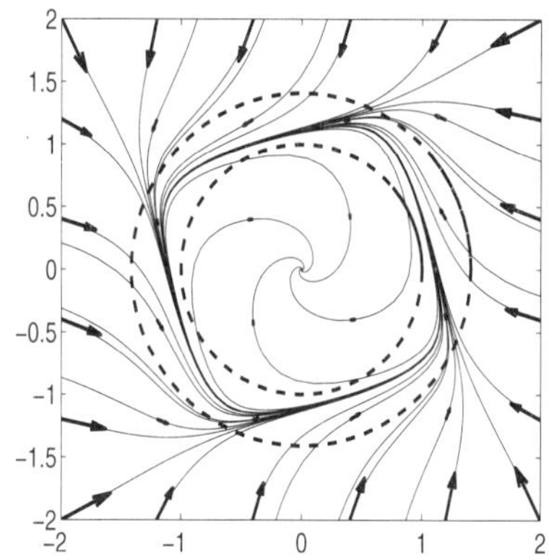

Figure 2.12. Phase portrait of (2.29) with the invariant region.

2.4.5. Gradient systems and Lyapunov functions. Another class of systems with restricted asymptotic behavior is given by gradient systems. They are of the form

$$\dot{u} = -\nabla V(u), \tag{2.30}$$

with the potential $V \in C^1(X, \mathbb{R})$, where $X = \mathbb{R}^d$ could be some general Banach space. For simplicity let $X = \mathbb{R}^d$. Obviously, fixed points of (2.30) satisfy $\nabla V(u) = 0$, i.e., they are critical points of V. Moreover, V decays along solutions of (2.30), see Figure 2.13 for a one-dimensional sketch.

Theorem 2.4.13. *The function $t \mapsto V(u(t))$ is strictly decaying for solutions $u = u(t)$ of* (2.30) *except in case that u is a fixed point. In particular, there are no non-trivial periodic solutions in gradient systems.*

Proof. We have

$$\frac{\mathrm{d}}{\mathrm{d}t} V(u(t)) = (\nabla V(u(t)))^T \left(\frac{\mathrm{d}}{\mathrm{d}t} u(t)\right) = -\|\nabla V(u(t))\|^2_{\mathbb{R}^d} < 0 \tag{2.31}$$

except in case that u is a fixed point. Suppose that there exists a non-trivial periodic solution with $u(t) = u(t+T)$ with minimal period $T > 0$. Then by (2.31) we have $V(u(t)) > V(u(t+T)) = V(u(t))$, which is a contradiction. □

Remark 2.4.14. The linearization $A \in \mathbb{R}^{d \times d}$ at a fixed point in a gradient system is a symmetric matrix. Therefore, all eigenvalues are real. ⌋

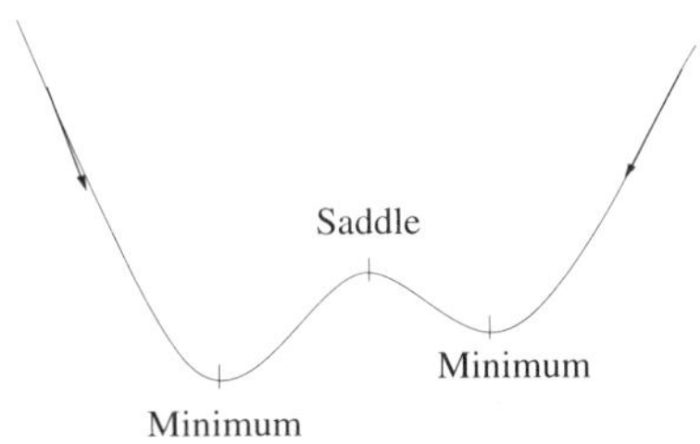

Figure 2.13. The solutions decay along the gradient of the potential.

It turns out that with only a few additional assumptions the ω-limit sets and the global attractor of gradient systems can be completely described.

Theorem 2.4.15. *Suppose that $V(u) \to \infty$ for $\|u\|_{\mathbb{R}^d} \to \infty$, that the set $\mathcal{E}$ of fixed points is finite, and that the fixed points are all hyperbolic. Then, for all $u_0 \in \mathbb{R}^d$ we have $\omega(u_0) = u^*$ for some fixed point u^*. Moreover, for* (2.30) *there exists a compact absorbing set B, and the attractor $\mathcal{A} = \omega(B)$ consists of finitely many fixed points and the connecting orbits between the fixed points.*

Proof. See [**Rob01**, §10.6.1]. □

Most properties of gradient systems are also true in case that a Lyapunov function exists for the dynamical system.

Definition 2.4.16. *A Lyapunov function for a dynamical system $\mathcal{S}_t : X \to X$ is a continuous function $\Phi : B \to \mathbb{R}$ on a positively invariant set $B \subset X$ such that*

(i) *given $u_0 \in B$ the function $t \mapsto \Phi(\mathcal{S}_t u_0)$ is non-increasing,*
(ii) *if $\Phi(\mathcal{S}_t u_0) = \Phi(u_0)$ for some $t > 0$ then u_0 is a fixed point.*

Obviously, for gradient systems $\dot{u} = -\nabla V(u)$ the potential V is a Lyapunov function on $\mathbb{R}^d$. Conversely, suppose that some dynamical system $\mathcal{S}_t$ has a Lyapunov function Φ defined on $X = \mathbb{R}^d$ with the properties that $\Phi(u) \to \infty$ as $\|u\|_{\mathbb{R}^d} \to \infty$ and $\frac{\mathrm{d}}{\mathrm{d}t}\Phi(u(t)) < 0$ outside some bounded set B. For such systems we have global existence of solutions and since $\overline{B}$ is a compact absorbing set the system is dissipative. Suppose further that the set $\mathcal{E}$ of fixed points is discrete. Then the assertions of Theorem 2.4.15 remain true. Therefore, such systems are often called gradient-like. Lyapunov functions are very often used to prove stability and instability of fixed points, cf. [**HK91**, §9.3-§9.4]. See also Example 2.6.3. The concepts of gradient systems and Lyapunov functions are used in the analysis of PDEs, too, cf. §5.3.

2.5. Chaotic dynamics

In contrast to the relatively simple dynamics which can be found for autonomous ODEs in one and two dimensions, for ODEs in dimensions three and higher very complicated behavior can occur. It is reasonable to expect that complicated dynamics occurs in very high-dimensional systems such as for instance the one describing positions and velocities of 1Mol $\approx 6.022\times 10^{23}$ particles of an ideal gas in some container. Statistical mechanics, which was initiated in 1870 by Ludwig Boltzmann, is based on the insight that a description of individual particles does not make any sense for such systems, and that a statistical description is more appropriate. However, it is surprising that very complicated dynamical behavior already occurs in low-dimensional systems. This was already observed by Henri Poincaré around 1890, when he studied the N-body problem, cf. Chapter 4. However, this fact only came apparent to a wider audience with the first computer simulations made in the early 1960s. The meteorologist Edward Lorenz [**Lor63**] found that already systems in $\mathbb{R}^3$ show a behavior which was later on called chaotic. As a consequence of the interesting pictures which were produced in the following years there was a big boom about chaos lasting for almost 30 years, cf. [**Gle88, Man91**]. According to [**Kel93**], chaos theory studies the behavior of dynamical systems that are highly sensitive to initial conditions, an effect which is popularly referred to as the butterfly effect. Small differences in initial conditions (such as those due to rounding errors in numerical computation) yield widely diverging outcomes for such dynamical systems, rendering long-term prediction impossible in general.

We will call a dynamical system chaotic if there is a subset of the phase space such that the flow restricted to this subset is conjugated to shift dynamics, which is a prototype of a chaotic dynamical system. The occurrence of shift dynamics in a dynamical system is proved with the help of an intermediate step which is called Smale's horseshoe. This is a geometric construction of a chaotic dynamical system which is easier to detect in a given dynamical system. Using this idea we will present with Silnikov chaos an example of a three-dimensional ODE which exhibits chaotic behavior. Routes to chaos in dissipative systems by sequences of local bifurcations will briefly be described subsequently in §3.4. The occurrence of chaotic behavior in Hamiltonian systems is discussed in §4.5. In the present book, chaos will not play a central role, but one should keep in mind its existence already in low-dimensional dynamical systems. Our presentation of this subject is based on [**GH83**].

2.5.1. Shift dynamics. We will use shift dynamics as a prototype of a chaotic dynamical system. On the set

$$\Sigma_2 = \{a : \mathbb{Z} \to \{0,1\} : a = (a_i)_{i\in\mathbb{Z}}\}$$

which is equipped with the distance

$$d(a,b) = \sum_{j\in\mathbb{Z}} 2^{-|j|}|a_j - b_j|.$$

we define the shift $\sigma : \Sigma_2 \to \Sigma_2$ by $(\sigma(a))_i = a_{i+1}$. Similarly, we define Σ_2^+ where the index set $\mathbb{Z}$ is replaced by $\mathbb{N}$. The subsequent theory will be formulated for the shift in Σ_2, but can also be formulated in Σ_2^+.

Theorem 2.5.1. *We have that $\sigma \in C(\Sigma_2, \Sigma_2)$ has the following properties:*

(i) There exist non-trivial periodic solutions to every minimal period;

(ii) There exists a dense orbit;

(iii) The sensitivity w.r.t. the initial conditions holds, i.e., for every $a \in \Sigma_2$ and every $\delta > 0$ there exist $b \in \Sigma_2$ and $j \geq 0$ such that $d(\sigma^j(a), \sigma^j(b)) \geq 1$, although $d(a,b) \leq \delta$.

Proof. (i) The 1-periodic solutions are $a = \ldots 00000 \ldots$ and $a = \ldots 11111 \ldots$. The 2-periodic solutions are generated by 00, 01, 10 and 11, and the 3-periodic ones by 000, 001, 010, 100, 110, 101, 011 and 111, etc.
(ii) Consider the orbit to the initial condition a, consisting of all finite sequences that generate the periodic solutions, i.e.,

$$a = \ldots 0000|0100011011000001010100110101011111000000010010 \ldots .$$

Right of $|$ we have the position $j = 0$ and left of $|$ the sequence is filled up with zeroes. For a given $\varepsilon > 0$ and $b \in \Sigma_2$ we have to find $n \in \mathbb{N}$ such that $d(b, \sigma^n(a)) \leq \varepsilon$. For $c \in \Sigma_2$ we have $d(b,c) \leq \varepsilon$, if at least $b_j = c_j$ for $|j| \leq j_0(\varepsilon)$. The other c_j for $|j| \geq j_0(\varepsilon)$ can be arbitrary. Since a contains all finite sequences the claim follows by shifting a until the finite sequence $(b_j)_{j=-j_0,\ldots,j_0}$ occurs at the positions between $-j_0$ and j_0.
(iii) Let $a \in \Sigma_2$ and $\delta > 0$, and set $b_j = a_j$ for $j \leq j_0(\delta)$ and $a_{j_0(\delta)+1} \neq b_{j_0(\delta)+1}$. Then $d(a,b) \leq \delta$, but $d(\sigma^{j_0(\delta)+1}(a), \sigma^{j_0(\delta)+1}(b)) \geq 1$. □

A general dynamical system is called chaotic if the flow is conjugated to shift dynamics on a subset of its phase space, i.e.,

Definition 2.5.2. *A discrete dynamical system $\Pi : \mathbb{R}^d \to \mathbb{R}^d$ is called chaotic, if there is a set $\Lambda \subset \mathbb{R}^d$ and a homeomorphism $h : \Lambda \to \Sigma_2$ such that on Λ the flows are conjugated, i.e., $\sigma \circ h|_\Lambda = h \circ \Pi|_\Lambda$.*

Remark 2.5.3. This is a very strict definition of a chaotic dynamical system. Chaos in the sense of [**Dev89**] for a map $f \in C(M,M)$, with M some metric space, is defined by

(1) the sensitive dependence on initial conditions;

(2) periodic points are dense in M;

(3) topological transitivity, i.e., for all open subsets U, V of M, there exists $k \in \mathbb{N}$ such that $f^k(U) \cap V \neq \emptyset$. ⌋

Remark 2.5.4. Another definition of chaotic behavior in a system where all forward orbits are bounded is the occurrence of a positive Lyapunov exponent [**Rue89**]. Lyapunov exponents describe how the distance of nearby solutions evolves in time. They are defined through

$$\lambda(u_0, \varphi) = \limsup_{t \to \infty} \left(\frac{1}{t} \ln \|Du(t, u_0)\varphi\| \right)$$

for $\varphi, u_0 \in \mathbb{R}^d$. For each initial condition u_0 there are d such Lyapunov exponents. Very often the Lyapunov exponents do not depend on u_0. A positive Lyapunov exponent implies a sensitive dependence on the initial conditions. ⌋

See Exercise 2.18 for an example of a map where conjugacy to the shift on Σ_2^+ can be shown explicitly. A related and famous 1D iteration for which chaotic behaviour can be shown for certain parameters is the logistic map, see, e.g., [**Dev89**], and §3.4.1.

2.5.2. Smale's horseshoe. The occurrence of shift dynamics in a general dynamical system is very often proved with the help of an intermediate step. There is a geometric construction of a chaotic dynamical system, called Smale's horseshoe, which is easier to detect in a given dynamical system than shift dynamics. The construction is as follows. Starting with $S = [0,1] \times [0,1] \subset \mathbb{R}^2$ we define via Figure 2.14 a map $f : S \to \mathbb{R}^2$ such that $f(S) \cap S$ consists of two components, namely the two vertical strips V_0 and V_1. There exist two horizontal strips H_0 and H_1 with $f(H_j) = V_j$.

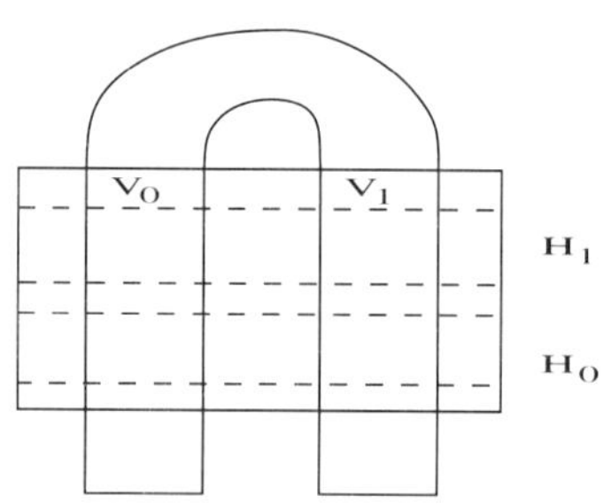

Figure 2.14. Smale's horseshoe.

Under iteration of f most of the points of S leave S. The points which stay in S under all iterations of f define a set

$$\Lambda = \{x : f^i(x) \in S, \ -\infty < i < \infty\}.$$

Remark 2.5.5. The invariant set $\Lambda \subset \mathbb{R}^d$ is called hyperbolic since there is a continuous invariant splitting of the tangent spaces $T_\Lambda \mathbb{R}^d = E^u_\Lambda \oplus E^s_\Lambda$ with the following property: There exist constants $C > 0$ and $\lambda \in (0,1)$ with

$$\text{a)} \quad \|Df^{-n}(x)v\|_{\mathbb{R}^d} \leq C\lambda^n \|v\|_{\mathbb{R}^d} \text{ if } v \in E^u_\Lambda(x),$$
$$\text{b)} \quad \|Df^{n}(x)v\|_{\mathbb{R}^d} \leq C\lambda^n \|v\|_{\mathbb{R}^d}, \text{ if } v \in E^s_\Lambda(x).$$

Hyperbolic dynamics, cf. [**KH97**, Part 4], is one branch in the description of chaos. ⌋

The set Λ has a complicated topological structure.

Lemma 2.5.6. *The set Λ is a Cantor set, i.e., an uncountable, compact, totally disconnected and nowhere dense set which consists entirely of limit points.*

Proof. Each horizontal strip H_i is mapped through f into the vertical strip $V_i = f(H_i)$. We consider $V_i \cap H_j$ which is the image of some thin horizontal strips H_{ij}. We obtain vertical strips $V_{ij} = f^2(H_{ij})$ by two iterations of f. See Figure 2.15.

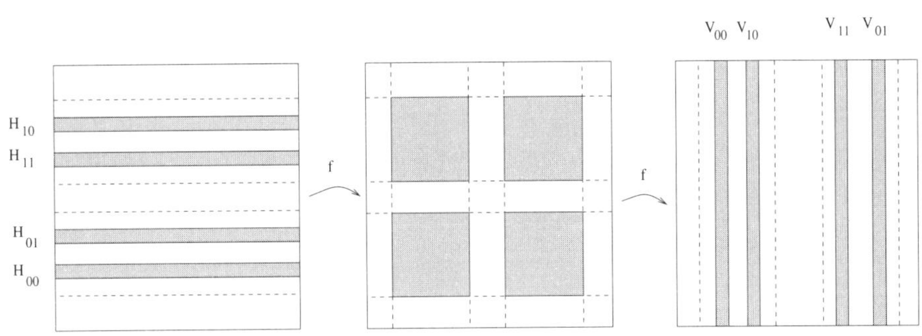

Figure 2.15. Iteration of the horseshoe map f.

By more forward and backward iterations of f and the intersection of all horizontal and vertical strips which are obtained in this way we get a closed, non-empty, completely disconnected set Λ. Each point in Λ is a limit point of Λ. Thus, Λ is a Cantor set. □

To each point $x \in \Lambda$ we associate an infinite sequence $a : \mathbb{Z} \to \{0,1\}$ via $\phi(x) = (a_i)_{i=-\infty}^{\infty}$ if $f^i(x) \in H_{a_i}$.

Theorem 2.5.7. *There is a one-to-one map ϕ between Λ and Σ_2 such that the sequence $b = \phi(f(x))$ can be computed from $a = \phi(x)$ by shifting the indices $b_i = a_{i+1}$. The map ϕ is a homeomorphism between the metric spaces (Σ_2, d) and $(\Lambda, \|\cdot\|_{\mathbb{R}^d})$.*

Proof. We consider $\phi(f(x)) = (b_i)_{i=-\infty}^{\infty}$ with $f^{i+1}(x) \in H_{b_i}$. Hence, $f^i(x) \in H_{b_{i-1}} = H_{a_i}$ and therefore $b_i = a_{i+1}$ which implies $\phi \circ f = \sigma \circ \phi$.

It remains to prove the continuity of $\phi : \Lambda \to \Sigma_2$ and $\phi^{-1} : \Sigma_2 \to \Lambda$. For $x \in \Lambda$ and all $\varepsilon > 0$ we have to show the existence of a $\delta > 0$ such that $d(\phi(x), \phi(y)) < \varepsilon$ if $\|x - y\| < \delta$. To a given $\varepsilon > 0$ there exists a $j_0 = j_0(\varepsilon)$ such that $d(a, b) \leq \varepsilon$ is equivalent to $a_j = b_j$ at least for $|j| \leq j_0$. a_j and b_j can be arbitrary for $|j| > j_0$. Therefore, the condition $d(\phi(x), \phi(y)) < \varepsilon$ uniquely defines two sequences $\overline{a}^+ = (a_i)_{j=0}^{j_0}$ and $\overline{a}^- = (a_i)_{j=-j_0-1}^{-1}$. Associated with these sequences there are strips $V_{\overline{a}^-}$ and $H_{\overline{a}^+}$. If we choose $\delta > 0$ so small that $y \in V_{\overline{a}^+} \cap H_{\overline{a}^-}$ if $\|y - x\|_{\mathbb{R}^d} \leq \delta$ we are done. The continuity of ϕ^{-1} follows in the same way. □

As a consequence it follows that if we find a Smale's horseshoe in a dynamical system, then chaotic behavior in the sense of Definition 2.5.2 is present in this system.

2.5.3. Silnikov chaos. Using the idea of Smale's horseshoe we present a first example of a chaotic dynamical system coming from an ODE, namely Silnikov chaos. The presentation is based on [**GH83**, §6.5.1]. However, we skip most analytic arguments and argue mostly by pictures. We consider an autonomous three-dimensional ODE with a homoclinic orbit γ at the origin which is assumed to be a hyperbolic fixed point with eigenvalues $\lambda \in \mathbb{R}, \omega, \overline{\omega} \in \mathbb{C}$, where $\operatorname{Im}\omega \neq 0$. See Figure 2.16(a). Silnikov [**Sil65**] proved in 1965 the following result.

Theorem 2.5.8. *If $|\operatorname{Re}\omega| < \lambda$, then the flow S_t can be perturbed in such a way that the perturbed flow $\widetilde{S}_t$ has a homoclinic orbit $\widetilde{\gamma}$ close to γ and that there exists a subset of $\mathbb{R}^3$ on which the first return map for the perturbed flow $\widetilde{S}_t$ is conjugated to Smale's horseshoe.*

Idea of the proof. By the Hartman-Grobman theorem 2.3.8 we may assume that the vector field is linear in a neighborhood of the origin, i.e., with $\alpha := \operatorname{Re}\omega$,

$$\begin{pmatrix} \dot{x} \\ \dot{y} \\ \dot{z} \end{pmatrix} = \begin{pmatrix} \alpha & -\beta & 0 \\ \beta & \alpha & 0 \\ 0 & 0 & \lambda \end{pmatrix} \begin{pmatrix} x \\ y \\ z \end{pmatrix}, \quad \omega = \alpha + \mathrm{i}\beta. \tag{2.32}$$

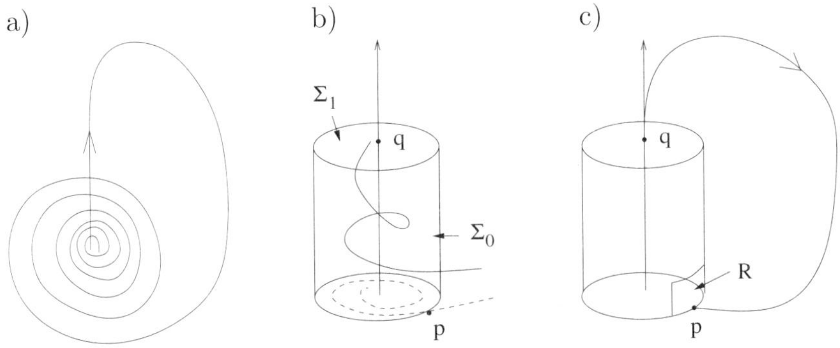

Figure 2.16. a) The homoclinic orbit in the example of Silnikov, b) the inner map $\psi_{int} : \Sigma_0 \to \Sigma_1$, and c) the outer map $\psi_{out} : \Sigma_1 \to \Sigma_0$.

The solutions of (2.32) are given by

$$\begin{pmatrix} x \\ y \\ z \end{pmatrix}(t) = \begin{pmatrix} e^{\alpha t}((\cos\beta t)x(0) - (\sin\beta t)y(0)) \\ e^{\alpha t}((\sin\beta t)x(0) + (\cos\beta t)y(0)) \\ e^{\lambda t}z(0) \end{pmatrix}. \tag{2.33}$$

We define two sets

$$\begin{aligned} \Sigma_0 =& \{(x, y, z) : x^2 + y^2 = r_0^2 \text{ and } 0 < z < z_1\}, \\ \Sigma_1 =& \{(x, y, z) : x^2 + y^2 < r_0^2 \text{ and } z = z_1 > 0\} \end{aligned}$$

and assume that these sets are contained in the previous neighborhood. The solutions go from Σ_0 to Σ_1 according to Figure 2.16 b). The inner map $\psi_{int} : \Sigma_0 \to \Sigma_1$, which maps a point $a \in \Sigma_0$ into the first intersection point of the associated solution with Σ_1, maps vertical vertical lines from Σ_0 into a logarithmic spiral in Σ_1. The outer map ψ_{out} transports a neighborhood of q through the homoclinic solution into

$$\widetilde{\Sigma}_0 = \{(x, y, z) : x^2 + y^2 = r_0^2,\ |z| < z_1\}.$$

See Figure 2.16 c). The map ψ is defined by $\psi = \psi_{out} \circ \psi_{int}$ for all points $X \in \Sigma_0$ with $\psi(X) \in \Sigma_0$ and has the same asymptotic behavior as ψ_{int} for $z \to 0$ since for $z \to 0$ the time needed by the solution to come from Σ_0 to Σ_1 becomes infinite, whereas the time needed by the solution to come from Σ_1 to Σ_0 stays finite. Hence, a rectangular set R around the entrance point p of the homoclinic orbit is mapped into a spiral like structure. See Figure 2.17. The assumption $|\operatorname{Re}\omega| < \lambda$ is necessary that this picture really occurs, for more details see [**GH83**, §6.5.1]. Therefore, graphically we have found a Smale's horseshoe for ψ. □

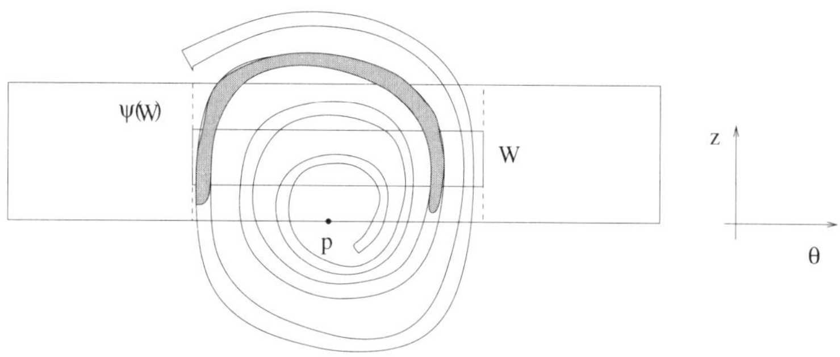

Figure 2.17. Smale's horseshoe in the Silnikov example.

2.6. Examples

The following series of examples is intended to give some familiarity with the notions and ideas introduced so far.

Example 2.6.1. a) For the potential $V(x, y) = (x^2 - 1)^2 + y^2$ we find $-\nabla V(x, y) = -(4x(x^2 - 1), 2y)$, leading to the fixed point $(x, y) = (0, 0)$, which is a saddle point of V, and to the fixed points $(\pm 1, 0)$, which are minima of V. For every $r > 1$ the set $\{(x, y) : x^2 + y^2 \leq r^2\}$ is absorbing. The attractor $\mathcal{A}$ is given by $[-1, 1] \times \{0\}$, consisting of the three fixed points and the heteroclinic connections between the unstable fixed point $(0, 0)$ and the stable fixed points $(\pm 1, 0)$.

b) For the potential $V(x, y) = (x^2 - 1)^2 + (y^2 - 1)^2$ we find that any neighborhood of the unit square $Q = [-1, 1] \times [-1, 1]$ is an absorbing set. The global attractor is $Q = \overline{W^u((0, 0))}$. ⌋

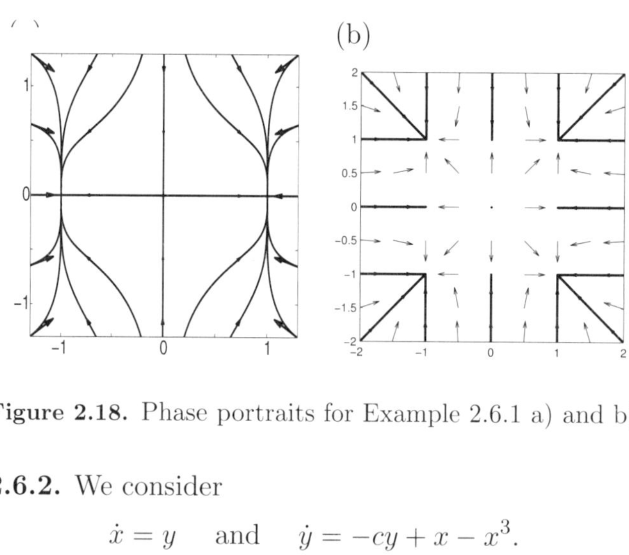

Figure 2.18. Phase portraits for Example 2.6.1 a) and b).

Example 2.6.2. We consider

$$\dot{x} = y \quad \text{and} \quad \dot{y} = -cy + x - x^3.$$

There is a simple mechanical interpretation of the orbits of this system as the orbits of a particle moving in the double-well potential $F(x) = -\frac{1}{2}x^2 + \frac{1}{4}x^4$ with friction $-cy$. In partiular, for $c = 0$ we have a similar situation as in Example 2.3.20, i.e., orbits are level lines of the energy $E(x, \dot{x}) = \frac{1}{2}\dot{x}^2 + F(x)$, see also Example 4.1.2 below for a general discussion.

The system possesses the fixed points $(x, y) = (\pm 1, 0)$ and $(x, y) = (0, 0)$. The linearization at the fixed point $(x, y) = (0, 0)$ yields the eigenvalues $\lambda_{1,2} = -c/2 \pm \sqrt{c^2/4 + 1}$. Hence, $(0, 0)$ is a saddle for all values of c. The unstable eigenspace is spanned by $\phi_1 = (1, 1)$, the stable one by $\phi_2 = (1, -1)$. The linearization at the fixed point $(x, y) = (1, 0)$ yields the eigenvalues $\lambda_{1,2} = -c/2 \pm \sqrt{c^2/4 - 1}$. Thus, $(1, 0)$ is a center for $c = 0$ (Figure 2.19(a)), a stable vortex for $0 < c < 2$ (Figure 2.19(b)), a stable node for $c \geq 2$ (Figure 2.19(c)), an unstable vortex for $-2 < c < 0$, and an unstable node for $c \leq -2$. The same classification holds for the fixed point $(-1, 0)$. The mechanical interpretation is that for $c > 2$ the friction is so large that the particle approaches the minima $x = \pm 1$ of the energy monotonically.

For $c > 0$ the system is dissipative, and the stable manifold $W_s((0, 0))$ separates the stable manifolds $W_s((1, 0))$ and $W_s((-1, 0))$, i.e., the domains of attraction of $(1, 0)$ and $(-1, 0)$. The global attractor consists of the three fixed points and the unstable manifold of $(0, 0)$.

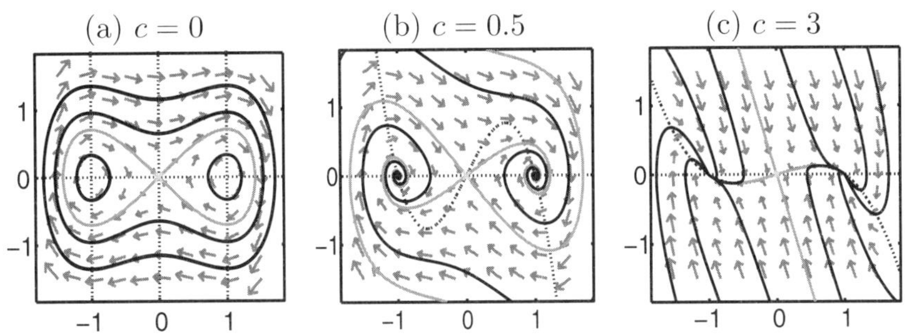

Figure 2.19. Phase portraits for Example 2.6.2; invariant manifolds in light grey, and nullclines as dashed lines.

For $c = 0$ we have homoclinic solutions and so $W_u((0, 0)) = W_s((0, 0))$, and the center manifolds of $(\pm 1, 0)$ can be defined as small disks around $(\pm 1, 0)$. To show the existence of the homoclinic solutions for $c = 0$, instead of the energy argument that $E(x, \dot{x})$ =const= 0 we may also use the symmetry (reversibility) $(t, x, y) \mapsto (-t, x, -y)$ under which the system is invariant. Hence, with $t \mapsto (x(t), y(t))$ also $t \mapsto (x(-t), -y(-t))$ is a solution. The unstable manifold of the origin intersects the fixed space $\Sigma = \{(x, 0) : x \in \mathbb{R}\}$ of reversibility transversally. W.l.o.g. taking this intersection at $t = 0$, the orbit can be extended to $t > 0$ by reflection at Σ. See Figure 2.20.

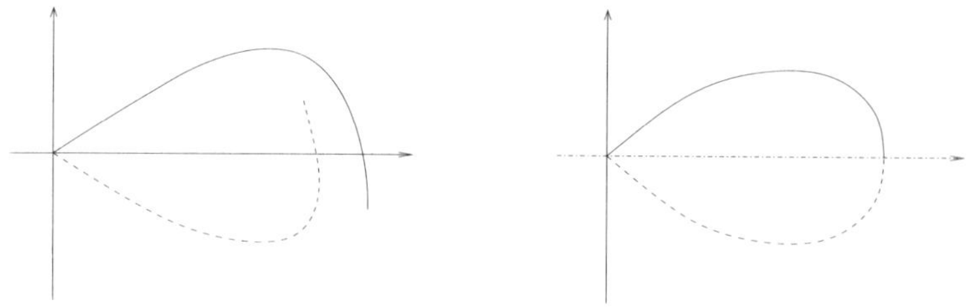

Figure 2.20. Persistence of homoclinic connections in planar systems. In non-reversible systems an additional parameter is needed for the intersection of the stable and unstable manifold. In reversible systems the fact that the stable manifold intersects the fixed space $\{y = 0\}$ of the reversibility operator transversally can be used for the persistence.

This homoclinic orbit persists under small perturbations respecting the reversibility of the vector field due to the transversal intersection. Such transversibility and symmetry arguments often also work when other arguments, such as the above energy argument, fail, see, e.g., Remark 13.3.1. For general perturbations ($c > 0$) the homoclinic orbit breaks up in accordance with the fact that the probability that two one-dimensional manifolds intersect in a two-dimensional phase space is zero. ⌋

The following examples are applications from mathematical biology; examples of this type, combined with diffusion lead to the important class of reaction diffusion systems, see Chapter 9.

Example 2.6.3. (Lotka-Volterra) We consider the predator-prey system

$$\dot{x} = x(a - y), \quad \dot{y} = y(x - c), \qquad a, c > 0 \tag{2.34}$$

with $x, y > 0$ ($x =$ prey, $y =$ predator). In case of no predators, i.e., $y = 0$, the prey will grow with some exponential rate according to $\dot{x} = ax$. In case of no prey, i.e., $x = 0$, the predators will die with some exponential rate according to $\dot{y} = -cy$. In case of predators, i.e., $y > 0$, the prey will be killed by the predator via the term $-xy$ with a rate proportional to the number of predators. On the other hand the term xy gives an exponential growth of the number of predators with a rate proportional to the number of preys.

The unique non-trivial fixed point is $(x_0, y_0) = (c, a)$. Its linearization

$$A = \begin{pmatrix} a - y & -x \\ y & x - c \end{pmatrix}\bigg|_{(x,y)=(c,a)} = \begin{pmatrix} 0 & -c \\ \gamma & 0 \end{pmatrix},$$

possesses the eigenvalues $\lambda = \pm i\sqrt{ca}$ such that (x_0, y_0) is a centre for the linearization. Thus, no stability result can be concluded from Theorem 2.3.4 a) for the nonlinear system.

However, $\phi(x, y) = x + y - c\ln(x) - a\ln(y)$ is conserved for (2.34), i.e., $\frac{\mathrm{d}}{\mathrm{d}t}\phi(x, y) = 0$. Since $(\dot{x}, \dot{y}) \neq (0, 0)$ for $(x, y) \neq (x_0, y_0)$ the solutions move on the level lines of ϕ. Calculus yields that ϕ has a unique critical point, namely a minimum in (x_0, y_0). Thus, (x_0, y_0) is nonlinearly stable and all other solutions move on periodic orbits around (x_0, y_0). This behavior agrees with observed data, for instance of canadian lynx and snowshoe hare pelt-trading records of the Hudson Bay Company between 1845 and 1935, cf. [**Mur89**].

The Lotka–Volterra model (2.34) as the oldest predator-prey model was partially motivated by the observation that during and shortly after world war I the fraction of predator fish caught in the mediterranean sea increased, when the total fishing decreased. Let $(x, y)(t)$ be a periodic solution of (2.34) with period T. Then

$$\overline{x} = \frac{1}{T}\int_0^T x(t)\,\mathrm{d}t = c, \qquad \overline{y} = \frac{1}{T}\int_0^T y(t)\,\mathrm{d}t = a.$$

This holds due to $\frac{1}{T}\int_0^T \frac{\dot{x}}{x}\,\mathrm{d}t = \frac{1}{T}\int_0^T a - y\,\mathrm{d}t = \frac{1}{T}(\ln x(T) - \ln x(0)) = 0$, hence $\overline{y} = a$, and similarly for $\overline{x} = c$. Fishing can be modeled by a simple modification of (2.34), namely to replace a, c by $a - p$ and $c + p$, where $0 \leq p \leq a$ denotes the "fishing pressure". Depending on p, the fraction of predators hence is

$$\frac{\overline{y}(p)}{\overline{x}(p) + \overline{y}(p)} = \frac{a - p}{a + c},$$

which is a decreasing function of p.

See also Exercise 2.19 for some modification of (2.34), for which the conserved quantity ϕ becomes a genuine Lyapunov function. ⌋

Example 2.6.4. (SI and SIS diseases) We are interested in the dynamics of some disease which proceeds on a time scale much shorter than the lifespan of its hosts. Thus, we assume that the size of the population is unchanged and that in the population of size N a fraction S of individuals is healthy, but susceptible to this disease, while a fraction I is infected. A general ODE model describing the evolution of the fractions S and I reads

$$\dot{S} = -f(S, I), \qquad \dot{I} = f(S, I),$$

where $f(S, I)$ is the rate of infections. The simplest model is $f(S, I) = \beta IS$, which in chemistry would be called the law of mass action, see also §9.1. Here β is called the pairwise infectious contact rate. Using $S + I = N$ yields

$$\dot{I} = \beta I(N - I).$$

It is easy to see that in this model with $\beta > 0$ the whole population becomes infected.

A slight extension is given by SIS models, where infected recover and become susceptible again with a rate γ. Thus

$$\dot{S} = -f(S,I) + \gamma I, \qquad \dot{I} = f(S,I) - \gamma I.$$

Introducing dimensionless variables $u = S/N, v = I/N, \tau = \gamma t$, and using $u + v = 1$ we obtain

$$v' = (R_0(1-v) - 1)v,$$

where $R_0 = \beta N/\gamma$ is called the reproductive ratio of the disease and where $'$ denotes the derivative w.r.t. the new time variable τ. For $R_0 < 1$ the disease dies out, but for $R_0 > 1$ it becomes endemic, i.e., it reaches the steady state $1 - 1/R_0$ as $t \to \infty$. Vaccination reduces the number of susceptible and hence R_0. Note that for desease control it is not necessary to vaccinate all, but sufficiently many to decrease R_0 below 1. ⌋

Example 2.6.5. (Mathematical ecology) The Kolmogorov form of the equations for 2-species interaction in mathematical ecology is

$$\dot{u} = uM(u,v), \qquad \dot{v} = vN(u,v), \tag{2.35}$$

where $(u,v) = (u,v)(x,t)$ are population densities and their respective growth rates M and N are smooth functions from $\mathbb{R}^2_+$ to $\mathbb{R}$. The models (2.35) are further classified as

$$\begin{aligned} \text{predator-prey (PP)} \quad & \partial_v M < 0 \text{ and } \partial_u N > 0 \text{ for } u,v > 0,\\ \text{competition (C)} \quad & \partial_v M < 0 \text{ and } \partial_u N < 0 \text{ for } u,v > 0,\\ \text{symbiosis (S)} \quad & \partial_v M > 0 \text{ and } \partial_u N > 0 \text{ for } u,v > 0. \end{aligned}$$

Usually, further conditions are imposed, namely

(PP1) $\exists k_0 > 0$ such that $M(u,0) < 0$ for $u > k_0$,

(PP2) $\exists$ a function l such that $N(u,v) < 0$ for $u > 0$ and $v > l(u)$.

(C1) $\exists k_0 > 0$ such that $M(u,0) < 0$ for $u > k_0$,

(C2) $\exists l_0 > 0$ such that $N(u,v) < 0$ for $v > l_0$.

(S1) $\exists$ a function k such that $M(u,0) < 0$ for $u > k(v)$,

(S2) $\exists$ a function l such that $N(u,v) < 0$ for $v > l(u)$,

(S3) $k(v) = o(v)_{v\to\infty}$ and $l(u) = o(u)_{u\to\infty}$.

Biologically, for instance (PP1) and (C1) essentially mean that even if there are no predators ($v = 0$), then the growth of the prey population saturates at k_0. By (PP2) and (C2), the predators saturate at $l(u)$. Finally, symbiosis means that each species thrives with the other, but (S3) ensures limits to this symbiotic growth. The condition (S3) should be complemented by demanding that there exists at least one non-trivial fixed point.

As examples consider

$$(1)\quad \dot{u} = u(1-u-v), \quad \dot{v} = v(u-v),$$

$$(2)\quad \dot{u} = u(1-u-v), \quad \dot{v} = v(1-u-\alpha v),$$

$$(3)\quad \dot{u} = u(2\arctan(2v)-u), \quad \dot{v} = v(3\arctan(2u)-v).$$

In (2), $\alpha > 0$ is some parameter. Clearly, (1)=(PP), (2)=(C), (3)=(S). After determining the unique fixed points $(u,v)^*$ with $uv > 0$ for (1) and (3), the phase portraits can be conveniently sketched by considering the signs of the growth rates M, N. For (1) we may additionally use the fact that, e.g., $[1/4, 3/4] \times [1/4, 3/4]$ is positively invariant.

For (2) we note that for $\alpha = 1$ we have $M = N$, and thus a line $\{u+v = 1\}$ of fixed points. For $\alpha \neq 1$ we again have a unique non-trivial fixed point. In particular, for $\alpha > 1$ ($\alpha < 1$) the v species (the u species) dies out. For $\alpha > 1$ the biological interpretation is that for $u = v$ the growth rate N of v is smaller than that of u, due to higher damping (faster saturation) of the growth of v by itself, hence u "wins". ⌋

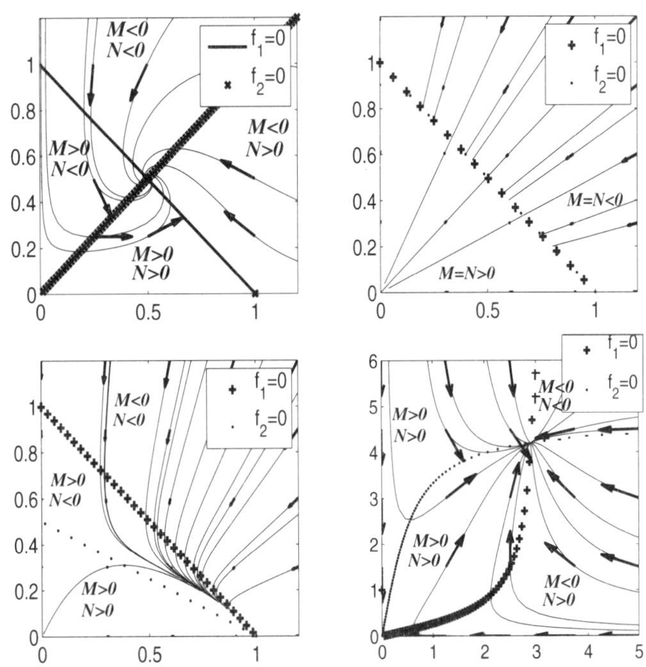

Figure 2.21. ODE phase portraits for (PP), (C) ($\alpha = 1$ and $\alpha = 2$) and (S). For $\alpha = 1$ in (C) we have a line $\{u+v=1\}$ of fixed points. For equations of the form (2.35) a convenient way to start the phase portrait is to consider the regions defined by the nullclines, i.e., $M = 0$ and $N = 0$.

Example 2.6.6. The van der Pol equation [**VdP26**] is given by

$$\ddot{u} + \varepsilon(u^2 - \alpha)\dot{u} + u = 0, \quad u(t) \in \mathbb{R}, \tag{2.36}$$

where $\alpha > 0$ and $0 \leq \varepsilon \ll 1$ are some parameters. As initial conditions we take $u(0) = a$ and $\dot{u}(0) = 0$. This describes some oscillator with small amplitude-dependent damping. It is known and might be expected from the form of the equation, that for every fixed $\alpha > 0$ and small $\varepsilon > 0$ there is a unique periodic solution. For $\varepsilon = 0$ we have solutions $u(t) = A\mathrm{e}^{\mathrm{i}t} + \mathrm{c.c.}$ with $A \in \mathbb{C}$ arbitrary, and thus for $\varepsilon > 0$ we try a two-scale ansatz of the form

$$u(t) = A(\varepsilon t)\mathrm{e}^{\mathrm{i}\omega t} + \mathrm{c.c.}. \tag{2.37}$$

Using $u^2 = A^2\mathrm{e}^{2\mathrm{i}\omega t} + 2|A|^2 + \overline{A}^2\mathrm{e}^{-2\mathrm{i}\omega t}$ this yields

$$\begin{aligned} \mathcal{O}(\varepsilon^0): &\quad -\omega^2 + 1 = 0, \quad \Rightarrow \omega = 1, \\ \mathcal{O}(\varepsilon^1): &\quad 0 = \mathrm{i}(-2\frac{\mathrm{d}}{\mathrm{d}\tau}A + \alpha A - A|A|^2)\mathrm{e}^{\mathrm{i}t} - \mathrm{i}A^3\mathrm{e}^{3\mathrm{i}t} + \mathrm{c.c.}, \end{aligned}$$

and thus equating the coefficient of $\varepsilon^1\mathrm{e}^{\mathrm{i}t}$ to zero yields

$$\frac{\mathrm{d}}{\mathrm{d}\tau}A = \frac{1}{2}(\alpha A - A|A|^2), \tag{2.38}$$

which is called the Landau equation for this problem. Introducing polar coordinates $A(\tau) = \rho(\tau)\mathrm{e}^{\mathrm{i}\phi(\tau)}$ gives $\rho' = \frac{1}{2}\rho(\alpha - \rho^2)$, $\phi' = 0$, with the abbreviation $' = \frac{\mathrm{d}}{\mathrm{d}\tau}$. From this, or directly from (2.38), we can see that $|A|$ converges to $\sqrt{\alpha}$, which predicts that u approaches the circle with radius $2\sqrt{\alpha}$ up to $\mathcal{O}(\varepsilon)$ terms. Incidentically, although nonlinear, (2.38) can be explicitly solved. Via $r = \rho^2$ and $r' = r(\alpha - r)$, and via $v = 1/r$ and $v' = -\alpha v + 1$, we find the solution $r(t) = \alpha r_0/(r_0 + (\alpha - r_0)\mathrm{e}^{-\tau})$, and hence

$$\rho(\tau) = \rho_0 \left(\frac{\alpha}{(\alpha - \rho_0^2)\mathrm{e}^{-\alpha\tau} + \rho_0^2} \right)^{1/2}, \tag{2.39}$$

with $\rho(0) = \rho_0 = a/2$, and $\phi(\tau) = \phi_0 = 0$. Figure 2.22 compares some numerical solutions to (2.36) with approximations via (2.37) and illustrates the distortion of the limit cycles of (2.36) from the circles described by (2.37) as ε becomes larger. Approximation results for this special problem can be found in [**Ver96**]. For PDEs on unbounded domains such a perturbation analysis is one of the most powerful tools. In Part IV of this book and such PDEs we prove error estimates for such formal approximations. ⌋

Exercises

Exercises 2.11 and 2.12 should be done with some software for ODE phase portraits, e.g., `xppaut` or `pplane`, and we also recommend to use such software for illustration *after* the analysis for the other planar ODEs, e.g., in Exercises 2.10, 2.15, 2.19, and 2.20.

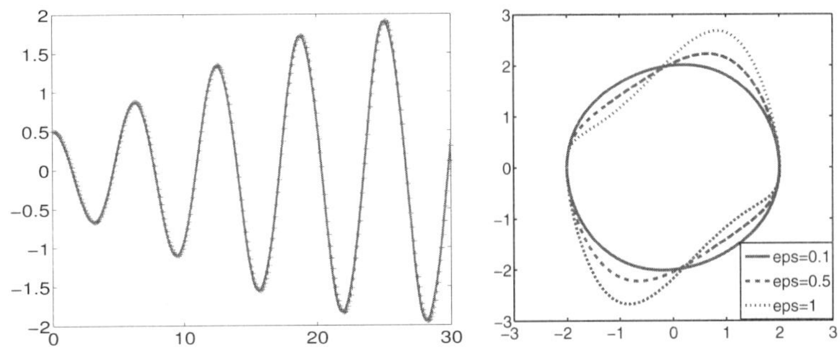

Figure 2.22. Left: numerical solution of (2.36) and approximation via (2.37), $\alpha = 1$, $\varepsilon = 0.2$. Right: Distortion of circle $\rho = 2\sqrt{\alpha}$ by higher-order terms.

2.1. Find the general solution of $\dot{x}(t) = Ax(t)$ with

$$\text{a) } A = \begin{pmatrix} 3 & 1 \\ 0 & 4 \end{pmatrix}, \qquad \text{b) } A = \begin{pmatrix} 2 & 1 \\ 0 & 2 \end{pmatrix}, \qquad \text{c) } A = \begin{pmatrix} 7 & 0 & 0 \\ 0 & 6 & 1 \\ 0 & 0 & 4 \end{pmatrix}.$$

2.2. Find the general solutions and the solutions of the initial value problems with $x(0) = \dot{x}(0) = 1$ for: a) $\ddot{x}(t)+3\dot{x}(t)+2x(t) = 0$, b) $\ddot{x}(t)+5\dot{x}(t)+4x(t) = \cos(3t)$.

2.3. Solve the initial value problems

$$\text{a) } \frac{\mathrm{d}}{\mathrm{d}x}y = xy, \quad y(0) = 1, \qquad \text{b) } \frac{\mathrm{d}}{\mathrm{d}x}y = (\cos x)y, \quad y(0) = 1.$$

2.4. Prove that $\mathrm{e}^{A+B} = \mathrm{e}^A\mathrm{e}^B$ for $d \times d$-matrices A, B, if $AB = BA$.

2.5. Compute real-valued logarithms of the matrices

$$\begin{pmatrix} 1 & 1 & 0 \\ 0 & 1 & 1 \\ 0 & 0 & 1 \end{pmatrix}, \quad \begin{pmatrix} -2 & 0 \\ 0 & -2 \end{pmatrix}, \quad \begin{pmatrix} 1 & 1 & 0 \\ 0 & 2 & 1 \\ 0 & 0 & 3 \end{pmatrix}, \quad \begin{pmatrix} a & a & 1 & 0 \\ -a & a & 0 & 1 \\ 0 & 0 & a & a \\ 0 & 0 & -a & a \end{pmatrix},$$

with $a > 0$. Are the solutions unique?

2.6. Consider $\ddot{y} + 2d\dot{y} + \left(k(t)^2 + d^2\right) y = 0$, with $k(t+T) = k(t)$, $T = r + \frac{\pi}{2}$, and $k(t) = \begin{cases} 0 & \text{for} \quad t \in [0, r), \\ 1 & \text{for} \quad t \in [r, T), \end{cases}$ with $r > 0$. Compute the evolution operator $U(T, 0) = U(T, r)U(r, 0)$. Show that the Floquet-multipliers are given by

$$\rho_{1,2} = \mathrm{e}^{-d(r+\frac{\pi}{2})}\left(-\frac{r}{2} \pm \sqrt{\frac{r^2}{4} - 1}\right).$$

Find the domain of stability in which $|\rho_{1,2}| < 1$.

2.7. i) Solve $x_{n+1} = Bx_n$ with $B = \begin{pmatrix} 3 & 4 \\ 4 & 3 \end{pmatrix}$. ii) Illustrate selected orbits of of $x_{n+1} = Bx_n$ for

$$\text{a) } \begin{pmatrix} 1 & 0 \\ 0 & 1 \end{pmatrix}, \quad \text{b) } \begin{pmatrix} 2 & 0 \\ 0 & \frac{1}{2} \end{pmatrix}, \quad \text{c) } \begin{pmatrix} -1 & 0 \\ 0 & 1 \end{pmatrix}, \quad \text{d) } \begin{pmatrix} \frac{1}{2} & 0 \\ 0 & -\frac{1}{2} \end{pmatrix}, \quad \text{e) } \begin{pmatrix} 0 & 1 \\ -1 & 0 \end{pmatrix}.$$

2.8. Consider the one-dimensional ODEs

a) $\dot{u} = u - u^2$, b) $\dot{u} = -u + 4u^3 - u^5$, c) $\dot{u} = \begin{cases} 0, & \text{if } u = 0, \\ -u^3 \sin(1/u), & \text{if } u \neq 0. \end{cases}$

Find the fixed points and compute their linearization. Which fixed points are stable and which fixed points are unstable? Sketch the phase portraits.

2.9. Consider $f : \mathbb{R}^3 \to \mathbb{R}^3$, $f(u) = Au + g(u)$ with

$$Au = (au_1, acu_2, cu_3), \quad g(u) = (0, acu_1u_3, 0),$$

where $a > 1 > c > a^{-1} > 0$. According to the discrete Hartman-Grobman theorem, cf. Remark 2.3.9, there exists a homeomorphism h such that $h^{-1} \circ f \circ h = A$. Show that h cannot be Lipschitz-continuous.

Hint: Clearly $h^{-1} \circ f^n \circ h = A^n$. Show that this implies

$$c^{-n}h_2(u_1, 0, c^n u_3) - a^n h_2(a^{-n}u_1, 0, u_3) = nh_1(u_1, 0, c^n u_3)h_3(a^{-n}u_1, 0, u_3).$$

Next show that $h_2(u_1, 0, 0) = 0$ and $h_2(0, 0, u_3) = 0$, if h is Lipschitz-continuous. Then obtain a contradiction for $n \to \infty$.

2.10. Consider $\dot{x} = y$, $\dot{y} = -cy - x + x^3$, with $c \in \mathbb{R}$ a parameter. Find the fixed points and compute their linearization. Which fixed points are stable and which fixed points are unstable? Sketch the phase portrait for different values of c. Find the stable, the unstable and the center manifolds for the fixed points. ⌋

2.11. Find the possible ω-limit sets for $\dot{x} = y$, $\dot{y} = x + \varepsilon y - x^3 + 0.1x^2y$, with $\varepsilon \in [-0.09, -0.07]$. Compute values of $\varepsilon \in [-0.09, -0.07]$ where a qualitative change of the periodic orbits occurs. *Hint:* Unstable objects can be found by $t \mapsto -t$. ⌋

2.12. Consider

$$\begin{pmatrix} \dot{x} \\ \dot{y} \end{pmatrix} = \begin{pmatrix} 1 & 1 \\ -1 & 1 \end{pmatrix} \begin{pmatrix} x \\ y \end{pmatrix} - \begin{pmatrix} (x^2 + \alpha y^2)x \\ (\alpha x^2 + y^2)y \end{pmatrix}.$$

Plot the phase portrait for $\alpha = 1, 5, 10$. Find the fixed points and the periodic solutions. Which of them are stable? Find the maximal $\alpha^* > 1$, such that there exists a non-trivial stable periodic solution for all $\alpha \in [1, \alpha^*]$. (Hint: Consider the phase portrait for $\alpha \in (10, 12)$ by computing the ω-limit set for the initial condition $(x, y) = (0.1, 0.1)$.) Let $\alpha = 12$. Find the fixed points and the associated stable and unstable manifolds. ⌋

2.13. Consider $\dot{u} = f(u)$ with $f \in C^1(\mathbb{R}^2, \mathbb{R}^2)$. Let $\Omega \subset \mathbb{R}^2$ be open and simply connected. Assume the existence of a $b \in C^1(\mathbb{R}^2, \mathbb{R})$ with $\operatorname{div}(bf) > 0$ in Ω. Use the integral law of Gauss to show the non-existence of a periodic orbit in Ω.

2.14. Use the idea from Example 2.4.12 to prove that $\dot{x} = x - y - x^3$, $\dot{y} = x + y - y^3$; has a periodic solution.

2.15. Discuss the stability of the fixed point $(x, y) = (0, 0)$ and sketch the phase portraits for the following systems; compare with $\dot{x} = y, \dot{y} = -x$ and explain the qualitative differences.

a) $\dot{x} = y$, $\dot{y} = -x^3$. b) $\dot{x} = y^{1999}$, $\dot{y} = -x^{1999}$.

Hint for a) Consider $V(x, y) = \alpha x^4 + y^2$ with suitable α.

2.16. Consider $\ddot{x} + \delta(x)\dot{x} + 25x = 0$ with $\delta(x) = 8$ for $|x| > 1$ and $\delta(x) = -6$ for $|x| \leq 1$. In order to show the existence of a periodic orbit consider the Poincaré map $G_1 : S_1 \to S_2$ and $G_2 : S_2 \to S_3$, where $S_1 = \{(x, \dot{x}) : x = -1,\ \dot{x} \geq 0\}$, $S_2 = \{(x, \dot{x}) : x = 1,\ \dot{x} \geq 0\}$ and $S_3 = \{(x, \dot{x}) : x = 1,\ \dot{x} \leq 0\}$. Use then the symmetry of the problem.

2.17. Let Σ' be the set of all $0-1$ sequences $(s_j)_{j\in\mathbb{N}}$ with the following property. If $s_j = 0$, then $s_{j+1} = 1$, i.e., Σ' consists of all sequences without two succeeding zeroes. Prove that:
a) the shift σ maps Σ' into itself;
b) there exists a dense orbit in Σ';
c) the set of periodic orbits is dense in Σ'.

2.18. Prove that the shift σ on Σ_2^+ is conjugated to the tent map $f : [0,1) \to [0,1)$ defined by

$$f(x) = \begin{cases} 2x, & \text{for } x \in [0, 1/2), \\ 2x-1, & \text{for } x \in [1/2, 1). \end{cases}$$

Hint: show that $\phi \circ f = \sigma \circ \phi$ where $\phi(x) = (x_j)_{j\in\mathbb{N}}$ for $x = \sum_{j=1}^{\infty} x_j 2^{-j}$.

2.19. The dynamics of the prey predator system (2.34) is not robust under perturbations. Consider (2.34) with the modification $\dot{x} = x(a(x) - y)$ with $a(x) = ae^{-\beta x}$ for a $\beta > 0$.
a) Give a biological interpretation of the modification.
b) Show the asymptotic stability of the fixed point $(x, y) = (c, a(c))$.
c) Use the Lyapunov function $\phi(x, y) = x + y - c\ln x - a(c)\ln y$ to prove that all solutions starting with $x(0) > 0$ and $y(0) > 0$ converge towards this fixed point.
d) Sketch the phase portrait.

2.20. Consider the 2-species interaction systems

$$(1)\ \begin{aligned} \dot{u} &= u\left(\tfrac{2}{1+v^2} - u\right), \\ \dot{v} &= v\,(u - v), \end{aligned} \qquad (2)\ \begin{aligned} \dot{u} &= u\,(v - u^2 - 1), \\ \dot{v} &= v\,(u - v + 3), \end{aligned} \qquad (3)\ \begin{aligned} \dot{u} &= u - u^2 + \tfrac{u}{1+v}, \\ \dot{v} &= 2v - v^2 + \tfrac{v}{1+u}, \end{aligned}$$

all on $u, v > 0$. For each system, compute the nontrivial fixed point and its linearized stability, and sketch the phase portrait. Classify the systems according to Example 2.6.5.

Chapter 3

Dissipative dynamics

In this chapter we provide the strategy and the tools to tackle dissipative systems, which are characterized by the existence of a compact absorbing set. In such systems very often through so called bifurcations complicated and eventually chaotic dynamics occur if some external parameter is varied. In applications such an external parameter can be for instance an external heating or the concentration of a chemical substance. A typical scenario is as follows. For small values of this parameter all solutions are attracted to some asymptotically stable fixed point. If the value of the parameter is increased the fixed point becomes unstable. Then more complicated dynamics can be expected in a neighborhood of the unstable fixed point, for instance new fixed points or time-periodic solutions may bifurcate, i.e., appear in a neighborhood of the first unstable fixed point. A further increase of the external parameter leads to instabilities of the bifurcating solutions. Then quasi-periodic solutions can occur. The next bifurcation may already lead to chaotic dynamics.

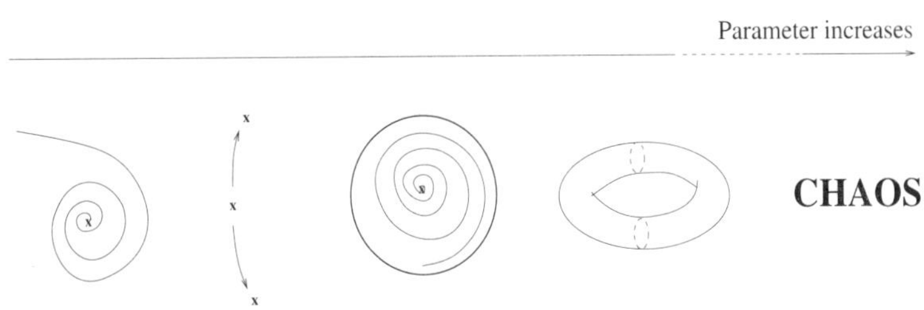

Figure 3.1. Complicated dynamics occurs in dissipative systems as a parameter is increased.

After introducing a number of elementary bifurcations for one- and two-dimensional systems we introduce with the Lyapunov-Schmidt reduction and the center manifold theorem two reduction methods which allow to find these elementary bifurcations in higher dimensional systems, too. Center manifold theory turns out to be a very powerful tool. Besides the construction of the bifurcating solutions it also yields information on their stability. If a fixed point changes from stable to unstable, then all nearby solutions are attracted with some exponential rate towards the center manifold, i.e., the interesting non-trivial dynamics happens on the center manifold of the fixed point. So called normal form transformations allow to analyze the dynamics on the center manifold. We will present this method in the context of the proof of the Hopf bifurcation theorem, i.e., we use it to prove the bifurcation of time-periodic solutions. The chapter is closed by sketching some routes of bifurcations to chaotic behavior in dissipative systems.

3.1. Bifurcations

We present a number of elementary bifurcations and explain how the implicit function theorem and the Lyapunov-Schmidt reduction can be used to prove their occurrence in more complicated systems.

3.1.1. Examples of elementary bifurcations. We start with a globally attracting fixed point which becomes unstable when an external parameter is changed. The following examples are the simplest ones which however turn out to be the 'generic' (see Remark 3.1.10) bifurcations occurring at a fixed point.

Example 3.1.1. (Pitchfork bifurcation of fixed points) Consider

$$\dot{x} = f(x, \mu) = \mu x - x^3,$$

with $x = x(t) \in \mathbb{R}$ and $\mu \in \mathbb{R}$. The linear stability analysis of $x = x_1^* = 0$ gives: $x_1^* = 0$ is asymptotically stable for $\mu < 0$ and unstable for $\mu > 0$. At $\mu = 0$ a real eigenvalue crosses the imaginary axis and two further fixed points $x_{2,3}^* = \pm\sqrt{\mu}$ bifurcate from $x_1^* = 0$. There is an exchange of stability: for $\mu < 0$, $x_1^* = 0$ is stable; for $\mu > 0$, $x_1^* = 0$ is unstable and $x_{2,3}^* = \pm\sqrt{\mu}$ are stable, since for $\mu > 0$ the linearization $A = (\mu - 3x^2)|_{x=x_{2,3}^*} = -2\mu$ has the negative eigenvalue -2μ. Since the fixed points only exist for $\mu > 0$, where $x_1^* = 0$ is unstable, this bifurcation is called a supercritical pitchfork bifurcation. In case $f(x, \mu) = \mu x + x^3 = 0$ we can explicitly compute the bifurcating unstable branches $x_{2,3}^* = \pm\sqrt{-\mu}$ for $\mu < 0$. These exist where the primary solution $x_1^* = 0$ is stable, and the bifurcation is called a subcritical pitchfork bifurcation. See Figure 3.2. ⌋

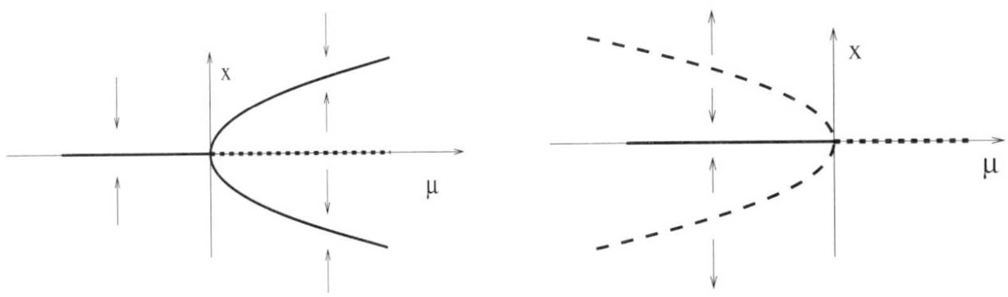

Figure 3.2. Super- and subcritical pitchfork bifurcation of fixed points.

There are two other elementary bifurcations of fixed points, namely the transcritical bifurcation and the saddle-node bifurcation.

Example 3.1.2. (Transcritical bifurcation of fixed points) Consider

$$\dot{x} = \mu x - x^2,$$

with $x = x(t) \in \mathbb{R}$ and $\mu \in \mathbb{R}$. The trivial fixed point $x = x_1^* = 0$ is asymptotically stable for $\mu < 0$ and unstable for $\mu > 0$. For $\mu = 0$ a real eigenvalue crosses the imaginary axis. There exists another fixed point $x_2^* = \mu$, which coincides with the trivial solution $x_1^* = 0$ for $\mu = 0$. Since in general we know a priori only the trivial solution, we say that the fixed point $x_2^* = \mu$ bifurcates from the trivial solution $x_1^* = 0$. For the transcritical bifurcation an exchange of stability takes place: for $\mu < 0$, $x_1^* = 0$ is stable and $x_2^* = \mu$ is unstable; for $\mu > 0$, $x_1^* = 0$ is unstable and $x_2^* = \mu$ is stable. See Figure 3.3. ⌋

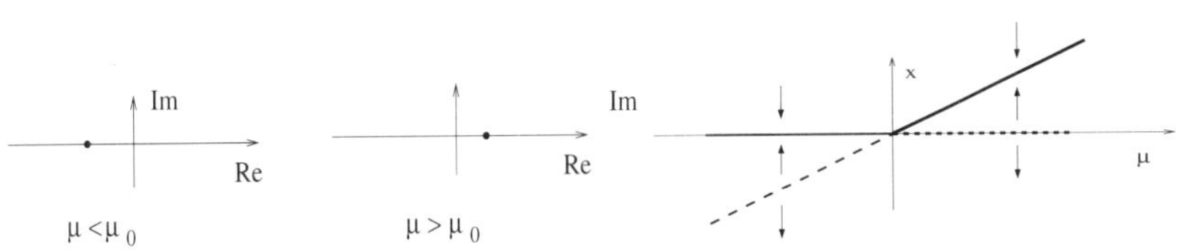

Figure 3.3. A real eigenvalue crosses the imaginary axis leading (here) to a transcritical bifurcation.

Example 3.1.3. (Saddle-node or flip bifurcation of fixed points) Consider

$$\dot{x} = \mu - x^2,$$

with $x = x(t) \in \mathbb{R}$ and $\mu \in \mathbb{R}$. Two fixed points $x_{1,2}^* = \pm\sqrt{\mu}$ appear at $\mu = 0$. The linearization around $x_{1,2}^*$ gives $\mp 2\sqrt{\mu}$. Thus, x_1^* is stable and x_2^* is unstable. See Figure 3.4. The origin of the name saddle-node bifurcation can be seen in $\dot{x} = \mu - x^2$, $\dot{y} = -y$: for this system, $(x_2^*, 0)$ is a saddle and $(x_1^*, 0)$ is a stable node. See Figure 3.4. ⌋

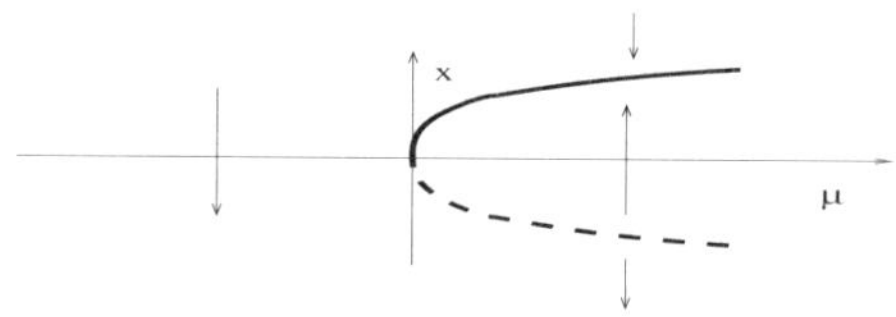

Figure 3.4. Saddle-node-bifurcation.

There is another elementary bifurcation which may occur when a globally attracting fixed point becomes unstable, namely the bifurcation of periodic solutions from a fixed point.

Example 3.1.4. (Hopf bifurcation) Consider

$$\dot{x}_1 = \mu x_1 + x_2 - x_1(x_1^2 + x_2^2) \quad \text{and} \quad \dot{x}_2 = -x_1 + \mu x_2 - x_2(x_1^2 + x_2^2), \tag{3.1}$$

with $x_j(t) \in \mathbb{R}$ and $\mu \in \mathbb{R}$. The linearization $A = \begin{pmatrix} \mu & 1 \\ -1 & \mu \end{pmatrix}$ around $x = 0$ possesses the eigenvalues $\lambda_{1,2} = \mu \pm \mathrm{i}$, i.e., two complex conjugate eigenvalues cross the imaginary axis for $\mu = 0$. See Figure 3.5. Introducing polar coordinates $x_1 = r \sin \phi$ and $x_2 = r \cos \phi$ with $r \geq 0$ and $\phi \in \mathbb{R}/(2\pi\mathbb{Z})$ gives

$$\dot{r} = \mu r - r^3 \quad \text{and} \quad \dot{\phi} = 1.$$

Hence, a family of periodic solutions

$$\{x = x_{\mathrm{per}}(t, \mu, \phi_0) : x_1 = \sqrt{\mu} \sin(t + \phi_0), \; x_2 = \sqrt{\mu} \cos(t + \phi_0)\}$$

bifurcates from the trivial solution $x = 0$ at $\mu = 0$. This is called a supercritical Hopf bifurcation. For fixed $\mu > 0$ the family attracts every solution with an exponential rate $\mathcal{O}(\exp(-2\mu t))$, see Figure 3.5.

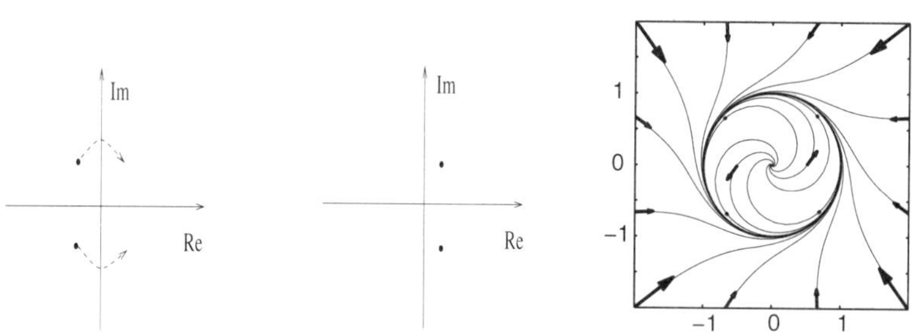

Figure 3.5. Two complex conjugate eigenvalues cross the imaginary axis and the phase portrait for (3.1) for $\mu > 0$.

In §3.3 we shall see that this bifurcation occurs generically when a fixed point loses stability due to two complex conjugate eigenvalues crossing the imaginary axis. ⌋

Example 3.1.5. (Subcritical Hopf bifurcation with turning point) Consider

$$\begin{pmatrix} \dot{x} \\ \dot{y} \end{pmatrix} = \begin{pmatrix} \varepsilon & -1 \\ 1 & \varepsilon \end{pmatrix} \begin{pmatrix} x \\ y \end{pmatrix} + (r^2 - r^4) \begin{pmatrix} x \\ y \end{pmatrix},$$

with $r^2 = x^2 + y^2$. The occurrence of $x^2 + y^2$ on the right-hand side suggests the use of polar coordinates which yields $\dot{r} = \varepsilon r + r^3 - r^5$, $\dot{\phi} = 1$. Thus, for $\varepsilon < 0$, with small $|\varepsilon|$, there exists an unstable periodic solution $r \equiv r_0 = \mathcal{O}(\sqrt{|\varepsilon|})$. The second (stable) fixed point $r_1 = \mathcal{O}(1)$ of $\dot{r} = \varepsilon r + r^3 - r^5$ yields a stable periodic solution, see Figure 3.6. The bifurcation is subcritical. The small amplitude non-trivial branch exists in the parameter regime ($\varepsilon < 0$) where the trivial solution is stable. Due to the turning point and the $\mathcal{O}(1)$ amplitude stable periodic orbits in the subcritical regime, this is also called a hard bifurcation since in systems described by such a model the solution may suddenly "jump" to the $\mathcal{O}(1)$ amplitude stable periodic orbit, because in applications noise will push the solution beyond the unstable periodic orbit. In contrast, supercritical bifurcations (stable non-trivial solutions only start to exist after the trivial solution becomes unstable) are soft, since then the bifurcating stable periodic orbits have $\mathcal{O}(\sqrt{\varepsilon})$ amplitude. ⌋

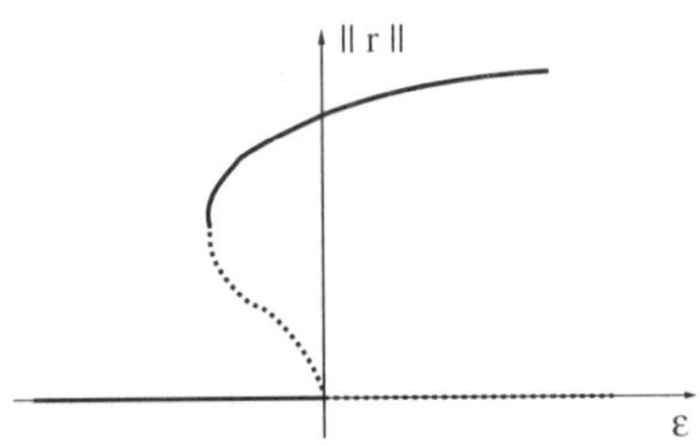

Figure 3.6. Subcritial Hopf bifurcation with turning point

3.1.2. Bifurcations of fixed points. It is the purpose of this section to prove the occurrence of transcritical and pitchfork bifurcations of fixed points from a fixed point in case when the branch of the bifurcating solutions cannot be computed explicitly. The following analysis is based on scaling arguments and the implicit function theorem. The detection of fixed points for the ODE $\dot{x} = f(x, \mu)$, where $\mu \in \mathbb{R}$, leads to the algebraic equation

$$f(x, \mu) = 0. \tag{3.2}$$

Throughout this section we restrict to analytic $f : \mathbb{R}^2 \to \mathbb{R}$. Assume that a solution (x_0, μ_0) of (3.2) is known, i.e., $f(x_0, \mu_0) = 0$, and assume that $\partial_x f(x_0, \mu_0) \neq 0$. Then, by the implicit function theorem there exists a unique smooth solution $x = x(\mu)$ of (3.2), i.e., $f(x(\mu), \mu) = 0$, in a neighborhood of (x_0, μ_0). This solution can be extended outside the neighborhood of (x_0, μ_0) with the same argument until the assumption $\partial_x f(x(\mu), \mu)) \neq 0$ is no longer satisfied. In such a point (x_0, μ_0) a new branch of solutions can bifurcate from this family of solutions $\eta \mapsto (x, \mu)(\eta)$. This so called bifurcation point can be analyzed with the Newton polygon which is explained subsequently. It turns out that generically only two situations for the bifurcations of fixed points can occur, namely the transcritical bifurcation from Example 3.1.2 or the pitchfork bifurcation from Example 3.1.1.

Scaling arguments. One way to establish the existence of bifurcating solutions in the general case are scaling arguments and the implicit function theorem.

Example 3.1.6. Let $f(x, \mu) = \mu x + x^2 + \sin x$. Then $f(0, \mu) = 0$ for all $\mu \in \mathbb{R}$. Hence, $x = x_1^* = 0$ is the trivial solution for all $\mu \in \mathbb{R}$. For all values of $\mu \in \mathbb{R}$ we have $\partial_\mu f|_{x=0} = 0$. Hence, it is sufficient to consider

$$\partial_x f|_{x=0} = (\mu - 2x + \cos x)|_{x=0} = \mu + 1.$$

Thus, a bifurcation can only take place when $\mu + 1 = 0$. Therefore, we introduce the small bifurcation parameter $\alpha = \mu + 1$. In order to find Example 3.1.2 in the present problem we rescale $x = \alpha y$ and introduce

$$F(y, \alpha) = \alpha^{-2} f(\alpha y, 1 + \alpha) = y + y^2 + \mathcal{O}(\alpha).$$

Thus, for $\alpha = 0$ we have the simple equation $F(y, 0) = y + y^2$ having the solutions $y_1^* = 0$ and $y_2^* = -1$. According to

$$\partial_y F|_{(y,\alpha)=(y_j^*,0)} = (1 + 2y)|_{(y,\alpha)=(y_j^*,0)} \neq 0$$

we can apply the implicit function theorem to solve $F = 0$ in a neighborhood of $(y, \alpha) = (y_j^*, 0)$ for y and and obtain $y_1^* = 0 + \mathcal{O}(\alpha)$ and $y_2^* = -1 + \mathcal{O}(\alpha)$. Hence, beside the trivial solution $x_1^* = 0$ we also found the bifurcating solution $x_2^* = \alpha + \mathcal{O}(\alpha^2)$. ⌋

Example 3.1.7. Consider $f(x, \mu) = \mu x + \sin x$. Again $x = x_1^* = 0$ becomes unstable at $\mu = -1$. Let $\alpha^2 = \mu + 1$ and $x = \alpha y$. The rescaled problem

$$F(y, \alpha) = \alpha^{-3} f(\alpha y, 1 + \alpha^2) = y - \frac{1}{6} y^3 + \mathcal{O}(\alpha^2) = 0$$

can be explicitly solved for $\alpha = 0$. Using the implicit function theorem we obtain $y_1^* = 0 + \mathcal{O}(\alpha^2)$ and $y_{2,3}^* = \pm\sqrt{6} + \mathcal{O}(\alpha^2)$, hence $x_1^* = 0$ and $x_{2,3}^* = \pm\sqrt{6}\alpha + \mathcal{O}(\alpha^3)$ for $\alpha > 0$. ⌋

Also more general situations can be handled by scaling arguments.

Example 3.1.8. Consider

$$f(x, \varepsilon) = x^2 + x\varepsilon + \varepsilon^3 = 0. \tag{3.3}$$

We have the trivial solution $(0,0)$ for which the assumptions of the implicit function theorem are not satisfied, i.e., $\partial_x f(0,0) = \partial_\varepsilon f(0,0) = 0$. Again we are interested in non-trivial solutions $x = x(\varepsilon)$ near the origin.

We make the ansatz $x(\varepsilon) = \varepsilon v(\varepsilon)$ and obtain

$$F(v,\varepsilon) = \varepsilon^{-2} f(\varepsilon v, \varepsilon) = v^2 + v + \varepsilon = 0.$$

For $\varepsilon = 0$ we find the non-trivial solution $v_1^* = -1$. We additionally have $\partial_v F(-1,0) = -1 \neq 0$ such that we can apply the implicit function theorem and obtain a smooth solution $v = v_1^*(\varepsilon) = -1 + \mathcal{O}(\varepsilon)$. Hence, we find a non-trivial solution $x_1^* = -\varepsilon + \mathcal{O}(\varepsilon^2)$ for $f = 0$. However, the ansatz $x(\varepsilon) = \varepsilon^2 v(\varepsilon)$ yields

$$F(v,\varepsilon) = \varepsilon^{-3} f(\varepsilon^2 v, \varepsilon) = \mathcal{O}(\varepsilon) + v + 1 = 0.$$

For $\varepsilon = 0$, we find the non-trivial solution $v_2^* = -1$ and $\partial_v F(-1,0) = 1 \neq 0$. Hence, we can apply the implicit function theorem and obtain a smooth solution $v = v_2^*(\varepsilon) = -1 + \mathcal{O}(\varepsilon)$. Therefore, we found a second curve of non-trivial solutions $x_2^* = -\varepsilon^2 + \mathcal{O}(\varepsilon^3)$ for $f = 0$. The expansions correspond to the solutions

$$x_{1,2}(\varepsilon) = -\frac{\varepsilon}{2} \pm \sqrt{\frac{\varepsilon^2}{4} - \varepsilon^3}$$

of (3.3), which only can be computed explicitly since (3.3) is a second order polynomial w.r.t. x. ⌋

The Newton-polygon. In the last example there exist at least two curves of non-trivial solutions. Since we have a polynomial in the example we can be sure that we found all solutions. For non-polynomial problems the scaling argument can be made rigorous with the help of the Weierstrass preparation theorem which allows to bring analytic f into a polynomial form w.r.t. one of the variables, cf. [**CH82**, §2.8]. With this preparation it is then clear that the solutions which we will find with the scaling arguments are the only non-trivial solutions near $(x,\varepsilon) = (0,0)$.

A systematic approach to find the scalings is as follows. Assume that f can be expanded in some convergent power series near the origin, i.e., $f(x,\varepsilon) = \sum_{m,n}^{\infty} a_{mn} x^m \varepsilon^n$. Whenever the coefficient $a_{mn} \in \mathbb{R}$ is nonzero, make a dot at $(m,n) \in \mathbb{N}_0 \times \mathbb{N}_0$. Then take the lower convex hull of all dots in the $\mathbb{N}_0 \times \mathbb{N}_0$-plane. This hull is the so called Newton-polygon with finitely many line segments with endpoints (m_i, n_i) and (m_{i+1}, n_{i+1}) and slopes $-\alpha_i$. Associated with each of these lines there are p_i solutions $x_i^*(\varepsilon) = \varepsilon^{\alpha_i} v_i^*(\varepsilon)$ of $f(x,\varepsilon) = 0$, where $p_i = m_i - m_{i-1}$.

Example 3.1.9. The Newton polygon for

$$f(x,\varepsilon) = x^3 + 3x^2\varepsilon + 2x\varepsilon^2 + \varepsilon^5 = 0$$

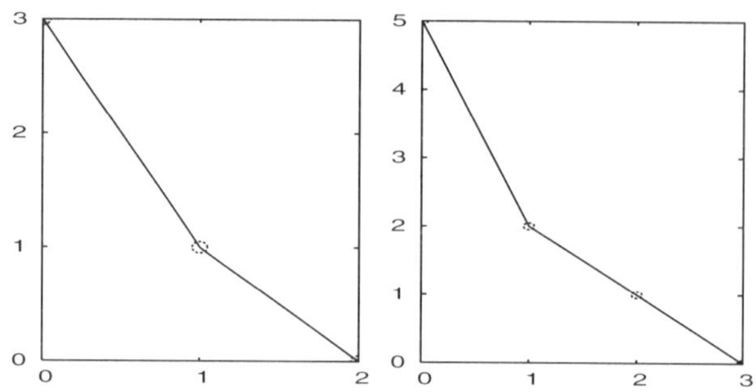

Figure 3.7. Newton polygons for $f(x,\varepsilon)=x^2+x\varepsilon+\varepsilon^3$ and for $f(x,\varepsilon)=x^3+3x^2\varepsilon+2x\varepsilon^2+\varepsilon^5$.

yields $\alpha_1=3$, $p_1=1$ and $\alpha_2=1$, $p_2=2$. With the ansatz $x(\varepsilon)=\varepsilon^\alpha v(\varepsilon)$ we obtain

$$\varepsilon^{3\alpha}v^3+3\varepsilon^{1+2\alpha}v^2+2\varepsilon^{2+\alpha}v+\varepsilon^5=0.$$

The first three terms are of the same leading order for $\alpha=1$. The third and the fourth term are of the same leading order for $\alpha=3$.

For $\alpha=1$ we obtain $F(v,\varepsilon)=v^3+3v^2+2v+\mathcal{O}(\varepsilon^2)=0$. For $\varepsilon=0$ we find the non-trivial solutions $v_1^*=-1$ and $v_2^*=-2$. Since $\partial_v F(-1,0)=(3v^2+6v+2)|_{v=-1}=-1\neq 0$ and $\partial_v F(-2,0)=2\neq 0$ by the implicit function theorem we find the non-trivial solutions $v_1^*(\varepsilon)=-1+\mathcal{O}(\varepsilon^2)$ and $v_2^*(\varepsilon)=-2+\mathcal{O}(\varepsilon^2)$ or equivalently $x_1^*(\varepsilon)=-\varepsilon+\mathcal{O}(\varepsilon^3)$ and $x_2^*(\varepsilon)=-2\varepsilon+\mathcal{O}(\varepsilon^3)$.

For $\alpha=3$, we obtain $F(v,\varepsilon)=\mathcal{O}(\varepsilon^2)+2v+1=0$. For $\varepsilon=0$, we find the non-trivial solution $v_3^*=-1/2$. Since $\partial_v F(-1/2,0)=2\neq 0$ by the implicit function theorem we find the non-trivial solution $v_3^*(\varepsilon)=-1/2+\mathcal{O}(\varepsilon^2)$ or equivalently $x_3^*(\varepsilon)=-\varepsilon^3/2+\mathcal{O}(\varepsilon^5)$. ⌋

Remark 3.1.10. (The genericity of transcritical and pitchfork bifurcations) If w.l.o.g. we assume that the trivial solution is given by $x=x^*=0$, then there exists a smooth function $g:\mathbb{R}^2\to\mathbb{R}$ with $f(x,\mu)=xg(x,\mu)$. If we further assume that the bifurcation point is given by $(x,\mu)=(0,0)$, then $\partial_x f(0,0)=\partial_\mu f(0,0)=0$. This gives the condition $g(0,0)=0$ such that

$$g(x,\mu)=g_{10}x+g_{01}\mu+g_{20}x^2+g_{11}x\mu+g_{02}\mu^2+\mathcal{O}(|x|^3+|\mu|^3),$$

with coefficients $g_{ij}\in\mathbb{R}$. Generically we have $g_{10}\neq 0$ and $g_{01}\neq 0$ such that we find a bifurcating branch with $x=-g_{10}^{-1}g_{01}\mu+\mathcal{O}(\mu^2)$, i.e., a transcritical bifurcation. However, symmetries such as $f(x,\mu)=-f(-x,\mu)$ can force $g_{10}=0$. Solving the equation $g(x,\mu)=0$ w.r.t. μ and using the Newton polygon we find by the ansatz $\mu(x)=x^2s(x)$ that $g_{01}\mu$ and $g_{20}x^2$ are of the same order. We obtain

$$G(x,s)=x^{-2}g(x,x^2s)=g_{01}s+g_{20}+\mathcal{O}(x)=0$$

and hence $s = -g_{01}^{-1} g_{20}$ such that $\mu = -g_{01}^{-1} g_{20} x^2 + \mathcal{O}(x^3)$, i.e., depending on the sign of $g_{01}^{-1} g_{20}$ a sub- or a supercritical pitchfork bifurcation occurs. More coefficients can vanish, but this is a degenerated situation, which requires more symmetries.

By a small perturbation the vanishing coefficients can be made nonzero. This is called unfolding of the bifurcation, cf. [**GS85**, Chapter III]. Such unfoldings are robust w.r.t. other small perturbations, i.e., additional parameters different from the unfolding parameters will not change the solution set qualitatively. ⌋

3.1.3. The Lyapunov-Schmidt reduction. We consider now an ODE

$$\dot{x} = f(x, \mu)$$

with $x(t) \in \mathbb{R}^d$ in case $d > 1$ under the assumption that at $\mu = \mu_0$ one simple eigenvalue crosses the imaginary axis and that all other eigenvalues have negative real part. See Figure 3.8. We remark that for the subsequent analysis it is sufficient that all eigenvalues except of one are bounded away from the imaginary axis, i.e., eigenvalues with positive real part are allowed, too.

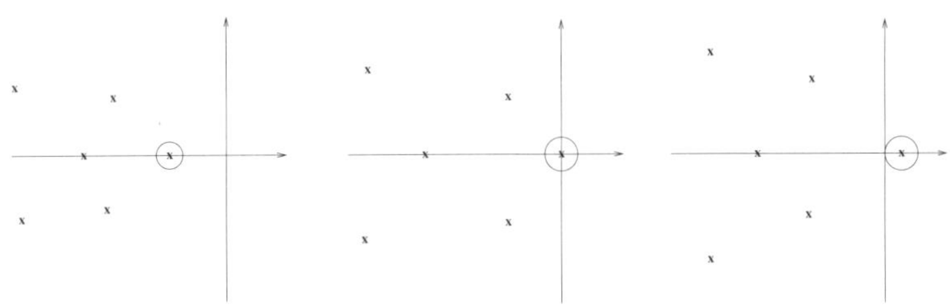

Figure 3.8. Spectral situation in case of a fixed point becoming unstable via a pitchfork or transcritical bifurcation.

In order to establish a pitchfork or a transcritical bifurcation we use the so called Lyapunov-Schmidt method, which allows to reduce the d-dimensional problem $f(x, \mu) = 0$ to a one-dimensional one. So far we have restricted ourselves to problems $f : \mathbb{R} \times \mathbb{R} \to \mathbb{R}$. In principle, the assertions from §3.1.2 remain valid also in the general case $f : B \times \mathbb{R} \to B$ with B a Banach space. This means that under the previous spectral assumption generically only transcritical and pitchfork bifurcations occur. In contrast to the examples above, x and μ are no longer equivalent. We distinguish between the variable x and the parameter μ.

Only to avoid a number of technicalities we restrict ourselves in the following to $B = \mathbb{R}^d$. By the implicit function theorem we can compute a solution $x = x(\mu)$ for growing μ until $M = \partial_x f(x(\mu), \mu) \in \mathbb{R}^{d \times d}$ is no longer invertible. We denote this point by (x_0^*, μ_0). Under the previous spectral assumption non-invertibility is equivalent to the fact that exactly one of the d eigenvalues of M is zero. W.l.o.g. we assume that the associated eigenvector is given by e_1. In order to apply the implicit function theorem in this situation, we split the system into two parts, namely into a part where the implicit function theorem can be applied and into a part where it cannot be applied. Denote by P_1 a projection on $\mathrm{span}\{\mathrm{e}_1\} = (1, 0, \ldots, 0)$ and let $P_2 = I - P_1$. Moreover, we set $x_j = P_j x$, denote by Q_2 a projection on the range of M, and let $Q_1 = I - Q_2$. Then, we consider

$$Q_1 f((x_1, x_2), \mu) = 0, \quad \text{and} \quad Q_2 f((x_1, x_2), \mu) = 0.$$

We find that $\partial_{x_2} Q_2 f(x_0, \mu_0) \in \mathbb{R}^{d-1\times d-1}$ is invertible since it possesses the $(d-1)$ non-zero eigenvalues of M. Hence, the second equation can be solved locally w.r.t. x_2, i.e., there exists a solution $x_2 = x_2(x_1, \mu)$. Inserting this solution into the first equation gives the so called reduced bifurcation problem

$$\widetilde{f}(x_1, \mu) = Q_1 f((x_1, x_2(x_1, \mu)), \mu) = 0.$$

After this so called Lyapunov-Schmidt reduction we are in the same situation as in the previous section. We have to find the zeroes of a smooth function $\widetilde{f} : \mathbb{R}^2 \to \mathbb{R}$. If symmetries are present in the problem, then the projections can be chosen in such a way that the symmetries are preserved by the reduction [**GS85**, Chapter VII.3].

Example 3.1.11. Consider

$$\begin{aligned} f_1(x, y, \varepsilon) =& \varepsilon x - yx - x^3 = 0, \\ f_2(x, y, \varepsilon) =& y + 2x^2 + y^2\varepsilon^2 = 0. \end{aligned}$$

The origin $(x, y) = (0, 0)$ is a solution for all $\varepsilon \in \mathbb{R}$, and we are interested in non-trivial solutions close to it. The linearization

$$\partial_{(x,y)}(f_1, f_2)|_{(x,y)=(0,0)} = \begin{pmatrix} \varepsilon & 0 \\ 0 & 1 \end{pmatrix}$$

has the eigenvalues ε and 1. Hence, a bifurcation is only possible for $\varepsilon = 0$. The kernel is given by $\{(x, 0) : x \in \mathbb{R}\}$ and the range by $\{(0, y) : y \in \mathbb{R}\}$. Thus, the above system is already in the form needed for the Lyapunov-Schmidt reduction, and the second equation can be solved w.r.t. y. In order to obtain an approximate solution we consider an iteration of the second equation, namely

$$y = -2x^2 - y^2\varepsilon^2 = -2x^2 - \left(-2x^2 - y^2\varepsilon^2\right)^2 \varepsilon^2 = -2x^2 + \mathcal{O}\left(\varepsilon^2 x^4\right).$$

Inserting the solution $y = -2x^2 + \mathcal{O}(\varepsilon^2 x^4)$ into the first equation gives the bifurcation equation

$$f_1(x, y(x, \varepsilon), \varepsilon) = \varepsilon x - \left(-2x^2 + \mathcal{O}\left(\varepsilon^2 x^4\right)\right) x - x^3 = \varepsilon x + x^3 + \mathcal{O}\left(\varepsilon^2 x^5\right) = 0.$$

Dividing this equation by x gives $\varepsilon+x^2+\mathcal{O}\left(\varepsilon^2x^4\right)=0$ which can be analyzed by the Newton polygon. We find a subcritical pitchfork bifurcation, i.e., non-trivial solutions $x^*(\varepsilon)=\pm\sqrt{-\varepsilon}+\mathcal{O}(\varepsilon)$ and $y^*(\varepsilon)=\mathcal{O}(\varepsilon)$ for $\varepsilon<0$. ⌋

Consequence. When a fixed point becomes unstable by a simple eigenvalue crossing the imaginary axis, generically a transcritical or a pitchfork bifurcation occurs. Hence, even in higher-dimensional phase spaces for this spectral situation no new bifurcations can occur.

Remark 3.1.12. The Lyapunov-Schmidt reduction has certain disadvantages. It does not provide information about the stability of bifurcating solutions. Treating Hopf bifurcations via Lyapunov-Schmidt reduction leads to an infinite-dimensional problem. In order to find $2\pi/\omega$-time-periodic solutions of the ODE $\dot{x}(t)=f(x,\mu)$ the problem is transferred by using Fourier series $x(t)=\sum_{j\in\mathbb{Z}}\widehat{x}_j e^{i\omega jt}$ to an infinite-dimensional stationary problem for the Fourier coefficients $\widehat{x}_j$. With the help of the Lyapunov-Schmidt reduction the problem can be reduced to a two-dimensional one, cf. [**CH82**, §1.4]. The construction of homoclinic and heteroclinic solutions with this method again leads to an infinite-dimensional problem and requires special properties of the underlying ODE, cf. [**PS07**]. ⌋

3.2. Center manifold theory

Center manifold theory is an alternative way to find the elementary bifurcations from above also in higher space dimensions. Additionally, it often yields information on the stability of the bifurcating solutions. Moreover, in contrast to the Lyapunov-Schmidt reduction with this method Hopf bifurcations and the occurrence of small amplitude homoclinic and heteroclinic solutions can be handled as finite-dimensional problems.

If a fixed point becomes unstable, all solutions are attracted with some exponential rate towards the center manifold, i.e., the interesting non-trivial dynamics happens on the center manifold of the unstable fixed point. In general, only polynomial approximations of the vector field on the center manifold are known. If the center manifold has two and more space dimensions, so called normal form transformations help us to analyze the dynamics on the center manifold. In the next section we use center manifold theory and normal forms to prove a general Hopf bifurcation theorem.

We already formulated in Theorem 2.3.19 the invariant manifold theorem. The part about the center manifold is now made more precise. We consider

$$\begin{aligned}\dot{u}_c &= B_c u_c + \widetilde{g}_c(u_c,u_s,u_u),\\ \dot{u}_s &= B_s u_s + \widetilde{g}_s(u_c,u_s,u_u),\\ \dot{u}_u &= B_u u_u + \widetilde{g}_u(u_c,u_s,u_u),\end{aligned}\tag{3.4}$$

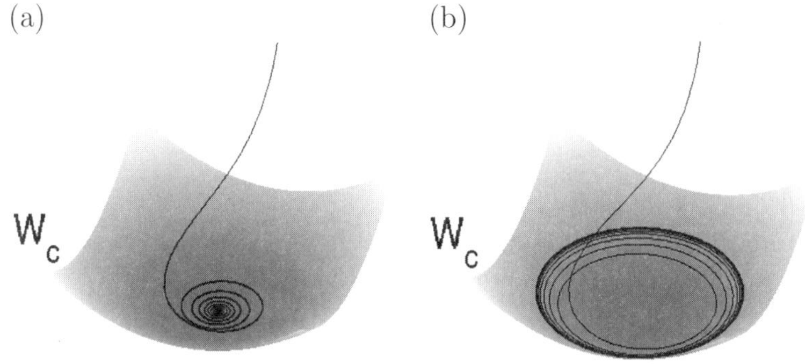

Figure 3.9. When a fixed point becomes unstable the bifurcating solutions can be found on the exponentially attracting center manifold. (a),(b) before and after the bifurcation of a stable periodic orbit on M_c.

with $u_c \in E_c = \mathbb{R}^{d_c}, u_s \in E_s = \mathbb{R}^{d_s}, u_u \in E_u = \mathbb{R}^{d_u}$ some finite-dimensional vectors, B_c a matrix with eigenvalues on the imaginary axis, B_s a matrix with eigenvalues with negative real part, B_u a matrix with eigenvalues with positive real part, and $\widetilde{g}_c : \mathbb{R}^d \to E_c, \widetilde{g}_s : \mathbb{R}^d \to E_s$ and $\widetilde{g}_u : \mathbb{R}^d \to E_u$, $d = d_c + d_s + d_u$, are C^{r+1}-maps without constant and linear terms.

Theorem 3.2.1. (Center manifold theorem) *There exists a neighborhood $U \subset E_c$ of $u_c = 0$ and a C^r-map $h : U \ni u_c \mapsto h(u_c)$ such that the manifold*

$$W_c = \{u = u_c \oplus h(u_c) : u_c \in U,\ (u_s, u_u) = h(u_c)\}$$

is invariant under the flow of (3.4). *W_c is called the center manifold. The reduced flow is determined by*

$$\dot{u}_c = B_c u_c + \widetilde{g}_c(u_c, h_s(u_c), h_u(u_c)). \tag{3.5}$$

The function h contains no constant and no linear terms w.r.t. u_c such that the center manifold W_c is tangential to the central subspace E_c associated to the eigenvalues with vanishing real part. In general the center manifold is not unique.

Some parts of the theorem, namely the invariance, the existence, and Lipschitz-continuity instead of r-times differentiability of the center manifold will be proven in §13.1. Here we will discuss some of the assertions of the theorem and concentrate on its application by giving a number of examples. With the first example we explain how center manifold theory can be used to handle bifurcation problems although the central eigenvalues are only on the imaginary axis for one particular value of the bifurcation parameter

Example 3.2.2. For μ close to zero consider the trivial decoupled system

$$\dot{x} = \mu x - x^3, \qquad \dot{y} = -y. \tag{3.6}$$

For $\mu < 0$, the origin $(x, y) = (0, 0)$ is stable. For $\mu = 0$, we find the one-dimensional center manifold $W_c = \{(x, y) \in \mathbb{R}^2 : y = 0\}$. In order to handle non-zero values of μ with the center manifold theorem the above system is extended to

$$\dot{x} = \mu x - x^3, \quad \dot{y} = -y, \quad \dot{\mu} = 0.$$

For this extended system we find the two-dimensional center manifold $W_c = \{(\mu, x, y) \in \mathbb{R}^3 : y = 0\}$. Note that after introducing $\dot{\mu} = 0$ the term μx is no longer a linear, but a nonlinear term. Since $\dot{\mu} = 0$ implies that μ is a constant, the two-dimensional center manifold is foliated by one-dimensional invariant manifolds. See Figure 3.10. Hence, the additional equation $\dot{\mu} = 0$ can be canceled again and on the two-dimensional center manifold μ can be considered again as a parameter. Therefore, by applying the center manifold theorem in a sloppy way the two-dimensional bifurcation problem (3.6) can be reduced to

$$\dot{x} = \mu x - x^3$$

in the one-dimensional center manifold $W_c = \{(x, y) \in \mathbb{R}^2 : y = 0\}$. Obviously, the reduction is trivial in this case, i.e., $h = 0$. In summary, bifurcation problems can be handled with the help of the center manifold theorem by introducing the equation $\dot{\mu} = 0$. ⌋

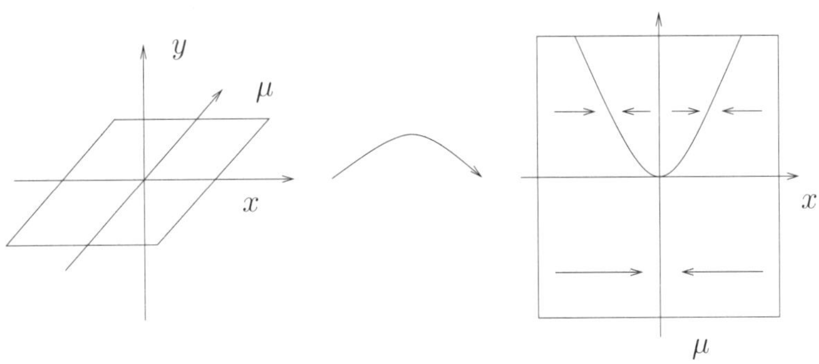

Figure 3.10. Reduction of the system $\dot{x} = \mu x - x^3$, $\dot{y} = -y$, $\dot{\mu} = 0$ to a two-dimensional center manifold which is foliated by invariant one-dimensional manifolds.

The next example shows how to compute approximations of the reduction function h and of the reduced system on the center manifold.

Example 3.2.3. For μ close to zero consider

$$\dot{x} = \mu x + x^3 - xy, \qquad \dot{y} = -y + 2x^2.$$

Like above we extend the system by the equation $\dot{\mu} = 0$. The linearized system is given by

$$\dot{x} = 0, \quad \dot{y} = -y, \quad \dot{\mu} = 0,$$

and hence $E_c = \{(\mu, x, y) \in \mathbb{R}^3 : y = 0\}$. Therefore, we make the ansatz

$$y = h(x, \mu) = ax^2 + b\mu x + c\mu^2 + \mathcal{O}(|\mu|^3 + |x|^3),$$

and from $\dot{y} = -y + 2x^2$ we obtain $2ax\dot{x} + \mu\dot{x} + \ldots = -(ax^2 + b\mu x + c\mu^2 + \ldots) + 2x^2$. Since $\dot{x} = \mu x + \ldots$, by comparing the coefficients it follows

$$x^2 : \ 0 = -a + 2, \quad x\mu : \ 0 = -b, \quad \mu^2 : \ 0 = -c, \quad \ldots$$

As a general rule, no powers μ^n without x can occur in h, and therefore

$$W_c = \{(\mu, x, y) \in \mathbb{R}^3 : y = 2x^2 + \mathcal{O}(|\mu|x^2 + |x|^3)\}.$$

Moreover, the function h cannot contribute to the quadratic terms of the reduced system which here is given by

$$\dot{x} = \mu x + x^3 - x(2x^2) + \text{h.o.t} = \mu x - x^3 + \text{h.o.t.},$$

i.e., the fixed point $(x, y) = (0, 0)$ is stable also for $\mu = 0$. We explain below that stability on the center manifold implies stability in the full system in such a situation. At $\mu = 0$ a supercritical pitchfork bifurcation occurs. ⌋

The following two examples are about the non-uniqueness and the non-smoothness of center manifolds.

Example 3.2.4. In order to illustrate the non-uniqueness of the center manifold we consider

$$\dot{x} = x^2, \qquad \dot{y} = -y. \tag{3.7}$$

Obviously, the central subspace E_c is given by the x-axis. The solutions of the ODEs are given by $x(t) = \frac{x_0}{1-tx_0}$ and $y(t) = y_0 e^{-t}$. Elimination of time t yields $y(x) = (y_0 e^{-1/x_0})e^{1/x}$. For $x < 0$ we have $\lim_{x\to 0, x<0} y^{(n)}(x) = 0$, i.e., every solution approaches the origin in a flat way. For $x > 0$ we find that $y = 0$ is the only solution which approaches the origin. Thus, we find infinitely many different C^∞-center manifolds which are tangential to E_c at the origin by glueing together the obits in the left half plane with the positive real axis. This shows that center manifolds are non-unique in general. The only analytic center manifold, i.e., with a convergent power series, is the x-axis. See Figure 3.11 for the phase portrait of (3.7). ⌋

Example 3.2.5. In order to illustrate the non-smoothness (and non-uniqueness) of the center manifold we consider

$$\dot{x} = -\mu x, \quad \dot{y} = -y, \quad \dot{\mu} = 0$$

with $0 < \mu < 1$. The vector field is C^∞ and E_c is given by the (x, μ)-plane. Obviously $W_c = E_c$ is a center manifold. However, the trajectories satisfy $\frac{\mathrm{d}}{\mathrm{d}x} y = \mu \frac{y}{x}$ and are given through $y(x) = C|x|^{1/\mu}$. If $r < 1/\mu < r + 1$ with $r \in \mathbb{N}$, then the trajectories for fixed μ are in C^r, but not in C^{r+1}. Each of

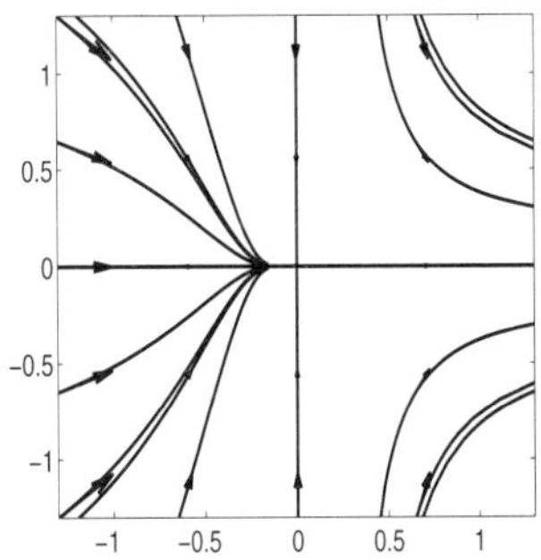

Figure 3.11. The phase portrait of $\dot{x} = x^2, \quad \dot{y} = -y$.

these curves is tangent to $y = 0$ and so the whole of these trajectories forms another center manifold. In a ball of radius $\mu < 1/r$ this center manifold is C^r, i.e., the larger r is chosen, the smaller is the center manifold. ⌋

Remark 3.2.6. In case of no eigenvalues with positive real part there is a neighborhood in which all solutions are attracted to the associated center manifold W_c with some exponential rate $\mathcal{O}(\mathrm{e}^{-\beta t})$ for a $\beta > 0$. More precisely, in [**Van89**, Theorem 5.17] it is shown that in this case there are strictly positive constants C and β, such that for all x_0 in a neighborhood of the center manifold there is a $t_0 \in \mathbb{R}$ and a $x_c \in W_c$ such that

$$\|x(t, x_0) - x(t - t_0, x_c)\| \leq C\mathrm{e}^{-\beta t}.$$

As a consequence the stability of bifurcating solutions is solely determined by the reduced ODE on the center manifold. ⌋

Similarly, center manifolds can be defined for discrete dynamical systems.

Example 3.2.7. We consider the discrete dynamical system

$$x_{n+1} = x_n + x_n y_n \qquad y_{n+1} = \lambda y_n - x_n^2$$

with $0 < \lambda < 1$. We find $E_c = \{(x, y) \in \mathbb{R}^2 : y = 0\}$. In order to compute the center manifold we make the ansatz

$$y = h(x) = ax^2 + bx^3 + \mathcal{O}(x^4)$$

and find

$$a(x + x(ax^2 + \ldots))^2 + b(x + x(ax^2 + \ldots))^3 + \ldots = \lambda(ax^2 + bx^3 + \ldots) - x^2$$

which yields $a = -\frac{1}{1-\lambda}$ and $b = 0$. Hence, we have

$$y = h(x) = -\frac{x^2}{1 - \lambda} + \mathcal{O}(x^4)$$

and find for the reduced equation

$$x_{n+1} = x_n - \frac{x_n^3}{1-\lambda} + \mathcal{O}(x_n^5) = x_n\left(1 - \frac{x_n^2}{1-\lambda} + \mathcal{O}(x_n^4)\right).$$

Therefore, $x = 0$ is asymptotically stable in the reduced equation which implies the asymptotic stability of the origin $(x, y) = (0, 0)$ in the full system, similarly to the previous Remark 3.2.6. ⌋

Example 3.2.8. (Saddle-Node bifurcation on center manifold) We consider

$$\dot{x} = \varepsilon + x^2 + y^2 \quad \text{and} \quad \dot{y} = -y + x^2 \tag{3.8}$$

with small ε. For $\varepsilon = 0$ we have the fixed point $(x, y) = (0, 0)$ with eigenvalues $0, -1$ with neutral direction $(1, 0)$. Thus we expand the center manifold as $y = h(x, \varepsilon) = ax^2 + bx\varepsilon + c\varepsilon^2 + \ldots$, which yields $a = 1$, $b = -2$, $c = 2$ and hence $\dot{x} = \varepsilon + x^2 + \mathcal{O}(x^4)$ as reduced equation. We have a saddle-node bifurcation with two fixed points $-\sqrt{-\varepsilon}$ (stable) and $\sqrt{-\varepsilon}$ (saddle) for $\varepsilon < 0$ and no fixed point for $\varepsilon > 0$. Thus we do not actually need h. See Figure 3.12 for the phase portrait. ⌋

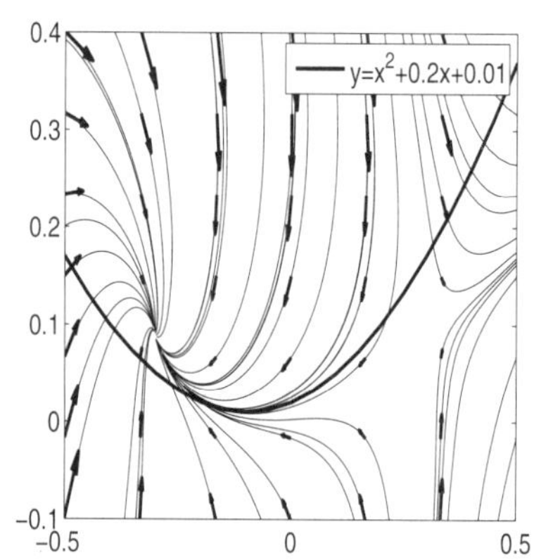

Figure 3.12. Saddle-node bifurcation on center manifold; $\varepsilon = -0.1$. .

As already said, some parts of the center manifold theorem, cf. Theorem 3.2.1, namely the invariance, the existence, and the Lipschitz-continuity of the center manifold will be proven in §13.1. There we will explain that center manifold theory is not restricted at all to the finite-dimensional situation. In Part IV it is used for the construction of bifurcating spatially periodic solutions of pattern forming systems, but also in the construction of traveling wave solutions in unbounded cylindrical domains.

3.3. The Hopf bifurcation

In case that two complex conjugate eigenvalues cross the imaginary axis, the analysis of the system can be reduced to the analysis of the associated two-dimensional center manifold. For the ODE on the center manifold, however, a large number of coefficients have to be computed, namely six coefficients for the quadratic terms and eight coefficients for the cubic terms. Hence, at a first glance a big zoo of possible dynamics can be expected. However, this is not true. By normal form transformations the problem can be reduced in polar coordinates to

$$\dot{r} = \nu_1 r + \nu_2 r^3 + \dots \quad \text{and} \quad \dot{\phi} = 1 + \dots, \tag{3.9}$$

with $\nu_1, \nu_2 \in \mathbb{R}$, i.e., to the computation of two efficient coefficients. Ignoring the higher order terms this system has already been discussed in Example 3.1.4. Therefore, in case that two complex conjugate eigenvalues cross the imaginary axis, under some non-degeneracy condition, always time-periodic solutions occur, either as sub- or supercritical bifurcation.

Theorem 3.3.1. *Consider* $\dot{x} = A_\mu x + g(x)$ *with* $x(t) \in \mathbb{R}^d$ *and* $\|g(x)\| = \mathcal{O}(\|x\|^2)$ *for* $x \to 0$. *Assume that for* $\mu = \mu_0$ *the matrix* A_μ *possesses two eigenvalues* $\lambda_\pm = \pm \mathrm{i}\omega$ *with* $\omega \neq 0$ *and that all other eigenvalues possess strictly negative real part. Furthermore assume that* $\frac{\mathrm{d}\,\mathrm{Re}\,\lambda_\pm}{d\mu}|_{\mu=\mu_0} \neq 0$. *If* $\nu_2 \neq 0$ *in* (3.9), *or more precisely* $\gamma_r \neq 0$ *in* (3.13) *below, then a one parametric family of periodic solutions bifurcates from* $x = 0$ *at* $\mu = \mu_0$. *The period of the bifurcating solutions is* $2\pi/\omega + \mathcal{O}(|\mu - \mu_0|)$ *and their amplitude is of order* $\mathcal{O}(|\mu - \mu_0|^{1/2})$.

Proof. For the somewhat lengthy proof we introduce the new bifurcation parameter $\varepsilon = \mu - \mu_0$ and extend the ODE system with $\dot{\varepsilon} = 0$. Then we apply the center manifold theorem and reduce the full system to a system on the three-dimensional center manifold associated to the eigenvalues $\lambda_\pm$ and the variable ε.

On the center manifold M_c for arbitrary coordinates $(y, z) \in \mathbb{R}^2$ the reduced system can be written as

$$\begin{aligned}
\dot{y} &= a_{11}y + a_{12}z + a_{120}y^2 + a_{111}yz + a_{102}z^2 + a_{130}y^3 + a_{121}y^2 z \\
&\qquad + a_{112}yz^2 + a_{003}z^3 + \mathcal{O}(\varepsilon^2(|x|+|y|) + |y|^4 + |z|^4), \\
\dot{z} &= a_{21}y + a_{22}z + a_{220}y^2 + a_{211}yz + a_{202}z^2 + a_{230}y^3 + a_{221}y^2 z \\
&\qquad + a_{212}yz^2 + a_{203}z^3 + \mathcal{O}(\varepsilon^2(|x|+|y|) + |y|^4 + |z|^4), \\
\dot{\varepsilon} &= 0,
\end{aligned} \tag{3.10}$$

where the values of the real-valued coefficients $a_\cdot = a_\cdot(\varepsilon)$ depends on our choice of basis. The only restriction so far is that the matrix $A = \begin{pmatrix} a_{11} & a_{12} \\ a_{21} & a_{22} \end{pmatrix}$

possesses the eigenvalues

$$\lambda_\pm(\mu) = \pm \mathrm{i}\omega + \mathcal{O}(\varepsilon).$$

At a first view, all kinds of dynamics seem to be possible. However, the system can be simplified heavily with a so called normal form transform. In order to do so we diagonalize (3.10) and obtain

$$\begin{aligned}
\dot{c}_1 &= \mathrm{i}\omega c_1 + \alpha^1_{110}\varepsilon c_1 + \alpha^1_{101}\varepsilon c_{-1} \\
&\quad + \alpha^1_{020} c_1^2 + \alpha^1_{011} c_1 c_{-1} + \alpha^1_{002} c_{-1}^2 + \ldots, \\
\dot{c}_{-1} &= -\mathrm{i}\omega c_{-1} + \alpha^2_{110}\varepsilon c_1 + \alpha^2_{101}\varepsilon c_{-1} \\
&\quad + \alpha^2_{020} c_1^2 + \alpha^2_{011} c_1 c_{-1} + \alpha^2_{002} c_{-1}^2 + \ldots, \\
\dot{\varepsilon} &= 0,
\end{aligned} \tag{3.11}$$

with $c_1 = \overline{c_{-1}}$ and coefficients $\alpha_{\cdot} \in \mathbb{C}$. The idea of the normal transform is to eliminate all terms which are not in resonance with the linear ones. As an example consider c_{-1}^2 in the equation for c_1. It oscillates as $\mathrm{e}^{-2i\omega t}$ and is therefore not in resonance with c_1, which oscillates as $\mathrm{e}^{i\omega t}$. Therefore, this c_{-1}^2 term be eliminated. With this heuristic argument the only terms which remain in the equation for c_1 are those of the form $\varepsilon^m c_1^n c_{-1}^{n-1}$, and in the equation for c_{-1} those of the form $\varepsilon^m c_1^{n-1} c_{-1}^n$. This heuristic argument can be made rigorous by making a number of near identity changes of variables.

3.3.1. Normal form transforms. We consider the autonomous system

$$\dot{x} = Ax + f(x)$$

for $x(t) \in \mathbb{R}^d$, with $A \in \mathbb{R}^{d\times d}$, and

$$f(x) = f_2(x) + f_3(x) + f_4(x) + \ldots,$$

with $f_m(kx) = k^m f_m(x)$ for all $k \geq 0$, i.e., f_m is a vector in $\mathbb{R}^d$ whose entries are homogeneous polynomials of degree m in the variables $x_1, \ldots, x_d$. Hence,

$$f_m = \begin{pmatrix} f_{m1} \\ \vdots \\ f_{md} \end{pmatrix}$$

is an element of the vector space

$$V_m = \left\{ u = \begin{pmatrix} \sum_{m_1+\ldots+m_d=m} \alpha^1_{m_1\ldots m_d} x_1^{m_1} \cdot \ldots \cdot x_d^{m_d} \\ \vdots \\ \sum_{m_1+\ldots+m_d=m} \alpha^d_{m_1\ldots m_d} x_1^{m_1} \cdot \ldots \cdot x_d^{m_d} \end{pmatrix} : \alpha^j_{m_1\ldots m_d} \in \mathbb{R} \right\},$$

the space of vector valued homogeneous polynomials of degree m in the variables $x_1, \ldots, x_d$.

We look for a near identity change of variables which allows us to eliminate as many terms as possible in order to make the system as simple as possible. Therefore, we make the ansatz

$$x = y + h(y),$$

where

$$h(y) = h_2(y) + h_3(y) + h_4(y) + \dots$$

with $h_m \in V_m$. We obtain

$$\dot{x} = \dot{y} + \frac{\partial h}{\partial y}\dot{y} = A(y + h(y)) + f(y + h(y)),$$

and therefore

$$\begin{aligned}\dot{y} =& \left(1 + \frac{\partial h}{\partial y}\right)^{-1} [A(y + h(y)) + f(y + h(y))] \\ =& Ay - \frac{\partial h_2}{\partial y}Ay + Ah_2(y) + f_2(y) + \mathcal{O}(\|y\|^3).\end{aligned}$$

In order to eliminate all quadratic terms in f_2 we have to find an h_2 such that

$$-\frac{\partial h_2}{\partial y}Ay + Ah_2(y) + f_2(y) = 0.$$

With the above interpretation of h_2 as an element of the vector space V_2,

$$(L_A h_2)(y) = -\frac{\partial h_2}{\partial y}Ay + Ah_2(y)$$

is a linear map of V_2 into itself which acts linearly on the coefficients $\alpha^m_{m_1\dots m_d}$.

Similarly, in order to eliminate terms of order m we have to solve the linear system

$$-\frac{\partial h_m}{\partial y}Ay + Ah_m(y) + \widetilde{f}_m(y) = 0,$$

where $\widetilde{f}_m$ represents the nonlinear terms of degree m after the application of the transformations h_2 to h_{m-1}.

For our purposes it is sufficient to restrict ourselves to the case of a diagonal A, i.e., $A = \operatorname{diag}(\lambda_1, \dots, \lambda_d)$. Then in the space V_m the linear map L_A possesses the eigenvectors $y_1^{m_1} \cdot \dots \cdot y_d^{m_d} \mathrm{e}_j$, where e_j is the jth unit vector of $\mathbb{R}^d$. The associated eigenvalues are given by $\mu = \sum_{k=1}^d m_k\lambda_k - \lambda_j$. In order to see this, we consider the jth component

$$\sum_{k=1}^{d} \frac{\partial h_{mj}}{\partial y_k}\lambda_k y_k - \lambda_j h_{mj} = \mu h_{mj}$$

of the eigenvalue equation $L_A h_m = \mu h_m$. Inserting the above eigenvectors shows the statement. Therefore, $L_A h_m = g_m$ can be solved w.r.t. h_m in all eigenspaces with eigenvalue $\mu \neq 0$. We found

term	non-resonance condition	can be eliminated
$\alpha_{20}c_1^2$	$-2\lambda_1 - 0\lambda_2 + \lambda_1 = -2\mathrm{i} + \mathrm{i} \neq 0$	yes
$\alpha_{11}c_1c_{-1}$	$-0\lambda_1 - 2\lambda_2 + \lambda_1 = -2(-\mathrm{i}) + \mathrm{i} \neq 0$	yes
$\alpha_{02}c_{-1}^2$	$-0\lambda_1 - 2\lambda_2 + \lambda_1 = -2(-\mathrm{i}) + \mathrm{i} \neq 0$	yes
$\alpha_{30}c_1^3$	$-3\lambda_1 - 0\lambda_2 + \lambda_1 = -3\mathrm{i} + \mathrm{i} \neq 0$	yes
$\alpha_{21}c_1^2c_{-1}$	$-2\lambda_1 - 1\lambda_2 + \lambda_1 = -2\mathrm{i} - (-\mathrm{i}) + \mathrm{i} = 0$	no
$\alpha_{12}c_1c_{-1}^2$	$-1\lambda_1 - 2\lambda_2 + \lambda_1 = -\mathrm{i} - 2(-\mathrm{i}) + \mathrm{i} \neq 0$	yes
$\alpha_{03}c_{-1}^3$	$-0\lambda_1 - 3\lambda_2 + \lambda_1 = -3(-\mathrm{i}) + \mathrm{i} \neq 0$	yes

Table 1. Non-resonance conditions for the terms in the first equation of (3.11).

Lemma 3.3.2. *Assume* $A = \operatorname{diag}(\lambda_1, \ldots, \lambda_d)$. *Then in order to eliminate the term* $y_1^{m_1} \cdot \ldots \cdot y_d^{m_d}$ *in the* j^{th} *equation we need the non-resonance condition*

$$\sum_{k=1}^{d} m_k \lambda_k - \lambda_j \neq 0. \tag{3.12}$$

3.3.2. Continuation of the proof of Theorem 3.3.1. We now make a normal form transform for (3.11). In Table 1 we list the various terms in the first equation of (3.11) and their non-resonance conditions. Hence, after the transformation we obtain for the new variables $b_j = c_j + \mathcal{O}(|\varepsilon|(|c_1| + |c_{-1}|) + |c_1|^2 + |c_{-1}|^2)$ that

$$\begin{aligned} \dot{b}_1 =& \mathrm{i}\omega b_1 + \beta_1 \varepsilon b_1 + \gamma_1 b_1^2 b_{-1} + \mathcal{O}(|\varepsilon^2|(|b_1| + |b_{-1}|) + |b_1|^4 + |b_{-1}|^4), \\ \dot{b}_{-1} =& -\mathrm{i}\omega b_{-1} + \beta_{-1}\varepsilon b_{-1} + \gamma_{-1} b_1 b_{-1}^2 + \mathcal{O}(|\varepsilon^2|(|b_1| + |b_{-1}|) + |b_1|^4 + |b_{-1}|^4), \end{aligned}$$

with $b_1 = \overline{b_{-1}}$, $\beta_1 = \overline{\beta_{-1}} = \beta_r + \mathrm{i}\beta_i$, and $\gamma_1 = \overline{\gamma_{-1}} = \gamma_r + i\gamma_i$, where β_r, β_i, γ_r, $\gamma_i \in \mathbb{R}$. Introducing polar coordinates $b_1 = re^{\mathrm{i}\phi}$ gives the system

$$\begin{aligned} \dot{r} &= \beta_r \varepsilon r + \gamma_r r^3 + \mathcal{O}(\varepsilon^2 r + r^4), \\ \dot{\phi} &= \omega + \beta_i \varepsilon + \gamma_i r^2 + \mathcal{O}(\varepsilon^2 + r^3). \end{aligned} \tag{3.13}$$

Hence, we have a system which can be analyzed for small ε. Ignoring the higher order terms we have $r = \mathcal{O}(\sqrt{\varepsilon})$ for the bifurcating time-periodic solutions. For the scaled variable $\widetilde{r}$ defined by $r = \sqrt{\varepsilon}\widetilde{r}$ we find the approximate time-periodic solution $\widetilde{r}_0^2 = -\beta_r/\gamma_r$. Depending on the sign of γ_r we have a sub- or supercritical bifurcation of time-periodic solutions. In order to prove the persistence of these solutions under the neglected $\mathcal{O}(\varepsilon^2)$-terms we construct the associated Poincaré map Π_ε for which the periodic solution is a fixed point. The fixed point is therefore a zero of

$$F(\widetilde{r}(0), \varepsilon) = \varepsilon^{-1/2}(\Pi_\varepsilon(\widetilde{r}(0)) - \widetilde{r}(0)).$$

We have $F(\widetilde{r}_0(0), 0) = 0$ and $\partial_1 F(\widetilde{r}_0(0), 0) = -4\pi\beta_r\omega^{-1} \neq 0$ such that the implicit function theorem can be applied and a fixed point $\widetilde{r}(0) = \widetilde{r}_\varepsilon(0)$ of

Π_ε exists for $\varepsilon > 0$, too. Associated to this fixed point are periodic solutions of (3.13), (3.11), and finally of (3.10). Therefore, we are done. □

3.3.3. An example and further remarks. In applications very often the parameter space has more than one dimension, i.e., the parameters are given by $p \in \mathbb{R}^m$. In principle, this situation can be handled in the same way as above by varying the parameters individually. Generically, the set $\{p \in \mathbb{R}^m : F(x(p), p) = 0, \partial_x F(x(p), p) = 0\}$ of possible bifurcation values becomes a $(m-1)$-dimensional manifold. Here we consider the following example of a 2-parameter bifurcation/stability diagram.

Example 3.3.3. Bifurcation diagram for a (toy) problem from chemistry. The system

$$\dot{\alpha} = \mu - \alpha(\kappa + \beta^2), \qquad \dot{\beta} = -\beta + \alpha(\kappa + \beta^2) \tag{3.14}$$

serves as a (drastically reduced) model for so called cubic autocatalysis in chemistry. Here, α, β are concentrations (hence $\alpha, \beta \geq 0$) of some substances and the parameters $\mu, \kappa > 0$ are some reaction rates. Unlike other reagents, here represented by α, that participate in the chemical reaction, a catalyst, here represented by β, is not consumed by the reaction itself. The above model describes a situation where the so called educt α is supplied into the system at constant rate μ and converted into β with rate $\kappa + \beta^2$. Thus, the catalyst β catalyses its own production, hence the name autocatalysis.

The unique fixed point of (3.14) is given by

$$(\alpha^*, \beta^*) = (\mu/(\kappa + \mu^2), \mu),$$

with the associated linearization

$$A = \begin{pmatrix} -(\mu^2 + \kappa) & -2\mu^2/(\kappa + \mu^2) \\ \mu^2 + \kappa & (\mu^2 - \kappa)/(\mu^2 + \kappa) \end{pmatrix}. \tag{3.15}$$

First, we discuss the eigenvalues of A which are given in terms of $p = \operatorname{trace} A$ and $q = \det A$ by

$$\lambda_{1,2} = \frac{p}{2} \pm \sqrt{\frac{p^2}{4} - q}, \tag{3.16}$$

This associated bifurcation/stability diagram is plotted in Figure 3.13. The meaning of this diagram is as follows. Starting, e.g., with p, q in the **sn** regime ($p < 0$ and $0 < q < p^2/4$) and crossing, e.g., the line $q = p^2/4$, the fixed point changes type from **sn** to **sf**, cf. (3.16). Next, depending on the nonlinearity, we may expect a Hopf bifurcation when crossing the line $p = 0, q > 0$, which is therefore called Hopf line. The point $(p, q) = (0, 0)$ is called Bogdanov-Takens or co-dimension-2 point since two parameters are needed to describe the possible bifurcations in its neighborhood.

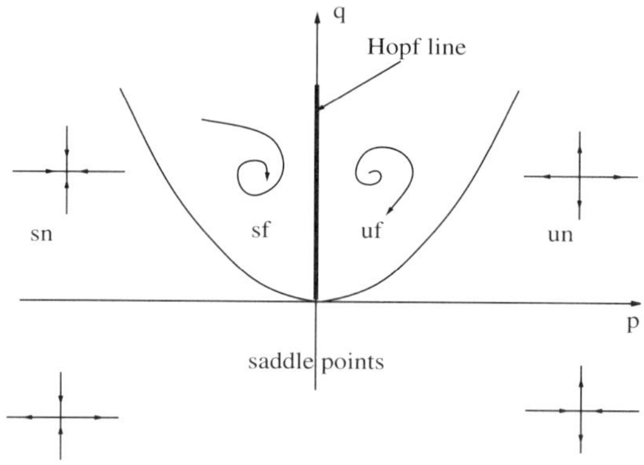

Figure 3.13. Bifurcation diagram for $\dot{x} = Ax$ in terms of the trace p and the determinant q of A. The abbreviations stand for: **sn**: stable node, two stable real eigenvalues; **sf**: stable focus, two stable conjugate complex eigenvalues; **uf**: unstable focus, two unstable conjugate complex eigenvalues; **un**: unstable node, two unstable real eigenvalues; **saddle points:** two real eigenvalues, one stable, the other unstable.

The transfer of Figure 3.13 to the fixed point (α_0, β_0) and to the parameters μ, κ yields to the solution of a number of algebraic equations. For instance, the Hopf line is given by solving the 4th order equation

$$\operatorname{trace} A = -(\mu^4 - (1-2\kappa)\mu^2 + \kappa/(1+\kappa))/(\mu^2+\kappa) = 0,$$

hence

$$\mu_{1,2}(\kappa) = \frac{1}{\sqrt{2}}\left((1-2\kappa) \pm (1-8\kappa)^{1/2}\right)^{1/2}.$$

In summary, we obtain the bifurcation diagram plotted in Figure 3.14, while Figure 3.15 shows two selected phase portraits. We will come back to such systems in Chapter 9. ⌋

Remark 3.3.4. Besides the analytical study of bifurcating branches close to bifurcation, i.e., the analysis of the reduced equation, there is the big field of numerical path following (or continuation) and bifurcation analysis. The basic idea of continuation is as follows. Given a solution $(x_0, \mu_0) \in \mathbb{R}^2$ of $f(x,\mu) = 0$ with $\partial_x f(x_0, \mu_0) \neq 0$ we choose a small $\delta > 0$, let $\mu = \mu_0 + \delta$, and use the Newton scheme to compute $x(\mu)$. In detail, we use the iteration

$$x_{n+1} = x_n - (\partial_x f(x_n, \mu))^{-1} f(x_n, \mu)$$

with starting point x_0. The scheme converges for $\delta > 0$ sufficiently small and we set $x(\mu) = \lim_{n\to\infty} x_n$. Replacing (x_0, μ_0) by $(x(\mu_0+\delta), \mu_0+\delta)$ we can start again and compute solutions $x = x(\mu)$ until $\partial_x f(x(\mu), \mu) = 0$. In case of $\partial_\mu f(x_0, \mu_0) \neq 0$ we can interchange the role of x and μ and obtain a solution

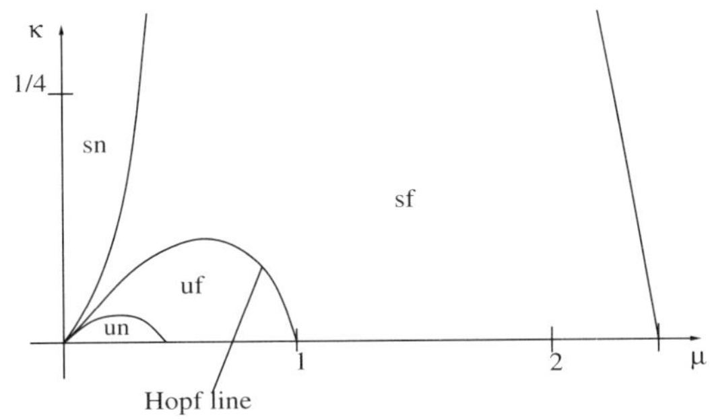

Figure 3.14. The bifurcation diagram for (3.14) is to be read as follows. If, e.g., we fix $\kappa = 0.1$ then we cross the Hopf line at approximately $\mu_1 \approx 0.41$ and $\mu_2 \approx 0.77$. At these lines Hopf bifurcations can be expected. Path following methods may allow us to follow the family of periodic solutions in the parameter plane.

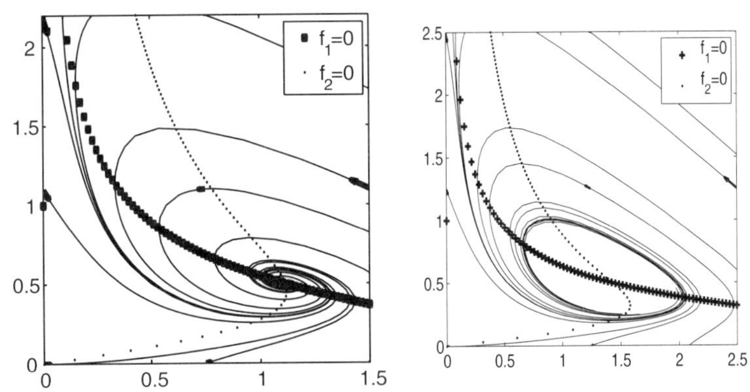

Figure 3.15. Phase portraits for $(\mu, \kappa) = (0.5, 0.2)$ (left) and $(\mu, \kappa) = (0.5, 0.1)$ (right)

$\mu = \mu(x)$, i.e., we have a smooth curve $(x, \mu) = (x, \mu)(s)$ parameterized with s until simultaneously $\partial_x f(x(s), \mu(s)) = \partial_\mu f(x(s), \mu(s)) = 0$.

There are variants of this idea [**Kel77, Kuz04, Doe07, Sey10**] which automatically allow the continuation of branches around folds and beyond bifurcation points, the detection and localization of bifurcation points, and branch switching at bifurcation points. One standard method is so called (pseudo-)arclength continuation, which is implemented in the package AUTO, [**Doe07, Dea16**], see also XPPAUT, [**Erm02**]. Many of these methods can also be applied to bifurcation problems in PDEs and are important tools there. A recent package specifically designed for elliptic systems in two space dimensions is `pde2path`, [**UWR14, DRUW14**]. ⌋

3.4. Routes to chaos

The chapter is closed by sketching two routes of bifurcations to chaotic behavior in dissipative systems, namely period-doubling, which is based on an infinite series of local bifurcations, and homoclinic explosion, which is a so called global bifurcation. There are many other routes to chaotic behavior, but we will only comment on one of them, namely the Ruelle-Takens scenario.

The theory of turbulence developed by Landau [**LL91**] in 1944 is based on the assumption that more and more pairs of complex conjugate eigenvalues cross the imaginary axis. This route to chaos is called the Landau-Hopf scenario. In 1971 Ruelle and Takens [**RT71**] showed that these infinitely many unstable eigenvalues are not necessary for the occurrence of chaos. The scenario starts with a stable fixed point and provides a very short route to chaotic behavior only using local bifurcations. The first bifurcation is a Hopf bifurcation leading to time-periodic solutions. Then the time-periodic solution becomes unstable via a pair of complex-conjugate Floquet multipliers crossing the unit circle leading to quasi-periodic solutions. The next bifurcation leads to a three-dimensional invariant torus where nearby chaotic behavior can be found. This route to chaos plays a certain role in hydrodynamical applications, but now we focus on the route with infinitely many period-doublings.

3.4.1. Period-doubling. Period-doubling is realized in nature in a number of systems, as cardiac diseases, leaking water-taps, laser dynamics, and various chemical reactions. It occurs if a periodic orbit becomes unstable and a stable periodic solution with roughly the double period occurs, and if this repeats under further increase (or decrease) of some parameter, see Figure 3.16.

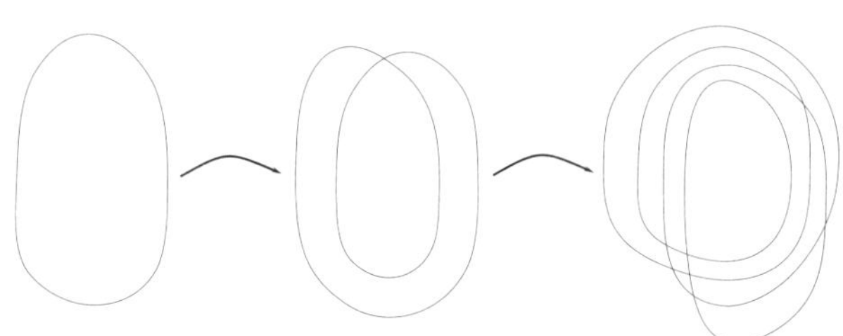

Figure 3.16. Sequence of period-doublings for a periodic solution

For the analysis of this phenomenon we consider the associated Poincaré map Π. The fixed point of Π, which is associated to the periodic orbit, becomes unstable via a real Floquet multiplier crossing the unit circle at

-1. The fixed point becomes also unstable for the second iterate of the Poincaré map Π^2, but now via a real Floquet multiplier crossing the unit circle at 1. On the center manifold associated to the Floquet multiplier -1 we have the following situation. If

$$\Pi x = -x + \alpha x^2 + \mathcal{O}(x^3), \quad \text{then} \quad \Pi^2 x = x + \beta x^3 + \mathcal{O}(x^4),$$

such that for Π^2 a pitchfork bifurcation occurs. The two stable bifurcating fixed points for Π^2 corresponds to a 2-periodic solution for Π itself, since for Π no bifurcation of fixed points occurs. Hence, a new periodic orbit with twice the period is bifurcating from the old one. Assuming that this new periodic orbit becomes unstable in the same way and that this procedure goes on and on we finally come to chaotic dynamics. A famous ODE example showing this behavior is by Rössler [**Rös76**].

Example 3.4.1. The Rössler system. Consider the ODE

$$\partial_t \begin{pmatrix} x \\ y \\ z \end{pmatrix} = \begin{pmatrix} -(y+z) \\ x + by \\ b + z(x-a) \end{pmatrix},$$

where typically $b = 0.1$ and $a \in \mathbb{R}$ serves as a bifurcation parameter. Starting from an asymptotically stable periodic orbit for a=4 we find for increasing a a period-doubling sequence, cf. Figure 3.17. See, e.g., [**PJS92**, §3.3] and the references therein for a more detailed introduction to the Rössler system. ⌋

3.4.2. The logistic map. There is a discrete model problem for the period-doubling route to chaos, namely the logistic map

$$x_{n+1} = \mu x_n(1 - x_n) = F(x_n)$$

with $\mu \geq 0$ and $x_n \in \mathbb{R}$. We have for the nth iteration $F^n(x) \to -\infty$ for $n \to \infty$ if $x < 0$ or $x > 1$. For $\mu \in [0,4]$ the map F maps the interval $[0,1]$ into itself. In the following we restrict ourselves to values μ and x_0 in these sets. More details can be found in [**Dev89**], including a discussion of chaos in the strict sense of Definition 2.5.2 in the logistic map for parameter values $\mu > 2 + \sqrt{5}$.

The condition $F(x) = \mu x(1-x) = x$ gives the fixed points $x_1^* = 0$ and $x_2^* = 1 - 1/\mu$. At $\mu_0 = 1$ a transcritical bifurcation of fixed points occurs. The linearization around the fixed point x^* is given by $y_{n+1} = F'(x^*)y_n$ where $F'(x^*) = \mu(1 - 2x^*)$. For $x^* = 1 - 1/\mu$ we obtain

$$F'(x^*) = \mu(1 - 2(1 - 1/\mu)) = 2 - \mu.$$

Hence, this fixed point is stable for $\mu \in (1,3)$ and becomes unstable at $\mu_1 = 3$ via some period-doubling. A stable two-periodic solution appears. See Figure 3.18.

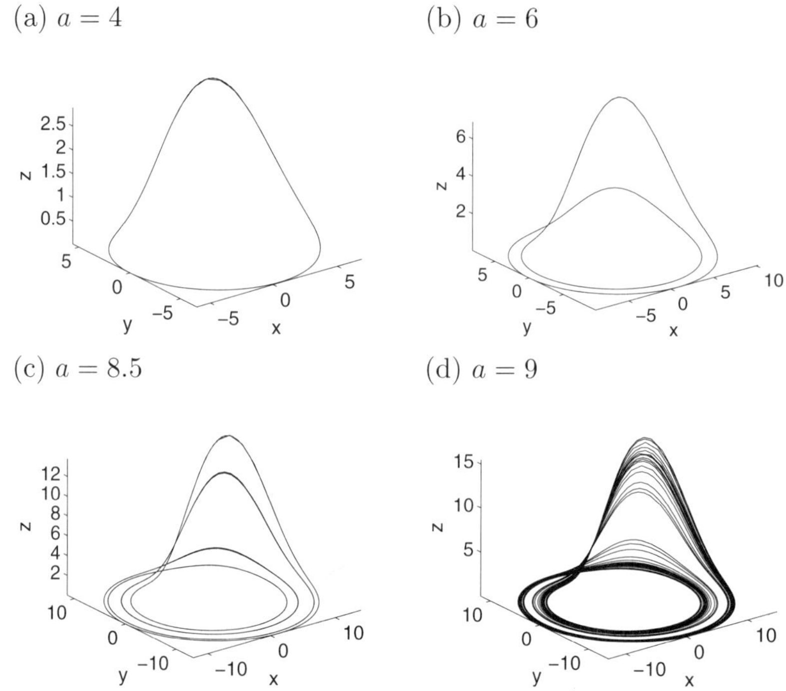

Figure 3.17. Period-doubling in the Rössler system, $b = 0.1$ fixed.

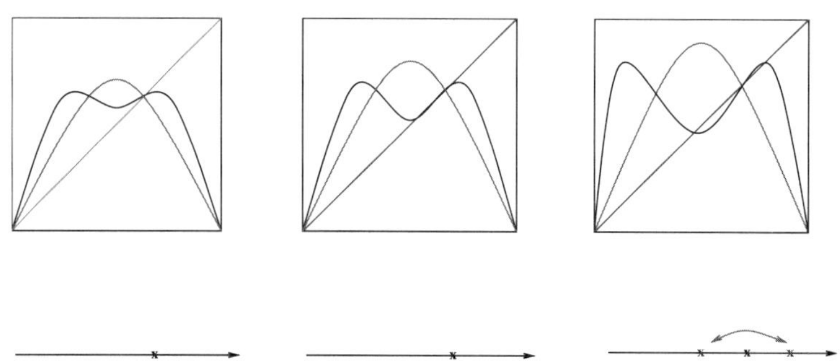

Figure 3.18. The map F^2 for $\mu = 2.9$, $\mu = 3$ and $\mu = 3.1$. A pitchfork bifurcation occurs which corresponds to the occurrence of a two-periodic solution for F.

A further increase of μ yields an instability of the two-periodic solution at a value $\mu = \mu_2$. We find a pitchfork bifurcation for F^4 and so the occurrence of a four-periodic solution for F. A further increase of μ yields an instability of the four-periodic solution at a value $\mu = \mu_3$. We find a pitchfork bifurcation of F^8 and so the occurrence of an eight-periodic solution for F. Interestingly, there is an infinite sequence of such bifurcations and so a further increase of μ yields an instability of the 2^{n-1}-periodic solution at a value $\mu = \mu_n$. We find a pitchfork bifurcation of F^{2^n} and so the occurrence of a 2^n-periodic solution for F.

Even more interestingly, the period-doublings show some asymptotic behavior. It can be proved rigorously by computer-assisted proofs that the limit

$$\lim_{n\to\infty} \frac{\mu_n - \mu_{n-1}}{\mu_{n+1} - \mu_n} \approx 4.6692,$$

called the Feigenbaum constant, exists, cf. [**CE80**]. As a consequence we have the existence of $\mu_\infty = \lim_{n\to\infty} \mu_n \approx 3.57$, too. A recent overview about the theoretical background of the occurrence of these limits is [**Avi11**]. For most values $\mu > \mu_\infty$ the system exhibits chaotic behavior. In Figure 3.19 the ω-limit set for starting point $x_0 = 1/2$ is plotted as a function over the bifurcation parameter μ. There are isolated regions on the μ-axis where no attractive chaotic behavior occurs, the so called windows of stability. Beginning at $1+\sqrt{8} \approx 3.83$ there is for instance a range of parameters μ with a stable 3-periodic solution. There is a general theory [**Dev89**, §1.10] that for maps from $\mathbb{R}$ to $\mathbb{R}$ solutions of period 3 imply the existence of periodic solutions of every period $m \in \mathbb{N}$, known as the Theorem of Sarkovskii.

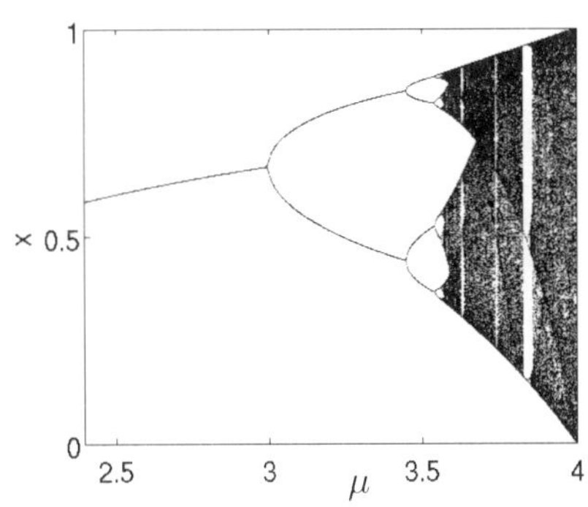

Figure 3.19. The ω-limit set for starting point $x_0 = 1/2$ is plotted as a function over the bifurcation parameter μ. For every fixed μ the iterates $x_N, \ldots x_{N+M}$ with N and M sufficiently large are plotted.

For ODEs the instability occurs in a one-dimensional center manifold where the dynamics can be described via a one-dimensional Poincaré map and hence it can be expected that the route to chaos via period-doublings may occur in ODE systems, too. Another example is the so called chemostat, see Exercise 3.6.

3.4.3. Homoclinic explosion – the Lorenz attractor. The Lorenz attractor [**Lor63, Spa82**] is a famous example of a three-dimensional ODE with chaotic dynamics. It was found by the meteorologist E. Lorenz in 1963 as a lowest order approximation for convection in fluids and is considered as a cartoon weather model. It is given by

$$
\begin{aligned}
\dot{x} &= \sigma(y-x), \\
\dot{y} &= \rho x - y - xz, \\
\dot{z} &= -\beta z + xy,
\end{aligned}
\tag{3.17}
$$

with $\sigma = 10$ and $\beta = 8/3$ fixed. Numerical simulations of this simple model for $\rho = 27$ show a complicated irregular dependence of the solutions on the initial conditions and the occurrence of a so called "strange" attractor. A rigorous proof for the occurrence of chaotic dynamics in the Lorenz model (similar to shift dynamics (Σ_2, σ) defined in §2.5.1) has been given [**Tuc02**]. We now explore the route to chaos in a bit more detail.

For any $\sigma, \beta, \rho > 0$ any large enough sphere around $(0, 0, \rho + \sigma)$ is absorbing. This can be shown with the Lyapunov function

$$V(x, y, z) = x^2 + y^2 + (z - \rho - \sigma)^2.$$

With $\alpha = \min\{2\sigma, 2, \beta\}$ we obtain

$$
\begin{aligned}
\frac{\mathrm{d}}{\mathrm{d}t} V &= -2\sigma x^2 - 2y^2 - 2\beta z^2 + 2\beta(\rho+\sigma)z \\
&= -2\sigma x^2 - 2y^2 - \beta(z - \rho - \sigma)^2 - \beta z^2 + \beta(\rho+\sigma)^2 \\
&\le -\alpha V + \beta(\rho+\sigma)^2.
\end{aligned}
$$

Hence, for t large enough by Gronwall's inequality we obtain

$$V(t) \le \frac{2\beta(\rho+\sigma)^2}{\alpha}.$$

By Theorem 2.4.4 there exists the global attractor $\mathcal{A} = \omega(B)$, for which numerical simulations show its geometric complexity. The attractor has a dimension less than three since the divergence of the vector field

$$\partial_{y_1}(\sigma(y_2 - y_1)) + \partial_{y_2}(\rho y_1 - y_2 - y_1 y_3) + \partial_{y_3}(-\beta y_3 + y_1 y_2) = -(\sigma + 1 + \beta)$$

is negative and, therefore, every test volume shrinks to zero for $t \to \infty$. For $\rho = 27$ numerical experiments show a non-integer Hausdorff-dimension of $\mathcal{A}$ of approximately 2.04.

The behavior occurs by a global bifurcation which is called homoclinic explosion. The route to chaos for the Lorenz system is as follows when ρ is increased from 0 to 27, for $\sigma = 10$ and $\beta = 8/3$ fixed. The z-axis is an invariant set, and the origin is a stable fixed point for $\rho < 1$. The linearization of (3.17) around 0 is

$$\nabla f(x, y, z) = \begin{pmatrix} -\sigma & -\sigma & 0 \\ \rho - z & -1 & -x \\ y & x & -\beta \end{pmatrix}.$$

At $(x, y, z) = (0, 0, 0)$ we find the eigenvalues

$$\lambda_{1,2} = -\frac{\sigma+1}{2} \pm \frac{1}{2}\sqrt{(\sigma+1)^2 + 4\sigma(\rho-1)}, \quad \lambda_3 = -\beta,$$

such that at $\rho = 1$ a bifurcation of fixed points occurs which turns out to be a supercritical pitchfork bifurcation. For $\rho > 1$ we have two non-trivial fixed points X_1^*, X_2^* with

$$z = \rho - 1 \quad \text{and} \quad x = y = \pm\sqrt{\beta(\rho-1)}. \tag{3.18}$$

The eigenvalues of $\nabla f(X_{1,2}^*)$ are the roots of

$$p(\lambda) = \lambda^3 + (\sigma + \beta + 1)\lambda^2 + \beta(\sigma + \rho)\lambda + 2\sigma\beta(\rho - 1),$$

From this we find that the fixed points $X_{1,2}^*$ are stable until $\rho = \rho_{\text{Hopf}} \approx 24.74$, where two complex conjugate eigenvalues cross the imaginary axis. It turns out that a subcritical Hopf bifurcation occurs. This means that for $\rho \lessapprox \rho_{\text{Hopf}}$ unstable periodic solutions near $X_{1,2}^*$ exist, which shrink to $X_{1,2}^*$ as $\rho \to \rho_{\text{Hopf}}$.

Figure 3.20 gives some numerical illustrations: In general for $\rho \in (1, 24]$ the two parts S_1 and S_2 of the one-dimensional unstable manifold of the origin are connected with the three-dimensional stable manifolds of the fixed points X_1^* and X_2^*. Though it is difficult to see, there is a value $\rho = \rho_{\text{global}} \approx$

(a) $S_{1,2}$ for $\rho = 13$.

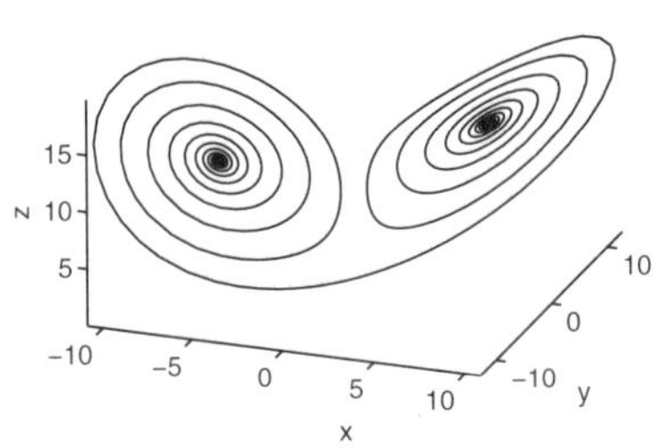

(b) $S_{1,2}$ for $\rho = 13.91$.

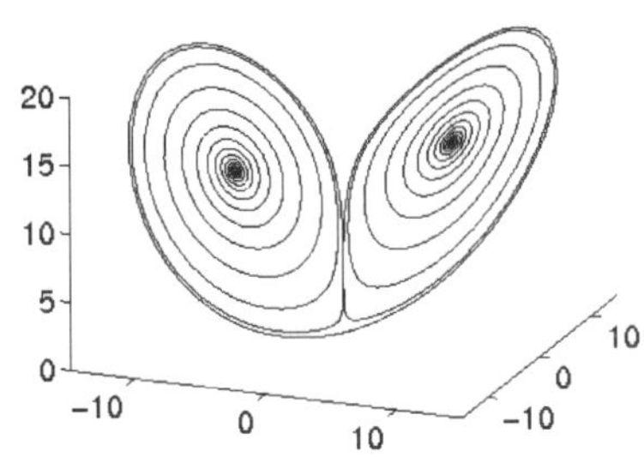

(c) $S_{1,2}$ for $\rho = 24$.

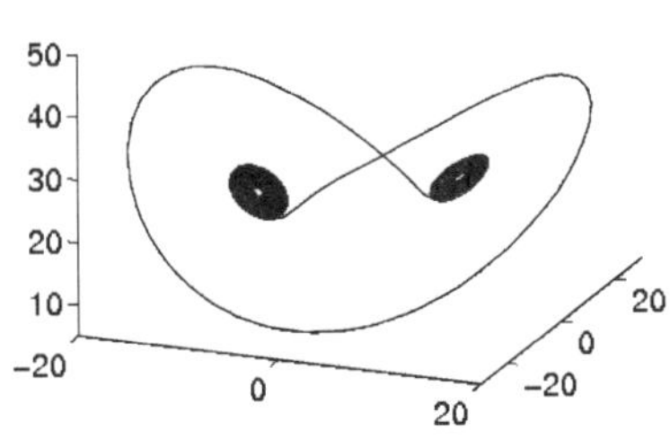

(d) A visualization of the attractor.

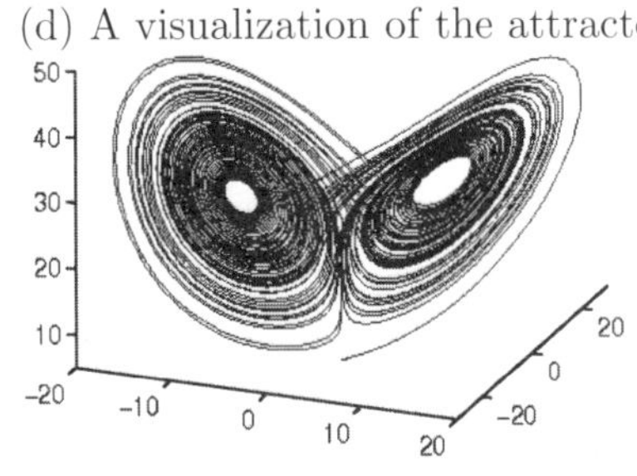

Figure 3.20. The unstable manifolds of the origin in the Lorenz system for different ρ (a)-(c), and the "Lorenz attractor" for $\rho = 27$ (d), visualized by one trajectory.

13.91 where the two parts S_1 and S_2 are connected with the two-dimensional stable manifold of the origin and form two homoclinic connections, see Figure 3.20(b). For $\rho < \rho_{\text{global}}$ the part S_1 connects to X_1^* and S_2 to X_2^*, see Figure 3.20(a), and vice versa for $\rho > \rho_{\text{glob}}$. This behavior is the origin of a so called homoclinic explosion, cf. [**Wig88**], and creates the chaotic behavior in the system. Therefore, chaotic behavior is present in the system already for ρ close to ρ_{global}, but becomes only attractive for larger values of ρ. Figures 3.20(c) and (d)) illustrate the different behavior for $\rho = 24 < \rho_{\text{Hopf}}$ and $\rho = 27$ (Figure 3.20(d)).

An elementary but more detailed introduction to the Lorenz system can be found in [**Str94**, §9], including explanations of simple mechanical and electronic systems able to simulate the Lorenz system, together with applications to send encrypted messages.

Exercises

3.1. For the following ODEs $\dot{x}=f(x)$ determine all fixed points and their stability in dependence of the parameter $\mu\in\mathbb{R}$. What bifurcations occur at what μ?
a) $\dot{x}=\mu+6+4x-x^2$, b) $\dot{x}=2-\mu+x(\mu-4)+3x^2-x^3$, c) $\dot{x}=\mu+x(\mu-1)-x^2$.

3.2. Compute the non-trivial solutions close to the origin for

$$f(x,y,\varepsilon) = \begin{pmatrix} \varepsilon x - yx^2 - x^4 \\ y + x^2 \end{pmatrix} = 0.$$

3.3. Check the stability of $(x,y) = (0,0)$ for the following systems by calculating the center manifold, cf. [**Wig03**].

(a) $\dot{x} = -xy - x^6$, $\dot{y} = -y + x^2$, (b) $\dot{x} = x^2y - x^5$, $\dot{y} = -y + x^2$, and

(c) $\begin{pmatrix} x_{n+1} \\ y_{n+1} \end{pmatrix} = \begin{pmatrix} 0 & 1 \\ -\frac{1}{2} & \frac{3}{2} \end{pmatrix}\begin{pmatrix} x_n \\ y_n \end{pmatrix} + \begin{pmatrix} 0 \\ -y_n^3 \end{pmatrix}$.

3.4. Which of the following systems has periodic orbits close to $z = (x,y) = (0,0)$? Does Theorem 3.3.1 apply?

$$\begin{aligned}
&\text{(a)} \quad \dot{z} = \begin{pmatrix} -\mu^2 & 1 \\ -1 & -\mu^2 \end{pmatrix} z - |z|^2 z, \\
&\text{(b)} \quad \dot{z} = \begin{pmatrix} -\mu^2 & 1 \\ -1 & -\mu \end{pmatrix} z - |z|^2 z, \qquad \mu \in \mathbb{R}, \text{ with small } |\mu|. \\
&\text{(c)} \quad \dot{z} = \begin{pmatrix} 0 & -1 \\ 1 & 0 \end{pmatrix} + \begin{pmatrix} x^2 - xy \\ xy \end{pmatrix}.
\end{aligned}$$

3.5. This exercise (from [**Str94**]) brings together a number of concepts treated above, namely bifurcation, center manifold calculations, and non-trivial gradient dynamics. The system

$$\begin{aligned}
\dot{\vartheta}_1 &= f_1(\vartheta) := k\sin(\vartheta_1 - \vartheta_2) - \sin\vartheta_1, \\
\dot{\vartheta}_2 &= f_2(\vartheta) := k\sin(\vartheta_2 - \vartheta_1) - \sin\vartheta_2,
\end{aligned}$$

with $\vartheta = (\vartheta_1, \vartheta_2)$ and parameter $k > 0$ describes two rotating magnets between two fixed magnets, in a geometry as follows:

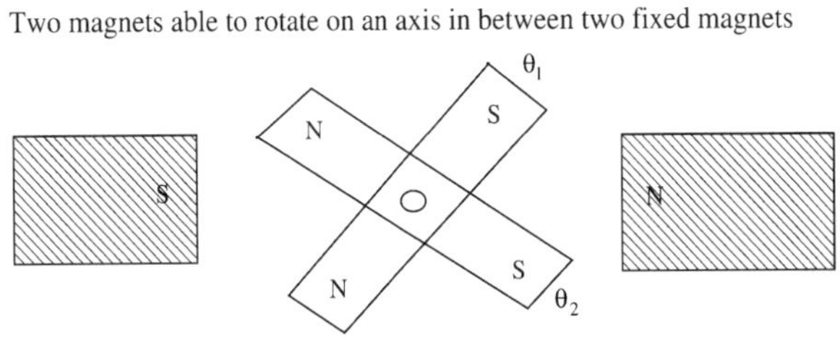

a) The system has exactly nine fixed points in $[-\pi,\pi]^2$ for $k < 1/2$. Determine these and their stability.
b) What bifurcation occurs at $k = 1/2$? What new fixed points emerge?
c) Find a potential V such that $\dot{\vartheta} = -\nabla V(\vartheta)$.
d) Sketch a phase portrait for $0 < k < 1/2$ and a phase portrait for $k > 1/2$.
Remark. Doing c) first and then the rest is a good idea.

3.6. The Chemostat. The chemostat is an industrially used "predator-prey system" to cultivate bacteria. In case of 3 species the system is modeled by

$$\dot{s}(t) = (1 - s(t)) - \frac{m_1 s(t)}{a_1 + s(t)} x_1(t),$$

$$\dot{x}_1(t) = x_1(t)\left(\frac{m_1 s(t)}{a_1 + s(t)} - 1 - \frac{m_2 x_2(t)}{a_2 + x_1(t)}\right),$$

$$\dot{x}_2(t) = x_2(t)\left(\frac{m_2 x_1(t)}{a_2 + x_1(t)} - 1 - \frac{m_3 x_3(t)}{a_3 + x_2(t)}\right),$$

$$\dot{x}_3(t) = x_3(t)\left(\frac{m_3 x_2(t)}{a_3 + x_2(t)} - 1\right).$$

a) Explain the modeling.
b) Let $\sigma(t) = 1 - s(t) - \sum_{k=1}^{3} x_k(t)$. Show that $\dot{\sigma}(t) = -\sigma(t)$ and use this to prove that the ω-limit set of any solution $(s(t), x_1(t), \ldots, x_3(t))$ is contained in

$$\Omega = \{(s, x_1, \ldots, x_3) : s + \sum_{k=1}^{3} x_k = 1\}.$$

c) Substitute $s = 1 - \sum_{k=1}^{3} x_k$ into the equations for $\dot{x}_k$ and try to reproduce the period-doubling shown in Figure 3.21.

3.7. Let $Q_c(x) = x^2 + c$. Prove that for all $c < \frac{1}{4}$ there exists a unique $\mu > 1$ such that Q_c is conjugated to $F_\mu(x) = \mu x(1-x)$ through the map $h(x) = \alpha x + \beta$.

3.8. Consider the iteration $x_{n+1} = T_\lambda(x_n)$, where

$$T_\lambda(x) = \lambda \begin{cases} 2x & \text{for } x \in [0, 1/2], \\ 2 - 2x, & \text{for } x \in (1/2, 1]. \end{cases}$$

a) Prove, $x^* = 0$ is an asymptotically stable fixed point for $\lambda \geq 0$ sufficiently small.

b) Compute $T_\lambda^2 = T_\lambda \circ T_\lambda$ and find graphically the 2-periodic solutions, i.e., solve $T_\lambda^2(x) = x$ by finding the intersection points of the functions $x \mapsto T_\lambda^2(x)$ and $x \mapsto x$.

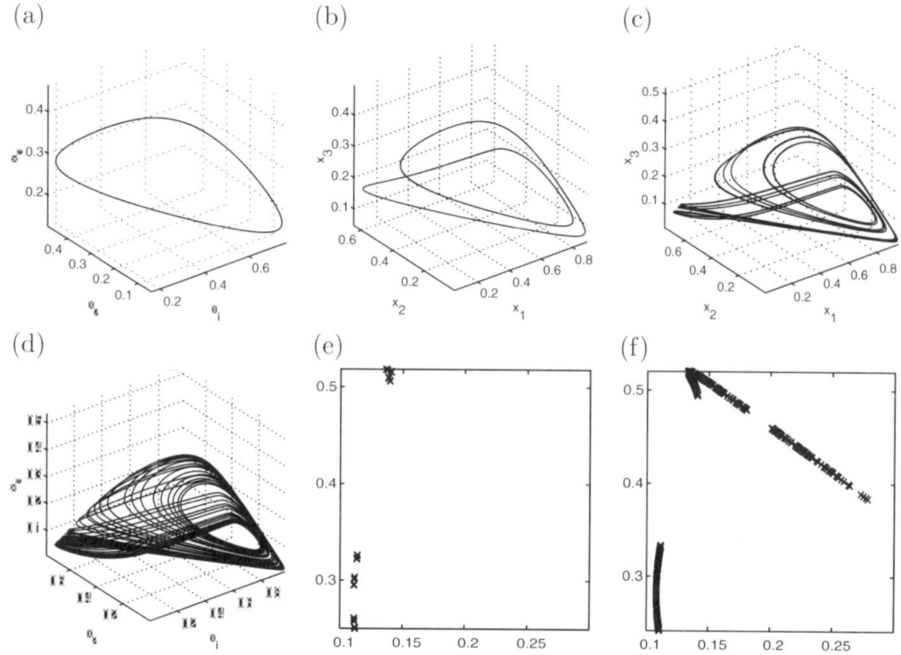

Figure 3.21. Period-doubling bifurcation in the 3 component chemostat. $(m_1, m_2, m_3) = (10, 4, 3.5)$, $(a_1, a_2) = (0.08, 0.23)$, and $a_3 = 0.4, 0.3, 0.225, 0.2$ in (a)-(d), respectively. (e,f) Poincaré section to $x_1 = 0.3$ for $a_3 = 0.225$ and $a_3 = 0.2$, respectively. For $a_3 = 0.2$ the time span is $t \in [0, 200]$.

c) Find for $\lambda = 4$ the structure of the set

$$S_\lambda = \{x : T_\lambda^n(x) \in [0,1]\ \forall n \in \mathbb{N}_0\}$$

What kind of dynamics do you expect in S_λ?

3.9. With $z = x + iy$ und $c = a + ib$ the discrete dynamical system

$$\pi_{a,b}(x,y) = (x^2 - y^2 + a, 2xy + b)$$

can be written in the complex form $f_c(z) = z^2 + c$. The Mandelbrot set is sketched in Figure 3.9, and is the set of all $c \in \mathbb{C}$ for which the sequence $(z_n)_{n\in\mathbb{N}}$, defined through $z_{n+1} = f_c(z_n)$, $z_0 = 0$, is bounded.

a) Prove that $\pi_{a,b}$ possesses an asymptotically stable fixed point (x_0, y_0) if $|f_c'(z_0)| < 1$. Express the fixed points of $D\pi_{a,b}(x_0, y_0)$ in terms of f'.

b) Find the fixed points of $\pi_{a,b}$ and compute their stability. Show that the set of all c, for which an asymptotically stable fixed point $z_0(c)$ exists is given by the interior of the cardioid

$$c(t) = \tfrac{1}{4} - \tfrac{1}{4}(1 - \mathrm{e}^{\mathrm{i}t})^2, \quad t \in \mathbb{R}.$$

c) Compute the fixed points of $\pi_{a,b}^2 = \pi_{a,b} \circ \pi_{a,b}$. Show that the non-trivial 2-periodic solutions of $\pi_{a,b}$ are asymptotically stable for $|c+1| < \frac{1}{4}$.

d) Which periods can be expected in the different parts of the Mandelbrot set? For this consider the eigenvalues of the linearization in the fixed point $z_0(c)$ for different values of c at the boundary of the cardioid.

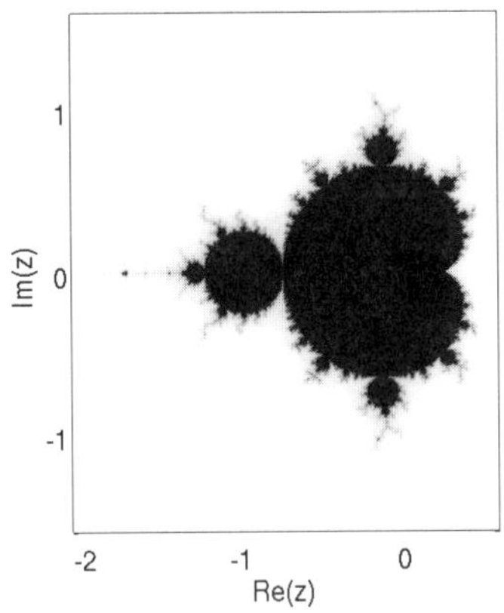

Figure 3.22. The Mandelbrot set, cf. [**Man91**].

Chapter 4

Hamiltonian dynamics

In the systems considered in Chapter 3, the evolution changes the volume of sets in phase space. However, many systems in nature conserve this volume, especially those of classical mechanics. For these systems we will discuss their dynamical properties, as stability and instability, and the occurrence of chaotic behavior. Our starting point of the bifurcation analysis of dissipative systems usually was a system with a globally attracting fixed point. In conservative systems such things cannot exist, and so we start from a completely integrable system, i.e., from a system in which all solutions can in principle be computed explicitly. It is the main purpose of this section to contrast the behavior of dissipative and conservative systems. Hence, essential parts in usual courses about Hamiltonian systems will be skipped. For an overview we refer to the textbooks [**Arn78, Thi88, MH92, HZ11**], or the selection of reprints [**MM87**].

4.1. Basic properties

The basic rule of classical mechanics is that the force $f : \mathbb{R}^d \to \mathbb{R}^d$ acting on a mass point at a position $q \in \mathbb{R}^d$ equals the product of mass m and acceleration $\ddot{q}$, i.e.,

$$m\ddot{q} = f(q). \tag{4.1}$$

For simplicity we set $m = 1$ in the following. In a conservative system, to a given force $f : \mathbb{R}^d \to \mathbb{R}^d$ there exists a potential $U : \mathbb{R}^d \to \mathbb{R}$ with $f = -\partial_q U$. Introducing the momentum variable $p = \dot{q} \in \mathbb{R}^d$ gives the first order system

$$\begin{pmatrix} \dot{q} \\ \dot{p} \end{pmatrix} = \begin{pmatrix} p \\ -\partial_q U \end{pmatrix} = \begin{pmatrix} \frac{1}{2}\partial_p(\|p\|^2) \\ -\partial_q U \end{pmatrix} = \begin{pmatrix} \partial_p H \\ -\partial_q H \end{pmatrix} = J\nabla H, \tag{4.2}$$

where $H : \mathbb{R}^{2d} \to \mathbb{R}$ given by $H = \frac{1}{2}\|p\|^2 + U$ is called the Hamiltonian and where

$$J = \begin{pmatrix} 0 & I \\ -I & 0 \end{pmatrix} \in \mathbb{R}^{2d \times 2d}$$

is a skew symmetric operator with I the identity matrix in $\mathbb{R}^{d\times d}$. All classical mechanical systems can be written as a Hamiltonian system

$$\dot{u} = J\nabla H(u), \tag{4.3}$$

with $u(t) \in \mathbb{R}^{2d}$, $H : \mathbb{R}^{2d} \to \mathbb{R}$ the Hamiltonian, and $J \in \mathbb{R}^{2d\times 2d}$ the skew symmetric operator from above. For ∇H locally Lipschitz-continuous, there exists a unique solution $u = u(t, u_0)$ of (4.3) with initial condition $u|_{t=0} = u_0$, cf. Theorem 2.2.1. An important property of (4.3) is the conservation of energy.

Theorem 4.1.1. *The Hamiltonian H is constant along a solution, i.e., $H(u(t, u_0)){=}H(u_0)$.*

Proof. Let $u = u(t)$ be a solution of the Hamiltonian system (4.3). Then

$$\frac{\mathrm{d}}{\mathrm{d}t}H(u(t)) = (\nabla H)^T \dot{u}(t) = (\nabla H)^T J\nabla H = 0,$$

due to the skew symmetry of J. □

In case $d = 1$ in (4.1) the full phase portrait can be constructed graphically.

Example 4.1.2. Consider an ODE

$$\ddot{x} = f(x) \tag{4.4}$$

with locally Lipschitz-continuous but otherwise arbitrary $f : \mathbb{R} \to \mathbb{R}$. In Figure 4.1 we explain how to draw the phase portrait without any formulas or calculations. Given f, without formulas, we may draw the potential energy $-F(x)$. In the top left we do this by first labeling the 4 zeros of f by $x_1, \ldots, x_4$. Setting, e.g., $-F(x) = -\int_0^x f(\xi)\,\mathrm{d}\xi$ we may plot $-F$ and find that $x_1, \ldots, x_4$ are stationary points of $-F$; here they are extrema since the roots of f are simple. Note that given $-F$ we obtain $E = \frac{1}{2}\dot{x}^2 - F$ by simply adding a parabola in $\dot{x}$ to $-F(u)$ at each x, see the bottom right for an illustration. Thus, minima of $-F$ are minima of E, while maxima of $-F$ are saddle points for E with stable direction $(0, 1)$ and unstable direction $(1, 0)$. To draw orbits we may think of small balls rolling around on the energy surface. For instance, consider a ball starting at $(x, \dot{x}) = (x_5, 0)$. It will slowly start to move to the right, thereby loosing potential energy $-F$ and taking up speed, hence gaining kinetic energy $\frac{1}{2}\dot{x}^2$. At $x = x_2$ it will have maximum speed, and then has to roll uphill, thus loosing kinetic energy but gaining potential energy. Thus, it will roll precisely until x_6, defined by $-F(x_6) = -F(x_5)$, where all kinetic energy has been transformed to potential energy again. The ball will now roll back, and all together we obtain a periodic orbit γ_1. Similar periodic orbits are obtained for all starting positions $(x, 0)$ with $x_2 < x < x_3$, or, equivalently for all $(x, \dot{x})$ inside the region bounded by the homoclinic orbit γ_2 to $(x_3, 0)$ and passing through $(x_7, 0)$. In a similar way all orbits can be constructed graphically. For instance, the orbit γ_3 corresponds to a ball coming from the far left with some large positive speed and rolling through the potential all the way to a position x_8 on the (far) right where it reaches some maximal potential energy $-F(x_8)$ and then rolls back. ⌋

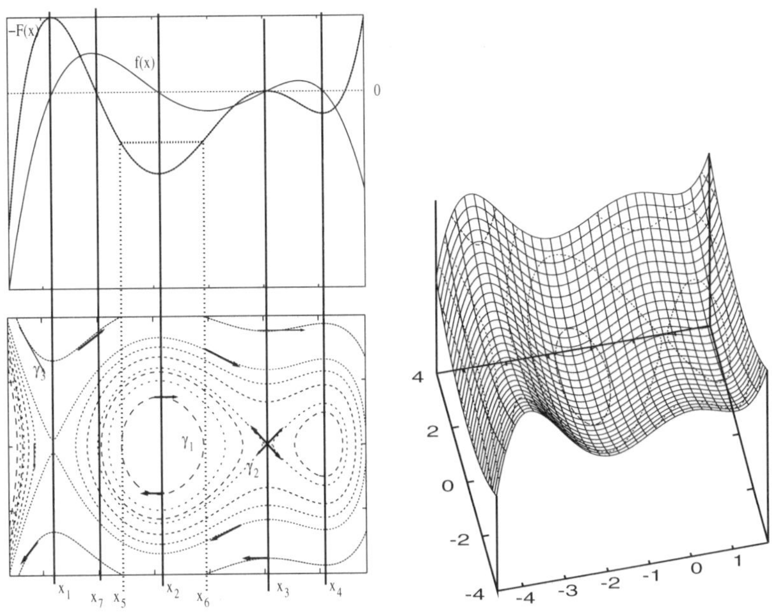

Figure 4.1. Phase portrait for a general scalar Newtonian system $\ddot{x} = f(x)$.

The volume of sets in the phase space of Hamiltonian systems is preserved.

Theorem 4.1.3. (Liouville) *Let*

$$u(t, \Omega) = \{v \in \mathbb{R}^{2d} : \exists u_0 \in \Omega : u(t, u_0) = v\}$$

be the image of a bounded and measurable $\Omega \subset \mathbb{R}^{2d}$ *under the map* $u(t, \cdot)$, *and let* μ *be the Lebesgue measure in phase space, i.e.,* $d\mu = 1du$. *For all* $t \in \mathbb{R}$ *we then have*

$$\int_{\Omega} \mathrm{d}\mu(u) = \int_{u(t,\Omega)} \mathrm{d}\mu(u).$$

Proof. A change of coordinates yields

$$\int_{u(t,\Omega)} d\mu(u) = \int_{\Omega} |D_{u_0}u(t,u_0)|\, d\mu(u_0).$$

The Jacobi matrix $Y(t) = D_{u_0}u(t,u_0)$ satisfies the differential equation

$$\frac{d}{dt}Y(t) = ((D(J\nabla H))(u(t,u_0)))Y(t).$$

Furthermore, the determinant $|Y(t)|$ satisfies

$$\frac{d}{dt}|Y(t)| = \mathrm{trace}(D(J\nabla H))|Y(t)|.$$

See Exercise 4.1 for a proof of this formula in $\mathbb{R}^2$. Using $Y(0) = I$, and

$$\mathrm{trace}(DJ\nabla H) = \sum_{i=1}^{2d} \delta_{ik} \sum_{k=1}^{2d} \partial_{u_k} \sum_{j=1}^{2d} J_{ij}\partial_{u_j} H = \sum_{i=1}^{2d}\sum_{j=1}^{2d} J_{ij}\partial_{u_i}\partial_{u_j} H = 0$$

due to the skew symmetry of J and due to the symmetry of the matrix $(\partial_{u_i}\partial_{u_j}H)_{i,j}$, we obtain the assertion. □

The invariant Lebesgue measure in phase space is called Liouville measure. The theory of measure preserving dynamical systems is the subject of ergodic theory, see, e.g., [**Wal82, Kre85**]. Complicated dynamical behaviour in Hamiltonian systems is described statistically in this theory.

4.1.1. Dynamics near a fixed point. As a direct consequence of the invariance of the phase space volume Hamiltonian systems cannot possess asymptotically stable fixed points. We start with the discussion of the linearization of the system at the fixed point. The Hamiltonian H must be quadratic in order to obtain a linear differential equation, i.e.,

$$H(u) = \frac{1}{2}\langle Mu, u\rangle = \frac{1}{2}\sum_{i,j=1}^{2d} m_{ij}u_i u_j.$$

Then

$$\partial_{u_k} H = \frac{1}{2}\sum_{i,j=1}^{2d}(m_{ij}\delta_{ik}u_j + m_{ij}u_i\delta_{jk}) = \frac{1}{2}\sum_{i=1}^{2d}(m_{ik} + m_{ki})u_i,$$

i.e., w.l.o.g. M can be considered as symmetric. Hence, a linear Hamiltonian system is of the form

$$\dot{u} = JMu$$

with $J = \begin{pmatrix} 0 & I \\ -I & 0 \end{pmatrix}$ and $M = M^T$.

Lemma 4.1.4. *Let λ be an eigenvalue of JM. Then also $-\lambda$, $\overline{\lambda}$ and $-\overline{\lambda}$ are eigenvalues of JM.*

Proof. We have $\det J = 1$, $J^2 = -I$ and $J^T = -J$. For the characteristic polynomial of JM we obtain

$$\begin{aligned} p(\lambda) &= \det(JM - \lambda I) = \det(J)\det(JM - \lambda I) = \det(J^2 M - \lambda J) \\ &= \det(-M - \lambda J) = \det(-M - \lambda J)\det(J) \\ &= \det(-MJ + \lambda I) = \det((-MJ + \lambda I)^T) \\ &= \det(-J^T M^T + \lambda I) = \det(JM + \lambda I) = p(-\lambda). \end{aligned}$$

Hence, with λ also $-\lambda$ is an eigenvalue. Since JM is a real-valued matrix, the complex conjugate of an eigenvalue must be an eigenvalue, too. □

A direct consequence is the following lemma, which again implies that in Hamiltonian systems no asymptotically stable fixed points can exist.

Lemma 4.1.5. *A fixed point of a Hamiltonian system can only be stable if all eigenvalues of the linearization lie on the imaginary axis with same geometric and algebraic multiplicity.*

For general ODEs in case that all eigenvalues lie on the imaginary axis the nonlinear terms decide about stability. For Hamiltonian systems the quadratic approximation of the Hamiltonian at the fixed point gives additional information.

Theorem 4.1.6. *Let $H(u) = \frac{1}{2}u^T A u + \mathcal{O}(\|u\|^3)$ with A strictly positive (or strictly negative) definite. Then $u = 0$ is stable.*

Proof. Let

$$\rho_0(r) = \min\{H(u) : |u| = r\}, \quad \rho_1(r) = \max\{H(u) : |u| = r\}.$$

Then

$$\rho_0(r) = \frac{1}{2}\lambda_{\min} r^2 + \mathcal{O}(r^3), \quad \rho_1(r) = \frac{1}{2}\lambda_{\max} r^2 + \mathcal{O}(r^3),$$

where $\lambda_{\min} > 0$, respectively $\lambda_{\max} > 0$, is the smallest, respectively the largest eigenvalue, of the positive definite matrix A. Then there exists an $r_0 > 0$, such that

$$\rho_0(r) \geq \frac{1}{4}\lambda_{\min} r^2 \quad \text{and} \quad \rho_1(r) \leq \lambda_{\max} r^2.$$

for all $r \in [0, r_0]$. Given $\varepsilon > 0$ we choose

$$0 \leq \delta \leq \sqrt{\frac{\lambda_{\min}}{4\lambda_{\max}}} \min(\varepsilon, r_0).$$

Then

$$\begin{aligned} |u(t, u_0)|^2 &\leq \frac{4}{\lambda_{\min}} \rho_0(u(t, u_0)) \leq \frac{4}{\lambda_{\min}} H(u(t, u_0)) \\ &= \frac{4}{\lambda_{\min}} H(u_0) \leq \frac{4}{\lambda_{\min}} \rho_1(u_0) \leq \frac{4\lambda_{\max}}{\lambda_{\min}} |u_0|^2 \leq \frac{4\lambda_{\max}}{\lambda_{\min}} \delta^2 \leq \varepsilon^2. \end{aligned}$$

Hence, the solution $u = u(t, u_0)$ cannot leave the ε-neighborhood, if the initial condition u_0 is contained in the δ-neighborhood. □

Example 4.1.7. Let $H = q^2 + p^2$. Then

$$\dot{q} = p, \quad \dot{p} = -q.$$

The orbits are circles with radius $\sqrt{H}$ and $u = (q, p) = (0, 0)$ is stable. ⌋

4.1.2. Lyapunov's subcenter theorem. Our next goal is the existence of periodic solutions. It turns that in each neighborhood of a fixed point with imaginary eigenvalues we can always find periodic solutions if the eigenvalues satisfy some non-resonance condition. Before we prove this result, we need two preparations.

First we introduce angle and action variables for linear systems. We consider the harmonic oscillator

$$\ddot{q} = -\omega^2 q,$$

which we write as Hamiltonian system

$$\dot{q} = \omega p = \partial_p H, \qquad \dot{p} = -\omega q = -\partial_q H,$$

with Hamiltonian $H = \frac{1}{2}\omega(p^2 + q^2)$. Introducing polar coordinates

$$q = \sqrt{2I}\cos(\phi), \qquad p = \sqrt{2I}\sin(\phi)$$

shows that the new variables ϕ and I satisfy the Hamiltonian system

$$\dot{\phi} = \omega = \partial_I H, \qquad \dot{I} = 0 = -\partial_\phi H$$

with Hamiltonian $H = \omega I$. The 2π-periodic variable ϕ is called angle variable and the variable I which is preserved under the flow is called action variable. Such variables play a fundamental role in the description of completely integrable Hamiltonian systems in the following.

Secondly we explain a reduction method for systems which are at least partly given in action angle variables. Let $H = H(\phi, q, I, p)$ be 2π-periodic w.r.t. ϕ. Moreover, assume that $\partial_I H > 0$ in some open subset of $\mathbb{R}^{2d}$. Then $H(\phi, q, I, p) = h$ can be solved for $I = -K(q, p, \phi, h)$.

From $\frac{\mathrm{d}}{\mathrm{d}t} H = 0$ and $H(\phi, q, -K(q, p, \phi, h), p) = h$ it follows that

$$\begin{aligned} \partial_{q_i} H + \partial_I H \cdot (-\partial_{q_i} K) &= 0, \\ \partial_{p_i} H + \partial_I H \cdot (-\partial_{p_i} K) &= 0, \end{aligned}$$

and hence using $\frac{\mathrm{d}}{\mathrm{d}t}\phi = \partial_I H$ and $\frac{\mathrm{d}}{\mathrm{d}t}\phi = (\frac{\mathrm{d}}{\mathrm{d}\phi} t)^{-1}$ yields

$$\partial_\phi q_i = \partial_t q_i \partial_\phi t = \partial_{p_i} H / \partial_I H = \partial_{p_i} K(q, p, \phi, h), \tag{4.5}$$

$$\partial_\phi p_i = -\partial_t p_i \partial_\phi t = \partial_{q_i} H / \partial_I H = -\partial_{q_i} K(q, p, \phi, h). \tag{4.6}$$

This is a $2(d-1)$-dimensional Hamiltonian system which is 2π-periodic w.r.t. the new time variable ϕ.

With these two preparations we are now going to prove

Theorem 4.1.8. (Lyapunovs subcenter theorem) *Let $u = 0$ be a fixed point of the Hamiltonian system $\dot{u} = J\nabla H(u)$. Let $\pm i\omega$, with $\omega \neq 0$, be simple eigenvalues of the linearization $JD^2H(0)$, and let all other eigenvalues λ_j fulfill $\lambda_j \neq in\omega$ for all $n \in \mathbb{Z}$. Then there is a neighborhood U of 0 and a two-dimensional manifold $M \subset U$ which is filled with periodic solutions with period close to $2\pi/\omega$. Moreover, M is tangential to the subspace spanned by the eigenvectors which are associated to the eigenvalues $\pm i\omega$.*

Proof. We seek small solutions $x = \varepsilon\widetilde{x}$, with $0 < \varepsilon \ll 1$ a small parameter, and consider the rescaled Hamiltonian

$$\widetilde{H}_\varepsilon(\widetilde{x}) = \frac{1}{\varepsilon^2}H(\varepsilon\widetilde{x}).$$

By this rescaling the quadratic part of the Hamiltonian stays independent of ε, whereas the higher order terms become small. W.l.o.g. assume that

$$D^2\widetilde{H}_\varepsilon(0) = D^2H(0) = \begin{pmatrix} \omega & 0 & 0 \\ 0 & \omega & 0 \\ 0 & 0 & A \end{pmatrix},$$

with $A \in \mathbb{R}^{2(d-1)\times 2(d-1)}$. This form can always be achieved by interchanging the second and the $(d+1)$th coordinate. The coordinates associated to the matrix A are denoted by $\widehat{x}$. For the first two coordinates $\widetilde{q}_1$ and $\widetilde{p}_1$ we introduce polar coordinates $\widetilde{q}_1 = \sqrt{2I}\cos(\phi)$ and $\widetilde{p}_1 = \sqrt{2I}\sin(\phi)$. In the new coordinates we have

$$H_\varepsilon(I,\phi,\widehat{x}) = \omega I + \frac{1}{2}\widehat{x}^T A\widehat{x} + \mathcal{O}(\varepsilon).$$

We look for solutions with $H_\varepsilon(I,\phi,\widehat{x}) = \omega$. For $\omega \neq 0$ we have $\partial_I H_\varepsilon \neq 0$ in a neighborhood of the origin such that $H_\varepsilon(I,\phi,\widehat{x}) = \omega$ can be solved w.r.t. I. We write this solution as $I = 1 - K_\varepsilon(\phi,\widehat{x})$. Inserting this ansatz in $H_\varepsilon(I,\phi,\widehat{x}) = \omega$ yields

$$K_\varepsilon(\phi,\widehat{x}) = \frac{1}{2\omega}\widehat{x}^T A\widehat{x} + \mathcal{O}(\varepsilon).$$

By the above reduction we obtain a 2π-periodic Hamiltonian system with new time variable ϕ and associated Poincaré map $\psi_\varepsilon(\widehat{x}_0) = \widehat{x}(2\pi,\widehat{x}_0)$. A fixed point $\widehat{x}^*$ of the Poincaré map ψ_ε yields a periodic solution $\widehat{x}(t,\widehat{x}^*) = \widehat{x}(t+2\pi,\widehat{x}^*)$. Thus, a periodic solution can be obtained via a zero of the function $F(\widehat{x}^*,\varepsilon) = \psi_\varepsilon(\widehat{x}^*) - \widehat{x}^*$. We have $F(0,0) = 0$ since $K_0(\phi,\widehat{x}) = \frac{1}{2\omega}\widehat{x}^T A\widehat{x}$ and thus $\widehat{x} = 0$ is a solution of the associated linear autonomous Hamiltonian system $\dot{\widehat{x}} = \frac{1}{\omega}JA\widehat{x}$. For the same reason we have $D_{\widehat{x}^*}F(0,0) =$

$e^{\frac{2\pi}{\omega}JA} - I$. Due to the non-resonance assumption, $D_{\widehat{x}^*}F(0,0)$ has only non-zero eigenvalues and hence is invertible. Then, by the implicit function theorem, $F(\widehat{x}^*, \varepsilon) = 0$ can be solved for $\widehat{x}^* = \widehat{x}^*(\varepsilon)$. The associated solution $(I, \phi, \widehat{x})$ is non-trivial since $I \neq 0$. The period is

$$\int_0^T \mathrm{d}t = \int_0^{2\pi} \frac{\mathrm{d}\phi}{\dot{\phi}} = \int_0^{2\pi} \frac{\mathrm{d}\phi}{\partial_I H} = \int_0^{2\pi} \frac{\mathrm{d}\phi}{\omega + \mathcal{O}(\varepsilon)} = \frac{2\pi}{\omega} + \mathcal{O}(\varepsilon).$$

This family of periodic solutions is tangential to $\operatorname{span}\{I, \phi\}$ since $\|\widehat{x}\| = \mathcal{O}(\varepsilon)$. □

4.2. Some celestial mechanics

To give some illustrations of computations in Hamiltonian dynamics we review some very basic celestial mechanics. A good reference is [**Gut94**].

4.2.1. The 1-body problem. Let $q \in \mathbb{R}^3$ be the position of a mass point, e.g., earth, that moves in a radially symmetric potential $U(q) = \widetilde{U}(\|q\|)$, i.e., $U(q)$ only depends on $\|q\|$, e.g. the gravitational potential of the sun. Then

$$\ddot{q} = -\nabla U(q) = -\widetilde{U}'(\|q\|)q/\|q\|,$$

or, in Hamiltonian form

$$\frac{\mathrm{d}}{\mathrm{d}t}\begin{pmatrix} q \\ p \end{pmatrix} = J\nabla H(q,p), \quad \text{with} \quad H(q,p) = U(q) + \frac{1}{2}\|p\|^2.$$

This is a 6-dimensional first order ODE, or, more precisely a Hamiltonian system with 3 degrees of freedom. Using the angular momentum, see Exercise 4.8,

$$C = q \times \dot{q}, \quad a \times b = \begin{pmatrix} a_2 b_3 - a_3 b_2 \\ a_3 b_1 - a_1 b_3 \\ a_1 b_2 - a_2 b_1 \end{pmatrix}$$

the dimension can be reduced. Let $C \neq 0$. Then C is orthogonal to the orbital plane $\{\alpha q + \beta \dot{q} : \alpha, \beta \in \mathbb{R}\}$. The area swept until t then is by Leibniz's sector formula

$$F(t) = \frac{1}{2}\int_0^t |q(s) \times \dot{q}(s)| \,\mathrm{d}s.$$

This yields **Kepler's second law**: the line segment joining a planet and the Sun sweeps out equal areas during equal intervals of time, or in modern formulation: $\frac{\mathrm{d}}{\mathrm{d}t}F(t) = \frac{1}{2}|q(t) \times \dot{q}(t)| = \frac{1}{2}C$.

Now assume that the orbital plane is the q_1-q_2 plane, let $q = (q_1, q_2) = (x, y)$, and introduce polar coordinates, i.e.

$$q = \begin{pmatrix} x \\ y \end{pmatrix} = r\begin{pmatrix} \cos\phi \\ \sin\phi \end{pmatrix}.$$

Then

$$p := \dot{q} = \dot{r}\begin{pmatrix}\cos\phi\\ \sin\phi\end{pmatrix} + r\dot{\phi}\begin{pmatrix}-\sin\phi\\ \cos\phi\end{pmatrix},$$

and C and H in polar coordinates become

$$C = r^2\dot{\phi} \neq 0, \quad H = \frac{1}{2}(\dot{r}^2 + r^2\dot{\phi}^2) + \widetilde{U}(r). \tag{4.7}$$

From (4.7) we obtain $\dot{\phi} \neq 0$ and $\dot{r}^2 = 2H - r^2\dot{\phi}^2 - 2\widetilde{U}(r)$. This yields a scalar first order equation as follows. Locally we can assume $r = r(\phi)$; then

$$\dot{r} = \frac{\mathrm{d}}{\mathrm{d}t}r(\phi) = (\frac{\mathrm{d}}{\mathrm{d}\phi}r)\dot{\phi} \Rightarrow r' := \frac{\mathrm{d}}{\mathrm{d}\phi}r = \frac{\dot{r}}{\dot{\phi}}.$$

In particular

$$(r')^2 = \frac{2(H-U)}{\dot{\phi}^2} - r^2 = \frac{2(H-U)r^4}{C^2} - r^2,$$

which yields the Clairaut ODE

$$r' = \pm g(r), \quad \text{with} \quad g(r) = \sqrt{2C^{-2}(H - U(r))r^4 - r^2}.$$

Instead of r we use the inverse radius $\sigma{=}1/r$, which yields the so called **fundamental equation of the 1-body problem**

$$\sigma' = g(1/\sigma)\sigma^2.$$

For the gravitational potential $U(r) = -Ar^{-1}$ we obtain

$$\sigma' = -\sqrt{-\sigma^2 + \alpha\sigma + \beta}, \quad \text{with} \quad \alpha = 2A/C^2, \quad \beta = 2H/C^2. \tag{4.8}$$

Lemma 4.2.1. *a) We have $H \geq -A^2/(2C^2)$ (lower energy bound).*
b) For $H = -A^2/(2C^2)$ the orbit is a circle with radius C^2/A.

Proof. Completing the square we write $-(\sigma')^2 = \sigma^2 - \alpha\sigma - \beta = (\sigma - \frac{\alpha}{2})^2 - \delta/4$, where $\delta = \alpha^2 + 4\beta$. Thus $\delta \geq 0$ and hence $H \geq -A^2/(2C)^2$. For $H = -A^2/(2C)^2$ we have $\delta = 0$ and hence $\sigma' = 0$ and $\sigma \equiv \alpha/2$. □

Henceforth let $\delta = \alpha^2 + 4\beta > 0$. We seek solutions of

$$\sigma' = -g(\sigma) \tag{4.9}$$

with $g(\sigma) = \sqrt{-\sigma^2 + \alpha\sigma + \beta}$ and $\sigma \in I := [\frac{\alpha - \sqrt{\delta}}{2}, \frac{\alpha + \sqrt{\delta}}{2}]$ to have a real radicand. In particular, an unbounded orbit is only possible for

$$0 \in I \Leftrightarrow \alpha \leq \sqrt{\delta} \Leftrightarrow \beta \geq 0 \Leftrightarrow H \geq 0,$$

while the orbit is always bounded if $H < 0$. Moreover, from $\sigma \leq \frac{\alpha+\sqrt{\delta}}{2}$ we obtain a minimal distance of the orbit to the origin, i.e.,

$$r \geq r_{\min} := \frac{2}{\alpha + \sqrt{\delta}}.$$

Let $\sigma(0) = \sigma_0$ with $\max\{0, \frac{\alpha-\sqrt{\delta}}{2}\} < \sigma_0 < \frac{\alpha+\sqrt{\delta}}{2}$. Then we have a local solution as g is locally Lipschitz near σ_0. Substituting $u = \frac{1}{2}\sqrt{\delta}x + \frac{\alpha}{2}$ we obtain

$$\phi = \int_{\sigma_0}^{\sigma} \frac{\mathrm{d}u}{-g(u)} = \int_{\sigma_0}^{\sigma} \frac{\mathrm{d}u}{-\sqrt{-u^2+\alpha u+\beta}} = \int_{\widehat{\sigma}_0}^{\widehat{\sigma}} \frac{\mathrm{d}x}{-\sqrt{1-x^2}}$$
$$= \arccos\widehat{\sigma} - \arccos\widehat{\sigma}_0,$$

with $\widehat{\sigma}_0 = \frac{2}{\sqrt{\delta}}(\sigma_0 - \frac{\alpha}{2})$ and $\widehat{\sigma} = \frac{2}{\sqrt{\delta}}(\sigma - \frac{\alpha}{2})$. From $-1 < \widehat{\sigma}_0 < 1$ we obtain $0 < \arccos(\widehat{\sigma}_0) < \pi$ and thus

$$\frac{2}{\sqrt{\sigma}}(\sigma - \alpha/2) = \widehat{\sigma} = \cos(\phi + b) \quad \text{with} \quad b = \arccos(\widehat{\sigma}_0). \tag{4.10}$$

W.l.o.g. we choose the initial condition $\sigma_0 = 1/r_{\min}$ and obtain

$$r(\phi) = \frac{p}{1 + e\cos(\phi)} \quad \text{with} \quad p = C^2/A, \quad e = \sqrt{1 + 2HC^2/A^2}. \tag{4.11}$$

We distinguish three cases.

(1) $0 \le e < 1 \Leftrightarrow H < 0$: Then $r(\phi)$ is defined for all $\phi \in \mathbb{R}$ and 2π-periodic in ϕ. Going back to cartesian coordinates we obtain

$$\begin{pmatrix} x \\ y \end{pmatrix} = \begin{pmatrix} q_1 \\ q_2 \end{pmatrix} = r \begin{pmatrix} \cos\phi \\ \sin\phi \end{pmatrix},$$

and $\cos\phi = (p/r - 1)/e$ yields $(p - ex)^2 = r^2 = x^2 + y^2$, and thus

$$\left(x + \frac{ep}{1-e^2}\right)^2 + \frac{y^2}{1-e^2} = \frac{p^2}{(1-e^2)^2}.$$

This is **Kepler's first law:** The body moves on an ellipse with focal points $(0,0)$ and $(-(2ep)/(1-e^2), 0)$, numerical eccentricity e and major semi-axis $a = -A/(2H)$. The point $(r_{\min}, 0)$ is called *peri center* (*perihel* for a planet in the solar system) and $(r_{\max}, \pi)$ is called *apo center* (*apohel*). Examples for numerical eccentricities e are $e = 0.0167$ for Earth, $e = 0.2056$ for Mercury, and $e = 0.9673$ for Halley's comet. The relatively large eccentricity of Mercury is of great importance historically since already in the 19th century it allowed the observation of the perihel precession of Mercury: after each elliptical orbit Mercury's perihel is shifted by a few angular seconds. This contradicts the above (newtonian) calculations, but could be explained by Einstein's relativity theory.

(2) $e = 1 \Leftrightarrow H = 0$: the existence interval is $-\pi < \phi < \pi$, and geometrically the orbit is a parabola opening to the left, $y^2 = -px + p^2$.

(3) $e > 1 \Leftrightarrow H > 0$: the orbit is the hyperbola $y^2 = (e^2-1)x^2 - 2epx + p^2$.

Thus we found the orbits for the 1-body problem in implicit form and without time dependence, determined by parameters H, A and C. Next, the orbits can be characterized via initial conditions and the time-dependence

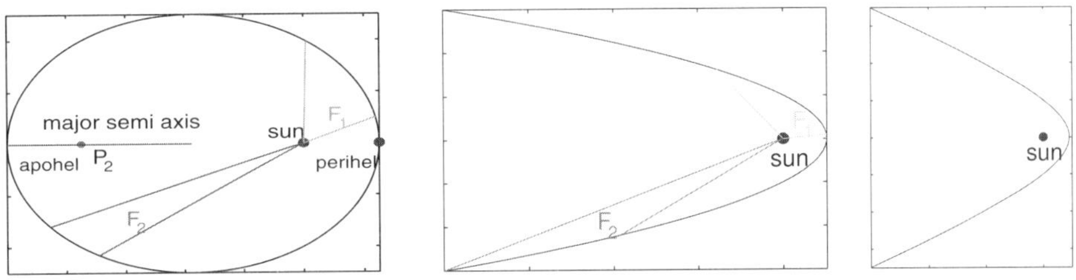

Figure 4.2. Kepler-ellipse, parabola and hyperbola. The areas F_1, F_2 are meant to illustrate Kepler's 2nd law.

can be reintroduced. This allows to derive **Kepler's 3rd law:** if a_γ is the length of an orbit's major semi-axis, and T_γ its period, then T_γ^2/a_γ^3 is independent of the orbit γ. More precisely, in this calculus we have $\frac{T^2}{a^3} = \frac{4\pi^2}{GM}$ for all orbits in the solar systems, where G is the gravitational constant and M the mass of the sun, which is a very good approximation to observations. This is only natural, as Kepler derived his laws from observations.

4.2.2. The restricted 3-body problem. N bodies which move under the influence of gravity have the Hamiltonian

$$H = \sum_{i=1}^{N} \frac{p_i^2}{2m_i} + \sum_{i \neq j} \frac{m_i m_j}{|q_i - q_j|}.$$

The solution of the associated differential equations, and the associated question about the mechanical stability of our solar system, have been considered as essential for mankind. However, it turned out that only the two body problem ($N = 2$), see above and Exercise 4.9, can be solved explicitly and already the three body problem shows chaotic behavior.

There is one intermediate problem, namely the so called restricted three body problem. There it is assumed that the third body K_3 has a very small mass compared to the other two bodies K_1 and K_2. The restricted three body problem is obtained by neglecting the forces of K_3 on K_1 and K_2, such that their motion is not affected by K_3, i.e., they move on Kepler ellipses around their center of mass.

In a coordinate system which moves with the two larger bodies of reduced masses $\mu = \frac{m_1}{m_1+m_2}$ and $1-\mu$, which lie fixed in $-\mu$ and $1-\mu$, the Hamiltonian for the motion of the third body is given by

$$H(q,p) = \frac{1}{2}(p_1^2 + p_2^2 + p_3^2) + q_2 p_1 - q_1 p_2 + U(q),$$

where

$$U(q) = -\frac{q_1^2}{2} - \frac{q_2^2}{2} - \frac{1-\mu}{\sqrt{(q_1+\mu)^2 + q_2^2 + q_3^2}} - \frac{\mu}{\sqrt{(q_1 - 1 + \mu)^2 + q_2^2 + q_3^2}}.$$

The second and third term in H and the first term in U come from the Coriolis force in the rotating coordinate system. There are five equilibria, called Lagrangian points, shown in Figure 4.3.

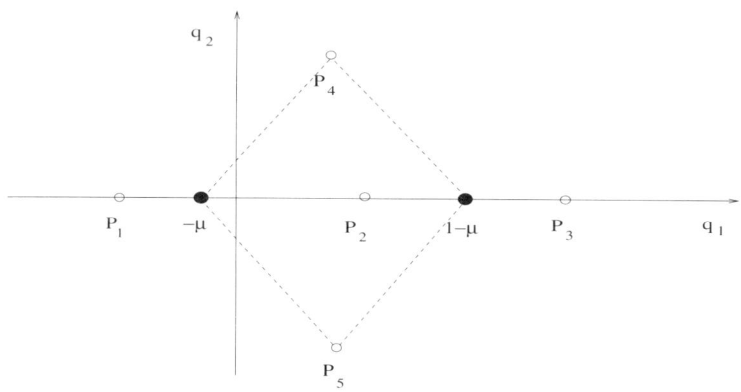

Figure 4.3. The equilibria in the restricted three body problem.

For the stability of these equilibria we first consider

$$M = D^2H(u) = \begin{pmatrix} \partial_{q_1}^2 U & \partial_{q_1}\partial_{q_2} U & 0 & 0 & -1 & 0 \\ \partial_{q_1}\partial_{q_2} U & \partial_{q_2}^2 U & 0 & 1 & 0 & 0 \\ 0 & 0 & \partial_{q_3}^2 U & 0 & 0 & 0 \\ 0 & 1 & 0 & 1 & 0 & 0 \\ -1 & 0 & 0 & 0 & 1 & 0 \\ 0 & 0 & 0 & 0 & 0 & 1 \end{pmatrix}.$$

For the linearisation JM in these points (especially $\partial_{q_j}\partial_{q_3} U|_{P_i} = 0$ for $j = 1, 2$ and $i = 1, \ldots, 5$) we thus obtain

$$JM = \begin{pmatrix} 0 & 1 & 0 & 1 & 0 & 0 \\ -1 & 0 & 0 & 0 & 1 & 0 \\ 0 & 0 & 0 & 0 & 0 & 1 \\ -\partial_{q_1}^2 U & -\partial_{q_1}\partial_{q_2} U & 0 & 0 & 1 & 0 \\ -\partial_{q_1}\partial_{q_2} U & -\partial_{q_2}^2 U & 0 & -1 & 0 & 0 \\ 0 & 0 & -\partial_{q_3}^2 U & 0 & 0 & 0 \end{pmatrix}.$$

We find that the q_3, p_3-part decouples and leads to the eigenvalue problem

$$0 = \lambda^2 + \partial_{q_3}^2 U.$$

Since $\partial_{q_3}^2 U > 0$ we have $\lambda_{1,2} \in i\mathbb{R}$. For the remaining eigenvalues we have

$$0 = \lambda^4 + \lambda^2(\partial_{q_1}^2 U + \partial_{q_2}^2 U + 4) + (\partial_{q_1}^2 U)(\partial_{q_2}^2 U) - (\partial_{q_1}\partial_{q_2} U)^2.$$

It turns out that the points P_1, P_2 and P_3 are saddles and therefore unstable. In the points P_4 and P_5 we find

$$q_1 = \frac{1}{2} - \mu,\ q_2 = \pm\frac{\sqrt{3}}{2},\ \partial^2_{q_1} U = -\frac{3}{4},\ \partial^2_{q_2} U = -\frac{9}{4},\ (\partial_{q_1}\partial_{q_2} U)^2 = -\frac{3\sqrt{3}}{2}(\frac{1}{2} - \mu).$$

For $4(\frac{1}{2}-\mu)^2 < 1$ the eigenvalues are purely imaginary, i.e., the points P_4 and P_5 are linearly stable. Unfortunately M is indefinite, such that we cannot conclude on the nonlinear stability of P_4 and P_5 with the above theorem. Nevertheless P_4 and P_5 are realized in nature and play an important role for space missions. For instance, Sun and Jupiter can be taken as the big bodies, and in an angle of 60 degrees before and after Jupiter on his orbit there are the so the called Greeks and Trojans, some families of asteroids.

4.3. Completely integrable systems

If there are several non-resonant eigenvalues on the imaginary axis, then there are several families of periodic solutions. In this and the following section we discuss situations with even more complex structures. We start with the linear Hamiltonian system

$$\dot{x} = JMx. \tag{4.12}$$

We assume that all eigenvalues $\mathrm{i}\omega_j$ of the matrix JM are semi-simple and on the imaginary axis. Then the system can be transformed into

$$\dot{q_j} = \omega_j p_j, \quad \dot{p_j} = -\omega_j q_j, \quad j = 1, \dots, d, \tag{4.13}$$

i.e., into a Hamiltonian system with Hamiltonian

$$H = \sum_{j=1}^{d} \frac{\omega_j}{2}(q_j^2 + p_j^2).$$

Clearly this system is the direct sum of d Hamiltonian systems with d independent Hamiltonians $H_j = \frac{\omega_j}{2}(q_j^2 + p_j^2)$. The $I_j = \frac{1}{2\omega_j} H_j$ are conserved also for the flow of (4.12), i.e., $\frac{\mathrm{d}}{\mathrm{d}t} I_j(x(t)) = 0$ for solutions $x = x(t)$ of (4.12). For the jth system the orbits are circles, i.e.,

$$q_j + ip_j = \sqrt{2I_j}\mathrm{e}^{i\phi_j} \quad \text{with} \quad \phi_j(t) = \phi_j(0) + \omega_j t \bmod 2\pi.$$

For (4.12) the phase space decomposes into d-dimensional tori

$$\{u \in \mathbb{R}^d : I_1 = c_1, \dots, I_d = c_d\}.$$

For one or more vanishing I_js we have dimensions of the tori between 1 and d. The d-dimensional tori contain so called quasi-periodic solutions

$$x(t) = g(\omega_1 t, \dots, \omega_d t)$$

with $g : S^1 \times \ldots \times S^1 \to \mathbb{R}^{2d}$. If the non-resonance condition

$$\omega \cdot n = \omega_1 n_1 + \ldots + \omega_d n_d \neq 0 \quad \text{for all} \quad (n_1, \ldots, n_d) \in \mathbb{Z} \times \ldots \times \mathbb{Z} \setminus \{0, \ldots, 0\}$$

holds, then the orbits are dense in the associated d-dimensional torus. If for instance $\omega_1 = 2\omega_2$ and all other ω_i are non-resonant, then the solutions are dense in $d-1$-dimensional tori.

In the following we study whether this situation persists under perturbations or not. We expect that it is more simple to destroy a torus filled with low-dimensional solutions associated to a set of resonant ωs than a torus with dense solutions associated to non-resonant ωs. Therefore, we expect non-resonance conditions to play an important role. Moreover, such tori are not only important for linear Hamiltonian systems but also for special nonlinear systems, which are called completely integrable, see below.

A coordinate transform $y = T(x)$ will in general destroy the Hamiltonian structure of a Hamiltonian system. Only so called symplectic transformations keep the Hamiltonian structure. For $\widetilde{H}(y) = H(x)$ we have

$$\partial_{x_j} H(T(x)) = \sum_{k=1}^{2d} \partial_{y_k} \widetilde{H}(y) \partial_{x_j} y_k,$$

and hence

$$\dot{y} = (DT)\dot{x} = (DT)J\nabla_x H(x) = (DT)J(DT)^T \nabla_y \widetilde{H}(y) = J\nabla_y \widetilde{H}(y),$$

if T is a so called symplectic transformation.

Definition 4.3.1. *Let $J^{-1} = -J = J^T$. The bilinear form*

$$\omega(v_1, v_2) = v_1^T J v_2$$

is called the symplectic structure induced by J. A transformation $y = T(x)$ is called canonical or symplectic if

$$((DT)(x))J(DT)^T(x) = J \quad \forall x \in \mathbb{R}^{2d}.$$

A Hamiltonian system is called completely integrable if it can be transformed into the form

$$(4.14) \qquad \dot{I}_j = -\partial_{\phi_j} H = 0, \quad \dot{\phi}_j = \partial_{I_j} H = \omega_j, \quad j = 1, \ldots, d$$

by a symplectic transformation. The Hamiltonian $H = H(I_1, \ldots, I_d)$ and the frequencies $\omega_j = \omega_j(I_1, \ldots, I_d)$ only depend on the conserved quantities $I_1, \ldots, I_d$. If the set

$$\{x \in \mathbb{R}^{2d} : I_1 = const_1, \ldots, I_d = const_d\}$$

is smooth and compact then it is a d-dimensional torus. The I_j and ϕ_j are called action and angle variables, respectively.

Lemma 4.3.2. *The map $x_0 \mapsto x(t, x_0)$ is symplectic for all t.*

Proof. Consider $R(t,\cdot) = (Dx(t,\cdot))J(Dx(t,\cdot))^T$. Then $R|_{t=0} = J$ and R solves the linear ODE

$$\begin{aligned}\dot{R} =&(D\dot{x})J(Dx)^T + (Dx)J(D\dot{x})^T\\ =&JD^2H(Dx)J(Dx)^T + DxJ(JD^2H(Dx))^T\\ =&JD^2HR + R(JD^2H)^T.\end{aligned}$$

Since $(D^2H)^T = D^2H$ and $J^T = -J$ we have

$$JD^2HJ + J(JD^2H)^T = JD^2HJ + J(D^2H)^T J^T = 0.$$

Hence, $R = J$ is the unique solution. □

Example 4.3.3. Let $F = F(\widetilde{I}, \phi)$. Then the map induced by $\widetilde{\phi} = \partial_{\widetilde{I}} F$ and $I = \partial_\phi F$ is symplectic, cf. [**Arn78**, §48]. ⌋

There are various sufficient conditions that ensure that a Hamiltonian system is completely integrable. Here we will only cite one major criterion.

Definition 4.3.4. *Let $F : \mathbb{R}^{2d} \to \mathbb{R}$ and $G : \mathbb{R}^{2d} \to \mathbb{R}$ be in C^1. Then*

$$\{F, G\} = (\nabla F)^T J \nabla G$$

is called the Poisson bracket of F and G.

Theorem 4.3.5. (Liouville's theorem) *Let $H : \mathbb{R}^{2d} \to \mathbb{R}$, and let $I_1 = H, I_2, \dots, I_d$ be independent integrals in involution, i.e. $\{I_j, I_k\} = 0$ for $i, j = 1, \dots, d$. Then the Hamiltonian system is completely integrable.*

Proof. See [**Arn78**, §49]. □

4.4. Perturbations of completely integrable systems

The question occurs how robust completely integrable systems are under perturbations. The answer turns out to be rather delicate and has to do with number theory. The starting situation is as follows. Consider a Hamiltonian of the form

$$H(\phi, I) = H_0(I) + \varepsilon H_1(\phi, I, \varepsilon), \tag{4.15}$$

with $I \in \mathbb{R}^d$, $\phi \in T^d$, and ε a small parameter. The associated Hamiltonian system reads

$$\dot{\phi} = \partial_I H_0 + \varepsilon \partial_I H_1(I, \phi), \qquad \dot{I} = -\varepsilon \partial_\phi H_1(I, \phi).$$

Hence, the action variable I only changes slowly in time. The idea is to find a change of coordinates such that the transformed system is of the original completely integrable form. According to Example 4.3.3 we obtain a symplectic transformation if we take a so called generating function $F : \mathbb{R}^{2d} \to \mathbb{R}$ and define a relation between the old and new variables through

$$\widetilde{\phi} = \partial_{\widetilde{I}} F(\phi, \widetilde{I}) \qquad \text{and} \qquad I = \partial_\phi F(\phi, \widetilde{I}).$$

We look for a transformation near the identity and therefore choose

$$F(\phi, \widetilde{I}) = \phi \widetilde{I} + \varepsilon f(\phi, \widetilde{I}), \tag{4.16}$$

which yields

$$\widetilde{\phi} = \partial_{\widetilde{I}} F = \phi + \varepsilon \partial_{\widetilde{I}} f, \qquad I = \partial_\phi F = \widetilde{I} + \varepsilon \partial_\phi f,$$

and therefore

$$\phi = \widetilde{\phi} - \varepsilon \partial_{\widetilde{I}} f + \mathcal{O}(\varepsilon^2), \qquad I = \widetilde{I} + \varepsilon \partial_{\widetilde{\phi}} f + \mathcal{O}(\varepsilon^2).$$

Plugging this into the Hamiltonian gives

$$\begin{aligned}\widetilde{H}(\widetilde{\phi}, \widetilde{I}) =& H_0(\widetilde{I} + \varepsilon \partial_{\widetilde{\phi}} f) + \varepsilon H_1(\widetilde{\phi}, \widetilde{I}, 0) + \mathcal{O}(\varepsilon^2) \\ =& H_0(\widetilde{I}) + \varepsilon[\partial_{\widetilde{I}} H_0 \cdot \partial_{\widetilde{\phi}} f + H_1(\widetilde{\phi}, \widetilde{I}, 0)] + \mathcal{O}(\varepsilon^2).\end{aligned}$$

The idea is to eliminate the terms of order $\mathcal{O}(\varepsilon)$ by finding f such that

$$\partial_{\widetilde{I}} H_0 \cdot \partial_{\widetilde{\phi}} f + H_1(\widetilde{\phi}, \widetilde{I}, 0) = 0.$$

If we find such an f, then we can go on and find in the next step another symplectic transformation which then eliminates the $\mathcal{O}(\varepsilon^2)$ terms, etc., until finally all perturbations are eliminated. Before we do so we look at the problem to eliminate the terms of order $\mathcal{O}(\varepsilon)$ in more detail. Given $H_1(\widetilde{\phi}, \widetilde{I}, 0)$ and $H_0(\widetilde{I})$ we seek $f : \mathbb{T}^d \times \mathbb{R}^d \to \mathbb{R}$, such that

$$\partial_{\widetilde{I}} H_0(\widetilde{I}) \cdot \partial_{\widetilde{\phi}} f(\widetilde{\phi}, \widetilde{I}) = -H_1(\widetilde{\phi}, \widetilde{I}, 0). \tag{4.17}$$

Since f is periodic in $\widetilde{\phi}$ we expand, for fixed $\widetilde{I}$, the problem into Fourier series, cf. §5.2.2. Thus,

$$f(\widetilde{\phi}) = \sum_{k \in \mathbb{Z}^d} a_k \mathrm{e}^{\mathrm{i}k \cdot \widetilde{\phi}}, \quad H_1(\widetilde{\phi}) = \sum_{k \in \mathbb{Z}^d} b_k \mathrm{e}^{\mathrm{i}k \cdot \widetilde{\phi}}, \quad k = (k_1, \dots, k_d).$$

Next we set

$$\omega(\widetilde{I}) = \partial_{\widetilde{I}} H_0(\widetilde{I}) = (\omega_1, \dots, \omega_d)(\widetilde{I}).$$

For $k \neq 0$ we obtain $\mathrm{i}(\omega \cdot k) a_k = -b_k$. Thus, if $\omega \cdot k \neq 0$, then

$$a_k = \frac{\mathrm{i} b_k}{\omega \cdot k},$$

is determined and the term $b_k \mathrm{e}^{\mathrm{i}k \cdot \widetilde{\phi}}$ can be removed from H_1. If there are no resonances at all, i.e., if $\omega \cdot k \neq 0$ for all $k \in \mathbb{Z} \setminus \{0\}$, then formally all $\mathcal{O}(\varepsilon)$ terms can be removed except of $b_0(\widetilde{I})$. However, this term can be included into $H_0(\widetilde{I})$ as a correction. Obviously the non-resonance condition for the elimination of the higher order terms is not changed since the left-hand side of (4.17) is not changed. If there are no resonances until the nth step, then the perturbation up to terms of order $\mathcal{O}(\varepsilon^n)$ can be removed. We have the following approximation theorem.

Theorem 4.4.1. *If the normal form transformations allow to remove all terms up to order $\mathcal{O}(\varepsilon^n)$, i.e., if $\partial_{\widetilde{\phi}}\widetilde{H} = \mathcal{O}(\varepsilon^{n+1})$, then there exist C_1 and $\varepsilon_0 > 0$ such that for all $\varepsilon \in (0, \varepsilon_0)$ we have in the original coordinates*

$$\sup_{t\in(-1/\varepsilon^n, 1/\varepsilon^n)} \|I(t) - I(0)\| \leq C_1\varepsilon.$$

Proof. We have $\|I - \widetilde{I}(\phi, I)\| \leq C_2\varepsilon$ for a $C_2 > 0$, and, since $\|\dot{\widetilde{I}}\| = \|\partial_{\widetilde{\phi}}\widetilde{H}\| \leq C_3\varepsilon^{n+1}$ for a $C_3 > 0$, we have

$$\|\widetilde{I}(t) - \widetilde{I}(0)\| \leq C_3|t|\varepsilon^{n+1} \leq C_3\varepsilon$$

for all $|t| \leq 1/\varepsilon^n$. This yields

$$\begin{aligned}\|I(t) - I(0)\| \leq& \|I(t) - \widetilde{I}(t)\| + \|\widetilde{I}(t) - \widetilde{I}(0)\| + \|I(0) - \widetilde{I}(0)\| \\ \leq& (2C_2 + C_3)\varepsilon =: C_1\varepsilon.\end{aligned}$$

So far we did not consider the convergence of the above Fourier series in the normal form transforms. This turns out to be complicated due to so called small divisor problems. This means that for given $\omega \in \mathbb{R}^d$ and (arbitrary small) $\delta > 0$ there always is a $k \in \mathbb{Z}^d$ such that

$$|k \cdot \omega| < \delta.$$

Hence, the divisors in the series for f become arbitrarily small and the convergence of the Fourier series is a serious problem. The problem is solved by restricting the set of possible frequencies.

Definition 4.4.2. *A vector $\omega \in \mathbb{R}^d$ is called of type (L, γ) if for all $k \in \mathbb{Z}^d \setminus \{(0, \ldots, 0)\}$ we have*

$$|k \cdot \omega| \geq L|k|^{-\gamma}$$

We remark that for given $\gamma > d$ and almost all $\omega \in \mathbb{R}^d$ there exists a $L > 0$ such that ω is of type (L, γ), cf. [**Arn88**, page 114]. To study the analytic properties of the generating function F we use the following functions spaces.

Definition 4.4.3. *For $n \in \mathbb{N}$ define the spaces*

$$\ell_{1,n} = \{a : \mathbb{Z}^d \to \mathbb{C} : \|a\|_{\ell_{1,n}} = |a_0| + \sum_{k\in\mathbb{Z}^d} |a_k||k|^n < \infty\}.$$

Remark 4.4.4. We have that $a \in \ell_{1,n}$ implies $\mathcal{F}^{-1}a \in C_b^n$, where

$$C_b^n = \{f : \mathbb{T}^d \to \mathbb{R} : f \text{ } n \text{ times continuously differentiable}\},$$

which is equipped with the norm $\|f\|_{C_b^n} = \sum_{|j|=0}^{n} \|\partial_\phi^j f\|_{C_b^0}$, where $\|f\|_{C_b^0} = \sup_{\phi\in\mathbb{T}^d} |f(\phi)|$, and where $\mathcal{F}^{-1} : a \mapsto f$ is defined by $f(\phi) = \sum_{k\in\mathbb{Z}^d} a_k e^{ik\cdot\phi}$, cf. §5.1. ⌋

Lemma 4.4.5. *Let $\omega \in \mathbb{R}^d$ be of type (L,γ) and $b \in \ell_{1,n}$ with $b_0 = 0$. Then a, defined by $a_k = \frac{ib_k}{k\cdot\omega}$ for $k \neq 0$, $a_0 = 0$, is in $\ell_{1,n-\gamma}$.*

Proof. We have

$$\|a\|_{\ell_{1,n-\gamma}} = \sum_{k\in\mathbb{Z}^d\setminus\{0\}} |a_k |k|^{n-\gamma}| = \sum_{k\in\mathbb{Z}^d\setminus\{0\}} \left| \frac{b_k}{k\cdot\omega} |k|^{n-\gamma} \right| \leq L^{-1}\|b\|_{\ell_{1,n}}. \quad \square$$

The correspondence of the exponent n in the weight of $\ell_{1,n}$ in Fourier space to regularity w.r.t. ϕ in physical space implies that in each iterative step we lose regularity of the Hamiltonian H. In order to eliminate the perturbation completely, infinitely many steps are necessary. Hence, there will be a loss of infinitely many derivatives. This problem can be solved by working in a space of analytic functions or by using some artificial smoothing in the so called hard implicit function theorem, cf. [**SR89**], when working with Hamiltonians of finite differentiability. The convergence is based on the quadratic convergence of the Newton scheme. This approach results in the famous **KAM**-theorem named after Kolmogorov, Arnold and Moser. We choose the analytic version and set

$$\mathcal{A}_{\sigma,\rho}(I^*) = \{(I,\phi) \in \mathbb{R}^n \times \mathbb{C}^n : |I - I^*| < \rho,\ |\mathrm{Im}(\phi_j)| < \sigma, j = 1,\dots,d\}.$$

We define the norm of a function f which is analytic w.r.t. ϕ on $\mathcal{A}_{\sigma,\rho}(I^*)$ by

$$\|f\|_{\sigma,\rho} = \sup_{(I,\phi)\in\mathcal{A}_{\sigma,\rho}(I^*)} |f(I,\phi)|.$$

In nonlinear Hamiltonian systems in general the frequencies vary in a nontrivial manner with I, i.e., $\partial_I\omega = \partial_I^2 H_0$ does not vanish. Since there is a dense set of resonant frequencies in $\mathbb{R}^d$ it cannot be expected that the phase space is completely filled with tori after the perturbation. Therefore, in any neighborhood of a torus $\mathbb{T}^d$ with a non-resonant ω there is a torus $\mathbb{T}^d$ with a resonant ω. This means that next to any torus for which the transformations can be carried through to arbitrary order there is a torus in which low order perturbations influence the dynamics in the torus and may destroy the torus. Nevertheless, almost all tori persist in the following sense.

Theorem 4.4.6. (KAM) *Let $\omega(I^*) = \omega^*$ be of type (L,γ) and let the Hessian $\partial_I^2 H_0$ be invertible in I^*. Then there exists an $\varepsilon_0 > 0$ such that for $\|f\|_{\sigma,\rho} < \varepsilon_0$ the Hamiltonian system has quasi-periodic solutions with frequencies ω^*, i.e., the torus to $I = I^*$ persists under the perturbation. Let $V \subset \mathbb{R}^d$ be an open set with finite Lebesgue measure, where the Hessian $\partial_I^2 H_0$*

is invertible. For all $\delta > 0$ there is an $\varepsilon_0 > 0$, such that for all $\varepsilon \in (0, \varepsilon_0)$ there is a set $P_\varepsilon \subset V \times T^d$ with the following properties. The Lebesgue measure μ of $(V \times T^d) \setminus P_\varepsilon$ is less than δ, and for all $(I_0, \phi_0) \in P_\varepsilon$ the orbit through (I_0, ϕ_0) is quasi-periodic.

Proof. See [**Arn78, Way96**] or [**KP03**, §2] for a review. □

Hence, for small perturbations most of the phase space is still filled with tori. In between the tori chaotic behavior may occur. This is explained subsequently in §4.5.

In $\mathbb{R}^4$ the KAM-theorem yields a stability theorem since the invariant tori form two-dimensional hypersurfaces in the three-dimensional energy surfaces. In higher space dimensions the d-dimensional tori cannot separate domains in the $2d-1$-dimensional energy surface, but we expect that solutions need a long time to wander around the tori. This is called Arnold-diffusion and is mathematically formulated in Nekhoroshov's Theorem below.

An important motivation of these investigations again comes from celestial mechanics, in particular the question of the stability of our solar system. If we ignore mutual gravitational forces between the planets then we obtain a completely integrable system. The forces between the planets compared to that of the sun have a ratio of $\varepsilon \approx 1/1000$. Hence, the interplanetary forces can be considered as small perturbations. Quite obviously it is impossible to say whether our solar system is resonant or non-resonant, i.e., whether the ratios between different rotation times are rational or irrational. In spite of the fact that the rotation times of Jupiter and Saturn have a ratio of about 2/5, our solar system seems to be remarkably stable. As said above, heuristically, even in the resonant case we expect the solutions to need a long time to wander around the remaining tori. This can be made precise for so called steep Hamiltonians, cf. [**AKN06**, §6.3.4].

Definition 4.4.7. *An analytic function is called steep if it has no real extrema and if all complex extrema are isolated.*

Theorem 4.4.8. (Nekhoroshov) *Let $H_0 = H_0(I)$ be a steep function. Then in the perturbed Hamiltonian system for a sufficiently small perturbation εH_1 we have*

$$\|I(t) - I(0)\|_{\mathbb{R}^d} < \varepsilon^b \tag{4.18}$$

for $0 \le t \le \frac{1}{\varepsilon}\exp(\frac{1}{c\varepsilon^a})$, where $a, b, c > 0$ only depend on H_0.

Remark 4.4.9. KAM-theory is used as an explanation for the so called Kirkwood gaps. Figure 4.4 shows the number of asteroids in the main asteroid belt as a function of their orbits major semi-axis in astronomical units (AU), where 1 AU is the length of the major semi-axis of the earth's orbit.

At certain values there are gaps in the distribution, and these correspond to low resonances between the periods of the asteroids and Jupiter. ⌋

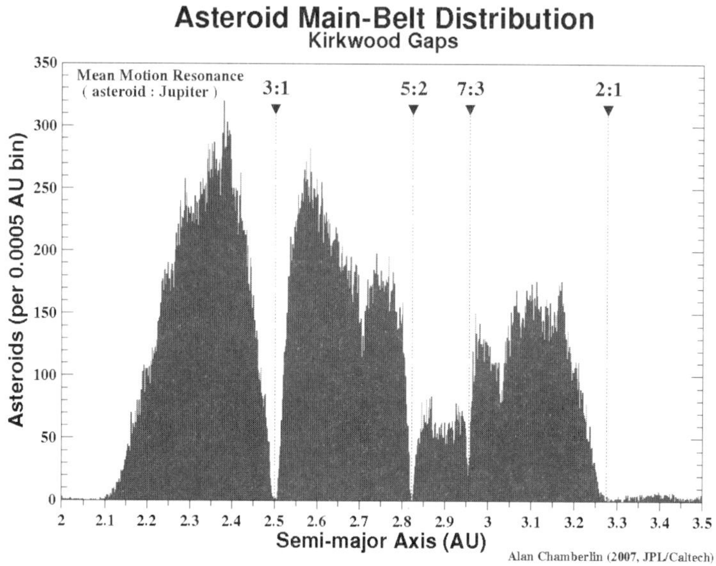

Figure 4.4. Kirkwood-gaps (ssd.jpl.nasa.gov/images/ast_histo.ps) Courtesy NASA/JPL-Caltech.

4.5. Homoclinic chaos

It is the purpose of this section to explain that in the part of the phase space which is not filled with invariant tori for small $\varepsilon > 0$ chaotic behavior can be expected. In between these tori there are periodic solutions and their stable and unstable manifolds. If there exists a heteroclinic connection with a transversal intersection of stable and unstable manifolds then a Smale horseshoe map and hence shift dynamics and chaotic behavior can be found. If for a map $f : \mathbb{R}^d \to \mathbb{R}^d$ the stable and unstable manifolds W_s and W_u of a hyperbolic fixed point p intersect transversally in a point q, then due to the invariance of the manifolds there must be infinitely many intersections. See Figure 4.5. Hence complicated dynamics can be expected. Recall that a fixed point p for an iteration $f : \mathbb{R}^d \to \mathbb{R}^d$ is called hyperbolic if the linearization $\partial_x f(p)$ possesses no eigenvalues on the unit circle.

Theorem 4.5.1. (The Smale-Birkhoff homoclinic orbit theorem) *Let $f : \mathbb{R}^d \to \mathbb{R}^d$ be a diffeomorphism such that p is a hyperbolic fixed point, and let $q \neq p$ be another point in which there is a transversal intersection of the stable manifold $W_s(p)$ and the unstable manifold $W_u(p)$. Then there*

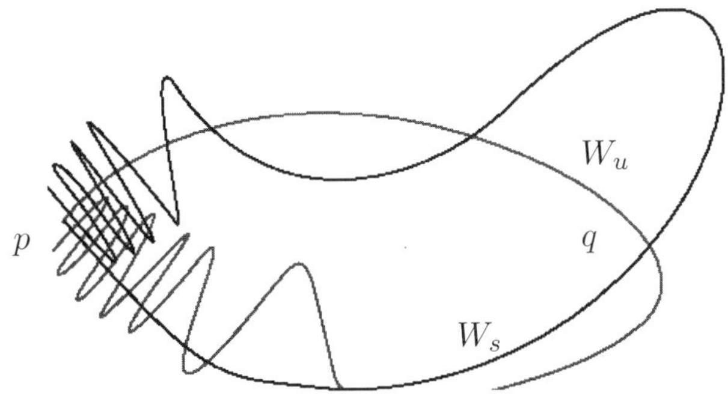

Figure 4.5. A transversal homoclinic point q implies infinitely many intersection points of the stable and unstable manifolds due to the invariance of the manifolds.

exists a (hyperbolic) set Λ on which an iteration of f is homeomorphic to shift dynamics.

Idea of the proof in $\mathbb{R}^2$: We are done if we find for an iteration of f a Smale's horseshoe, cf. Figure 2.14. W.l.o.g. let the saddle p be in the origin. By the Hartman-Grobman theorem, cf. Theorem 2.3.8, the saddle $(x, y) = (0, 0)$ has a neighborhood in which after some change of coordinates the dynamics is given by

$$x_{n+1} = \lambda x_n \quad \text{and} \quad y_{n+1} = \mu y_n$$

with $|\mu| > 1 > |\lambda|$. W.l.o.g. we can assume $\mu, \lambda > 0$. If this is not the case we consider the second iteration f^2. Then we consider

$$S = \{(x, y) \in \mathbb{R}^2 : 0 \leq x \leq \delta, \ |y| \leq \delta\}$$

for $\delta > 0$ sufficiently small. The kth iteration of f applied to S for k sufficiently large looks as sketched in Figure 4.6. Hence, we found a horseshoe in case of a homoclinic transversal point. □

This idea can be applied to 2π-time-periodic systems by considering the time 2π-map Π_ε. We derived from a $2d$-dimensional Hamiltonian system the $2(d-1)$-dimensional 2π-time-periodic Hamiltonian system (4.5)-(4.6). The d-dimensional tori break up and periodic solutions occur which are fixed points for the associated time 2π-map Π_ε. For a variety of systems numerical experiments indicate a transversal intersection of the associated stable and unstable manifolds and the occurrence of chaotic behavior between the persisting invariant tori, cf. [**Wig03**].

Remark 4.5.2. For time-periodic perturbations of an autonomous system with a homoclinic orbit the occurrence of a transversal intersection of the

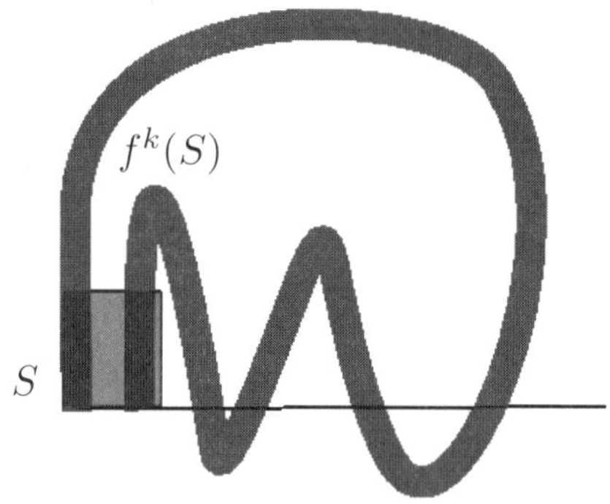

Figure 4.6. Smale's horseshoe in case of a homoclinic transversal point. The light gray rectangle is the set S and the dark gray set is $f^k(S)$.

stable and unstable manifolds can be established by finding single zeroes of the associated so called Melnikov function, cf. [**GH83**, §4.5]. ⌋

Exercises

4.1. Prove that $\dot{y} = (\text{trace}M)y$ for $y = \det Y$, where $Y(t) \in \mathbb{R}^{2\times 2}$ satisfies $\dot{Y} = MY$ for $M = M(t) \in \mathbb{R}^{2\times 2}$.

4.2. The "$6-12$ Lennart-Jones potential" models the forces between two neutral particles (atoms or molecules), namely an attractive van der Waals force at long ranges and a repulsive force at short ranges due to overlapping electron orbitals. In a simple (dimensionless) form it is given by $F(u) = au^{-12} - bu^{-6}$ where u is the distance between the particles and $a, b > 0$ are suitable constants. Choose $a = 0.001$ and $b = 1$ and discuss the phase portrait of the system $\ddot{u} = -F'(u)$.

4.3. Consider the pair $\ddot{x} + \omega^2 x = 0$, $\ddot{y} + \mu^2 y = 0$ of (uncoupled) harmonic oscillators. Write this as a Hamiltonian system. Find two integrals in polar coordinates. Discuss the cases (i) ω/μ rational and (ii) ω/μ irrational.

4.4. Given $r, \mu > 0$, write down explicitly a circular solution of the 1-body problem $\ddot{q} = -\mu q/\|q\|^3$, $q \in \mathbb{R}^2$, i.e., find initial conditions $q_0, \dot{q}_0$ such that the solution satisfies $\|q(t)\| = r$ for all $t \in \mathbb{R}$.

4.5. In dimensionless form, the first "Post-Newtonian" approximation for the orbit of a planet around the sun is

$$\partial_\theta^2 u + u = \alpha + \varepsilon u^2,$$

where $u = 1/r$ and (r, θ) are the polar coordinates of the planet and $\alpha, \varepsilon > 0$ are parameters. Discuss the phase portrait of this system.

4.6. Let $M \in \mathbb{R}^{n\times n}$ be nonsingular and symmetric and $F : \mathbb{R}^n \to \mathbb{R}$ be smooth. Write the Newtonian equation $M\ddot{x} + \nabla F(x) = 0$ as a Hamiltonian system.

4.7. Write the 4th order ODE $u'''' + qu'' + f(u) = 0$ as a Hamiltonian system for (u, u', u'', u'''). *Hint.* Let $z=(u, u'')$ and derive a system $Tz''+\nabla V(z)=0$ with non-singular $T\in\mathbb{R}^{2\times 2}$.

4.8. Consider the 1-body problem $\frac{\mathrm{d}}{\mathrm{d}t}\begin{pmatrix} q \\ p \end{pmatrix} = J\nabla H(q,p), H(q,p) = U(q) + \frac{1}{2}|p|^2$. Show that the angular momentum $C = q \times \dot{q}$ is constant.

4.9. (The 2-body problem) Consider two mass points with positions $q_j \in \mathbb{R}^3$ and masses m_j that move under mutual gravitational attraction. The equations are

$$m_1\ddot{q}_1 = F_{21}, \quad m_2\ddot{q}_2 = F_{12},$$

with

$$F_{ij} = m_i m_j g(|q_i - q_j|)(q_i - q_j), \quad g(r) = G/r^3.$$

This problem can be completely reduced to the 1 body problem. For this consider the center of mass $q_s = (m_1 q_1 + m_2 q_2)/m_s$, with $m_s = m_1 + m_2$. Find the ODE for q_s and express the orbits $q_{1,2}$ via q_s and orbits of the 1 body problem for the distance $q = q_2 - q_1$.

4.10. Let $F, G, H : \mathbb{R}^{2n} \to \mathbb{R}$ be smooth. Show that (a)

$$\{F, \{G, H\}\} + \{G, \{H, F\}\} + \{H, \{F, G\}\} = 0.$$

(b) F is an integral of $\dot{u} = J\nabla H$ iff $\{F, H\} = 0$. (c) $\frac{\mathrm{d}}{\mathrm{d}t}F(u) = \{F, H\}$.

4.11. Let $F = F(\widetilde{I}, \phi)$. Show that the map induced by $\widetilde{\phi} = \partial_{\widetilde{I}}F$ and $I = \partial_\phi F$ is symplectic.

Chapter 5

PDEs on an interval

The second part of this book is about nonlinear dynamics in countably many dimensions. It contains this chapter about PDEs on an interval and Chapter 6 about the Navier-Stokes equations.

We start with ordinary differential equations in $\mathbb{R}^\infty$, where $\mathbb{R}^\infty$ stands for the spaces $\mathbb{R}^\mathbb{N}$ or $\mathbb{R}^\mathbb{Z}$, i.e., for the spaces of real or complex (identifying $\mathbb{C}$ with $\mathbb{R}^2$) sequences $(a_j)_{j\in\mathbb{N}}$ or $(a_j)_{j\in\mathbb{Z}}$. In this book these countably many ODEs arise from PDEs, for which the spatial variable lives in a bounded domain. For function spaces on such domains very often a countable basis exists. By an expansion of the PDE w.r.t. this basis, for instance by an expansion into Fourier series in case of rectangular domains and suitable boundary conditions, the PDE can be transformed into an ODE in $\mathbb{R}^\infty$

Example. Consider the linear heat equation $\partial_t u = \partial_x^2 u$ for $x\in[0,\pi]$ with boundary condition $u(0,t)=u(\pi,t)=0$. Expanding

$$u(x,t)=\sum_{k\in\mathbb{N}} \widehat{u}_k(t)\sin(kx),$$

the PDE is formally equivalent to the countably many (uncoupled) ODEs $\frac{d}{dt}\widehat{u}_k=-k^2\widehat{u}_k$ for the sequence of Fourier coefficients $(\widehat{u}_k)_{k\in\mathbb{N}}$. ⌋

There are major differences between finitely and infinitely many dimensions due to the non-equivalence of norms in infinite-dimensional spaces and due to the loss of compactness of bounded closed sets. As a consequence, in infinite dimensions there can be stability w.r.t. one norm, but instability w.r.t. another norm. On the other hand there is a large class of equations, namely dissipative systems with smoothing properties, where the choice of the phase space does not matter. We however do not aim at a complete functional analytic treatment of such PDEs posed on bounded domains with dynamical systems concepts. For this we refer to the textbooks [**Hen81, Hal88, Rob01**].

Most dynamical system questions addressed in this part will be more involved when one considers PDEs on unbounded domains, as we do in Parts III and IV of this book. Hence, one of the main purposes of this section is to prepare for the additional difficulties as they appear in transferring the dynamical systems concept to PDEs posed on unbounded domains.

In §5.1 we consider the non-equivalence of norms, the loss and regain of compactness, and the local existence and uniqueness theory for countable many linear and nonlinear differential equations in $\mathbb{R}^\infty$. In §5.2 we discuss a number of basic function spaces, in particular those that are isomorphic via Fourier series to some sequence spaces. We explain local existence and uniqueness of solutions for some prototype linear and nonlinear PDEs, most of which will later be considered also over unbounded domains, and, moreover, explain how to prove global existence results. For these, the main tools are energy estimates and Gronwall type inequalities. We also give a characterization of the attractor of the so called Chafee-Infante problem, the scalar equation $\partial_t u = \partial_x^2 u + \alpha u - u^3$ on an interval $(0,\pi)$ with Dirichlet boundary conditions $u|_{x=0,\pi} = 0$, where $\alpha \in \mathbb{R}$ is a parameter.

5.1. From finitely to infinitely many dimensions

We consider systems of countably many linear and nonlinear differential equations. We discuss continuity of solutions w.r.t. time and some abstract local existence and uniqueness theory for ODEs in $\mathbb{R}^\infty$. Moreover, we explain how to differentiate and integrate in spaces of infinitely many dimensions and very briefly recall some basic facts from functional analysis, in particular compactness, which plays a crucial role for the dynamical systems point of view for PDEs

5.1.1. Non-equivalent norms. Concepts such as convergence in $\mathbb{R}^d$ or stability and instability for ODEs in $\mathbb{R}^d$ are independent of the chosen norm in $\mathbb{R}^d$. The reason for this is the equivalence of norms in finite-dimensional vector spaces, cf. Theorem 2.1.1. Setting $u = (u_1, \ldots, u_d)$, examples for norms in $\mathbb{R}^d$ have been

$$\|u\|_1 = \sum_{j=1}^d |u_j|, \qquad \|u\|_2 = \left(\sum_{j=1}^d |u_j|^2 \right)^{1/2},$$

and more generally $\|u\|_p := (\sum_{j=1}^d |u_j|^p)^{1/p}$, $1 \le p \le \infty$, and finally $\|u\|_\infty = \max_{j=1,\dots,d} |u_j|$. We have for instance

$$\|u\|_\infty \le \|u\|_p \le d^{1/p}\|u\|_\infty.$$

In infinite dimensions there are infinitely many non-equivalent norms. The norms which we use in this section are as follows.

Definition 5.1.1. *For $p \in [1,\infty)$ and $\theta \in \mathbb{R}$ let*

$$\|u\|_{\ell_{p,\theta}(\mathbb{R}^\mathbb{Z})} = \left(\sum_{n\in\mathbb{Z}} |u_n|^p \max(1,|n|)^{p\theta}\right)^{1/p}.$$

For $p = \infty$ and $\theta \in \mathbb{R}$ let

$$\|u\|_{\ell_{\infty,\theta}(\mathbb{R}^\mathbb{Z})} = \sup_{n\in\mathbb{Z}} |u_n| \max(1,|n|)^\theta.$$

We set

$$\ell_{p,\theta}(\mathbb{R}^\mathbb{Z}) = \{u : \mathbb{Z} \to \mathbb{R} : \|u\|_{\ell_{p,\theta}(\mathbb{R}^\mathbb{Z})} < \infty\}.$$

Similarly, we define $\|\cdot\|_{\ell_{p,\theta}(\mathbb{R}^\mathbb{N})}$ and $\ell_{p,\theta}(\mathbb{R}^\mathbb{N})$. We use the abbreviations $\|\cdot\|_{\ell_{p,\theta}}$ and $\ell_{p,\theta}$ for $\|\cdot\|_{\ell_{p,\theta}(\mathbb{R}^\infty)}$ and $\ell_{p,\theta}(\mathbb{R}^\infty)$.

The norms for different p or different θ are not equivalent. As a consequence a sequence can converge in one norm towards 0 while it diverges to ∞ in another norm.

Example 5.1.2. For the sequence $(u^m)_{m\in\mathbb{N}}$, with $u^m \in \ell_{1,2}(\mathbb{R}^\mathbb{N})$ for fixed m defined through $u_n^m = \delta_{nm}/n$, we have $\|u^m\|_{\ell_{1,0}} = 1/m \to 0$ for $m \to \infty$, while $\|u^m\|_{\ell_{1,2}} = m \to \infty$ for $m \to \infty$. ⌋

Remark 5.1.3. The spaces $\ell_{p,\theta}$ are Banach spaces, i.e., complete normed vector spaces. We recall that a metric space M is called complete, if every Cauchy sequence in M possesses a limit in M. The spaces $\ell_{2,\theta}$ are Hilbert spaces, i.e., complete normed vector spaces where the norm is induced by a scalar product. The space $c_{00} = \{u : \mathbb{Z} \to \mathbb{R} : u_n \ne 0 \text{ for finitely many } n\}$ equipped with the ℓ_1 norm is not complete. See Exercise 5.1. ⌋

5.1.2. Linear differential equations in $\mathbb{R}^\infty$. For notational simplicity in the following we work with equations in $\mathbb{R}^\mathbb{N}$. The results for $\mathbb{R}^\mathbb{Z}$ are exactly the same. We consider linear differential equations

$$\frac{\mathrm{d}}{\mathrm{d}t}u = Au, \qquad \text{i.e.,} \qquad \frac{\mathrm{d}}{\mathrm{d}t}u_k = \sum_{j\in\mathbb{N}} a_{kj}u_j.$$

We briefly recall the basic notions of semigroup theory which is the abstract version of the subsequent analysis.

We are not interested in such equations in greatest generality and therefore restrict ourselves mainly to equations having to do with PDEs, i.e., we consider A in diagonal form or with Jordan blocks of finite size. In this situation the equation can be solved explicitly, but the analytic properties of the solutions still turn out to be rather subtle. In case of A in diagonal form

$$\frac{\mathrm{d}}{\mathrm{d}t}u_k = \lambda_k u_k$$

we find the solutions

$$u_k(t) = \mathrm{e}^{\lambda_k t}u_k(0).$$

Solutions $u(t)$ to linear differential equations in $\mathbb{R}^d$ with constant coefficients are arbitrarily smooth w.r.t. t. In $\mathbb{R}^{\mathbb{N}}$ this is no longer true. Even for the boundedness additional conditions are necessary.

Lemma 5.1.4. *Let* $\sup_{k\in\mathbb{N}} \mathrm{Re}\lambda_k = \alpha < \infty$. *Then for every* $\theta \in \mathbb{R}$, $T_0 > 0$, *and* $p \in [1,\infty]$ *the curve* $t \mapsto u(t)$ *is bounded in* $\ell_{p,\theta}$ *for* $t \in [0, T_0]$.

Proof. We have $\|u(t)\|_{\ell_{p,\theta}} \leq (\sup_{k\in\mathbb{N}} |\mathrm{e}^{\lambda_k t}|)\|u(0)\|_{\ell_{p,\theta}} \leq \mathrm{e}^{\alpha t}\|u(0)\|_{\ell_{p,\theta}}$. □

The next question is the continuity of the curve $t \mapsto u(t)$ in the spaces $\ell_{p,\theta}$. Which conditions do we have to impose on the eigenvalues λ_k to have continuity? We put this question into a bigger framework, namely the theory of semigroups. The solution operator $T(t) = \mathrm{diag}(\mathrm{e}^{\lambda_1 t}, \mathrm{e}^{\lambda_2 t}, \ldots)$ defined through $T(t)u(0) = u(t)$ is an example of a semigroup of bounded linear operators, here from $\ell_{p,\theta}$ to $\ell_{p,\theta}$.

Definition 5.1.5. *Let* $(X, \|\cdot\|)$ *be a Banach space. A one-parameter family* $T(t)$, $0 \leq t < \infty$, *of bounded linear operators from* X *to* X *is called semigroup of bounded linear operators on* X, *if*

(i) $T(0) = I$,

(ii) $T(t+s) = T(t)T(s)$ *for all* $t, s \geq 0$.

The linear operator $A : D(A) \to X$, *defined by*

$$D(A) = \{u \in X : \lim_{t\downarrow 0} \frac{T(t)u - u}{t} \text{ exists }\},$$

$$Au = \lim_{t\downarrow 0} \frac{T(t)u - u}{t}, \quad \textit{for } u \in D(A),$$

is called the infinitesimal generator of $T(t)$.

According to the semigroup property, for the continuity of the maps $t \mapsto T(t)$ or $t \mapsto T(t)u$ the continuity at $t = 0$ is sufficient, cf. Remark 5.1.11. There are different concepts of continuity for semigroups. The first one is as follows.

Definition 5.1.6. *The semigroup is called operator-continuous if*

$$\lim_{t\downarrow 0} \|T(t) - I\| = 0$$

where $\|\cdot\|$ denotes the operator norm.

Example 5.1.7. Consider $\dot{u} = Au$ with $u(t) \in \mathbb{R}^d$ and A a $d \times d$-matrix with constant coefficients. Then $T(t) = \mathrm{e}^{tA}$ defines an operator-continuous semigroup in $\mathbb{R}^d$ since

$$\|\mathrm{e}^{tA} - I\| = \|\sum_{k=1}^{\infty}(At)^k/k!\| \le |t|\|A\|\mathrm{e}^{|t|\|A\|} \to 0 \text{ for } t \to 0. \qquad \lrcorner$$

Theorem 5.1.8. *Let $\sup_{k\in\mathbb{N}} |\lambda_k| = \alpha < \infty$. Then for every $\theta \in \mathbb{R}$ and $p \in [1,\infty]$ the associated semigroup is operator-continuous in $\ell_{p,\theta}$ for all $t \in \mathbb{R}$.*

Proof. We have $\|u(t) - u(0)\|_{\ell_{p,\theta}} \le (\sup_{k\in\mathbb{N}} |\mathrm{e}^{\lambda_k t} - 1|)\|u(0)\|_{\ell_{p,\theta}} \le |\mathrm{e}^{\alpha t} - 1|\|u(0)\|_{\ell_{p,\theta}} \to 0$ for $t \to 0$. This implies $\|T(t) - I\|_{\ell_{p,\theta}\to\ell_{p,\theta}} \le |\mathrm{e}^{\alpha t} - 1| \to 0$ for $t \to 0$ and so continuity holds. $\square$

For completeness we remark that a semigroup $T(t)$ of bounded linear operators on X is operator-continuous if and only if the generator $A : X \to X$ is bounded, cf. [**Paz83**, §1, Theorem 1.2]. Hence, as seen in the above example, the solutions of finite-dimensional ODEs always define an operator-continuous semigroup.

Since linearized operators in PDEs are usually unbounded, the generated semigroups are in general only strongly continuous.

Definition 5.1.9. *A semigroup $T(t)$, $0 \le t < \infty$, of bounded linear operators in X is called strongly continuous semigroup, or C_0-semigroup, if*

$$\lim_{t\downarrow 0} \|T(t)u - u\| = 0 \qquad \textit{for each} \quad u \in X.$$

Theorem 5.1.10. *For every $\theta \in \mathbb{R}$, $p \in [1,\infty)$, and $u(0) \in \ell_{p,\theta}$, the curve $t \mapsto u(t)$ is continuous in $\ell_{p,\theta}$ for $t \ge 0$ if and only if $\sup_{j\in\mathbb{N}} \mathrm{Re}\lambda_j = \alpha < \infty$.*

Proof. Let $\varepsilon > 0$. Using the triangle inequality in $\ell_{p,\theta}$ we have that

$$\begin{aligned}
&\|u(t) - u(0)\|_{\ell_{p,\theta}} \\
&= (\sum_{n=1}^{N} |(\mathrm{e}^{\lambda_n t} - 1)u_n(0)|^p |n|^{p\theta})^{1/p} + (\sum_{n=N+1}^{\infty} |(\mathrm{e}^{\lambda_n t} - 1)u_n(0)|^p |n|^{p\theta})^{1/p} \\
&= s_1 + s_2
\end{aligned}$$

for a $N \in \mathbb{N}$ suitably chosen in the following. In order to prove that $s_1+s_2 < \varepsilon$ for $t > 0$ sufficiently small we first estimate s_2 by choosing N so big that

$$s_2 \leq (\mathrm{e}^{\alpha t}+1)(\sum_{n=N+1}^{\infty} |u_n(0)|^p|n|^{p\theta})^{1/p} < \varepsilon/2.$$

For this N we then find a $t_0 > 0$ such that for all $t \in (0, t_0)$ we can estimate

$$s_1 \leq (\max_{n=1,\dots,N} |\mathrm{e}^{\lambda_n t} - 1|)\|u(0)\|_{\ell^{p,\theta}} < \varepsilon/2.$$

Therefore, we are done. □

Remark 5.1.11. Since $T(t_0+h)u - T(t_0)u = (T(h)-I)T(t_0)u \to 0$ and $T(t_0-h)u - T(t_0)u = -T(t_0-h)(T(h)-I)u \to 0$ for $h \downarrow 0$ the right-continuity in $t_0 = 0$ implies the continuity in every $t_0 > 0$ if the semigroup is uniformly bounded on every compact interval, cf. Lemma 5.1.4. In fact the assumption of the uniform boundedness on every compact interval is satisfied for C_0-semigroups due to a deep result from functional analysis, namely the uniform boundedness principle, cf. [**Paz83**, §1.2, Theorem 2.2]. ⌋

For $\dot{u} = Au$, with A in diagonal form, for $p \in [1, \infty)$ every bounded trajectory is also continuous in t. However, there is no uniformity w.r.t. the initial conditions $u(0)$. In $\ell_{\infty,\theta}$ the assumption about the boundedness of the eigenvalues λ_j is also necessary for continuity as the following example shows.

Example 5.1.12. Let $\lambda_k = -k^2$. Then there exists an $u(0) \in \ell_{\infty,0}$, for instance $u(0) = (1, 1, 1, \dots)$, such that $\|u(t) - u(0)\|_{\ell_{\infty,0}} = 1$ for every $t > 0$ such that continuity cannot hold. ⌋

Solutions of ODEs in $\mathbb{R}^d$ are smooth if the data is smooth. In $\mathbb{R}^\infty$ for $u(t) = (\mathrm{e}^{\lambda_n t}u_n(0))_{n\in\mathbb{N}}$ additional conditions about the eigenvalues λ_n are necessary.

- The mth derivative $u^{(m)}(t)$ is given by $u_n^{(m)}(t) = \lambda_n^m \mathrm{e}^{\lambda_n t}u_n(0)$. For $u(0) \in \ell_{p,\theta}$ we can guarantee the m-times differentiability of $t \mapsto u(t)$ in $\ell_{p,\theta}$ if the eigenvalues are in a set

$$\{\lambda \in \mathbb{C} : t\mathrm{Re}\lambda \leq a - m\ln|\mathrm{Im}\lambda|\}$$

for some constant $a \in \mathbb{R}$, cf. [**Paz83**, §2.4, Theorem 4.8]. As an example we consider $\lambda_n = -\ln n + in$. In order to have the m-times differentiability we need that $u^{(m)}(t) \in \ell_{p,\theta}$ which follows if

$$\sup_{n\in\mathbb{N}} |(-\ln n + in)^m \mathrm{e}^{-(\ln n)t}| = \sup_{n\in\mathbb{N}} |(-\ln n + in)^m n^{-t}| < \infty.$$

This means that the curve is one time differentiable for $t \in (1, 2]$, two times differentiable for $t \in (2, 3]$, etc.

- For $u(0) \in \ell_{p,\theta}$ we have the analyticity of $t \mapsto u(t)$ if the eigenvalues are in a sector
$$S_{a,b} = \{\lambda \in \mathbb{C} : \mathrm{Re}\lambda \leq a - b \ln |\mathrm{Im}\lambda|\}$$
for some constants $a \in \mathbb{R}$ and $b \geq 0$, cf. [**Paz83**, §2.5, Theorem 5.2]. As an example we consider $\lambda_n = -n + (-1)^n in$. The function $t \mapsto \mathrm{e}^{\lambda_n t} u_n(0)$ can be extended analytically into a sector of the complex plane. For $t = t_r + it_i$ we find $u(t_r + it_i) \in \ell_{p,\theta}$ if
$$\sup_{n\in\mathbb{N}} |\mathrm{e}^{(-n+(-1)^n in)(t_r+it_i)}| \leq \sup_{n\in\mathbb{N}} |\mathrm{e}^{n(-t_r+|t_i|)}| < \infty$$
which holds if $|t_i| < t_r$. Such generators are called sectorial and play a major role in the analysis of dissipative systems. The associated semigroup $(\mathrm{e}^{\lambda_n t})_{n\in\mathbb{N}}$ is called analytic. We come back to this in §6.3.

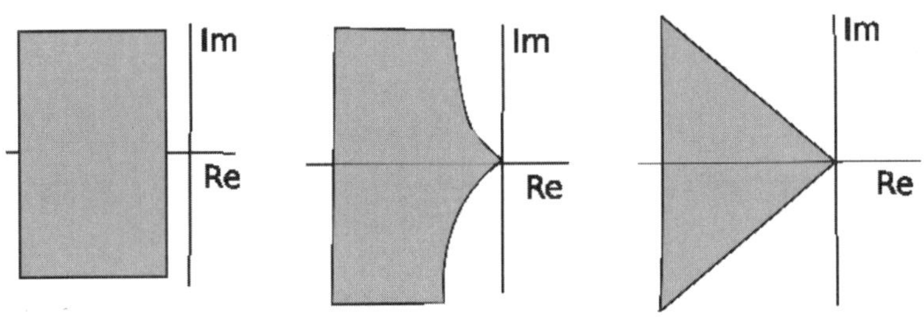

Figure 5.1. The picture shows the regions where the spectrum of the generators must be contained in to have a continuous (left panel), a differentiable (middle panel), or an analytic (right panel) semigroup.

We refer to the textbook [**Paz83**] for a thorough introduction to semigroup theory. Generators of C_0-semigroups are characterized by the theorem of Hille-Yosida, cf. [**Paz83**, §1.3, Theorem 3.1] or the Lumer-Phillips theorem, cf. [**Paz83**, §1.4, Theorem 4.3].

5.1.3. Differentiation and integration in Banach spaces. Before we proceed with the consideration of nonlinear infinite-dimensional ODEs, we need some additional functional analytic tools. For the stability of fixed points in ODEs the linearization, i.e., the derivative $A = Df \in \mathbb{R}^{d\times d}$ of a vector field $f : \mathbb{R}^d \to \mathbb{R}^d$ plays a central role. Hence, the concept of derivatives has to be generalized from $\mathbb{R}^d$ to general Banach spaces X. The same is true for integration. In the iteration scheme used in the proof of the local existence and uniqueness theorem for ODEs, a continuous function on an interval with values in $\mathbb{R}^d$ is integrated. If this iteration scheme is transferred to PDEs, then a continuous function on an interval with values in some infinite-dimensional Banach space has to be integrated. Such integrations occur in other iteration schemes used for PDEs, too. Hence, we have to define the integral of a continuous function on an interval with values in some Banach space. It turns out that the usual definition with Riemann sums is sufficient for our purposes.

Let us start with the derivatives. The Gateaux derivative is a generalization of the concept of the directional derivative. In Banach spaces the derivative is also called Fréchet derivative, cf. [**AA11**].

Definition 5.1.13. *Suppose that X and Y are Banach spaces, that $U \subset X$ is open, and consider $F : X \to Y$. The Gâteaux derivative $DF(u)[v]$ of F at $u \in U$ in the direction $v \in X$ is defined as*

$$DF(u)[v] = \lim_{\tau\to 0} \frac{F(u+\tau v) - F(u)}{\tau} = \frac{\mathrm{d}}{\mathrm{d}\tau} F(u+\tau v)\Big|_{\tau=0}.$$

If the limit exists for all $v \in X$, then F is called Gâteaux differentiable at u. $F : X \to Y$ is called differentiable in $u \in U$ if there exists a bounded linear operator $A = A(u) : X \to Y$ such that

$$\lim_{h\to 0} \frac{\|F(u+h) - F(u) - A(u)h\|_Y}{\|h\|_X} = 0.$$

The function $f : \mathbb{R}^2 \to \mathbb{R}$ with

$$f(x) = \begin{cases} 1, & \text{if } y = x^2,\ x \neq 0 \\ 0, & \text{elsewhere} \end{cases}$$

is a finite-dimensional example of a function for which every directional derivative exists, but which is not differentiable. In infinite-dimensional spaces less 'exotic' examples are possible.

Example 5.1.14. Consider $X = Y = L^2(0,1)$ and $F(u)(x) = \sin(u(x))$. We show that F is Gâteaux differentiable, but not differentiable at $u = 0$. We have

$$\lim_{\tau\to 0} \frac{F(u+\tau v) - F(u)}{\tau} = \lim_{\tau\to 0} \frac{\sin(\tau v(x))}{\tau} = \cos(0)v(x) = v(x)$$

due to the differentiability of $\sin : \mathbb{R} \to \mathbb{R}$. For the Fréchet differentiability we can vary v not only along lines. Due to the above computed Gâteaux derivative the only possible candidate for the derivative $A(0)$ is the identity. We set

$$v_n(x) = \begin{cases} n\pi, & \text{if } x \in (0, 1/n^4) \\ 0, & \text{elsewhere} \end{cases}$$

and find

$$\frac{\|F(v_n) - F(0) - A(0)v_n\|_Y}{\|v_n\|_X} = \frac{\|v_n\|_Y}{\|v_n\|_X} = 1 \not\to 0$$

although $\|v_n\|_{L^2} = \pi/n \to 0$ for $n \to \infty$. We remark that with the choice $X = Y = C_b^0([0,1])$ equipped with the sup-norm the map F would be analytic. ⌋

However, in spaces with additional algebra properties, virtually all results from complex power series carry over.

Example 5.1.15. Let $(X, \|\cdot\|)$ be a Banach space. Then $Y = \mathcal{L}(X, X) = \{F : X \to X : \text{linear, continuous}\}$ is a Banach space equipped with the operator norm

$$\|A\| = \sup\{\|Au\|_X \; : \; \|u\|_X = 1\}.$$

For $A, B \in Y$ we have that AB and A^n are in Y with $\|AB\| \leq \|A\|\|B\|$, and $\|A^n\| \leq \|A\|^n$. Therefore, the series $F(A) = \exp(A) = \sum_{k=0}^{\infty} \frac{1}{k!} A^k$ is convergent in Y. We have the continuity and the Fréchet differentiability of $F : Y \to Y$ with $DF(A) = F(A)$. The analyticity of F follows like for real-valued power series, cf. Exercise 5.3. ⌋

Next we come to the integration of continuous functions $f : [a, b] \to X$ with values in a Banach space X. Let $P = \{x_0, \ldots, x_n\}$ with $a = x_0 < x_1 < \ldots < x_{n-1} < x_n = b$ be a partition of the interval. Its fineness is defined by

$$\|P\| = \max\{|x_{j+1} - x_j| : j = 0, \ldots, n-1\}.$$

Let $\xi = (\xi_1, \ldots, \xi_n)$ with $\xi_j \in [x_{j-1}, x_j]$. Then define the Riemann sum

$$S(\xi, P) = \sum_{j=1}^{n} f(\xi_j)(x_j - x_{j-1}).$$

Definition 5.1.16. *A function $f : [a, b] \to X$ is called Riemann integrable if the limit*

$$\lim_{n\to\infty} S(\xi(n), P(n))$$

exists for every sequence $(\xi(n), P(n))$ with $\lim_{n\to\infty} \|P(n)\| = 0$. If the limit exists, then we define the Riemann integral by

$$\int_a^b f(x)\,\mathrm{d}x = \lim_{\|P\|\to 0} S(\xi, P).$$

Theorem 5.1.17. *Continuous functions $f : [a, b] \to X$ are Riemann integrable.*

Proof. We have to show that for every $\varepsilon > 0$ there exists a $\delta > 0$ such that for all (ξ^1, P_1), (ξ^2, P_2) with $\|P_1\| \leq \delta$ and $\|P_2\| \leq \delta$ we have $\|S(\xi^1, P_1) - S(\xi^2, P_2)\|_X \leq \varepsilon$.

We set $P_3 = P_1 \cup P_2$ and choose an arbitrary ξ^3. Then by the triangle inequality we have

$$\begin{aligned} \|S(\xi^1, P_1) - S(\xi^2, P_2)\|_X \leq& \|S(\xi^1, P_1) - S(\xi^3, P_3)\|_X + \|S(\xi^3, P_3 - S(\xi^2, P_2)\|_X \\ \leq& \varepsilon/2 + \varepsilon/2 \leq \varepsilon. \end{aligned}$$

where we used

$$\|S(\xi^1,P_1)-S(\xi^3,P_3)\|_X = \|\sum_{j=1}^{N_1} f(\xi_j^1)(\widetilde{x}_j-\widetilde{x}_{j-1}) - \sum_{j=1}^{N_3} f(\xi_j^3)(x_j-x_{j-1})\|_X$$

$$\leq \sum_{j=1}^{N_1}\sum_{k=\alpha(j-1)}^{\alpha(j)-1} \left\|f(\xi_j^1)-f(\xi_{k+1}^3)\right\|_X |x_{k+1}-x_k|, \tag{5.1}$$

where $[\widetilde{x}_{j-1},\widetilde{x}_j]=\bigcup_{k=\alpha(j-1)}^{\alpha(j)-1}[x_k,x_{k+1}]$. See Figure 5.2.

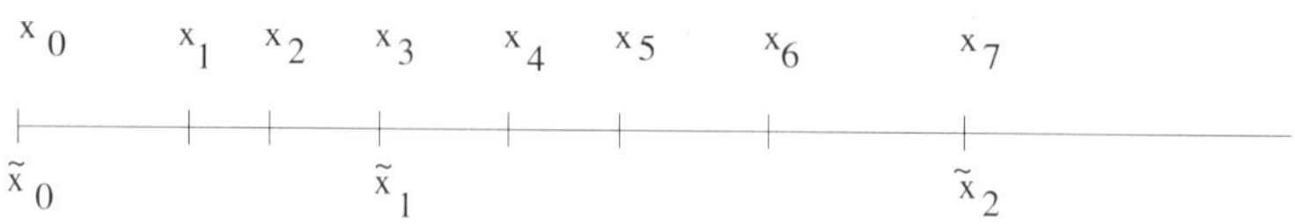

Figure 5.2. The partition P_1 is drawn below the line and P_3 above the line. In this example we have $\alpha(0)=0$, $\alpha(1)=3$, $\alpha(2)=7,\dots$.

By uniform continuity of f, which follows from the continuity of f on the compact interval $[a,b]$, we have that for all $\varepsilon>0$ there exists a $\delta>0$ such that $|y-\widetilde{y}|<\delta$ implies $\|f(y)-f(\widetilde{y})\|_X<\frac{\varepsilon}{2(b-a)}$. Hence, if $\|P_3\|\leq\|P_1\|<\delta$ is chosen sufficiently small, (5.1) can be estimated by

$$\leq \sum_{k=1}^{N_3}\frac{\varepsilon}{2(b-a)}|x_k-x_{k-1}|\leq\frac{\varepsilon}{2}. \qquad \square$$

Remark 5.1.18. Not only the Riemann integral can be generalized to functions $u:\mathbb{R}\to X$, with X some Banach space, but also the Lebesgue integral, cf. [**Alt16**, §A1]. ⌋

5.1.4. Nonlinear differential equations in $\mathbb{R}^\infty$. Since in general the ODEs in $\mathbb{R}^\infty$ obtained from PDEs have unbounded λ_ks the right-hand side is no longer Lipschitz-continuous from $\ell_{p,\theta}$ to $\ell_{p,\theta}$. Thus, the Picard-Lindelöf theorem no longer applies and has to be replaced. The simplest idea to obtain a contraction as in the proof of the Picard-Lindelöf theorem is the use of the variation of constant formula, cf. §2.1.3.

For simplicity we first consider

$$\frac{\mathrm{d}}{\mathrm{d}t}u=\Lambda u+B(u,u), \tag{5.2}$$

where $u(t)\in\ell_{p,\theta}$, where Λ is a diagonal matrix with entries λ_k satisfying

$$\sup_k \mathrm{Re}\lambda_k=\beta<\infty, \tag{5.3}$$

and where B is a bilinear symmetric map from $\ell_{p,\theta}$ into $\ell_{p,\theta}$ satisfying

$$\|B(u,v)\|_{\ell_{p,\theta}}\leq C_B\|u\|_{\ell_{p,\theta}}\|v\|_{\ell_{p,\theta}}. \tag{5.4}$$

In order to prove the local existence and uniqueness of solutions of (5.2) on an interval $[0, T_0]$ we use the variation of constant formula to rewrite (5.2) into

$$u(t) = \mathrm{e}^{t\Lambda} u(0) + \int_0^t \mathrm{e}^{(t-\tau)\Lambda} B(u(\tau), u(\tau))\,\mathrm{d}\tau =: F(u)(t). \tag{5.5}$$

Definition 5.1.19. *a) A function $u \in C^0([0, T_0], \ell_{p,\theta})$ which satisfies* (5.5) *is called a mild solution of* (5.2).
b) A function $u \in C^1([0, T_0], \ell_{p,\theta})$, with $\Lambda u \in C([0, T_0], \ell_{p,\theta})$, is called a strong solution of (5.2), *if* (5.2) *holds in $\ell_{p,\theta}$ for every $t \in (0, T_0)$.*

Clearly, every strong solution is a mild solution. Conversely, a mild solution which satisfies $u \in C^1([0, T_0], \ell_{p,\theta})$ and $\Lambda u \in C([0, T_0], \ell_{p,\theta})$ is a strong solution. In the following until further notice solutions will always mean mild solutions.

Theorem 5.1.20. *Assume* (5.3) *and* (5.4). *For all $C_1 > 0$ there exists a $T_0 > 0$ such that for all $w \in \ell_{p,\theta}$ with $\|w\|_{\ell_{p,\theta}} \leq C_1$ we have a unique solution $u \in C([0, T_0], \ell_{p,\theta})$ of* (5.2) *with initial condition $u(0) = w$.*

Proof. We fix a $C_2 > 0$ and show that for $T_0 \in (0, 1)$ sufficiently small the right-hand side of (5.5) is a contraction in the set

$$M = C([0, T_0], \{u(t) \in \ell_{p,\theta} : \|u(t) - \mathrm{e}^{t\Lambda} u(0)\|_{\ell_{p,\theta}} \leq C_2\}),$$

and apply the contraction mapping theorem. M is a complete metric space, but since the metric is induced by a norm we will use the norm notation in the following. We use the abbreviation $C_3 = \sup_{u \in M} \|u\|_M \leq C_1 \mathrm{e}^{\beta} + C_2$.

In a first step we prove that F maps M into itself. We have

$$\begin{aligned}
\|F(u) - (\mathrm{e}^{t\Lambda} u(0))_{t \geq 0}\|_M &= \sup_{t \in [0, T_0]} \|F(u)(t) - \mathrm{e}^{\Lambda t} u(0)\|_{\ell_{p,\theta}} \\
&\leq \sup_{t \in [0, T_0]} \Big\| \int_0^t \mathrm{e}^{(t-\tau)\Lambda} B(u(\tau), u(\tau))\,\mathrm{d}\tau \Big\|_{\ell_{p,\theta}} \\
&\leq \sup_{t \in [0, T_0]} \int_0^t \mathrm{e}^{\beta(t-\tau)} \|B(u(\tau), u(\tau))\|_{\ell_{p,\theta}}\,\mathrm{d}\tau \\
&\leq T_0 \mathrm{e}^{\beta T_0} C_B C_3^2 \leq C_1
\end{aligned}$$

for $T_0 > 0$ sufficiently small.

Secondly, we find that F is a contraction since

$$\begin{aligned}\|F(u)-F(v)\|_M &= \sup_{t\in[0,T_0]} \|F(u)(t)-F(v)(t)\|_{\ell_{p,\theta}}\\ &\leq \sup_{t\in[0,T_0]} \|\int_0^t \mathrm{e}^{(t-\tau)\Lambda}(B(u(\tau),u(\tau))-B(v(\tau),v(\tau)))\,\mathrm{d}\tau\|_{\ell_{p,\theta}}\\ &\leq \sup_{t\in[0,T_0]} \int_0^t \mathrm{e}^{\beta(t-\tau)}\|B(u(\tau),u(\tau))-B(v(\tau),v(\tau))\|_{\ell_{p,\theta}}\,\mathrm{d}\tau\\ &\leq T_0\mathrm{e}^{\beta T_0} \sup_{\tau\in[0,T_0]} \|B(u(\tau)+v(\tau),u(\tau)-v(\tau))\|_{\ell_{p,\theta}}\\ &\leq 2T_0\mathrm{e}^{\beta T_0}C_BC_3\|u-v\|_M \leq \|u-v\|_M/2\end{aligned}$$

for $T_0>0$ sufficiently small. □

This procedure of constructing solutions to ODEs in $\mathbb{R}^\infty$ can be extended to a wider class of problems. We consider again

$$\frac{\mathrm{d}}{\mathrm{d}t}u = \Lambda u + B(u,u), \tag{5.6}$$

where $u(t)\in\ell_{p,\theta}$, but now with the following assumptions:

- Λ a diagonal matrix satisfying

$$\|\mathrm{e}^{t\Lambda}u\|_{\ell_{p,\theta}} \leq C_{\theta-r}\mathrm{e}^{\beta t}t^{-\alpha}\|u\|_{\ell_{p,r}} \tag{5.7}$$

  for an $\alpha\in[0,1)$, a constant $C_{\theta-r}$, and $\theta-r\geq 0$.

- B a bilinear symmetric map from $\ell_{p,\theta}$ into $\ell_{p,r}$ satisfying

$$\|B(u,v)\|_{\ell_{p,r}} \leq C_B\|u\|_{\ell_{p,\theta}}\|v\|_{\ell_{p,\theta}}. \tag{5.8}$$

The property described by equation (5.7) is called smoothing since the evolution operator maps for $t>0$ the space $\ell_{p,r}$ into $\ell_{p,\theta}$ and since functions whose Fourier coefficients are in $\ell_{p,\theta}$ are smoother than functions whose Fourier coefficients are only $\ell_{p,r}$. Many of the subsequent examples will satisfy estimates like (5.7).

Example 5.1.21. Consider $\lambda_n=-n^2$. We have the decay estimate

$$\|(\mathrm{e}^{\lambda_n t}u_n)_{n\in\mathbb{N}}\|_{\ell_{p,\theta}} \leq \sup_{n\in\mathbb{N}}|\mathrm{e}^{-n^2t}n^\theta|\|u\|_{\ell_{p,0}} \leq Ct^{-\theta/2}\|u\|_{\ell_{p,0}},$$

which corresponds to smoothing of functions in physical space, see Example 5.2.19. ⌋

Remark 5.1.22. Smoothing is not directly related to regularity w.r.t. time t, as the following examples show. In case $\lambda_n=0$ for all $n\in\mathbb{N}$ all eigenvalues are identical and contained in a sector. However, the associated semigroup is the identity which is not smoothing from $\ell_{p,r}$ into $\ell_{p,\theta}$ for $r<\theta$ although we have an analytic semigroup (w.r.t. time). In case $\lambda_n=-n^2+\mathrm{i}(-1)^n n^3$

obviously the eigenvalues are not contained in a sector and the semigroup is not analytic (w.r.t. time $t = t_r + it_i$) since

$$\sup_{n\in\mathbb{N}} |e^{(-n^2+i(-1)^n n^3)(t_r+it_i)}| = \sup_{n\in\mathbb{N}} |e^{-n^2 t_r + (-1)^{n+1} n^3 t_i}| = \infty$$

for $t_i \neq 0$. However, we have the decay estimate

$$\|(e^{\lambda_n t} u_n)_{n\in\mathbb{N}}\|_{\ell_{p,\theta}} \le \sup_{n\in\mathbb{N}} |e^{(-n^2 \pm in^3)t} n^\theta| \|u\|_{\ell_{p,0}} \le Ct^{-\theta/2} \|u\|_{\ell_{p,0}}. \qquad \rfloor$$

In order to prove the local existence and uniqueness of solutions of (5.6) on an interval $[0, T_0]$, we again use the variation of constant formula

$$u(t) = e^{t\Lambda} u(0) + \int_0^t e^{(t-\tau)\Lambda} B(u(\tau), u(\tau))\, d\tau =: F(u)(t). \tag{5.9}$$

We show that for $T_0 \in (0, 1)$ sufficiently small the right-hand side of (5.9) is a contraction in the set

$$M = C([0, T_0], \{u(t) \in \ell_{p,\theta} : \|u(t) - e^{t\Lambda} u(0)\|_{\ell_{p,\theta}} \le C_2\}),$$

where $\|u(0)\|_{\ell_{p,\theta}} \le C_1$ and $C_2 > 0$ is a fixed constant. In a first step we prove that F maps M into itself. With C_3 as above we have

$$\begin{aligned}
\|F(u) - (e^{t\Lambda} u(0))_{t\ge 0}\|_M &= \sup_{t\in[0,T_0]} \|F(u)(t) - e^{t\Lambda} u(0)\|_{\ell_{p,\theta}} \\
&\le \sup_{t\in[0,T_0]} \| \int_0^t e^{(t-\tau)\Lambda} B(u(\tau), u(\tau))\, d\tau \|_{\ell_{p,\theta}} \\
&\le \sup_{t\in[0,T_0]} C_{\theta-r} \int_0^t (t-\tau)^{-\alpha} e^{\beta(t-\tau)} \|B(u(\tau), u(\tau))\|_{\ell_{p,r}}\, d\tau \\
&\le C_{\theta-r} (1-\alpha)^{-1} T_0^{1-\alpha} e^{\beta T_0} C_B C_3^2 \le C_1
\end{aligned}$$

for $T_0 > 0$ sufficiently small.

Secondly, we find

$$\begin{aligned}
\|F(u) - F(v)\|_M &= \sup_{t\in[0,T_0]} \|F(u)(t) - F(v)(t)\|_{\ell_{p,\theta}} \\
&\le \sup_{t\in[0,T_0]} \| \int_0^t e^{(t-\tau)\Lambda} (B(u(\tau), u(\tau)) - B(v(\tau), v(\tau)))\, d\tau \|_{\ell_{p,\theta}} \\
&\le \sup_{t\in[0,T_0]} \int_0^t C_{\theta-r} (t-\tau)^{-\alpha} e^{\beta(t-\tau)} \|B(u(\tau), u(\tau)) - B(v(\tau), v(\tau))\|_{\ell_{p,r}}\, d\tau \\
&\le C_{\theta-r} (1-\alpha)^{-1} T_0^{1-\alpha} e^{\beta T_0} \sup_{t\in[0,T_0]} \|B(u(\tau) + v(\tau), u(\tau) - v(\tau))\|_{\ell_{p,\theta}} \\
&\le 2C_{\theta-r} (1-\alpha)^{-1} T_0^{1-\alpha} e^{\beta T_0} C_B C_3 \|u - v\|_M \le \|u - v\|_M / 2
\end{aligned}$$

for $T_0 > 0$ sufficiently small. Hence, the contraction F possesses a unique fixed point in M and so we have proved

Theorem 5.1.23. *Assume* (5.7) *and* (5.8)*. For all* $C_1 > 0$ *there exists a* $T_0 > 0$ *such that for all* $w \in \ell_{p,\theta}$ *with* $\|w\|_{\ell_{p,\theta}} \leq C_1$ *we have a unique solution* $u \in C([0,T_0], \ell_{p,\theta})$ *of* (5.6) *with initial condition* $u(0) = w$.

Remark 5.1.24. Both theorems obviously also hold if B is replaced by a general locally Lipschitz-continuous map from $\ell_{p,\theta}$ into $\ell_{p,\theta}^{\cdot}$ or $\ell_{p,r}$ respectively, i.e., for instance in the latter case that for all C_1 there exists an L such that $\max(\|u\|_{\ell_{p,\theta}}, \|v\|_{\ell_{p,\theta}}) \leq C_1$ implies

$$\|N(u) - N(v)\|_{\ell_{p,r}} \leq L\|u - v\|_{\ell_{p,\theta}}.$$

Every polynomial nonlinearity is locally Lipschitz-continuous in this sense.⌋

Moreover, Theorem 5.1.20 and Theorem 5.1.23 are prototypes for other local existence and uniqueness theorems for semi-linear evolutionary PDEs below.

5.1.5. A first look on Fourier series. PDEs with periodic boundary conditions for the spatial coordinates can be transferred to ODEs in $\mathbb{R}^\infty$ with the help of Fourier series. Here we give the definition and some elementary properties. For later purposes we consider here the d-dimensional situation.

Definition 5.1.25. *A series of the form*

$$u(x) = \sum_{k \in \mathbb{Z}^d} \widehat{u}_k \mathrm{e}^{\mathrm{i}k\cdot x}, \tag{5.10}$$

is called Fourier series, its partial sums $u(x) = \sum_{|k|\leq N} \widehat{u}_k \mathrm{e}^{\mathrm{i}k\cdot x}$, *are called Fourier polynomials of order* N, *and* $\widehat{u}_k$ *is called the* k^{th} *Fourier coefficient.*

See §5.2.2 for more details, in particular a number of convergence results for (5.10). Since our main interest is in nonlinear PDEs we also have to handle products of functions in physical space with Fourier series. The point-wise multiplication in physical space correspond in Fourier space to convolution. That is,

$$u(x)v(x) = \left(\sum_{k\in\mathbb{Z}^d} \widehat{u}_k e^{ik\cdot x}\right)\left(\sum_{m\in\mathbb{Z}^d} \widehat{v}_m e^{im\cdot x}\right) = \sum_{k\in\mathbb{Z}^d}\left(\sum_{m\in\mathbb{Z}^d} \widehat{u}_{k-m}\widehat{v}_m\right) e^{ik\cdot x}.$$

This motivates the definition of the convolution

$$(\widehat{u} * \widehat{v})_k = \sum_{m\in\mathbb{Z}^d} \widehat{u}_{k-m}\widehat{v}_m.$$

For the control of the nonlinear terms in Fourier space we need

Lemma 5.1.26. (Young's inequality for convolutions) *For* $p \in [1,\infty]$ *we have*

$$\|\widehat{u} * \widehat{v}\|_{\ell_p} \leq \|\widehat{u}\|_{\ell_p}\|\widehat{v}\|_{\ell_1}.$$

Proof. For $p \in [1,\infty]$ we find

$$\begin{aligned}\|\widehat{u} * \widehat{v}\|_{\ell_p} =& (\sum_{k\in\mathbb{Z}} |\sum_{m\in\mathbb{Z}} \widehat{u}_{k-m}\widehat{v}_m|^p)^{1/p} \leq (\sum_{l\in\mathbb{Z}}(\sum_{m\in\mathbb{Z}} |\widehat{u}_l\widehat{v}_m|)^p)^{1/p} \\ =& (\sum_{l\in\mathbb{Z}} |\widehat{u}_l|^p(\sum_{m\in\mathbb{Z}} |\widehat{v}_m|)^p)^{1/p} \leq \|\widehat{u}\|_{\ell_p}\|\widehat{v}\|_{\ell_1}.\end{aligned}$$

The case $p = \infty$ is obvious. □

Young's inequality for convolutions allows us to prove that the $\ell_{2,\theta}$-spaces are closed under convolution if θ is sufficiently big. In order to do so we prove the following version of Sobolev's embedding theorem

Lemma 5.1.27. *For $m - d/2 > n$ there exists a $C > 0$, such that*

$$\|\widehat{u}\|_{\ell_{1,n}(\mathbb{R}^d)} \leq C\|\widehat{u}\|_{\ell_{2,m}(\mathbb{R}^d)}.$$

Proof. With $\widehat{\rho}_k = \max(1,|k|)$ the estimate follows from

$$\begin{aligned}\|\widehat{u}\|_{\ell_{1,n}} &= \sum_{k\in\mathbb{Z}^d} |\widehat{u}_k|\widehat{\rho}_k^n = \sum_{k\in\mathbb{Z}^d} |\widehat{u}_k|\widehat{\rho}_k^m\widehat{\rho}_k^{(n-m)} \\ &\leq \left(\sum_{k\in\mathbb{Z}^d} |\widehat{u}_k|^2\widehat{\rho}_k^{2m}\right)^{1/2}\left(\sum_{k\in\mathbb{Z}^d} \widehat{\rho}_k^{2(n-m)}\right)^{1/2} \leq C\|\widehat{u}\|_{\ell_{2,m}},\end{aligned}$$

since $\sum_{k\in\mathbb{Z}^d} \widehat{\rho}_k^{2(n-m)} < \infty$, due to $m - d/2 > n$ by assumption. □

We use this embedding to establish

Lemma 5.1.28. *a) For all $m \geq 0$ there exists a $C > 0$, such that for all $\widehat{u}, \widehat{v} \in \ell_{1,m}$ we have*

$$\|\widehat{u} * \widehat{v}\|_{\ell_{1,m}} \leq C\|\widehat{u}\|_{\ell_{1,m}}\|\widehat{v}\|_{\ell_{1,m}}.$$

b) For all $m > d/2$ there exists a $C > 0$, such that for all $\widehat{u}, \widehat{v} \in \ell_{2,m}$ we have

$$\|\widehat{u} * \widehat{v}\|_{\ell_{2,m}} \leq C\|\widehat{u}\|_{\ell_{2,m}}\|\widehat{v}\|_{\ell_{2,m}}.$$

Proof. a) Since $\widehat{\rho}_k^m \leq C(\widehat{\rho}_{k-l}^m + \widehat{\rho}_l^m)$ with $C = 2^m$ for $\widehat{\rho}_k = \max(1,|k|)$ using Lemma 5.1.26 it follows that

$$\begin{aligned}\|\widehat{u} * \widehat{v}\|_{\ell_{1,m}} =& \sum_{k\in\mathbb{Z}^d} |\sum_{l\in\mathbb{Z}^d} \widehat{u}_{k-l}\widehat{v}_l\widehat{\rho}_k^m| \leq \sum_{k\in\mathbb{Z}^d}\sum_{l\in\mathbb{Z}^d} |\widehat{u}_{k-l}\widehat{v}_l\widehat{\rho}_k^m| \\ \leq& C\sum_{k\in\mathbb{Z}^d}\sum_{l\in\mathbb{Z}^d}(|\widehat{u}_k\widehat{v}_l|\widehat{\rho}_k^m + |\widehat{u}_k\widehat{v}_l|\widehat{\rho}_l^m) \\ \leq& C(\|\widehat{u}\|_{\ell_{1,0}}\|\widehat{v}\|_{\ell_{1,m}} + \|\widehat{u}\|_{\ell_{1,m}}\|\widehat{v}\|_{\ell_{1,0}}) \leq 2C\|\widehat{u}\|_{\ell_{1,m}}\|\widehat{v}\|_{\ell_{1,m}}.\end{aligned}$$

b) With Lemma 5.1.26 we have

$$\begin{aligned}
\|\widehat{u} * \widehat{v}\|_{\ell_{2,m}} =& (\sum_{k\in\mathbb{Z}^d} |\sum_{l\in\mathbb{Z}^d} \widehat{u}_{k-l}\widehat{v}_l \widehat{\rho}_k^m|^2)^{1/2} \\
\leq& (\sum_{k\in\mathbb{Z}^d} (\sum_{l\in\mathbb{Z}^d} |\widehat{u}_{k-l}\widehat{v}_l|(\widehat{\rho}_{k-l}^m + \widehat{\rho}_l^m))^2)^{1/2} \\
\leq& C(\||\widehat{u}|\widehat{\rho}^m * |\widehat{v}|\|_{\ell_{2,0}} + \||\widehat{u}| * |\widehat{v}|\widehat{\rho}^m\|_{\ell_{2,0}}) \\
\leq& C(\|\widehat{u}\|_{\ell_{2,m}}\|\widehat{v}\|_{\ell_{1,0}} + \|\widehat{u}\|_{\ell_{1,0}}\|\widehat{v}\|_{\ell_{2,m}}).
\end{aligned}$$

The final assertion follows from Sobolev's embedding theorem $\|\widehat{v}\|_{\ell_{1,0}} \leq C\|\widehat{v}\|_{\ell_{2,m}}$ for $m > d/2$. □

We now give a number of classical examples of nonlinear PDEs over intervals with periodic boundary conditions. In fact, over unbounded domains each of these equations will play an important role in this book. For the modeling and physical background of the equations we refer in particular to Part III.

Example 5.1.29. Let $u(x,t) = u(x+2\pi,t) \in \mathbb{R}$, $A(X,T) = A(X+2\pi,T) \in \mathbb{C}$, and $\widehat{u}_k(t)$, $\widehat{A}_k(T)$ be the associated Fourier coefficients.

a) The Kolmogorov, Petrovsky, Piskounov (KPP) equation $\partial_t u = \partial_x^2 u + u - u^2$ transforms into

$$\partial_t \widehat{u}_k = -k^2\widehat{u}_k + \widehat{u}_k - \sum_{m\in\mathbb{Z}} \widehat{u}_{k-m}\widehat{u}_m.$$

b) The Allen-Cahn equation $\partial_t u = \partial_x^2 u + u - u^3$ transforms into

$$\partial_t \widehat{u}_k = -k^2\widehat{u}_k + \widehat{u}_k - \sum_{m\in\mathbb{Z}}\sum_{l\in\mathbb{Z}} \widehat{u}_{k-m}\widehat{u}_{m-l}\widehat{u}_l.$$

c) The Burgers equation $\partial_t u = \partial_x^2 u + \partial_x(u^2)$ transforms into

$$\partial_t \widehat{u}_k = -k^2\widehat{u}_k + \mathrm{i}k\sum_{m\in\mathbb{Z}} \widehat{u}_{k-m}\widehat{u}_m.$$

d) The Korteweg-deVries (KdV) equation $\partial_t u = \partial_x^3 u + \partial_x(u^2)$ transforms into

$$\partial_t \widehat{u}_k = -\mathrm{i}k^3\widehat{u}_k + \mathrm{i}k\sum_{m\in\mathbb{Z}} \widehat{u}_{k-m}\widehat{u}_m.$$

e) Using $(\mathcal{F}\overline{A})_j = \overline{\widehat{A}}_{-j}$ the Nonlinear Schrödinger (NLS) equation $\partial_T A = \mathrm{i}\partial_X^2 A + \mathrm{i}|A|^2 A$ transforms into

$$\partial_T \widehat{A}_k = -\mathrm{i}k^2\widehat{A}_k + \mathrm{i}\sum_{m\in\mathbb{Z}}\sum_{l\in\mathbb{Z}} \widehat{A}_{k+l-m}\overline{\widehat{A}}_l\widehat{A}_m.$$

f) The complex Ginzburg-Landau (GL) equation $\partial_T A = (1+\mathrm{i}\alpha)\partial_X^2 A + RA - (1+\mathrm{i}\beta)|A|^2 A$, with $\alpha, \beta, R \in \mathbb{R}$, transforms into

$$\partial_T \widehat{A}_k = -(1+\mathrm{i}\alpha)k^2 \widehat{A}_k + \widehat{A}_k - (1+\mathrm{i}\beta) \sum_{m\in\mathbb{Z}} \sum_{l\in\mathbb{Z}} \widehat{A}_{k+l-m} \overline{\widehat{A}_l} \widehat{A}_m.$$

Except for the KdV equation, for all equations from above the local existence and uniqueness theory can be handled with Theorem 5.1.20 and Theorem 5.1.23. The linear parts are given by the eigenvalues λ_k, $k \in \mathbb{Z}$, where

$$\text{a), b) } \lambda_k = -k^2 + 1, \quad \text{c) } \lambda_k = -k^2,$$
$$\text{d) } \lambda_k = -\mathrm{i}k^3, \quad \text{e)}\lambda_k = -\mathrm{i}k^2, \quad \text{f) } \lambda_k = -(1+\mathrm{i}\mu)k^2 + 1.$$

In a), b), e) and f) we only need that $(\mathrm{e}^{\lambda_k t})_{k\in\mathbb{Z}} : \ell_{2,\theta} \to \ell_{2,\theta}$ is bounded for fixed t. Since the nonlinear terms in a), b), e) and f) are bi- and trilinear maps from $\ell_{2,\theta} \to \ell_{2,\theta}$ for $\theta > 1/2$ we have the local existence and uniqueness for these equations in $\ell_{2,\theta}$ for $\theta > 1/2$ according to Theorem 5.1.20. Since for c) and d) the nonlinear terms are only bilinear maps from $\ell_{p,\theta+1}$ into $\ell_{p,\theta}$ we need an estimate

$$\|(\mathrm{e}^{\lambda_k t})_{k\in\mathbb{Z}}\|_{\ell_{p,\theta}\to\ell_{p,s+1}} \le C\max(1, t^{-\alpha}) \tag{5.11}$$

with $\alpha \in [0,1)$ for the semigroup in order to apply our local existence and uniqueness result from Theorem 5.1.23. According to Example 5.1.21 such an estimate is true for c) with $\alpha = 1/2$, but not for d). The KdV equation is a so called a quasilinear (hyperbolic) equation. There is local existence in $\ell_{2,\theta}$ for $\theta = 3$, for instance. However, the proof is more involved, cf. [**Paz83**, §8, Theorem 5.6] or §8.2 for further remarks. ⌋

5.1.6. Loss and regain of compactness. We close this section with a number of comments on compactness, which is a crucial concept to define attractors in dynamical systems. In metric spaces there are the following equivalent characterizations of compact sets, cf. [**Alt16**, §2.5].

Definition 5.1.30. *Let (M,d) be a complete metric space.*

a) A set $A \subset M$ is compact if every covering of A by open sets contains a finite subcovering.

b) A set $A \subset M$ is (sequentially) compact if every sequence in A has a convergent subsequence with limit in A.

c) A set $A \subset M$ is compact if A is closed and pre-compact, where a set $A \subset M$ is said to be pre-compact if for every $\varepsilon > 0$, there exists a finite subset $\{s_1, s_2, \ldots, s_n\}$ of A such that $A \subset \bigcup_{k=1}^n B(s_k, \varepsilon)$, where $B(s_k, \varepsilon)$ denotes the open ball around s_k with radius ε.

Compactness arguments in the sense of b) were used a number of times in Part I. Examples are the existence, respectively the non-emptyness, of ω-limit sets and attractors. For the term 'pre-compact' also the term 'totally bounded' is used in the literature. In $\mathbb{R}^d$ compact sets can be characterized by the theorem of Heine-Borel.

Theorem 5.1.31. *In $\mathbb{R}^d$ a set is compact if and only if it is closed and bounded.*

In infinite-dimensional spaces compactness is more restrictive due to the fact that the theorem of Heine-Borel is no longer true as the following example shows.

Example 5.1.32. Consider the closed unit ball in $(\mathbb{R}^{\mathbb{N}}, \|\cdot\|_\infty)$. The sequence $(u_n)_{n\in\mathbb{N}}$ with $u_n = e_n$ satisfies $\|u_n - u_m\|_\infty = \delta_{nm}$ such that no convergent subsequence can exist. Variants of this example works in all norms from above such that the closed unit ball is not compact in any of the norms from Definition 5.1.1. ⌋

The equivalence of compactness to boundedness and closedness is a precise distinction between finite- and infinite-dimensional Banach spaces [**Alt16**, Satz 2.9]. There are famous theorems about the characterization of pre-compact subsets of function spaces. These are the Arzela-Ascoli theorem [**Alt16**, Satz 2.11], the theorem of Riesz [**Alt16**, Satz 2.15], and Sobolev's embedding theorem [**Alt16**, Satz 8.9].

Compactness in infinite-dimensional spaces will be regained by smoothing properties of the evolution operators. For instance the evolution operator of Example 5.1.21 maps bounded balls of $\ell_{2,0}$ into bounded balls of $\ell_{2,1}$ for every fixed $t > 0$. Since the subsequent version of Sobolev's embedding theorem 5.1.33 guarantees that bounded balls of $\ell_{2,1}$ are pre-compact sets of $\ell_{2,0}$, the evolution operator of Example 5.1.21 maps bounded balls of $\ell_{2,0}$ into pre-compact sets of $\ell_{2,0}$. This property will be used for showing that ω-limit sets and attractors for such systems are non-empty. The following theorem is also known under the name Rellich's embedding theorem.

Theorem 5.1.33. *The space $\ell_{p,\theta}$ can be compactly embedded into the space $\ell_{p,r}$ for all $p \geq 1$ and $\theta > r$.*

Proof. For notational simplicity we restrict to the index set $\mathbb{N}$. Compactly embedded means that every bounded set of $\ell_{p,\theta}$ is pre-compact in $\ell_{p,r}$. Due to the homogeneity of the spaces it is sufficient to prove that the unit ball of $\ell_{p,\theta}$ can be covered by finitely many balls of $\ell_{p,r}$ with radius ε. In order to do so we consider the first n_0 coordinates. The restriction of the unit ball of $\ell_{p,\theta}$ to these coordinates is a pre-compact set in $\mathbb{R}^{n_0}$. Hence, for every $\varepsilon > 0$ it can be covered by finitely many balls $B_\varepsilon(z_j)$ of $\mathbb{R}^{n_0}$ w.r.t. the $\ell_{p,r}$-norm and with $z_j \in \mathbb{R}^{n_0}$ for $j = 1, \ldots, N$. We claim that the unit ball of $\ell_{p,\theta}$ is contained in the union of the balls $B_{\varepsilon/2}((z_j, 0))$ of $\ell_{p,r}(\mathbb{R}^{\mathbb{N}})$ if we choose $(n_0+1)^{r-\theta} \leq \varepsilon/2$. This follows since for $u = ((u_k)_{k=1,\ldots,n_0}, u_\infty)$ in the unit ball of $\ell_{p,\theta}(\mathbb{R}^{\mathbb{N}})$ we have a $j \in \{1, \ldots, N\}$ such that

$$\|u-(z_j,0)\|_{\ell_{p,r}} \leq \|((u_k)_{k=1,\ldots,n_0},0)-(z_j,0)\|_{\ell_{p,r}}+\|(0,u_\infty)\|_{\ell_{p,r}} \leq \varepsilon/2+\varepsilon/2=\varepsilon,$$

where $\|((u_k)_{k=1,\ldots,n_0}, 0) - (z_j, 0)\|_{\ell_{p,r}} \leq \varepsilon/2$ due to the construction of the points z_j and where $\|(0, u_\infty)\|_{\ell_{p,r}} \leq \varepsilon/2$ due to

$$\begin{aligned}\|(0,u_\infty)\|_{\ell_{p,r}} \leq & \Big(\sum_{n_0+1}^{\infty} |u_n|^p |n|^{pr}\Big)^{1/p} \leq \sup_{n=n_0+1,\ldots,\infty} |n|^{r-\theta} \Big(\sum_{n_0+1}^{\infty} |u_n|^p |n|^{p\theta}\Big)^{1/p} \\ \leq & (n_0+1)^{r-\theta} \|(0,u_\infty)\|_{\ell_{p,\theta}} \leq \varepsilon/2\end{aligned}$$

for n_0 sufficiently large since $\|(0, u_\infty)\|_{\ell_{p,\theta}} \leq 1$. □

5.2. Basic function spaces and Fourier series

PDEs posed on spatially bounded domains are very often isomorphic to ODEs in $\mathbb{R}^\infty$. Thus, the abstract set-up from the last section can often be applied to solve PDEs posed on spatially bounded domains. However, a big part of PDE theory is concerned with problems coming from the boundary of the considered domains. These play almost no role in this book, i.e., they are circumvented by considering almost all systems subsequently with periodic boundary conditions or on the real line. This allows us to concentrate on phenomena coming from the equations. In other words, a complete functional analytic treatment of PDEs posed on bounded domains with dynamical systems concepts is beyond the scope of this book. For this we refer to the textbooks [**Hen81, Hal88, Tem97**]. However, in Part IV of this book some of the methods to handle problems posed on cylindrical domains $\mathbb{R} \times \Sigma$, with $\Sigma \subset \mathbb{R}^d$ a bounded domain, are explained. In this section we concentrate on PDEs where the spatial coordinate lives on a bounded interval with periodic boundary conditions. Such problems can easily be related to ODEs in $\mathbb{R}^\infty$ with the help of Fourier series. These explanations are embedded in some theoretical background about basic function spaces and Fourier series.

5.2.1. Basic function spaces. The solution $u = u(\cdot, t)$ of a PDE is for fixed t in some function space. Here, we introduce some basic function spaces following [**Alt16, Wlo87**]. In the following let $\Omega \subset \mathbb{R}^d$ be an open set, $x = (x_1, \ldots, x_d) \in \Omega$, $n = (n_1, \ldots, n_d)$ a multi-index, $|n| = n_1 + \ldots + n_d$, and $\partial_x^n = \partial_{x_1}^{n_1} \ldots \partial_{x_d}^{n_d}$.

Continuous and differentiable functions. The space of continuous functions in $\overline{\Omega}$ is

$$C^0(\overline{\Omega}, \mathbb{R}) = \{u : \overline{\Omega} \to \mathbb{R} : u \text{ is continuous}\},$$

equipped with the norm

$$\|u\|_{C_b^0} = \sup_{x \in \overline{\Omega}} |u(x)|.$$

The space of m-times continuously differentiable functions in $\overline{\Omega}$ is

$$C^m(\overline{\Omega}, \mathbb{R}) = \{u : \overline{\Omega} \to \mathbb{R} : \partial_x^j u \text{ is continuous for } |j| = 0, \ldots, m\},$$

equipped with the norm

$$\|u\|_{C_b^m} = \sum_{0 \le |j| \le m} \|\partial_x^j u\|_{C_b^0}.$$

From the definition it is clear that for $u \in C^m(\overline{\Omega}, \mathbb{R})$ we have $\|u\|_{C_b^m} < \infty$, if Ω is bounded. More generally, we define

$$C_b^m(\overline{\Omega}, \mathbb{R}) = \{u \in C^m(\overline{\Omega}, \mathbb{R}) : \|u\|_{C_b^m} < \infty\}.$$

For $\Omega = \overline{\Omega} = \mathbb{R}$ the function $u(x) = x$ is in C^0, but not in C_b^0. For Ω bounded, C_b^m is dense in C_b^0. For the treatment of unbounded Ω we define

$$\begin{aligned} C_{b,unif}^m(\overline{\Omega}, \mathbb{R}) = \{u : \overline{\Omega} \to \mathbb{R} :&\partial_x^j u \text{ is uniformly continuous for} \\ & |j| = 0, \ldots, m, \|u\|_{C_b^m} < \infty\}. \end{aligned}$$

For $\Omega = \mathbb{R}$ the function $u(x) = \sin(x^2)$ is in C_b^0, but not in $C_{b,unif}^0$. For similar reasons $C_b^n(\mathbb{R}, \mathbb{R})$ is not dense in $C_b^0(\mathbb{R}, \mathbb{R})$, but $C_{b,unif}^n(\mathbb{R}, \mathbb{R})$ in $C_{b,unif}^0(\mathbb{R}, \mathbb{R})$. All these spaces are Banach spaces.

Hölder spaces. The spaces C_b^0 and C_b^m are not the optimal choice for solving linear PDEs. Even for arbitrarily smooth boundary $\partial\Omega$ the boundary value problem

$$\Delta u = f \qquad \text{in } \Omega, \qquad u|_{\partial\Omega} = 0, \tag{5.12}$$

for $f \in C_b^0$ in general does not possess a solution u with optimal regularity, cf. Example 5.2.4 on page 157. Optimal regularity holds for the subsequently defined Hölder-continuous functions and Sobolev functions, i.e., for instance for (5.12) from $f \in C^{0,\alpha}$ it follows $u \in C^{2,\alpha}$. For $\alpha \in (0, 1]$ we define

$$C^{0,\alpha}(\overline{\Omega}, \mathbb{R}) = \{u : \overline{\Omega} \to \mathbb{R} : u \ \text{ is } \alpha\text{ -Hölder-continuous}, \ \|u\|_{C^{0,\alpha}} < \infty\}$$

equipped with the norm

$$\|u\|_{C^{0,\alpha}} = \|u\|_{C_b^0} + \sup_{x,y \in \overline{\Omega}, x \ne y, |x-y| \le 1} \frac{|u(x) - u(y)|}{|x - y|^\alpha},$$

and, for $k \in \mathbb{N}$ and $\alpha \in (0,1]$,

$$C^{k,\alpha}(\overline{\Omega},\mathbb{R}) = \{u : \overline{\Omega} \to \mathbb{R} : \partial_x^j u \in C^0 \text{ for } |j| = 0,\dots,k, \\ \partial_x^k u \in C^{0,\alpha},\ \|u\|_{C^{k,\alpha}} < \infty\}$$

equipped with the norm

$$\|u\|_{C^{k,\alpha}} = \|u\|_{C_b^{k-1}} + \sum_{|j|=k} \|\partial_x^j u\|_{C^{0,\alpha}}.$$

All these function spaces are Banach spaces, cf. Exercise 5.9. $C^{0,1}(\overline{\Omega},\mathbb{R})$ is the space of Lipschitz-continuous functions.

Lebesgue and Sobolev spaces. Unfortunately, the above spaces are not equipped with a scalar product and so tools from linear algebra related to orthogonality are not available. A natural choice of a scalar product for functions would be

$$\langle u, v\rangle_{L^2} = \int_\Omega u(x)\overline{v(x)}\,\mathrm{d}x. \tag{5.13}$$

However, if the above spaces are equipped with the above scalar product they are not complete w.r.t. the induced norm. For instance the sequence $(u_n)_{n\in\mathbb{N}}$ with

$$u_n(x) = \begin{cases} 1, & \text{for } |x| \le 1 - 1/n, \\ 0, & \text{for } |x| \ge 1, \\ n(1-|x|), & \text{for } |x| \in (1-1/n, 1), \end{cases}$$

is a Cauchy sequence w.r.t. the norm induced by the L^2-scalar product. However, the limit function is not in C_b^0 although $u_n \in C_b^0$ for all $n \in \mathbb{N}$.

Since the limit of a Cauchy sequence of Riemann integrable functions is in general no longer Riemann integrable the Riemann integral has to be replaced by the Lebesgue integral in order to define complete function spaces [**Alt16**, §A1]. In order to define the Lebesgue and Sobolev spaces we introduce

$$C^\infty(\Omega,\mathbb{R}) = \{u : \Omega \to \mathbb{R} : u \text{ is arbitrarily many times differentiable}\}$$

and

$$C_0^\infty(\Omega,\mathbb{R}) = \{u \in C^\infty(\Omega,\mathbb{R}) : u \text{ has compact support in } \Omega\}$$

where the support of a function is defined by $\mathrm{supp}(u) = \mathrm{cl}_{\mathbb{R}^d}\{x \in \Omega : u(x) \neq 0\}$. The Lebesgue spaces are defined by

$$L^p(\Omega,\mathbb{R}) = \mathrm{cl}_{\|\cdot\|_{L^p}}(C_0^\infty(\Omega,\mathbb{R})), \quad \text{where} \quad \|u\|_{L^p} = \left(\int_\Omega |u(x)|^p\,\mathrm{d}x\right)^{1/p}$$

for all $p \in [1,\infty)$. By construction all these spaces are Banach spaces consisting of equivalence classes of Cauchy sequences, with two Cauchy sequences in the same class if their difference converges to zero. The L^p-spaces constructed in this way coincide with the spaces known from measure theory. The space $L^2(\Omega,\mathbb{R})$, respectively $L^2(\Omega,\mathbb{C})$, is a Hilbert space equipped with the scalar product (5.13).

For the solution of PDEs so called Sobolev spaces turn out to be useful. For $p \in [1,\infty)$ and Ω bounded we define

$$W^{m,p}(\Omega,\mathbb{R}) = \mathrm{cl}_{\|\cdot\|_{W^{m,p}}}(C^\infty(\Omega,\mathbb{R})),$$

where

$$\|u\|_{W^{m,p}} = \Big(\sum_{|j|\le m} \|\partial_x^j u\|_{L^p}^p\Big)^{1/p}$$

and

$$W_0^{m,p}(\Omega,\mathbb{R}) = \mathrm{cl}_{\|\cdot\|_{W^{m,p}}}(C_0^\infty(\Omega,\mathbb{R})),$$

for general Ω. Since the sum in the definition of $\|\cdot\|_{W^{m,p}}$ is finite there are various equivalent norms such as $\|u\|_{W^{m,p}} = \sum_{|j|\le m}\|\partial_x^j u\|_{L^p}$. By construction these spaces are Banach spaces, too. The spaces $H^m(\Omega,\mathbb{R}) = W^{m,2}(\Omega,\mathbb{R})$ and $H_0^m(\Omega,\mathbb{R}) = W_0^{m,2}(\Omega,\mathbb{R})$ are Hilbert spaces equipped with the scalar product

$$\langle u,v\rangle_{H^m} = \sum_{|j|\le m} \langle \partial_x^j u, \partial_x^j v\rangle_{L^2}.$$

By Sobolev's embedding theorem [**Alt16**, Satz 8.8.], Sobolev spaces can be embedded continuously into classical function spaces. We have

$$W^{m,p}(\Omega,\mathbb{R}) \hookrightarrow C^{n,\alpha}(\overline{\Omega},\mathbb{R}) \quad \text{if} \quad m - d/p > n+\alpha,$$

i.e., there exists a $C>0$, such that for all $u \in W^{m,p}$

$$\|u\|_{C^{n,\alpha}} \le C\|u\|_{W^{m,p}}$$

and in the equivalence class of $u \in W^{m,p}$ there is a representative $u \in C^{n,\alpha}(\overline{\Omega},\mathbb{R})$. For the proof of special cases see Lemma 5.1.27 and Lemma 5.2.3.

A different characterization of these spaces is (e.g., [**Alt16**, §1.25])

$$W^{m,p}(\Omega,\mathbb{R}) = \{u : \Omega \to \mathbb{R} : \partial_x^\alpha u \in L^p \text{ for } |\alpha| = 0,\dots,m,\ \|u\|_{W^{m,p}} < \infty\},$$

where $\partial_x^\alpha u$ denotes the αth weak derivative of u. For $\Omega \subset \mathbb{R}^d$ the function $\partial_x^\alpha u \in L^p(\Omega,\mathbb{R})$ is called α^{th} weak derivative of $u \in L^p(\Omega,\mathbb{R})$ if for all $\phi \in C_0^\infty(\Omega,\mathbb{R})$ we have

$$\int_\Omega (\partial_x^\alpha u(x))\phi(x)\,\mathrm{d}x = (-1)^{|\alpha|}\int_\Omega u(x)(\partial_x^\alpha \phi(x))\,\mathrm{d}x.$$

We define $L^\infty(\Omega, \mathbb{R})$ as the space of all measurable functions $u : \Omega \to \mathbb{R}$ for which

$$\sup_{x\in\Omega\setminus N} |f(x)| < \infty \text{ for a null set } N.$$

This space is equipped with the norm

$$\|u\|_{L^\infty} = \inf_{N \text{ is a null set}} \sup_{x\in\Omega\setminus N} |f(x)| .$$

We introduce

$$\|u\|_{W^{m,\infty}} = \sum_{|\alpha|=0}^{m} \|\partial_x^\alpha u\|_{L^\infty}$$

and the space $W^{m,\infty}(\Omega, \mathbb{R})$ as the space of all functions $u : \Omega \to \mathbb{R}$ for which the weak derivatives $\partial_x^\alpha u$ exist for $|\alpha| = 0, \dots, m$ and for which $\|u\|_{W^{m,\infty}} < \infty$.

Remark 5.2.1. The concept of weak derivatives can be generalized to the concept of distributional derivatives [**RR04**, Chapter 5]. A priori, the sets C^∞ and C_0^∞ are just vector spaces. There is no norm for which these spaces are complete. However, using inductive limits of semi-norms, $C_0^\infty(\Omega, \mathbb{R})$ can be made to be a complete metric space $\mathcal{D}(\Omega)$, called space of test functions, where convergence $u_n \to u$ in $\mathcal{D}(\Omega)$ means: **a)** There exists a compact $K \subset \Omega$ such that $\operatorname{supp}(u_n), \operatorname{supp}(u) \subset K$, **b)** $\lim_{n\to\infty} \partial_x^\alpha u_n(x) = \partial_x^\alpha u(x)$ uniformly in K for all $\alpha \in \mathbb{N}^d$. However, this convergence is not induced by a norm.

The elements of the dual space of $\mathcal{D}(\Omega, \mathbb{R}) = C_0^\infty(\Omega, \mathbb{R})$ are called distributions, i.e., a distribution T is a continuous linear map from $\mathcal{D}$ into the real or complex numbers. This means that $u_n \to u$ in $\mathcal{D}$ implies $Tu_n \to Tu$, which is equivalent to the formulation that for all open bounded sets D there is a constant C and a number $m \in \mathbb{N}$ such that

$$|T(\phi)| \le C\|\phi\|_{C_b^m} \quad \text{for all} \quad \phi \in C_0^\infty(D, \mathbb{R}). \tag{5.14}$$

For a continuous function $u \in C_b^0(\mathbb{R}^d, \mathbb{R})$, or for u in one of the above other spaces

$$T_u(\phi) = \int_{\mathbb{R}^d} u(x)\phi(x)\,\mathrm{d}x$$

defines the so called associated distribution, which is then called regular. For the distribution associated to $\partial_x^\alpha u$ we find

$$\begin{aligned} T_{\partial_x^\alpha u}(\phi) =& \int_{\mathbb{R}^d} (\partial_x^\alpha u(x))\phi(x)\,\mathrm{d}x \\ =&(-1)^{|\alpha|} \int_{\mathbb{R}^d} u(x)(\partial_x^\alpha \phi(x))\,\mathrm{d}x = (-1)^{|\alpha|} T_u(\partial_x^\alpha \phi). \end{aligned}$$

This property is taken to define the αth derivative of an arbitrary distribution by

$$(\partial_x^\alpha T)(\phi) = (-1)^{|\alpha|} T(\partial_x^\alpha \phi).$$

The distributional derivative of a function u is not necessarily again a function, as the next example shows. If $(\partial_x^\alpha T_u)$ can be represented by a function g, i.e., $\partial_x^\alpha T_u = T_g$, then g is the αth weak derivative of u. ⌋

Example 5.2.2. For $u(x) = |x|$ we show that $u \in H^1((-1,1))$ but $u \notin H^2((-1,1))$ by computing the weak derivatives $\partial_x u, \partial_x^2 u$. For $\phi \in C_0^\infty((-1,1))$ we have $T_u(\partial_x \phi) = \int_{-1}^0 -x\partial_x\phi \,\mathrm{d}x + \int_0^1 x\partial_x \phi \,\mathrm{d}x = \int_{-1}^0 \phi \,\mathrm{d}x + \int_0^1 -\phi \,\mathrm{d}x = -T_g(\phi)$ with

$$g(x) = \partial_x u(x) = \begin{cases} -1, & x < 0, \\ 1, & x > 0. \end{cases}$$

Similarly, $T_u(\partial_x^2 \phi) = \int_{-1}^0 \partial_x \phi(x) \,\mathrm{d}x + \int_0^1 -\partial_x\phi(x)\,\mathrm{d}x = 2\phi(0) = 2\delta_0(\phi)$ where δ_0 is called the Dirac δ distribution. Thus, $\partial_x^2 u \notin L^2$ as there is no function g such that $\int_{-1}^1 g(x)\phi(x)\,\mathrm{d}x = \phi(0)$. ⌋

In order to solve nonlinear PDEs two additional properties have to be satisfied by the function spaces in use. First, the values of the functions on the boundary have to be well defined. For $X = L^2(\Omega)$ a function $u \in L^2(\Omega)$ is only unique up to a null set in $\Omega \subset \mathbb{R}^d$. Since a smooth boundary $\partial\Omega$ is a null set, boundary conditions in L^2 are not well defined. Secondly, in the nonlinearity we have products of functions of X, i.e., with $u \in X$, also u^2 should be in X, i.e., X should be an algebra. For $u \in L^2$ in general we do not have $u^2 \in L^2$. However, for m sufficiently large (depending on p and the space dimension d), H^m or more generally $W^{m,p}$ are algebras, and point-wise values are defined, or at least the boundary conditions can be fulfilled in a generalized sense.

We close this subsection with the proof of a very simple version of Sobolev's embedding theorem and an example indicating which of the function spaces are suitable for solving PDEs and which are not.

Lemma 5.2.3. *Let* $-\infty < a < b < \infty$. *Then* $H^1((a,b)) \subset C^{0,1/2}((a,b))$ *and*

$$\|u\|_{L^\infty}^2 \le 2\|u\|_{L^2}\left(\frac{1}{b-a}\|u\|_{L^2} + \|\partial_x u\|_{L^2}\right), \tag{5.15}$$

$$|u(x) - u(y)| \le \sqrt{x-y}\,\|\partial_x u\|_{L^2}^2. \tag{5.16}$$

Proof. Since $C^1((a,b))$ is dense in $H^1((a,b))$ w.r.t. the $\|\cdot\|_{H^1}$-norm, it is sufficient to prove (5.15) and (5.16) for $u \in C^1((a,b))$. We have

$$\begin{aligned} u^2(x) &= \int_a^x \frac{\mathrm{d}}{\mathrm{d}s}\left(\frac{s-a}{x-a}u^2(s)\right)\mathrm{d}s \\ &= \int_a^x \frac{1}{x-a}u^2(s)\,\mathrm{d}s + \int_a^x \frac{s-a}{x-a}2u(s)\partial_x u(s)\,\mathrm{d}s \\ &\le \frac{1}{x-a}\|u\|_{L^2}^2 + 2\|u\|_{L^2}\|\partial_x u\|_{L^2} \end{aligned}$$

and similarly

$$u^2(x) = \int_x^b \frac{\mathrm{d}}{\mathrm{d}s}\left(\frac{s-x}{b-x}u^2(s)\right)\mathrm{d}s \le \frac{1}{b-x}\|u\|_{L^2}^2 + 2\|u\|_{L^2}\|\partial_x u\|_{L^2}.$$

Hence, $u^2(x) \le \min\{\frac{1}{x-a}, \frac{1}{b-x}\}\|u\|_{L^2}^2 + 2\|u\|_{L^2}\|\partial_x u\|_{L^2}$. For the second estimate we use the Cauchy-Schwarz inequality, namely

$$|u(x) - u(y)| = \left|\int_x^y \partial_s u(s)\,\mathrm{d}s\right| \le \int_x^y 1|\partial_s u(s)|\,\mathrm{d}s \le \sqrt{|x-y|}\|\partial_x u\|_{L^2}. \qquad \square$$

We already stated that C^k-spaces are in general not optimal concerning the regularity of solutions of PDEs. More life is given to this statement by the following example [**Sal08**, Example 8.2]. This gives a motivation for the use of Sobolev spaces, in particular for the use of H^m-spaces for which Hilbert space methods are available.

Example 5.2.4. For $0 < \alpha < 2\pi$ let $\Omega_\alpha := \{(r,\phi) : 0 < r < 1, -\alpha/2 < \theta < \alpha/2\}$ be the two-dimensional sector with opening angle α. Consider the Dirichlet boundary value problem

$$(5.17) \qquad -\Delta u = 0 \text{ in } \Omega_\alpha, \quad u|_{\partial\Omega} = g_\alpha(r,\phi) \text{ on } \partial\Omega_\alpha,$$

with $g_\alpha(r,\phi) = \cos(\pi\phi/\alpha)$ for $r = 1$, $g(r,\phi) = 0$ else, where (r,ϕ) are polar coordinates. Identifying $\mathbb{R}^2$ with $\mathbb{C}$ we find that $f(z) = z^{\pi/\alpha}$ is holomorphic in Ω_α, and thus

$$u(r,\phi) = \mathrm{Re}(f(z)) = r^{\pi/\alpha}\cos(\pi\phi/\alpha)$$

is harmonic in Ω_α and satisfies the boundary conditions. Thus, it is the unique solution of (5.17).

Clearly, $u \in C^\infty(\Omega)$, and we now consider the regularity of u up to the boundary and compare it with Sobolev regularity. Let $\alpha \neq \pi$, otherwise $u(x,y) = x$. We find

$$|\nabla u|^2 = (\partial_r u)^2 + \frac{1}{r^2}(\partial_\phi u)^2 = \frac{\pi^2}{\alpha^2}r^{2(\pi/\alpha - 1)},$$

and this is in $C^1(\overline{\Omega})$ only for $\alpha \le \pi$. But $\int_{\Omega_\alpha} |\nabla u|^2\,\mathrm{d}x = \frac{\pi^2}{\alpha}\int_0^1 r^{2\pi/\alpha - 1}\,\mathrm{d}r = \pi/2$ independent of α, and thus $u \in H^1(\Omega)$ for all α. Next, $|\partial_{x_i}^2 u| \sim$

$r^{\pi/\alpha-2}$ for $r \to 0$, thus, $u \in C^2(\overline{\Omega})$ for $\alpha \leq \pi/2$, and $u \in H^2(\Omega)$ for $\alpha \leq \pi$, i.e., if the sector is convex. By setting $\widetilde{g}(r,\phi) = r^j g(\phi)$ with $j \geq 2$ and $v = u - \widetilde{g}$ we find $\widetilde{g} \in C^\infty(\overline{\Omega_\alpha})$ and $-\Delta v = -\Delta u + \Delta\widetilde{g} = \Delta\widetilde{g} =: f$ in Ω_α, and $v|_{\partial\Omega_\alpha} = 0$. For the last system the Lax-Milgram theorem [**Eva98**, §6.2.1] guarantees $u \in H^1$ for general spatial domains and regularity theory [**Eva98**, §6.3] guarantees $u \in H^2$ for convex spatial domains. ⌋

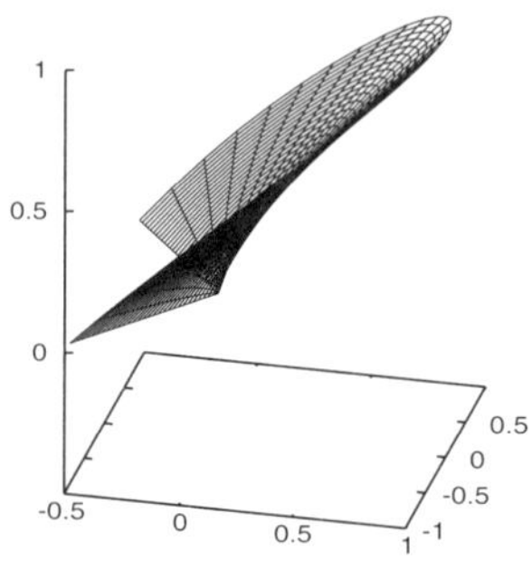

Figure 5.3. $z \mapsto (\mathrm{Re} z)^{3/4}$ solves the boundary value problem (5.17) if $\alpha = 4\pi/3$. The derivative is unbounded at the origin.

5.2.2. Fourier series. PDEs with periodic boundary conditions for the spatial coordinates can be transferred to ODEs in $\mathbb{R}^\infty$ with the help of Fourier series. For notational simplicity we restrict ourselves first to the torus $\mathbb{T}^d = \mathbb{R}^d/(2\pi\mathbb{Z})^d$. Let C^∞_{per} be the space of functions $u : \mathbb{T}^d \to \mathbb{R}^d$, with a C^∞ periodic extension $\widetilde{u} : \mathbb{R}^d \to \mathbb{R}^d$ satisfying

$$\begin{aligned}\widetilde{u}(x_1, x_2, \ldots, x_d) &= \widetilde{u}(x_1 + 2\pi, x_2, \ldots, x_d) = \widetilde{u}(x_1, x_2 + 2\pi, \ldots, x_d)\\ &= \ldots = \widetilde{u}(x_1, x_2, \ldots, x_d + 2\pi).\end{aligned}$$

We define

$$H^m_{\mathrm{per}} = \mathrm{clos}_{\|\cdot\|_{H^m(\mathbb{T}^d)}}(C^\infty_{\mathrm{per}}).$$

The question is if and in what sense a function can be represented by its Fourier series, or equivalently, in which norms Fourier series converge. In L^2 we have a simple answer which follows from the general theory of orthonormal systems, and which for convenience we summarize here.

Definition 5.2.5. *Let H be a Hilbert space with scalar product $\langle\cdot,\cdot\rangle : H \times H \to \mathbb{C}$. A (finite or infinite) system (ϕ_j) in H, $j = 1, \ldots, N$ or $j \in \mathbb{N}$, is called orthogonal system if $\langle\phi_i, \phi_j\rangle = 0$ for $i \neq j$. It is called orthonormal if additionally $\langle\phi_j, \phi_j\rangle = 1$. It is called a complete orthonormal system (complete ONS) or Hilbert basis if $\langle u, \phi_j\rangle = 0$ for all j implies $u = 0$ for $u \in H$.*

In Hilbert spaces H the following holds.

Lemma 5.2.6. *a) If $u_n \to u$ and $v_n \to v$ in H, then $\langle u_n, v_n\rangle \to \langle u, v\rangle$.*
b) (Pythargoras) For $\phi_1, \dots, \phi_n \in H$ with $\langle \phi_i, \phi_j\rangle = 0$ for $i \neq j$ we have

$$\|\phi_1 + \ldots + \phi_n\|^2 = \|\phi_1\|^2 + \ldots + \|\phi_n\|^2.$$

c) If $(\phi_j)_{j\in\mathbb{N}}$ is an orthonormal system in H, and u_j a sequence in $\mathbb{C}$, then $\sum_{j=1}^{\infty} u_j\phi_j$ converges if and only if $\sum_{j=1}^{\infty} |u_j|^2$ converges, i.e., if $(u_j)_{j\in\mathbb{N}} \in \ell^2$.
d) (Bessel's inequality) If $(\phi_j)_{j\in\mathbb{N}}$ is an orthonormal sequence in H, then for all $u \in H$ we have $\sum_{j=1}^{\infty} |\langle u, \phi_j\rangle|^2 \leq \|u\|^2$.

Proof. a) By Cauchy-Schwarz we have

$$\begin{aligned}|\langle u_n, v_n\rangle - \langle u, v\rangle| &= |\langle u_n, v_n - v\rangle + \langle u_n - u, v\rangle| \\ &\leq \|u_n\|\|v_n - v\| + \|u_n - u\|\|v\|,\end{aligned}$$

where $\sup_{n\in\mathbb{N}} \|u_n\| < \infty$ for convergent series $(u_n)_{n\in\mathbb{N}}$.
b) Direct calculation for the finite sums.
c) For all $m \leq n$ we have $\|\sum_{j=m}^{n} u_j\phi_j\|^2 = \sum_{j=m}^{n} |u_j|^2$ by b). Thus, $\sum_{j=1}^{\infty} u_j\phi_j$ is a Cauchy sequence if and only if $\sum_{j=1}^{\infty} |u_j|^2$ is a Cauchy sequence.
d) For $N \in \mathbb{N}$ we have

$$0 \leq \left\langle u - \sum_{j=1}^{N} \langle u, \phi_j\rangle \phi_j, u - \sum_{j=1}^{N} \langle u, \phi_j\rangle \phi_j \right\rangle = \|u\|^2 - \sum_{j=1}^{N} |\langle u, \phi_j\rangle|^2,$$

and hence $\sum_{j=1}^{N} |\langle u, \phi_j\rangle|^2 \leq \|u\|^2$, which implies convergence of the series and Bessel's inequality. □

Lemma 5.2.7. *The following statements are equivalent:*
(i) $(\phi_j)_{j\in\mathbb{N}}$ is a complete ONS.
(ii) For all $u \in H$ we have $u = \sum_{j=1}^{\infty} \langle u, \phi_j\rangle \phi_j$.
(iii) For all $u, v \in H$ we have Parseval's identity

$$\langle u, v\rangle = \sum_{j=1}^{\infty} \langle u, \phi_j\rangle \langle v, \phi_j\rangle. \tag{5.18}$$

(iv) For all $u \in H$ we have Bessel's equality

$$\|u\|^2 = \sum_{j=1}^{\infty} |\langle u, \phi_j\rangle|^2. \tag{5.19}$$

Proof. (i)⇒(ii). For $u \in H$ we have convergence of $\sum_{j=1}^{\infty} |\langle u, \phi_j\rangle|^2$ by d) and convergence of $\sum_{j=1}^{\infty} \langle u, \phi_j\rangle \phi_j$ to some $v \in H$ by Lemma 5.2.6 c). By Lemma 5.2.6 a) we have

$$\langle u - v, \phi_j\rangle = \langle u, \phi_j\rangle - \sum_{n=1}^{\infty} \langle u, \phi_n\rangle \langle \phi_n, \phi_j\rangle = \langle u, \phi_j\rangle - \langle u, \phi_j\rangle = 0,$$

and since (ϕ_j) is complete this implies $v = u$. (ii)$\Rightarrow$(iii) again follows from Lemma 5.2.6 a), and (iv) follows from (iii) with $v = u$. Finally, (iv)$\Rightarrow$(i) since $\langle u, \phi_j\rangle = 0$ for all j and (iv) imply $\|u\| = 0$, hence $u = 0$. $\square$

Due to the equivalence of (5.18) and (5.19), often both are called Parseval's identity. Clearly, Lemma 5.2.6 and Lemma 5.2.7 also holds if sequences $(\phi_j)_{j\in\mathbb{N}}$ are replaced by sequences $(\phi_j)_{j\in\mathbb{Z}}$, $(\phi_j)_{j\in\mathbb{N}^d}$, $(\phi_j)_{j\in\mathbb{Z}^d}$, with the respective replacements in the sums. The most important example are classical Fourier series.

Theorem 5.2.8. *a) The functions $\phi_k = \frac{1}{2\pi}\mathrm{e}^{\mathrm{i}k\cdot x}$ with $k \in \mathbb{Z}^d$ are a complete ONS in $L^2(\mathbb{T}^d)$ w.r.t. the inner product*

$$\langle u, v\rangle_{L^2} = \frac{1}{(2\pi)^d}\int_{\mathbb{T}^d} u(x)\overline{v(x)}\,\mathrm{d}x.$$

For $u \in L^2(\mathbb{T}^d)$ we have L^2-convergence of the Fourier series, i.e., for

$$S_N(x) = \sum_{|k|\le N} \widehat{u}_k \mathrm{e}^{\mathrm{i}k\cdot x}, \qquad \text{with} \quad \widehat{u}_k = \langle \mathrm{e}^{\mathrm{i}k\cdot x}, u\rangle_{L^2} = \frac{1}{(2\pi)^d}\int_{\mathbb{T}^d} u(x)\mathrm{e}^{-\mathrm{i}k\cdot x}\,\mathrm{d}x$$

we have $\|u - S_N\|_{L^2} \to 0$ as $N \to \infty$. This convergence is abbreviated as $u(x) = \sum_{k\in\mathbb{Z}^d} \widehat{u}_k \mathrm{e}^{\mathrm{i}k\cdot x}$.

b) For all $\phi \in T_N = \operatorname{span}\{\mathrm{e}^{\mathrm{i}k\cdot x} : |k| \le N\}$ we have $\|u - S_N\|_{L^2} \le \|u - \phi\|_{L^2}$, i.e., S_N is the best approximation of u in T_N in the quadratic mean.

c) We have Parseval's identity

$$\sum_{k\in\mathbb{Z}^d} |\widehat{u}_k|^2 = \frac{1}{(2\pi)^d}\|u\|_{L^2}^2.$$

d) There exists a $C > 0$ such that if $u \in C^m$ is 2π-periodic in each direction, then $|\widehat{u}_k| \le C|k|^{-m}$.

Proof. a) By direct calculation we find $\langle \phi_k, \phi_m\rangle = \begin{cases} 1, & \text{if } k = m, \\ 0, & \text{else,}\end{cases}$ i.e., the (ϕ_k) are an ONS. The completeness of this ONS can be shown with the Weierstraß approximation theorem, see [**Alt16**, Satz 7.10].

b) follows since S_N is the orthogonal projection of u on T_N.

c) Parseval's identity can be computed directly for finite sums. Going to the limits shows the assertion.

d) Through integration by parts we find

$$\begin{aligned}\widehat{u}_k = \langle \mathrm{e}^{\mathrm{i}k\cdot x}, u\rangle_{L^2} &= \frac{1}{(2\pi)^d}\int_{\mathbb{T}^d} u(x)\mathrm{e}^{-\mathrm{i}k\cdot x}\,\mathrm{d}x \\ &= (-1)^n \frac{1}{(2\pi)^d k^m}\int_{\mathbb{T}^d} (\partial_x^m u(x))\mathrm{e}^{-\mathrm{i}k\cdot x}\,\mathrm{d}x.\end{aligned}$$

□

The map $u \mapsto (\widehat{u})_{k\in\mathbb{Z}^d}$ will be abbreviated with $\mathcal{F}$. By c), $\mathcal{F}$ is an isometric isomorphism from L^2 to ℓ_2. Its inverse $(\widehat{u}_k)_{k\in\mathbb{Z}} \mapsto u$ is denoted by $\mathcal{F}^{-1}$. By d) the smoothness of u is related to the decay of its Fourier coefficients.

Formally we have $\partial_x u(x) = \sum_{k\in\mathbb{Z}} \mathrm{i}k\widehat{u}_k \mathrm{e}^{\mathrm{i}kx}$, or equivalently $\mathcal{F}(\partial_x u) = (\mathrm{i}k\widehat{u}_k)_{k\in\mathbb{Z}}$. It follows, that $\mathcal{F}$ is in fact an isomorphism between the Sobolev spaces H^m_{per} and the spaces of sequences $\ell_{2,m}$ which have been introduced in §5.1. Moreover, $\mathcal{F}^{-1}$ maps $\ell_{1,m}$ to C_b^m.

Lemma 5.2.9. *Let $m \in \mathbb{N}_0$. a) There exists a $C > 0$, such that for all $\widehat{u} \in \ell_{1,m}$*

$$\|u\|_{C_b^m} \le C\|\widehat{u}\|_{\ell_{1,m}}.$$

b) There exist $C_1, C_2 > 0$, such that for all $\widehat{u} \in \ell_{2,m}$

$$C_1\|\widehat{u}\|_{\ell_{2,m}} \le \|u\|_{H^m} \le C_2\|\widehat{u}\|_{\ell_{2,m}}.$$

c) There exist $C_1, C_2 > 0$, such that for all $u \in H^m_{per}$

$$C_1\|u\|_{H^m} \le \|\widehat{u}\|_{\ell_{2,m}} \le C_2\|u\|_{H^m}.$$

Proof. For notational simplicity we consider $d = 1$ and u with $\widehat{u}_0 = 0$. Moreover, we first consider $\mathcal{F}$ and $\mathcal{F}^{-1}$ on the dense subspaces C^∞_{per} respectively the space of finite sequences. The results then follow by continuous extension, see the subsequent Lemma 5.2.10.

a) We have

$$\begin{aligned}\|u\|_{C_b^m} &\le C\sup_{x\in\mathbb{R}}\sup_{0\le j\le m}\left|\partial_x^j \sum_{k\in\mathbb{Z}} \widehat{u}_k \mathrm{e}^{\mathrm{i}kx}\right| \le C\sup_{x\in\mathbb{R}}\sup_{0\le j\le m}\sum_{k\in\mathbb{Z}} |k|^j|\widehat{u}_k|\,|\mathrm{e}^{\mathrm{i}kx}| \\ &\le C\|\widehat{u}\|_{\ell_{1,m}}.\end{aligned}$$

Continuity in respect to the differentiability of u follows from the uniform and absolute convergence of the series.

b) and c) The second estimate in b) and the first estimate in c) follow from

$$\|u\|_{H^m}^2 = \sum_{j=0}^{m} \|\partial_x^j u\|_{L^2}^2 = \sum_{j=0}^{m} \int_{\mathbb{T}^d} \left| \partial_x^j \sum_{k\in\mathbb{Z}} \widehat{u}_k \mathrm{e}^{\mathrm{i}kx} \right|^2 dx$$

$$= \sum_{j=0}^{m} \int_{\mathbb{T}^d} \left| \sum_{k\in\mathbb{Z}} \widehat{u}_k (ik)^j \mathrm{e}^{\mathrm{i}kx} \right|^2 dx \le 2\pi \sum_{j=0}^{m} \sum_{k\in\mathbb{Z}} |\widehat{u}_k|^2 |k|^{2j} \le C\|\widehat{u}\|_{\ell_{2,m}}^2 .$$

The first estimate in b) and the second estimate in c) follow from

$$\|\widehat{u}\|_{\ell_{2,m}}^2 = \sum_{k\in\mathbb{Z}} |\widehat{u}_k|^2 |k|^{2m} = \sum_{k\in\mathbb{Z}} \left| \int_{\mathbb{T}^d} \mathrm{e}^{-ikx} u(x)dx \right|^2 |k|^{2m}$$

$$\overset{\int \text{by parts}}{=} \sum_{k\in\mathbb{Z}} \left| \int_{\mathbb{T}^d} \mathrm{e}^{-ikx} \partial_x u(x)dx \right|^2 |k|^{2(m-1)} = \dots$$

$$= \sum_{k\in\mathbb{Z}} \left| \int_{\mathbb{T}^d} \mathrm{e}^{-ikx} \partial_x^m u(x)dx \right|^2$$

$$= \|\mathcal{F}(\partial_x^m u)\|_{\ell_{2,0}}^2 \overset{\text{Parseval}}{=} \frac{1}{2\pi} \|\partial_x^m u\|_{L^2}^2 \le \|u\|_{H^m}^2 .$$

□

Lemma 5.2.10. *Let X be a metric space, $A \subset X$ a dense set, and Y a complete metric space. Then every uniformly continuous function $f : A \to Y$ possesses a unique uniformly continuous extension $\widetilde{f} : X \to Y$.*

Proof. The condition that the extension must be continuous leads to the only possible extension of f, namely

$$\widetilde{f}(x) = \lim_{x' \in A, x' \to x} f(x').$$

It remains to prove the existence of this limit, i.e., to prove that $\widetilde{f}$ is well defined. In order to do so, let $(x_n)_{n\in\mathbb{N}}$ be a sequence with $x_n \in A$ and $\lim_{n\to\infty} x_n = x$. Hence, $(x_n)_{n\in\mathbb{N}}$ is a Cauchy sequence in X and from the uniform continuity it follows that the image sequence $(f(x_n))_{n\in\mathbb{N}}$ is a Cauchy sequence in Y. Since Y is complete, we have the existence of

$$y = \lim_{n\to\infty} f(x_n)$$

in Y. Obviously the limit y is independent of the chosen sequence. It is an easy exercise to prove the uniform continuity of $\widetilde{f}$. □

We give a number of remarks and further results about Fourier series which will be useful later.

Remark 5.2.11. (Hausdorff-Young) ßThe discrete Fourier transform $\mathcal{F}$ is continuous from L^p to ℓ_q with $1/p+1/q=1$ for $p \le 2$. The discrete inverse Fourier transform $\mathcal{F}^{-1}$ is continuous from ℓ_q to L^p with $1/p+1/q=1$ for $q \le 2$, but not for $q > 2$, cf. [**Duo01**, Corollary 1.20]. This can be shown with the so called Riesz-Thorin interpolation between the inequalities from Theorem 5.2.8 c) and Lemma 5.2.9 a).

Remark 5.2.12. Lemma 5.2.9 suggests to define non-integer Sobolev spaces by Fourier series, i.e., for $\theta \in \mathbb{R}$ let

$$H^\theta_{per} = \mathcal{F}^{-1}\ell_{2,\theta} \quad \text{with} \quad \|u\|_{H^\theta} = \|\widehat{u}\|_{\ell_{2,\theta}}.$$

We will come back to this definition in §6.2.1. ⌋

Remark 5.2.13. (Real Fourier series) Besides the complex Fourier expansion also real Fourier polynomials and series of the form

$$u(x) = \frac{a_0}{2} + \sum_{k\in\mathbb{N}^d}^{\infty} [a_k \cos(k\cdot x) + b_k \sin(k \cdot x)],$$

with $a_k, b_k \in \mathbb{R}$, are in use, where

$$\begin{aligned} a_k =& \frac{2}{(2\pi)^d}\int_{\mathbb{T}^d} u(x)\cos(k\cdot x)\,\mathrm{d}x, \quad k \ge 0,\\ b_k =& \frac{2}{(2\pi)^d}\int_{\mathbb{T}^d} u(x)\sin(k\cdot x)\,\mathrm{d}x, \quad k \ge 1. \end{aligned}$$

The relations between $\widehat{u}_k$ and (a_k, b_k) are

$$\widehat{u}_0 = \frac{1}{2}a_0, \quad \widehat{u}_k = \frac{1}{2}(a_k - \mathrm{i}b_k), \quad \widehat{u}_{-k} = \frac{1}{2}(a_k + \mathrm{i}b_k),$$

$$a_k = \widehat{u}_k + \widehat{u}_{-k}, \quad b_k = (\widehat{u}_k - \widehat{u}_{-k})\mathrm{i}, \quad k \in \mathbb{N}.$$

For $u(x) \in \mathbb{R}$ we have $\widehat{u}_k = \overline{\widehat{u}}_{-k}$. In this book we prefer the concise complex notation. ⌋

Remark 5.2.14. (General periodic boxes) Let $L_1, \ldots, L_d > 0$, $\Omega = (0, L_1) \times \cdots \times (0, L_d)$. As in Theorem 5.2.8 we may expand $u \in L^2(\Omega)$ as

$$u(x) = \sum_{k\in\mathbb{Z}^d}^{\infty} \widehat{u}_k \mathrm{e}^{\mathrm{i}\omega_k\cdot x} \quad \text{with} \quad \omega_k = \left(\frac{2\pi k_1}{L_1}, \ldots, \frac{2\pi k_d}{L_d}\right),$$

where

$$\widehat{u}_k = \frac{1}{L_1 L_2 \cdots L_d}\int_\Omega u(x)\mathrm{e}^{-\mathrm{i}\omega_k\cdot x}\,\mathrm{d}x = (\mathcal{F}u)_k.$$

Again, $\mathcal{F}$ is an isomorphism between $H^m_{\mathrm{per}}(\Omega)$ and $\ell_{2,m}$. ⌋

Remark 5.2.15. Let $\Omega \subset \mathbb{R}^d$ be a bounded domain and let $(\varphi_j)_{j\in\mathbb{N}}$ be a complete orthonormal system in $L^2(\Omega)$, i.e., every $u \in L^2(\Omega)$ possesses a unique representation as convergent series in $L^2(\Omega)$, i.e., $u = \sum_{j\in\mathbb{N}} c_j\varphi_j$ with $c_j \in \mathbb{C}$. Then $\mathcal{F}$, defined by $(\mathcal{F}u)_j = c_j$, is an isomorphism between $L^2(\Omega)$ and ℓ_2. However, in general $\mathcal{F}$ is not an isomorphism between $H^m(\Omega)$ and $\ell_{2,m}$. The set $\{\sin nx : n \in \mathbb{N}\}$ is a basis of $L^2((0,\pi))$, but not a basis of $H^1((0,\pi))$. Since $H^1 \subset C_b^0$ in H^1, only functions u with $u(0) = u(\pi) = 0$ can be approximated. In L^2 the two points $x = 0, \pi$ are a null set. ⌋

Remark 5.2.16. Point-wise convergence of Fourier series is a rather delicate issue. For instance, the Fourier series of $u \in L^1$ may diverge almost everywhere [**Kol27**], while for $u \in L^2$ we have convergence almost everywhere [**Car66**]. No necessary and sufficient conditions are known for the point-wise convergence of the Fourier series of a function u. However, there are various sufficient conditions, for instance if u is piecewise C^1, then $S_n(x) \to u(x)$ at points of continuity. More generally,

$$S_n(x) \to \frac{1}{2}(u(x+) + u(x-)),$$

where $u(x+)$ and $u(x-)$ denote the right and the left limit of u in x. Thus, $S_n(x_0)$ converges to the mean of u at jump points x_0. This convergence comes with notable oscillations ($\approx 19\%$) to the left and right of x_0, which is known as Gibbs phenomenon, see Figure 5.4. ⌋

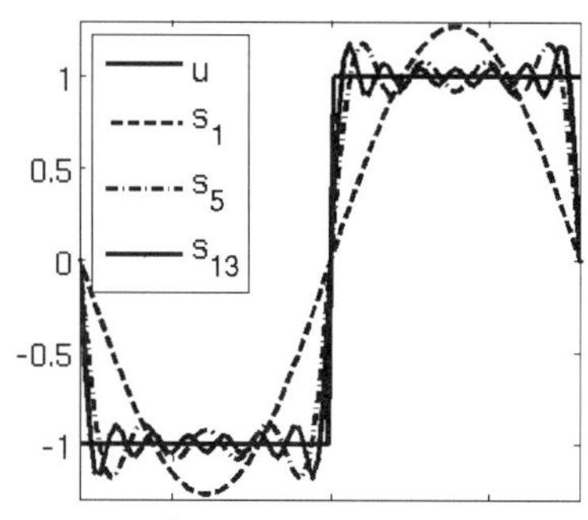

Figure 5.4. Fourier expansion of $u(x){=}-1$, $x \in (-\pi, 0)$, $u(x){=}1$, $x \in (0,\pi)$ yields $u(x) = \frac{4}{\pi}\sum_{n=1}^{\infty} \frac{1}{2n-1}\sin(2n-1)x$. The figure shows the partial sums s_1, s_5, s_{13} and thus illustrates the Gibbs phenomenon.

Fourier series allow giving simple proofs of classical inequalities when the functions involved are spatially periodic.

Lemma 5.2.17. (Poincaré's inequality) *For* $u \in H^1_{\text{per}}(\mathbb{T}^d, \mathbb{R})$ *with* $\int_{\mathbb{T}^d} u \,\mathrm{d}x = 0$ *we have*

$$\int_{\mathbb{T}^d} |u|^2 \,\mathrm{d}x \le \int_{\mathbb{T}^d} |\nabla u|^2 \,\mathrm{d}x. \tag{5.20}$$

Proof. Parseval's identity gives

$$\int_{\mathbb{T}^d} |u|^2 \, \mathrm{d}x = 2\pi \sum_{k\in\mathbb{Z}^d\setminus\{0\}} |\widehat{u}_k|^2 \le 2\pi \sum_{k\in\mathbb{Z}^d\setminus\{0\}} |k|^2 |\widehat{u}_k|^2 = \int_{\mathbb{T}^d} |\nabla u|^2 \, \mathrm{d}x. \qquad \square$$

Next we have the following version of Sobolev's embedding theorem in d space dimensions.

Lemma 5.2.18. *For $m - d/2 > n$ there exists a $C > 0$, such that*

$$\|u\|_{C^n(\mathbb{T}^d)} \le C\|u\|_{H^m(\mathbb{T}^d)}.$$

Proof. The assertion follows from Lemma 5.2.9 and Lemma 5.1.27. $\square$

Analytic properties of the solution operator of a linear evolution equation can be established with the help of Fourier series.

Example 5.2.19. We consider the solution operator $T(t)$ defined via the solution $u(x,t) = T(t)u_0)(x)$ of the linear heat equation $\partial_t u = \partial_x^2 u$, with $x \in [0,\pi]$, under Dirichlet boundary condition $u(0,t) = u(\pi,t) = 0$ to the initial value $u(x,0) = u_0(x)$. In order to prove that $(T(t))_{t\ge 0}$ is a C_0-semigroup in $L^2((0,\pi))$ and in $H^m((0,\pi))\cap H_0^1((0,\pi))$ for every $m \in \mathbb{N}$ we make an odd 2π-periodic extension of the functions with $u(0,t) = u(\pi,t) = 0$. The semigroup in the space of 2π-periodic functions is denoted again by $T(t)$. We proved in §5.1.2 that $\widehat{T}(t) = \mathcal{F}T(t)\mathcal{F}^{-1}$ defined by $(\widehat{T}(t)\widehat{u}(0))_{k\in\mathbb{Z}} = (\mathrm{e}^{-k^2 t}\widehat{u}_k(0))_{k\in\mathbb{Z}}$ is continuous in $\ell_{2,m}$, i.e., for every $\widehat{u}(0) \in \ell_{2,m}$ we have

$$\|\widehat{T}(t)\widehat{u}(0) - \widehat{u}(0)\|_{\ell_{2,m}} \to 0 \quad \text{for} \quad t \to 0\,.$$

Due to the isomorphism property of $\mathcal{F}$ between H^m_{per} and $\ell_{2,m}$, cf. Lemma 5.2.9 b) and c), it follows that

$$\begin{aligned} \|T(t)u(0) - u(0)\|_{H^m_{per}} &\le C_1\|\mathcal{F}^{-1}T(t)u(0) - \widehat{u}(0)\|_{\ell_{2,m}} \\ &= C_1\|\widehat{T}(t)\widehat{u}(0) - \widehat{u}(0)\|_{\ell_{2,m}} \to 0 \text{ for } t \to 0\,, \end{aligned}$$

i.e., $T(t)$ is a C_0-semigroup in H^m_{per}. The restriction of x to $[0,\pi]$ gives the result. Moreover, from Example 5.1.21 it is known that for $r \ge 0$ the semigroup $\widehat{T}(t)$ can be estimated by

$$\|\widehat{T}(t)\widehat{u}(0)\|_{\ell_{2,m+r}} \le \widetilde{C}\max(1, t^{-r/2})\|\widehat{u}(0)\|_{\ell_{2,m}}\,.$$

Using again that $\mathcal{F}$ is an isomorphism between H^m_{per} and $\ell_{2,m}$ shows

$$\begin{aligned} \|T(t)u_0\|_{H^{m+r}_{per}} \le& C_1\|\widehat{T}(t)\widehat{u}(0)\|_{\ell_{2,m+r}} \le C_1\widetilde{C}\max(1, t^{-r/2})\|\widehat{u}(0)\|_{\ell_{2,m}} \\ \le& C\max(1, t^{-r/2})\|u_0\|_{H^m_{per}}\,, \end{aligned}$$

with $C = C_1\widetilde{C}C_2$. ⌋

Example 5.2.20. We consider $\partial_t^2 u = \partial_x^2 u$ with $x \in (0, 2\pi)$ and periodic boundary conditions. We rewrite this equation as first order system for $z = (\partial_t u, \partial_x u)$ and obtain

$$\frac{\mathrm{d}}{\mathrm{d}t} z(x,t) = Az(x,t), \quad \text{with} \quad A = \begin{pmatrix} 0 & \partial_x \\ \partial_x & 0 \end{pmatrix},$$

or in Fourier space

$$\frac{\mathrm{d}}{\mathrm{d}t} \widehat{z}(k,t) = \widehat{A}\widehat{z}(k,t), \quad \text{with} \quad \widehat{A} = \begin{pmatrix} 0 & \mathrm{i}k \\ \mathrm{i}k & 0 \end{pmatrix}.$$

The general solution is given by

$$z(x,t) = \sum_{k \in \mathbb{Z}} c_1(k) \mathrm{e}^{\mathrm{i}k(t+x)} \widehat{z}_1 + c_2(k) \mathrm{e}^{\mathrm{i}k(-t+x)} \widehat{z}_2 =: \mathrm{e}^{tA} z(\cdot, 0)(x),$$

where

$$\begin{pmatrix} c_1(k) \\ c_2(k) \end{pmatrix} = \frac{1}{2} \begin{pmatrix} 1 & 1 \\ 1 & -1 \end{pmatrix} \widehat{z}(k,0).$$

Hence, we have a uniformly bounded C_0-semigroup for z in, e.g., $H^m \times H^m$, which however is not smoothing. ⌋

5.2.3. Some nonlinear PDE examples. We start this section with a version of the local existence and uniqueness theorem, Theorem 5.1.23, in physical space. In order to use the results from §5.1 we refrain from greatest generality and restrict ourselves to Sobolev spaces H^θ_{per} in accordance with Lemma 5.2.9. We consider

$$\frac{\mathrm{d}}{\mathrm{d}t} u = \Lambda u + N(u), \quad u|_{t=0} = u_0, \tag{5.21}$$

with Λ generating a C_0-semigroup which satisfies

$$\|\mathrm{e}^{t\Lambda} u\|_{H^\theta_{per}} \leq C_{\theta - r} \mathrm{e}^{\beta t} t^{-\alpha} \|u\|_{H^r_{per}} \tag{5.22}$$

for $\theta \geq r$ with constants $C_{\theta-r}$, β, and $\alpha \in [0,1)$. Moreover, let N be a locally Lipschitz-continuous map from H^θ into H^r.

Similar to Definition 5.1.19 we define

Definition 5.2.21. *a) A function $u \in C([0,T_0], H^\theta_{per})$ which satisfies*

$$u(t) = \mathrm{e}^{t\Lambda} u_0 + \int_0^t \mathrm{e}^{(t-\tau)\Lambda} N(u(\tau))\, \mathrm{d}\tau \tag{5.23}$$

is called a mild solution of (5.21).
b) A function $u \in C^1([0,T_0], H^\theta_{per})$, with $\Lambda u \in C([0,T_0], H^\theta_{per})$, is called a strong solution of (5.21), *if* (5.21) *holds in H^θ_{per} for each $t \in [0,T_0]$.*

Theorem 5.2.22. *For all $C_1 > 0$ there exists a $T_0 > 0$ such the following holds. For $u_0 \in H^\theta$ with $\|u_0\|_{H^\theta} \leq C_1$ there exists a unique solution $u \in C([0,T_0], H^\theta)$ of* (5.21) *with $u|_{t=0} = u_0$.*

Proof. This a direct consequence of Theorem 5.1.23, Remark 5.1.24, and Lemma 5.2.9. □

There are other straightforward generalizations from the finite- to the infinite-dimensional situation.

Theorem 5.2.23. *Consider* (5.21), *where* (5.22) *is satisfied for a* $\beta < 0$. *Then the fixed point* $u^* = 0$ *is asymptotically stable.*

Proof. The proof goes line for line as the proof of Theorem 2.3.4 a). □

In order to check the assumptions for system (5.21) for a concrete nonlinear PDE we have to handle products of functions in physical space. The H^θ_{per}-spaces are closed under multiplication if θ is sufficiently big.

Lemma 5.2.24. *For all* $\theta > d/2$ *there exists a* $C > 0$, *such that for all* $u, v \in H^\theta_{per}$ *we have*

$$\|uv\|_{H^\theta} \leq C\|u\|_{H^\theta}\|v\|_{H^\theta}.$$

Proof. This follows from Lemma 5.1.28 by using the isomorphism $\mathcal{F} : H^\theta_{per} \to \ell_{2,\theta}$. □

We come back to the PDEs introduced in Example 5.1.29. We have already proved the local existence and uniqueness of solutions of the Fourier transformed versions. The isomorphism property between $\ell_{2,\theta}$ and H^θ_{per} gives the following result.

Theorem 5.2.25. *For the KPP equation, the Allen-Cahn equation, the NLS equation, the Burgers equation, and the GL equation with* 2π*-periodic boundary conditions we have the local existence and uniqueness of solutions in* H^θ_{per} *if* $\theta > 1/2$, *i.e., for all* $C > 0$ *there exists a* $T_0 > 0$ *such the following holds. For* $u_0 \in H^\theta_{per}$ *with* $\|u_0\|_{H^\theta} \leq C$ *there exists a unique solution* $u \in C([0, T_0], H^\theta_{per})$ *with* $u|_{t=0} = u_0$, *respectively,* $A \in C([0, T_0], H^\theta_{per})$ *with* $A|_{T=0} = A_0$.

The θ in the last theorem can be made smaller by using the smoothing properties of the semigroup, cf. §6.2.1. Moreover, the smoothing estimate (5.11) can be used to show that solutions to the KPP equation, the Allen-Cahn equation, and the GL equation become arbitrary smooth and even analytic for $t > 0$. This is done for instance in §5.3.3 or §6.2.2.

5.3. The Chafee-Infante problem

After having discussed the local existence and uniqueness theory of PDEs on an interval we now consider the qualitative behavior of solutions in a specific example, namely the Chafee-Infante problem [**CI75**]. The presentation is based on [**Hen81**, §5.3].

The Chafee-Infante problem is to find the attractor of a semi-linear parabolic PDE, the Allen-Cahn equation,

$$\partial_t u = \partial_x^2 u + \alpha u - u^3, \tag{5.24}$$

with $\alpha \in \mathbb{R}$, $u = u(x,t) \in \mathbb{R}$, $t \geq 0$, and $x \in (0,\pi)$, under Dirichlet boundary conditions $u(0,t) = u(\pi,t) = 0$. Our goal is to characterize the attractor of this system for different values of α. This PDE can be interpreted as an infinite-dimensional gradient system. Similar to finite-dimensional gradient systems, see §2.4.5, this fact restricts the elements of the attractor in the following to fixed points and heteroclinic connections.

5.3.1. Local and global existence of solutions. As phase space we use $H_0^1 = H_0^1(0,\pi)$. Solving the Allen-Cahn equation with 2π-periodic boundary conditions and restricting to the invariant subspace of odd functions is the same as solving the Allen-Cahn equation with Dirichlet boundary conditions. Hence, Theorem 5.2.25 applies and we have the local existence and uniqueness of solutions in H^θ if $\theta > 1/2$.

Theorem 5.3.1. *For all $C > 0$ there exists a $T_0 > 0$ such that for all $u_0 \in H_0^1$ with $\|u_0\|_{H^1} \leq C$ there exists a unique solution $u \in C([0,T_0], H_0^1)$ of the Allen-Cahn equation (5.24) with $u|_{t=0} = u_0$.*

To prove the global existence of solutions it is sufficient to bound the H^1-norm. We prove more, namely the existence of an absorbing set for (5.24).

Theorem 5.3.2. (Global existence and existence of an absorbing set) *For all $\alpha \in \mathbb{R}$ there exists a $R > 0$ such that for all $C_1 \geq 0$ we have a $T > 0$ such that the followings holds. If $u_0 \in H_0^1$ satisfies $\|u_0\|_{H^1} \leq C_1$, then the associated solution satisfies $u(t) \in B = \{u \in H_0^1 : \|u\|_{H^1} \leq R\}$ for all $t \geq T$.*

Proof. Again the solutions are extended to odd 2π-spatially periodic solutions. Then we have

$$\frac{\mathrm{d}}{\mathrm{d}t}\int_0^{2\pi} u^2 \,\mathrm{d}x = 2\int_0^{2\pi} -(\partial_x u)^2 + \alpha u^2 - u^4 \,\mathrm{d}x \tag{5.25}$$

$$\frac{\mathrm{d}}{\mathrm{d}t}\int_0^{2\pi} (\partial_x u)^2 \,\mathrm{d}x = 2\int_0^{2\pi} -(\partial_x^2 u)^2 + \alpha(\partial_x u)^2 - 3u^2(\partial_x u)^2 \,\mathrm{d}x. \tag{5.26}$$

If $\alpha < 0$ all terms on the right-hand side are negative and we have

$$\limsup_{t\to\infty} \|u(t)\|_{H^1} = 0.$$

In order to obtain estimates which are also good for small $\alpha \geq 0$ we split the parameter regime $\alpha \geq 0$ in two parts. First let $\alpha \in [0, 1/2]$. Adding (5.25) and (5.26) yields

$$
\begin{aligned}
\frac{\mathrm{d}}{\mathrm{d}t}\int_0^{2\pi} & u^2 + (\partial_x u)^2 \,\mathrm{d}x \\
&= 2\int_0^{2\pi} -(\partial_x^2 u)^2 + (\alpha-1)(\partial_x u)^2 - 3u^2(\partial_x u)^2 + \alpha u^2 - u^4 \,\mathrm{d}x \\
&\le 2\int_0^{2\pi} -(\partial_x u)^2/2 + \alpha u^2 - u^4 \,\mathrm{d}x \\
&\le 2\int_0^{2\pi} -(\partial_x u)^2/2 - u^2/2 + 1/8 + \alpha^2/2 \,\mathrm{d}x \\
&\le -\int_0^{2\pi} u^2 + (\partial_x u)^2 \,\mathrm{d}x + \pi.
\end{aligned}
$$

This immediately shows that

$$\limsup_{t\to\infty} \|u(t)\|_{H^1}^2 \le \pi.$$

For $\alpha > 1/2$ we consider

$$
\begin{aligned}
\frac{\mathrm{d}}{\mathrm{d}t}\int_0^{2\pi} & 2\alpha u^2 + (\partial_x u)^2 \,\mathrm{d}x \\
&= 2\int_0^{2\pi} -(\partial_x^2 u)^2 - \alpha(\partial_x u)^2 - 3u^2(\partial_x u)^2 + 2\alpha^2 u^2 - 2\alpha u^4 \,\mathrm{d}x \\
&\le 2\int_0^{2\pi} -\alpha(\partial_x u)^2 + 2\alpha^2 u^2 - 2\alpha u^4 \,\mathrm{d}x \\
&\le 2\int_0^{2\pi} -\alpha(\partial_x u)^2 - 2\alpha^2 u^2 + 2\alpha^3 \,\mathrm{d}x.
\end{aligned}
$$

Hence, for $E = \int_0^{2\pi} 2\alpha u^2 + (\partial_x u)^2 \,\mathrm{d}x$ we have $E' \le -2\alpha E + 8\pi\alpha^3$ and thus

$$E(t) \le \mathrm{e}^{-2\alpha t} E(0) + \frac{8\pi\alpha^3}{2\alpha}(1 - \mathrm{e}^{-\alpha t}).$$

Hence, $\limsup_{t\to\infty} E(t) \le 4\pi\alpha^2$. Since $\|u\|_{H^1} \le (\int 2\alpha u^2 + (\partial_x u)^2 \,\mathrm{d}x)^{1/2}$ for $\alpha > 1/2$, we are done. $\square$

5.3.2. Existence of the attractor. The existence proof of attractors in finite dimensions uses the argument that a bounded sequence contains a convergent subsequence. In infinite dimensions this is in general no longer true. Hence, this compactness argument has to be recovered by using the smoothing properties of the solution operator $\mathcal{S}_t$ with $\mathcal{S}_t u_0 = u(\cdot, t)$.

Lemma 5.3.3. *For $t > 0$ fixed the solution operator $\mathcal{S}_t$ maps bounded balls of H_0^1 into bounded balls of $H^2 \cap H_0^1$.*

Proof. We consider the variation of constant formula and estimate

$$\begin{aligned}\|u(t)\|_{H^2} \le& \|T(t)u_0\|_{H^2} + \int_0^t \|T(t-\tau)u^3(\tau)\|_{H^2}\,\mathrm{d}\tau \\ \le& Ct^{-1/2}\|u_0\|_{H^1} + \int_0^t C(t-\tau)^{-1/2}\,\mathrm{d}\tau \sup_{\tau\in[0,t]} \|u(\tau)\|_{H^1}^3 < \infty,\end{aligned} \tag{5.27}$$

where $T(t)$ is the semigroup from Example 5.2.19. □

Since $H^2 \cap H_0^1$ is compactly embedded in H_0^1 we have compactness of the operator $\mathcal{S}_t$ in H_0^1. Theorem 5.3.2 thus shows that the Chafee-Infante problem (5.24) defines a dissipative dynamical system such that Theorem 2.4.4 applies.

Theorem 5.3.4. *For the Chafee-Infante problem* (5.24) *there exists a non-empty, compact, time-invariant set* $\mathcal{A} = \omega(B) \subset H_0^1$*, the global attractor, for which*

$$\operatorname{dist}(u(t,B),\mathcal{A}) = \sup_{b\in B}\inf_{a\in\mathcal{A}} \|u(t,b)-a\|_{H^1} \to 0 \quad \textit{for} \quad t\to\infty,$$

where B is the absorbing set from Theorem 5.3.2.

Proof. For convenience we repeat the main steps from the proof of Theorem 2.4.4. The attractor is defined by

$$\mathcal{A} = \bigcap_{t\ge 0} A_t$$

with $A_t = \operatorname{clos}_{H^1}(\mathcal{S}_t(B))$. Since B is positively invariant, the family $(A_t)_{t\ge0}$, satisfies $A_{t_1} \subset A_{t_2}$ for $t_1 > t_2$. Hence, $\mathcal{A} \subset A_0$ is bounded. Since $\mathcal{S}_t$ is a compact operator for $t > 0$, the set A_t is compact for $t > 0$. Since $(A_t)_{t\ge0}$ is a decreasing family of compact non-empty sets, the attractor $\mathcal{A} = \cap_{t\ge0} A_t$ is non-empty and compact.

We skip the proof of the time invariance and restrict ourselves to the attractivity which is proved by contradiction. We assume that B is not attracted by $\mathcal{A}$. Then there exists a $\delta > 0$, sequences $t_n \to \infty$ and $u_n \in B$, such that $\operatorname{dist}(\mathcal{S}_{t_n}(u_n),\mathcal{A}) > \delta > 0$ for all $n \in \mathbb{N}$. For a small $t > 0$ the sequence $\mathcal{S}_{t_n - t}(u_n)$, $(n \in \mathbb{N})$ is bounded. Since $\mathcal{S}_t$ is a compact operator there exists a subsequence such that $v_j = \mathcal{S}_{t_{n_j}}(u_{n_j})$ converges towards a w for $j \to \infty$. Therefore, $w \in \mathcal{A}$ which contradicts the above assumption that the sequence is bounded away from $\mathcal{A}$. □

5.3.3. The choice of regularity does not matter (much).

As already said, major differences between finite-dimensional systems and infinite-dimensional systems are due to the fact that in finite dimensions all norms are equivalent, whereas in infinite dimensions there are infinitely many non-equivalent norms and so infinitely many possible non-equivalent phase spaces. More or less all definitions in the theory of dynamical systems, such as continuity of solutions w.r.t. time, stability of solutions, etc. depend on the chosen norm. Therefore, we expect that the choice of a suitable phase space in infinitely many dimensions in general plays a crucial role. It is the purpose of this section to explain that for systems with smoothing properties this often is not the case. If there is a global bound in one H^θ-space, then it does not matter which H^θ-space is chosen as long as these spaces are connected with a smoothing estimate.

The estimate (5.27) can be generalized to

$$\begin{aligned}\|u(t)\|_{H^{\theta+1}} \leq& \|T(t-\tau)u(\tau)\|_{H^{\theta+1}} + \int_\tau^t \|T(t-s)u^3(s)\|_{H^{\theta+1}}\,\mathrm{d}s \\ \leq& C(t-\tau)^{-1/2}\|u(\tau)\|_{H^\theta} + \int_\tau^t C(t-s)^{-1/2}\,\mathrm{d}s \sup_{s\in[\tau,t]} \|u(s)\|_{H^\theta}^3 \\ \leq& C(t-\tau)^{-1/2}C_\theta(\tau) + 2C(t-\tau)^{1/2}C_\theta(\tau)^3,\end{aligned}$$

where $C_\theta(\tau) = \sup_{s\in[\tau,\infty]} \|u(s)\|_{H^\theta}$. Hence

$$C_{\theta+1}(t) \leq C(t-\tau)^{-1/2}C_\theta(\tau) + 2C(t-\tau)^{1/2}C_\theta(\tau)^3. \tag{5.28}$$

From Theorem 5.3.2 we know that $C_1(0) < \infty$ and that $\limsup_{\tau\to\infty} C_1(\tau) \leq R$. In H^2 we have the local existence and uniqueness of solutions, i.e., for $u_0 \in H^2$ with $\|u_0\|_{H^2} \leq \widetilde{C}$ there is a $T_2 > 0$ and a $\widetilde{C}_2 < \infty$ such that the solutions exist for all $t \in [0, T_2]$ and $\sup_{t\in[0,T_2]} \|u(t)\|_{H^2} \leq \widetilde{C}_2$. Moreover, we get

$$C_2(T_2+\tau) \leq CT_2^{-1/2}C_1(\tau) + 2CT_2^{1/2}C_1(\tau)^3 < \infty.$$

Combining the last two estimates shows that

$$\sup_{t\in[0,\infty)} \|u(t)\|_{H^2} \leq \max(\widetilde{C}_2, C_2(T_2)) < \infty.$$

Moreover, choosing $t-\tau = 1$ in (5.28) yields

$$\limsup_{t\to\infty} \|u(t)\|_{H^2} \leq CR + 2CR^3.$$

Hence, we have established an absorbing set in H^2. With the same procedure we show the existence of an absorbing set in H^3, etc.. As a consequence the attractor exists in each H^θ and the choice of phase space is not crucial with this respect.

5.3.4. Characterization and bifurcation of the attractor. Here we give a characterization of the attractor for different values of α. In a first step we write (5.24) as a gradient system, cf. §2.4.5. We have

$$\partial_t u = \partial_x^2 u + \alpha u - u^3 = -\beta\partial_u V(u)$$

with potential

$$V(u) = \int_0^\pi \frac{1}{2}(\partial_x u(x))^2 - \frac{\alpha}{2}u(x)^2 + \frac{1}{4}u(x)^4 \, \mathrm{d}x$$

and β a linear map defined below. In order to justify this formula, first recall that for a function $V : \mathbb{R}^d \to \mathbb{R}$ we have

$$V(u + \varepsilon v) = V(u) + \varepsilon a^T v + \mathcal{O}(\varepsilon^2) = V(u) + \varepsilon \langle a, v \rangle + \mathcal{O}(\varepsilon^2) \text{ for all } v \in \mathbb{R}^d,$$

where $\langle u, v \rangle = u^T v$ is the scalar product between the vectors u and v, i.e., the derivative is defined as an element of the dual space of $\mathbb{R}^d$. However, it can be identified with $\mathbb{R}^d$ through the map

$$\beta : \mathrm{Lin}(\mathbb{R}^\mathrm{d}, \mathbb{R}) \to \mathbb{R}^d, \qquad \langle a, \cdot \rangle \mapsto a.$$

For a map $V : X \to \mathbb{R}$ where the function space X is equipped with the scalar product

$$\langle u, v \rangle = \int_0^\pi u(x)\overline{v(x)} \, \mathrm{d}x$$

we define the map

$$\beta : \mathrm{Lin}(\mathrm{X}, \mathbb{R}) \to X, \qquad \langle a, \cdot \rangle \mapsto a.$$

This is well defined since in Hilbert spaces the dual space $\mathrm{Lin}(\mathrm{X}, \mathbb{R})$ can be identified with X by the Riesz representation theorem [**Alt16**, Satz 4.1]. Using the boundary conditions and integration by parts we find

$$\begin{aligned}
&\lim_{\varepsilon \to 0} \frac{V(u + \varepsilon v) - V(u)}{\varepsilon} \\
&\quad = \lim_{\varepsilon \to 0} \varepsilon^{-1} \int_0^\pi \Big(\frac{1}{2}(\partial_x (u + \varepsilon v))^2 - \frac{\alpha}{2}(u + \varepsilon v)^2 + \frac{1}{4}(u + \varepsilon v)^4\Big) \\
&\qquad\qquad - \Big(\frac{1}{2}(\partial_x u)^2 - \frac{\alpha}{2}u^2 + \frac{1}{4}u^4\Big) \, \mathrm{d}x \\
&\quad = - \int_0^\pi (\partial_x^2 u + \alpha u - u^3) v \, \mathrm{d}x
\end{aligned}$$

and so by comparison

$$\beta \partial_u V(u) = -(\partial_x^2 u + \alpha u - u^3).$$

Therefore, (5.24) is a gradient system in H_0^1, and hence the function $t \mapsto V(u(t))$ decreases along solutions $u = u(t)$, i.e., $\frac{\mathrm{d}}{\mathrm{d}t} V(u(t)) \le 0$, where equality only holds in fixed points. Consequently, no non-trivial periodic solution can occur. Moreover, V is bounded from below, since

$$V(u) = \int_0^\pi \frac{1}{2}(\partial_x u(x))^2 - \frac{\alpha}{2}u(x)^2 + \frac{1}{4}u(x)^4 \, \mathrm{d}x \ge - \int_0^\pi \frac{\alpha^2}{4} \, \mathrm{d}x = - \frac{\pi \alpha^2}{4}.$$

Similar to the finite-dimensional situation, cf. Theorem 2.4.15, the attractor consists of the fixed points and their unstable manifolds, cf. [**Rob01**, Theorem 10.13]. In case that only finitely many fixed points exist, the attractor consists of these fixed points and their heteroclinic connections. This can be seen directly. In a gradient system every solution must end in a fixed point. Solutions in the attractor must also start in one of the finitely many fixed points. This follows from the fact that backwards in time the system in the attractor is a gradient system, too. The potential is given by $-V$ and it is bounded on the attractor.

We compute the fixed points, or stationary solutions, of the PDE, which satisfy

$$\partial_x^2 u + \alpha u - u^3 = 0.$$

Due to the boundary conditions $u(0) = u(\pi) = 0$ in the (u, u')-plane we have to find solutions which start from the $v = u'$-axis, end on this axis, and need for this part of the orbit the 'time' $x = \pi$. For all $\alpha > 0$ the phase portrait looks qualitatively the same. The periodic orbits around the origin have a periodicity which is i) minimal at the origin, namely the periodicity of the linearization, $2\pi/\sqrt{\alpha}$, ii) infinity at the heteroclinic orbits, and iii) which increases strictly monotonic with the distance from the origin. Thus, non-trivial equilibria of (5.24) can only exist for $\alpha > 1$ since for $\alpha \leq 1$ the solutions are too slow to make half of the periodic orbit in a time π.

Using i)-iii) the complete bifurcation picture can be established in a rigorous way. The number of solutions with $u(0) = u(\pi) = 0$ changes for $m\pi/\sqrt{\alpha} = \pi$ with $m \in \mathbb{N}$, an integer multiple of half the minimal period. As a consequence, for $\alpha \in (-\infty, 1]$ we have one equilibrium, the origin; for $\alpha \in (1, 4]$ we have 3 equilibria, the origin, and two equilibria called $u_{\pm 1}$; for $\alpha \in (4, 9]$ we have 5 equilibria, $\ldots$; and for $\alpha \in (m^2, (m+1)^2]$ we have $2m + 1$ equilibria, the origin, $u_{\pm 1}, \ldots$, and $u_{\pm m}$.

Hence, for fixed α there are only finitely many fixed points which are elements of the attractor. In order to understand the dynamics in the attractor, i.e., to find the heteroclinic connections between the fixed points, we analyze the linearization at the fixed point $u \equiv 0$, i.e.,

$$\partial_t u = \partial_x^2 u + \alpha u,$$

with $u(0, t) = u(\pi, t) = 0$, or equivalently, with $u(x, t) = \sum_{n \in \mathbb{N}} \widehat{u}_n(t) \sin(nx)$,

$$\frac{\mathrm{d}}{\mathrm{d}t} \widehat{u}_n = (\alpha - n^2) \widehat{u}_n.$$

Therefore, the linear operator

$$\Lambda \cdot = \partial_x^2 \cdot + \alpha \cdot$$

with Dirichlet boundary conditions has eigenvectors $u(x) = \sin mx$ with associated eigenvalues $\lambda = \alpha - m^2$ for $m \in \mathbb{N}$. Equivalently the infinite-dimensional diagonal matrix

$$(\widehat{\Lambda}_{nm})_{n,m \in \mathbb{N}} = ((\alpha - m^2)\delta_{nm})_{n,m \in \mathbb{N}}$$

has the eigenvectors e_m defined by $(e_m)_n = \delta_{mn}$ with associated eigenvalues $\lambda = \alpha - m^2$.

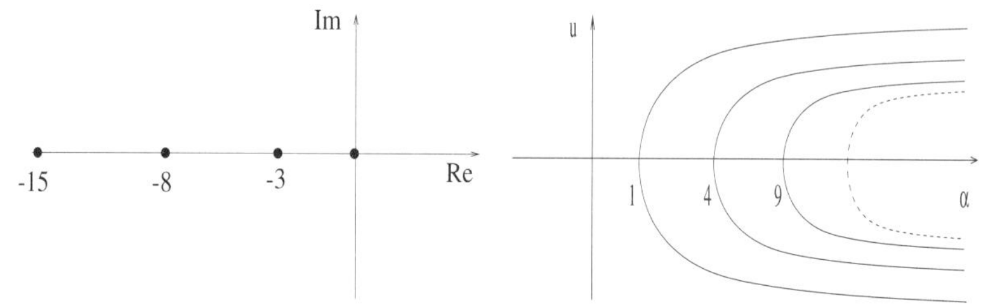

Figure 5.5. Left: the spectrum of the operator $\Lambda \cdot = \partial_x^2 \cdot + \alpha \cdot$ under Dirichlet boundary conditions for $\alpha = 1$. Right: the bifurcation diagram. At the parameter values $\alpha = n^2$ unstable equilibria bifurcate via a pitchfork bifurcation from the trivial branch $u(\alpha) \equiv 0$.

Hence, for $\alpha \in (-\infty, 1)$ the origin is asymptotically stable. For $\alpha > 1$ the origin is unstable, with a one-dimensional unstable manifold for $\alpha \in (1, 4]$, with a two-dimensional unstable manifold for $\alpha \in (4, 9]$, and with an m-dimensional unstable manifold for $\alpha \in (m^2, (m+1)^2]$.

For $\alpha \in (1, 4)$ the one-dimensional unstable manifold of the origin ends in the stable equilibria $u_{\pm 1}$. For $\alpha \in (4, 9)$ the equilibria $u_{\pm 1}$, $u_{\pm 2}$ lie on the two-dimensional unstable manifold of the origin. Since $u_{\pm 2}$ bifurcates from the unstable origin, these fixed points are also unstable and their one-dimensional unstable manifold ends in $u_{\pm 1}$.

The reasons are as follows. Since for fixed α the fixed points u_j are isolated and since the linearization only has real eigenvalues due to the gradient structure, no eigenvalue of the linearization around the equilibria u_j crosses the imaginary axis after the bifurcation when α is increased. Therefore, the dimension of the unstable manifold of u_j is the same as at their bifurcation point from the trivial branch. The fixed points $u_{\pm 1}$ bifurcating at $\alpha = 1$ are always stable. The fixed points $u_{\pm 2}$ bifurcating at $\alpha = 4$ have a one-dimensional unstable manifold which ends in the fixed points $u_{\pm 1}$. The fixed points $u_{\pm 3}$ bifurcating at $\alpha = 9$ have a two-dimensional unstable manifold and so heteroclinic connections to the fixed points to $u_{\pm 1}$ and $u_{\pm 2}$ exist. Figure 5.6 sketches the dynamics in the attractor of (5.24). For $\alpha \in (n^2, (n+1)^2)$ we have an attractor of dimension n consisting of finitely many fixed points and heteroclinic orbits between these fixed points, in particular, it contains the n-dimensional unstable manifold of the origin. A local bifurcation analysis via center manifold reduction can be found in §13.2.1.

Further Reading. Our point of view of PDEs over bounded sets as countably many ODEs is similar to [**Hal88, Rob01, KP13**], while [**Paz83**] gives

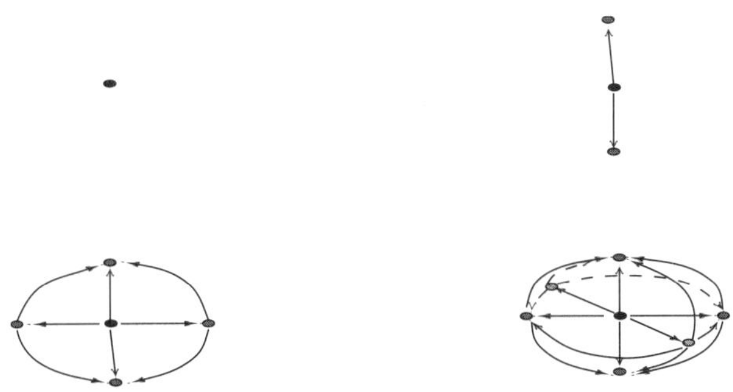

Figure 5.6. The finite-dimensional attractors consisting of fixed points and connecting orbits for $\alpha < 1$, $1 < \alpha < 4$, $4 < \alpha < 9$, and $9 < \alpha < 16$. Only selected heteroclinic orbits have been plotted. The attractor always contains the n-dimensional unstable manifold of the origin.

an excellent and concise account of the general semigroup approach, see also [**RR04**, Chapter 11]. Fourier series can be found in most textbooks on analysis and functional analysis, and in many books on PDE; we recommend [**Olv14**, Chapter 3] for an introduction with a PDE point of view, and [**Duo01**] for a concise but comprehensive treatment. Classical books on linear functional analysis, covering much wider ground than what is used here and in the following chapters are [**Yos71, RS75a, Kat95**]; our favorites are [**Rud73, Wer00, Alt16**]. A concise introduction to nonlinear functional analysis is [**AA11**]. Our presentation of function spaces follows [**Alt16, Wlo87**], but the same material can be found in many textbooks, for instance, from a PDE perspective, in [**Str92, RR04, Eva98, Sal08**]. Comprehensive treatments of distributions and Sobolev spaces, including various versions of Poincaré inequalities and Sobolev imbeding theorems, are given in [**Hör83, Maz11**], and [**Tay96**, Chapters 3,4,13]. See also [**Geo15**] for a concise introduction aimed at graduate students, and Section 7.3 of this book for the case of unbounded domains.

Exercises

5.1. Prove that the space $c_{00} = \{u : \mathbb{Z} \to \mathbb{R} : u_n \neq 0 \text{ for finitely many } n\}$ equipped with the ℓ_1-norm is not complete.

5.2. Consider $\frac{\mathrm{d}^2}{\mathrm{d}t^2}u_n = -\omega_n^2 u_n$, $n \in \mathbb{N}$, with $u_n(t) \in \mathbb{R}$, and $\omega_n \in \mathbb{R}$. Write the equation as first order system and find some phase space where the infinitely many first order ODEs define a C_0-semigroup. Under which additional assumptions on the ω_n is the semigroup uniformly continuous, differentiable, or analytic?

5.3. Work out the details for Example 5.1.15.

5.4. Let $(H, \langle\cdot,\cdot\rangle)$ be a Hilbert space and (e_n) an orthonormal basis of H. Let $\lambda_n > 0$, $\lambda_n \le \lambda_{n+1}$ and $\lim_{n\to\infty} \lambda_n = \infty$. Finally, let

$$V := \{u \in H : \sum_{n\in\mathbb{N}} \lambda_n \langle u, e_n\rangle^2 < \infty\}.$$

Show that $a(u,v) = \sum_{n\in\mathbb{N}} \lambda_n \langle u, e_n\rangle \langle v, e_n\rangle$ defines a scalar product in V such that V is a Hilbert space, and $V \subset H$ compact.

5.5. Let $(H, \langle\cdot,\cdot\rangle)$ be a Hilbert space with an orthonormal basis $(e_n)_{n\in\mathbb{N}}$. Let $\lambda_n > 0$, $\lambda_n \le \lambda_{n+1}$ and $\lim_{n\to\infty} \lambda_n = \infty$. For $t \ge 0$ let $T(t) : H \to H$ be defined by $T(t)u = \sum_{n\in\mathbb{N}} \mathrm{e}^{-\lambda_n t} \langle u, e_n\rangle e_n$. Show that $(T(t))$ is a C_0-semigroup, and that $\lim_{t\searrow 0} \frac{1}{t}(T(t)u - u) =: Au$ exists iff $\sum_{n\in\mathbb{N}} \lambda_n^2 \langle u, e_n\rangle^2 < \infty$.
Finally, for $v = \sum_{n\in\mathbb{N}} \lambda_n \langle u, e_n\rangle e_n$ show that $T(t)u - u = \int_0^t T(s)v\,\mathrm{d}s$.

5.6. Prove local existence and uniqueness of solutions for $\partial_t u_n = -n^4 u_n + u_n^3$, with $n \in \mathbb{Z}$, in spaces $\ell_{p,\theta}$. Do we have global existence and uniqueness of solutions?

5.7. Consider the discrete NLS equation

$$\mathrm{i}\partial_t u_n = \varepsilon(u_{n+1} - 2u_n + u_{n-1}) + |u_n|^2 u_n,$$

with $u_n(t) \in \mathbb{C}$ and $n \in \mathbb{Z}$ for $0 \le \varepsilon \le 1$.

a) Show the local existence and uniqueness of solutions in ℓ_2. Note that the linear part is not diagonal, but bounded. Show that the system conserves the ℓ_2-norm. Conclude the global existence and uniqueness of solutions in ℓ_2 from this fact.

b) For $\varepsilon = 0$ find non-trivial solutions U_0 of the form $u_0(t) = r_0 \mathrm{e}^{\mathrm{i}\omega t}$ with $r_0 \in \mathbb{R}$ and $u_n = 0$ for all other $n \in \mathbb{Z}$. Use the implicit function theorem for instance in ℓ_∞ to prove that for $\varepsilon > 0$ there are solutions U_ε of the form $u_n(t) = r_n \mathrm{e}^{\mathrm{i}\omega t}$ nearby U_0. What additional information is gained if $\ell_{\infty,\theta}$ instead of ℓ_∞ is used?

5.8. Show that any $f \in C^1(\mathbb{R}^d, \mathbb{R}^d)$ is locally Lipschitz continuous.

5.9. a) Let $I \subset \mathbb{R}$ be a compact interval and $\alpha \in (0,1)$. Show that $C^{0,\alpha}(I)$ is a Banach space.
b) Let $I = [0,1]$ and $0 < \beta < \alpha < 1$. Show that $C^{0,\alpha}(I) \subset C^{0,\beta}(I)$ as a proper subset.

5.10. Let $\Omega = (0,1)$. Find sequences (u_n), $u_n : \Omega \to \mathbb{R}$ such that:
(a) (u_n) bounded in $H^1(\Omega)$, but (u_n) does not converge in $L^2(\Omega)$.
(b) $u_n \to 0$ in $L^2(\Omega)$ but $u_n(x) \not\to 0$ for all $x \in \Omega$.
(c) $\|u_n\|_{L^2} = 1$ for all $n \in \mathbb{N}$ and $u_n \to 0$ weakly in $L^2(\Omega)$.
Remark. Since $H^1(\Omega) \subset L^2(\Omega)$ compact, in (a) we always have $u_{n_k} \to v$ in $L^2(\Omega)$ for a subsequence u_{n_k}.

5.11. Let $\Omega = B_{1/2}^{\mathbb{R}^2}(0)$. Show that $u(x) = \ln|\ln|x|| \in H^1(\Omega) \setminus L^\infty(\Omega)$.

5.12. Let $d \in \mathbb{N}$ and $\Omega = B_1(0)$ in $\mathbb{R}^d$. For which α do we have (a) $|x|^\alpha \in H^1(\Omega)$; (b) $(\sin|x|)^\alpha \in H^1(\Omega)$; (c) $(\ln|x|)^\alpha \in H^1(\Omega)$?

5.13. For $H = L^2(0,1)$ define $F : H \to \mathbb{R}$ by $F(u) = \int_0^{1/2} u(x)\,\mathrm{d}x$. Do we have $F \in H'$? If so, find a representation $\langle\cdot, v\rangle_{L^2}$ of F with $v \in H$.

5.14. For the following PDEs with $x \in (0, 2\pi)$ and periodic boundary conditions investigate whether the solution operator defines a C_0-semigroup in $L^2_{per}((0, 2\pi), \mathbb{R})$ with smoothing properties

$$a)\ \partial_t u = \partial_x^4 u, \qquad b)\ \partial_t u = -\partial_x^4 u, \qquad c)\ \partial_t u = \partial_x^3 u.$$

5.15. Consider the complex GL equation

$$\partial_t u = (1 + \mathrm{i}\alpha)\partial_x^2 u + Ru - (1 + \mathrm{i}\beta)\,|u|^2\, u$$

with 2π-periodic boundary conditions, $u(x,t) \in \mathbb{C}$, and $\alpha, \beta \in \mathbb{R}$. In case $|\beta| < 1/3$ prove the global existence of solutions in H^1 for all $R \in \mathbb{R}$.

5.16. Consider $\partial_t u = \partial_x^2 u + u^3$ for $t \geq 0$, $x \in (0, \pi)$ and $u(x,t) \in \mathbb{R}$ with boundary condition $u(0,t) = u(\pi,t) = 0$ and initial condition $u(x,0) = \phi(x)$, cf. [**Hen81**, Page 49]. Prove that there are solutions which converge in finite time towards ∞. Hint: Derive a differential inequality for $s(t) = \int_0^\pi \sin(x)u(x,t)dx$. With Hölder's inequality we obtain $s(t) \leq 2^{2/3}(\int_0^\pi \sin(x)u^3(x,t)dx)^{1/3}$ and so $\frac{\mathrm{d}}{\mathrm{d}t}s \geq -s + s^3/4$.

5.17. Write $\partial_t u = -\partial_x^4 u + \sin(u)$, with $u(x,t) \in \mathbb{R}$, $x \in \mathbb{R}$, and $t \geq 0$, subject to periodic boundary conditions $u(x,t) = u(x + 2\pi, t)$ as a gradient system.

5.18. Consider the Cahn-Hilliard equation $\partial_t u = \partial_x^2(-\gamma\partial_x^2 u - u + u^3)$, with $u(x,t) \in \mathbb{R}$, $\gamma > 0$, and 2π-periodic boundary conditions.

a) Prove that $\frac{\mathrm{d}}{\mathrm{d}t}C = 0$, where $C(t) = \int_0^{2\pi} u(x,t)\,\mathrm{d}x$.

b) Let $F(u) = \int_0^{2\pi} \frac{1}{4}(u^2 - 1)^2 + \frac{\gamma}{2}\,|\partial_x u|^2\ \mathrm{d}x$ and show that $\frac{\mathrm{d}}{\mathrm{d}t}F = -\int |\nabla w|^2\ \mathrm{d}x$ with $w = u^3 - u - \gamma\partial_x^2 u$.

c) Find the possible ω-limit sets.

Chapter 6

The Navier-Stokes equations

6.1. Introduction

In this chapter we give an introduction to the Euler and Navier-Stokes equations, which over unbounded domains will also play a role in subsequent chapters. The global existence and uniqueness of solutions of the three-dimensional (3D) Navier-Stokes equations is one of the seven so called 'one million dollar' or millennium problems in mathematics presented by the Clay Mathematics Institute in the year 2000. There are a number of reasons for this choice. On the one hand, the solution of this problem would allow us to understand and simulate the motion of fluids more rigorously. On the other hand, the 3D Navier-Stokes equations are interesting PDEs which resisted so far all attempts to prove the global existence and uniqueness of solutions.

Their history goes back a long way. The equations describing the motion of non-viscous fluids are called Euler equations and have been derived by Leonhard Euler (1707–1783). The Navier-Stokes equations generalize the Euler equations and include the case of viscous fluids. They have been derived independently by a number of people, including Claude-Louis Navier (1785–1839), George Stokes (1819–1903), Simeon Poisson (1781–1840) and Jean Claude Saint-Venant (1797–1886).

First we recall the derivation of the Navier-Stokes equations, following [**Fow97**, §6]. Then we focus on the analysis of the Navier-Stokes equations in $\Omega = \mathbb{T}^d = \mathbb{R}^d/(2\pi\mathbb{Z})^d$, i.e., $\Omega = [0, 2\pi)^d$ with periodic boundary conditions. After the presentation of some local existence and uniqueness results we formulate the global existence question. The local existence and uniqueness of solutions in some phase space X is obtained by a fixed point argument in $C([0,T_0],X)$ for a $T_0>0$, using the variation of constant formula. A good choice of the phase space X turns out to be essential in this construction. The background of the millennium problem is the fact that in infinite-dimensional spaces there are infinitely many non-equivalent norms. For the 3D Navier-Stokes equations so far in no phase space simultaneously the global existence of solutions and their uniqueness can be shown.

6.1.1. Derivation of the Navier-Stokes equations. The Navier-Stokes equations describe the velocity and the pressure field of an incompressible fluid. By Newton's law (force=mass×acceleration) the N molecules of the fluid satisfy the system of ODEs

$$m\ddot{x}_j = F_j(x_1,\dots,x_N)$$

for $j=1,\dots,N$. The motion of the fluid is completely determined by the evolution of this system. However, the system is pretty useless due to the very large number N. Therefore, the fluid is modeled as a continuum. In doing so we have to guarantee that no molecules are lost, i.e., that mass is conserved.

The velocity field of the continuum at a position $x\in\mathbb{R}^d$ at a time t is denoted by $u(x,t)\in\mathbb{R}^d$ for $d=2,3$. With $\rho=\rho(x,t)\in\mathbb{R}$ we denote similarly the density of the fluid. The Navier-Stokes equations consist of two equations, a scalar one for the conservation of mass and a second equation with d components for the conservation of the momentum. In general, by the internal friction of the fluid heat will be produced which leads to a coupling of the Navier-Stokes equations with a heat equation. However, here we will neglect this aspect.

Conservation of mass. We consider a fixed test volume V with surface S. The total mass in V can only change by the flow through the boundary S, i.e.,

$$\begin{aligned}\frac{\mathrm{d}}{\mathrm{d}t}\int_V \rho\,\mathrm{d}V &= \int_V \partial_t\rho\,\mathrm{d}V = -\int_S \rho u\cdot n\,\mathrm{d}S = -\int_S\sum_{j=1}^d \rho u_j n_j\,\mathrm{d}S\\ &= -\int_V \mathrm{div}(\rho u)\,\mathrm{d}V = -\int_V\sum_{j=1}^d \partial_{x_j}(\rho u_j)\,\mathrm{d}V,\end{aligned}$$

where we used the Gauss integral theorem and where $n(x)=(n_1,\dots,n_d)(x)$ is the outer unit normal in the point x at the boundary S. Since this relation holds for all test volumes V the integrands must be equal, i.e.,

$$\partial_t\rho + \mathrm{div}(\rho u) = 0. \tag{6.1}$$

Conservation of momentum. Similarly, the momentum of a test volume V can only change by the flow through the boundary and by forces, for instance friction forces, on the surface of the test volume. For these forces f we assume the existence of a matrix $\sigma = (\sigma_{ij})_{i,j=1,\dots,d}$, the so called stress tensor, which relates the direction of the outer normal n with the direction and magnitude of the resulting force f, i.e.,

$$f_i = \sum_{j=1,\dots,d} \sigma_{ij} n_j.$$

For examples see below. Hence, we obtain for the change of the momentum

$$\frac{\mathrm{d}}{\mathrm{d}t} \int_V \rho u_i \,\mathrm{d}V = -\int_S \sum_{j=1}^{d} (\rho u_i) u_j n_j \,\mathrm{d}S + \int_S \sum_{j=1}^{d} \sigma_{ij} n_j \,\mathrm{d}S.$$

Application of the Gauss integral theorem and the above arguments yield

$$\partial_t(\rho u_i) + \operatorname{div}(\rho u_i u) = \operatorname{div}(\sigma_{i\cdot}).$$

Using conservation of mass gives

$$\partial_t(\rho u_i) = \rho \partial_t u_i + u_i \partial_t \rho = \rho \partial_t u_i - u_i \sum_{j=1,\dots,d} \partial_{x_j}(\rho u_j),$$

thus, in vector notation,

$$\rho[\partial_t u_i + (u \cdot \nabla) u_i] = \nabla \cdot \sigma_{i\cdot} \tag{6.2}$$

or, in coordinates,

$$\rho \left[\partial_t u_i + \sum_{j=1}^{d} (u_j \cdot \partial_{x_j}) u_i \right] = \sum_{j=1}^{d} \partial_{x_j} \sigma_{ij}.$$

Constitutive laws. In order to obtain a closed set of equations from (6.1) und (6.2) we need to know how the stress tensor σ depends on the velocity u and the density ρ. Such a relation $\sigma = \sigma(u, \rho)$ is called a constitutive law and depends on the fluid under consideration, i.e., the function differs strongly for instance between water and honey. It is possible that $\sigma(t)$ is not only a function of $(u, \rho)(t)$, but depends on the whole history of (u, ρ), cf. [**Ren00**].

i) For a non-viscous fluid, i.e., for a fluid without internal friction, the surface forces f are only due to pressure. Hence, the forces f are parallel to the outer normals n and their magnitude does not depend on the direction, i.e.,

$$f_i = \sum_{j=1}^{d} \sigma_{ij} n_j = -\sum_{j=1}^{d} p \delta_{ij} n_j = -p n_i, \tag{6.3}$$

which yields the constitutive law

$$\sigma_{ij} = -p\delta_{ij},$$

where $\delta_{ij} = 1$ for $i = j$ and $\delta_{ij} = 0$ else. Since

$$\sum_{j=1}^{d} \partial_{x_j}\sigma_{ij} = -\sum_{j=1}^{d} \partial_{x_j}(p\delta_{ij}) = -\partial_{x_i}p$$

we obtain for non-viscous fluids

$$\partial_t u + (u \cdot \nabla)u = -\frac{1}{\rho}\nabla p. \tag{6.4}$$

The system consisting of (6.1) and (6.4) is still not closed. We have to relate p with the velocity u and density ρ. A typical choice would be $p = c\rho^\gamma$ with some constants $c > 0$ and $\gamma \geq 1$. In case of incompressible fluids, i.e., $\rho = const.$, we obtain Euler's equations

$$\begin{aligned} \partial_t u + (u \cdot \nabla)u &= -\frac{1}{\rho}\nabla p, \\ \nabla \cdot u &= 0. \end{aligned} \tag{6.5}$$

ii) For a viscous fluid the constitutive law is given by

$$\sigma_{ij} = -p\delta_{ij} + \tau_{ij},$$

where τ_{ij} will model the internal friction. In order to find a model for τ_{ij} as simple as possible we consider a stationary constant density shear flow with velocity field $(u_1(x_2), 0)$, see Figure 6.1. The friction forces act on the top surface and bottom surface, and they are proportional to the difference of these velocities. Hence, for an infinitesimal small test volume we find that the surface force f is proportional to the so called strain $\partial_{x_2}u_1$. Since the friction forces are perpendicular to the top surface we find

$$\tau_{12} = \mu\partial_{x_2}u_1,$$

where $\mu > 0$ is called the dynamic viscosity. Due to the isotropy of the fluid σ must be symmetric. This finally leads to

$$\tau_{ij} = 2\mu\dot{\varepsilon}_{ij} + \lambda\sum_{k=1}^{d}\dot{\varepsilon}_{kk}\delta_{ij}$$

where

$$\dot{\varepsilon}_{ij} = \frac{1}{2}(\partial_{x_j}u_i + \partial_{x_i}u_j).$$

The constants μ for the shear and λ for the compressions are called the Lamé-constants.

Remark 6.1.1. There is no reason why in our modeling σ should only depend linearly on the first derivatives of u. An answer to the millennium problem can lead to some corrections at this point of the modeling. ⌋

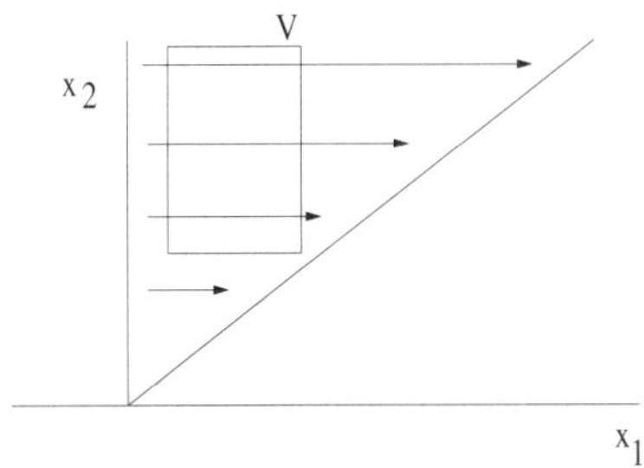

Figure 6.1. Shear flow: the internal friction is proportional to $\partial_{x_2} u_1$.

The Navier-Stokes equations. For air, compressibility is an important issue. However, for fluids such as water, ρ can be considered to be a constant, i.e., $\partial_t \rho = 0$. As in (6.5) the conservation of mass then simplifies to

$$\sum_{j=1}^{d} \partial_{x_j} u_j = \nabla \cdot u = \mathrm{div} u = 0.$$

From $\sum_{k=1}^{d} \dot{\varepsilon}_{kk} = \sum_{k=1}^{d} \partial_{x_k} u_k = 0$ we obtain $\tau_{ij} = 2\mu \dot{\varepsilon}_{ij}$. Moreover, we find

$$\sum_{j=1}^{d} \partial_{x_j} \tau_{ij} = \sum_{j=1}^{d} \partial_{x_j} (\partial_{x_j} u_i + \partial_{x_i} u_j) = \sum_{j=1}^{d} \partial_{x_j}^2 u_i + \partial_{x_i} (\sum_{j=1}^{d} \partial_{x_j} u_j) = \sum_{j=1}^{d} \partial_{x_j}^2 u_i,$$

such that

$$\partial_t u + (u \cdot \nabla) u = -\frac{1}{\rho} \nabla p + \nu \Delta u,$$

where $\nu = \mu / \rho$ is called the kinematic viscosity. The Navier-Stokes equations are then given by

$$\begin{aligned} \partial_t u + (u \cdot \nabla) u &= -\frac{1}{\rho} \nabla p + \nu \Delta u, \\ \nabla \cdot u &= 0. \end{aligned} \tag{6.6}$$

In order to handle these equations as a dynamical system boundary conditions have to be added. At rigid boundaries, a viscous fluid satisfies $u = 0$. At free surfaces, boundary conditions involve for instance the prescription of stresses. In this section we consider the Navier-Stokes equations with periodic boundary conditions. These have no physical meaning, but allow us to focus on the equations itself.

In order to eliminate the physical units from the Navier-Stokes equations, let U be a typical velocity and l be a typical length of the flow. We set

$$u = U u^*, \quad x = l x^*, \quad p = \rho U^2 p^*, \quad t = l t^* / U,$$

and obtain after dropping the * the Navier-Stokes equations in dimensionless form

$$\partial_t u + (u \cdot \nabla)u = \qquad -\nabla p + \frac{1}{\mathcal{R}}\Delta u, \qquad \nabla \cdot u = 0,$$

where $\mathcal{R} = Ul/\nu$ is called the Reynolds number. The larger $\mathcal{R}$, the more complex the flow. Until further notice we assume w.l.o.g. for our purposes that $\mathcal{R} = 1$.

6.1.2. The vorticity, and some explicit solutions. The vorticity ω of the 3D velocity field u is defined by

$$\omega = \nabla \times u = \begin{pmatrix} \partial_{x_2} u_3 - \partial_{x_3} u_2 \\ \partial_{x_3} u_1 - \partial_{x_1} u_3 \\ \partial_{x_1} u_2 - \partial_{x_2} u_1 \end{pmatrix},$$

while for 2D flows the vorticity is the scalar $\omega = \partial_{x_1} u_2 - \partial_{x_2} u_1$. Applying the curl-operator $\nabla\times$ to the Navier-Stokes equations gives in $\mathbb{R}^3$ that

$$\partial_t \omega = \nu \Delta \omega - (u \cdot \nabla)\omega + (\omega \cdot \nabla)u. \tag{6.7}$$

Specializing (6.7) to two-dimensional flows shows $\omega \perp \nabla u$, and so in $\mathbb{R}^2$ we have

$$\partial_t \omega = \nu \Delta \omega - (u \cdot \nabla)\omega. \tag{6.8}$$

The pressure gradient has vanished from (6.7) and (6.8). On the other hand, the velocity u still appears and has to be reconstructed from ω by solving the PDEs

$$\omega = \nabla \times u \qquad \text{and} \qquad \nabla \cdot u = 0.$$

If $\Omega = \mathbb{R}^d$ the solution is given by the Biot-Savart law [**FLS64**, II-14-10] from magnetostatics, cf. Exercise 6.3.

There are major differences between the 2D case and the 3D case. Beside the diffusion and transport of vorticity which appear in 2D and 3D, in 3D there is also the production term $(\omega\cdot\nabla)u$ for vorticity. Hence, it is likely that the differences in the global existence and uniqueness question in 2D and 3D are not only an artificial functional analytic problem. In fact, experiments and simulations, cf. the discussion in [**GW06a**], show that in 3D smaller and smaller vortices are created, whereas in 2D the smaller vortices vanish and are eaten up by the larger ones.

In $\mathbb{R}^d$ or $\mathbb{T}^d$ we have that $\omega = 0$ is a solution of the vorticity equation. The fact that vorticity is preserved by the motion of the fluid together with the incompressibility of the fluid allows us to construct a number of non-trivial solutions for the Navier-Stokes equations. The existence of a potential $\Phi : \mathbb{R}^d \to \mathbb{R}$ with $u = \nabla\Phi$ follows from $\omega = \nabla \times u = 0$. The potential Φ satisfies $\Delta\Phi = 0$ due to the incompressibility of the fluid. Such flows are called potential flows. However, in general the boundaries of the physical domain will create vorticity.

Remark 6.1.2. The conditions

$$\nabla \times u = 0, \qquad \text{and} \qquad \nabla \cdot u = 0 \tag{6.9}$$

can be interpreted in two space dimensions, i.e. $u = (u_1, u_2)$, as the Cauchy-Riemann differential equations of a complex valued function $z \mapsto w(z)$ defined through $w(x_1 + \mathrm{i}x_2) = u_1(x_1, x_2) - \mathrm{i}u_2(x_1, x_2)$. From complex analysis it is well known that the complex differentiability of w, together with (6.9), implies the analyticity of w, respectively u. ⌋

Here are a number of examples for potential flows. Further interesting and more complicated exact solutions of the Navier-Stokes equations can for instance be found in [**MB02**].

Example 6.1.3. We already encountered the explicit constant shear flow solution (2D)

$$u(x,t) = \begin{pmatrix} u_1(x_2) \\ 0 \end{pmatrix} = \begin{pmatrix} cx_2 \\ 0 \end{pmatrix}, \qquad p(x,t) = p_0,$$

with p_0 some constant. This is an example of a parallel or laminar flow with non-vanishing vorticity. Further examples are Couette and Poiseuille flow. The latter describes the flow in an infinitely long pipe $\Omega = \mathbb{R} \times \Sigma$, where $\Sigma \subset \mathbb{R}^d$ is a bounded cross-section with rigid boundary conditions $u|_{\partial\Sigma\subset\mathbb{R}^d} = 0$. As an explicit example we again consider the 2D case and set $\Sigma = (-1, 1)$. Then

$$u(x,t) = \begin{pmatrix} u_1(x_2) \\ 0 \end{pmatrix}, \quad \text{with} \quad u_1(x_2) = c(x_2^2 - 1), \quad c \in \mathbb{R},$$

is an exact solution, see Figure 6.2a), with the pressure given by $p(x,t) = p_0 - \frac{c}{\mathrm{R}} x_1$. ⌋

An important property of all parallel flows is that the nonlinear term $(u \cdot \nabla)u$ of the Navier-Stokes equations drops out. As a consequence, in $\mathbb{R}^d$ parallel flows can always be superimposed. This also works in general domains if the boundary conditions permit it.

Example 6.1.4. In 2D so called irrotational strain flows are given by

$$u(x,t) = \gamma \begin{pmatrix} -x_1 \\ x_2 \end{pmatrix}, \quad p(x,t) = p_0 - \frac{\gamma^2}{2}(x_1^2 + x_2^2),$$

see Figure 6.2b), while vortices are given by

$$u(x,t) = \omega_0 \begin{pmatrix} -x_2 \\ x_1 \end{pmatrix}, \quad p = p_0 + \frac{\omega_0^2}{2}(x_1^2 + x_2^2),$$

see Figure 6.2c). A 3D flow generalizing the 2D strain flow is the irrotational stationary jet

$$u(x,t) = \begin{pmatrix} -\gamma_1 x_1 \\ -\gamma_2 x_2 \\ (\gamma_1+\gamma_2)x_3 \end{pmatrix}, \quad p(x,t) = p_0 - \frac{1}{2}(\gamma_1^2 x_1^2 + \gamma_2^2 x_2^2 + (\gamma_1+\gamma_2)^2 x_3^2).$$

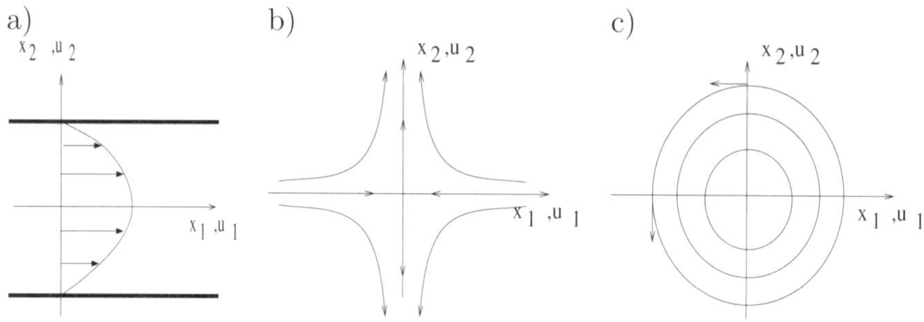

Figure 6.2. Three planar exact solutions: Poiseuille flow, strain flow, a vortex.

⌋

To get a feeling for the behavior of general solutions of the Navier-Stokes equation we recommend to do some numerical experiments. For this we refer to [**Uec09**] and the `matlab` scripts provided and explained therein, including some brief discussion of turbulence.

6.2. The equations on a torus

6.2.1. Local existence and uniqueness. We remind the reader of basic problems with existence and uniqueness of solutions of ODEs and PDEs. In Example 2.2.4 we saw that the scalar equation $\dot{u} = \sqrt{|u|}$ with the initial condition $u(0) = 0$ has infinitely many different solutions. Two examples are $u(t) = 0$ and $u(t) = t^2/4$. In Example 2.2.5 we saw that the unique solution $u(t) = \tan(t)$ of $\dot{u} = (1 + u^2)$ with the initial condition $u(0) = 0$ becomes unbounded for $t = \pi/2$.

We study the local existence and uniqueness of solutions of the Navier-Stokes equations

$$\partial_t u + (u \cdot \nabla)u = -\nabla p + \Delta u, \qquad \nabla \cdot u = 0. \tag{6.10}$$

We follow the formulation of the millennium problem [**Fef06**] and consider the Navier-Stokes equations in $\mathbb{T}^d$, i.e., in $[0, 2\pi)^d$ with periodic boundary conditions. The phase space is chosen in such a way that its elements satisfy the boundary conditions. The Navier-Stokes equations have the special difficulty that the second equation, $\nabla \cdot u = 0$, is without a time derivative, and that the variable p occurs without time derivative at all. This problem is solved by prescribing the equation $\nabla \cdot u = 0$ as additional condition in the definition of the phase space. The term $-\nabla p$ in the first equation will be interpreted as projection P onto the divergence free vector fields such that the Navier-Stokes equations can be written as

$$\partial_t u = P\Delta u - P(u \cdot \nabla)u.$$

For periodic boundary conditions we will have $P\Delta = \Delta P$ such that we finally have to consider $\partial_t u = \Delta u - P(u \cdot \nabla)u$ in the space of divergence free vector fields $\{u : u = Pu\}$.

The Navier-Stokes equations are semi-linear parabolic differential equations such that for the construction of local solutions in time we use again the variation of constant formula

$$u(t) = \mathrm{e}^{t\Delta} u(0) - \int_0^t \mathrm{e}^{(t-s)\Delta} (P(u \cdot \nabla)u)(s)\,\mathrm{d}s \tag{6.11}$$

and the scheme introduced in §5.1.4. The semigroup $\mathrm{e}^{t\Delta}$ generated by the linear part is smoothing, i.e., $u_0 \in L^2$ implies that $t^{m/2}\partial_x^m u(t)$ is bounded in $L^2(\mathbb{T}^d)$ for every $t > 0$ and $m \in \mathbb{N}$. Semi-linear means here that the nonlinearity only contains terms with less derivatives than in the linear part, i.e., for the Navier-Stokes equations first derivatives in $(u \cdot \nabla)u$ compared with the second order derivatives in Δu. More precisely, we will prove that $\mathrm{e}^{t\Delta}$ maps H^m into H^{m+1} with a singularity $t^{-1/2}$ and that $P(u \cdot \nabla)u$ is a bilinear map from $H^{m+1} \times H^{m+1} \to H^m$. Then all assumptions following (5.21) are satisfied and the local existence and uniqueness Theorem 5.2.22 will apply. Hence, the major step is to give a precise definition of P and to investigate its analytic properties.

The projection on the divergence free vector fields. We define the projection P via the solution $v = Pf$ of the system of PDEs

$$v + \nabla p = f, \quad \nabla \cdot v = 0, \tag{6.12}$$

where $f : \mathbb{R}^d \to \mathbb{R}^d$ should be thought as a placeholder for the nonlinear terms $(u \cdot \nabla)u$. For notational simplicity we carry out the analysis only in case $x \in \mathbb{R}^2$ with 2π-periodic boundary conditions. In order to solve (6.12) we make an expansion in Fourier series

$$v_j(x) = \sum_{k \in \mathbb{Z}^2} \widehat{v}_{j,k} \mathrm{e}^{\mathrm{i}k \cdot x}, \quad p(x) = \sum_{k \in \mathbb{Z}^2} \widehat{p}_k \mathrm{e}^{\mathrm{i}k \cdot x}, \quad f_j(x) = \sum_{k \in \mathbb{Z}^2} \widehat{f}_{j,k} \mathrm{e}^{\mathrm{i}k \cdot x},$$

with $\widehat{v}_{j,k}$, $\widehat{f}_{j,k}$, $\widehat{p}_k \in \mathbb{C}$ for $j = 1, 2$. Plugging this into (6.12) yields

$$\widehat{v}_{1,k} + \mathrm{i}k_1 \widehat{p}_k = \widehat{f}_{1,k}, \quad \widehat{v}_{2,k} + \mathrm{i}k_2 \widehat{p}_k = \widehat{f}_{2,k}, \quad \mathrm{i}k_1 \widehat{v}_{1,k} + \mathrm{i}k_2 \widehat{v}_{2,k} = 0. \tag{6.13}$$

In case $|k| \neq 0$ we find the solution

$$\begin{pmatrix} \widehat{v}_{1,k} \\ \widehat{v}_{2,k} \\ \widehat{p}_k \end{pmatrix} = \frac{1}{k_1^2 + k_2^2} \begin{pmatrix} k_2^2 & -k_1 k_2 & -\mathrm{i}k_1 \\ -k_1 k_2 & k_1^2 & -\mathrm{i}k_2 \\ -\mathrm{i}k_1 & -\mathrm{i}k_2 & 1 \end{pmatrix} \begin{pmatrix} \widehat{f}_{1,k} \\ \widehat{f}_{2,k} \\ 0 \end{pmatrix}.$$

For the subspace $k_1 = k_2 = 0$ there are two possibilities.

Case i). We prescribe the periodicity of the pressure p. Then we have that $\widehat{v}_{j,0} = \widehat{f}_{j,0}$ in (6.13) and that $\widehat{p}_0$ is arbitrary which is no problem since only ∇p occurs in the Navier-Stokes equations. This choice can lead to a non-vanishing mean flow.

Case ii). We require that the mean flows $\int_\Omega v_j(x_1, x_2)\,\mathrm{d}x$ vanish for $j = 1, 2$, i.e., $\widehat{v}_{1,0} = \widehat{v}_{2,0} = 0$. In order to do so we consider a pressure

$$p(x, t) = \sum_{j=1}^{2} \alpha_j x_j + \widetilde{p}(x, t),$$

where $\widetilde{p}(x, t)$ is 2π-periodic w.r.t. the x_j. Then $\partial_{x_j} p(x, t) = \alpha_j + \partial_{x_j} \widetilde{p}(x, t)$ and so $\widehat{v}_{j,0} + \alpha_j = \widehat{f}_{j,0}$. Thus, to a $\widehat{f}_{j,0}$ we always find an α_j such that $\widehat{v}_{j,0} = 0$.

Example 6.2.1. To illustrate the difference between Case i) and Case ii), we consider the vector field

$$f(x_1, x_2) = \mathrm{e}^{-2(x_1-\pi)^2 - 2(x_2-\pi)^2} \begin{pmatrix} 2 + \tanh(x_2 - \pi) \\ 0 \end{pmatrix}.$$

In Figure 6.3 we show the different effects of choosing i) or ii). ⌋

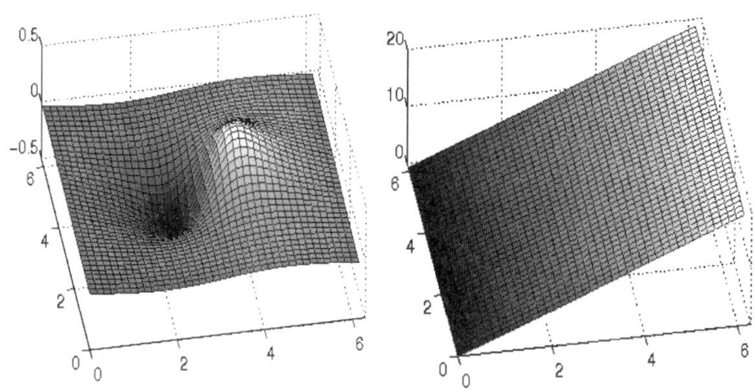

Figure 6.3. Illustration of the difference between cases i) and ii), concerning the boundary conditions for the pressure in the definition of the projection $u = Pf$ for f from Remark 6.2.1, via a plot of the pressure function. In the left panel we require a periodic pressure, giving a mean flow in Pf, while in the right panel we require zero mean flow, giving a linear growth of the pressure p.

Choosing between i) and ii) is a question of modeling. i) has the disadvantage that if a constant mean force is added to the Navier-Stokes equations then this choice leads to unbounded growth of (laminar) mean flows. Therefore, in the following we opt for ii), i.e., $\widehat{v}_{j,0} = 0$ for $j = 1, 2$ and define the projection $\widehat{P}$ as direct sum of the projections $\widehat{P}_k$, i.e., $\widehat{v}_k = (\widehat{P}\widehat{f})_k = \widehat{P}_k \widehat{f}_k$, where

$$\widehat{v}_0 = 0 \text{ and } \begin{pmatrix} \widehat{v}_{1,k} \\ \widehat{v}_{2,k} \end{pmatrix} = \frac{1}{k_1^2 + k_2^2} \begin{pmatrix} k_2^2 & -k_1 k_2 \\ -k_1 k_2 & k_1^2 \end{pmatrix} \begin{pmatrix} \widehat{f}_{1,k} \\ \widehat{f}_{2,k} \end{pmatrix} \tag{6.14}$$

for $k \neq 0$. In physical space we define P by $P = \mathcal{F}^{-1}\widehat{P}\mathcal{F}$.

Lemma 6.2.2. *The projection $\widehat{P}$ is a bounded linear map in $\ell_{1,m}$ and in $\ell_{2,m}$, i.e., for all $m \in \mathbb{R}$ there exists a $C > 0$ such that*

$$\|\widehat{P}\widehat{f}\|_{\ell_{1,m}} \leq C\|\widehat{f}\|_{\ell_{1,m}} \quad \text{and} \quad \|\widehat{P}\widehat{f}\|_{\ell_{2,m}} \leq C\|\widehat{f}\|_{\ell_{2,m}}.$$

Hence, the projection P is also a bounded linear map in H^m_{per}, i.e., for all $m \in \mathbb{R}$ there exists a $C > 0$ such that

$$\|Pf\|_{H^m} \leq C\|f\|_{H^m}.$$

Proof. We find

$$\|\widehat{P}\widehat{f}\|_{\ell_{1,m}} = \|(\widehat{P}_k\widehat{f}_k)_{k\in\mathbb{Z}^2}\|_{\ell_{1,m}} \leq \sup_{k\in\mathbb{Z}^2} \|\widehat{P}_k\| \ \|(\widehat{f}_k)_{k\in\mathbb{Z}^2}\|_{\ell_{1,m}} \leq C\|\widehat{f}\|_{\ell_{1,m}}.$$

The proof for $\ell_{2,m}$ works exactly the same. Using that Fourier transform is an isomorphism between $\ell_{2,m}$ and H^m_{per}, cf. Lemma 5.2.9, yields

$$\|Pf\|_{H^m} \leq C_1\|\widehat{P}\widehat{f}\|_{\ell_{2,m}} \leq C_1 C\|\widehat{f}\|_{\ell_{2,m}} \leq C_1 C C_2\|f\|_{H^m}.$$ □

Remark 6.2.3. a) Lemma 6.2.2 is valid in arbitrary dimensions $d \geq 2$.
b) In Fourier space we have $\widehat{v}_k \in V_k := \{\widehat{v}_k \in \mathbb{C}^d : k \cdot \widehat{v}_k = 0\}$, due to $\nabla \cdot v = 0$. The pressure gradient ∇p defines in each $\mathbb{C}^d$ a vector $\mathrm{i}\widehat{p}_k k$, with $\widehat{p}_k \in \mathbb{C}$, which is orthogonal to V_k. This property can be generalized to general domains $\Omega \subset \mathbb{R}^d$, cf. §6.3. ⌋

The phase space and the fixed point argument. In the following we will solve the Fourier transformed Navier-Stokes equations in two classes of phase spaces, namely

$$\ell^{div}_{1,m} = \{\widehat{u} \in (\ell_{1,m})^d : \widehat{u} = \widehat{P}\widehat{u}\}, \qquad \ell^{div}_{2,m} = \{\widehat{u} \in (\ell_{2,m})^d : \widehat{u} = \widehat{P}\widehat{u}\},$$

and the Navier-Stokes equations in physical space in the class of phase spaces

$$H^{div,m}_{\mathrm{per}} = \{u \in (H^m_{\mathrm{per}})^d : u = Pu\}.$$

In case of periodic boundary conditions we have $P\Delta = \Delta P$. Hence, for u with $u = Pu$ the solution operator of the linearized Navier-Stokes equations is given coordinate-wise by the solution operator $\mathrm{e}^{t\Delta}$ of the linear diffusion equation $\partial_t u = \Delta u$. In one space dimension this operator has been discussed a number of times, cf. Example 5.1.21 and Example 5.2.19. The statements made above about this operator transfer line to line from $\mathbb{R}^1$ to $\mathbb{R}^d$.

Theorem 6.2.4. *The solution operator* $(\mathrm{e}^{-|k|^2 t})_{k\in\mathbb{Z}^d}$ *of the linearized Navier-Stokes equations* $\partial_t \widehat{u}_k = -|k|^2 \widehat{u}_k$ *in Fourier space defines a* C_0*-semigroup in* $\ell^{div}_{1,m}$ *and* $\ell^{div}_{2,m}$ *for all* $m \in \mathbb{R}$. *The associated solution operator* $\mathrm{e}^{t\Delta} = \mathcal{F}^{-1}(\mathrm{e}^{-|k|^2 t})_{k\in\mathbb{Z}^d}\mathcal{F}$ *of the linearized Navier-Stokes equations in physical space defines a* C_0*-semigroup in* $H^{div,m}_{\mathrm{per}}$. *Moreover, for all* $m \in \mathbb{R}$ *and* $r \geq 0$ *there exists a* $C > 0$*, such that*

$$\begin{aligned}
\|(\mathrm{e}^{-|k|^2 t})_{k\in\mathbb{Z}^d}\widehat{u}\|_{\ell^{div}_{1,m+r}} &\leq C\max(1, t^{-r/2})\|\widehat{u}\|_{\ell^{div}_{1,m}}, \\
\|(\mathrm{e}^{-|k|^2 t})_{k\in\mathbb{Z}^d}\widehat{u}\|_{\ell^{div}_{2,m+r}} &\leq C\max(1, t^{-r/2})\|\widehat{u}\|_{\ell^{div}_{2,m}}, \\
\|\mathrm{e}^{t\Delta}u\|_{H^{div,m+r}_{\mathrm{per}}} &\leq C\max(1, t^{-r/2})\|u\|_{H^{div,m}_{\mathrm{per}}}.
\end{aligned}$$

Since we already analyzed P in §6.2.1 it remains to bound the nonlinear term $(u\cdot\nabla)u)$. It is easy to see that it is smooth from $H^{div,m}_{\mathrm{per}}$ to H^{m-1}_{per} if m is sufficiently large. Hence, with the previous estimates the local existence and uniqueness of solutions easily follows in every $H^{div,m}_{\mathrm{per}}$ if m is sufficiently large. However, since we are also interested in the global existence and uniqueness of solutions, and since it is more easy to obtain a priori estimates on the solutions in $H^{div,m}_{\mathrm{per}}$-spaces for small m, we would like to have m as small as possible. Hence, we spend a little bit more time at this point to optimize the estimates.

We use the variation of constant formula (6.11) to prove local existence and uniqueness of solutions. The key ingredients are the smoothing properties of the semigroup $\mathrm{e}^{t\Delta}$ summarized in Theorem 6.2.4 and the Lipschitz-continuity of the nonlinear terms $P(u\cdot\nabla)u)$. Since a singularity $t^{-1+\delta}$, with $\delta > 0$, is integrable, our approach also works if the nonlinearity $P((u\cdot\nabla)u)$ is Lipschitz-continuous from $H^{div,m}_{\mathrm{per}}$ into $H^{div,m-2+\delta}_{\mathrm{per}}$. By using the incompressibility $\sum_{j=1}^d \partial_{x_j} u_j = 0$ the lth component of $(u\cdot\nabla)u$ can be written as

$$\sum_{j=1}^d u_j \partial_{x_j} u_l = \sum_{j=1}^d u_j \partial_{x_j} u_l + \sum_{j=1}^d u_l \partial_{x_j} u_j = \sum_{j=1}^d \partial_{x_j}(u_j u_l),$$

or equivalently in vector notation as

$$(u\cdot\nabla)u = \nabla\cdot(uu^T). \tag{6.15}$$

Using this representation we have to establish

$$\|\nabla\cdot uv^T\|_{H^{m-2+\delta}} \leq C\|u\|_{H^m}\|v\|_{H^m},$$

or equivalently

$$\|uv^T\|_{H^{m-1+\delta}} \le C\|u\|_{H^m}\|v\|_{H^m} \tag{6.16}$$

for a $\delta > 0$ and all $m > m^*$ for a $m^* \in \mathbb{R}$. To find the minimal m^* we start with the following lemma.

Lemma 6.2.5. *For all $\widetilde{m} \in \mathbb{R}$, $r \ge 0$, and $\delta > 0$ there exists a $C > 0$ such that*

$$\|\widehat{u} * \widehat{v}\|_{\ell_{2,\widetilde{m}}} \le C\ (\|\widehat{u}\|_{\ell_{2,\widetilde{m}+r}}\|\widehat{v}\|_{\ell_{2,\frac{d}{2}-r+\delta}} + \|\widehat{u}\|_{\ell_{2,\frac{d}{2}-r+\delta}}\|\widehat{v}\|_{\ell_{2,\widetilde{m}+r}}).$$

Proof. We define $\widehat{\varrho}(k) = (1+|k|)^{\widetilde{m}}$, $\widehat{\varrho}_1(k) = (1+|k|)^{\widetilde{m}+r}$, and $\widehat{\varrho}_2(k) = (1+|k|)^{-r}$, for which we have the inequality

$$\begin{aligned}\widehat{\varrho}(k) \le&(1+|k|)^{\widetilde{m}} \le C((1+|k-l|)^{\widetilde{m}+r}(1+|l|)^{-r}+(1+|l|)^{\widetilde{m}+r}(1+|k-l|)^{-r})\\ \le&C(\widehat{\varrho}_1(k-\ell)\widehat{\varrho}_2(\ell)+\widehat{\varrho}_2(k-\ell)\widehat{\varrho}_1(\ell)).\end{aligned}$$

Using this and Lemma 5.1.26 yields

$$\begin{aligned}\|\widehat{u} * \widehat{v}\|_{\ell_{2,\widetilde{m}}} =&\|\widehat{\varrho}(\widehat{u} * \widehat{v})\|_{\ell_2} = 2\|\,(\widehat{u}\widehat{\varrho}_1) * (\widehat{v}\widehat{\varrho}_2)\,\|_{\ell_2} + 2\|\,(\widehat{u}\widehat{\varrho}_2) * (\widehat{v}\widehat{\varrho}_1)\,\|_{\ell_2}\\ \le&2\|\widehat{u}\widehat{\varrho}_1\|_{\ell_2}\|\widehat{v}\widehat{\varrho}_2\|_{\ell_1} + 2\|\widehat{u}\widehat{\varrho}_2\|_{\ell_1}\|\widehat{v}\widehat{\varrho}_1\|_{\ell_2}\\ =&2C\left(\|\widehat{u}\|_{\ell_{2,\widetilde{m}+r}}\|\widehat{v}\|_{\ell_{1,-r}} + 2\|\widehat{u}\|_{\ell_{1,-r}}\|\widehat{v}\|_{\ell_{2,\widetilde{m}+r}}\right)\\ \le&2C\widetilde{C}\left(\|\widehat{u}\|_{\ell_{2,\widetilde{m}+r}}\|\widehat{v}\|_{\ell_{2,\frac{d}{2}-r+\delta}} + \|\widehat{u}\|_{\ell_{2,\frac{d}{2}-r+\delta}}\|\widehat{v}\|_{\ell_{2,\widetilde{m}+r}}\right)\end{aligned}$$

for a $\delta > 0$ according to Sobolev's embedding, cf. Lemma 5.1.27. □

For the validity of (6.16) we have to choose $r = 1 - \delta$ for a $\delta > 0$ such that $m^* = d/2 - 1$ and use that Fourier transform is an isomorphism between $\ell_{2,m}$ and H^m_{per}.

Remark 6.2.6. (The critical Sobolev number m^* for $\ell_{1,m}$) For all $m \ge 0$ there exists a constant C such that $(1+|k+l|)^m \le C((1+|k|)^m + (1+|l|)^m)$ for all k and l. Thus, we have $\|\widehat{u} * \widehat{v}\|_{\ell_{1,m}} \le C\|\widehat{u}\|_{\ell_{1,m}}\|\widehat{v}\|_{\ell_{1,m}}$ which yields $m \ge m^* = 0$. ⌋

Therefore, we obtain the following local existence and uniqueness result.

Theorem 6.2.7. *a) Let $\widehat{u}_0 \in \ell^{div}_{2,m}$ for $m > m^* = d/2 - 1$. Then there exists a $T_0 = T_0(\|\widehat{u}_0\|_{\ell^{div}_{2,m}}) > 0$ and unique mild solution $\widehat{u} \in C([0,T_0], \ell^{div}_{2,m})$ of the Fourier transformed Navier-Stokes equations* (6.11) *with $\widehat{u}|_{t=0} = \widehat{u}_0$. The same is true if $\ell^{div}_{2,m}$ is replaced by $\ell^{div}_{1,m}$ and $m \ge 0$.*

b) Let $u_0 \in H^{div,m}_{\text{per}}$ for $m > m^* = d/2 - 1$. Then there exists a $T_0 = T_0(\|u_0\|_{H^{div,m}_{\text{per}}}) > 0$ and unique mild solution $u \in C([0,T_0], H^{div,m}_{\text{per}})$ of the Navier-Stokes equations (6.11) *with $u|_{t=0} = u_0$.*

6.2.2. Analyticity of solutions. As pointed out in §5.3.3 the solutions of semi-linear equations where the semigroup is smoothing are infinitely often differentiable for every $t > 0$. The scheme which has been explained in §5.3.3 can also be applied to the Navier-Stokes equations if $m > m^*$. However, the step size for increasing the differentiability has to be decreased from 1 to $\delta/2$. Actually, the solution becomes analytic in a strip along the real axis in the complex plane if $m > m^*$. In order to prove this we define

Definition 6.2.8. *For $\beta \geq 0$ let*

$$\ell_{1,\beta}^{\omega} = \{\widehat{u} : \mathbb{Z}^d \to \mathbb{C} : \|\widehat{u}\|_{\ell_{1,\beta}^{\omega}} < \infty\} \quad \textit{and} \quad \|\widehat{u}\|_{\ell_{1,\beta}^{\omega}} = \sum_{k\in\mathbb{Z}^d} |\widehat{u}_k| \mathrm{e}^{\beta|k|}.$$

Lemma 6.2.9. *If $\widehat{u} \in \ell_{1,\beta}^{\omega}$ then $u = \mathcal{F}^{-1}\widehat{u}$ is analytic in a strip*

$$\mathcal{S}_\beta = \{z = (z_1, \ldots, z_d) \in \mathbb{C}^d : \max_{j=1,\ldots,d} |\mathrm{Im} z_j| < \beta\}$$

in $\mathbb{C}^d$ and $\sup_{z\in\mathcal{S}_\beta} |u(z)| \leq \|\widehat{u}\|_{\ell_{1,\beta}^{\omega}}$.

Proof. The estimate

$$\sup_{z\in\mathcal{S}_\beta} |u(z)| \leq \sum_{k\in\mathbb{Z}^d} (|\widehat{u}_k| \sup_{z\in\mathcal{S}_\beta} |\mathrm{e}^{ikz}|) \leq \sum_{k\in\mathbb{Z}^d} (|\widehat{u}_k| \mathrm{e}^{\beta|k|}) \leq \|\widehat{u}\|_{\ell_{1,\beta}^{\omega}} < \infty \tag{6.17}$$

shows that the function $u(z) = \sum_{k\in\mathbb{Z}^d} \widehat{u}_k \mathrm{e}^{ikz}$ is the uniform limit of the analytic functions $z \mapsto s_N(z) = \sum_{|k|\leq N} \widehat{u}_k \mathrm{e}^{ikz}$ in their domain of convergence $\mathcal{S}_\beta$ and so the function u is analytic in $\mathcal{S}_\beta$. □

For every $t > 0$ and $\beta \geq 0$ the linear Navier-Stokes semigroup $\widehat{T}(t)$ defined by $(\widehat{T}(t)\widehat{u})_k = \mathrm{e}^{-|k|^2 t}\widehat{u}_k$ maps ℓ_1 into $\ell_{1,\beta}^{\omega}$ due to

$$\|\widehat{T}(t)\widehat{u}\|_{\ell_{1,\beta}^{\omega}} \leq \sum_{k\in\mathbb{Z}^d} |\mathrm{e}^{-|k|^2 t}\widehat{u}_k \mathrm{e}^{\beta|k|}| \leq \sup_{k\in\mathbb{Z}^d} |\mathrm{e}^{-|k|^2 t}\mathrm{e}^{\beta|k|}| \sum_{k\in\mathbb{Z}^d} |\widehat{u}_k| \leq C(\beta, t)\|\widehat{u}\|_{\ell_1},$$

where $\sup_{k\in\mathbb{Z}^d} |\mathrm{e}^{-|k|^2 t}\mathrm{e}^{\beta|k|}| \leq C(\beta, t) < \infty$, since $\mathrm{e}^{-|k|^2 t}\mathrm{e}^{\beta|k|} \to 0$ for $|k| \to \infty$ if $t > 0$. The constant $C = C(\beta, t)$ satisfies $C(\beta, t) \to \infty$ for $t \to 0$ with a non-integrable singularity which makes this estimate useless for nonlinear problems. However, this singularity can be avoided if we choose β proportional to $\sqrt{t}$.

Lemma 6.2.10. *There exists a constant $C < \infty$ such that the semigroup $\widehat{T}(t)$ defined by $(\widehat{T}(t)\widehat{u})_k = \mathrm{e}^{-|k|^2 t}\widehat{u}_k$ satisfies for all $t \geq 0$ that*

$$\|\widehat{T}(t)\widehat{u}\|_{\ell_{1,\sqrt{t}}^{\omega}} \leq C\|\widehat{u}\|_{\ell_1}.$$

Proof. The assertion follows since

$$\sum_{k\in\mathbb{Z}^d} |\mathrm{e}^{-|k|^2 t}\mathrm{e}^{\sqrt{t}|k|}| \leq \sum_{s\in\mathbb{R}^d} |\mathrm{e}^{-|s|^2}\mathrm{e}^{|s|}| \leq C < \infty$$

can be bounded independently of t. □

This estimate can be used to prove the analyticity of the solutions of Navier-Stokes equations w.r.t. x in $\mathcal{S}_{\sqrt{t}} \subset \mathbb{C}$ for $t > 0$. According to Lemma 6.2.9 this assertion follows if $\widehat{u}(t) \in \ell^{\omega}_{1,\sqrt{t}}$ for $t > 0$.

Lemma 6.2.11. *For every $M > 0$ there exists a $T_0 > 0$ such that the map*

$$F(\widehat{u})(k,t) = \mathrm{e}^{-k^2 t}\widehat{u}(k,0) + \int_0^t \mathrm{e}^{-k^2(t-\tau)}\widehat{P}(k)\mathrm{i}k(\widehat{u} * \widehat{u}^T)(k,\tau)\,\mathrm{d}\tau$$

is a contraction in

$$\mathcal{X} = \{\widehat{u} : \mathbb{Z} \times [0, T_0] \to \mathbb{C} : \|\widehat{u} - \widehat{u}_{lin}\|_{\mathcal{X}} \leq M\},$$

where $\|\widehat{u}\|_{\mathcal{X}} = \sup_{t\in[0,T_0]} \|\widehat{u}(t)\|_{\ell^{\omega}_{1,\sqrt{t}}}$ *and where* $\widehat{u}_{lin}(k,t) = \mathrm{e}^{-k^2 t}\widehat{u}(k,0)$.

Proof. It is easy to see, cf. Exercise 6.4, that $\|\widehat{u} * \widehat{v}\|_{\ell^{\omega}_{1,\sqrt{t}}} \leq \|\widehat{u}\|_{\ell^{\omega}_{1,\sqrt{t}}}\|\widehat{v}\|_{\ell^{\omega}_{1,\sqrt{t}}}$ which implies

$$\|u * v\|_{\mathcal{X}} \leq \|u\|_{\mathcal{X}}\|v\|_{\mathcal{X}}. \tag{6.18}$$

Using this estimate and $\sqrt{t-s} \geq \sqrt{t} - \sqrt{s}$ shows that

$$\begin{aligned}
\|F(\widehat{u}) - \widehat{u}_{lin}\|_{\mathcal{X}} \leq & \sup_{t\in[0,T_0]} \sum_{k\in\mathbb{Z}} \Big| \int_0^t \mathrm{e}^{-k^2(t-\tau)}\widehat{P}(k)\mathrm{i}k(\widehat{u} * \widehat{u}^T)(k,\tau)\,\mathrm{d}\tau \Big| \mathrm{e}^{\sqrt{t}|k|} \\
\leq & \sup_{t\in[0,T_0]} \sum_{k\in\mathbb{Z}} \int_0^t |\mathrm{e}^{-k^2(t-\tau)}\widehat{P}(k)\mathrm{i}k(\widehat{u} * \widehat{u}^T)(k,\tau)\mathrm{e}^{\sqrt{t}|k|}\mathrm{e}^{-\sqrt{\tau}|k|}\mathrm{e}^{\sqrt{\tau}|k|}|\,\mathrm{d}\tau \\
\leq & \sup_{t\in[0,T_0]} \int_0^t \sum_{k\in\mathbb{Z}} |\mathrm{e}^{-k^2(t-\tau)}\widehat{P}(k)\mathrm{i}k\mathrm{e}^{\sqrt{t-\tau}|k|}(\widehat{u} * \widehat{u}^T)(k,\tau)\mathrm{e}^{\sqrt{\tau}|k|}|\,\mathrm{d}\tau \\
\leq & C \sup_{t\in[0,T_0]} \int_0^t \sup_{k\in\mathbb{Z}} |\mathrm{e}^{-k^2(t-\tau)}\mathrm{i}k\mathrm{e}^{\sqrt{t-\tau}|k|}|\,\mathrm{d}\tau \sup_{s\in[0,T_0]} \sum_{k\in\mathbb{Z}} |(\widehat{u} * \widehat{u}^T)(k,\tau)\mathrm{e}^{\sqrt{\tau}|k|}| \\
\leq & C T_0^{1/2}\|\widehat{u}\|^2_{\mathcal{X}} < \infty.
\end{aligned}$$

Thus, we proved that F maps the space $\mathcal{X}$ into itself if $T_0 > 0$ is sufficiently small. The proof of the contraction property works the same way. □

Corollary 6.2.12. *For all $C > 0$ there exists a $T_0 > 0$ such that the solutions u of the Navier-Stokes equations are analytic w.r.t. x in $\mathcal{S}_{\sqrt{t}} \subset \mathbb{C}$ for all $t \in [0, T_0]$ if $\|\widehat{u}(0)\|_{\ell_1} \leq C$.*

Remark 6.2.13. It is easy to see that Theorem 5.1.23 can be generalized to $\ell^{\omega}_{1,\beta}$-spaces and the Fourier transformed Navier-Stokes equations such that there is local existence and uniqueness in $\ell^{\omega}_{1,\beta}$, too. ⌋

Remark 6.2.14. In general it cannot be expected that in nonlinear problems the strip of analyticity is arbitrarily wide. An explicit, but typical, example is the $z \mapsto \tanh(z)$ equilibrium of the Allen-Cahn equation

$\partial_t u = \partial_x^2 u + u - u^3$ in the subsequent §7.2. The function tanh has singularities in the complex plane, due to $\tanh(iy) = \tan(y)$, for $z = i\pi/2 + ik\pi$ with $k \in \mathbb{Z}$. ⌋

6.2.3. Global existence in 2D. In order to prove the global existence and uniqueness of solutions of the Navier-Stokes equations in a phase space X we need a local existence and uniqueness result in X and a priori bounds for the solutions in X. Then as explained already a number of times the local existence and uniqueness result can be applied again and again to construct a solution for all $t \geq 0$.

Bounds for the L^2-norm. The L^2-norm of the solutions u of the Navier-Stokes equations can be bound in every space dimension. By using integration by parts and the incompressibility $\sum_{j=1}^d \partial_j u_j = 0$ we obtain

$$\begin{aligned}\frac{1}{2}\frac{\mathrm{d}}{\mathrm{d}t}\int_{\mathbb{T}^d}\sum_{j=1}^d u_j u_j \,\mathrm{d}x &= \int_{\mathbb{T}^d}\sum_{j=1}^d u_j \partial_t u_j \,\mathrm{d}x \\ &= \int_{\mathbb{T}^d}\sum_{j=1}^d\sum_{\ell=1}^d u_j(\partial_{x_\ell}\partial_{x_\ell} u_j - \partial_j p - \sum_{\ell=1}^d u_\ell \partial_{x_\ell} u_j)\,\mathrm{d}x \\ &= \int_{\mathbb{T}^d} - (\partial_{x_\ell} u_j)(\partial_{x_\ell} u_j)\,\mathrm{d}x.\end{aligned}$$

Poincaré's inequality, cf. Lemma 5.2.17, implies

Lemma 6.2.15. *For all $d \geq 2$ we have*

$$\frac{1}{2}\frac{\mathrm{d}}{\mathrm{d}t}\|u\|_{L^2}^2 \leq -\|u\|_{L^2}^2, \quad \textit{and so} \quad \|u(t)\|_{L^2} \leq \mathrm{e}^{-t}\|u(0)\|_{L^2}.$$

Bounds for the H^1-norm. In $\mathbb{R}^2$ with periodic boundary conditions also the H^1-norm can be bound. By using integration by parts and the incompressibility $\sum_{j=1}^2 \partial_j u_j = 0$, in $\mathbb{R}^2$ we find after some explicit calculation, cf. Exercise 6.5, that

$$\int_{\mathbb{T}^2}\sum_{j=1}^2\sum_{\ell=1}^2\sum_{m=1}^2 (\partial_{x_j} u_\ell)\partial_{x_j}(u_m \partial_{x_m} u_\ell) = 0. \tag{6.19}$$

Therefore, we find as above

$$\frac{1}{2}\frac{\mathrm{d}}{\mathrm{d}t}\int_{\mathbb{T}^2}\sum_{j=1}^2\sum_{\ell=1}^2(\partial_{x_j} u_\ell)(\partial_{x_j} u_\ell)\,\mathrm{d}x = -\int_{\mathbb{T}^2}\sum_{j=1}^2\sum_{k=1}^2\sum_{\ell=1}^2(\partial_{x_j}\partial_{x_k} u_\ell)(\partial_{x_j}\partial_{x_k} u_\ell)\,\mathrm{d}x.$$

Again with Poincaré's inequality, cf. Lemma 5.2.17, we find

$$\frac{1}{2}\frac{\mathrm{d}}{\mathrm{d}t}\int_{\mathbb{T}^2}\sum_{j=1}^2\sum_{\ell=1}^2(\partial_{x_j} u_\ell)(\partial_{x_j} u_\ell)\,\mathrm{d}x = -\int_{\mathbb{T}^2}\sum_{j=1}^2\sum_{\ell=1}^2(\partial_{x_j} u_\ell)(\partial_{x_j} u_\ell)\,\mathrm{d}x.$$

Combining this estimate with the L^2-estimate yields

Lemma 6.2.16. *For $d = 2$ (and only for $d = 2$ and periodic boundary conditions) we have*

$$\frac{1}{2}\frac{\mathrm{d}}{\mathrm{d}t}\|u(t)\|_{H^1}^2 \leq -\|u(t)\|_{H^1}^2 \quad \textit{and so} \quad \|u(t)\|_{H^1} \leq \mathrm{e}^{-t}\|u(0)\|_{H^1}.$$

Combining this a priori estimate with the previous local existence and uniqueness result in $\mathbb{R}^2$ yields

Theorem 6.2.17. *Let $u_0 \in H_{\text{per}}^{div,1}(\mathbb{T}^2)$. Then there exists a unique mild solution $u \in C([0,\infty), H_{\text{per}}^{div,1}(\mathbb{T}^2))$ of the Navier-Stokes equations* (6.11) *with $u|_{t=0} = u_0$. Moreover, for $t > 0$ the solution is an analytic function.*

The same is true for every $H_{\text{per}}^{div,m}(\mathbb{T}^2)$ with $m > m^* = 0$. For $m \in (m^*, 1)$ the result follows from local existence and uniqueness of solutions in such $H_{\text{per}}^{div,m}(\mathbb{R}^2)$ and due to $\|u(t)\|_{H_{\text{per}}^{div,m}(\mathbb{T}^2)} \leq \|u(t)\|_{H_{\text{per}}^{div,1}(\mathbb{T}^2)}$ for such m. For $m > 1$ we have

$$\|uv^T\|_{H^m} \leq C\|u\|_{H^m}\|v\|_{H^m} \tag{6.20}$$

and the local existence and uniqueness in $H_{\text{per}}^{div,m}(\mathbb{T}^2)$. Arguing as in §5.3.3 yields an a priori-bound for $\|u(t)\|_{H_{\text{per}}^{div,m}(\mathbb{T}^2)}$ in terms of $\|u(t-T_0)\|_{H_{\text{per}}^{div,1}(\mathbb{T}^2)}$ for all $t \geq T_0$. Since $\|u(t-T_0)\|_{H_{\text{per}}^{div,1}(\mathbb{T}^2)}$ is globally bounded and decays to zero, the local existence and uniqueness result in $H_{\text{per}}^{div,m}(\mathbb{T}^2)$ can be applied again and again to construct a solution for all $t \geq 0$. Therefore, we have the following theorem about the asymptotic stability of the zero solution.

Theorem 6.2.18. *Let $u_0 \in H_{\text{per}}^{div,m}(\mathbb{T}^2)$ with $m > m^* = 0$. Then there exists a unique mild solution $u \in C([0,\infty), H_{\text{per}}^{div,m}(\mathbb{T}^2))$ of the Navier-Stokes equations* (6.11) *with $u|_{t=0} = u_0$. Moreover, for $t > 0$ the solution is an analytic function and satisfies*

$$\lim_{t\to\infty} \|u(t)\|_{H_{\text{per}}^{div,m}(\mathbb{T}^2)} = 0$$

with some exponential rate.

6.2.4. The millennium problem.

Figure 6.4 displays the Sobolev numbers for the local existence and uniqueness and the a priori estimates.

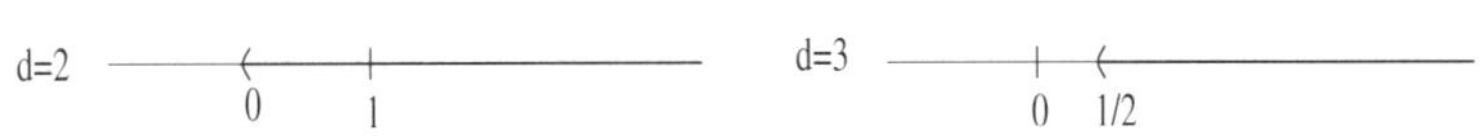

Figure 6.4. A priori estimates in L^2 for $d = 3$ and H^1 for $d = 2$. Local existence and uniqueness holds in H^m for $m > d/2 - 1$. For $d = 2$ there is no gap between the exponents m for which we have a priori estimates and the ones for which we have local existence and uniqueness of solutions. For $d = 3$ a gap remains, and global existence of smooth solutions cannot be concluded.

In $\mathbb{R}^3$ we have no a priori estimate for the spaces where we have local existence and uniqueness. Hence, global existence of smooth solutions is unclear. This question is exactly the content of the millennium problem formulated in [**Fef06**]:

Millennium problem of the Clay-foundation. *Prove (or disprove) the global existence and uniqueness of solutions of the Navier-Stokes equations in three space dimensions. For instance, show $T_0 = \infty$ in Theorem 6.2.7, i.e., close the gap between the a priori estimates and the local existence- and uniqueness theorem, respectively.*

Partial results are already known, from which we list only the two absolute basic ones.

- Jean Leray [**Ler34**] proved the global existence of so called weak solutions, cf. §7.4.2, of the Navier-Stokes equations. These solutions are very rough and they are not unique.
- For small initial conditions due to the linear stability of the origin one easily obtains an a priori estimate and thus it follows $\lim_{t\to\infty} \|u(t)\|_{H^{div,m}_{\text{per}}(\mathbb{T}^3)} = 0$ with some exponential rate for every $m > m^* = 1/2$.

6.2.5. Some qualitative theory. The 2D Navier-Stokes equations without forcing are a bit boring since $\|u(t)\|_{H^1} \to 0$ for $t \to \infty$, cf. Lemma 6.2.16. Thus, we show the existence of a global attractor for the two-dimensional Navier-Stokes equations with forcing, i.e.,

$$\partial_t u = \Delta u - \nabla p - (u \cdot \nabla)u + f, \quad \nabla \cdot u = 0, \tag{6.21}$$

with 2π-periodic boundary conditions and external (time-independent) force $f \in L^2$ with $\widehat{f}_0 = 0$. As above we find

$$\frac{1}{2}\frac{\mathrm{d}}{\mathrm{d}t}\|u\|_{H^1}^2 \leq -\|u\|_{H^1}^2 + \|u\|_{L^2}\|f\|_{L^2} \leq -\frac{1}{2}\|u\|_{H^1}^2 + \frac{1}{2}\|f\|_{L^2}^2$$

using $\|u\|_{L^2}\|f\|_{L^2} \leq \frac{1}{2}\|u\|_{L^2}^2 + \frac{1}{2}\|f\|_{L^2}^2$, and so

$$\|u(t)\|_{H^1} \leq \mathrm{e}^{-t}\|u_0\|_{H^1} + \|f\|_{L^2}(1 - \mathrm{e}^{-t}).$$

Hence, the set

$$\mathcal{B} = \{u \in H^1 : \|u\|_{H^1} \leq 2\|f\|_{H^1}\}$$

is absorbing, i.e., attracts balls of finite size in finite times. Moreover, it is (positively) invariant under the flow of (6.21), i.e., $u_0 \in \mathcal{B}$ implies $u(t, u_0) \in \mathcal{B}$ for all $t \geq 0$.

Since the embedding $H^1 \hookrightarrow L^2$ is compact (cf. Theorem 5.1.33), as in Theorem 5.3.4 we thus obtain the existence of the global attractor (in L^2), given by

$$\mathcal{A} = \bigcap_{t \geq 0} A_t, \qquad \text{with} \quad A_t = \overline{\mathcal{S}_t(\mathcal{B})}.$$

Theorem 6.2.19. *The 2D Navier-Stokes equations* (6.21) *have a non-empty, compact, time-invariant set $\mathcal{A} \subset L^2$, the global attractor, with*

$$\operatorname{dist}_{L^2}(S_t(\mathcal{B}), \mathcal{A}) = \sup_{b \in S_t(\mathcal{B})} \inf_{a \in \mathcal{A}} \|a - b\|_{L^2} \to 0 \text{ as } t \to \infty.$$

Remark 6.2.20. With a little more work it can be shown that $\mathcal{A}$ is a H^1_{per} attractor for (6.21), i.e., $\mathcal{A}$ is compact in H^1 and attracts in H^1, i.e.,

$$\operatorname{dist}_{H^1}(S_t(\mathcal{B}), \mathcal{A}) = \sup_{b \in S_t(\mathcal{B})} \inf_{a \in \mathcal{A}} \|a - b\|_{H^1} \to 0 \text{ as } t \to \infty.$$

To show this, use a priori estimates to obtain an absorbing set in H^2_{per}. ⌋

6.3. Other boundary conditions and more general domains

In this section we consider the Navier-Stokes equations with Dirichlet boundary conditions $u|_{\partial\Omega} = 0$ in an open domain $\Omega \subset \mathbb{R}^d$ with smooth boundary. In order to prove the local existence and uniqueness of solutions we generalize our previous approach and recall the basics of analytic semigroup theory, cf. [**Hen81**]. We recall that the resolvent set of a (bounded or unbounded) linear operator $A : D(A) \subset X \to X$ is defined as the set of all $\lambda \in \mathbb{C}$ for which $(\lambda - A)$ has a bounded inverse $(\lambda - A)^{-1} : X \to X$, the resolvent.

Definition 6.3.1. *A closed and densely defined operator A in a Banach space X is called sectorial if there exists an $a \in \mathbb{R}$, a $\phi \in (0, \frac{\pi}{2})$ and an $M \geq 1$, such that the sector*

$$S_{a,\phi} = \{\ \lambda : \phi \leq |\arg(\lambda - a)| \leq \pi,\ \lambda \neq a\ \}$$

is a part of the resolvent set of A, and such that for all $\lambda \in S_{a,\phi}$ we have the estimate

$$\|(\lambda - A)^{-1}\|_{X \to X} \leq \frac{M}{|\lambda - a|}.$$

The negative of a sectorial operator generates an analytic semigroup.

Definition 6.3.2. *A C_0-semigroup $T(t)$ of bounded linear operators is called analytic if $t \mapsto T(t)u$ is analytic for $0 < t < \infty$ and all $u \in X$.*

The following theorem gives an explicit construction of the semigroup generated by the negative of a sectorial operator.

Theorem 6.3.3. *Let A be a sectorial operator. Then $-A$ generates an analytic semigroup with the representation*

$$\mathrm{e}^{-tA} := \frac{1}{2\pi i}\int_\Gamma (\lambda + A)^{-1}\mathrm{e}^{\lambda t}\,\mathrm{d}\lambda,$$

where Γ is a curve in the resolvent set $\rho(-A)$ with $\arg\lambda \to \pm\theta$ for $|\lambda| \to \infty$ and a $\theta \in (\frac{\pi}{2}, \pi)$. The semigroup can be extended analytically into the sector $\{t \neq 0 : |\arg t| \leq \varepsilon\}$ for a $\varepsilon > 0$. If $\mathrm{Re}\,\lambda > a$ for λ in the spectrum $\sigma(A)$ then

$$\|\mathrm{e}^{-tA}\| \leq C\mathrm{e}^{-at} \qquad \textit{und} \qquad \|A\mathrm{e}^{-tA}\| \leq \frac{C}{t}\mathrm{e}^{-at}$$

for $t > 0$ and a constant C. Moreover,

$$\frac{\mathrm{d}}{\mathrm{d}t}\mathrm{e}^{-tA} = -A\mathrm{e}^{-tA}.$$

Remark 6.3.4. a) For $t > 0$ the integral

$$\frac{1}{2\pi i}\int_\Gamma (\lambda + A)^{-1}\mathrm{e}^{\lambda t}\,\mathrm{d}\lambda$$

is well defined since $\mathrm{Re}\,\lambda \to -\infty$ for $|\lambda| \to \infty$ and $\|(\lambda + A)^{-1}\|_{X\to X} \leq \frac{M}{|\lambda - a|}$. Since the resolvent is holomorphic on the resolvent set, due to Cauchy's theorem of complex analysis the value of the integral is independent of the special choice of Γ.

b) The estimate $\|A\mathrm{e}^{-tA}\|_{X\to X} \leq \frac{C}{t}\mathrm{e}^{-at}$ implies that e^{-tA} maps the space X into the domain of definition $D(A)$. Since $A^n\mathrm{e}^{-tA} = (A\mathrm{e}^{-A\frac{t}{n}})^n$ we even have

$$\|A^n\mathrm{e}^{-tA}\|_{X\to X} \leq \|A\mathrm{e}^{-A\frac{t}{n}}\|^n_{X\to X} \leq \left(\frac{C}{t}\mathrm{e}^{-a\frac{t}{n}}\right)^n \leq \frac{\widetilde{C}}{t^n}\mathrm{e}^{-at},$$

such that $\mathrm{e}^{-tA} : X \to D(A^n)$ for $t > 0$, and the semigroup is smoothing. ⌋

Remark 6.3.5. a) The concept of sectorial operators is very robust under perturbations. Let A be a sectorial operator with $\|A(\lambda - A)^{-1}\| \leq C$ for all λ in a chosen sector. Moreover, let B be a linear operator with $D(B) \supset D(A)$ satisfying

$$\|Bx\| \leq \varepsilon\|Ax\| + K\|x\|$$

with ε, K some constants. If $\varepsilon C < 1$, then also $A + B$ is a sectorial operator. For a self-adjoint A it is sufficient that $\varepsilon < 1$. Hence, it is sufficient to check the assumptions for the principal part of a given operator. Such estimates can often be found in the existing literature.

b) The most essential remark is that, due to Parseval's identity, in a Hilbert space every self-adjoint operator which is bounded from below is a sectorial operator. ⌋

In order to apply the previous ideas to the Navier-Stokes equations we first have to get rid of the pressure term and of the equation $\nabla \cdot u = 0$. Therefore, let $u \in C^1(\Omega, \mathbb{R}^d)$ with $\nabla \cdot u = 0$ and $u \cdot n|_{\partial\Omega} = 0$. Then for $\phi \in C^1(\Omega, \mathbb{R})$ we have $\int_\Omega u \cdot \nabla\phi \, dx = 0$. On the other hand a vector field u which is orthogonal to $\{\nabla\phi : \phi \in C^1(\Omega, \mathbb{R})\}$ satisfies $\nabla \cdot u = 0$ and $u \cdot n|_{\partial\Omega} = 0$.

We define L^2_p to be the closure of $\{\nabla\phi : \phi \in C^1(\Omega, \mathbb{R})\}$ and L^2_{div} to be the closure of $\{u \in C^1(\Omega, \mathbb{R}^d) : \nabla \cdot u = 0, \quad u \cdot n|_{\partial\Omega} = 0\}$. Then L^2_p and L^2_{div} are orthogonal subspaces of L^2 with $L^2 = L^2_P \oplus L^2_{div}$. We introduce P to be the orthogonal projection on the subspace L^2_{div}.

As before we write the Navier Stokes equations as $\partial_t u = -Au + N(u)$ with $Au = -P\Delta$ under Dirichlet boundary conditions and $N(u) = -P((u \cdot \nabla)u)$. It is easy to see that A is a self adjoint and positive definite operator which immediately implies that A is a sectorial operator, too. Hence, $-A$ is the generator of an analytic semigroup, cf. Remark 6.3.5 b).

Again u is called mild solution if u satisfies the variation of constant formula

$$u(t) = \mathrm{e}^{-tA} u_0 + \int_0^t \mathrm{e}^{-(t-\tau)A} N(u(\tau)) \, \mathrm{d}\tau.$$

In order to control the nonlinear terms we need so called X^α-spaces. We consider a sectorial operator A with $\operatorname{Re}\sigma(A) > \delta > 0$. For a given sectorial operator A this can always be achieved by considering $\widetilde{A} = A + \beta I$ for a suitable chosen $\beta > 0$.

Definition 6.3.6. *For $\alpha > 0$ define*

$$A^{-\alpha} = -\frac{1}{2\pi i} \int_\Gamma \lambda^{-\alpha} (\lambda I - A)^{-1} \, \mathrm{d}\lambda,$$

where Γ is a curve asymptotically coming from $\mathrm{e}^{-i\theta}\infty$ and asymptotically going to $\mathrm{e}^{i\theta}\infty$ with $\frac{\pi}{2} - \widetilde{\delta} < \theta < \pi$ running between the origin and $\sigma(A)$. The branch of the function $\lambda \mapsto \lambda^{-\alpha}$ is chosen in such a way that the slit in the complex plane where $\lambda \mapsto \lambda^{-\alpha}$ is not analytic coincides with the negative real axis, see [**Gam01**] *for an introduction to complex analysis.*

Since $\|(\lambda I - A)^{-1}\| \leq \frac{C}{|\lambda|}$ we have the convergence of the integral for $\alpha > 0$. There exists a $C \geq 0$, such that $\|A^{-\alpha}\| \leq C$ for $0 < \alpha \leq 1$. Moreover, we have $A^{-\alpha}A^{-\beta} = A^{-(\alpha+\beta)}$ if $\alpha, \beta \in (0, 1]$.

Since $\operatorname{Re}\sigma(A) > \delta > 0$ we have the injectivity of A^{-n}. Since $A^{-n} = A^{-n+\alpha}A^{-\alpha}$ for $n > \alpha$ we have the injectivity of $A^{-\alpha}$. Therefore, $A^{-\alpha} : X \to \mathbb{R}(A^{-\alpha})$ is bijective and we have a (non bounded) inverse. Other representation formulas for A^α can be found for instance in [**Hen81**, §1.4]

Definition 6.3.7. *We set $A^\alpha = (A^{-\alpha})^{-1}$ for $\alpha > 0$. The domain of definition is given by $D(A^\alpha) = R(A^{-\alpha})$. We introduce $X^\alpha = D(A^\alpha)$ equipped with the norm*

$$\|u\|_{X^\alpha} = \|A^\alpha u\|_X.$$

For $\alpha \in (0,1)$ we have $D(A) \subset D(A^\alpha)$ since $R(A^{-1}) \subset R(A^{-\alpha})$, and so $D(A^\alpha)$ is dense in X. Due to the construction of the operators we have that

$$\|e^{-tA}u\|_{X^\alpha} \leq M_\alpha t^{-\alpha} e^{-\delta t}\|u\|_X.$$

In order to proceed with the local existence uniqueness of solutions as above we need the Lipschitz-continuity of the nonlinear terms $N(u)$ from X^α to X. The following lemma reduces this proof to the proof of the Lipschitz-continuity from $W^{k,q}$ to L^p, cf. [**Hen81**, Theorem 1.6.1].

Lemma 6.3.8. *Let $\Omega \subset \mathbb{R}^d$ be an open set with smooth boundary, let $1 \leq p < \infty$, and let A be a sectorial operator in $X = L^p(\Omega)$ with $D(A) = X^1 \subset W^{m,p}(\Omega)$ for a $m \geq 1$. Then*

$$X^\alpha \subset W^{k,q} \qquad \text{or} \qquad X^\alpha \subset C^\nu(\Omega)$$

for $\alpha \in [0,1]$ if $k - d/q < m\alpha - d/p$, $q \geq p$ or $0 \leq \nu < m\alpha - d/p$.

In order to prove that $D(A) = X^1 \subset W^{2,2}$, for $f \in L^2$ one has to find solutions $u \in W^{2,2}$ of the elliptic problem

$$-\Delta u + \nabla p = f, \qquad \nabla \cdot u = 0$$

in Ω with Dirichlet boundary conditions for u. The existence of such solutions is implied by elliptic regularity theory [**ADN59, ADN64**]. Since $D(A) \subset W^{2,2}$ it follows for $d = 3$ by Lemma 6.3.8 that for $\alpha \in (1/2, 1)$ that $X^\alpha \subset W^{1,q}$ provided $1/q > (5 - 4\alpha)/6$ and that for $\alpha \in (3/4, 1)$ that $X^\alpha \subset L^\infty$. Therefore

$$\|N(u)\|_X = \|N(u)\|_{L^2} \leq \|u\|_{L^\infty}\|\nabla u\|_{L^2} \leq C\|u\|_{X^\alpha}^2$$

for $\alpha \in (3/4, 1)$. For $d = 2$ we find $\alpha \in (1/2, 1)$. Hence, we have the Lipschitz-continuity of the polynomial $N(u)$ from X^α to X for $d = 3$ if $\alpha \in (3/4, 1)$ and for $d = 2$ if $\alpha \in (1/2, 1)$.

Theorem 6.3.9. *Let $\alpha \in (3/4, 1)$ if $d = 3$ or $\alpha \in (1/2, 1)$ if $d = 2$. For $u_0 \in X^\alpha$ there exists a $T_0 > 0$ such that the Navier-Stokes equations possess a unique mild solution $u \in C([0, T_0], X^\alpha)$ with $u|_{t=0} = u_0$.*

In the proof of the global existence of solutions $u \in C([0, T_0], H^{1,div}_{\text{per}})$ of the Navier-Stokes equations in $d = 2$ space dimensions in §6.2.3 we used (6.19) which only holds in $\mathbb{R}^2$ and periodic boundary conditions. In this section we prove the global existence without using (6.19). The method gives weaker estimates but it is more general and periodic boundary conditions are not needed; nevertheless we keep them for simplicity.

With the Gagliardo-Nirenberg inequality L^p-norms can be estimated by L^q-Norms and gradients, for $p > q$. Such inequalities are called interpolation estimates since a norm in the middle (L^p) is interpolated with the help of a weaker norm (L^q) and a stronger norm (gradient). Here we give a simple version. See, e.g., [**Hen81**, Page 37] for a general version, and Exercise 6.6 for the proof of an even weaker version.

Lemma 6.3.10. (A simple Gagliardo-Nirenberg inequality) *For $d = 1, 2, 3, 4$ we have*

$$\|u\|_{L^4(\mathbb{T}^d)} \le C\|u\|_{L^2(\mathbb{T}^d)}^{1-d/4}\|u\|_{H^1(\mathbb{T}^d)}^{d/4}.$$

With the help of Lemma 6.3.10 we can proceed as follows. The L^2-estimate in domains with general Lipschitz-continuous boundary and Dirichlet boundary conditions works exactly the same as before. For the H^1-norm we estimate again as before

$$\frac{1}{2}\frac{\mathrm{d}}{\mathrm{d}t}\int_{\mathbb{T}^d}|\nabla u|^2\,\mathrm{d}x \le -\int_{\mathbb{T}^d}|\Delta u|^2\,\mathrm{d}x + g(u) \tag{6.22}$$

with $g(u) = \left|\int_{\mathbb{T}^d}(\Delta u)\cdot((u\cdot\nabla)u)\,\mathrm{d}x\right|$. For periodic boundary conditions and $d = 2$ we have $g(u) = 0$. For general boundary conditions and/or $d = 3$ the best estimate known is

$$\begin{aligned} g(u) &\le \|\Delta u\|_{L^2}\|(u\cdot\nabla)u\|_{L^2} \le \|\Delta u\|_{L^2}\left(\int_{\mathbb{T}^d}|\nabla u|^2|u|^2\,\mathrm{d}x\right)^{1/2} \\ &\le \|\Delta u\|_{L^2}\|u\|_{L^4}\|\nabla u\|_{L^4} \end{aligned}$$

where we used the Cauchy-Schwarz inequality. The Gagliardo-Nirenberg estimate and the Poincaré-inequality $\|u\|_{H^m} \le C\sum_{|\alpha|=m}\|\partial_x^\alpha u\|_{L^2}$, see §5.2.2, give

$$\begin{aligned} g(u) \le& \|\Delta u\|_{L^2}\left(C\|u\|_{L^2}^{1-d/4}\|u\|_{H^1}^{d/4}\right)\left(C\|\nabla u\|_{L^2}^{1-d/4}\|\nabla u\|_{H^1}^{d/4}\right) \\ \le& C\|\Delta u\|_{L^2}^{1+d/4}\|u\|_{L^2}^{1-d/4}\|\nabla u\|_{L^2}. \end{aligned}$$

In order to balance the factor $\|\Delta u\|_{L^2}^2$ on the right-hand side against $-\|\Delta u\|_{L^2}^2$ in (6.22) we use Young's inequality

$$ab \le \frac{a^p}{p} + \frac{b^q}{q} \quad \text{for } a, b \ge 0 \text{ and } p, q > 1 \text{ with } 1/p + 1/q = 1. \tag{6.23}$$

Case $d = 2$: We choose $\varepsilon = (4/3)^{3/4}$ and obtain

$$g(u) \le \frac{C}{\varepsilon}\|u\|_{L^2}^{1/2}\|\nabla u\|_{L^2}\;\varepsilon\|\Delta u\|_{L^2}^{3/2}$$

$$\leq \frac{1}{4}\left(\frac{C}{\varepsilon}\|u\|_{L^2}^{1/2}\|\nabla u\|_{L^2}\right)^4 + \frac{3\varepsilon^{4/3}}{4}\|\Delta u\|_{L^2}^2$$
$$\leq \|\Delta u\|_{L^2}^2 + C\|u\|_{L^2}^2\|\nabla u\|_{L^2}^4$$

using (6.23) with $q = 4/3$ and $p = 4$. Hence, we obtain

$$\frac{\mathrm{d}}{\mathrm{d}t}\|\nabla u\|_{L^2}^2 \leq a(t)\|\nabla u\|_{L^2}^2, \qquad \text{where} \qquad a(t) = 2C\|u\|_{L^2}^2\|\nabla u\|_{L^2}^2.$$

From

$$\frac{1}{2}\frac{\mathrm{d}}{\mathrm{d}t}\|u\|_{L^2}^2 = -\|\nabla u\|_{L^2}^2,$$

cf. §6.2.3, follows $\|u(t)\|_{L^2} = \mathrm{e}^{-t}\|u_0\|_{L^2}$, and then

$$\int_0^t \|\nabla u(\tau)\|_{L^2}^2\,\mathrm{d}\tau = \|u_0\|_{L^2}^2 - \|u(t)\|_{L^2}^2.$$

Note, that this estimate does not imply a uniform bound for $h(t) = \|\nabla u(t)\|_{L^2}$, but its square-integrability. As a consequence,

$$0 \leq \int_0^t a(t) \leq C \sup_{\tau\in(0,t)} \|u(\tau)\|_{L^2}^2 \int_0^t \|\nabla u\|_{L^2}^2\,\mathrm{d}\tau < \infty$$

uniformly for all $t \geq 0$ and so

$$\|\nabla u(t)\|_{L^2}^2 \leq \mathrm{e}^{\int_0^t a(\tau)\,\mathrm{d}\tau}\|\nabla u_0\|_{L^2}^2 \leq M\|\nabla u_0\|_{L^2}^2$$

for a $M \geq 0$ independent of $t \geq 0$. Therefore, the local solution can be extended to a global solution, i.e., we obtain the global existence in 2D.

Case $d = 3$: Similar to the 2D case we use the Gagliardo-Nirenberg estimate and Young's inequality with $q = 8/7$ and $p = 8$ and obtain

$$g(u) = \left|\int_{\mathbb{T}^d}(\Delta u)\cdot((u\cdot\nabla)u)\,\mathrm{d}x\right| \leq \|\Delta u\|_{L^2}\|u\|_{L^4}\|\nabla u\|_{L^4}$$
$$\leq C\|\Delta u\|_{L^2}^{7/4}\|\nabla u\|_{L^2}\|u\|_{L^2}^{1/4} \leq C\left(\|\nabla u\|_{L^2}\|u\|_{L^2}^{1/4}\right)^8 + \|\Delta u\|_{L^2}^2.$$

If we proceed as above it follows

$$\frac{\mathrm{d}}{\mathrm{d}t}\|\nabla u\|_{L^2}^2 \leq a(t)\|\nabla u\|_{L^2}^2, \qquad \text{where} \qquad a(t) = 2C\|u\|_{L^2}^2\|\nabla u\|_{L^2}^6.$$

The equation for the dissipation only guarantees that the function $h(t) = \|\nabla u(t)\|_{L^2}$ is in $L^2((0,T_0))$, but not in $L^6((0,T_0))$, and therefore we cannot proceed as above. For a further discussion we refer to [**Wie99, Con01**].

Remark 6.3.11. Since so far it cannot be proved that for $d = 3$ unique global solutions exist, it appears to be nonsense to discuss their long time dynamics. Nevertheless, if one assumes the existence of global strong solutions, then the concept of attractor is again quite useful. In particular, there are a number of estimates for the dimension of attractors (which is finite) for the Navier-Stokes equations in two and three space dimensions, cf. [**Tem97**]. However, a finite-dimensional attractor by no means implies that the dynamics is "simple". In fact there is a lot of theory on turbulent flows and also on so called fully developed turbulence which is mainly based on methods from statistical physics, cf. [**FRMT01**]. ⌋

Further Reading. Classical books about local existence and uniqueness of the solutions of the Navier-Stokes equations are [**vW85, Tem01**]. More background on the derivation and applications of the Navier-Stokes equations and related equations can be found in [**Fow97**]. Concise treatments of these equations as a dynamical system, embedded in the general existence theory of semilinear parabolic equations can be found for instance in [**Hen81, DG95, Tem97, Rob01**]; the latter three are also recommended for the so called Galerkin method as an alternative to semigroup methods for proving local existence in the Navier-Stokes equations and general parabolic systems.

An excellent textbook, going way beyond the brief summary given here, is [**MB02**]. An essential reference for steady problems in exterior domains, including the necessary function spaces and inequalities, is [**Gal11**]. The dynamics and stability of vortices is treated in [**MB02, GW05, GW06b**]. See [**WW15**, Chapter 1] for a very accessible account on metastable states and the finite dimensionality of the global attractor for the 2D Navier-Stokes equations. An enlightening essay about the Navier-Stokes Millenium problem is [**Tao09**], emphazising the scale invariance, see Exercise 6.7. Finally, [**Lem16**] gives an impressive overview about the state of the art of the mathematical analysis of the Navier-Stokes equations, and an excellent review of the Millennium problems, with focus on the Navier-Stokes equations.

Exercises

6.1. (a) Show that $u : \mathbb{R}^2 \to \mathbb{R}^2$ with $\nabla \times u = 0$ and $\nabla \cdot u = 0$ is equivalent to the Cauchy-Riemann differential equations for $w(z) = u_1(x,y) - \mathrm{i}u_2(x,y)$, $z = x + \mathrm{i}y$.
(b) Sketch the flow belonging to $w(z) = z^2$ and calculate the associated pressure.

6.2. (d'Alembert Paradox) Let $a > 0$, $U \in \mathbb{R}^3$, and $\phi(x) = \left(\frac{a^3}{2\|x\|^3} + 1\right)\langle U, x\rangle$ for $x \in \Omega = \mathbb{R}^3 \setminus B_a(0)$. Sketch $u = \nabla\phi$, show that $\operatorname{div} u = 0$, and calculate the drag $f = -\int_{B_a(0)} \phi n \, \mathrm{d}S$.

6.3. For given $\omega = \omega(x)$, $x \in \mathbb{T}^2$, find in Fourier space an explicit solution u of

$$\nabla \times u = \omega \quad \text{and} \quad \nabla \cdot u = 0.$$

6.4. Show that $\|\widehat{u} * \widehat{v}\|_{\ell^\omega_{1,\sqrt{t}}} \le \|\widehat{u}\|_{\ell^\omega_{1,\sqrt{t}}} \|\widehat{v}\|_{\ell^\omega_{1,\sqrt{t}}}$.

6.5. Show the enstrophy identity (6.19).

6.6. Prove the following weak form of the Gagliardo-Nirenberg inequality for $(2\pi)^d$-periodic functions. For $d = 1, 2, 3$ and $\delta > 0$ we have

$$\|u\|_{L^4(\mathbb{T}^d)} \leq C\|u\|_{L^2(\mathbb{T}^d)}^{1-d/4-\delta}\|u\|_{H^1(\mathbb{T}^d)}^{d/4+\delta}.$$

Hint: Use $\|u\|_{L^p} \leq C\|\hat{u}\|_{L^q}$ for $1/p+1/q = 1$ and $q \in [1,2]$ and the Hölder inequality

$$\sum_{k\in\mathbb{Z}} |a_k b_k c_k| \leq (\sum_{k\in\mathbb{Z}} |a_k|^{p_1})^{1/p_1}(\sum_{k\in\mathbb{Z}} |b_k|^{p_2})^{1/p_2}(\sum_{k\in\mathbb{Z}} |c_k|^{p_3})^{1/p_3}$$

with $1/p_1 + 1/p_2 + 1/p_3 = 1$.

6.7. Let $(u, p) : \mathbb{R}^d \times [0, T) \to \mathbb{R}^d \times \mathbb{R}$ be a solution of the d-dimensional Navier-Stokes equations over $\mathbb{R}^d$. For $l > 0$ set $u^l(x, t) = l^{\alpha_1}u(l^\beta x, l^\gamma t)$ and $p^l(x, t) = l^{\alpha_2}p(l^\beta x, l^\gamma t)$. Find $\alpha_{1,2}, \beta, \gamma$, and δ such that (u^l, p^l) is again a solution of the Navier-Stokes equation. Discuss how the energy and energy dissipation behave under this scaling.

Chapter 7

Some dissipative PDE models

This is the first chapter of Part III of this book. Here and in the remainder of this book we consider PDEs on unbounded domains. In order to avoid dealing with far away boundaries, whose influence on the solutions in the interior of the domain is small at least for a long time, we idealize the large domain to an unbounded domain. For instance, instead of x from some large interval $(-L, L)$ we consider $x \in \mathbb{R}$. From a didactic point of view the consideration of unbounded domains has certain advantages. Since we do not have to deal with boundary conditions which are often a source of functional analytic difficulties, this idealization allows to explain genuine PDE phenomena such as transport, diffusion and dispersion. Hence, it allows us to keep the functional analytic tools at a minimum. Unbounded domains are easy in this respect.

On the other hand, compared to PDEs over bounded domains there are new fundamental and challenging open questions, mainly due to the fact that PDEs on unbounded domains define dynamical systems with uncountably many modes (degrees of freedom). In contrast to the situation of countably many modes considered in Chapters 5 and 6, a separation of the uncountably many modes into single modes is a highly singular action from a functional analytic point of view, and therefore in general of little use. The recovery of compactness by smoothing properties is no longer true, and therefore finite-dimensional attractors in general cannot be expected.

To illustrate our point of view, the following example shows that also for PDEs defined on a very large domain in space the interpretation as countably many ODEs is no longer a big help.

Example. Consider the linear wave equation $\partial_t^2 u = \partial_x^2 u$ for $t \in \mathbb{R}$, $x \in (-L, L)$, $L > 0$ very large, $u(x,t) \in \mathbb{R}$, with Dirichlet boundary conditions $u(-L,t) = u(L,t) = 0$. We consider two special classes of solutions, first the oscillations of the eigenfunctions,

$$u(x,t) = \sin(n\pi t/(2L)) \sin(n\pi(x-L)/(2L)), \quad n \in \mathbb{N},$$

and secondly the traveling wave solutions

$$u(x,t) = f(x-t) + g(x+t)$$

with f and g arbitrary smooth functions with compact support

$$\mathrm{supp}(f) = \{x : f(x) \neq 0\} \subset [-1,1] \qquad \text{and} \qquad \mathrm{supp}(g) \subset [-1,1].$$

As long as $|t| < L - 1$ this is a solution of the PDE, i.e., for a very large time interval traveling wave solutions play a role. An expansion of these solutions in eigenfunctions is of no use. ⌋

In this Chapter 7 we start with some scalar model problems. These are the Kolmogorov-Petrovsky-Piskounov (KPP) or Fisher equation in §7.1, the Allen-Cahn equation in §7.2, and the Burgers equation in §7.4. Moreover, there is the method oriented §7.3 about Fourier transform. We keep the exposition rather brief and aim for a basic understanding of the various models. We are interested in the local existence and uniqueness of solutions and in the existence and stability of special solutions, which are important for the underlying physical processes which are described by the models. To construct these special solutions we often use the ODE methods from Part I. In order to make this part more self-contained we recall a number of definitions and constructions which are only small adaptions of respective concepts from Part I. In Chapter 8 we consider with the NLS, KdV, and the GL equation the three canonical modulation equations whose dynamics we will recover in more complicated PDEs in Part IV of this book. Part III of this book is closed with Chapter 9 about reaction-diffusion systems.

For each of the equations considered in this Part there already exists much literature on various levels, see the "further reading" at the end of each chapter.

7.1. The KPP equation

The Kolmogorov-Petrovsky-Piskounov (KPP) equation [**KPP37**] or Fisher equation [**Fis37**]

$$\partial_t u = \partial_x^2 u + u - u^2, \tag{7.1}$$

with $t \geq 0$, $x \in \mathbb{R}$, and $u = u(x,t) \in \mathbb{R}$, occurs as a model for various systems in nature, for instance for chemical reactions or population dynamics. The equation consists of two parts, namely the diffusion term $\partial_x^2 u$ and the nonlinear reaction term $u - u^2$. Therefore, it brings together PDE with ODE dynamics.

Inserting $u(x,t) = v(t)$ into (7.1) gives the one-dimensional ODE

$$\dot{v} = v - v^2. \tag{7.2}$$

The 1D phase portrait shows that the fixed point $v = 0$ is unstable and that the fixed point $v = 1$ is asymptotically stable. The term $+v$ in the KPP equation describes exponential growth for small v and the term $-v^2$ represents saturation. For instance, a population of animals or a chemical reaction initially increases with some exponential rate until the growth is saturated by the available food or the missing reactant. If $v(0) > 0$, then $\lim_{t\to\infty} v(t) = 1$.

Before we combine the ODE dynamics coming from the reaction term $u - u^2$ with the dynamics coming from the diffusion term $\partial_x^2 u$ we discuss the modeling and the properties of linear diffusion in the next two subsections.

7.1.1. The modeling of diffusion.

Diffusion occurs in various situations. We explain three such situations, namely Brownian motion, a discrete random walk, and Fourier's law.

Brownian motion and diffusion. The term Brownian motion is named after Robert Brown who in 1827 described the irregular motion of pollen particles suspended in water. In [**Ein05**] Einstein studied Brownian motion the following way. Consider a long, thin tube filled with clear water, into which we inject at time $t = 0$ a unit amount of ink, at the location $x = 0$. Let $u(x,t)$ denote the density of ink at position $x \in \mathbb{R}$ and time $t \geq 0$. Suppose that the probability that an ink particle moves from x to $x + y$ in a time τ is translational invariant, i.e., does not depend on x. This probability is denoted by $\rho(y,\tau)$. Then

$$\begin{aligned} u(x, t+\tau) &= \int_{\mathbb{R}} u(x-y,t)\rho(y,\tau)\,\mathrm{d}y \\ &= \int_{\mathbb{R}} \Big(u(x,t) - (\partial_x u(x,t))y + \frac{1}{2}(\partial_x^2 u(x,t))y^2 + \dots \Big)\, \rho(y,\tau)\,\mathrm{d}y. \end{aligned}$$

Now $\rho(-y,\tau) = \rho(y,\tau)$ by symmetry such that $\int_{\mathbb{R}} y\rho(y,\tau)\,\mathrm{d}y = 0$. Next assume that the variance is linear in τ, i.e.,

$$\int_{\mathbb{R}} y^2 \rho(y,\tau)\,\mathrm{d}y = 2D\tau \tag{7.3}$$

for a $D > 0$. Then

$$\frac{1}{\tau}(u(x,t+\tau) - u(x,t)) = D\partial_x^2 u(x,t) + \text{h.o.t.}$$

Under the assumption that all higher moments of ρ decay faster than τ for $\tau \to 0$, in the limit $\tau \to 0$ we obtain the linear diffusion equation

$$\partial_t u = D\partial_x^2 u. \tag{7.4}$$

Einstein derived the relation $2D = RT/(N_A\nu)$ where R is the gas constant, T the temperature, N_A the Avogadro number, and ν a friction coefficient.

A discrete random walk. We consider a two-dimensional rectangular lattice, comprising the sites $\{(m\delta x, n\delta t) : m = 0, \pm 1, \pm 2, \ldots ; n = 0, 1, 2, \ldots\}$. A particle starting in $x = 0$ at a time $t = 0$ decides at each time $n\delta t$ to move an amount δx to the left or to move an amount δx to the right, both possibilities with probability 1/2. Denote by $p(m, n)$ the probability that the particle is at the position $m\delta x$ at the time $n\delta t$. Then $p(0, 0) = 1$ and $p(m, 0) = 0$ for $m \neq 0$. Also, $p(m, n + 1) = \frac{1}{2}(p(m - 1, n) + p(m + 1, n))$, hence

$$p(m, n + 1) - p(m, n) = \frac{1}{2}(p(m - 1, n) - 2p(m, n) + p(m + 1, n)).$$

Now assume that $(\delta x)^2/(\delta t) = 2D$ which corresponds to (7.3) above. Then

$$\frac{1}{\delta t}(p(m, n + 1) - p(m, n)) = \frac{D}{(\delta x)^2}(p(m - 1, n) - 2p(m, n) + p(m + 1, n)),$$

and sending $\delta \to 0$ again yields the linear diffusion equation (7.4).

Fourier's law. Let $u : \mathbb{R}^3 \to \mathbb{R}$ be the temperature inside a material body and let $V \subset \mathbb{R}^3$ be a test volume with surface S. Then

$$\frac{\mathrm{d}}{\mathrm{d}t}\int_V u\,\mathrm{d}V = -\int_S j \cdot n\,\mathrm{d}S = -\int_V \operatorname{div} j\,\mathrm{d}V\,,$$

where $j = j(x, t) \in \mathbb{R}^3$ is the heat flow. Since this is true for all test volumes V we find

$$\partial_t u + \operatorname{div} j = 0\,.$$

It is reasonable to assume that the heat flow from warm to cold is proportional to the negative temperature gradient, i.e., $j = -D\nabla u$. This so called Fourier's law again yields the heat equation

$$\partial_t u = D \operatorname{div} \nabla u = D\Delta u.$$

7.1.2. Diffusion on the real line. Throughout this subsection we consider the linear diffusion equation with diffusion coefficient $D = 1$, i.e.,

$$\partial_t u = \partial_x^2 u, \tag{7.5}$$

which always can be achieved by a rescaling of time or space. We already observed the dissipative character of this equation in previous sections where we found the solutions $u(x, t) = \mathrm{e}^{-k^2 t}\sin(kx)$ which decay to zero for $|k| > 0$ with some exponential rate.

For $t > 0$ the general formula for the solutions of the linear diffusion equation (7.5) is given by

$$u(x,t) = \frac{1}{\sqrt{4\pi t}} \int_{-\infty}^{\infty} \mathrm{e}^{-\frac{(x-y)^2}{4t}} u(y,0)\,\mathrm{d}y. \tag{7.6}$$

The derivation of this formula is given subsequently, but also by a different method in §7.3. The existence of this integral is guaranteed for $t > 0$ if for instance $\sup_{y\in\mathbb{R}} |u(y,0)| < \infty$.

From (7.6) we immediately obtain the estimate

$$\sup_{x\in\mathbb{R}} |u(x,t)| \leq \frac{1}{\sqrt{4\pi t}} \int_{-\infty}^{\infty} |u(x,0)|\,\mathrm{d}x,$$

i.e., solutions to spatially localized initial conditions decay uniformly towards zero with a rate $t^{-1/2}$. Since mass is conserved, i.e.,

$$\int_{-\infty}^{\infty} u(x,t)\,\mathrm{d}x = \int_{-\infty}^{\infty} u(x,0)\,\mathrm{d}x$$

for all $t \geq 0$, this is how diffusion is expected to work. The conservation of mass follows for instance with the use of the solution formula from

$$\begin{aligned} \int_{-\infty}^{\infty} u(x,t)\,\mathrm{d}x &= \int_{-\infty}^{\infty} \left(\int_{-\infty}^{\infty} \frac{1}{\sqrt{4\pi t}} \mathrm{e}^{-\frac{(x-y)^2}{4t}}\,\mathrm{d}x \right) u(y,0)\,\mathrm{d}y \\ &= \int_{-\infty}^{\infty} 1 \cdot u(y,0)\,\mathrm{d}y. \end{aligned}$$

The decay happens in a universal manner. The initial condition

$$u(y,0) = \delta_0(y),$$

with δ_0 the "δ-distribution in $x = 0$", cf. Example 5.2.2, leads to the self-similar solution

$$u(x,t) = \frac{1}{\sqrt{4\pi t}} \mathrm{e}^{-\frac{x^2}{4t}}. \tag{7.7}$$

The solution only exists for $t > 0$. This is a general rule. For an arbitrary initial condition the diffusion equation cannot be solved backwards in time. Since this solution is the starting point of the construction of the general solution formula it is also called fundamental solution. In order to derive (7.7) we make the ansatz $u(x,t) = v\left(\frac{x}{\sqrt{t}}\right)$ and find $v'' = -\frac{1}{2}\xi v'$ which is solved by $v' = c\mathrm{e}^{-\frac{\xi^2}{4}}$ with a constant $c \in \mathbb{R}$. Since with $u(x,t) = v\left(\frac{x}{\sqrt{t}}\right)$ also $\partial_x u = \frac{1}{\sqrt{t}} v'\left(\frac{x}{\sqrt{t}}\right)$ is a solution of (7.5) we find (7.7).

In lowest order self-similar behavior is also observed for general spatially localized initial conditions, namely,

$$u(x,t) = \frac{A^*}{\sqrt{t}} v\left(\frac{x}{\sqrt{t}}\right) + \mathcal{O}(1/t) \quad \text{with} \quad v(\xi) = \mathrm{e}^{-\xi^2/4} \tag{7.8}$$

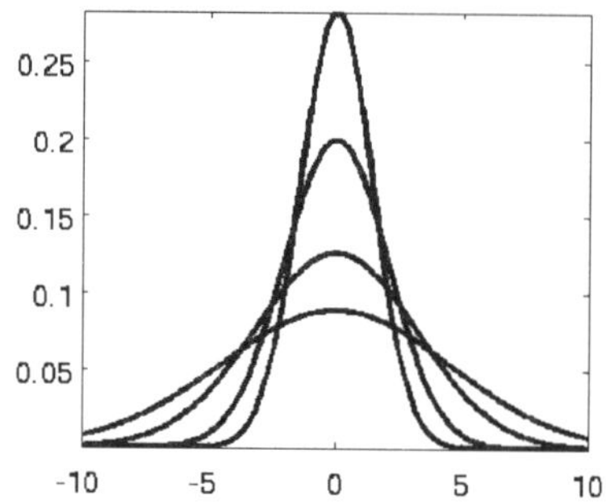

Figure 7.1. Self-similar decay of the special solution $u(x,t) = \frac{1}{\sqrt{4\pi t}}\mathrm{e}^{-\frac{x^2}{4t}}$ for $t \in \{1,2,5,10\}$.

and with a constant $A^* \in \mathbb{R}$ depending on the initial conditions. This is explained in detail in §14.

Diffusion is smoothing. We obtain the estimate

$$\begin{aligned}\sup_{x\in\mathbb{R}} |\partial_x u(x,t)| &= \sup_{x\in\mathbb{R}} \left| \frac{1}{\sqrt{4\pi t}} \int_{-\infty}^{\infty} -\frac{(x-y)}{2t} \mathrm{e}^{-\frac{(x-y)^2}{4t}} u(y,0)\,\mathrm{d}y \right| \\ &= \sup_{x\in\mathbb{R}} \left| \frac{1}{\sqrt{\pi}} \int_{-\infty}^{\infty} -\frac{s}{\sqrt{t}} \mathrm{e}^{-s^2} u\left(x - 2\sqrt{t}s, 0\right) \mathrm{d}s \right| \\ &\leq \frac{1}{\sqrt{t}} \frac{2}{\sqrt{\pi}} \int_0^{\infty} s\mathrm{e}^{-s^2}\,\mathrm{d}s \sup_{x\in\mathbb{R}} |u(x,0)| \leq \frac{C}{\sqrt{t}} \sup_{x\in\mathbb{R}} |u(x,0)|,\end{aligned}$$

with a constant C independent of t and of $u(\cdot,0)$, where we made the transformation $s = (x-y)/(2\sqrt{t})$, i.e., $\mathrm{d}s = -\,\mathrm{d}y/(2\sqrt{t})$. This can easily be generalized.

Theorem 7.1.1. *Let $u = u(\cdot,t)$ be a solution of the linear diffusion equation. Then for all $n \in \mathbb{N}$ there exists a $C > 0$ such that for all $t > 0$*

$$\|\partial_x^n u(\cdot,t)\|_{C_b^0} \leq Ct^{-n/2} \|u(\cdot,0)\|_{C_b^0}.$$

Finally for every $t_0 > 0$ and $x_0 \in \mathbb{R}$ the function $(x,t) \mapsto u(x,t)$ can be expanded in a convergent power series around (x_0,t_0), i.e., u is an analytic function and can be extended into the complex plane. See §6.2.2.

In order to handle the linear diffusion equation $\partial_t u = \partial_x^2 u$ with dynamical systems methods we have to choose a suitable phase space. We already know that in infinite dimensions the choice of phase space and associated norm is fundamental. A first choice is $X = C_{b,unif}^0(\mathbb{R},\mathbb{R})$, the space of uniformly bounded and uniformly continuous functions $u : \mathbb{R} \to \mathbb{R}$, equipped with the norm

$$\|u(t)\|_{C_{b,unif}^0} = \sup_{x\in\mathbb{R}} |u(x,t)|\,.$$

Lemma 7.1.2. *The curve $t \mapsto u(t, u_0)$ is continuous in X if $u_0 \in X$.*

Proof. Since $u(t+s, u_0) = u(t, u(s, u_0))$ it is sufficient to prove the continuity of the orbit $t \mapsto u(t)$ for $t \searrow 0$ in the space X. With $H(z) = \frac{1}{\sqrt{4\pi}} e^{-z^2/4}$ we estimate

$$\begin{aligned}
\|u(t,u_0) - u_0\|_{C_b^0} &= \sup_{x\in\mathbb{R}} |u(x,t) - u(x,0)| \\
&= \sup_{x\in\mathbb{R}} \left| \int_{-\infty}^{\infty} \frac{1}{\sqrt{t}} H\left(\frac{x-y}{\sqrt{t}}\right) (u(y,0) - u(x,0)) \mathrm{d}y \right| \\
&= \sup_{x\in\mathbb{R}} \left| \int_{-\infty}^{\infty} H(z) \left(u(x - \sqrt{t}z, 0) - u(x,0) \right) \mathrm{d}z \right| \\
&\leq \sup_{x\in\mathbb{R}} \int_{|z|\leq R} |\ldots| \mathrm{d}z + \sup_{x\in\mathbb{R}} \int_{|z|\geq R} |\ldots| \mathrm{d}z = s_1 + s_2.
\end{aligned}$$

For a given $\varepsilon > 0$ we have to find a $t_0 > 0$ such that for all $t \in (0, t_0)$ we have $s_1 + s_2 < \varepsilon$. We can estimate

$$s_2 \leq 2 \int_{|z|\geq R} H(z) \mathrm{d}z \sup_{x\in\mathbb{R}} |u(x,0)| < \varepsilon/2$$

by choosing an $R > 0$ sufficiently large due to the definition of H. Next, we estimate

$$\begin{aligned}
s_1 &\leq \int_{-\infty}^{\infty} H(z) \mathrm{d}z \sup_{x\in\mathbb{R}, |z|\leq R} |u(x + \sqrt{t}z, 0) - u(x,0)| \\
&= \sup_{x\in\mathbb{R}, |z|\leq R} |u(x + \sqrt{t}z, 0) - u(x,0)|.
\end{aligned}$$

Since $x \mapsto u(x, 0)$ is uniformly continuous for all $\varepsilon > 0$ there exists a $\delta > 0$ such that for all $y \in \mathbb{R}$ with $|y| < \delta$ we have $|u(x + y, 0) - u(x, 0)| < \varepsilon/2$. Choosing $t_0 > 0$ so small that $t_0 R < \delta$ we are done. $\square$

The deeper reason why $X = C_b^0$ would not be a good choice is explained in the following remark, but for a slightly simpler PDE.

Remark 7.1.3. If we consider the translation semigroup $T(t) : u(\cdot) \mapsto u(\cdot + t)$ which is the solution operator of the transport equation $\partial_t u = \partial_x u$ in the space $X = C_b^0(\mathbb{R}, \mathbb{R})$, we have that $T(t)$ is not a C_0-semigroup, cf. Definition 5.1.9, since for $u(x) = \sin(x^2)$ which is an element of C_b^0, but not of $C_{b,unif}^0$ we always have $\|u(\cdot) - u(\cdot + t)\|_{C_b^0} = 2$, if $t > 0$. The same is true for the linear diffusion equation. The deeper reason for the non-continuity of both semigroups in C_b^0 is the fact that the domain of definition C_b^1, respectively C_b^2, is not dense in C_b^0, cf. the theorem of Hille-Yosida, cf. [**Paz83**, Section 1.3, Theorem 3.1]. In order to have a C_0-semigroup we have to restrict to $X = C_{b,unif}^0(\mathbb{R}, \mathbb{R})$, which excludes the counter-example

$u(x) = \sin(x^2)$, where the faster and faster oscillations for $|x| \to \infty$ destroy the uniform continuity w.r.t. x and t. ⌋

Remark 7.1.4. With $(x,t) \mapsto u(x,t)$ a solution of (7.5) also $(x,t) \mapsto u(x+y,t)$ is a solution of (7.5), i.e., every solution shifted by y is a solution, too. More abstractly, the solution operator and the translation operator commute. With $t \mapsto u(x,t)$ a solution, obviously every derivative $t \mapsto \partial_x^n u(x,t)$ and every integral is a solution, too.

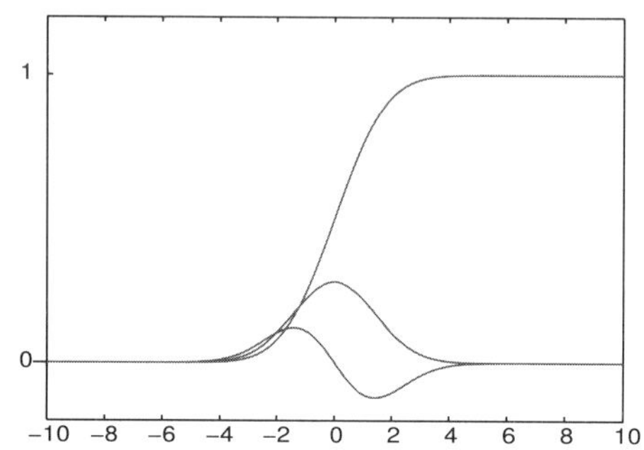

Figure 7.2. The solutions $u(x,t) = \frac{1}{\sqrt{4\pi t}} e^{-\frac{x^2}{4t}}$, $\partial_x u$ and its first integral for $t = 1$.

Moreover, every linear combination and every convergent series of solutions, or every convergent integral over a set of solutions are again solutions. These properties are used in the following paragraph. ⌋

Our starting point to derive the solution formula (7.6) is the explicit solution (7.7). With this solution also

$$u(x,t) = \frac{1}{\sqrt{t}} e^{-\frac{(x-y)^2}{4t}} \qquad \text{and} \qquad u(x,t) = \int_{-\infty}^{\infty} \frac{c(y)}{\sqrt{t}} e^{-\frac{(x-y)^2}{4t}} dy$$

are solutions of (7.5). From the limit $t \to 0$ and Lemma 7.1.2 we find that $c(y) = \frac{u(y,0)}{\sqrt{4\pi}}$. The solution formula can be interpreted as linear combination of fundamental solutions, i.e., of diffusion processes starting in every point $x \in \mathbb{R}$ with a δ-distribution as initial condition.

7.1.3. The reaction-diffusion equation. The KPP equation is obtained by adding the diffusion term to the ODE (7.2). Thus, the KPP equation describes for instance a chemical reaction or the evolution of a disease in a large, here in an infinitely extended, domain, where the concentration $u = u(x,t)$ spreads into space by diffusion.

In order to handle the KPP equation as an abstract dynamical system we choose the same phase space as for the linear diffusion equation, namely $X = C^0_{b,\text{unif}}(\mathbb{R}, \mathbb{R})$. Solutions $u = u(t) \in X$ satisfy the KPP equation only in a weak sense.

Definition 7.1.5. *Solutions of* (7.5) *are called mild solutions of* (7.5) *if they satisfy the variation of constant formula*

$$u(t) = T(t)u_0 + \int_0^t T(t-\tau)N(u)(\tau)\,\mathrm{d}\tau, \tag{7.9}$$

where $T(t)$ is the solution operator of the linear diffusion equation, i.e.,

$$(T(t)u_0)(x) = \frac{1}{\sqrt{4\pi t}}\int_{-\infty}^{\infty} \mathrm{e}^{-\frac{(x-y)^2}{4t}} u_0(y)dy,$$

and where $N(u)(\tau)$ represents the reaction term $u(x,\tau) - u(x,\tau)^2$.

Lemma 7.1.6. *If $u \in C([0,T_0], C^2_{b,\mathrm{unif}}) \cap C^1([0,T_0], C^0_{b,\mathrm{unif}})$ solves the variation of constant formula* (7.9) *then u is also a classical solution of the KPP equation, i.e., $u = u(x,t)$ can be inserted into the KPP equation.*

Proof. See Exercise 7.5. □

There is local existence and uniqueness of mild solutions for the KPP equation in X.

Theorem 7.1.7. *For all $C_0 > 0$ there exists a $T_0 > 0$ such that for $u_0 \in X$ with $\|u_0\|_X \leq C_0$ there exists a unique solution $u \in C([0,T_0], X)$ of the KPP equation* (7.5) *with $u|_{t=0} = u_0$ and $X = C^0_{b,\mathrm{unif}}(\mathbb{R},\mathbb{R})$.*

Proof. The proof of the local existence and uniqueness of solutions is based on the contraction mapping principle and on the variation of constant formula. Fix $C_1 > 0$. For sufficiently small $T_0 > 0$ the right-hand side of the variation of constant formula (7.9) defines a contraction F in the complete metric space

$$M = C^0([0,T_0], \{u(t) \in X : \|u(t) - T(t)u_0\|_X \leq C_1\})$$

with the metric $d(u,v) = \|u - v\|_M$ induced by the norm

$$\|u\|_M = \sup_{t\in[0,T_0]} \|u(t)\|_X.$$

We have $C_2 := \sup_{u\in M} \|u\|_M \leq C_0 + C_1$. In order to prove the contraction mapping property we need three estimates. For $t \geq 0$, $u_0 \in X$ and $u, v \in M$ we have

$$\textbf{(E1)} \|T(t)u_0\|_X \leq \|u_0\|_X,$$

$$\textbf{(E2)} \|N(u)\|_M \leq C_2 + C_2^2,$$

$$\textbf{(E3)} \|N(u) - N(v)\|_M \leq (1 + 2C_2)\|u - v\|_M.$$

Estimate **(E1)** follows from

$$\begin{aligned}\|T(t)u_0\|_X &= \sup_{x\in\mathbb{R}}\left|\frac{1}{\sqrt{4\pi t}}\int_{-\infty}^{\infty}\mathrm{e}^{-\frac{(x-y)^2}{4t}}u_0(y)dy\right|\\ &\le \sup_{x\in\mathbb{R}}\left|\frac{1}{\sqrt{4\pi t}}\int_{-\infty}^{\infty}\mathrm{e}^{-\frac{(x-y)^2}{4t}}dy\right|\cdot\sup_{y\in\mathbb{R}}|u_0(y)|\\ &= \sup_{y\in\mathbb{R}}|u_0(y)| = \|u_0\|_X\,.\end{aligned}$$

Estimate **(E2)** follows from

$$\begin{aligned}\|N(u)\|_M &= \sup_{t\in[0,T_0]}\sup_{x\in\mathbb{R}}\left|u(x,t)-u(x,t)^2\right|\\ &\le \sup_{t\in[0,T_0]}\sup_{x\in\mathbb{R}}|u(x,t)| + \sup_{t\in[0,T_0]}\sup_{x\in\mathbb{R}}|u(x,t)|^2 \le C_2 + C_2^2.\end{aligned}$$

Estimate **(E3)** follows from

$$\begin{aligned}\|N(u)-N(v)\|_M &= \sup_{t\in[0,T_0]}\sup_{x\in\mathbb{R}}\left|u(x,t)-u(x,t)^2-v(x,t)+v(x,t)^2\right|\\ &\le \sup_{t\in[0,T_0]}\sup_{x\in\mathbb{R}}\left(|u(x,t)-v(x,t)|+|u(x,t)+v(x,t)|\,|u(x,t)-v(x,t)|\right)\\ &\le \|u-v\|_M + (\|u\|_M+\|v\|_M)\|u-v\|_M \le (1+2C_2)\|u-v\|_M\,.\end{aligned}$$

i) F maps M into M, since

$$\begin{aligned}\|F(u)-(T(t)u_0)_{t\ge 0}\|_M &= \sup_{t\in[0,T_0]}\|\int_0^t T(t-\tau)N(u)(\tau)\,\mathrm{d}\tau\|_X\\ &\le \sup_{t\in[0,T_0]}\int_0^t\|T(t-\tau)N(u)(\tau)\|_X\,\mathrm{d}\tau\\ &\le \sup_{t\in[0,T_0]}\int_0^t\|N(u)(\tau)\|_X\,\mathrm{d}\tau\\ &\le \sup_{t\in[0,T_0]}\int_0^t\mathrm{d}\tau\|N(u)\|_M \le \left(C_2+C_2^2\right)T_0 \le C_1\end{aligned}$$

for $T_0 > 0$ sufficiently small.

ii) F is a contraction in M, since

$$\begin{aligned}\|F(u)-F(v)\|_M &= \sup_{t\in[0,T_0]}\left\|\int_0^t T(t-\tau)(N(u)-N(v))(\tau)\,\mathrm{d}\tau\right\|_X\\ &\le \sup_{t\in[0,T_0]}\int_0^t\|T(t-\tau)(N(u)-N(v))(\tau)\|_X\,\mathrm{d}\tau\\ &\le \sup_{t\in[0,T_0]}\int_0^t\|(N(u)-N(v))(\tau)\|_X\,\mathrm{d}\tau\end{aligned}$$

$$\leq \sup_{t\in[0,T_0]} \int_0^t d\tau \|(N(u) - N(v))\|_M$$
$$\leq T_0(1 + 2C_2)\|u - v\|_M \leq \frac{1}{2}\|u - v\|_M$$

for $T_0(1 + 2C_2) \leq 1/2$. Thus, for $T_0 > 0$ sufficiently small there is a fixed point $u = F(u)$, which is a mild solution of the KPP equation. □

Remark 7.1.8. The KPP equation is a semi-linear parabolic equation [**Hen81**], i.e., the semigroup $T(t)$ generated by the operator $\Lambda = \partial_x^2$ is smoothing and the nonlinearity N contains only derivatives of strictly lower order than in Λ. Like for the diffusion equation the solutions of the KPP equation or more general of a semi-linear parabolic equation on the real line are analytic for all $t > 0$. The proof is similar to the one given in §6.2.2. ⌋

7.1.4. The maximum principle. Second order scalar parabolic and elliptic PDEs have a special property which helps a lot in their analysis, namely the maximum principle.

Theorem 7.1.9. *Let u_1, u_2 be bounded solutions of the KPP equation* (7.1) *in X for $t \in [t_0, t_1]$. From $u_1(x, t_0) \leq u_2(x, t_0)$ for all $x \in \mathbb{R}$ it follows that $u_1(x, t) \leq u_2(x, t)$ for all $x \in \mathbb{R}$ and all $t \in [t_0, t_1]$.*

Idea of the proof. We show a slightly modified statement, namely that $u_1(x, t_0) < u_2(x, t_0)$ for all $x \in \mathbb{R}$ implies $u_1(x, t) < u_2(x, t)$ for all $x \in \mathbb{R}$ and all $t \in [t_0, t_1]$.

If this is wrong, then there exists a $t_* \in (t_0, t_1)$ and a $x_0 \in \mathbb{R}$ such that $u_1(x_0, t_*) = u_2(x_0, t_*)$. Since the solutions change smoothly in x and t in this point we necessarily have $\partial_x u_1(x_0, t_*) = \partial_x u_2(x_0, t_*)$ and generically $\partial_x^2 u_1(x_0, t_*) < \partial_x^2 u_2(x_0, t_*)$. Then it follows

$$\begin{aligned}\partial_t(u_2(x_0, t_*) - u_1(x_0, t_*)) &= \partial_x^2(u_2(x_0, t_*) - u_1(x_0, t_*)) \\ &\quad + (u_2(x_0, t_*) - u_1(x_0, t_*)) - (u_2^2(x_0, t_*) - u_1^2(x_0, t_*)) \\ &= \partial_x^2(u_2(x_0, t_*) - u_1(x_0, t_*)).\end{aligned}$$

In the generic case $\partial_x^2(u_2(x_0, t_*) - u_1(x_0, t_*)) > 0$, we have again $u_2(x_0, t) > u_1(x_0, t)$ for $t \in [t_0, t_1]$, i.e., u_1 can never be larger than u_2. For a complete proof see for instance [**RR04**, Theorem 4.26]. □

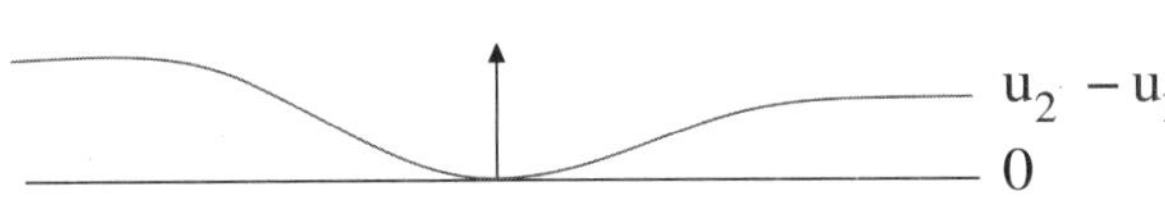

Figure 7.3. The maximum principle: the difference of the solutions for $t = t_*$. For $t > t_*$ we have $u_2 - u_1 > 0$ again.

A direct consequence is the following lemma.

Lemma 7.1.10. *Solutions that start at t_0 with values in $[0,1]$ will stay with their values in $[0,1]$ for $t \geq t_0$, i.e., $0 \leq u(x,t_0) \leq 1$ for all $x \in \mathbb{R}$ implies $0 \leq u(x,t) \leq 1$ for all $x \in \mathbb{R}$ and all $t \geq t_0$.*

Proof. Since $u \equiv 0$ and $u \equiv 1$ are solutions of the KPP equation, the assertion follows from the maximum principle. □

Hence, it makes sense to talk about concentrations. A second consequence of the maximum principle is the global existence and uniqueness of mild solutions of the KPP equation. The estimate from Lemma 7.1.10 is called a priori estimate. A priori estimates and local existence and uniqueness of solutions gives global existence and uniqueness of solutions. See e.g. §2.2, §5.3.1, or §6.2.3.

Theorem 7.1.11. *For $u_0 \in X$ with $0 \leq u_0(x) \leq 1$ there exists a unique solution $u \in C([0,\infty),X)$ of the KPP equation with $u|_{t=0} = u_0$ and $X = C^0_{b,\mathrm{unif}}(\mathbb{R},\mathbb{R})$.*

Proof. For all initial conditions with values in $[0,1]$ the local existence and uniqueness theorem 7.1.7 guarantees the existence for all $t \in [0,T_0]$ with for instance $T_0 = 1/6$. By the last lemma we have $0 \leq u(x,T_0) \leq 1$. Then we start the KPP equation again, but now with initial condition $u(T_0)$. The local existence and uniqueness theorem guarantees for the same reason solutions for all $t \in [T_0, 2T_0]$. Repeating the argument shows the assertion. □

Theorem 7.1.12. *For $u_0 \in X$ with $0 \leq u_0(x) \leq 1$ there exists a unique solution $u \in C([0,\infty),X)$ of the KPP equation with $u|_{t=0} = u_0$ and $X = C^0_{b,\mathrm{unif}}(\mathbb{R},\mathbb{R})$.*

Proof. For all initial conditions with values in $[0,1]$ the local existence and uniqueness theorem 7.1.7 guarantees the existence for all $t \in [0,T_0]$ with for instance $T_0 = 1/10$. By the last lemma we have $0 \leq u(x,T_0) \leq 1$. Then we start the KPP equation again, but now with initial condition $u(T_0)$. The local existence and uniqueness theorem guarantees for the same reason solutions for all $t \in [T_0, 2T_0]$. Repeating the argument shows the assertion. □

7.1.5. Stability, instability, and fronts. We are interested in the qualitative behavior of solutions of the KPP equation. Due to the chemical or biological motivation of the KPP equation we restrict to solutions u with $0 \leq u(x,t) \leq 1$. As usual we start with the fixed points, i.e., with the stationary solutions of the KPP equation. They satisfy

$$0 = \partial_x^2 u + u - u^2.$$

By having a look at the phase portrait in Figure 7.4 we see that the only equilibria having only values in $[0,1]$ are the fixed points $u=0$ and $u=1$.

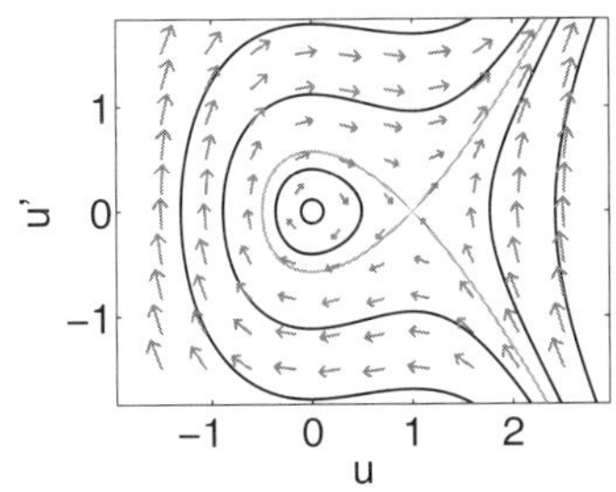

Figure 7.4. The phase portrait for the stationary solutions of the KPP equation.

Next we are interested in the stability of these stationary solutions. We have exactly the same definition as for ODEs, cf. Definition 2.3.3, except that the ODE phase space $\mathbb{R}^d$ has to be replaced by the PDE phase space X.

Definition 7.1.13. *A fixed point u^* is called stable for the KPP equation in X if for any $\varepsilon > 0$ there is a $\delta > 0$ such that $\|u_0 - u^*\|_X < \delta$ implies $\|u(t,u_0) - u^*\|_X < \varepsilon$ for all $t \geq 0$. Otherwise, it is called unstable. A stable fixed point is called asymptotically stable in X if additional* $\lim_{t\to\infty} u(t,u_0) = u^*$ *holds.*

For ODEs, in most cases the eigenvalues of the linearized system determine stability or instability, cf. Theorem 2.3.4. For $u^*=0$ we find the linear operator $(\partial_x^2+1)\cdot$ which has the eigenfunctions e^{ikx} for $k\in\mathbb{R}$ and the spectrum $(-\infty,1]$. For $u^*=1$ we find the linear operator $(\partial_x^2-1)\cdot$ which has the eigenfunctions e^{ikx} for $k\in\mathbb{R}$ and the spectrum $(-\infty,-1]$. Hence, we expect that $u^*=0$ is unstable, whereas $u^*=1$ is expected to be asymptotically stable. However, instead of using these spectra, here we use the maximum principle to study stability and instability.

Theorem 7.1.14. *The fixed point $u^*=0$ is unstable in X.*

Proof. Let $u_0 = \delta/2$ for $0<\delta\ll 1$. Since u_0 is constant in space, no diffusion occurs and the dynamics is determined by the ODE $\frac{\mathrm{d}}{\mathrm{d}t}v = v - v^2$. Therefore, we have $\lim_{t\to\infty} u(t,u_0) = 1$ and hence $u(t,u_0)$ leaves every ε-neighborhood of $u^*=0$ independent of how small $\delta>0$ was at the beginning. □

Theorem 7.1.15. *The fixed point $u^*=1$ is asymptotically stable in X.*

Proof. Choose $\delta = \min\{\varepsilon, 1/10\}$ and let $u_0 \in X$ with $\|u_0 - u^*\|_X < \delta$. Let $v_- = 1-\delta$ and $v^+ = 1+\delta$ such that $v_- \le u_0(x) \le v_+$ for all $x \in \mathbb{R}$. By the maximum principle we have for the associated solutions

$$v_-(t) \le u(x,t) \le v_+(t)$$

for all $x \in \mathbb{R}$ and $t \ge 0$. Since $\lim_{t\to\infty} v_-(t) = \lim_{t\to\infty} v_+(t) = 1$ we can conclude

$$\lim_{t\to\infty} \|u(t,u_0) - u^*\|_X = 0,$$

i.e., the asymptotic stability of u^*. □

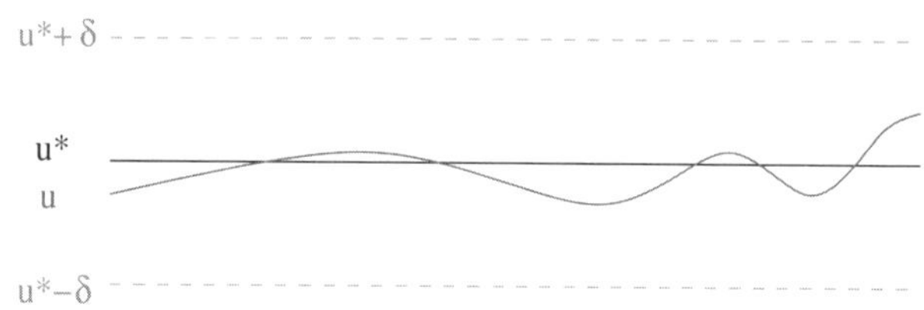

Figure 7.5. By the maximum principle the solution u is confined by the spatial constants.

If a chemical reaction (which is described by the KPP equation) is started locally in space, i.e., if we have a small positive perturbation with compact support of the unstable equilibrium $u^* = 0$, then we expect two reaction fronts to be created, one moving to the left and one moving to the right. In the following we are interested in the velocities of these reaction fronts.

Thus, we consider solutions of permanent form, i.e. $u(x,t) = v(x-ct) = v(\xi)$, with velocity c, satisfying $\lim_{\xi\to-\infty} v(\xi) = 1$ and $\lim_{\xi\to\infty} v(\xi) = 0$. Inserting this into the KPP equation gives the ODE

$$-c\partial_\xi v = \partial_\xi^2 v + v - v^2$$

which we write as first order system $(' = \partial_\xi)$

$$\begin{aligned} v' &= w, \\ w' &= -cw - v + v^2. \end{aligned}$$

We find that the qualitative form of the phase portraits in Figure 7.6 changes when c is increased.

For all values of $c \neq 0$ we find a heteroclinic connection between the fixed points $(v,w) = (0,0)$ and $(v,w) = (1,0)$. Since the heteroclinic connections are in the intersection of the one-dimensional unstable manifold of $(1,0)$ and the two-dimensional stable manifold of $(0,0)$ this connection is very robust under perturbations. However, for $|c| < 2$ these heteroclinic connections spiral into $(v,w) = (0,0)$ which can be seen by considering the linearization

$$v' = w, \quad w' = -cw - v. \tag{7.10}$$

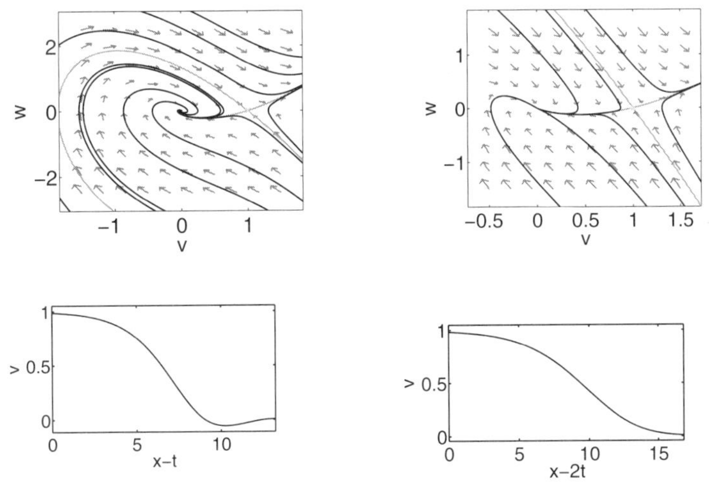

Figure 7.6. The phase portraits for the front solutions for the KPP equation for $c = 1$ (left) and $c = 2$ (right), with the associated fronts.

We find the eigenvalues

$$\lambda_{1,2} = \frac{-c \pm \sqrt{c^2 - 4}}{2}.$$

For $c \in (0, 2)$ we have complex conjugate eigenvalues with $\operatorname{Re}\lambda_{1,2} < 0$ and $\operatorname{Im}\lambda_{1,2} \neq 0$, and for $c \geq 2$ we have real negative eigenvalues. Thus, monotonic fronts which only take values in $[0, 1]$ can only exist for $c \geq 2$. The goal of Exercise 7.6 is an analytical existence proof of these fronts.

Remark 7.1.16. There is a very general result that heteroclinic connections for the ODE which is satisfied by the spatially homogeneous solutions imply the existence of fronts with large c in the PDE, cf. [**AJMR95**]. In detail, for the KPP equation the front satisfies

$$\partial_\xi^2 u + c\partial_\xi u + u - u^2 = 0.$$

We rescale time $\xi = c\zeta$, such that $\partial_\xi = \varepsilon\partial_\zeta$ where $\varepsilon = 1/c$ is small. Then

$$\varepsilon^2\partial_\zeta^2 v + \partial_\zeta v + v - v^2 = 0 \tag{7.11}$$

which, for $\varepsilon \to 0$, converges towards the ODE of the pure reaction, i.e.,

$$\partial_\zeta v + v - v^2 = 0. \tag{7.12}$$

The theory of singularly perturbed ODEs [**Fen79**] guarantees that the heteroclinic connection between the fixed points of (7.12) persists for small $\varepsilon > 0$ in (7.11). ⌋

In order to control the velocity of the reaction fronts created by small positive spatially localized initial conditions we use the maximum principle. To an initial condition u_0 with compact support and values in $[0, 1)$ we find an exact front $u_- = u_-(x)$ moving to the left and an exact front $u_+ = u_+(x)$ moving to the right with minimal velocity $c = 2$ satisfying

$$u_-(x) \geq u_0(x) \leq u_+(x)$$

for all $x \in \mathbb{R}$. From the maximum principle we obtain

$$u_-(x + 2t) \geq u(x, t) \leq u_+(x - 2t).$$

Therefore, the reaction fronts cannot move faster than the exact fronts with the minimal velocities $c = \pm 2$. In [**Bra83**] it has been shown that the upper bound is a good approximation and that e.g. the reaction front moving to the right moves like

$$x_f(t) = 2t - c_1 \ln t + c_2(t),$$

with $\lim_{t\to\infty} c_2(t) = c_2^* < \infty$, and c_1, $c_2^* \in \mathbb{R}$ some constants. See Figure 7.7.

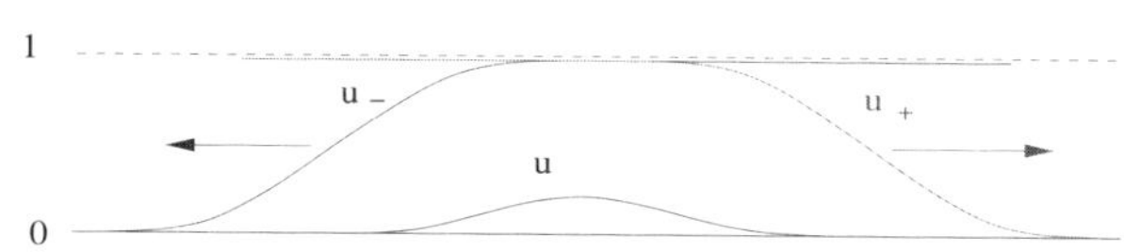

Figure 7.7. Solutions to localized initial conditions are confined by two fronts with minimal velocity.

In the following we restrict to front solutions with velocities $|c| \geq 2$ which allows us to argue with the maximum principle.

Theorem 7.1.17. *The front solutions are unstable in X.*

Proof. Let u_f be a front solution with velocity c, and $u_0(x) = \min(u_f(x), \delta/2)$. Then $\|u_f - u_0\|_X < \delta$ with $u(t, u_f)[x] = u_f(x - ct)$, but $\lim_{t\to\infty} u(t, u_0) = 1$ due to the maximum principle, since $\delta/2 \leq u_0(x) \leq 1$, and hence

$$\lim_{t\to\infty} \|u(t, u_f) - u(t, u_0)\|_X = 1 > \varepsilon$$

independent of how small $\delta > 0$ is chosen. See Figure 7.8. □

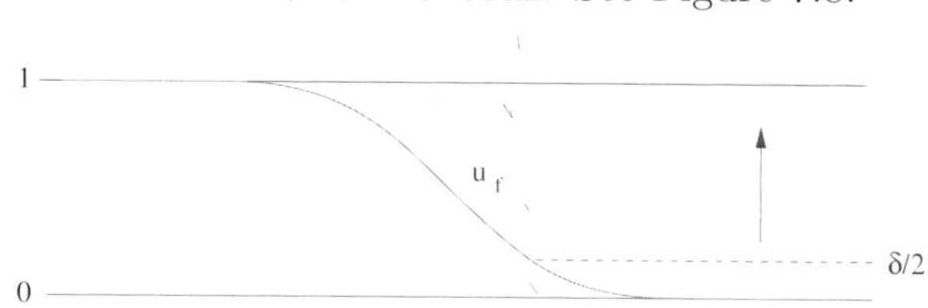

Figure 7.8. Instability of front solutions w.r.t. perturbations in $C^0_{b,\mathrm{unif}}$.

Remark 7.1.18. On the other hand, by the maximum principle the front solutions are stable w.r.t. smaller sets of perturbations. If the small perturbation v of the front solution u_f is contained between two translates of the same front then by the maximum principle it will stay there for all times. In detail, if

$$u_f(x - x_-) \leq u_f(x) + v(x,0) \leq u_f(x + x_+)$$

for all $x \in \mathbb{R}$, then

$$u_f(x - ct - x_-) \leq u_f(x - ct) + v(x,t) \leq u_f(x - ct + x_+)$$

for all $t \geq 0$ and all $x \in \mathbb{R}$. See Figure 7.9. ⌋

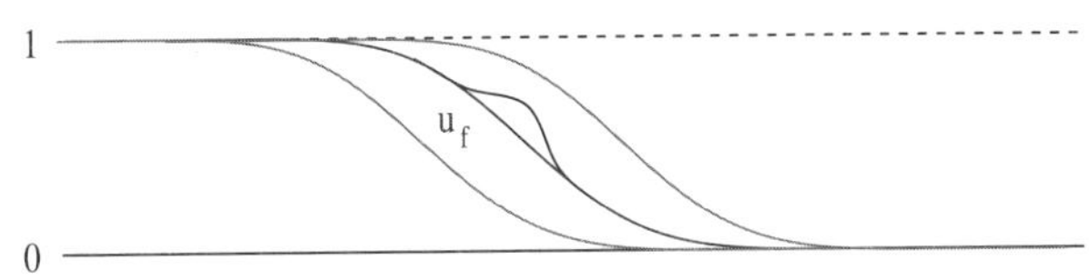

Figure 7.9. Stability of the fronts w.r.t. smaller sets of perturbations. The perturbations are confined by parallel fronts with the same velocity.

This leads to the following generalization of the definition of stability.

Definition 7.1.19. *A fixed point u^* is called (X_1, X_2)-stable, if for all $\varepsilon > 0$ there exists a $\delta > 0$ such that $\|u_0 - u^*\|_{X_1} < \delta$ implies $\|u(t, u_0) - u^*\|_{X_2} < \varepsilon$ for all $t \geq 0$.*

Definition 7.1.19 makes sense due to the fact that in infinite dimensions there are infinitely many non-equivalent norms. For instance, the fronts are $(X_1, C^0_{b,\text{unif}})$ stable, where $\|u\|_{X_1} = \sup_{x\in\mathbb{R}} |u(x)e^{x^2}|$, see, e.g., [**BK00**, §9.3]. Note that e^{-x^2} decays much faster than the difference of two fronts which converge towards each other with some exponential rate.

Example 7.1.20. Consider

$$\partial_t u = \partial_x^2 u + c\partial_x u + u.$$

Obviously $u = 0$ is unstable in $C^0_{b,\text{unif}}$. However, in a weighted space stability can be obtained. The transform $u(x,t) = v(x,t)e^{\beta x}$ yields

$$\begin{aligned}\partial_t v =&\partial_x^2 v + 2\beta\partial_x v + \beta^2 v + c\partial_x v + c\beta v + v\\ =&\partial_x^2 v + (2\beta + c)\partial_x v + (\beta^2 + c\beta + 1)v\ .\end{aligned}$$

Hence, $v = 0$ may expected to be stable if $\beta^2 + c\beta + 1 < 0$ which can be satisfied for $|c| > 2$ by choosing $\beta \in (\beta_-, \beta_+)$ with $\beta_\pm = -c/2 \pm \sqrt{c^2/4 - 1}$. Indeed, setting $\gamma_1 = 2\beta + c$, $\gamma_2 = \beta^2 + c\beta + 1$ and $v(x,t) = w(x + \gamma_1 t, t)$ we obtain $\partial_t w(x,t) = \partial_x^2 w(x,t) + \gamma_2 w(x,t)$ and from this $w(x,t) = e^{\gamma_2 t} e^{t\partial_x^2} w_0(x) \to 0$

in $C^0_{b,\text{unif}}$ as $t \to \infty$. In summary, we have asymptotic (X, X) stability of $u = 0$, where $\|u\|_X = \sup_x |u(x)e^{-\beta x}|$, i.e., $\|u_0\|_X \le \delta \Rightarrow \|u(t)\|_X \to 0$ as $t \to \infty$. Note that this does not mean that u stays bounded. What happens is that for $c > 2$ and hence $\beta < 0$ the (growing) mass is transported to $-\infty$ where it vanishes in the X-norm. This is sometimes called convective stability. ⌋

Remark 7.1.21. The idea of weighted variables from Example 7.1.20 can be used to prove the nonlinear stability of the front solutions u_f for $|c| > 2$. In a co-moving frame $\xi = x - ct$ the deviation $v = u - u_f$ from the front satisfies

$$\partial_t v(\xi, t) = \partial_\xi^2 v(\xi, t) + c\partial_\xi v(\xi, t) + (1 - 2u_f(\xi))v(\xi, t) - v(\xi, t)^2.$$

The weighted variable $w(\xi, t) = v(\xi, t)e^{-\beta\xi}$ satisfies

$$\partial_t w(\xi, t) = L_w w(\xi, t) - e^{\beta\xi} w(\xi, t)^2,$$

with

$$L_w w(\xi, t) = \partial_\xi^2 w(\xi, t) + (2\beta + c)\partial_\xi w(\xi, t) + (\beta^2 + c\beta + 1 - 2u_f(\xi))w(\xi, t).$$

Since $u_f \ge 0$ we can find a $\beta < 0$ such that $w = 0$ is linearly stable, e.g., in H^1. However, the nonlinearity is not well defined in H^1 due to the factor $e^{\beta\xi}$. Therefore, the equation for v and w have to be combined, namely

$$\begin{aligned}\partial_t v(\xi, t) =& L_v v(\xi, t) + (2 - 2u_f(\xi))e^{\beta\xi} w(\xi, t) - v(\xi, t)^2,\\ \partial_t w(\xi, t) =& L_w w(\xi, t) - v(\xi, t) w(\xi, t),\end{aligned}$$

with

$$L_v v(\xi, t) = \partial_\xi^2 v(\xi, t) + c\partial_\xi v(\xi, t) - v(\xi, t).$$

Since the semigroups generated by L_v and L_w decay with some exponential rate and since $(2-2u_f(\xi))e^{\beta\xi}$ is bounded, the asymptotic stability of $v{=}w{=}0$ follows, i.e., u_f is asymptotically stable w.r.t. perturbations v which are small w.r.t. the norm $\|v\|_{H^1} + \|ve^{\beta x}\|_{H^1}$. This idea goes back to [**Sat76**]. ⌋

For an extensive discussion of traveling wave solutions in parabolic systems see [**VVV94**]. An overview about tools for investigating the stability of pulses and fronts can be found in [**San02**].

7.2. The Allen-Cahn equation

Closely related to the KPP equation is the Allen-Cahn equation

$$\partial_t u = \partial_x^2 u + u - u^3, \tag{7.13}$$

with $t \ge 0$, $x \in \mathbb{R}$, and $u = u(x, t) \in \mathbb{R}$, a prototype of a reaction-diffusion equation with bistable behavior, cf. also §5.3. The equation again consists of two parts, namely the diffusion term $\partial_x^2 u$ and the nonlinear reaction term $u - u^3$. Inserting $u(x,t) = v(t)$ into (7.13) gives the one-dimensional ODE

$$\dot{v} = v - v^3. \tag{7.14}$$

The phase portrait shows that the fixed point $v = 0$ is unstable and that the two other fixed points $v = \pm 1$ are asymptotically stable. The last two fixed points can be interpreted as two stable phases. Hence, in the Allen-Cahn equation both phases have the tendency to grow. At their interfaces both phases try to move the interfaces towards the other phase. The Allen-Cahn equation therefore describes interface motion between two stable phases with applications in material science, image processing, biology and geology.

The local existence and uniqueness of solutions follows exactly as for the KPP equation. As a consequence of the dynamics of (7.14) and the maximum principle we have the following global existence und uniqueness result for mild solutions.

Theorem 7.2.1. *For $u_0 \in C^0_{b,\mathrm{unif}}$ there exists a unique mild solution $u \in C([0,\infty), C^0_{b,\mathrm{unif}})$ of the Allen-Cahn equation with $u|_{t=0} = u_0$.*

Time-independent solutions $u(x,t) = v(x)$ satisfy $0 = \partial_x^2 v + v - v^3$. In the (v,v')-phase plane, we find a family of periodic solutions around $v = (0,0)$ and two heteroclinic connections, cf. Exercise 2.10 and §5.3.4. We do not further discuss the periodic solutions since they turn out to be unstable as stationary solutions of the Allen-Cahn equation (7.13). The heteroclinic connection between $(-1,0)$ and $(1,0)$ is called kink and the connection between $(1,0)$ and $(-1,0)$ is called anti-kink. In contrast to the stable/unstable steady state connection the stable/stable steady state connection exists exactly for one velocity, here $c = 0$. The reason for this is that the velocity is needed as a parameter to get an intersection of the one-dimensional unstable manifold of $(-1,0)$ and the one-dimensional stable manifold of $(1,0)$.

The stability of these kink solutions w.r.t. small spatially localized perturbations can again be seen with the help of the maximum principle. If the small perturbation v of the kink solution u_{kink} is contained between two translates of the kink then by the maximum principle it will stay there for all times. In detail, if

$$u_{kink}(x - x_-) \le u_{kink}(x) + v(x,0) \le u_{kink}(x + x_+)$$

for all $x \in \mathbb{R}$, then

$$u_{kink}(x - x_-) \le u_{kink}(x) + v(x,t) \le u_{kink}(x + x_+)$$

for all $t \ge 0$ and all $x \in \mathbb{R}$. See Figure 7.10.

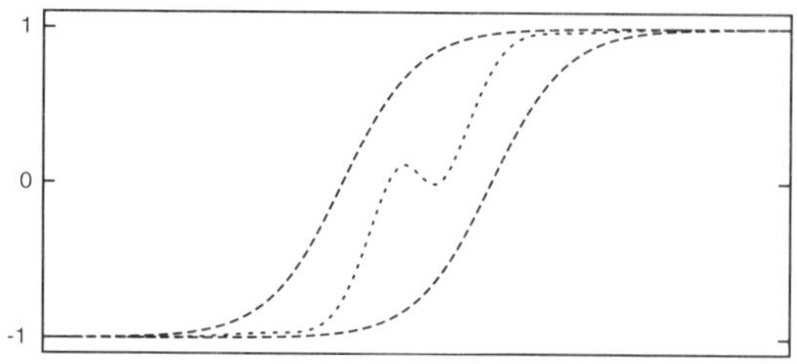

Figure 7.10. Stability of the kink w.r.t. to small spatially localized perturbations. The perturbations are confined by parallel kinks.

Further remarks. We expect that due to asymptotic stability of the states for $x \to \pm\infty$ the kink solutions are stable w.r.t. to a much larger class of perturbations. In the following we state a number of results concerning the long time behavior of solutions, but refer to the literature for proofs and more details, in particular to [**ER99**].

Linearizing the Allen-Cahn equation around the kink-solution u_{kink} gives

$$\partial_t v = \partial_x^2 v + v - 3u_{kink}^2 v = Lv + Bv$$

where $Lv = \partial_x^2 v - 2v$ and $Bv = -3(u_{kink}^2 - 1)v$. The operator B is a relatively compact perturbation of the operator L since it contains less derivates than the elliptic operator $-L$ and since it decays to 0 with some exponential rate for $|x| \to \infty$. As a consequence, the essential spectra in the sense of [**Hen81**, Chapter 5] of $L + B$ and L coincide, i.e., they only differ by discrete eigenvalues, cf. [**Hen81**, Chapter 5, Theorem A.1]. We find $\sigma(L) = \{-k^2 - 2 : k \in \mathbb{R}\}$ for the spectrum of L in various spaces, such as $C^0_{b,\mathrm{unif}}$ or L^2.

For the computation of the discrete spectrum we remark that $L + B$ is self-adjoint in L^2, i.e., $((L + B)u, v)_{L^2} = (u, (L + B)v)_{L^2}$. Hence, we only have real eigenvalues. There is one eigenvalue zero corresponding to the derivative $\partial_x u_{kink}$ of the kink-solution. Since this eigenfunction possesses no zeroes and since the eigenvalue problem comes from the linearization of a scalar translational invariant reaction-diffusion problem it must be the eigenfunction to the largest eigenvalue. Hence, all other eigenvalues must be strictly negative. The nonlinear stability of the kink-solution follows with the help of the center manifold theorem [**Hen81**, Ch. 6]. Since the operator $L + B$ is sectorial both in $C^0_{b,\mathrm{unif}}$ or L^2 the position of the spectrum gives the growth rates of the associated semigroups, cf. [**Hen81**, Theorem 1.3.4]. Thus, we have

Theorem 7.2.2. *Let $X = C^0_{b,\mathrm{unif}}$ or L^2. The family of translates of the kink solutions is asymptotically stable. In detail, there are $\delta_0 > 0$, $\beta > 0$ and $C > 0$ such that the following holds. For $u_0 \in X$ with $\|u_0 - u_{kink}\|_X \leq \delta_0$*

there is a $\gamma \in \mathbb{R}$ *such that*

$$\|u(t, u_0) - u_{kink}(\cdot + \gamma)\|_X \leq C\mathrm{e}^{-\beta t}.$$

Solutions of the Allen-Cahn equation with more than one interface are the next non-trivial set of solutions which can be understood completely. These solutions consist of almost flat regions, where they are close to the exponentially stable states ± 1. In between there are the interface regions, where the solutions connect the almost flat regions.

We define classes of functions with m interfaces

$$I_m = \{u \in C_b^0 \text{ is an interface function with } m \text{ zeros, } \lim_{x\to\pm\infty} u(x) = 1 \text{ or } -1\}.$$

Then from the analysis in [**CP89, CP90**] it is known that the interfaces annihilate. Moreover, it is known that this happens in a very robust way, i.e., small perturbations do not change the dynamics. With the help of the maximum principle the following global result can be obtained

$$\lim_{t\to\infty} u(\cdot,t) \in I_0, \text{ if } u(\cdot,0) \in I_{2m}, \quad \lim_{t\to\infty} u(\cdot,t) \in I_1, \text{ if } u(\cdot,0) \in I_{2m+1},$$

with $I_0 = \{-1, 1\}$. The time for the annihilation of the interfaces goes exponentially ($\mathcal{O}(\mathrm{e}^{CL})$) with the distance L of the interfaces. It has been shown that by separating the interfaces in the right way, i.e., making the distances bigger and bigger for $x \to \pm\infty$ it is possible to have an annihilation of interfaces for all times, i.e., infinitely many annihilations.

Comparing to the Allen-Cahn equation on a bounded domain where it has a finite dimensional attractor, cf. §5.3, we see that large time asymptotics over unbounded domains can be fundamentally different from the bounded domain case.

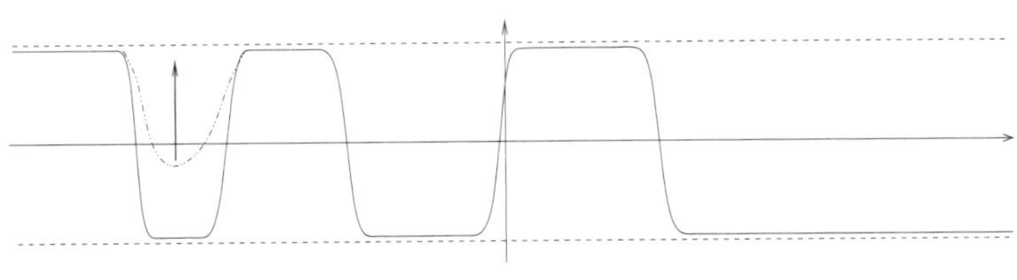

Figure 7.11. Annihilation of two interfaces to a solution with 3 interfaces

7.3. Intermezzo: Fourier transform

The analysis of the KPP and AC equations and of many more equations of similar form, heavily relies on the explicit solution formula for the linear diffusion equation. In general, explicit solution formulas of linear partial differential equations are not available. A powerful tool to analyze linear partial differential equations with constant coefficients on the real line is Fourier transform. We first define it for functions in $L^1(\mathbb{R}^d)$. We shall find that Fourier transform extends to an unitary operator in $L^2(\mathbb{R}^d)$ and finally to tempered distributions. We use Fourier transform to characterize the Sobolev spaces $H^s(\mathbb{R}^d)$ and introduce some further function spaces needed in the sequel. In applications we will mostly focus on the case of one unbounded direction, but the generalization to $\mathbb{R}^d$ will be straightforward. Fourier transform is treated in most books on analysis, and in most introductory books on PDEs. There are also a number of books specialized on Fourier transform, see for instance [**Duo01**]. We recall the main concepts and adapt them to our purposes.

7.3.1. Fourier transform in L^1. Before the definition of Fourier transform we recall without proof some important theorems from integration theory and basic facts about L^p-spaces.

Definition 7.3.1. *For $1 \leq p < \infty$ let $L^p(\mathbb{R}^d, Y)$ be the space of measurable functions with values in a Banach space Y such that the L^p-norm which defined by*

$$\|f\|_{L^p} = \left(\int_{\mathbb{R}^d} \|f(x)\|_Y^p \, \mathrm{d}x \right)^{1/p} \tag{7.15}$$

is finite.

If $f_1, f_2 \in L^p$ and $f_1 = f_2$ almost everywhere (a.e.), i.e., except on a set of measure zero, then $\|f_1 - f_2\|_p = 0$. We start with two convergence results in case $Y = \mathbb{R}$, namely the lemma of Beppo-Levi [**Alt16**, Folgerung 1.12 c)] and Lebesgue's dominated convergence theorem [**Alt16**, Satz 1.21].

Lemma 7.3.2. *Assume that $f_n \in L^1(\mathbb{R}^d, \mathbb{R})$ for $n \in \mathbb{N}$ with $0 \leq f_n \nearrow f$ for $n \to \infty$ a.e., and that $\limsup_{n\to\infty} \int f_n(x)\,\mathrm{d}x < \infty$. Then $f \in L^1(\mathbb{R}^d, \mathbb{R})$ and $\|f_n - f\|_{L^1} \to 0$ for $n \to \infty$.*

Theorem 7.3.3. *Let $f_n, f : \mathbb{R}^d \to \mathbb{R}$ be measurable functions and let $g \in L^1(\mathbb{R}^d, \mathbb{R})$. If $|f_n| \leq g$ for all $n \in \mathbb{N}$ a.e. and if $f_n \to f$ a.e. for $n \to \infty$, then $f \in L^1(\mathbb{R}^d, \mathbb{R})$ and $\|f_n - f\|_{L^1} \to 0$ for $n \to \infty$.*

We further need Fubini's theorem about iterated integrals [**Alt16**, Satz 4.10], and the fundamental lemma of variations [**Alt16**, Lemma 2.21].

Theorem 7.3.4. *Let Y be a Banach space and $1 \leq p < \infty$. For $f \in L^p(\mathbb{R}^{d_1} \times \mathbb{R}^{d_2}, Y)$ we have the existence of*

$$F(x_1) = \int_{\mathbb{R}^{d_2}} f(x_1, x_2)\,\mathrm{d}x_2,$$

for almost all $x_1 \in \mathbb{R}^{d_1}$, *and* $F \in L^p(\mathbb{R}^{d_1}, Y)$. *Moreover, by symmetry*

$$\int_{\mathbb{R}^{d_1}\times\mathbb{R}^{d_2}} f(x_1,x_2)\,\mathrm{d}(x_1,x_2) = \int_{\mathbb{R}^{d_1}}\int_{\mathbb{R}^{d_2}} f(x_1,x_2)\,\mathrm{d}x_2\,\mathrm{d}x_1$$
$$= \int_{\mathbb{R}^{d_2}}\int_{\mathbb{R}^{d_1}} f(x_1,x_2)\,\mathrm{d}x_1\,\mathrm{d}x_2.$$

Lemma 7.3.5. *Let* $\Omega \subset \mathbb{R}^d$ *be an open set and* Y *a Banach space. For* $g \in L^1(\Omega, Y)$ *we have the equivalence of (i)* $\int_\Omega \zeta(x)g(x)dx = 0$ *for all* $\zeta \in C_0^\infty(\Omega)$ *and (ii)* $g = 0$ *a.e. in* Ω.

For the rest of the section we restrict to $Y = \mathbb{R}$. Allthough $f \in L^p(\mathbb{R}^d)$ in general is not defined point-wise, it is clear that to have the existence of the integral in (7.15), in some sense $|f(x)|$ should decay to zero sufficiently fast as $|x| \to \infty$. This can be made precise as follows.

Lemma 7.3.6. *If* $f \in L^1(\mathbb{R})$, *then* $\lim_{n\to\infty} f(x+n) = 0$ *for almost all* $x \in \mathbb{R}$.

Proof. Since for all $x \in \mathbb{R}$ there exists a $z \in \mathbb{Z}$ with $x + z \in [0,1]$ we may assume $x \in [0,1]$. Let $f_k(x) = f(x+k)$, $g_n(x) = \sum_{k=1}^n |f_k(x)|$. Then

$$\int_0^1 g_n(x)\,\mathrm{d}x = \int_0^1 \sum_{k=1}^n |f_k(x)|\,\mathrm{d}x = \sum_{k=1}^n \int_0^1 |f(x+k)|\,\mathrm{d}x$$
$$= \int_1^n |f(x)|\,\mathrm{d}x \le \int_{\mathbb{R}} |f(x)|\,\mathrm{d}x$$

is bounded by a constant $M < \infty$. Thus, $g_n \le g_{n+1}$ and $\int_0^1 g_n(x)\,\mathrm{d}x \le M$, i.e., $(g_n)_{n\in\mathbb{N}}$ is monotonically increasing with a uniformly bounded L^1-norm. By the Beppo Levi Lemma 7.3.2 there exists a $g \in L([0,1])$ with $g_n \to g$ a.e., and hence $|f_k| \to 0$ a.e.. □

Fourier transform can be interpreted as an expansion into the eigenfunctions $\mathrm{e}^{\mathrm{i}k\cdot x}$, $k \in \mathbb{R}^d$, of spatially homogeneous linear partial differential operators on $\mathbb{R}^d$. We start with a definition for L^1 functions which is later on extended to L^2 and tempered distributions.

Definition 7.3.7. (Fourier transform) *For* $u \in L^1(\mathbb{R}^d)$ *we define* $\widehat{u} : \mathbb{R}^d \to \mathbb{C}$ *by*

$$\widehat{u}(k) = \frac{1}{(2\pi)^d}\int_{\mathbb{R}^d} \mathrm{e}^{-\mathrm{i}k\cdot x} u(x)\,\mathrm{d}x, \tag{7.16}$$

and call $\widehat{u} = \mathcal{F}u$ *the Fourier transform of* u.

Clearly, $\widehat{u}$ is bounded with the estimate

$$\sup_{k\in\mathbb{R}} |\widehat{u}(k)| \le \frac{1}{(2\pi)^d}\int_{\mathbb{R}^d} |u(x)|\,\mathrm{d}x. \tag{7.17}$$

The proof of the continuity of $\widehat{u}$ is the goal of Exercise 7.11. Similarly, it is easy to prove that decay of u corresponds to smoothness of $\widehat{u}$. In the following $\alpha = (\alpha_1, \dots, \alpha_d) \in \mathbb{N}_0^d$ is some multi-index, $|\alpha| = \alpha_1 + \dots + \alpha_d$, and

$$x^\alpha = x_1^{\alpha_1} \cdots x_d^{\alpha_d} \quad \text{and} \quad \partial_x^\alpha = \partial_{x_1}^{\alpha_1} \cdots \partial_{x_1}^{\alpha_1}.$$

Theorem 7.3.8. (Smoothness and decay) *a) If $u \in C_b^m$ fulfills $\partial_x^\alpha u \in L^1$ for $|\alpha| \leq m$ then $\mathcal{F}(\partial_x^\alpha u)(k) = (\mathrm{i}k)^\alpha \widehat{u}(k)$ for $|\alpha| \leq m$. Moreover, there exists a constant C such that for all $k \in \mathbb{R}^d$:*

$$|\widehat{u}(k)| \leq C(1 + |k|)^{-m}.$$

b) If $v_\alpha(x) = x^\alpha u(x)$ satisfies $v_\alpha \in L^1$ for $|\alpha| \leq m$, then $\widehat{u} \in C_b^m$ and $\mathcal{F}(v_\alpha) = (-\mathrm{i}\partial_k)^\alpha \widehat{u}$.

Proof. Using integration by parts the estimate in a) is straightforward. □

Using the estimate (7.17) immediately yields the so called roll off formula which subsequently will turn out to be very useful.

Lemma 7.3.9. *For $u, v \in L^1$ we have $\widehat{u}v, u\widehat{v} \in L^1$ and $\int_{\mathbb{R}^d} \widehat{u}(x)v(x)\,\mathrm{d}x = \int_{\mathbb{R}^d} u(x)\widehat{v}(x)\,\mathrm{d}x$.*

We will also use the following formulas which again follow directly from the definition of Fourier transform.

Lemma 7.3.10. *a) Let $u \in L^1$. For $a \in \mathbb{R}^d$ and $u_a(x) = u(x - a)$ we have*

$$\widehat{u}_a(k) = \mathrm{e}^{-\mathrm{i}k \cdot a}\widehat{u}(k).$$

b) For $r > 0$ and $v(x) = u(x/r)$ we have $\widehat{v}(k) = r^d \widehat{u}(rk)$.

Above we noted that Fourier transform can be interpreted as an expansion into the eigenfunctions $\mathrm{e}^{\mathrm{i}k \cdot x}$, $k \in \mathbb{R}^d$, of spatially homogeneous linear partial differential operators on $\mathbb{R}^d$. The coefficients of this expansion define the Fourier transform. Hence, we expect that the reconstruction formula for u from $\widehat{u}$ is as follows.

Theorem 7.3.11. (Inverse Fourier transform) *Assume $u \in L^1(\mathbb{R}^d)$ and $\widehat{u} \in L^1(\mathbb{R}^d)$. Then*

$$u(x) = \int_{\mathbb{R}^d} \mathrm{e}^{\mathrm{i}k \cdot x}\widehat{u}(k)\,\mathrm{d}k.$$

The map $\widehat{u} \mapsto u$ is called inverse Fourier transform and is denoted by $u = \mathcal{F}^{-1}\widehat{u}$.

The proof is given below. Note that $u \in L^1$ does not imply $\widehat{u} \in L^1$. The inversion formula means that $\mathcal{F}$ and $\mathcal{F}^{-1}$ map the subspace $L^1 \cap \mathcal{F}(L^1) \subset C_b^0$ bijectively into itself. However, since a characterization of $\mathcal{F}(L^1) \subset C_b^0$ is a non-trivial task, this is not very satisfactory.

7.3.2. Fourier transform in $\mathcal{S}$, L^2, and general L^p-spaces. The goal is to find spaces where the analytic properties of Fourier transform are better than in L^1. Our starting point are the relations

$$\mathcal{F}(\partial_x^\alpha u)(k) = (\mathrm{i}k)^\alpha \widehat{u}(k) \quad \text{and} \quad \mathcal{F}^{-1}((\mathrm{i}\partial_k)^\alpha \widehat{u})(x) = x^\alpha u(x).$$

We look for a subspace of L^1 which is mapped into itself by $\mathcal{F}$, differentiation, and multiplication with polynomials. Such a subspace is the Schwartz space of quickly decaying functions.

Definition 7.3.12. *The Schwartz space $\mathcal{S} = \mathcal{S}(\mathbb{R}^d)$ is defined by*

$$\mathcal{S} = \{u \in C^\infty(\mathbb{R}^d) : x^\alpha \partial_x^\beta u(x) \text{ is bounded for any } \alpha, \beta \in \mathbb{N}_0^d\}.$$

Theorem 7.3.13. *For $u \in \mathcal{S}$ we have $\widehat{u} \in \mathcal{S}$ and vice versa.*

Proof. Let $u \in \mathcal{S}$, and let $\alpha, \beta \in \mathbb{N}_0^d$ be some multi-indices. Clearly $v_\beta(x) = x^\beta u(x)$ satisfies $v_\beta \in \mathcal{S}$, hence $\widehat{u} \in C^{|\beta|}(\mathbb{R}^d)$ and $\partial_k^\beta \widehat{u} = (-1)^{|\beta|}\mathcal{F}(v_\beta)$ by Theorem 7.3.8. Since also $\partial_x^\alpha(v_\beta) \in \mathcal{S} \subset L^1$ we have $\mathcal{F}(\partial_x^\alpha(v_\beta)) \in C_b^0$, i.e., $k \mapsto k^\alpha \partial_k^\beta \widehat{u}(k)$ is in C_b^0. □

Theorem 7.3.14. *$\mathcal{F}$ is a bijection on $\mathcal{S}$, i.e., $\mathcal{F}^{-1}\mathcal{F}|_{\mathcal{S}} = \mathcal{F}\mathcal{F}^{-1}|_{\mathcal{S}} = \mathrm{I}$.*

Proof. It will turn out in §7.3.4 that the Fourier transform of the δ-distribution is the constant function with value 1. With this knowledge the formal argument is as follows.

$$\begin{aligned}(\mathcal{F}^{-1}\mathcal{F}u)(x) &= \int_{\mathbb{R}^d} \mathrm{e}^{\mathrm{i}k\cdot x}\left(\frac{1}{(2\pi)^d}\int_{\mathbb{R}^d} \mathrm{e}^{-\mathrm{i}k\cdot y} u(y)\,\mathrm{d}y\right)\mathrm{d}k \\ &= \int_{\mathbb{R}^d}\left(\frac{1}{(2\pi)^d}\int_{\mathbb{R}^d} \mathrm{e}^{\mathrm{i}k\cdot(x-y)}\,\mathrm{d}k\right) u(y)\,\mathrm{d}y \\ &= \int_{\mathbb{R}^d} \delta(x-y)u(y)\,\mathrm{d}y = u(x).\end{aligned}$$

In order to make this formal argument rigorous we use an approximation of the δ-distribution and the roll-off formula. Hence, we set $v(k) = \mathrm{e}^{-|k|^2/2}$ and let $v_r(k) = v(k/r)$ for $r > 0$. For $r \to \infty$ and fixed $k \in \mathbb{R}^d$ we have $v_r(k) \to 1$ monotonically from below.

For $u \in \mathcal{S}$ we set $u_x(k) = u(k-x)$ and find

$$\begin{aligned}\int_{\mathbb{R}^d} \mathrm{e}^{\mathrm{i}k\cdot x}\widehat{u}(k)v_r(k)\,\mathrm{d}k &= \int_{\mathbb{R}^d} \widehat{u_{-x}}(k)v_r(k)\,\mathrm{d}k = \int_{\mathbb{R}^d} u_{-x}(k)\widehat{v}_r(k)\,\mathrm{d}k \\ &= \int_{\mathbb{R}^d} u_{-x}(k)r^d\widehat{v}(rk)\,\mathrm{d}k = \int_{\mathbb{R}^d} u_{-x}(m/r)\widehat{v}(m)\,\mathrm{d}m \\ &= (2\pi)^{-d/2}\int_{\mathbb{R}^d} u_{-x}(y/r)v(y)\,\mathrm{d}y\end{aligned}$$

by Lemma 7.3.10 and since $\mathcal{F}(e^{-|x|^2/2})(k) = (2\pi)^{-d/2}e^{-|k|^2/2}$, cf. Exercise 7.12. Now

$$\int_{\mathbb{R}^d} \mathrm{e}^{\mathrm{i}k\cdot x}\widehat{u}(k)v_r(k)\,\mathrm{d}k \to \int_{\mathbb{R}^d} \mathrm{e}^{\mathrm{i}k\cdot x}\widehat{u}(k)\,\mathrm{d}k \quad \text{as} \quad r \to \infty$$

by the Lebesgue dominated convergence theorem 7.3.3 since the integrand is dominated by $|\widehat{u}| \in L^1$. Similarly,

$$(2\pi)^{-d/2}\int_{\mathbb{R}^d} u_{-x}(y/r)v(y)\,\mathrm{d}y \to (2\pi)^{-d/2}u(x)\int_{\mathbb{R}^d} v(y)\,\mathrm{d}y = u(x)$$

as $r \to \infty$ since the integrand is dominated by $\|u\|_\infty v \in L^1$. Hence $\int_{\mathbb{R}^d} \mathrm{e}^{\mathrm{i}k\cdot x}\widehat{u}(k)\,\mathrm{d}k = u(x)$. □

The inversion formula is true in much larger sets. It is first extended from $\mathcal{S}$ to L^1.

Proof of Theorem 7.3.11. Let $u, \widehat{u} \in L^1$ and set $\widetilde{u}(x) = \int_{\mathbb{R}^d} \mathrm{e}^{\mathrm{i}k\cdot x}\widehat{u}(k)\,\mathrm{d}k$. By Theorem 7.3.8 we have $\widetilde{u} \in C_b^0$ and we want to show that $\widetilde{u} = u$. By Theorem 7.3.14, for any $\phi \in \mathcal{S}$ there exists a $v \in \mathcal{S}$ with $\widehat{v} = \phi$. Then, by Lemma 7.3.9 and again by Theorem 7.3.14 we find

$$\begin{aligned}
\int_{\mathbb{R}^d} u(y)\widehat{v}(y)\,\mathrm{d}y &= \int_{\mathbb{R}^d} \widehat{u}(y)v(y)\,\mathrm{d}y \\
&= \int_{\mathbb{R}^d} \widehat{u}(y)\left(\int_{\mathbb{R}^d} \mathrm{e}^{\mathrm{i}k\cdot y}\widehat{v}(k)\,\mathrm{d}k\right)\mathrm{d}y = \int_{\mathbb{R}^d} \widehat{v}(k)\widetilde{u}(k)\,\mathrm{d}k.
\end{aligned}$$

Hence

$$\int_{\mathbb{R}^d} (u - \widetilde{u})\widehat{v}\,\mathrm{d}x = \int_{\mathbb{R}^d} (u - \widetilde{u})\phi\,\mathrm{d}x = 0,$$

and since ϕ was arbitrary we find $\widetilde{u} = \mathcal{F}^{-1}\widehat{u} = u$ by Lemma 7.3.5. □

The following crucial result allows us to bound the norm of the Fourier transform and of its inverse simultaneously. This will allow us to go from physical space to Fourier space and back, and to subsequently extend $\mathcal{F}$ to $L^2(\mathbb{R}^d)$.

Theorem 7.3.15. (Parseval's identity for Schwartz functions) *For $u, v \in \mathcal{S}$ we have*

$$\langle u, v\rangle_{L^2} = (2\pi)^d \langle \widehat{u}, \widehat{v}\rangle_{L^2},$$

where $\langle u, v\rangle_{L^2} = \int_{\mathbb{R}^d} u(x)\overline{v}(x)\,\mathrm{d}x$, and in particular

$$\|\widehat{u}\|_{L^2} = (2\pi)^{-d/2}\|u\|_{L^2}. \tag{7.18}$$

Proof. Let $u, v \in \mathcal{S}$ and set $g = \overline{\widehat{v}}$ which implies $(2\pi)^d\widehat{g} = \overline{v}$. Then, by Lemma 7.3.9,

$$\langle u, v\rangle_{L^2} = \int_{\mathbb{R}^d} u(x)\overline{v}(x)\,\mathrm{d}x = (2\pi)^d\int_{\mathbb{R}^d} u(x)\widehat{g}(x)\,\mathrm{d}x$$

$$=(2\pi)^d \int_{\mathbb{R}^d} \widehat{u}(x) g(x) \, \mathrm{d}x = (2\pi)^d \langle \widehat{u}, \widehat{v} \rangle_{L^2} .$$

□

Since $\mathcal{S}$ is dense in L^2 we find that $\mathcal{F}$ can be extended to an isomorphism in L^2 which is isometric up to a factor $(2\pi)^d$.

Theorem 7.3.16. *There exists a unique linear bounded invertible map $\mathcal{F} : L^2 \to L^2$ which equals $\mathcal{F}$ on $\mathcal{S}$. Its inverse $\mathcal{F}^{-1}$ equals $\mathcal{F}^{-1}$ on $\mathcal{S}$.*

Remark 7.3.17. There exist $u \in L^2(\mathbb{R})$ which are not in $L^1(\mathbb{R})$, for instance $u(x) = (1+|x|)^{-1}$. For these u the Fourier transform $\mathcal{F}u$ cannot be computed by the formula (7.16). However, with

$$\widehat{u}_n(k) = \frac{1}{(2\pi)^d} \int_{-n}^{n} \mathrm{e}^{-\mathrm{i}k\cdot x} u(x) \, \mathrm{d}x$$

we always have $\|\widehat{u} - \widehat{u}_n\|_{L^2} \to 0$ again by the Lebesgue dominated convergence theorem 7.3.3. ⌋

Similar to Remark 5.2.11 from $\|\widehat{u}\|_{L^\infty} \leq \|u\|_{L^1}$ and $\|\widehat{u}\|_{L^2} = (2\pi)^{-d/2}\|u\|_{L^2}$, by the so called Riesz-Thorin interpolation $\mathcal{F}$ can be estimated on $L^p(\mathbb{R}^d)$ for $1 \leq p \leq 2$, cf. [**Duo01**, Corollary 1.20].

Theorem 7.3.18. (Hausdorff-Young) *For $1 \leq p \leq 2$ there exists a $C_p > 0$ such that for $f \in L^p(\mathbb{R}^d)$ we have $\widehat{u} \in L^{p'}(\mathbb{R}^d)$, $1/p + 1/p' = 1$, and $\|\widehat{u}\|_{L^{p'}} \leq C_p \|u\|_{L^p}$.*

7.3.3. Convolution. The convolution of two functions is formally defined by

$$(u * v)(x) = \int_{\mathbb{R}^d} u(x-y) v(y) \, \mathrm{d}y.$$

It occurs as the Fourier transform of the product of two functions, in detail

$$\begin{aligned}
\mathcal{F}(uv)(k) &= (2\pi)^{-d} \int_{\mathbb{R}^d} \mathrm{e}^{-\mathrm{i}k\cdot x} u(x) v(x) \, \mathrm{d}x \\
&= (2\pi)^{-d} \int_{\mathbb{R}^d} u(x) \mathrm{e}^{-\mathrm{i}k\cdot x} \left(\int_{\mathbb{R}^d} \widehat{v}(\ell) \mathrm{e}^{\mathrm{i}l\cdot x} \, \mathrm{d}l \right) \mathrm{d}x \\
&= \int_{\mathbb{R}^d} \widehat{v}(\ell) \frac{1}{(2\pi)^d} \int_{\mathbb{R}^d} u(x) \mathrm{e}^{-\mathrm{i}(k-\ell)\cdot x} \, \mathrm{d}x \, \mathrm{d}\ell \\
&= \int_{\mathbb{R}^d} \widehat{u}(k-\ell) \widehat{v}(\ell) \, \mathrm{d}l = (\widehat{u} * \widehat{v})(k)
\end{aligned}$$

by Fubini's theorem 7.3.4 if for instance $u, v \in \mathcal{S}$. The identity can be extended to $u, v \in L^1$ by the dominated convergence theorem, similar to the proof of Theorem 7.3.11. For later purposes we also need the formula

$$\mathcal{F}^{-1}(\widehat{u}\widehat{v}) = (2\pi)^{-d} \, u * v \tag{7.19}$$

which can be obtained in the same way. We have the following estimate

Lemma 7.3.19. (Young's inequality for convolutions) *For* $1 \le p \le \infty$ *and* $f \in L^p(\mathbb{R}^d)$, $g \in L^1(\mathbb{R}^d)$ *we have* $f*g \in L^p$ *and* $\|f*g\|_{L^p} \le \|f\|_{L^p}\|g\|_{L^1}$.

Proof. For $p = \infty$ we have $\int_{\mathbb{R}^d} |f(x-y)g(y)|\,\mathrm{d}y \le \|f\|_{L^\infty}\|g\|_{L^1} < \infty$ a.e.. Thus let $1 \le p < \infty$. The map $(x,y) \mapsto f(x-y)g(y)$ is measurable as a product of measurable functions. By Hölder's inequality we obtain

$$\begin{aligned}\int_{\mathbb{R}^d} |f(x-y)g(y)|\,\mathrm{d}y &= \int_{\mathbb{R}^d} |f(x-y)||g(y)|^{1/p}|g(y)|^{1/p'}\,\mathrm{d}y \\ &\le \left(\int_{\mathbb{R}^d} |f(x-y)|^p|g(y)|\,\mathrm{d}y\right)^{1/p}\left(\int_{\mathbb{R}^d} |g(y)|\,\mathrm{d}y\right)^{1/p'}\end{aligned}$$

for $1/p + 1/p' = 1$, which implies

$$\begin{aligned}\int_{\mathbb{R}^d}\left(\int_{\mathbb{R}^d} |f(x-y)g(y)|\,\mathrm{d}y\right)^p \mathrm{d}x &\le \|g\|_1^{p/p'}\int_{\mathbb{R}^d}\int_{\mathbb{R}^d} |f(x-y)|^p\,\mathrm{d}x|g(y)|\,\mathrm{d}y \\ &= \|g\|_{L^1}^{1+p/p'}\|f\|_{L^p}^p < \infty.\end{aligned}$$

Taking the pth root and using $1 + p/p' = p$ yields the result. □

Example 7.3.20. For the linear diffusion equation $\partial_t u = \partial_x^2 u$ we once more derive the solution formula (7.6), i.e.,

$$u(x,t) = \frac{1}{\sqrt{4\pi t}}\int_{-\infty}^{\infty} \mathrm{e}^{-\frac{(x-y)^2}{4t}}u(y,0)\,\mathrm{d}y,$$

now using the Fourier transform. Inserting the approach

$$u(x,t) = \int_{-\infty}^{\infty} \widehat{u}(k,t)e^{\mathrm{i}kx}\,\mathrm{d}k,$$

into the linear diffusion equation $\partial_t u = \partial_x^2 u$ gives

$$\int_{-\infty}^{\infty} (\partial_t\widehat{u}(k,t) + k^2\widehat{u}(k,t))\mathrm{e}^{\mathrm{i}kx}\,\mathrm{d}k = 0,$$

and so for each $k \in \mathbb{R}$ the ODE $\partial_t\widehat{u} = -k^2\widehat{u}$, which is solved by $\widehat{u}(k,t) = \widehat{G}(k,t)\widehat{u}(k,0)$ with $\widehat{G}(k,t) = e^{-k^2t}$. Using (7.19) we then have

$$\begin{aligned}u(x,t) =&(\mathcal{F}^{-1}\widehat{u})(x,t) = (\mathcal{F}^{-1}\widehat{G}(\cdot,t)\widehat{u}(\cdot,0))(x,t) \\ =&\frac{1}{2\pi}\int_{-\infty}^{\infty} G(x-y,t)u(y,0)\,\mathrm{d}y.\end{aligned}$$

The inverse Fourier transform of $\widehat{G}$ is given by

$$G(x,t) = \sqrt{\frac{\pi}{t}}\mathrm{e}^{-\frac{x^2}{4t}},$$

see Exercise 7.12, which yields the assertion. We have the following commutative diagram

$$\begin{array}{ccc} u_0 & \stackrel{G(t)}{\longrightarrow} & u(\cdot,t) \\ \downarrow^{\mathcal{F}} & & \uparrow^{\mathcal{F}^{-1}} \\ \widehat{u}_0 & \stackrel{\widehat{G}(t)}{\longrightarrow} & \widehat{u}(\cdot,t) \end{array}$$

⌋

7.3.4. Fourier transform of tempered distributions. So far we defined Fourier transform for $u \in L^p$ with $p \in [1,2]$. These spaces are still too small for many applications since one wants to use Fourier transform for instance also in spaces as $C^0_{b,unif}$. Since any bounded function defines a (regular) distribution, the basic idea is to define the Fourier transform of distributions by rolling off the transform to the test function. In detail, recall that a classical function u can be interpreted as a regular distribution T_u by

$$T_u(\phi) = \int_{\mathbb{R}} u(x)\phi(x)\,\mathrm{d}x \qquad \text{for all } \phi \in \mathcal{D},$$

cf. Remark 5.2.1. We defined derivatives $\partial^j T$ for $T \in \mathcal{D}' = L(\mathcal{D},\mathbb{K})$ via

$$\partial^j T(\phi) = (-1)^j T(\partial_x^j \phi),$$

which corresponds to the formula of integration by parts. Similarly, for $T \in \mathcal{D}'$ we might want to define $\mathcal{F}T$ by

$$(\mathcal{F}T)(\phi) = T(\widehat{\phi}) \quad \text{for all } \phi \in \mathcal{D}.$$

However, for $\phi \in \mathcal{D}$ the problem is that $\widehat{\phi}$ does not have compact support and hence is not a test function, since first the Paley-Wiener theorem, cf. Lemma 6.2.9, implies that a compact support of $\widehat{\phi}$ implies the analyticity of ϕ, and secondly that the identity theorem for complex power series implies that an analytic function with compact support must be identical zero. Therefore, we shall restrict the definition of Fourier transform to the subspace $\mathcal{S}' \subset \mathcal{D}'$. The Schwartz space $\mathcal{S}$ can be equipped with some topology, namely it can be made a Fréchet-space with the semi-norms $p_{\beta,m}$, where

$$p_{\beta,m}(\phi) := \sup_{x\in\mathbb{R}^d} |\partial_x^\beta \phi(x)(1+|x|)^m| < \infty.$$

An equivalent characterization is

$$\phi_k \to \phi \text{ in } \mathcal{S} \Leftrightarrow x^\alpha \partial_x^\beta \phi_k(x) \to x^\alpha \partial_x^\beta \phi(x) \text{ uniformly on } \mathbb{R}^d \text{ for any } \alpha,\beta \in \mathbb{R}^d.$$

For details see [**Wlo87**]. The space of tempered distributions $\mathcal{S}'$ is defined as the dual space of $\mathcal{S}$ consisting of all continuous linear functionals $T : \mathcal{S} \to \mathbb{C}$.

Definition 7.3.21. *$T \in \mathcal{S}'$ is called a tempered distribution if there exists $m \in \mathbb{N}, \beta \in \mathbb{N}^d$, and a $C > 0$ such that*

$$|T(\phi)| \leq C p_{\beta,m}(\phi) \textit{ for all } \phi \in \mathcal{S}.$$

Since $\mathcal{D} \subset \mathcal{S}$ we have $\mathcal{S}' \subset \mathcal{D}'$ as proper subsets.

Example 7.3.22. A continuous function u which satisfies $\sup_{x\in\mathbb{R}} |u(x)| = C_u < \infty$ can be interpreted as tempered distribution T_u since

$$\begin{aligned}|T_u(\phi)| &\leq \int_{\mathbb{R}} |u(x)\phi(x)|\,\mathrm{d}x \leq C_u \int_{\mathbb{R}} |\phi(x)|\,\mathrm{d}x \\ &\leq C_u \int_{\mathbb{R}} (1+x^2)^{-1}\,\mathrm{d}x\; p_{0,2}(\phi) \leq C p_{0,2}(\phi)\end{aligned}$$

for all $\phi \in \mathcal{S}$. More general, $T_u : \mathcal{S} \to \mathbb{C}$ with $\phi \mapsto \int_{\mathbb{R}^d} u(x)\phi(x)\,\mathrm{d}x$ defines a tempered distribution if for instance for $u \in L^1_{\text{loc}}$ there exists an $N \in \mathbb{N}$ such that $\int_{\mathbb{R}} |u(x)|(1+\|x\|)^{-N}\,\mathrm{d}x < \infty$ or if $u \in L^p(\mathbb{R}^d)$ for some $p \geq 1$. On the other hand, $u(x) = \mathrm{e}^{a|x|}$ with $a > 0$ gives a $T_u \in \mathcal{D}' \setminus \mathcal{S}'$. ⌋

Since $\mathcal{F}$ maps $\mathcal{S}$ to $\mathcal{S}$ we may now define

Definition 7.3.23. *The Fourier transform $\mathcal{F}$ is defined for tempered distributions T via*

$$(\mathcal{F}T)(\phi) = T(\mathcal{F}\phi) \quad \text{for all } \phi \in \mathcal{S}.$$

This extended Fourier transform $\mathcal{F} : \mathcal{S}' \to \mathcal{S}'$ is the adjoint of the classical Fourier transform $\mathcal{F} : \mathcal{S} \to \mathcal{S}$. Again $\mathcal{F} : \mathcal{S}' \to \mathcal{S}'$ is an isomorphism.

Example 7.3.24. For $u(x) = e^{ik\cdot x}$ we have $\mathcal{F}T_u = \delta_k$, the δ-distribution in $k \in \mathbb{R}$, since $(\mathcal{F}T_u)(\phi) = \int e^{ik\cdot\xi}\widehat{\phi}(\xi)\,\mathrm{d}\xi = \phi(k)$. ⌋

Remark 7.3.25. Associated to the δ-distribution sometimes the δ-function formally satisfying $\int_{\mathbb{R}} \delta_0(x)\phi(x)\,\mathrm{d}x = \phi(0)$ is introduced. To show that δ_0 is not a regular distribution assume that $\delta_0 = T_u$ for some $u \in L^p_{\text{loc}}$. If $\phi \in \mathcal{D}$ then also $\psi = \|x\|^2\phi$ is a test function and we have

$$0 = \psi(0) = \delta_0\psi = \int_{\mathbb{R}} u(x)\|x\|^2\phi(x)^2\,\mathrm{d}x$$

for all ϕ, hence $u(x)\|x\|^2 = 0$ a.e., hence $u = 0$ a.e., which contradicts $\delta_0 \neq 0$.

Nevertheless the notation $\int_{\mathbb{R}} \delta_0(x)\phi(x)\,\mathrm{d}x = \phi(0)$ for $\delta_0\phi$ is often used. The motivation is that δ_0 can be approximated by regular distributions. For instance the fundamental solution

$$G(x,t) = \frac{1}{(4\pi t)^{1/2}} e^{-|x|^2/4t}$$

of the linear heat equation $\partial_t u = \partial_x^2 u$ approximates the δ_0-distribution in the sense of distributions for $t \to 0$, i.e.,

$$\int_{\mathbb{R}} G(x,t)\phi(x)\,\mathrm{d}x \to \phi(0)$$

for all $\phi \in \mathcal{D}$ and $t \to 0$. To show this let $\varepsilon > 0$ and $\phi \in \mathcal{D}$. Then there exists a $\delta > 0$ with $|\phi(x) - \phi(0)| < \varepsilon/2$ for $|x| < \delta$, and thus

$$\left|\int_{\mathbb{R}} G(x,t)\phi(x)\,\mathrm{d}x - \phi(0)\right| = \left|\int_{\mathbb{R}} G(x,t)(\phi(x) - \phi(0))\,\mathrm{d}x\right|$$
$$\leq \int_{|x|<\delta} \ldots\,\mathrm{d}x + \int_{|x|>\delta} \ldots\,\mathrm{d}x < \varepsilon/2 + \varepsilon/2$$

for $t > 0$ sufficiently small, where we used in the first integral that $\int_{\mathbb{R}} G(x,t)\,\mathrm{d}x = 1$. ⌋

7.3.5. Sobolev spaces. For the handling of nonlinear PDEs, L^p-spaces are of no use since they are not closed under multiplication. In §5.2.1 we defined the Sobolev spaces $H^m(\Omega)$ over some bounded domain Ω, and in §5.2.2 we showed that for suitable boundary conditions there is an easy alternative characterization using Fourier series, see in particular Lemma 5.2.9. Here we use Fourier transform to study the Sobolev spaces over $\mathbb{R}^d$. Again we usually do not distinguish real-valued from complex-valued or vector-valued functions.

Definition 7.3.26. *For $m \in \mathbb{N}$ define the Sobolev space*

$$W^{m,p}(\mathbb{R}^d) := \{u \in L^p(\mathbb{R}^d) : \partial_x^\alpha u \in L^p(\mathbb{R}^d) \text{ for } |\alpha| \leq m\},$$

where ∂_x^α with multi-index α denotes the distributional derivative of order α. It is equipped with the norm

$$\|u\|_{W^{m,p}} = \left(\sum_{|\alpha|\leq m} \|\partial_x^\alpha u\|_{L^p}^p\right)^{1/p}.$$

Of particular importance are the Hilbert spaces $H^m(\mathbb{R}^d) = W^{m,2}(\mathbb{R}^d)$ with scalar product $\langle u, v\rangle_{H^m} = \sum_{|\alpha|\leq m} \langle \partial_x^\alpha u, \partial_x^\alpha v\rangle_{L^2}$.

In Theorem 7.3.8 we have seen that smoothness of u is related to decay of $\widehat{u}$ and vice versa. As in Remark 5.2.12, this can be used to generalize Sobolev spaces H^m to non-integer m.

Definition 7.3.27. *Let $\rho(x) = (1 + |x|^2)^{1/2}$. For $m \geq 0$ define the weighted L^2-space*

$$L_m^2 := \{u \in L^2 : \|u\|_{L_m^2} := \|u\rho^m\|_{L^2} < \infty\}$$

and the Sobolev space $H^m := \{u \in L^2 : \widehat{u} \in L_m^2\}$ with norm $\|u\|_{H^m} = \|\widehat{u}\|_{L_m^2}$.

For $m \in \mathbb{N}_0$ this definition coincides with the previous one. Next, similar to Lemmas 5.2.3 and 5.2.24 we obtain the following embedding and multiplication results.

Lemma 7.3.28. *We have* $H^1(\mathbb{R}) \subset C^{0,1/2}(\mathbb{R})$ *and*

$$\|u\|_\infty^2 \le 2\|u\|_{L^2}\|\partial_x u\|_{L^2}, \tag{7.20}$$

$$|u(x)-u(y)| \le \sqrt{|x-y|}\|\partial_x u\|_{L^2}. \tag{7.21}$$

Proof. Since $C^1(\mathbb{R})\cap H^1(\mathbb{R})$ is dense in $H^1(\mathbb{R})$ it is sufficient to prove (7.20) and (7.21) for $u \in C^1(\mathbb{R}) \cap H^1(\mathbb{R})$. We have

$$u^2(x) = \int_{-\infty}^x \frac{\mathrm{d}}{\mathrm{d}s} u^2(s)\,\mathrm{d}s = 2\int_{-\infty}^x u(s)\partial_s u(s)\,\mathrm{d}s \le 2\|u\|_{L^2}\|\partial_x u\|_{L^2}.$$

Using the Cauchy-Schwarz inequality yields the second estimate

$$|u(x)-u(y)| = \left|\int_x^y \partial_s u(s)\,\mathrm{d}s\right| \le \int_x^y 1|\partial_s u(s)|\,\mathrm{d}s \le \sqrt{|x-y|}\|\partial_x u\|_{L^2},$$

which immediately implies the embedding $H^1(\mathbb{R}) \subset C^{0,1/2}(\mathbb{R})$. □

Lemma 7.3.29. *For* $u, v \in H^m(\mathbb{R}^d)$, $m > d/2$, *we have* $u, v \in L^\infty$ *and* $\|uv\|_{H^m} \le C\|u\|_{H^m}\|v\|_{H^m}$.

Proof. For $\rho(k) = (1+|k|^2)^{1/2}$ we have $\rho^{-m} \in L^2$ if $m > d/2$, and thus

$$\|\widehat{u}\|_{L^1} = C\|\rho^{-m}\rho^m\widehat{u}\|_{L^1} \le C\|\rho^{-m}\|_{L^2}\|\widehat{u}\|_{L^2_m} \le C\|\widehat{u}\|_{L^2_m}, \tag{7.22}$$

which implies

$$\|u\|_\infty \le C\|\widehat{u}\|_{L^1} \le C\|\widehat{u}\|_{L^2_m} \le C\|u\|_{H^m}.$$

Next $\rho^m(k) \le C(\rho^m(k-\ell) + \rho^m(\ell))$ for a constant C, and thus by Lemma 7.3.19 we obtain

$$\begin{aligned}\|uv\|_{H^m} \le& C\|\widehat{u} * \widehat{v}\|_{L^2_m} \le C(\|\rho^m\widehat{u}\|_{L^2}\|\widehat{v}\|_{L^1} + \|\widehat{u}\|_{L^1}\|\rho^m\widehat{v}\|_{L^2})\\ \le& 2C\|\widehat{u}\|_{L^2_m}\|\widehat{v}\|_{L^2_m} \le 2C\|u\|_{H^m}\|v\|_{H^m},\end{aligned}$$

where we used again (7.22). □

For later purposes we also need weighted Sobolev spaces.

Definition 7.3.30. *For* $r, m \ge 0$ *and* $\rho(x) = (1+x^2)^{1/2}$ *let*

$$H^r_m := \{u \in L^2 : \|u\rho^m\|_{H^r} < \infty\}.$$

The fact that decay in Fourier space corresponds to smoothness in x-space can now be quantified as follows.

Lemma 7.3.31. *Fourier transform is an isomorphism between* H^r_m *and* H^m_r.

7.4. The Burgers equation

The (inviscid) Burgers equation

$$\partial_t u = -\frac{1}{2}\partial_x(u^2), \tag{7.23}$$

with $t \geq 0$, $x \in \mathbb{R}$, and $u(x,t) \in \mathbb{R}$, appears as a model for the velocity field of (an inviscid) compressible one-dimensional fluid or gas. Traffic problems are very often modeled as a one-dimensional gas with the cars taking the role of the gas particles. Therefore, the Burgers equation also occurs as a model for traffic flows.

Example 7.4.1. Consider traffic flow with the following modeling assumptions: **a1)** There is only one lane. **a2)** The local (car)-density

$$\rho(x,t) = \frac{\text{number of cars at position } x \text{ and time } t}{\text{unit length}}$$

makes sense. In particular, we consider a long stretch of the road. **a3)** The local speed of traffic $v(x,t)$ is a function of $\rho(x,t)$ only. (This is typically revised if viscous or other effects are to be taken into account.)

Let $x_1 < x_2$ be arbitrary. With the flux function $f(\rho) = \rho v(\rho)$, where v is the speed, we obtain the integral form of conservation of mass (cars)

$$\begin{aligned}\int_{x_1}^{x_2} \partial_t \rho(x,t)\,\mathrm{d}x = \partial_t \int_{x_1}^{x_2} \rho(x,t)\,\mathrm{d}x &= f(\rho(x_1,t)) - f(\rho(x_2,t)) \\ &= -\int_{x_1}^{x_2} \partial_x f(\rho(x,t))\,\mathrm{d}x.\end{aligned}$$

Since x_1 and x_2 are arbitrary we obtain

$$\partial_t \rho + \partial_x(f(\rho)) = 0. \tag{7.24}$$

To close the equation we need a constitutive law $v(\rho)$ for the dependence of v on ρ. The following assumptions seem reasonable: There is a maximal density $\rho_{\max}$ with $v(\rho_{\max}) = 0$ (traffic jam), and there is a maximal speed at low density $v(0) = v_{\max}$. W.l.o.g. we assume that $\rho_{\max} = 1$ and $v_{\max} = 1$, and set $v(\rho) = 1 - \rho$. Thus, $f(\rho) = \rho(1-\rho)$. ⌋

The inviscid Burgers equation (7.23) is a one-dimensional scalar conservation law, which in general form reads

$$\partial_t u + \partial_x(f(u)) = 0, \tag{7.25}$$

where $f : \mathbb{R} \to \mathbb{R}$ is called the flux function. This is an important class of PDEs, e.g. [**LeV92, Ser99, Ser00**], [**Deb05**, §3-§7], or [**Eva98**, §3.4]. We do not consider this class in this book, except for the brief section 7.4.2, which however only serves to explain the nonlinear transport phenomena associated with (7.25).

If viscous effects are taken into account, one obtains the viscous Burgers equation

$$\partial_t u = \nu \partial_x^2 u - \frac{1}{2}\partial_x(u^2), \tag{7.26}$$

where $\nu > 0$ is the viscosity parameter. This equation also occurs as a modulation equation describing the behavior of, e.g., the free surface in viscous flows [**Uec07**], or the evolution of wave numbers in stable periodic pattern [**DSSS09**]. On the real line the parameter $\nu > 0$ can be scaled to 1, hence we consider $\nu = 1$ in the following.

Remark 7.4.2. a) Like for the KPP equation the maximum principle, cf. Theorem 7.1.9, holds for the viscous Burgers equation, too. With the notation from page 215 we have

$$\begin{aligned}
&\partial_t \big[u_2(x_0,t_1) - u_1(x_0,t_1)\big] \\
&\quad = \partial_x^2 \big[u_2(x_0,t_1) - u_1(x_0,t_1)\big] \\
&\qquad - \partial_x \big[u_2(x_0,t_1) - u_1(x_0,t_1)\big]\big(u_2(x_0,t_1) + u_1(x_0,t_1)\big) \\
&\qquad - \big[u_2(x_0,t_1) - u_1(x_0,t_1)\big]\partial_x\big(u_2(x_0,t_1) + u_1(x_0,t_1)\big) \\
&\quad = \partial_x^2 \big[u_2(x_0,t_1) - u_1(x_0,t_1)\big],
\end{aligned}$$

and so the same argument applies.
b) The Burgers equation possesses the so called boost invariance $u \mapsto u + c$ and $x \mapsto x - ct$, i.e., if we add a constant to a solution and go into a co-moving frame, then we again obtain a solution. Therefore, for a number of purposes it is sufficient to restrict to, e.g., solutions with zero mean, i.e. $\int_{\mathbb{R}} u(x,t)\,\mathrm{d}x = 0$. ⌋

7.4.1. A local existence and uniqueness result.

In order to handle the viscous Burgers equation (7.26) or simply the Burgers equation as a dynamical system we again choose the phase space

$$X = C^0_{b,\text{unif}}(\mathbb{R},\mathbb{R}).$$

We have the local existence and uniqueness of mild solutions in X, and by the maximum principle also the global existence and uniqueness of solutions in X of Burgers equation.

Definition 7.4.3. *$u \in C([0,T_0], C^0_{b,unif})$ is called mild solution with initial condition u_0 of the Burgers equation (7.26) if u satisfies*

$$u(t) = T(t)u_0 + \frac{1}{2}\int_0^T T(t-s)\partial_x\left(-(u(s)^2\right)\,\mathrm{d}s$$

where $T(t)$ is the solution operator of the linear diffusion equation which has been introduced in Definition 7.1.5.

Theorem 7.4.4. *For all $C_0 > 0$ there exists a $T_0 > 0$ such that the following holds. For $u_0 \in X$ with $\|u_0\|_X \leq C_0$. Then there exists a unique solution $u \in C([0,T_0],X)$ of the Burgers equation (7.26) with initial datum u_0.*

Proof. The proof of Theorem 7.1.7 has to be modified as follows. As before we set

$$M = C^0([0,T_0], \{u(t) \in X : \|u(t) - T(t)u_0\|_X \leq C_1\}),$$

but the variation of constant formula is then given by

$$u(t) = T(t)u_0 + \int_0^t T(t-\tau)\partial_x N(u)(\tau)\,\mathrm{d}\tau \tag{7.27}$$

where $N(u)$ represents the term $-u^2(x,t)/2$. Additionally to $T(t)\partial_x = \partial_x T(t)$, see Remark 7.4.5, we need the estimates

$$\begin{aligned}
\|T(t)u_0\|_X &\leq \|u_0\|_X\,, \\
\|T(t)\partial_x u_0\|_X &\leq C(1+t^{-1/2})\|u_0\|_X, \\
\|N(u)\|_M &\leq \frac{1}{2}\|u\|_M^2 \leq C_2^2, \\
\|N(u)-N(v)\|_M &\leq \|u+v\|_M\|u-v\|_M/2 \leq C_2\|u-v\|_M,
\end{aligned}$$

which follow almost line for line as for the KPP equation. The second estimate is not sharp. The one on the right-hand side is kept to have the estimates which occur in the general case, cf. (5.7) and Remark 7.4.6.

With these estimates we obtain that the map F defined by the right-hand side of (7.27) maps M into M, since

$$\begin{aligned}
\|F(u) - (T(t)u_0)_{t\geq 0}\|_M &= \sup_{t\in[0,T_0]} \|\int_0^t T(t-\tau)\partial_x N(u)(\tau)\,\mathrm{d}\tau\|_X \\
&\leq \sup_{t\in[0,T_0]} \int_0^t \|T(t-\tau)\partial_x N(u)(\tau)\|_X\,\mathrm{d}\tau \\
&\leq \sup_{t\in[0,T_0]} \int_0^t \left(1+(t-\tau)^{-1/2}\right)\|N(u)(\tau)\|_X\,\mathrm{d}\tau \\
&\leq \sup_{t\in[0,T_0]} \int_0^t \left(1+(t-\tau)^{-1/2}\right)\mathrm{d}\tau\|N(u)\|_M \\
&\leq C_2^2\left(T_0+2T_0^{1/2}\right)/2 \leq C_1
\end{aligned}$$

for $T_0 > 0$ sufficiently small. The map F is a contraction in M, since

$$\begin{aligned}
\|F(u)-F(v)\|_M &= \sup_{t\in[0,T_0]} \|\int_0^t T(t-\tau)(N(u)-N(v))(\tau)\,\mathrm{d}\tau\|_X \\
&\leq \sup_{t\in[0,T_0]} \int_0^t \|T(t-\tau)(N(u)-N(v))(\tau)\|_X\,\mathrm{d}\tau
\end{aligned}$$

$$
\begin{aligned}
&\leq \sup_{t\in[0,T_0]} \int_0^t \left(1+(t-\tau)^{-1/2}\right) \|(N(u)-N(v))(\tau)\|_X \, d\tau \\
&\leq \sup_{t\in[0,T_0]} \int_0^t \left(1+(t-\tau)^{-1/2}\right) d\tau \|(N(u)-N(v))\|_M \\
&\leq \left(T_0 + 2T_0^{1/2}\right) C_2 \|u-v\|_M \leq \frac{1}{2}\|u-v\|_M
\end{aligned}
$$

for $T_0 > 0$ sufficiently small. Hence, there is a unique fixed point u of the map F which is a mild solution of the Burgers equation. $\square$

Remark 7.4.5. The operator $T(t)\partial_x$ is applied to functions in $C^0_{b,unif}$ although it is a priori only defined for functions in $C^1_{b,unif}$. However, there is a continuous extension of this operator to a bounded linear operator on $C^0_{b,unif}$. The reason for this is as follows. On $C^1_{b,unif}$ we have $T(t)\partial_x = \partial_x T(t)$ such that on the dense subspace $(C^1_{b,unif}, \| \ \|_{C^0_{b,unif}})$ we have

$$
\|T(t)\partial_x u\|_{C^0_{b,unif}} = \|\partial_x T(t) u\|_{C^0_{b,unif}} \leq C t^{-1/2} \|u\|_{C^0_{b,unif}} .
$$

Therefore, due to Lemma 5.2.10 this bounded linear operator for $t > 0$ can be extended to a bounded linear operator from the dense subspace to the full space by

$$
T(t)\partial_x u = \lim_{n\to\infty} \partial_x T(t) u_n
$$

for $u \in C^0_{b,unif}$, and $u_n \in C^1_{b,unif}$ with $u_n \to u$ in $C^0_{b,unif}$ for $n \to \infty$. ⌋

Remark 7.4.6. Like the KPP equation, the Burgers equation is a semilinear parabolic equation [**Hen81**], i.e., the semigroup $T(t)$ generated by the operator $-A = \partial_x^2$ is smoothing and the nonlinearity N contains only derivatives of lower order than A. As already explained in §5.2.3 the scheme and the estimates of the last proof work for equations of the form

$$
\partial_t u = \partial_x^2 u + f(u, \partial_x u),
$$

with f a smooth function. It is also the same scheme and the same estimates as for the Navier-Stokes equations in §6.2.1. It does not work for

$$
\partial_t u = \partial_x^2 u + f(u, \partial_x u, \partial_x^2 u)
$$

since then $(1+(t-\tau)^{-1/2})$ in the proof of Theorem 7.4.4 has to be replaced by $(1+(t-\tau)^{-1})$ which is not integrable from 0 to t. As pointed out in §5.3.3, the solutions of semi-linear parabolic equations are infinitely often differentiable for every $t > 0$, and as pointed out in §6.2.2, they even become analytic in a strip along the real axis in the complex plane for every $t > 0$. ⌋

7.4.2. Characteristics. In case of small diffusion we may expect that solutions of the viscous Burgers equation behave similar to the solutions of the inviscid Burgers equation. Thus, to get a first understanding of the dynamical behavior of (7.26) we first consider the inviscid case (7.23) which can be viewed as a nonlinear transport equation where the speed at (x,t) is given by $u(x,t)$. Therefore, we look for curves $(t, x(t, x_0))$ in $\mathbb{R}^2$ along which u is constant. Differentiating $u(x(t,x_0),t) = u(x_0,0)$ w.r.t. t gives $(\partial_x u)\dot{x} + \partial_t u = 0$. Comparison with (7.23) shows that $\dot{x}(t,x_0) = u(x(t,x_0),t) = u(x_0,0)$. The solution curves $x(t,x_0) = x_0 + u(0,x_0)t$ of this ODE are called the characteristics of (7.23). Along the characteristics the values of u are constant. Obviously, even for smooth initial data $x \mapsto u_0(x)$, characteristics can intersect.

Example 7.4.7. Consider

$$u_0(x) = \begin{cases} 1, & \text{for } x \le 0, \\ 1-x, & \text{for } x \in [0,1], \\ 0, & \text{for } x \ge 1. \end{cases}$$

The point $(x,t) = (1,1)$ is the intersection point of the characteristics starting in $[0,1] \times \{0\}$, see Figure 7.12. ⌋

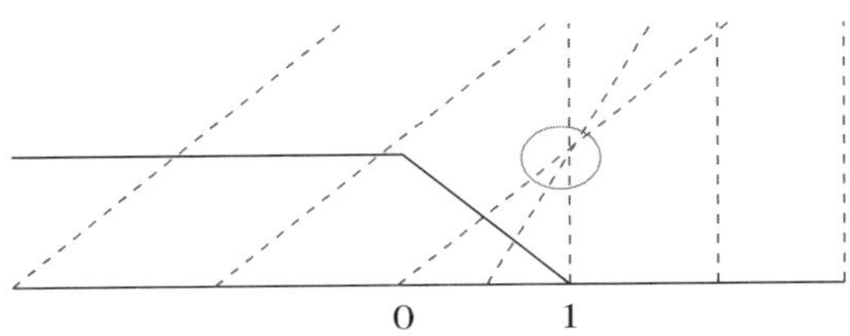

Figure 7.12. Characteristics intersect and create a shock.

Once more, Example 7.4.7 shows that properties of dynamical systems in infinite dimensions strongly depend on the phase space. The solutions stay bounded in $C^0_{b,\mathrm{unif}}$, but explode in $C^1_{b,\mathrm{unif}}$. Such solutions describe the creation of a shock.

After the characteristics have intersected a new understanding of what is meant by a solution is needed. We briefly review the concept of weak solutions, see, e.g., [**Eva98**, §7] or [**RR04**, §3.4]. If the Burgers equation is tested with a $\phi \in C_0^\infty(\mathbb{R} \times \mathbb{R}_+)$ we obtain via integration by parts

$$\begin{aligned} &\int_{\mathbb{R}_+} \int_{\mathbb{R}} \Big(\partial_t u(x,t) + \frac{1}{2} \partial_x (u(x,t)^2) \Big) \phi(x,t) \, \mathrm{d}x \, \mathrm{d}t \\ &= - \int_{\mathbb{R}_+} \int_{\mathbb{R}} \Big(u(x,t) \partial_t \phi(x,t) + \frac{1}{2} u(x,t)^2 \partial_x \phi(x,t) \Big) \, \mathrm{d}x \, \mathrm{d}t = 0. \end{aligned}$$

Hence, all derivatives fall on the smooth so called test function ϕ.

Definition 7.4.8. *u is called weak solution of the Burgers equation if*

$$\int_{\mathbb{R}_+}\int_{\mathbb{R}}\left(u(x,t)\partial_t\phi(x,t)+\frac{1}{2}u(x,t)^2\partial_x\phi(x,t)\right)\mathrm{d}x\,\mathrm{d}t=0$$

for all $\phi\in C_0^\infty(\mathbb{R}\times\mathbb{R}_+)$.

Example 7.4.9. The shock

$$u(x,t)=\begin{cases}1, & \text{for} \quad x\geq t/2,\\ 0, & \text{for} \quad x<t/2,\end{cases}$$

is a weak solution of the inviscid Burgers equation (7.23) since

$$\begin{aligned}\int_0^\infty\int_{\mathbb{R}}\left(u\partial_t\phi+\frac{1}{2}u^2\partial_x\phi\right)\mathrm{d}x\,\mathrm{d}t&=\int_0^\infty\int_{t/2}^\infty\partial_t\phi+\frac{1}{2}\partial_x\phi\,\mathrm{d}x\,\mathrm{d}t\\&=\int_0^\infty\int_0^{2x}\partial_t\phi(x,t)\,\mathrm{d}t\,\mathrm{d}x-\frac{1}{2}\int_0^\infty\phi(t/2,t)\,\mathrm{d}t=0\end{aligned}$$

for all $\phi\in C_0^\infty(\mathbb{R}\times\mathbb{R}_+)$. ⌋

As the following example together with Exercise 7.4.9 shows, weak solutions are in general not unique.

Example 7.4.10. Consider $u_0(x)=\begin{cases}1, & \text{for } x\geq 0,\\ 0, & \text{for } x\leq 0.\end{cases}$ For this initial condition the characteristics leave a complete region in the (x,t)-plane empty. See Figure 7.13. There are many possibilities to fill this region. A physically realistic solution will be a so called rarefaction wave

$$u(x,t)=\begin{cases}1, & \text{for } x\geq t,\\ x/t, & \text{for } x\in[0,t],\\ 0 & \text{for } x\leq 0.\end{cases}$$

We remark that $u(x,t)=\lim_{\varepsilon\to 0}u_\varepsilon(x,t)$, where u_ε solves $\partial_t u_\varepsilon=\varepsilon\partial_x^2u_\varepsilon-u_\varepsilon\partial_x u_\varepsilon$. Thus, u is called viscosity solution. ⌋

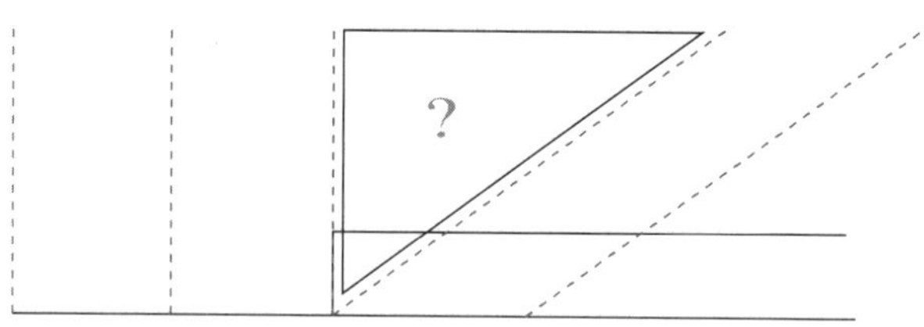

Figure 7.13. The characteristics do not enter parts of the (x,t)-plane and create a rarefaction wave.

7.4.3. Special solutions and Cole-Hopf transformation. We are interested in special solutions such as stationary or traveling wave solutions of the viscous Burgers equation

$$\partial_t u = \nu \partial_x^2 u - u \partial_x u.$$

Obviously every constant in space is a stationary solution. In order to find the traveling waves we make the ansatz $u(x,t) = v(x-ct) = v(\xi)$ and obtain

$$-cv' = \nu v'' - vv'.$$

Integration yields

$$cv + \nu v' - v^2/2 + d = 0$$

with a constant d. This one-dimensional ODE has two fixed points v_- and v_+ satisfying $cv - v^2/2 + d = 0$. To given v_- for $\xi \to -\infty$ and v_+ for $\xi \to \infty$ the system of linear equations

$$cv_+ + d = v_+^2/2 \quad \text{and} \quad cv_- + d = v_-^2/2$$

always has unique solutions c and d if $v_- \neq v_+$. However, looking at the phase portrait of the one-dimensional ODE shows a heteroclinic connection $v_{h,\nu}$ between the fixed points with $\lim_{\xi \to \pm\infty} v_h(\xi) = v_\pm$ only if $(v_+ - v_-)/c < 0$. This heteroclinic connection gives a monotonic front $u(x,t) = v_h(x-ct)$ in the Burgers equation, which is a diffusively smoothed shock. These solutions describe for instance a front of gas with constant high density moving into a domain with constant low density. In case $(v_+ - v_-)/c > 0$ there exist no shocks, and the diffusively smoothed rarefaction wave from Example 7.4.10 occurs. In case $v_- = 1$ and $v_+ = 0$ we find $d = 0$ and $c = 1/2$. Since $\nu v' = -v/2 + v^2/2$ we have $v_{h,\nu}(\xi) = v_{h,1}(\xi/\nu)$ such that for the vanishing viscosity limit

$$\lim_{\nu \to 0} v_{h,\nu}(\xi) = \begin{cases} 0, & \text{for} \quad x \geq 0, \\ 1, & \text{for} \quad x < 0, \end{cases}$$

i.e., the weak shock from Example 7.4.9 occurs as a vanishing viscosity limit.

With the help of the maximum principle a number of stability results can be established. Recall that we have chosen $X = C^0_{b,\mathrm{unif}}(\mathbb{R}, \mathbb{R})$ as phase space.

Theorem 7.4.11. *The fixed points $u^* \in X$ defined by $u(x,t) = u^* \in \mathbb{R}$ are stable, but not asymptotically stable.*

Proof. For given $\varepsilon > 0$ choose $\delta = \varepsilon > 0$. Then by the maximum principle from $\|u_0 - u^*\|_X < \delta$, i.e., $u^* - \delta < u_0(x) < u^* + \delta$ for all $x \in \mathbb{R}$, we have that $u^* - \delta < u(x,t) < u^* + \delta$ for all $x \in \mathbb{R}$ and $t > 0$, i.e., $\|u(t, u_0) - u^*\|_X < \varepsilon$ which shows the stability. Since there are infinitely many fixed points $\{u^* + \delta : |\delta| < \delta_0\}$ in every δ_0-neighborhood of u^* in X, the fixed point u^* cannot be asymptotically stable. □

Remark 7.4.12. Like for the KPP equation again the maximum principle can be used to prove the stability of the traveling waves under small localized perturbations by confining the perturbation between two translates of the traveling wave. The instability in $C^0_{b,unif}$ comes from the fact that there are nearby traveling waves with a different velocity. ⌋

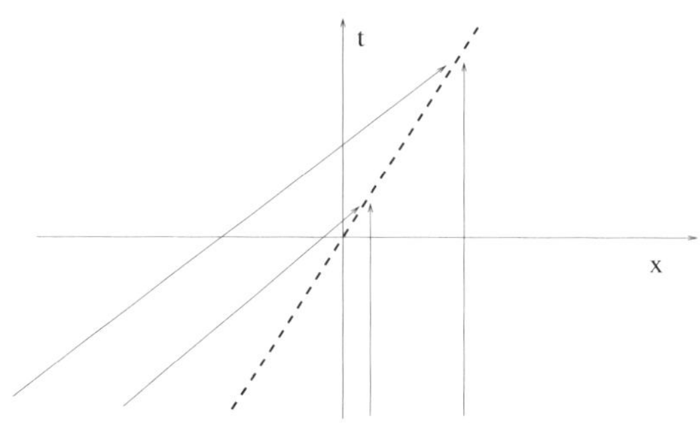

Figure 7.14. Shocks are robust against small spatially localized perturbations. For the shock from Example 7.4.9 these perturbations are transported by the characteristics into the shock with relative velocities $\pm 1/2$, where they finally vanish. This idea can also be used for analytic stability proofs of viscous shocks, cf. [**Kap94b**].

With the help of the Cole-Hopf transformation the nonlinear viscous Burgers equation (7.26) is transformed into the linear heat equation $\partial_t \psi = \partial_x^2 \psi$. The idea behind this remarkable transformation is as follows. Substituting $u = \partial_x \phi$ into (7.26) yields the so called integrated Burgers equation

$$\partial_t \phi = \nu \partial_x^2 \phi - \frac{1}{2}(\partial_x \phi)^2. \tag{7.28}$$

Next we let $\phi = g(\psi)$ with g to be chosen such that the quadratic term in (7.28) vanishes. We obtain

$$g'(\psi)\partial_t \psi = \nu g''(\psi)(\partial_x \psi)^2 + \nu g'(\psi)\partial_x^2 \psi - \frac{1}{2}(g'(\psi)\partial_x \psi)^2$$

which we rewrite as

$$g'(\psi)(\partial_t \psi - \nu \partial_x^2 \psi) = (\nu g''(\psi) - \frac{1}{2} g'(\psi)^2)(\partial_x \psi)^2.$$

The condition $\nu g''(\psi) - \frac{1}{2} g'(\psi)^2 = 0$ leads to $g(\psi) = -2\nu \log \psi$. Thus,

$$\psi(x,t) = \exp\left(-\frac{1}{2\nu}\int_a^x u(y,t)\,\mathrm{d}y\right) \quad \text{and} \quad u(x,t) = -2\nu \frac{\partial_x \psi(x,t)}{\psi(x,t)},$$

where $a \in \mathbb{R}$ is arbitrary. Moreover, we have by construction $\partial_t \psi = \nu \partial_x^2 \psi$. Denoting the Cole-Hopf transformation by $\mathcal{T}$ we thus have the commutative diagram

$$\begin{array}{ccc} u_0 & \xrightarrow{\text{nonlin. evolution in the Burgers eq.}} & u(\cdot,t) \\ \downarrow\mathcal{T} & & \uparrow\mathcal{T}^{-1} \\ \psi_0 & \xrightarrow{\text{linear evolution in the heat equation}} & \psi(\cdot,t) \end{array}$$

Example 7.4.13. We consider the explicit solution

$$\psi(x,t) = 1 + z\mathrm{Erf}(x/\sqrt{\nu t}) \quad \text{where} \quad \mathrm{Erf}(x) = \frac{1}{\sqrt{4\pi}}\int_{-\infty}^{x} \mathrm{e}^{-y^2/4}\,\mathrm{d}y, \tag{7.29}$$

which describes the diffusive mixing of the asymptotic states 1 for $x < 0$ and $1 + z$ for $x \geq 0$. Here Erf is related to the more standard definition of the so called error function

$$\mathrm{erf}(x) = \frac{2}{\sqrt{\pi}}\int_0^x \mathrm{e}^{-t^2}\,\mathrm{d}t \quad \text{via} \quad \mathrm{Erf}(x) = \frac{1}{2}(1 + \mathrm{erf}(x/2)).$$

We obtain

$$u(x,t) = -2\nu\frac{\partial_x\psi}{\psi} = -\frac{\sqrt{\nu}z}{\sqrt{\pi t}}\frac{\mathrm{e}^{-x^2/4\nu t}}{1 + z\mathrm{erf}(x/\sqrt{\nu t})}. \tag{7.30}$$

This is illustrated in Figure 7.15 with the small diffusion coefficient $\nu = 0.01$. Initially the wave steepening by nonlinear transport dominates. However, for $t > 0$ the solution is always smooth, and as $t \to \infty$ the solution decays to zero in L^∞ with a rate $\mathcal{O}(t^{-1/2})$. Figure 7.16 shows the so called renormalized asymptotic profiles

$$\sqrt{t}u(x\sqrt{t},t) = -\frac{\sqrt{\nu}z}{\sqrt{\pi}}\frac{\mathrm{e}^{-x^2/4\nu}}{1 + z\mathrm{erf}(x/\sqrt{\nu})} =: f_z^*(x) \tag{7.31}$$

which are independent of t. ⌋

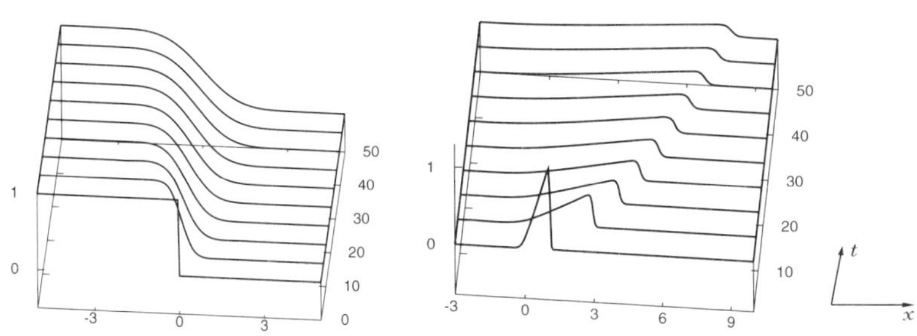

Figure 7.15. Diffusive mixing (7.29) in the linear heat equation (left) and associated exact solution (7.30) of Burgers equation (right), $M = \int_{-\infty}^{\infty} u(y,t)\,\mathrm{d}y = 1$, $\nu = 0.01$. Note that for u we start plotting at $t = 1$.

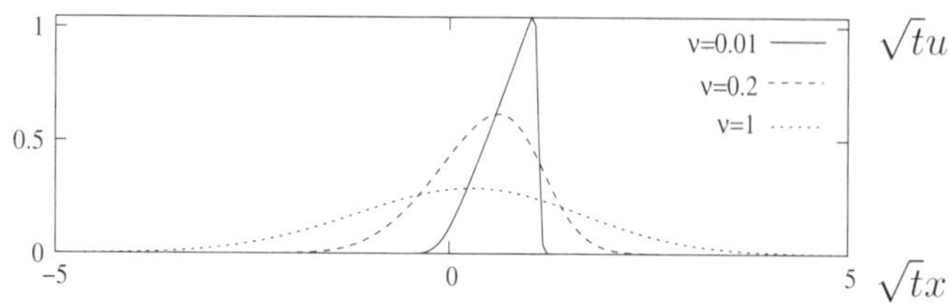

Figure 7.16. Asymptotic Burgers profiles (7.31), normalized to $M = 1$, i.e., $z = \mathrm{e}^{-1/2\nu} - 1$.

More generally, solutions of the linear heat equation $\partial_t \psi = \partial_x^2 \psi$ to initial conditions

$$\lim_{x\to-\infty} \psi(x,0) = 1 \qquad \text{and} \qquad \lim_{x\to\infty} \psi(x,0) = 1 + z < 1$$

satisfy under some additional assumptions, see §14, that

$$\lim_{t\to\infty} \psi(x\sqrt{t},t) = 1 + z\mathrm{erf}(x/\sqrt{\nu})$$

with a rate $\mathcal{O}(1/\sqrt{t})$. This shows that solutions u to spatially localized initial conditions in the Burgers equation satisfy

$$\lim_{t\to\infty} \sqrt{t}u(x\sqrt{t},t) = f_z^*(x)$$

with a rate $\mathcal{O}(1/\sqrt{t})$. Therefore, the renormalized solutions converge to a non-Gaussian limit, see again Figure 7.16. This will be reconsidered in more detail in §14.

Further Reading. A standard reference for the maximum principle is [**PW84**]. The long time dynamics in the KPP equation is dominated by front solutions, and the KPP attractor is fully characterized in [**HN01**]. An extensive review about front propagation, containing a wealth of references, is [**vS03**], and an encyclopaedic book about traveling waves for parabolic problems is [**VVV94**].

For more results on Fourier transform we again refer to [**Duo01**] and [**RS75b**]. See also the "further reading" of Chapter 5.

The method of characteristics can be extended to general scalar first order PDEs of the form

$$F(x_1,\ldots,x_d,u,\partial_{x_1}u,\ldots,\partial_{x_d}u) = 0,$$

with $F : \mathbb{R}^{2d+1} \to \mathbb{R}$ a smooth function. It can be shown that the solution of this PDE is equivalent to the solution of an ODE in $\mathbb{R}^{2d+1}$. See [**CH89**, §2]. The shock speed in conservation laws is determined by the the Rankine-Hugoniot conditions; see Exercises 7.22 and 7.23 for the scalar case, and, e.g., [**Whi99, LeV92, Ser00**] for the case of systems.

Exercises

7.1. Derive the two-dimensional diffusion equation $\partial_t u = \Delta u, u = u(x, y, t)$, from a two-dimensional random walk.

7.2. Compute $u = u(z, t)$ with $z = x_r + \mathrm{i}x_i$ using the solution formula for linear diffusion. Show the convergence of the integral for every $t > 0$ and $z \in \mathbb{C}$.

7.3. Prove that the solutions $u(t, u_0)$ of the linear diffusion equation satisfy the fundamental property of a dynamical system $u(t + s, u_0) = u(t, u(s, u_0))$.

7.4. Use the maximum principle to give an alternative proof of **(E1)** on page 213.

7.5. Prove that if $u \in C([0, T_0], C^2_{b,\mathrm{unif}}) \cap C^1([0, T_0], C^0_{b,\mathrm{unif}})$ solves the variation of constant formula, then u is also a classical solution of the KPP equation.

7.6. Let $R = \{(v, w) : 0 < v < 1, -kv < w < 0\}$. Show that for suitable k the vector field $f = (w, -cw - v + v^2)$ for (7.10) points inwards on ∂R. Use this and the Poincaré-Bendixon theorem to prove that for $c = 2$ there exists a monotonic front connecting $u = 1$ and $u = 0$ for the KPP equation.

7.7. a) Discuss for which α, β there exist similarity solutions $u(x, t) = t^\alpha v(x/t^\beta)$ of $\partial_t u = \partial_x^2 u + f(x)$, $x \in \mathbb{R}$, where (i) $f(x) = 0$, (ii) $f(x) = 1$, (iii) $f(x) = x$.
b) Find α, β such that $\partial_t u + u^p \partial_x u + \mu \partial_x^3 u = 0$ is invariant under $u \mapsto au$, $x \mapsto a^\alpha x$, $t \mapsto a^\beta t$.

7.8. a) Solve explicitly $\partial_t u = \partial_x^2 u$ with IC $u_0(x) = x^2$.
b) Use a) to calculate explicitly $\int s^2 \mathrm{e}^{-s^2}\,\mathrm{d}s$.
c) Compare the solution in a) with the general solution formula.
Hints: For a) make an ansatz $u(x, t) = a(t)x^2 + b(t)x + c(t)$.

7.9. For $d = 1, 2, 3$ and $\delta > 0$ prove the Gagliardo-Nirenberg inequality $\|u\|_{L^4(\mathbb{R}^d)} \leq C\|u\|^{1-d/4-\delta}_{L^2(\mathbb{R}^d)} \|u\|^{d/4+\delta}_{H^1(\mathbb{R}^d)}$. *Hint:* cf. Exercise 6.6

7.10. a) For $u(x) = e^{-a|x|}$ with $a > 0$ calculate $\widehat{u}(k)$.
b) Let $u(x) = \chi_{[-a,a]}$ with $a > 0$. Show that $\widehat{u}(k) = \frac{1}{\pi}\frac{\sin ak}{k}$ for $k \neq 0$ and $\widehat{u}(0) = \frac{a}{\pi}$.

7.11. Show that $u \in L^1$ implies $\widehat{u} \in C^0_b$.

7.12. a) For $u(x) = e^{-x^2/2}$ show $\widehat{u}(k) = \frac{1}{\sqrt{2\pi}} e^{-k^2/2}$ by solving the Fourier transformed ODE which is satisfied by u. Calculate $u(x)$ for $\widehat{u}(k) = e^{-k^2 t}$, $t > 0$.

7.13. Consider the linear heat equation with initial conditions

$$u_0(x) = \begin{cases} a, & \text{for } x < 0, \\ b, & \text{for } x \geq 0. \end{cases}$$

Prove that $u(x, t) = a + (b - a)\mathrm{erf}(x/\sqrt{t})$, where $\mathrm{erf}(x) = \frac{1}{\sqrt{4\pi}} \int_{-\infty}^x \mathrm{e}^{-y^2/4}\,\mathrm{d}y$.

7.14. Prove that Fourier transform is an isomorphism between H^r_m and H^m_r.

7.15. Let $g(x) = \mathrm{e}^{-x^2/2}$. Find $g * g$.

7.16. For $f \in \mathcal{S}$, prove the Heisenberg uncertainty principle $\|xf\|_{L^2}\|k\widehat{f}\|_{L^2} \geq \frac{1}{2}\|f\|^2_{L^2}$. *Hint.* Use Parseval, the relation between differentiation and Fourier transform, and the Cauchy-Schwarz inequality.

7.17. The scaling of a distribution is, as usual, defined by applying the scaling to the test function. I.e., for $k \neq 0$, $T \in D'$ and $\phi \in D$ we set $S_k^1 T(\phi) = T(\phi(k\cdot))$, $S_k^2 T(\phi) = \frac{1}{k} T(\phi(\cdot/k))$. Show that $S_k^1 \delta = S_k^2 \delta$.

7.18. Prove that (a) If $u_n \in C^0(\mathbb{R})$ and $u_n \to u$ uniformly in x, then $u_n \stackrel{D'}{\to} u$ for $n \to \infty$.
(b) If $u_n \in L^2(\mathbb{R})$ and $\|u_n - u\|_{L^2} \to 0$, then $u_n \stackrel{D'}{\to} u$ for $n \to \infty$.
(c) $\sin nx \stackrel{D'}{\to} 0$ and $n^2 \sin nx \stackrel{D'}{\to} 0$ for $n \to \infty$.

7.19. The Hermite functions are $\psi_n(x) = (-1)^n \mathrm{e}^{x^2/2} \frac{\mathrm{d}^n}{\mathrm{d}x^n} \mathrm{e}^{-x^2}$, $n \in \mathbb{N}_0$.
i) Sketch ψ_0, ψ_1 and ψ_2.
ii) Show the recursions $\psi_n'(x) = x\psi_n(x) - \psi_{n+1}(x)$ and $\mathcal{F}(\psi_n)'(k) = k\widehat{\psi}_n(k) - \mathrm{i}\widehat{\psi}_{n+1}(k)$.
iii) Use i) to inductively show $\widehat{\psi}_n = \lambda_n \psi_n$ with $\lambda_n = (-\mathrm{i})^n$.

Remark. iii) means that all the ψ_n are eigenfunctions of $\mathcal{F}$. They form a orthogonal system in $L^2(\mathbb{R})$, $\langle \psi_n, \psi_m \rangle = 2^n n! \sqrt{\pi} \delta_{nm}$, and the normalized $h_n = \psi_n / \|\psi_n\|$ form a complete orthonormal system in $L^2(\mathbb{R})$, which is quite useful in various fields, e.g., quantum mechanics.

7.20. Consider the scalar first order PDE

$$a(x,t,u)\partial_t u + b(x,t,u)\partial_x u = c(x,t,u).$$

Make the ansatz $v(s) = u(x(s), t(s))$ and find a system for $(x(s), t(s), v(s))$.

7.21. Show that $u(x,t) = u_0(x_0) = u_0(x - tu(x,t))$ is an implicit representation of the solutions of $\partial_t u + u\partial_x u = 0$. Show by differentiation of this representation w.r.t. x that for $t = t^* = -1/(\inf_{x_0 \in \mathbb{R}} u_0'(x_0))$ a shock occurs, i.e., that $u = u(x,t)$ is no longer differentiable w.r.t. x.

7.22. Given $u_l > u_r$ and $u_0(x) = u_l$ for $x \leq 0$ and $u(x) = u_r$ for $x > 0$ calculate the shock speed α of the weak solution $u(x,t) = \begin{cases} u_l, & \text{for} \quad x \leq \alpha t, \\ u_r, & \text{for} \quad x > \alpha t \end{cases}$ of the inviscid Burgers equation.

7.23. Consider for a general scalar conservation law $\partial_t u = -\partial_x(f(u))$ the weak solution from Exercise 7.22. Derive the so called Rankine-Hugoniot jump condition $\alpha = [f]/[u] := (f(u_r) - f(u_l))/(u_r - u_l)$ for the velocity of the shock.

7.24. Discuss the phase portrait of $-cv' = v'' - vv'$ in the (v, v')-plane. For $v_- = 1$ and $v_+ = 0$ show that $c = 1/2$ and that $v(\xi) = (1 + \mathrm{e}^{\xi})^{-1}$ is the explicit solution.

7.25. Consider the scalar conservation law $\partial_t u + u^3 \partial_x u = 0$, $x \in \mathbb{R}$, $t > 0$. Find the solution for $u(x,0) = g(x) = \begin{cases} 1, & 0 < x < 1, \\ 0, & \text{else.} \end{cases}$

7.26. Solve the viscous Burgers equation $\partial_t u = \nu \partial_x^2 u - \frac{1}{2}\partial_x(u^2)$ with $u(x,0) = H(x) = \begin{cases} -1, & \text{for } x < 0, \\ 1, & \text{for } x \geq 0. \end{cases}$ Give a sketch of the solution.

Chapter 8

Three canonical modulation equations

In this chapter we consider another three scalar equations on the real line, namely the Nonlinear Schrödinger (NLS) equation, the Korteveg-de Vries (KdV) equation, and the Ginzburg-Landau (GL) equation. These equations are the three most important so called modulation equations, sometimes also called amplitude or envelope equations. By multiple scaling analysis the NLS equation can be derived in order to describe the evolution of an envelope of a spatially and temporarily oscillating wave packet. This will be discussed in Chapter 11. Similarly, the KdV equation can be derived for the description of long waves in dispersive media, see Chapter 12, and the GL equation can be derived to describe the evolution of the envelope of spatio-temporal pattern in dissipative systems, see Chapters 10.

Besides the basic theory and the occurring phenomena for each equation, we introduce some general concepts, which are useful for the analysis of these equations but also in other circumstances. These are for instance the method of stationary phase in §8.1.4, the use of uniformly local Sobolev spaces in §8.3.1, and a concept of attractors on unbounded domains in §8.3.4.

As already said, each of the equations considered in this Chapter plays a very important role in the description of various phenomena in more complicated systems, and thus this chapter is partly a preparation for the remainder of this book. On the other hand, as for the equations from Chapter 7, for each of these equations there exists many papers, and often whole books entirely devoted to the particular equation. Again we keep the exposition rather brief, and give a list of further reading at the end of the chapter.

8.1. The NLS equation

The Nonlinear Schödinger (NLS) equation

$$\partial_t u = \mathrm{i}\nu_1 \partial_x^2 u + \mathrm{i}\nu_2 |u|^2 u, \tag{8.1}$$

with $t \in \mathbb{R}$, $x \in \mathbb{R}$, $u = u(x,t) \in \mathbb{C}$, and coefficients $\nu_1, \nu_2 \in \mathbb{R}$, can be derived by multiple scaling analysis in order to describe the evolution of the envelope of a spatially and temporarily oscillating wave packet, as will be discussed in Chapter 11. By rescaling $u \mapsto \beta_1 u$, $x \mapsto \beta_2 x$, and $t \mapsto \beta_3 t$ with $\beta_j \in \mathbb{R}$ for $j \in \{1,2,3\}$ it can be transformed into

$$\partial_t u = -\mathrm{i}\partial_x^2 u + \alpha \mathrm{i}|u|^2 u, \qquad \alpha = \pm 1. \tag{8.2}$$

See Exercise 8.1. Two cases remain, namely $\alpha = -1$ and $\alpha = 1$. The case $\alpha = -1$ is called focusing and the case $\alpha = +1$ is called defocusing. In particular, the focusing NLS equation is widely used in nonlinear optics to describe the evolution and interaction of optical pulses. The NLS equation also plays a role in some theories about the occurrence of so called freak or rogue waves, cf. [**Osb10**]. The variant

$$\partial_t u = -\mathrm{i}\partial_x^2 u + \mathrm{i}Vu + \alpha \mathrm{i}|u|^2 u, \tag{8.3}$$

with $V = V(x)$ some potential, is called the Gross-Pitaevsky equation and can be derived for the description of Bose-Einstein condensates, cf. [**Pel11**]. In summary, the NLS equation is a widely used model in nonlinear physics.

The evolution of the NLS equation is not smoothing in the sense of the KPP or the Burgers equation. Hence, the phase spaces are not connected by smoothing. We can have global existence or stability in one space, but explosion or instability in another space.

8.1.1. Nonlinear oscillations and pulse solutions. The NLS equation consists of two parts, namely the dispersion part $-\mathrm{i}\partial_x^2 u$ and the nonlinear oscillation part $\alpha \mathrm{i} u|u|^2$.

Making the ansatz $u(x,t) = v(t) = r(t)\mathrm{e}^{\mathrm{i}\phi(t)}$ for x-independent solutions we find

$$\partial_t r = 0 \quad \text{and} \quad \partial_t \phi = \alpha r^2,$$

with the solutions $r(t) = r(0)$ and $\phi(t) = \phi(0) + \alpha r^2(0)t$, i.e., oscillations where the frequency increases with r. Next, we search for solutions of the form $u(x,t) = v(t)\mathrm{e}^{\mathrm{i}kx} = r(t)\mathrm{e}^{\mathrm{i}\phi(t)}\mathrm{e}^{\mathrm{i}kx}$. We find

$$\partial_t r = 0 \quad \text{and} \quad \partial_t \phi = k^2 + \alpha r^2,$$

with the solutions $r(t) = r(0)$ and $\phi(t) = \phi(0) + \omega(k,r)t$, where $\omega(k,r) = (k^2 + \alpha r^2)$. Thus, we have solutions which are periodic in time and space, namely

$$u(x,t) = u(x,t;k,r,\phi_0) = r\mathrm{e}^{(\mathrm{i}(kx + \phi_0 + (k^2 + \alpha r^2)t))}.$$

In the defocusing case, $\alpha = 1$, where always $\omega(k, r) = k^2 + \alpha r^2 > 0$, all periodic waves travel left, whereas in the focusing case, $\alpha = -1$, where $\omega(k, r) = k^2 + \alpha r^2$ can have either sign, the periodic waves can travel left or right. For illustration, we sketch some of these nonlinear oscillations in Figure 8.1. Interestingly, in the focussing case they are all unstable, which is known as Benjamin–Feir instability. See Exercise 8.23 where we consider the related stability question for spatially periodic solutions in the GL equation.

a) $\alpha = 1$, $k = 1$ b) $\alpha = -1$, $k = 1$ c) $\alpha = -1$, $k = 2$

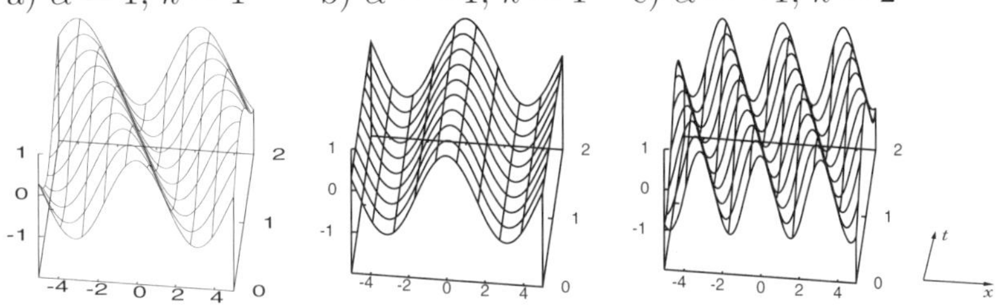

Figure 8.1. Nonlinear oscillations (real part) for the NLS equation, with $r = 1$. a) shows the defocusing case, where all waves travel left, while b), c) show the focusing case, where waves can travel left or right.

However, from the point of applications of the NLS equation, in particular due to its derivation for the description of modulations of electromagnetic waves, pulse solutions are more interesting. In order to find them, we make the ansatz

$$u(x, t) = B(x - ct)e^{i(qx - \omega t + \phi_0)},$$

with $B(\xi) \in \mathbb{R}$, which yields

$$-i\omega B - cB' = -iB'' + 2qB' + iq^2 B + \alpha i B^3.$$

Separating real and imaginary part gives the relations $c = -2q$ and

$$0 = B'' - (\omega + q^2)B - \alpha B^3. \tag{8.4}$$

This ODE can be discussed using the methods from §2.3.3, see in particular Remark 2.3.21. One finds that for $\omega + q^2 > 0$ and $\alpha = -1$ there exist solutions of (8.4) which are homoclinic to the origin. Figure 8.2 shows the phase portrait. In fact, there exist explicit formulas for homoclinic solutions of (8.4). Before computing these in Exercise 8.4, it is useful to have a look at the symmetries of the NLS equation in Exercise 8.3.

8.1.2. Dispersion. The linear Schrödinger equation

$$\partial_t u = -i\partial_x^2 u \tag{8.5}$$

is a prototype example of a linear dispersive equation. Dispersion means that the phase velocity $c_p(k) = \omega(k)/k$ of harmonic waves $e^{i(kx - \omega(k)t)}$ depends

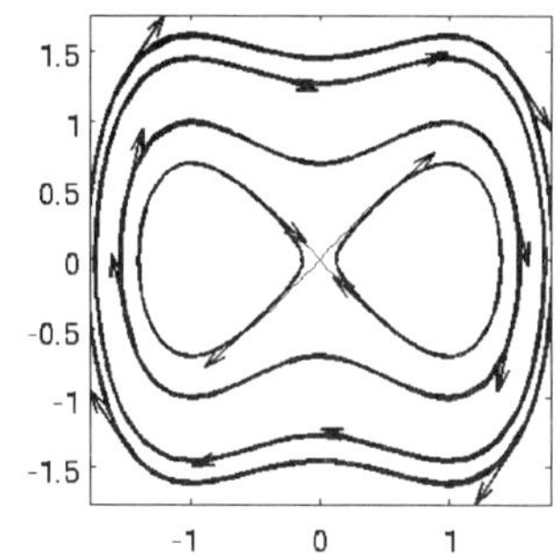

Figure 8.2. Phase portrait for (8.4), $\omega + q^2 = 1$ and $\alpha = -1$.

in a non-trivial way on k. This has a number of consequences which we explain in the following. For (8.5) we have solutions $u(x,t) = \mathrm{e}^{\mathrm{i}(kx-\omega t)}$ where $\omega = -k^2$. Hence, $c_p(k) = k$ and the amplitude of the harmonic waves is conserved. This behavior is in contrast to the linear diffusion equation $\partial_t u = \partial_x^2 u$ where we have $u(x,t) = \mathrm{e}^{\mathrm{i}kx+\lambda t}$, with $\lambda = -k^2$, and thus all spatially harmonic waves are damped with some exponential rate, except for $k = 0$.

Like for the diffusion equation there exists an explicit solution formula which can be derived by Fourier transform and which at first looks very similar to the one of the diffusion equation.

Lemma 8.1.1. *The initial value problem for the linear Schrödinger equation* $\partial_t u = -\mathrm{i}\partial_x^2 u$, $u|_{t=0} = u_0$, *is solved by*

$$u(x,t) = \frac{1}{\sqrt{-4\pi \mathrm{i}t}} \int_{-\infty}^{\infty} \mathrm{e}^{-\frac{\mathrm{i}(x-y)^2}{4t}} u_0(y)\,\mathrm{d}y. \tag{8.6}$$

The integral exists for $u_0 \in L^1(\mathbb{R})$ *and extends to an isometry in* $L^2(\mathbb{R})$, *i.e.*, $\|u(\cdot,t)\|_{L^2} = \|u_0\|_{L^2}$.

Proof. Proceeding exactly as in Example 7.3.20 it remains to calculate $G(x,t) = \int_{\mathbb{R}} \mathrm{e}^{\mathrm{i}k^2 t}\mathrm{e}^{\mathrm{i}kx}\,\mathrm{d}k$. We have

$$\partial_x G(x,t) = \int_{\mathbb{R}} \mathrm{i}k\mathrm{e}^{\mathrm{i}k^2 t}\mathrm{e}^{\mathrm{i}kx}\,\mathrm{d}k = -\frac{\mathrm{i}x}{2t} G(x,t).$$

This differential equation is solved by $G(x,t) = C\mathrm{e}^{-\mathrm{i}x^2/4t}$. The constant C can be computed through

$$C = G(0,t) = \int_{\mathbb{R}} \mathrm{e}^{\mathrm{i}k^2 t}\,\mathrm{d}k = \sqrt{\frac{1}{t}} \int_{\mathbb{R}} \mathrm{e}^{\mathrm{i}y^2}\,\mathrm{d}y = \sqrt{\frac{\pi}{t}}\mathrm{e}^{\mathrm{i}\pi/4} = \sqrt{\frac{\pi}{-\mathrm{i}t}}.$$

The integral $\int_{\mathbb{R}} \mathrm{e}^{\mathrm{i}y^2}\,\mathrm{d}y$ exists as improper integral due to faster and faster oscillations of $\mathrm{e}^{\mathrm{i}y^2}$ for $|y| \to \infty$. The isometry follows immediately from

$$\|u(\cdot,t)\|_{L^2} = (2\pi)^{1/2}\|\widehat{u}(\cdot,t)\|_{L^2} = (2\pi)^{1/2}\|\widehat{u}(\cdot,0)\|_{L^2} = \|u_0\|_{L^2}. \qquad \square$$

Like for diffusion, from (8.6) we immediately obtain the estimate

$$\sup_{x\in\mathbb{R}} |u(x,t)| \leq \frac{C}{\sqrt{t}} \int_{-\infty}^{\infty} |u(x,0)|\,\mathrm{d}x, \tag{8.7}$$

i.e., solutions to spatially localized initial conditions decay uniformly towards zero with a rate $t^{-1/2}$. But there are major differences: by diffusion energy is lost, i.e., $\frac{\mathrm{d}}{\mathrm{d}t}\int |u(x,t)|^2\,\mathrm{d}x \leq 0$, while dispersion conserves energy, i.e., $\frac{\mathrm{d}}{\mathrm{d}t}\int |u(x,t)|^2\,\mathrm{d}x = 0$, but spreads it all over the real line.

Dispersion smoothes solutions locally in space. We can compute

$$\partial_x u(x,t) = \frac{1}{\sqrt{4\pi \mathrm{i} t}} \int_{-\infty}^{\infty} \frac{-2\mathrm{i}(x-y)}{4t} \mathrm{e}^{-\frac{\mathrm{i}(x-y)^2}{4t}} u(y,0)\,\mathrm{d}y < \infty$$

for all $x \in \mathbb{R}$ and all $t > 0$ if $\int_{-\infty}^{\infty} |y||u(y,0)|\,\mathrm{d}y < \infty$, i.e., $\partial_x u(x,t)$ can be computed point-wise, although the initial condition may only be continuous. More generally, if the moment $\int_{-\infty}^{\infty} |y|^n |u(y,0)|\,\mathrm{d}y$ is finite, then $\partial_x^n u(x,t)$ is finite for all $x \in \mathbb{R}$ and all $t > 0$.

The reason for this behavior is as follows: Consider initial conditions which are slow modulations in space of an underlying carrier wave, i.e., initial conditions of the form

$$u(x,0) = u_0(\varepsilon x)\mathrm{e}^{\mathrm{i}kx}$$

where $0 < \varepsilon \ll 1$ is a small parameter and u_0 a smooth spatially localized function. The right-hand side of the Schrödinger equation applied to this initial condition yields

$$\mathrm{i}\partial_x^2(u_0(\varepsilon x)\mathrm{e}^{\mathrm{i}kx}) = \mathrm{i}\mathrm{e}^{\mathrm{i}kx}\left(-k^2 u_0(\varepsilon x) + \varepsilon 2\mathrm{i}k u_0'(\varepsilon x) + \varepsilon^2 u_0''(\varepsilon x)\right).$$

This motivates us to make the ansatz

$$u(x,t) = B\left(\varepsilon(x - c_g t), \varepsilon^2 t\right) \mathrm{e}^{\mathrm{i}k(x - c_p t)},$$

with constants c_p and c_g, and a function $B = B(\xi, \tau)$ with $\xi = \varepsilon(x - c_g t)$ and $\tau = \varepsilon^2 t$. Inserting this ansatz and computing the coefficients in front of ε^0, ε^1, and ε^2 shows that

$$-\mathrm{i}kc_p = -\mathrm{i}k^2, \qquad -c_g\partial_\xi B = -2k\partial_\xi B, \qquad \text{and} \qquad \partial_\tau B = \mathrm{i}\partial_\xi^2 B.$$

The constant $c_p = \omega(k)/k = k$ with frequency $\omega = k^2$ is called phase velocity. The constant $c_g = \omega'(k) = 2k$ is called the group velocity. This calculation shows that wave packets with carrier wave $\mathrm{e}^{\mathrm{i}kx}$ with k large are transported with large group velocity $c_g = 2k$ towards infinity. If the initial condition is spatially localized, then nothing can come from infinity. Therefore, at fixed $x \in \mathbb{R}$ only wave packets with low derivatives remain, and hence a local smoothing occurs.

On the other hand this also has the consequence that if the initial condition is not spatially localized, then packets with high derivatives can come from infinity. In fact, there is no local existence and uniqueness of solutions in $C^0_{b,\mathrm{unif}}(\mathbb{R},\mathbb{R})$.

Remark 8.1.2. For the transport equation and the linear wave equation we obtain $\omega(k) = ck$ with a constant $c \in \mathbb{R}$ and hence the group velocity $\omega'(k) = c$ is bounded and the transport equation and the linear wave equation can be solved in $C^0_{b,\mathrm{unif}}(\mathbb{R},\mathbb{R})$. ⌋

It turns out that $X = L^2(\mathbb{R},\mathbb{C})$ is a good choice for the handling of the linear Schrödinger equation as a dynamical system.

Lemma 8.1.3. *The curve of solutions $t \mapsto U(t,u_0)$ with $U(t,u_0)(x) = u(x,t)$ and $u(x,0) = u_0$ is continuous in X if $u_0 \in X$. Moreover, $U(t+s,u_0) = U(t,U(s,u_0))$.*

Proof. First of all we have the semigroup property

$$\begin{aligned} U(t+s,u_0) =& \mathcal{F}^{-1}(k \mapsto \mathrm{e}^{\mathrm{i}k^2(t+s)}\widehat{u}_0(k)) \\ =& \mathcal{F}^{-1}(k \mapsto \mathrm{e}^{\mathrm{i}k^2t}(\mathrm{e}^{\mathrm{i}k^2s}\widehat{u}_0(k))) = U(t,U(s,u_0)) \end{aligned}$$

such that is sufficient to prove continuity in $t = 0$. For all $\varepsilon > 0$ we have to find a $t_0 > 0$ such that for all $t \in (0,t_0)$ we have

$$\begin{aligned} \|U(t,u_0) - u_0\|^2_{L^2} =& \int_{\mathbb{R}} |u(x,t) - u(x,0)|^2 \,\mathrm{d}x = 2\pi \int_{\mathbb{R}} |\widehat{u}(k,t) - \widehat{u}(k,0)|^2 \,\mathrm{d}k \\ =& 2\pi \int_{\mathbb{R}} |(\mathrm{e}^{\mathrm{i}k^2t} - 1)\widehat{u}(k,0)|^2 \,\mathrm{d}k = 2\pi \int_{|k|<L} \ldots \,\mathrm{d}k + 2\pi \int_{|k|\geq L} \ldots \,\mathrm{d}k < \varepsilon^2, \end{aligned}$$

where L is chosen so large that

$$2\pi \int_{|k|\geq L} \ldots \,\mathrm{d}k \leq 4\pi \int_{|k|\geq L} |\widehat{u}(k,0)|^2 \,\mathrm{d}k < \varepsilon^2/2.$$

For this L we find for the first integral

$$\begin{aligned} 2\pi \int_{|k|<L} \ldots \,\mathrm{d}k \leq& 2\pi \sup_{|k|<L} |\mathrm{e}^{\mathrm{i}k^2t} - 1|^2 \int_{\mathbb{R}} \widehat{u}(k,0)|^2 \,\mathrm{d}k \\ \leq& 2\pi |\mathrm{e}^{\mathrm{i}L^2t} - 1|^2 \int_{\mathbb{R}} |\widehat{u}(k,0)|^2 \,\mathrm{d}k < \varepsilon^2/2 \end{aligned}$$

if $t_0 > 0$ is sufficiently small. Therefore, we are done. □

The choice $X = L^2(\mathbb{R},\mathbb{C})$ is only a good choice for the handling of the linear Schrödinger equation. For the NLS equation Sobolev spaces $H^m(\mathbb{R},\mathbb{C})$ with $m > 1/2$ are more adequate.

8.1.3. Local existence and uniqueness, and the Hamiltonian. In the following we choose $X = H^m(\mathbb{R}, \mathbb{C})$ with $m > 1/2$ as phase space for the NLS equation.

Theorem 8.1.4. *Let $u_0 \in X$. Then there exist a $T_0 > 0$ and a unique local solution $u \in C([0, T_0], X)$ of the NLS equation with $u|_{t=0} = u_0$.*

Proof. According to Exercise 8.6 the operator $-\mathrm{i}\partial_x^2$ generates a C_0-semigroup $\mathrm{e}^{-\mathrm{i}t\partial_x^2} : X \to X$. Moreover, by Lemma 7.3.29 the nonlinearity $u \mapsto \mathrm{i}\alpha|u|^2 u$ is locally Lipschitz-continuous in X. The local existence thus follows by the variation of constant formula and the contraction mapping theorem, cf. the proof of Theorem 7.1.7. □

Remark 8.1.5. Like for the wave equation or the transport equation the solution also exists backwards in time, i.e., $u \in C([-T_0, T_0], X)$). ⌋

This local solution can be continued as long as the H^m-norm of the solution stays bounded. First we show that the L^2-norm is conserved along solutions. We have

$$\begin{aligned} \frac{\mathrm{d}}{\mathrm{d}t}\|u\|_{L^2}^2 &= \frac{\mathrm{d}}{\mathrm{d}t}\int_{\mathbb{R}} u\overline{u}\,\mathrm{d}x = \int_{\mathbb{R}} (\partial_t u)\overline{u} + u(\partial_t \overline{u})\,\mathrm{d}x \\ &= 2\,\mathrm{Re}\int_{\mathbb{R}} \overline{u}\left(-\mathrm{i}(\partial_x^2 u - \alpha|u|^2 u)\right)\,\mathrm{d}x = 0. \end{aligned}$$

The estimate for the H^1-norm is related to the Hamiltonian structure of the NLS equation. It turns out that the NLS equation, like the KdV equation in the next section, is a completely integrable Hamiltonian system which can be solved explicitly. See [**DJ89**, §6] for an overview. Here, we only show that the NLS equation is a Hamiltonian system. With

$$H(u) = \int_{\mathbb{R}} \frac{1}{2}|\partial_x u(x)|^2 + \frac{1}{4}\alpha|u(x)|^4\,\mathrm{d}x \tag{8.8}$$

we find

$$\begin{aligned} \partial_u H[v] &= \lim_{\varepsilon\to 0} \varepsilon^{-1}(H(u+\varepsilon v) - H(u)) \\ &= \lim_{\varepsilon\to 0} \varepsilon^{-1}\int_{\mathbb{R}} |\partial_x(u+\varepsilon v)|^2/2 + \alpha|u+\varepsilon v|^4/4 - |\partial_x u|^2/2 - |u|^4/4\,\mathrm{d}x \\ &= \mathrm{Re}\int_{\mathbb{R}} (-\partial_x^2\overline{u} + \alpha\overline{u}|u|^2)v\,\mathrm{d}x \end{aligned}$$

which maps $v \in X = L^2(\mathbb{R}, \mathbb{C})$ linearly into $\mathbb{R}$, i.e., $\partial_u H$ is a linear map from X to $\mathbb{R}$ and hence an element of the dual space. According to the Riesz representation theorem [**Alt16**, Satz 4.1], in Hilbert spaces the dual space can be identified with X by defining a map (the canonical isomorphism)

$$\beta : \mathrm{Lin}(X, \mathbb{R}) \to X, \qquad \left(v \mapsto \langle u, v\rangle_{L^2} = \mathrm{Re}\int_{\mathbb{R}} u(x)\overline{v(x)}\,\mathrm{d}x\right) \mapsto u \tag{8.9}$$

and therefore $\beta\partial_u H = -\partial_x^2 u + \alpha u|u|^2$. We finally have

$$\partial_t u = -\mathrm{i}\partial_x^2 u + \alpha \mathrm{i} u|u|^2 = \mathrm{i}\beta\partial_u H(u) = J\beta\partial_u H(u)$$

where the operator $Ju = \mathrm{i}u$ is skew symmetric in X since

$$\langle Ju, v\rangle_{L^2} = \operatorname{Re}\int_{\mathbb{R}} \overline{\mathrm{i}u(x)}v(x)\,\mathrm{d}x = -\operatorname{Re}\int_{\mathbb{R}} \overline{u(x)}\mathrm{i}v(x)\,\mathrm{d}x = -\langle u, Jv\rangle_{L^2}.$$

The Hamiltonian is well defined on H^1 since $H \le \|\partial_x u\|_{L^2}^2 + \|u\|_{C^0}^2\|u\|_{L^2}^2$. In the defocusing case the Hamiltonian is positive definite, i.e., $H(u) > 0$ for $u \neq 0$, whereas in the focusing case it is indefinite. Hence, in the defocusing case we get the H^1-estimate for free since then

$$\|u(t)\|_{H^1}^2 \le H(t) + \|u(t)\|_{L^2}^2 = H(0) + \|u_0\|_{L^2}^2$$

for all $t \in \mathbb{R}$ and thus the solution exists globally with a uniform bound in $H^1(\mathbb{R})$, and we already proved half of the following theorem.

Theorem 8.1.6. *For $u_0 \in H^1(\mathbb{R})$ the local solution of the NLS equation exists globally in time and stays uniformly bounded in $H^1(\mathbb{R})$.*

Proof. It remains to consider the focusing case. We have $\|u(t)\|_{L^2}^2 = \|u(0)\|_{L^2}^2$ and $H(u(t)) = H(u(0))$ and thus

$$\frac{1}{2}\int_{\mathbb{R}} |\partial_x u|^2\,\mathrm{d}x = H(u(0)) + \frac{1}{4}\int_{\mathbb{R}} |u|^4\,\mathrm{d}x. \tag{8.10}$$

To estimate the last term we use the Gagliardo-Nirenberg estimate, cf. Lemma 6.3.10,

$$\int_{\mathbb{R}} |u|^{q+1}\,\mathrm{d}x \le \left(\int_{\mathbb{R}} |u|^2\,\mathrm{d}x\right)^{(2(q+1)-d(q-1))/4}\left(\int_{\mathbb{R}} |\nabla u|^2\,\mathrm{d}x\right)^{d(q-1)/4}, \tag{8.11}$$

which for $q = 3$ and $d = 1$ yields

$$\int_{\mathbb{R}} |u|^4\,\mathrm{d}x \le \left(\int_{\mathbb{R}} |u|^2\,\mathrm{d}x\right)^{3/2}\left(\int_{\mathbb{R}} |\partial_x u|^2\,\mathrm{d}x\right)^{1/2}. \tag{8.12}$$

Therefore, with $\phi(u) = \frac{1}{2}\int_{\mathbb{R}} |\partial_x u|^2\,\mathrm{d}x$ we have

$$0 \le \phi(u(t)) \le H(u(0)) + \|u(0)\|_{L^2}^{3/2}\sqrt{\phi(u(t))}$$

since the L^2-norm and the Hamiltonian are conserved. This inequality immediately implies a uniform bound for $\phi(u(t))$ and so also the H^1-norm stays bounded. □

Remark 8.1.7. Note that (8.11) depends on d. In fact, for $x \in \mathbb{R}^d$ with $d \ge 2$ the solutions of $\partial_t u = -\mathrm{i}\Delta u - \mathrm{i}|u|^2 u$ can blow up in finite time T, i.e., $\|u(t)\|_{H^1} \to \infty$ as $t \to T$, cf. [**SS99b, Fib15**]. ⌋

As already said, we come back to the NLS equation in Part IV, while here we close with some bounds for oscillatory integrals which are used in the derivation of dispersive estimates.

8.1.4. The method of stationary phase. We explain a method which gives more insight into dispersion and which especially allows to compute decay rates like (8.7) from the Fourier representation $\widehat{u}(k,t) = \widehat{G}(k,t)\widehat{u}(k,0)$ of the solution u also in situations where in physical space no explicit representation formula for G such as (8.6) is known. Our simple approach is based on the subsequent Lemma of van der Corput, cf. [**Ste93**].

Lemma 8.1.8. (Lemma of van der Corput) *Let $\phi : \mathbb{R} \to \mathbb{R}$ in C^ℓ with $|\phi^{(\ell)}| \geq 1$ for all $\theta \in [a,b]$. For $\ell = 1$ assume the monotonicity of ϕ'. Then there exists a $C_\ell > 0$ such that*

$$\left| \int_a^b e^{it\phi(\theta)d\theta} \right| \leq C_\ell t^{-1/\ell}.$$

Proof. For $\ell = 1$ we obtain

$$\begin{aligned} I(t) = \int_a^b e^{it\phi(\theta)}\, d\theta = & \int_a^b \frac{1}{it\phi'(\theta)} \left(\frac{d}{d\theta}(e^{it\phi(\theta)}) \right) d\theta \\ = & \frac{1}{it\phi'(\theta)} e^{it\phi(\theta)}|_a^b - \frac{1}{it} \int_a^b \left(\frac{d}{d\theta} \frac{1}{\phi'(\theta)} \right) e^{it\phi(\theta)} d\theta. \end{aligned}$$

This can be estimated by

$$\begin{aligned} |tI(t)| \leq & \frac{1}{|\phi'(b)|} + \frac{1}{|\phi'(a)|} + \int_a^b \left| \frac{d}{d\theta} \frac{1}{\phi'(\theta)} \right| d\theta \\ \leq & \frac{1}{|\phi'(b)|} + \frac{1}{|\phi'(a)|} + \frac{1}{|\phi'(b)^{-1} - \phi'(a)^{-1}|} \leq 4 \end{aligned}$$

where we used the monotonicity of ϕ'.

For all other ℓ we use induction. Assume that $|\phi^{(\ell+1)}(\theta)| \geq 1$ for all $\theta \in [a,b]$. Then there exists at most one point $\theta_0 \in [a,b]$ with $\phi^{(\ell)}(\theta_0) = 0$ and we have $|\phi^{(\ell)}(\theta)| \geq \delta$ for $|\theta - \theta_0| \geq \delta$. If this point does not exist we apply the Lemma after a possible rescaling for ℓ. Thus, we assume the existence of such a θ_0 and write $I(t) = I_1(t) + I_2(t)$ where $I_1(t)$ is the integral over $(a, \theta_0 - \delta) \cup (\theta_0 + \delta, b)$ and $I_2(t)$ over $(\theta_0 - \delta, \theta_0 + \delta)$. We find

$$|I_1(t)| \leq 2C_\ell(\delta t)^{-1/\ell}$$

since for $\theta \in (a, \theta_0 - \delta) \cup (\theta_0 + \delta, b)$ we have $|\phi^{(\ell)}(\theta)| \geq \delta$ such that the induction for ℓ can be applied. Moreover, we have

$$|I_2(t)| \leq | \int_{\theta_0 - \delta}^{\theta_0 + \delta} e^{it\phi(\theta)}\, d\theta| \leq 2\delta.$$

Choose $\delta = t^{-1/(\ell+1)}$. Since $(\delta t)^{1/\ell} = (t^{-1/(\ell+1)}t)^{1/\ell} = t^{-1/(\ell+1)}$ we have $I(t) \leq C_{\ell+1}t^{-1/(\ell+1)}$, with $C_{\ell+1} = 2(1 + C_\ell)$. □

Remark 8.1.9. Since no compactness argument has been used in the proof of Lemma 8.1.8 values $a = -\infty$ and $b = \infty$ are allowed if the integrals exist.

We use Lemma 8.1.8 and Remark 8.1.9 to confirm the dispersive estimate (8.7).

Example 8.1.10. The linear Schrödinger equation $\partial_t u = \mathrm{i}\partial_x^2 u$ with $t, x \in \mathbb{R}$, $u(x,t) \in \mathbb{C}$ is solved by $u(x,t) = \int_{-\infty}^{\infty} G(x-y,t)u(y,0)\,\mathrm{d}y$ with

$$G(x,t) = \int_{-\infty}^{\infty} \mathrm{e}^{-\mathrm{i}k^2 t}\mathrm{e}^{\mathrm{i}kx}\,\mathrm{d}k = \int_{-\infty}^{\infty} \mathrm{e}^{\mathrm{i}t\phi(k)}\,\mathrm{d}k$$

such that $\phi(k) = -k^2 + kx/t$. The integral is estimated for every fixed $\xi = x/t$. Since $\phi''(k) = -2$ we thus have $\sup_{\xi\in\mathbb{R}} |G(\xi t, t)| \leq Ct^{-1/2}$. From $\sup_{\xi\in\mathbb{R}} |G(\xi t, t)| = \sup_{x\in\mathbb{R}} |G(x,t)|$ it follows

$$\|u(\cdot,t)\|_{L^\infty} \leq C\|G(\cdot,t)\|_{L^\infty}\|u(\cdot,0)\|_{L^1} \leq Ct^{-1/2}\|u(\cdot,0)\|_{L^1}$$

due to Lemma 7.3.19. ⌋

Example 8.1.11. The Airy equation $\partial_t u = \partial_x^3 u$ with $t, x, u(x,t) \in \mathbb{R}$ is solved by $u(x,t) = \int_{-\infty}^{\infty} G(x-y,t)u(y,0)dy$ with

$$G(x,t) = \int_{-\infty}^{\infty} \mathrm{e}^{-\mathrm{i}k^3 t}\mathrm{e}^{\mathrm{i}kx}\,\mathrm{d}k = \int_{-\infty}^{\infty} \mathrm{e}^{\mathrm{i}t\phi(k)}\,\mathrm{d}k$$

such that $\phi(k) = -k^3 + kx/t$. Since $\phi'''(k) = -6$ we have like in Example 8.1.10 that $\|G(\cdot,t)\|_{L^\infty} \leq Ct^{-1/3}$ and therefore

$$\|u(\cdot,t)\|_{L^\infty} \leq C\|G(\cdot,t)\|_{L^\infty}\|u(\cdot,0)\|_{L^1} \leq Ct^{-1/3}\|u(\cdot,0)\|_{L^1}.$$

⌋

The handling of oscillatory integrals by the stationary phase method is a well developed theory, cf. [**Ste93**]. Even more complicated dispersive estimates can be obtained with this approach, cf. [**LP09**]. Dispersive estimates can be transferred into so called Strichartz estimates. In case of the linear Schrödinger semigroup, for $2 \leq q \leq \infty$, $\frac{2}{q} + \frac{d}{p} = \frac{d}{2}$, and $1 \leq p < \infty$, we have for instance

$$\|e^{it\Delta/2}u_0\|_{L^q((0,t),L^p)} \leq C\|u_0\|_{L^2}.$$

Such estimates allow to prove local and global existence in lower regularity spaces, cf. [**Tao06**].

8.2. The KdV equation

The Korteweg-deVries (KdV) equation [**KdV95**]

$$\partial_t u = -\partial_x^3 u + 6u\partial_x u, \tag{8.13}$$

with $t \in \mathbb{R}$, $x \in \mathbb{R}$, and $u = u(x,t) \in \mathbb{R}$, can be derived as a modulation equation from various models for the description of long unidirectional waves, such as long wave length surface water waves, see Chapter 12. It consists of two parts, namely the dispersive term $-\partial_x^3 u$ and the nonlinear transport term $6u\partial_x u$. By rescaling u, x, and t every other value for the coefficients can be obtained. The choice in (8.13) is the one which is most often used in the literature and is motivated by its derivation from the water wave problem. Our presentation follows in big parts the textbook [**DJ89**].

The linearized KdV equation or Airy equation is given by

$$\partial_t u = -\partial_x^3 u.$$

It possesses solutions $u(x,t) = \mathrm{e}^{ikx+\lambda t}$ with $\lambda = ik^3$, i.e., the amplitude of harmonic waves is preserved. Moreover, the total energy is conserved by the Airy equation since

$$\frac{\mathrm{d}}{\mathrm{d}t}\int_{\mathbb{R}} u^2(x,t)\,\mathrm{d}x = -2\int_{\mathbb{R}} u(x,t)\partial_x^3 u(x,t)\,\mathrm{d}x = \int_{\mathbb{R}} \partial_x(\partial_x u(x,t))^2\,\mathrm{d}x = 0.$$

The Airy equation shows dispersion, i.e., energy is spread over the real line. The estimate

$$\sup_{x\in\mathbb{R}} |u(x,t)| \leq Ct^{-1/3}\int_{\mathbb{R}} |u(y,0)|\,\mathrm{d}y,$$

can be established with the stationary phase method, cf. Example 8.1.11. The decay rate $t^{-1/3}$ is a consequence of $\lambda''(0) = 0$, but $\lambda'''(0) \neq 0$. Alternatively, it can be obtained by an explicit solution formula which can be build with the help of self-similar solutions of the form $u(x,t) = t^{-1/3}\widetilde{u}(xt^{-1/3})$ as fundamental solutions, cf. [**Rau91**].

The nonlinear transport term

$$\partial_t u = 6u\partial_x u$$

already appeared in the Burgers equation. We found that shocks may be created in finite time. Like in the Burgers equation, the linear semigroup, here generated by ∂_x^3, inhibits the creation of shocks in the full KdV equation.

The local existence and uniqueness theory, cf. Theorem 8.2.1, for the solutions of the KdV equation is more advanced due to the fact that the KdV equation is a quasilinear hyperbolic equation for which the previous approach with the variation of constant formula does not work. The solutions are constructed with the help of the iteration scheme

$$\partial_t u_{n+1} = -\partial_x^3 u_{n+1} + 6u_n\partial_x u_{n+1}. \tag{8.14}$$

For the existence of solutions u_{n+1} of the linear equation (8.14) for a given function u_n, see, e.g., [**Paz83**, Chapter 5]. Next the convergence of the sequence $(v_n)_{n\in\mathbb{N}}$, with $v_n = u_{n+1} - u_n$, towards zero and the boundedness of the sequence $(u_n)_{n\in\mathbb{N}}$ is shown. Then we have, cf. [**Kat81, Paz83, KPV91**],

Theorem 8.2.1. *Let $u_0 \in X = H^s(\mathbb{R}, \mathbb{C})$ with $s > 3/2$. There exist a $T_0 > 0$ and a unique local solution $u \in C([0, T_0], X)$ of the KdV equation* (8.13) *with $u|_{t=0} = u_0$.*

With a more clever approach solutions can be constructed in Sobolev spaces of smaller s, cf. [**Tao06**, §4] and [**LP09**, Ch. 7], which is a prerequisite for the study of statistical aspects of the behavior of ensembles of solutions of these equations with the help of invariant measures.

8.2.1. The solitary wave. The starting point of the KdV history is the observation of a solitary wave in 1834 by the engineer Scott Russell in a canal between Edinburgh and Glasgow. The solitary wave was created by a stopping boat and was of 40 cm height and 10 m length. It traveled with a speed of 16 km/h for many kilometers through the canal without changing significantly its shape. Russell was the first to realize the importance of this phenomenon. Motivated by his observation, in the following years he made a number of experiments. After the publication of his results [**Rus44**], a vigorous scientific debate followed on whether or not such a wave of permanent form could exist. Airy [**Air45**] argued that even if dissipation, i.e., the loss of energy, is neglected, dispersion, i.e., the spreading of energy, which is concentrated in the middle of the solitary wave, in the linear problem will destroy the solitary wave. It was finally accepted that such waves exist when Boussinesq [**Bou77**] and Rayleigh [**Ray76**] found approximations to such a wave by deriving the stationary KdV equation (8.15) with some perturbation analysis which takes into account the nonlinear nature of the problem. In 1895 Korteweg and de Vries [**KdV95**] derived the time-dependent KdV equation. A rigorous proof that the full water wave problem possesses such solitary waves remained open for another few decades and was given by Friedrichs and Hyers [**FH54**] in 1954.

What already Boussinesq [**Bou77**] observed is that in the KdV equation there is a balance between linear dispersion and nonlinear transport. This balance creates a wave of permanent form, i.e., $u(x,t) = v(x-ct) = v(\xi)$. Inserting this ansatz into the KdV equation yields

$$-cv' = -v''' + 6vv'.$$

Integration w.r.t. ξ yields

$$-cv + v'' - 3v^2 = D$$

with a constant $D \in \mathbb{R}$. We are interested in localized waves and therefore we have $\lim_{|\xi|\to\infty} v(\xi) = 0$ and $\lim_{|\xi|\to\infty} v''(\xi) = 0$ which together yield $D = 0$. By rescaling $\xi = \sqrt{c}\zeta$ and $v = c\widetilde{v}$ we obtain the first order system

$$\widetilde{v}' = w \quad \text{and} \quad w' = \widetilde{v} + 3\widetilde{v}^2, \tag{8.15}$$

which can be discussed in phase plane. In the phase portrait, Figure 8.3, we find a homoclinic orbit $\widetilde{v}_{\text{hom}} = \widetilde{v}_{\text{hom}}(\zeta)$. Undoing the scalings yields in the original variables a family of solitary waves

$$u(x,t) = c\widetilde{v}_{\text{hom}}\left(\sqrt{c}(x-ct)\right), \tag{8.16}$$

i.e., the higher the solitary wave, the smaller the width and the faster the solitary wave. For an explicit formula see Exercise 8.12.

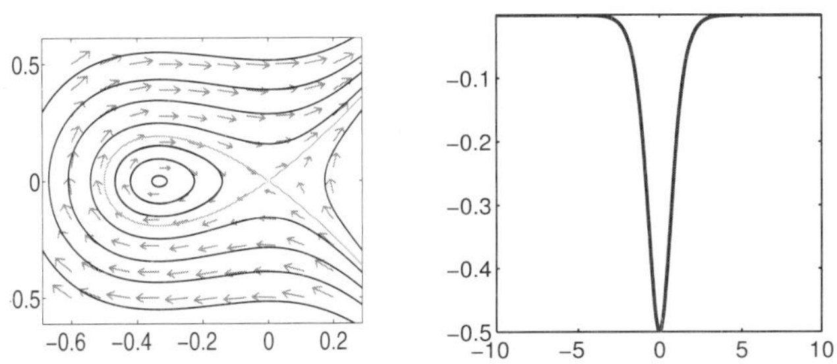

Figure 8.3. The phase portrait for the stationary KdV equation with $D = 0$, and the solitary wave.

8.2.2. The KdV equation as a completely integrable system. The KdV equation is a completely integrable Hamiltonian system which can be solved explicitly. After showing that the KdV equation can be written in Hamiltonian form we prove that the KdV equation possesses infinitely many independent conservation laws, which is a necessary prerequisite for complete integrability.

With

$$H(u(t)) = \int_{\mathbb{R}} \frac{1}{2}(\partial_x u(x,t))^2 + u(x,t)^3 \, \mathrm{d}x$$

we find

$$\begin{aligned} \partial_u H[v] &= \lim_{\varepsilon\to 0} \varepsilon^{-1}(H(u+\varepsilon v) - H(u)) \\ &= \lim_{\varepsilon\to 0} \varepsilon^{-1} \int_{\mathbb{R}} \frac{1}{2}(\partial_x(u(x,t)+\varepsilon v(x,t)))^2 + (u(x,t)+\varepsilon v(x,t))^3 \\ &\qquad - \frac{1}{2}(\partial_x u(x,t))^2 - u(x,t)^3 \, \mathrm{d}x \end{aligned}$$

$$= \int_{\mathbb{R}} (-\partial_x^2 u(x,t) + 3u(x,t)^2) v(x,t)\,\mathrm{d}x,$$

which maps $v \in X = L^2(\mathbb{R}, \mathbb{C})$ linearly into $\mathbb{R}$. Hence, $\partial_u H$ is a linear map from X to $\mathbb{R}$, i.e., an element of the dual space which again can be identified with $-\partial_x^2 u + 3u^2$ by the canonical isomorphism β from (8.9) and therefore $\beta\partial_u H = -\partial_x^2 u + 3u^2$. We finally have

$$\partial_t u = -\partial_x^3 u + 3\partial_x(u^2) = \partial_x \beta \partial_u H(u) = J\beta\partial_u H(u),$$

where the operator $(Ju) = \partial_x u$ is skew symmetric in X since

$$\langle Ju, v\rangle_{L^2} = \int_{\mathbb{R}} (\partial_x u(x)) v(x)\,\mathrm{d}x = -\int_{\mathbb{R}} u(x)\partial_x v(x)\,\mathrm{d}x = -\langle u, Jv\rangle_{L^2}.$$

If the KdV equation is a completely integrable Hamiltonian system, then there must be infinitely many independent conserved quantities for the KdV equation. An equation

$$\partial_t T(x,t) + \partial_x X(x,t) = 0$$

is called a conservation law. We have

$$\partial_t \int_{\mathbb{R}} T\,\mathrm{d}x = \int_{\mathbb{R}} \partial_t T\,\mathrm{d}x = -\int_{\mathbb{R}} \partial_x X\,\mathrm{d}x = -X|_{-\infty}^{\infty} = 0,$$

and hence $\int_{\mathbb{R}} T\,\mathrm{d}x$ is a conserved quantity, i.e., it does not change in time. For the KdV equation we find

$$\partial_t u + \partial_x(\partial_x^2 u - 3u^2) = 0$$

and therefore $\int_{\mathbb{R}} u(x,t)\,\mathrm{d}x$ is independent of time. Next, we find that

$$\begin{aligned}&\partial_t \left(u^3 + \frac{1}{2}(\partial_x u)^2\right) \\ &\quad + \partial_x \left(-\frac{9}{2}u^4 + 3u^2\partial_x^2 u - 6u(\partial_x u)^2 + (\partial_x u)(\partial_x^3 u) - \frac{1}{2}(\partial_x^2 u)^2\right) = 0,\end{aligned}$$

and hence $T_3 = u^3 + \frac{1}{2}(\partial_x u)^2$ yields the conserved quantity $\int_{\mathbb{R}} T_3(x,t)\,\mathrm{d}x$ which is the Hamiltonian which we already know to be conserved. It turns out that there are infinitely many conserved quantities:

$$\begin{aligned}T_4 &= 5u^4 + 10u(\partial_x u)^2 + (\partial_x^2 u)^2, \\ T_5 &= 21u^5 + 105u^2(\partial_x u)^2 + 21u(\partial_x^2 u)^2 + (\partial_x^3 u)^2, \\ &\vdots\end{aligned}$$

Remark 8.2.2. For the linearized system we have that $T_{n+2} = (\partial_x^n u)^2$ since

$$\partial_t(\partial_x^n u)^2 + \partial_x\left(\partial_x^{n+2} u \partial_x^n u - (\partial_x^{n+1} u)^2/2\right).$$

In Fourier space we find the solutions $\widehat{u}(k,t) = \mathrm{e}^{\mathrm{i}k^3 t}\widehat{u}(k,0)$. Thus, by introducing polar coordinates $\widehat{u}(k,t) = r(k,t)\mathrm{e}^{\mathrm{i}\phi(k,t)}$ we find that $r(k,t) = r(k,0)$ and $\phi(k,t) = \phi(k,0) + k^3 t \bmod 2\pi$ or, equivalently,

$$\partial_t r(k,t) = 0 \quad \text{and} \quad \partial_t \phi(k,t) = k^3 = \text{const}.$$

Thus, for the linearized system we found uncountably many action variables $r(k,\cdot)$ and uncountably many angle variables $\phi(k,\cdot)$, and consequently tori of any dimension. ⌋

In the following we describe how a transformation discovered by Miura [**Miu68**] and then generalized in [**GGKM74**] leads very easily to the conclusion that there are infinitely many conserved quantities for the KdV equation. The basic idea is that, given a transformation which maps solutions of one equation to solutions of a second one, the existence of simple conserved quantities for the first equation leads, via the transformation, to more complicated conserved quantities for the second equation.

To a given $u = u(x,t)$ we implicitly define $w(x,t)$ via the formula

$$u(x,t) = w(x,t) + \varepsilon\partial_x w(x,t) + \varepsilon^2 (w(x,t))^2. \tag{8.17}$$

If w is smooth enough and ε is small, then we can invert this relation recursively to obtain w in terms of u via the formula

$$\begin{aligned} w =& u - \varepsilon\partial_x u - \varepsilon^2(u^2 + \partial_x^2 u) + \varepsilon^3(\partial_x^3 u + 4u\partial_x^2 u) \\ &+ \varepsilon^4\left(2u^3 + 5(\partial_x u)^2 + 6u\partial_x^2 u + \partial_x^4 u\right) + \mathcal{O}(\varepsilon^5). \end{aligned} \tag{8.18}$$

Now we compute

$$\begin{aligned} \partial_t u + \partial_x^3 u - 6u\partial_x u =& (\partial_t w - 6w\partial_x w - 6\varepsilon^2 w^2 \partial_x w + \partial_x^3 w) \\ &+ 2\varepsilon^2 w(\partial_t w - 6w\partial_x w - 6\varepsilon^2 w^2\partial_x w + \partial_x^3 w) \\ &+ \varepsilon\partial_x(\partial_t w - 6w\partial_x w - 6\varepsilon^2 w^2\partial_x w + \partial_x^3 w). \end{aligned} \tag{8.19}$$

From this we immediately see that, if w satisfies the modified KdV equation

$$\partial_t w - 6w\partial_x w + 6\varepsilon^2 w^2\partial_x w - \partial_x^3 w = 0, \tag{8.20}$$

then u, defined by (8.17), satisfies the KdV equation. However, one also sees immediately that the integral of w is a conserved quantity of (8.20) for all values of ε, i.e., if we define $I_\varepsilon(t) = \int_{\mathbb{R}} w(x,t)\,\mathrm{d}x$, then I_ε is a constant for all values of ε. (We assume here that w is defined on the real line, and that w and its derivatives go to zero as $|x|$ tends to infinity. Similar results hold for x running over a finite interval with periodic boundary conditions.) But this immediately implies that, if we use (8.18) to expand I_ε in powers of ε, then the coefficients in this expansion must also be constant in time. Since these coefficients will be expressed as integrals of u and its derivatives they will give us (infinitely many) conserved quantities for the KdV equation. For the first few of these we find:

- $K_0 = \int_{\mathbb{R}} u(x,t)\,\mathrm{d}x$. The conservation of this quantity follows immediately from the form of the KdV equation.
- $K_1 = \int_{\mathbb{R}} \partial_x u(x,t)\,\mathrm{d}x = 0$, if we assume that u and its derivatives tend to zero as $|x|$ tend to infinity. Thus, we gain no new information from this quantity and in fact, all the integrals coming from the odd powers of ε turn out to be "trivial". Therefore, we ignore them and focus just on the even powers of ε.
- $K_2 = \int_{\mathbb{R}}(u^2 - \partial_x^2 u)\,\mathrm{d}x = \int_{\mathbb{R}} u^2\,\mathrm{d}x$. That this is a conserved quantity is again easy to see directly from the KdV equation, just by multiplying the equation by u and integrating w.r.t. x.
- $K_4 = \int_{\mathbb{R}} \partial_x(2u\partial_x - \partial_x(\partial_x^2 u - u^2)) + 2u(\partial_x^2 u - u^2) + (\partial_x u)^2\,\mathrm{d}x = \int (3u^2 + (\partial_x u)^2)\,\mathrm{d}x$.

In [**ZF71**] it has been proved that the KdV equation is a completely integrable Hamiltonian system. In particular, there exists a canonical transformation such that, w.r.t. the new coordinates, the Hamiltonian is a function only of the action variables and hence the action variables remain constant in time.

8.2.3. Solitons. In the experiments made by Russell from 1834 to 1844 he was also interested in the interaction of solitary waves. He made some sketches indicating that solitary waves after some nonlinear interaction seem to have gone through each other as if there had been no interaction, i.e., the interaction of two solitary waves leads to two solitary waves with the same shapes and velocities as before the interaction. By some first computer experiments in the 1950s and 1960s this behavior was observed for a number of systems. The first one was the Fermi-Pasta-Ulam system [**FPU55**], a nonlinear lattice differential equation which in contrast to the observed behavior was created to understand why heat conductivity in solids is finite. Zabusky and Kruskal [**ZK65**] observed that the KdV equation shows this behavior, too. Although the solitary waves A_{β_j} interact nonlinearly, after the interaction they seem to have gone through each other without changing their shape, i.e., asymptotically

$$A_{N-\mathrm{Sol.}}(x,t) \sim \sum_{j=1}^{N} A_{\beta_j}\left(x - \beta_j^2 t \pm \gamma_j + \delta_j\right),$$

for $t \to \pm\infty$, with $\beta_1 > \ldots > \beta_N$ and $\gamma_j, \delta_j \in \mathbb{R}$ some phase-shifts. It is the purpose of this section to explain where this behavior comes from.

The Lax pair formulation. The key observation for the understanding of all the remarkable properties of the KdV equation is the formulation of Lax [**Lax68**] which allows to transfer the KdV equation into a system of explicitly solvable equations. We write the KdV equation as an abstract evolutionary system

$$\partial_t u = N(u),$$

and interpret this equation as an evolution equation for operators acting in some Hilbert space. Both sides are interpreted as multiplication operators, namely

$$\partial_t u : \begin{cases} L^2(\mathbb{R}) \to L^2(\mathbb{R}), \\ \psi \mapsto (\partial_t u)\psi, \end{cases} \qquad \text{and} \qquad N(u) : \begin{cases} L^2(\mathbb{R}) \to L^2(\mathbb{R}), \\ \psi \mapsto (N(u))\psi. \end{cases}$$

In a first step we complicate the situation heavily and assume that the evolutionary system can be written as

$$\partial_t L(t) = [M, L](t) := M(t)L(t) - L(t)M(t)$$

with self-adjoint $L(t) = L^*(t)$ and skew-adjoint $M(t) = -M^*(t)$.

Example 8.2.3. Consider $L(t)\psi(x) = -\partial_x^2\psi(x)+u(x,t)\psi(x)$. Then for each fixed t we have $\partial_t L(t)\psi(x) = \partial_t u(\cdot,t)\psi(x)$. The operator $L(t)$ is self-adjoint in $L^2(\mathbb{R})$ since

$$\begin{aligned}(L(t)\psi,\phi)_{L^2} &= \int_{\mathbb{R}} (-\partial_x^2\psi(x) + u(x,t)\psi(x))\phi(x)\,\mathrm{d}x \\ &= \int_{\mathbb{R}} \partial_x\psi(x)\partial_x\phi(x) + u(x,t)\psi(x)\phi(x)\,\mathrm{d}x = (\psi, L(t)\phi)_{L^2},\end{aligned}$$

i.e., t is considered as fixed parameter in this calculation. ⌋

Hence, with $L(t)$ from the example the left-hand side of the KdV equation and the left-hand side of the Lax pair formulation coincide. Before we explain how to find operators $M(t)$ such that also the right-hand sides coincide we explain first consequences of a Lax pair representation.

Lemma 8.2.4. *If the self-adjoint operator* $L{=}L(t)$ *satisfies* $\partial_t L(t){=}[M,L](t)$, *with* $M(t) = -M^*(t)$ *a skew-adjoint operator, then the eigenvalues of* $L(t)$ *are independent of* t.

Proof. In the following $\psi(t)$ denotes a L^2-normalized eigenfunctions of $L(t)$ associated to the eigenvalues $\lambda(t)$. Differentiating the eigenvalue problem

$$L(t)\psi(t) = \lambda(t)\psi(t)$$

w.r.t. t, where $\lambda(t) \in \mathbb{R}$, yields

$$\partial_t L(t)\psi(t) + L(t)\partial_t\psi(t) = \partial_t\lambda(t)\psi(t) + \lambda(t)\partial_t\psi(t).$$

Using the right-hand side of the Lax pair formulation gives

$$\begin{aligned}\partial_t\lambda(t)\psi(t) =&(M(t)L(t) - L(t)M(t))\psi(t) + L(t)\partial_t\psi(t) - \lambda(t)\partial_t\psi(t) \\ =&M(t)\lambda(t)\psi(t) - L(t)M(t)\psi(t) + L(t)\partial_t\psi(t) - \lambda(t)\partial_t\psi(t) \qquad (8.21)\end{aligned}$$

$$=(L(t)-\lambda(t))(\partial_t\psi(t)-M(t)\psi(t)).$$

Using the fact that L is self-adjoint and the normalization $\langle\psi(t),\psi(t)\rangle_{L^2}=1$ we find

$$\begin{aligned}\partial_t\lambda(t) &= \langle\psi(t),(L(t)-\lambda(t))(\partial_t\psi(t)-M(t)\psi(t))\rangle_{L^2}\\ &= \langle(L(t)-\lambda(t))\psi(t),\partial_t\psi(t)-M(t)\psi(t)\rangle_{L^2}\\ &= \langle 0,\partial_t\psi(t)-M(t)\psi(t)\rangle_{L^2}=0.\end{aligned}$$

□

Lemma 8.2.5. *If the self-adjoint operator $L=L(t)$ satisfies $\partial_t L(t)=[M,L](t)$, with $M(t)=-M^*(t)$ a skew adjoint operator, and if the eigenvalues of $L(t)$ are semi-simple, then there exists a real-valued function $c=c(t)$ such that*

$$\partial_t\psi(t)-M(t)\psi(t)=c(t)\psi(t). \tag{8.22}$$

Proof. From (8.21) it follows that $(L(t)-\lambda(t))(\partial_t\psi(t)-M(t)\psi(t))=0$. If the eigenvalues of $L(t)$ are semi-simple, then $\partial_t\psi(t)-M(t)\psi(t)$ must be a multiple of $\psi(t)$. Hence, there exists a $c=c(t)$ such that (8.22) holds. □

The eigenvalues of $L(t)$ from Example 8.2.3 are semi-simple. Since

$$M(t)L(t)-L(t)M(t)=(M(t)-c(t))L(t)-L(t)(M(t)-c(t))$$

we can always redefine $M(t)$ such that $\psi(t)$ satisfies $\partial_t\psi(t)=M(t)\psi(t)$.

We summarize what we found so far. If the KdV equation can be written as a Lax pair formulation then the eigenvalues of $L(t)$ are conserved quantities for the KdV equation although $L(t)$ contains the solution u explicitly. Moreover, the associated eigenfunctions $\psi(t)$ satisfy $\partial_t\psi(t)=M(t)\psi(t)$. It will turn out subsequently that although $M(t)$ also contains u explicitly this equation can be solved without the knowledge of u. Finally, we will be able to reconstruct $u(\cdot,t)$ out of the λs and the eigenfunctions $\psi(t)$.

The KdV hierarchy. In order to find the Lax pair formulation for the KdV equation we have to find operators $L(t)$ and $M(t)$ with

$$\partial_t L(t)\psi=(\partial_t u)\psi \qquad \text{and} \qquad (M(t)L(t)-L(t)M(t))\psi=(-\partial_x^3 u+6u\partial_x u)\psi.$$

Especially, all derivatives ∂_x of $M(t)$ and $L(t)$ which fall on ψ have to cancel such that the commutator of $M(t)$ and $L(t)$ is a pure multiplication operator.

Example 8.2.6. We consider the symmetric operator

$$L(t)\cdot=-\partial_x^2\cdot+u(x,t)\cdot$$

and the skew symmetric operator $M=\partial_x$. We find

$$\begin{aligned}(M(t)L(t)-L(t)M(t))\psi(x) &= \partial_x(-\partial_x^2+u(x,t))\psi(x)+(\partial_x^2-u(x,t))\partial_x\psi(x)\\ &= (\partial_x u(x,t))\psi(x).\end{aligned}$$

Hence, the PDE

$$\partial_t u = \partial_x u,$$

which possesses the solution $u(x,t) = u(x+t,0)$ can be solved alternatively with a Lax pair formulation. The eigenfunctions $\psi = \psi(x,t)$ of

$$-\partial_x^2\psi(x,t) + u(x,t)\psi(x,t) = -\partial_x^2\psi(x,t) + u(x+t,0)\psi(x,t) = \lambda(t)\psi(x,t)$$

evolve according to the above theory as a solution of

$$\partial_t\psi(x,t) = \partial_x\psi(x,t).$$

which is solved by $\psi(x,t) = \psi(x+t,0)$. This result is consistent with the above theory. The eigenvalues of the Schrödinger operator $L(t)$ are independent of t since the potential u is only translated in time. Hence, the eigenfunctions are also translated in time. For the reconstruction of u from the λs and the ψs see §8.2.3. ⌋

It turns out that for the operator $L(t)$ from Example 8.2.6 an operator $M(t)$ can be found such that L and M give a Lax pair formulation of the KdV equation. In order to derive the KdV equation we make the ansatz

$$M(t)\cdot = -\alpha\partial_x^3 \cdot + U\partial_x \cdot + \partial_x(U\cdot) + A$$

with $\alpha \in \mathbb{R}$, $U = U(x,t)$ and $A = A(x,t)$. We find

$$\begin{aligned}(M(t)&L(t) - L(t)M(t))\\ =&(-\alpha\partial_x^3 + \partial_x^3 U + \partial_x^2 A + 2(\partial_x u)U) - (3\alpha\partial_x^2 u + 4\partial_x^2 U + 2\partial_x A)\partial_x\\ &- (-3\alpha\partial_x u + 4\partial_x U)\partial_x^2.\end{aligned}$$

In order to have a multiplication operator the coefficients in front of ∂_x and ∂_x^2 in the second and third line have to vanish identically. From the third line we obtain $U = 3\alpha u/4$ and then from the second line that $A = A(t)$ is independent of x, such that finally

$$M(t)L(t) - L(t)M(t) = \frac{\alpha}{4}(-\partial_x^3 u(x,t) + 6u(x,t)\partial_x u(x,t)).$$

Choosing $\alpha = 4$ gives the right-hand side of our KdV equation.

Summary. If $u = u(x,t)$ solves the KdV equation (8.13), then the eigenvalues $\lambda = \lambda(t)$ of the eigenvalue problem

$$-\partial_x^2\psi(x,t) + u(x,t)\psi(x,t) = \lambda(t)\psi(x,t)$$

are independent of time and the associated eigenfunctions evolve according to

$$\partial_t\psi(x,t) = -4\partial_x^3\psi(x,t) + 3u(x,t)\partial_x\psi(x,t) + 3\partial_x(u(x,t)\psi(x,t)) + A(t)\psi(x,t).$$

For the same $L(t)$, beside the two $M(t)$s from above there are infinitely many skew symmetric operators $M(t)$ which lead to a multiplication operator for $M(t)L(t) - L(t)M(t)$ giving a "hierarchy of KdV equations". The next one would be

$$\partial_t u = -\partial_x^5 u + 10u\partial_x^3 u + 20\partial_x u\partial_x^2 u - 30u^2\partial_x u.$$

The Hamiltonian of the nth equation in KdV hierarchy is given by the nth conserved quantity of the KdV equation obtained via the Miura transformation, cf. [**New85**].

The evolutionary problem $\partial_t\psi(t) = M(t)\psi(t)$. As the next step on our way to solve the KdV equation with the Lax pair formulation we have to solve the evolutionary problem $\partial_t\psi(t) = M(t)\psi(t)$ for the eigenfunctions $\psi(t)$. It is given by

$$\begin{aligned}\partial_t\psi(x,t) =& -4\partial_x^3\psi(x,t)+3u(x,t)\partial_x\psi(x,t)+3\partial_x(u(x,t)\psi(x,t))+A(t)\psi(x,t)\\ =&2(u(x,t)+2\lambda)\partial_x\psi(x,t) - \partial_x u(x,t)\psi(x,t)+A(t)\psi(x,t),\end{aligned} \tag{8.23}$$

where we used the eigenvalue problem to simplify the system slightly. At first sight it seems that we need to know the solution $u = u(x,t)$ of the KdV equation to solve this problem which would make the approach pretty useless. However, it turns out that $\partial_t\psi(t) = M(t)\psi(t)$ can be solved explicitly without the knowledge of u. In order to explain why this is the case we have to look more closely at the eigenfunctions $\psi(t)$ of the self-adjoint linear operator $L(t)$ defined by

$$L(t)\psi(x,t) = -\partial_x^2\psi(x,t) + u(x,t)\psi(x,t),$$

where $u = u(x,t)$ is a solution of the KdV equation. There are finitely many, say N, discrete eigenvalues, $0 > \lambda_1 > \lambda_2 > \ldots > \lambda_N$ and a continuum $[0,\infty)$ of spectral values which are denoted by $\lambda(k) = k^2$ for $k \in \mathbb{R}$. Corresponding to the eigenvalues there are eigenfunctions $\psi_k \in L^2$ for $k = 1,\ldots,N$ and $\psi(k) \in L^\infty \setminus L^2$ for $k \in \mathbb{R}$.

Example 8.2.7. We compute the eigenvalues and eigenfunctions of $L\psi = \psi'' - u(x)\psi$ in case $u(x) = -U_0\delta(x)$, with U_0 a positive constant and δ the Dirac delta-distribution, cf. Example 5.2.2. Integration of $-\partial_x^2\psi - U_0\delta(x)\psi = \lambda\psi$ gives

$$-\partial_x\psi|_{x=-\varrho}^{\varrho} - \int_{-\varrho}^{\varrho} U_0\delta(x)\psi(x)\,\mathrm{d}x = \int_{-\varrho}^{\varrho}\lambda\psi(x)\,\mathrm{d}x,$$

which, since $\psi \in C^0$, for $\rho \to 0$ yields the jump condition

$$-[\partial_x\psi] - U_0\psi(0) = 0, \qquad [u] := \lim_{\varepsilon\to 0} u(\varepsilon) - \lim_{\varepsilon\to 0} u(-\varepsilon).$$

For $x \neq 0$ we have to solve $-\partial_x^2 \psi = \lambda\psi$. For $\lambda > 0$, i.e., $\lambda = k^2$ we have $\psi \sim \mathrm{e}^{\pm \mathrm{i}kx}$. We make the ansatz

$$\widehat{\psi}(x,k) = \begin{cases} \mathrm{e}^{-\mathrm{i}kx} + b(k)\mathrm{e}^{\mathrm{i}kx} & \text{for } x > 0, \\ a(k)\mathrm{e}^{-\mathrm{i}kx} & \text{for } x < 0. \end{cases}$$

Continuity in $x = 0$ leads to $1 + b = a$ and the jump condition to

$$-\mathrm{i}k + b\mathrm{i}k - (-\mathrm{i}ka) = -U_0(1+b) = U_0 a(k)$$

which finally leads to

$$b(k) = -\frac{U_0}{U_0 + 2\mathrm{i}k}.$$

For $\lambda < 0$, i.e., $\lambda = -\kappa_n^2$ we have $\psi \sim \mathrm{e}^{\pm\kappa_n x}$. Therefore, we make the ansatz

$$\psi_n(x) = \begin{cases} \alpha_n \mathrm{e}^{-\kappa_n x} & \text{for } x > 0, \\ \beta_n \mathrm{e}^{+\kappa_n x} & \text{for } x < 0. \end{cases}$$

Continuity yields $\alpha_n = \beta_n$. Normalisation gives

$$\int_{-\infty}^0 \alpha_n^2 \mathrm{e}^{2\kappa_n x}\, \mathrm{d}x + \int_0^\infty \alpha_n^2 \mathrm{e}^{-2\kappa_n x}\, \mathrm{d}x = 1,$$

and hence $\alpha_n = \kappa_n^2$. The jump condition then gives

$$[\psi_n'] = -\,\kappa_n\alpha_n - \kappa_n\alpha_n = -\,U_0\alpha_n,$$

and therefore $\kappa_1 = \frac{1}{2}U_0$, i.e., we have only negative eigenvalues. ⌋

Since the eigenfunctions satisfy a second order scalar ODE they are completely determined by their asymptotic behavior for $x \to \pm\infty$. For ψ_k normalized by $\int |\psi_k(x)|^2\, \mathrm{d}x = 1$ we have

$$\psi_k \sim c_k \begin{cases} \mathrm{e}^{-\kappa_k x} & x \to \infty, \\ \mathrm{e}^{\kappa_k x} & x \to -\infty, \end{cases}$$

where $\kappa_k^2 = -\lambda_k$ and $\kappa_k > 0$. For $\psi(k)$ we have

$$\psi(k) \sim \begin{cases} \mathrm{e}^{-\mathrm{i}kx} + b(k)\mathrm{e}^{\mathrm{i}kx} & x \to \infty, \\ a(k)\mathrm{e}^{-\mathrm{i}kx} & x \to -\infty. \end{cases}$$

The question is if it possible to derive some ODEs for the so called scattering data $(\lambda_k(t), c_k(t))$ for $k = 1, \ldots, N$ and $(\lambda(k,t), a(k,t), b(k,t))$ for $k \in \mathbb{R}$ from the PDEs $\partial_t\psi(t) = M(t)\psi(t)$ for the eigenfunctions $\psi(t)$. It turns out that this is possible, and that these ODEs can be solved explicitly without knowing the evolution of the KdV solution $u = u(x,t)$.

i) We start with the discrete spectrum. Multiplying (8.23) by $\psi_k(x,t)$ gives that $A(t)\psi_k(x,t)^2$ equals

$$\frac{1}{2}\partial_t(\psi_k(x,t)^2) + \partial_x(u(x,t)\psi_k(x,t)^2 - 2(\partial_x\psi_k(x,t))^2 - 4\lambda_k\psi_k(x,t)^2),$$

by straightforward calculation. Performing the integration $\int_{\mathbb{R}} \ldots \,\mathrm{d}x$ and using the normalization $\int_{\mathbb{R}} \psi_k(x,t)^2 \,\mathrm{d}x = 1$ yields

$$\frac{1}{2}\frac{\mathrm{d}}{\mathrm{d}t}\int_{\mathbb{R}} \psi_k(x,t)^2 \,\mathrm{d}x = 0 = A(t)\int_{\mathbb{R}} \psi_k(x,t)^2 \,\mathrm{d}x = A(t),$$

i.e., $A(t) = 0$, such that $\psi_k(t)$ satisfies

$$\partial_t \psi_k(x,t) = 2(u(x,t) + 2\lambda_k)\partial_x \psi_k(x,t) - (\partial_x u(x,t))\psi_k(x,t). \tag{8.24}$$

As already said, $\psi_k(x,t)$ is fully determined by its asymptotics $c_k(t)\mathrm{e}^{-\kappa_k x}$ for $x \to \infty$. Since $u(x,t) \to 0$ for $x \to \infty$ we find by multiplying (8.24) with $\mathrm{e}^{\kappa_k x}$ in the limit $x \to \infty$ that

$$\dot{c}_k(t) = -4\lambda_k \kappa_k c_k(t) = 4\kappa_k^3 c_k(t).$$

This equation is independent of the evolution of $u = u(x,t)$ and can be solved explicitly, namely

$$c_k(t) = \mathrm{e}^{4\kappa_k^3 t} c_k(0),$$

where $c_k(0)$ has to be computed from the KdV initial condition $u = u(x,0)$. The asymptotics

$$\psi_k(x,t) \sim c_k(t)\mathrm{e}^{-\kappa_k x} = c_k(0)\mathrm{e}^{4\kappa_k^3 t - \kappa_k x} = c_k(0)\mathrm{e}^{-\kappa_k(x - 4\kappa_k^2 t)}$$

for $x \to \infty$ describes a translation of the eigenfunction $\psi_k(x,t)$ with velocity $4\kappa_k^2$.

ii) We come now to the continuous spectrum. Inserting the asymptotic behavior

$$\psi(k;x,t) \sim \mathrm{e}^{-\mathrm{i}kx} + b(k,t)\mathrm{e}^{\mathrm{i}kx}$$

for $x \to \infty$ into (8.23), and using again $u(x,t) \to 0$ for $x \to \infty$, gives for the coefficient in front of $\mathrm{e}^{-\mathrm{i}kx}$ that $0 = 4(-\mathrm{i}k)\lambda + A(t) = -4\mathrm{i}k^3 + A(t)$ and for the coefficient in front of $\mathrm{e}^{\mathrm{i}kx}$ that $\dot{b}(k,t) = 4\mathrm{i}k\lambda b(k,t) + A(t)b(k,t) = (4\mathrm{i}k^3 + A(t))b(k,t)$ such that finally

$$\dot{b}(k,t) = 8\mathrm{i}k^3 b(k,t).$$

Again this equation can be solved explicitly, namely

$$b(k,t) = \mathrm{e}^{8\mathrm{i}k^3 t} b(k,0),$$

where $b(k,0)$ has to be computed from the KdV initial condition $u = u(x,0)$. With the same argument the evolution equation for $a(k,t)$ can be derived. Inserting the asymptotic behavior

$$\psi(k;x,t) \sim a(k,t)\mathrm{e}^{-\mathrm{i}kx}$$

for $x \to -\infty$ into (8.23), using $u(x,t) \to 0$ for $x \to -\infty$, and $A(t) = 4\mathrm{i}k^3$ gives for the coefficient in front of $\mathrm{e}^{-\mathrm{i}kx}$ that

$$\dot{a}(k,t) = -4\mathrm{i}k\lambda a(k,t) + A(t)a(k,t) = (-4\mathrm{i}k^3 + A(t))a(k,t) = 0,$$

such that $a(k,t) = a(k,0)$.

Remark 8.2.8. The scattering data variables can be interpreted as action and angle variables of a completely integrable Hamiltonian system. The eigenvalues λ_k and $\lambda(k)$ are the conserved quantities, i.e., the action variables, whereas the $c_k(t)$ and $b(k,t)$ behave linearly, i.e., are the angle variables. ⌋

The inverse scattering transform. Two steps remain. First we have to compute the initial scattering data

$$(\lambda_k(0), c_k(0), b(k,0), a(k,0))$$

from the initial condition $u = u(x,0)$ of the KdV equation. Secondly, we have to reconstruct the solution $u = u(x,t)$ of the KdV equation from the scattering data

$$(\lambda_k(t), c_k(t), b(k,t), a(k,t)) = (\lambda_k(0), \mathrm{e}^{4\kappa_k^3 t} c_k(0), \mathrm{e}^{8ik^3 t} b(k,0), a(k,0)).$$

Both problems are well studied in quantum mechanics since the operator $L(t)$ is the Schrödinger operator with potential $u = u(x,t)$ describing the wave function of a non-relativistic electron. We recall the parts of scattering and inverse scattering theory which are necessary for the construction of $u = u(x,t)$ from the data $(\lambda_k, c_k(t), \lambda(k), a(k,t), b(k,t))$ by the Marchenko- or Gelfand-Levitan equation [**GL51, Mar55**]. For each fixed time t the solution $u = u(x,t)$ is given by

$$u(x) = -2\frac{\mathrm{d}}{\mathrm{d}x} K(x,x),$$

with $K(x,z)$ satisfying the linear integral equation

$$K(x,z) + F(x+z) + \int_x^\infty K(x,y) F(y+z)\,\mathrm{d}y = 0$$

where

$$F(x) = \sum_{j=1}^n c_j^2 \mathrm{e}^{-\kappa_j x} + \frac{1}{2\pi} \int_{\mathbb{R}} \mathrm{e}^{\mathrm{i}kx} b(k)\,\mathrm{d}k.$$

If we denote the calculation of the scattering data by $\mathcal{T}$ and the solution of the Marchenko or Gelfand-Levitan equation, i.e., the inverse scattering transform, by $\mathcal{T}^{-1}$ then we obtain the following commutative diagram.

u_0	$\xrightarrow{\text{nonlinear evolution in KdV}}$	$u(\cdot,t)$
$\downarrow \mathcal{T}$		$\uparrow \mathcal{T}^{-1}$
scattering data $(\lambda_k(0), c_k(0))_{k=1,\dots,N}$ $(\lambda(k,0), b(k,0))_{k\in\mathbb{R}}$	$\xrightarrow{\text{linear evolution of scattering data}}$	scattering data $(\lambda_k(t), c_k(t))_{k=1,\dots,N}$ $(\lambda(k,t), b(k,t))_{k\in\mathbb{R}}$

That this procedure is really useful is explained in the next example where we compute 2-soliton solutions of the KdV equation.

Example 8.2.9. We consider the situation of two negative eigenvalues $\lambda_1 = -1$ and $\lambda_2 = -4$, such that $\kappa_1 = 1$ and $\kappa_2 = 2$. Next we choose the initial scattering data

$$c_1(0) = \sqrt{6}, \quad c_2(0) = 2\sqrt{3}, \quad \text{and} \quad b(k,0) = 0$$

for all $k \in \mathbb{R}$. Such a potential is called reflectionless. We find

$$c_1(t) = \sqrt{6}\mathrm{e}^{4t}, \quad c_2 = 2\sqrt{3}\mathrm{e}^{32t}, \quad \text{and} \quad b(k,t) = 0$$

for all $k \in \mathbb{R}$ and $t \in \mathbb{R}$. The function F in the Marchenko or Gelfand-Levitan equation is then given by

$$F(x) = 6\mathrm{e}^{8t-x} + 12\mathrm{e}^{64t-2x}.$$

In order to solve this equation we make the ansatz

$$K(x,z) = \ell_1(x)\mathrm{e}^{-z} + \ell_2(x)\mathrm{e}^{-2z}.$$

With $F_j(x) = c_j^2 \mathrm{e}^{-\kappa_j x}$ we obtain

$$\begin{aligned} 0 = & \ell_1 \mathrm{e}^{-z} + \ell_2 \mathrm{e}^{-2z} + F_1(x)\mathrm{e}^{-z} + F_2(x)\mathrm{e}^{-2z} \\ & + \int_x^\infty \left(\ell_1(x)\mathrm{e}^{-y} + \ell_2(x)\mathrm{e}^{-2y}\right)\left(F_1(z)\mathrm{e}^{-y} + F_2(z)\mathrm{e}^{-2y}\right) \mathrm{d}y. \end{aligned}$$

The coefficients in front of e^{-z} and e^{-2z} are then given by

$$\begin{aligned} \ell_1 + 6\mathrm{e}^{8t-x} + 3\ell_1\mathrm{e}^{8t-2x} + 2\ell_2\mathrm{e}^{8t-3x} &= 0, \\ \ell_2 + 12\mathrm{e}^{64t-2x} + 4\ell_1\mathrm{e}^{64t-3x} + 3\ell_2\mathrm{e}^{64t-4x} &= 0. \end{aligned}$$

The solution of this linear system is

$$\ell_1 = 6\left(\mathrm{e}^{72t-5x} - \mathrm{e}^{8t-x}\right)/D \quad \text{and} \quad \ell_2 = -12\left(\mathrm{e}^{64t-2x} + \mathrm{e}^{72t-4x}\right)/D,$$

where $D = 1 + 3\mathrm{e}^{8t-2x} + 3\mathrm{e}^{64t-4x} + \mathrm{e}^{72t-6x}$. Finally, we obtain the remarkable explicit solution

$$\begin{aligned} u(x,t) &= -2\partial_x\left(\ell_1\mathrm{e}^{-x} + \ell_2\mathrm{e}^{-2x}\right) \\ &= -12\frac{3 + 4\cosh(2x-8t) + \cosh(4x-64t)}{(3\cosh(x-28t) + \cosh(3x-36t))^2}. \end{aligned} \tag{8.25}$$

This 2-soliton solution describes the reflectionless interaction of two solitons, cf. Exercise 8.14 and Figure 8.4. ⌋

The previous computation can easily be generalized to N negative eigenvalues in case of a reflectionless potential, i.e., $b(k,0) = 0$. The explicit solution formula obtained in this way describes a so called N-soliton. For

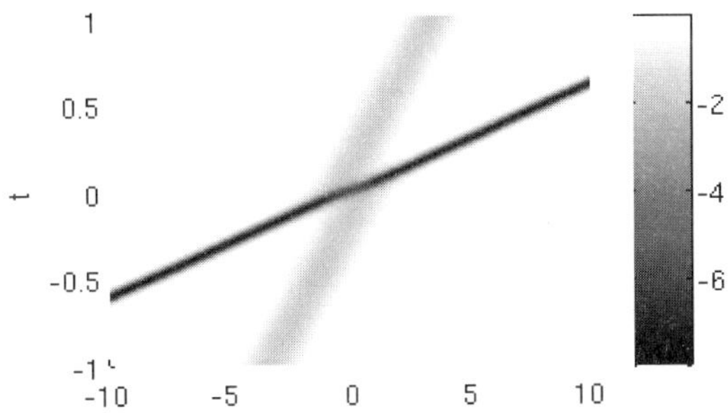

Figure 8.4. Density plot of the 2-soliton (8.25).

$t \to \pm\infty$ the complicated solution separates in N individual solitary waves, cf. Exercise 8.14.

Remark 8.2.10. The procedure can also be used to characterize the asymptotic behavior for $t \to \infty$ for arbitrary initial conditions u_0 satisfying $\int_{\mathbb{R}} u_0(x)(1+|x|)\,\mathrm{d}x < \infty$, cf. [**EvH81**]. Suppose that we have N negative eigenvalues. Each eigenvalue corresponds to a soliton in the KdV equation. Then for large t the solution u consists of N-solitons which are ordered w.r.t. their height which as we have seen is proportional to the velocity of the solitons. Behind the solitons we find a dispersive rest decaying with a rate $t^{-1/3}$ which is described by the coefficients $b(k,t)$. ⌋

8.2.4. Stability of the solitary wave. We cannot expect naive stability of a solitary wave u_c since according to formula (8.16) for $t = 0$ there will always be a nearby solitary wave u_{c^*} with different velocity c^* which will run away from u_c for $t \to \infty$. However, using the Hamiltonian structure of the KdV equation so called orbital stability can be established [**Ben72**]. Orbital stability means that for all $\varepsilon > 0$ there exists a $\delta > 0$ such that

$$\|u_0(\cdot) - u_c(\cdot,0)\|_{H^1} \leq \delta \quad \Rightarrow \quad \inf_{x_0\in\mathbb{R}} \|u(\cdot,t) - u_c(x+x_0,t)\|_{H^1} \leq \varepsilon, \tag{8.26}$$

i.e., the solution $u(\cdot,t)$ stays close to a translate of u_c for all times.

The argument is similar to the one in Theorem 4.1.6 where the stability of a fixed point has been shown using the positive definiteness of the second derivative of the Hamiltonian at the fixed point. See Figure 8.5.

In case of orbital stability the argument is a little bit more complicated. First of all the solitary waves u_c are only constrained minima, i.e., they minimize H in the invariant subspaces of the functions with the same L^2-norm. Moreover, in each of these subspaces there is a one-dimensional family of minima due to the translation invariance of the problem. In the following we sketch the idea to overcome these difficulties.

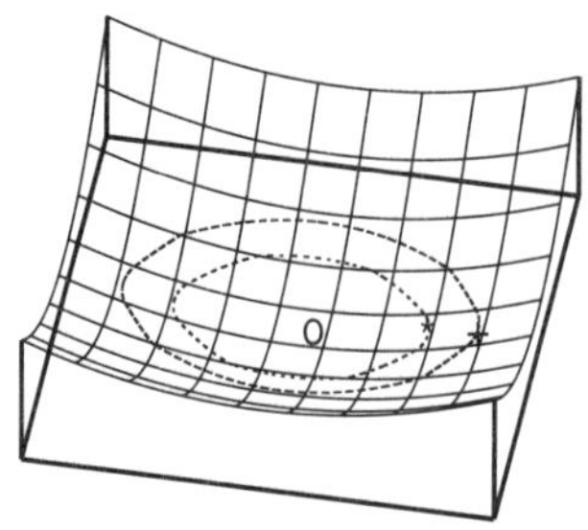

Figure 8.5. Cartoon picture for stability. o is the minimum of a conserved quantity M, hence perturbations $*$ and $+$ can't get away from o.

We use Lagrange multipliers for the constraint and consider

$$M(u) = H(u) + cE(u) = \int_{\mathbb{R}} \frac{1}{2}(\partial_x u)^2 + u^3 + \frac{c}{2}u^2 \, \mathrm{d}x$$

which is a constant of motion, i.e., $\frac{\mathrm{d}}{\mathrm{d}t}M(u(t)) = 0$. We find

$$\begin{aligned} M(u_c + \varepsilon v) &= \int_{\mathbb{R}} \frac{1}{2}(u_c' + \varepsilon v')^2 + (u_c + \varepsilon v)^3 + \frac{c}{2}(u_c + \varepsilon v)^2 \, \mathrm{d}x \\ &= M(u_c) + \varepsilon \int_{\mathbb{R}} (-u_c'' + 3u_c^2 + cu_c) v \, \mathrm{d}x \\ &\quad + \varepsilon^2 \int_{\mathbb{R}} \frac{1}{2}v'^2 + 3u_c v^2 + \frac{c}{2}v^2 \, \mathrm{d}x + \varepsilon^3 \int_{\mathbb{R}} v^3 \, \mathrm{d}x \\ &= M(u_c) + \partial_u M(u_c)[\varepsilon v] + \partial_u^2 M(u_c)[\varepsilon v, \varepsilon v] + \partial_u^3 M(u_c)[\varepsilon v, \varepsilon v, \varepsilon v]. \end{aligned} \tag{8.27}$$

Since u_c satisfies the traveling wave equation $u_c'' - cu_c - 3u_c^2 = 0$ for the KdV equation we see that $\partial_u M(u_c) = 0$ and that u_c is a stationary point of M.

Next we restrict u_0 to the energy surface of the traveling wave u_c, i.e., we restrict ourselves to the consideration of initial conditions u_0 with $E(u_0) = E(u_c) =: E_c$. It can then be shown that u_c is a minimum of M in this surface, and moreover that there exist $C_1, C_2 > 0$ such that

$$0 \leq C_1 \inf_{x_0 \in \mathbb{R}} \|u(\cdot + x_0) - u_c(\cdot)\|_{H^1}^2 \leq M(u) - M(u_c) \leq C_2 \|u - u_c\|_{H^1}^2 \tag{8.28}$$

for all sufficiently small $\|u - u_c\|_{H^1}$, see [**Ben72**]. Next we have the positive definiteness of the second variation

$$\partial_u^2 M(u_c)[v, v] = \int_{\mathbb{R}} \frac{1}{2}v'^2 + 3u_c' v^2 + \frac{c}{2}v^2 \, \mathrm{d}x$$

of $M(u_c)$, cf.(8.27), in the subspace of constant L^2-norm. From (8.28) we get orbital stability of u_c in the energy surface $E(u_0) = E_c$. This is called in the following conditional orbital stability.

From $E(u_c) = 8c^{3/2}/3$ it follows that for $|E(u_0) - E_c| < \delta$ there exists a c^* with $|c - c^*| \leq C\delta$ and $E(u_0) = E_{c^*}$. Hence, $\|u_c - u_{c^*}\|_{H^1} < \varepsilon/2$ if $\delta > 0$ is sufficiently small. Using this and the conditional orbital stability of u_{c^*} yields

$$\begin{aligned} &\inf_{x_0\in\mathbb{R}} \|u(\cdot,t)-u_c(\cdot+x_0,t)\|_{H^1} \\ \leq &\inf_{x_0\in\mathbb{R}} \|u(\cdot,t) - u_{c^*}(\cdot + x_0,t)\|_{H^1} + \inf_{x_0\in\mathbb{R}} \|u_{c^*}(\cdot,t) - u_c(\cdot + x_0,t)\|_{H^1} \\ \leq &\varepsilon/2 + \varepsilon/2 \end{aligned}$$

for $\delta > 0$ sufficiently small, i.e., the orbital stability of u_c. The orbital stability of N-solitons of the KdV equation has been established in [**MS93**].

8.3. The GL equation

The Ginzburg-Landau (GL) equation

$$\partial_t u = (1 + i\alpha)\partial_x^2 u + u - (1 + i\beta)|u|^2 u, \tag{8.29}$$

with $\alpha, \beta, x \in \mathbb{R}$, $t \geq 0$, and $u(x,t) \in \mathbb{C}$, appears as a universal modulation equation describing bifurcating solutions in spatially extended pattern forming systems close to the first instability. Typical examples are reaction-diffusion systems such as the Brusselator, or hydrodynamical stability problems such as the Couette-Taylor problem or Bénard's problem. This connection and more examples are discussed in Part IV of this book, where the mathematical justification of this so called GL approximation is one of the major topics.

However, the name of this equation comes from a completely different application. It has been derived in a multi-dimensional real version

$$\partial_t u = \Delta u + u - |u|^2 u, \tag{8.30}$$

with $x \in \mathbb{R}^3$, by Ginzburg and Landau as a model for the description of superconductivity [**GL50**]. In this theory u is an order parameter which describes in each point the fraction of the superconducting phase. The most prominent solutions of (8.30) are the so called GL vortices [**BBH94**]. Due to the applications which we have in mind we focus on the 1D GL equation over the real line and completely ignore any connection to superconductivity.

The dynamics of (8.29) differs strongly depending on the coefficients α and β. It spans from very simple dynamics to spatio-temporal chaos [**TZ10**], but for all α and β the origin is unstable with an (uncountable) infinite-dimensional unstable manifold. There are already a number of overview articles, cf. [**vS95, AK02**], about the GL equation itself. We concentrate on topics which are of interest for Part IV of this book. These are the local and global existence of solutions in uniformly local Sobolev spaces which will be introduced in a first step. Then we will discuss the existence of an attractor and the linear stability analysis of spatially periodic pattern. On unbounded domains there is another source of loss of compactness, namely the transport to infinity and so the analysis of PART II of this book about the existence of attractors has to be modified in several directions.

As we have seen, Sobolev spaces on $\mathbb{R}^d$ and Fourier transform fit very well together. However, for many systems over unbounded domains Sobolev spaces are not rich enough to contain all interesting solutions since for instance they do not contain spatially periodic functions or fronts, or, more generally, solutions which do not decay to zero for $|x| \to \infty$. Therefore, an alternative choice could be $C^0_{b,unif}$ and generalizations. On the one hand Fourier transform is available in $C^0_{b,unif}$-spaces, but on the other hand its analytic properties turn out to be rather complicated, see §5.2.1. Therefore, we look for spaces which contain all these non-decaying functions and for which additionally Fourier transform can be used to control the solution operators of the linearized equations. The so called uniformly local Sobolev spaces H^θ_{ul} turn out to be a good choice.

Thus, before discussing special solutions of the GL equation, and local (and global) existence of general solutions, here we start with some functional analysis to set up the spaces where the interesting solutions of the GL equation, such as periodic solutions and fronts, live in. Then we consider spatially periodic solutions, but we postpone a more comprehensive discussion of special solutions to Part IV, in particular §10.2.

Remark 8.3.1. a) If $\alpha = \beta = 0$, then (8.29) is called real GL equation. If $\alpha \neq 0$ or $\beta \neq 0$, then (8.29) is called complex GL equation or just GL equation. For the real GL equation we have the invariant subspace of real-valued functions $u : \mathbb{R} \to \mathbb{R}$, which then fulfill the Allen-Cahn equation

$$\partial_t u = \partial_x^2 u + u - u^3.$$

Thus, the dynamics which we found for the Allen-Cahn equation in §7.2 is also present in the real GL equation.

b) Like the NLS equation the GL equation is S^1-symmetric, i.e., if $u : \mathbb{R} \to \mathbb{C}$ is a solution of (8.29), then so is $v = e^{i\phi}u$ with arbitrary $\phi \in \mathbb{R}$.

c) The GL equation (8.29) can also be considered as a two component real-valued system either for the real and imaginary parts of u, cf. Exercise 8.17, or for the polar coordinates, see §8.3.6. There is no maximum principle for (8.29). ⌋

8.3.1. Uniformly local Sobolev spaces H^θ_{ul}. We choose a positive weight function $\rho : \mathbb{R} \to (0, \infty)$ which is continuous, bounded, and has a finite integral $\int_{\mathbb{R}} \rho(x)\, dx$. For later purposes we also impose $\rho \in C^2(\mathbb{R}, \mathbb{R})$ and assume $|\rho'(x)|, |\rho''(x)| \leq \rho(x)$ for all x. It turns out that the subsequent definition of the uniformly local Sobolev spaces H^θ_{ul} is rather independent of the particular choice of ρ. Therefore, we may fix ρ once and for all to

$$\rho(x) = 1/\cosh(x) \quad \text{or} \quad \rho(x) = 2/(2+x^2),$$

which both satisfy the conditions on ρ. We let

$$\widetilde{L^2_{\mathrm{ul}}}(\mathbb{R}) = \{\, u \in L^2_{\mathrm{loc}}(\mathbb{R}) \; : \; \|u\|_{L^2_{\mathrm{ul}}} < \infty \,\},$$

where

$$\|u\|^2_{L^2_{\mathrm{ul}}} = \sup_{y\in\mathbb{R}} \int_{\mathbb{R}} \rho(y+x) u(x)^2 \, \mathrm{d}x.$$

The statement that different weight functions lead to the same uniform space with equivalent norms is made rigorous with the following lemma.

Lemma 8.3.2. *The norm $\|\cdot\|_{L_{\mathrm{ul}}}$ is equivalent to the norm $\|\cdot\|_*$ defined by*

$$\|u\|_* = \sup_{y\in\mathbb{R}} \left(\int_{y-1/2}^{y+1/2} |u(x)|^2 \, \mathrm{d}x \right)^{1/2}.$$

Proof. Let $\alpha = \min_{x\in[-1/2,1/2]} \rho(x)$. Then

$$\begin{aligned} \alpha \left(\int_{y-1/2}^{y+1/2} |u(x)|^2 \, \mathrm{d}x \right)^{1/2} &\leq \left(\int_{y-1/2}^{y+1/2} \rho(y+x) |u(x)|^2 \, \mathrm{d}x \right)^{1/2} \\ &\leq \left(\int_{\mathbb{R}} \rho(y+x) |u(x)|^2 \, \mathrm{d}x \right)^{1/2}. \end{aligned}$$

Taking the supremum w.r.t. y shows $\alpha\|u\|_* \leq \|u\|_{L_{\mathrm{ul}}}$. On the other hand we have

$$\begin{aligned} \left(\int_{\mathbb{R}} \rho(y+x) |u(x)|^2 \, \mathrm{d}x \right)^{1/2} &= \left(\sum_{y\in\mathbb{Z}} \int_{y-1/2}^{y+1/2} \rho(y+x) |u(x)|^2 \, \mathrm{d}x \right)^{1/2} \\ &\leq \left(\sum_{y\in\mathbb{Z}} \sup_{x\in[-1/2,1/2]} \rho(y+x) \int_{y-1/2}^{y+1/2} |u(x)|^2 \, \mathrm{d}x \right)^{1/2} \\ &\leq \left(\sup_{y\in\mathbb{R}} \int_{y-1/2}^{y+1/2} |u(x)|^2 \, \mathrm{d}x \right)^{1/2} \left(\sum_{y\in\mathbb{Z}} \sup_{x\in[-1/2,1/2]} \rho(y+x) \right)^{1/2} \\ &\leq C \sup_{y\in\mathbb{R}} \left(\int_{y-1/2}^{y+1/2} |u(x)|^2 \, \mathrm{d}x \right)^{1/2} \end{aligned}$$

since $C = (\sum_{y\in\mathbb{Z}} \sup_{x\in[-1/2,1/2]} \rho(y+x))^{1/2} < \infty$ due to the assumptions on ρ. This shows $\|u\|_{L_{\mathrm{ul}}} \leq \alpha\|u\|_*$. □

In the rest of this section we omit the domain $\mathbb{R}$ as all function spaces considered here are defined over the real line. With the translation operator $T_y : \widetilde{L^2_{\mathrm{ul}}} \to \widetilde{L^2_{\mathrm{ul}}}$, $(T_y u)(\cdot) \mapsto u(\cdot + y)$, our final space of uniformly local L^2 functions is given as

$$L^2_{\mathrm{ul}} = \{ u \in \widetilde{L^2_{\mathrm{ul}}} : \|T_y u - u\|_{L^2_{\mathrm{ul}}} \to 0 \text{ as } y \to 0 \}.$$

This construction is very similar to the construction of the space of uniformly continuous functions, cf. Exercise 8.20. For $\theta \in \mathbb{N}$ we define the associated Sobolev spaces $\widetilde{H^\theta_{\mathrm{ul}}}$ and H^θ_{ul} by requiring that the first m distributional derivatives lie in the spaces as well. Some first properties of these spaces are summarized in the following lemma, cf. [**MS95**, Lemma 3.1].

Lemma 8.3.3. *a) L^2_{ul} is a closed subspace of $\widetilde{L^2_{\mathrm{ul}}}$.*
b) The spaces $\widetilde{H^\theta_{\mathrm{ul}}}$ and H^θ_{ul} are dense in L^2_{ul}.
c) L^2_{ul} is a proper subset of $\widetilde{L^2_{\mathrm{ul}}}$.

Proof. a) To show that L^2_{ul} is a closed subspace let $(u_n)_{n\in\mathbb{N}}$ be a Cauchy sequence in L^2_{ul} with limit $u \in \widetilde{L^2_{\mathrm{ul}}}$. Then

$$\|T_y u - u\|_{L^2_{\mathrm{ul}}} \le \|T_y u - T_y u_n\|_{L^2_{\mathrm{ul}}} + \|T_y u_n - u_n\|_{L^2_{\mathrm{ul}}} + \|u_n - u\|_{L^2_{\mathrm{ul}}} = s_1 + s_2 + s_3.$$

Since u_n approximates u there exists an n such that $s_1 = s_3 < \varepsilon/3$. For the corresponding $u_n \in L^2_{\mathrm{ul}}$ we find a $\delta > 0$ such that for all $y \in (-\delta, \delta)$ we have $s_2 < \varepsilon/3$.

b) Let $u \in \widetilde{H^1_{\mathrm{ul}}}$. With part a) we conclude

$$\|T_y u - u\|^2_{L^2_{\mathrm{ul}}} = \sup\{ \int_{\mathbb{R}} \rho(x)(u(x+y+z) - u(x+z))^2 \, dx : z \in \mathbb{R} \} \le C|y| \|u\|^2_{H^1_{\mathrm{ul}}}.$$

Thus, $\widetilde{H^1_{\mathrm{ul}}} \subset L^2_{\mathrm{ul}}$ and hence $H^\theta_{\mathrm{ul}} \subset \widetilde{H^\theta_{\mathrm{ul}}} \subset L^2_{\mathrm{ul}}$ for $\theta \in \mathbb{N}$. To see that H^1_{ul} is dense in L^2_{ul} choose any $u \in L^2_{\mathrm{ul}}$. As $y \mapsto T_y u \in L^2_{\mathrm{ul}}$ is continuous the Riemann integral $v_h = \frac{1}{h} \int_0^h T_y u \, dy$ exists and $\|v_h - u\|_{L^2_{\mathrm{ul}}} \to 0$ for $h \to 0$. Moreover, $v_h \in H^1_{\mathrm{ul}}$ as $v_h(x) = \frac{1}{h} \int_0^h u(x+y) \, dy = \frac{1}{h} \int_x^{x+h} u(z) \, dz$. Iterating this process it follows that H^θ_{ul} is dense in L^2_{ul}.

c) It suffices to consider the example $u(x) = n$ for $x \in [n, n + 1/n^2]$, $n \in \mathbb{N}$, and $u(x) = 0$ else. This u lies in $\widetilde{L^2_{\mathrm{ul}}} \setminus L^2_{\mathrm{ul}}$. Hence, L^2_{ul} is a proper subspace of $\widetilde{L^2_{\mathrm{ul}}}$. □

Compactness. Over bounded domains, smoothness yields (pre)compactness by Sobolev embeddings, see §5.3. This is not true over unbounded domains.

Example 8.3.4. The embedding $H^1(\mathbb{R}) \hookrightarrow L^2(\mathbb{R})$ is not compact, since the sequence $(u_n)_{n\in\mathbb{N}}$, with $u_n(x) = u_0(x+n)$ and $u_0(x) = e^{-x^2}$ is bounded in H^1, but does not contain a convergent subsequence. ⌋

Instead, we have a form of localized compactness which plays a role in §8.3.4 where we construct an attractor for the GL equation on the real line based on H^θ_{ul} spaces, cf. [**MS95**, Theorem 3.2].

Theorem 8.3.5. *Let $\theta > 1$ and $B \subset H^1_{\mathrm{ul}}$ be any set which is bounded in H^θ_{ul}. Then B is pre-compact in*

$$H^1_\rho = \{u : \mathbb{R} \to \mathbb{R} : u, \partial_x u \in L^2_{loc},\ \|u\|_{H^1_\rho} = \|u\rho\|_{H^1} < \infty\}.$$

Proof. For every $\varepsilon > 0$ we have to show that B admits in H^1_ρ a finite covering by balls of radius less than ε. We decompose every $u \in B$ into $u = v + w$ with $v = u\chi_\beta$ and $w = u(1-\chi_\beta)$, where the smooth cut-off function χ_β vanishes for $|x| \le \beta$ and equals 1 for $|x| \ge \beta + 1$. Then $\|v\|_{H^1_\rho} < \varepsilon$ for β sufficiently large, as ρ decays for $|x| \to \infty$ and u varies in a bounded set. Moreover, $w \in H^\theta([-\beta-1, \beta+1])$ which can be embedded compactly into $H^1([-\beta-1, \beta+1])$. Since for functions with support $[-\beta-1, \beta+1]$ the norms in H^θ and H^θ_ρ are equivalent, there is a finite covering of this set by balls $B_{H^1_\rho}(w_i, \varepsilon)$ with $i = 1, \ldots, m < \infty$. Thus, $\cup_{i=1,\ldots,m}\{B_{H^1_\rho}(w_i, 2\varepsilon)\}$ is a finite covering of B, since $\|u - w_i\|_{H^1_\rho} \le \|v\|_{H^1_\rho} + \|w - w_i\|_{H^1_\rho}$. □

Multiplier theory in H^θ_{ul}-spaces. An important tool for studying translational invariant operators on the spaces H^θ_{ul} is multiplier theory which uses Fourier transform methods. An operator $M : H^q_{\mathrm{ul}} \to H^\theta_{\mathrm{ul}}$ is called multiplier if there exists a function $\widehat{M} : \mathbb{R} \to \mathbb{C}$ such that $Mu = \mathcal{F}^{-1}(\widehat{M}\mathcal{F}u)$, i.e., if the associated operator in Fourier space is a multiplication operator. Taking the Fourier transform of a function $u \in L^2_{l,u}$ gives a tempered distribution. This allows easily to define operators via their action in Fourier space. However, it is rather complicated to estimate the norm of the operators in physical space. In order to do so we proceed as follows. We will make use of the fact that Fourier transform is an isomorphism between H^m_n and H^n_m, cf. Definition 7.3.30 and Lemma 7.3.31. This is the case for almost all operators considered so far in this book. We have the following classical result of multiplier theory.

Lemma 8.3.6. *Let $q, \theta \ge 0$ and $w_{\theta-q}(k) = (1+k^2)^{(\theta-q)/2}\widehat{M}(k) \in C^0_b(\mathbb{R}, \mathbb{C})$. Then $M : H^q \to H^\theta$, $u \mapsto \mathcal{F}^{-1}(\widehat{M}\mathcal{F}u)$ is well defined with the estimate*

$$\|Mu\|_{H^\theta} \le C(q,\theta)\|w_{\theta-q}\|_{C^0_b(\mathbb{R},\mathbb{C})}\|u\|_{H^q},$$

where $C(q, \theta)$ does not depend on $\widehat{M}$.

Proof. The statement follows immediately from

$$\|Mu\|_{H^\theta} \leq C\|\widehat{M}\widehat{u}\|_{H^0_\theta} \leq C\|w_{\theta-q}\|_{C^0_b(\mathbb{R},\mathbb{C})}\|\widehat{u}\|_{H^0_q} \leq C\|w_{\theta-q}\|_{C^0_b(\mathbb{R},\mathbb{C})}\|u\|_{H^q}.$$

□

The similar result in H^θ_{ul}-spaces is as follows, cf. [**Sch94a**, Lemma 5], which also contains an extension of the result from $L_{\mathrm{ul}}(\mathbb{R},\mathbb{C})$ to $L_{\mathrm{ul}}(\mathbb{R},H)$, with H some Hilbert space.

Lemma 8.3.7. *Let $q,\theta \geq 0$ and $w_{\theta-q}(k) = (1+k^2)^{(\theta-q)/2}\widehat{M}(k) \in C^2_b(\mathbb{R},\mathbb{C})$. Then $M_{\mathrm{ul}} : H^q_{\mathrm{ul}} \to H^\theta_{\mathrm{ul}}$, $u \mapsto \mathcal{F}^{-1}(\widehat{M}\mathcal{F}u)$ is well defined with the estimate*

$$\|M_{\mathrm{ul}}u\|_{H^\theta_{\mathrm{ul}}} \leq C(q,\theta)\|w_{\theta-q}\|_{C^2_b(\mathbb{R},\mathbb{C})}\|u\|_{H^q_{\mathrm{ul}}},$$

where $C(q,\theta)$ does not depend on $\widehat{M}$.

Proof. We choose $\chi \in C^\infty_0$ such that its support is contained in $[-1,1]$ and $\sum_{j\in\mathbb{Z}}\chi(x+j) \equiv 1$. The operator M_{ul} then can be defined as follows. For $u \in H^q_{\mathrm{ul}}$ we set $u_j(x) = u(x-j)\chi(x)$. Since $u_j \in H^q_2$ we find $v_j = Mu_j \in H^\theta_2$ according to

$$\|Mu_j\|_{H^\theta_2} \leq C\|\widehat{M}\widehat{u}_j\|_{H^2_\theta} \leq C\|w_{\theta-q}\|_{C^2_b}\|\widehat{u}_j\|_{H^2_q} \leq \|u_j\|_{H^q_2}.$$

Now let $Mu = \sum_{j\in\mathbb{Z}} T_j v_j$. Clearly this sum does not converge in H^θ_{ul}. But, since $T_j v_j$ is concentrated around $x = j$ and decays like $1/(1+(x-j)^2)$ it is easy to see that $\sum_{j\in\mathbb{Z}} T_j v_j$ converges locally to a function in H^θ_{ul} with norm $\leq C(q,\theta)\|w_{\theta-q}\|_{C^2_b(\mathbb{R},\mathbb{C})}\|u\|_{H^q_{\mathrm{ul}}}$. □

8.3.2. Local existence and uniqueness of solutions. We already considered the local existence and uniqueness of solutions of the GL equation in §5.2.3 in case of periodic boundary conditions. In exactly the same way we prove here the local existence and uniqueness of solutions in Sobolev spaces H^θ with $\theta > 1/2$ on the real line.

Theorem 8.3.8. *Let $\theta > 1/2$ and $u_0 \in H^\theta$. Then there exists a $T_0 = T_0(\|u_0\|_{H^\theta}) > 0$ and unique mild solution $u \in C([0,T_0],H^\theta)$ of the GL equation with $u|_{t=0} = u_0$.*

Proof. The semigroup $T(t) : H^\theta \to H^\theta$ generated by $(1+\mathrm{i}\alpha)\partial_x^2$ is defined by $T(t) = \mathcal{F}\widehat{T}(t)\mathcal{F}^{-1}$, where $\widehat{T}(t)[k] = \mathrm{e}^{-(1+\mathrm{i}\alpha)k^2 t}$. We have the estimate

$$\|T(t)u\|_{H^\theta} = \|\widehat{T}(t)\widehat{u}\|_{L^2_\theta} \leq \|\widehat{T}(t)\|_{C^0_b}\|\widehat{u}\|_{L^2_\theta} \leq \|\widehat{u}\|_{L^2_\theta} = \|u\|_{H^\theta}.$$

The nonlinear term $N(u) = u - (1+\mathrm{i}\beta)|u|^2u$ is locally Lipschitz-continuous from H^θ to H^θ if $\theta > 1/2$. Thus, the right-hand side of the variation of constant formula

$$u(t) = T(t)u(0) + \int_0^t T(t-\tau)N(u)(\tau)\,\mathrm{d}\tau$$

is a contraction in the space $C([0,T_0],H^\theta)$ if $T_0>0$ is sufficiently small. $\square$

Remark 8.3.9. Following the proof of Lemma 6.2.5 we have the estimate

$$\begin{aligned}\|\widehat{u}*\widehat{v}*\widehat{w}\|_{L^2_\theta} \leq C(&\|\widehat{u}\|_{L^2_{\theta+2r}}\|\widehat{v}\|_{L^1_{-r}}\|\widehat{w}\|_{L^1_{-r}}\\ &+\|\widehat{u}\|_{L^1_{-r}}\|\widehat{v}\|_{L^2_{\theta+2r}}\|\widehat{w}\|_{L^1_{-r}}+\|\widehat{u}\|_{L^1_{-r}}\|\widehat{v}\|_{L^1_{-r}}\|\widehat{w}\|_{L^2_{\theta+r}}).\end{aligned}$$

Since $\|\widehat{u}\|_{L^1_{-r}} \leq C\|\widehat{u}\|_{L^2_{1/2-r+\delta}}$ for every $\delta>0$ and since Fourier transform is an isomorphism between L^2_s and H^s we can choose $\theta=-1-2\delta$, $r=1/2+\delta$ to establish the local Lipschitz-continuity of the nonlinear term $N(u)=u-(1+\mathrm{i}\beta)|u|^2u$ from L^2 into $H^{-1-2\delta}$ for every $\delta>0$. Since the semigroup $T(t)$ is smoothing with a singularity $t^{-1/2-\delta}$ from $H^{-1-2\delta}$ to L^2 the right-hand side of the variation of constant formula is a contraction in the space $C([0,T_0],L^2)$, too, if $T_0>0$ is sufficiently small. Hence, Theorem 8.3.8 is already true for every $\theta\geq 0$. ⌋

However, H^θ as a phase space does not contain spatially periodic functions or fronts and is therefore too small for our purposes. Thus, we next choose $C^0_{b,unif}$ as phase space. In case of $\alpha=0$ the semigroup $T(t)$ of the GL equation coincides with the semigroup of the linear diffusion equation. Since the nonlinear term $N(u)=u-(1+\mathrm{i}\beta)|u|^2u$ is locally Lipschitz-continuous in $C^0_{b,unif}$ the right-hand side of the variation of constant formula is a contraction in the space $C([0,T_0],C^0_{b,unif})$ if $T_0>0$ is sufficiently small. Therefore, we have

Theorem 8.3.10. *Let $\alpha=0$ and $u_0\in C^0_{b,unif}$. Then there exists a $T_0=T_0(\|u_0\|_{C^0_{b,unif}})>0$ and unique mild solution $u\in C([0,T_0],C^0_{b,unif})$ of the GL equation with $u|_{t=0}=u_0$.*

We refrain from generalizing the result to the case $\alpha\neq 0$ since $C^0_{b,unif}$ is not suitable for establishing the existence of an absorbing set. Instead we choose the uniformly local Sobolev spaces H^θ_{ul}. For the local existence and uniqueness theory we need

Lemma 8.3.11. *The space H^θ_{ul} is an algebra and can be continuously embedded in $C^0_{b,unif}$ if $\theta>1/2$.*

Proof. We write $u,v\in H^\theta_{\mathrm{ul}}$ as $u(x)=\sum_{n\in\mathbb{Z}}u_n(x)$ where $\mathrm{supp}\,u_n(x)\in(n-1,n+2)$, and similarly for v. For the product we find $u(x)v(x)=\sum_{l\in\mathbb{Z}}\sum_{n\in\mathbb{Z}}u_l(x)v_n(x)$ where $u_l(x)v_n(x)=0$ if $|l-n|>2$ due to disjoint supports. Hence

$$\begin{aligned}\|uv\|_{H^\theta[k,k+1]} =&\|(u_{k-1}+u_k+u_{k+1})(v_{k-1}+v_k+v_{k+1})\|_{H^\theta[k,k+1]}\\ \leq&(\|u_{k-1}\|_{H^\theta[k,k+1]}+\|u_k\|_{H^\theta[k,k+1]}+\|u_{k+1}\|_{H^\theta[k,k+1]})\\ &\times(\|v_{k-1}\|_{H^\theta[k,k+1]}+\|v_k\|_{H^\theta[k,k+1]}+\|v_{k+1}\|_{H^\theta[k,k+1]})\end{aligned}$$

$$\leq 9\|u\|_{H^\theta_{\mathrm{ul}}}\|v\|_{H^\theta_{\mathrm{ul}}}$$

for $\theta > 1/2$ and as a consequence $\|uv\|_{H^\theta_{\mathrm{ul}}} \leq 9\|u\|_{H^\theta_{\mathrm{ul}}}\|v\|_{H^\theta_{\mathrm{ul}}}$. Similarly, the embedding follows. □

Then we have the following local existence and uniqueness result.

Theorem 8.3.12. *Let $\theta > 1/2$ and $u_0 \in H^\theta_{\mathrm{ul}}$. Then there exists a $T_0 = T_0(\|u_0\|_{H^\theta_{\mathrm{ul}}}) > 0$ and unique mild solution $u \in C([0,T_0], H^\theta_{\mathrm{ul}})$ of the GL equation with $u|_{t=0} = u_0$.*

Proof. The nonlinear term $N(u) = u - (1+\mathrm{i}\beta)|u|^2u$ is locally Lipschitz-continuous in H^θ_{ul}. The linear operator $(1+\mathrm{i}\alpha)\partial_x^2$ is the generator of a semigroup which is given in Fourier space by $\widehat{T}(t)[k] = \mathrm{e}^{-(1+\mathrm{i}\alpha)k^2t}$. In order to get a bound for the semigroup $T(t) = \mathcal{F}\widehat{T}(t)\mathcal{F}^{-1}$ in H^θ_{ul} we use Lemma 8.3.7. We have $\sup_{k\in\mathbb{R}} |\mathrm{e}^{-(1+\mathrm{i}\alpha)k^2t}| \leq 1$ and for the second derivative

$$\sup_{k\in\mathbb{R}} |(-4(1+\mathrm{i}\alpha)^2k^2t^2 - 2(1+\mathrm{i}\alpha)t)\mathrm{e}^{-(1+\mathrm{i}\alpha)k^2t}| = \mathcal{O}(t)$$

such that $\|\widehat{T}(t)\|_{C_b^2}$ is bounded for finite t. Hence, the right-hand side of the variation of constant formula is a contraction in the space $C([0,T_0], H^\theta_{\mathrm{ul}})$ if $T_0 > 0$ is sufficiently small. Therefore, we are done. □

The local existence and uniqueness theorem can be improved from $\theta > 1/2$ to $\theta \geq 0$. In order to do so we combine Remark 8.3.9 and Lemma 8.3.11. Using the inequality of Remark 8.3.9 yields

$$\begin{aligned}
\|uvw\|_{H^\theta[k,k+1]} &= \left\| \Big(\sum_{j=-1}^{1} u_{k+j}\Big)\Big(\sum_{j=-1}^{1} v_{k+j}\Big)\Big(\sum_{j=-1}^{1} w_{k+j}\Big)\right\|_{H^\theta[k,k+1]} \\
&\leq \sum_{j_1=-1}^{1}\sum_{j_2=-1}^{1}\sum_{j_3=-1}^{1} \|u_{k+j_1}v_{k+j_2}w_{k+j_3}\|_{H^\theta[k,k+1]} \\
&\leq \sum_{j_1=-1}^{1}\sum_{j_2=-1}^{1}\sum_{j_3=-1}^{1} \|u_{k+j_1}v_{k+j_2}w_{k+j_3}\|_{H^\theta} \\
&\leq C\sum_{j_1=-1}^{1}\sum_{j_2=-1}^{1}\sum_{j_3=-1}^{1} \|u_{k+j_1}\|_{L^2}\|v_{k+j_2}\|_{L^2}\|w_{k+j_3}\|_{L^2} \\
&\leq 9C\|u\|_{L^2_{\mathrm{ul}}}\|v\|_{L^2_{\mathrm{ul}}}\|w\|_{L^2_{\mathrm{ul}}},
\end{aligned}$$

and so $\|uvw\|_{H^\theta_{\mathrm{ul}}} \leq 9C\|u\|_{L^2_{\mathrm{ul}}}\|v\|_{L^2_{\mathrm{ul}}}\|w\|_{L^2_{\mathrm{ul}}}$ if $\theta = -1-2\delta$ for every $\delta > 0$. As a consequence, it follows

Theorem 8.3.13. *Let $\theta \geq 0$ and $u_0 \in H^\theta_{\rm ul}$. Then there exists a $T_0 = T_0(\|u_0\|_{H^\theta_{\rm ul}}) > 0$ and unique mild solution $u \in C([0,T_0], H^\theta_{\rm ul})$ of the GL equation with $u|_{t=0} = u_0$.*

Remark 8.3.14. The smoothing properties of the semigroup allows us to establish as before that the solutions are analytic for every $t > 0$. Again the global existence of solutions or the existence of an absorbing set in $C^0_{b,unif}$, respectively $L^2_{\rm ul}$, implies the global existence of solutions or the existence of an absorbing set in $C^m_{b,unif}$, respectively $H^\theta_{\rm ul}$, also for every $m \in \mathbb{N}$, respectively $\theta > 0$. ⌋

8.3.3. Global existence for the GL equation. In order to establish the existence of an attractor of a dynamical system one needs the existence of an absorbing set in the chosen phase space X, cf. §2.4.2. If such a set exists the solutions must exist globally in time, which follows by combining the local existence and uniqueness of solutions with a priori estimates for the norm $\|u(t)\|_X$. According to Remark 8.3.14 it is sufficient to establish these a priori estimates in $C^0_{b,unif}$, respectively $L^2_{\rm ul}$.

The real case in $C^0_{b,unif}$. Let $\alpha = \beta = 0$ and assume that $u_0 \in C^0_{b,unif}$. We introduce polar coordinates $u(x,t) = r(x,t)e^{i\phi(x,t)}$ and find

$$\partial_t \phi = \partial_x^2 \phi + 2\frac{\partial_x r}{r} + \partial_x \phi, \qquad \partial_t r = \partial_x^2 r + r - r^3 - r(\partial_x \phi)^2. \tag{8.31}$$

Since $\sup_{x\in\mathbb{R}} |u(x,t)| = \sup_{x\in\mathbb{R}} r(x,t) = R(t)$ it is sufficient to get a bound for r. Since $(\partial_x \phi)^2 \geq 0$ we obtain

$$\partial_t r \leq \partial_x^2 r + r - r^3.$$

The maximum principle then implies $\partial_t R \leq R - R^3$ such that

$$\limsup_{t\to\infty} \sup_{x\in\mathbb{R}} |u(x,t)| = \limsup_{t\to\infty} R(t) \leq 1.$$

Hence, we have the following global existence and uniqueness result

Theorem 8.3.15. *Let $\alpha = \beta = 0$ and $u_0 \in C^0_{b,unif}$. Then there exists a unique mild solution $u \in C([0,\infty), C^0_{b,unif})$ of the GL equation with $u|_{t=0} = u_0$ and $\limsup_{t\to\infty} \|u(\cdot,t)\|_{C^0_{b,unif}} \leq 1$.*

The real case in $L^2_{\rm ul}$. Let $\alpha = \beta = 0$ and assume that $u_0 \in L^2_{\rm ul}$. In order to find the bound in $L^2_{\rm ul}$ we consider weighted energy estimates for $E_y(t) = \int_{\mathbb{R}} u(x,t)\overline{u(x,t)}\rho_y(x)\,dx$ with $\rho(x) = 2(2+x^2)^{-1}$ and $\rho_y(x) = \rho(x+y)$. We find through integration by parts and by using $|\rho''| \leq \rho$ that

$$\begin{aligned}\frac{\mathrm{d}}{\mathrm{d}t}E_y(t) =& \int_{\mathbb{R}} (\partial_x^2 u + u - |u|^2 u)\overline{u}\rho_y \,\mathrm{d}x + \int_{\mathbb{R}} \overline{(\partial_x^2 u + u - |u|^2 u)} u\rho_y \,\mathrm{d}x \\ =& 2\int_{\mathbb{R}} (-|\partial_x u|^2 + |u|^2 - |u|^4)\rho_y \,\mathrm{d}x + 2\int_{\mathbb{R}} |u|^2 \rho_y'' \,\mathrm{d}x \\ \leq& 2\int_{\mathbb{R}} (2|u|^2 - |u|^4)\rho_y \,\mathrm{d}x \leq 2\int_{\mathbb{R}} (4 - 2|u|^2)\rho_y \,\mathrm{d}x \\ \leq& 8\int_{\mathbb{R}} \rho \,\mathrm{d}x - 4E_y(t) = 16\pi\sqrt{2} - 4E_y(t).\end{aligned}$$

Thus, $\limsup_{t\to\infty} E_y(t) \leq 4\pi\sqrt{2}$, and since $y \in \mathbb{R}$ was arbitrary we finally have

$$\limsup_{t\to\infty} \|u(t)\|_{L^2_{\mathrm{ul}}} = \limsup_{t\to\infty} \sup_{y\in\mathbb{R}} E_y(t)^{1/2} \leq \sqrt{4\pi\sqrt{2}}.$$

The general case in L^2_{ul}. Now let $\alpha, \beta \in \mathbb{R}$ be arbitrary. Again it is sufficient to bound $E_y(t)$, for which we find

$$\begin{aligned}\frac{\mathrm{d}}{\mathrm{d}t}E_y(t) =& \int_{\mathbb{R}} ((1+\mathrm{i}\alpha)\partial_x^2 u + u - (1+\mathrm{i}\beta)|u|^2 u)\overline{u}\rho_y \,\mathrm{d}x \\ &+ \int_{\mathbb{R}} \overline{((1+\mathrm{i}\alpha)\partial_x^2 u + u - (1+\mathrm{i}\beta)|u|^2 u)} u\rho_y \,\mathrm{d}x \\ =& 2\int_{\mathbb{R}} (-|\partial_x u|^2 + |u|^2 - |u|^4)\rho_y \,\mathrm{d}x + 2\int_{\mathbb{R}} |u|^2 \rho_y'' \,\mathrm{d}x \leq 16\pi\sqrt{2} - 4E_y(t)\end{aligned}$$

as before since the imaginary parts cancel. Therefore, as before

$$\limsup_{t\to\infty} \|u(t)\|_{L^2_{\mathrm{ul}}} = \limsup_{t\to\infty} \sup_{y\in\mathbb{R}} E_y(t)^{1/2} \leq \sqrt{4\pi\sqrt{2}}.$$

Using the variation of constant formula and the smoothing properties of the semigroup we find that

$$\limsup_{t\to\infty} \|u(t)\|_{H^1_{\mathrm{ul}}} = \mathcal{O}(|\beta|)$$

for $|\beta| \to \infty$. See Exercise 8.21.

Remark 8.3.16. First a priori estimates for the size of the solutions have been established in [**BCD⁺90**] leading to the terminology of soft and hard turbulence regimes in the (α, β)-plane. Later on the estimates have been improved in [**Mie98**] where also estimates for higher space dimensions can be found. ⌋

8.3.4. Attractors on unbounded domains. In §2.4 we introduced the notion of an attractor and studied simple ODE cases. We then transferred this to the case of (dissipative) PDEs over spatially bounded domains, see §5.3. A key property was compactness, which for ODEs with finite-dimensional phase space simply follows from boundedness. For dissipative PDEs over spatially bounded domains the compactness is obtained from boundedness of orbits, smoothing properties of the semigroup, and compactness of embeddings. This no longer works for PDEs over spatially unbounded domains. In particular, for translational invariant problems we cannot expect a compact embedding, cf. Example 8.3.4, and as a consequence no compact attractor exists in any translational invariant norm.

Example 8.3.17. Consider the linear heat equation $\partial_t u = \partial_x^2 u$ in the space $C^0_{b,unif}(\mathbb{R},\mathbb{R})$. We are interested in the attractor $\mathcal{A}$ in the positively invariant set $\mathcal{B} = \{u \in Z : \|u\|_\infty \leq R\}$ for an $R > 0$. The only solutions which stay in $\mathcal{B}$ for all negative times are the constants, thus $\mathcal{A} = \{u \equiv \theta : \theta \in [-R, R]\}$ is the only candidate for an attractor. Consider now the solution $\mathcal{S}_t(u_0)$ with the initial condition $u_0(x) = (2/\pi)\arctan(x)$. Using the explicit solution formula $u(x,t) = \int_{\mathbb{R}} G(x-y,t)u_0(y)dy$ with $G(x,t) = \frac{1}{2\sqrt{\pi t}}\mathrm{e}^{(x-y)^2/(4t)}$ we find that for all t we have $u(x,t) \to \pm 1$ for $x \to \pm\infty$. Hence, we always have $\operatorname{dist}_{C^0_b}(\mathcal{S}_t(u_0), \mathcal{A}) = 1$, and attractivity of $\mathcal{A}$ cannot hold in the uniform norm of Z. However, we have convergence to 0 on every finite interval I, i.e., $\sup_{x\in I} |\mathcal{S}_t(u_0) - 0| \to 0$ for $t \to \infty$. ⌋

Hence, we can only expect attractivity in a spatially localized norm. We closely follow the setup introduced in [**MS95**] to show the existence of such an attractor. We start with some abstract theory which will be applied to the GL equation and later in Part IV to more complicated pattern forming systems.

We are interested in attractors for semigroups $(\mathcal{S}_t)_{t\geq 0}$ on a Banach space $(Z, \|\cdot\|)$. Our basic assumption is that $\mathcal{S}_t(u) \in Z$ depends continuously on $(t,u) \in [0,\infty) \times Z$. As before a subset $\mathcal{B} \subset Z$ is called positively invariant for $(\mathcal{S}_t)$ if $\mathcal{S}_t(\mathcal{B}) \subset \mathcal{B}$ for all $t > 0$, and it is called an absorbing set for $(\mathcal{S}_t)$, if it is bounded, positively invariant and every bounded set $B \subset Z$ is absorbed into $\mathcal{B}$ in finite time, i.e., there exists $t > 0$ such that $\mathcal{S}_t(B) \subset \mathcal{B}$.

Beside the 'uniform' norm $\|\cdot\|$ we work with a 'spatially localized' norm $\|\cdot\|_\rho$ such that $\|u\|_\rho \leq \|u\|$. We denote by Z_ρ the set Z, but equipped with the weaker topology induced by the norm $\|\cdot\|_\rho$.

Definition 8.3.18. *Let $\mathcal{B} \subset Z$ be positively invariant. A subset $\mathcal{A} \subset \mathcal{B}$ is called an (Z, Z_ρ)-attractor for $\mathcal{S}_t$ in $\mathcal{B}$ if the following conditions hold:*
a) $\mathcal{A}$ is nonempty, closed, bounded in Z, and compact in Z_ρ.
b) $\mathcal{A}$ is invariant under $\mathcal{S}_t$, i.e., $\mathcal{S}_t(\mathcal{A}) = \mathcal{A}$ for all $t > 0$.

c) Every $B \subset \mathcal{B}$ which is bounded in Z is attracted to $\mathcal{A}$ in the distance induced by the norm of Z_ρ, i.e.,

$$\mathrm{dist}_{Z_\rho}(\mathcal{S}_t(B), \mathcal{A}) := \sup_{b\in B} \inf_{a\in\mathcal{A}} \|\mathcal{S}_t(b) - a\|_\rho \to 0 \quad \textit{for } t \to \infty.$$

If $\mathcal{A}$ is an (Z, Z_ρ)-attractor in $\mathcal{B} = Z$, then it is called the (global) attractor of $\mathcal{S}_t$.

If $Z = Z_\rho$ in the sense of topological spaces this definition coincides with our previous Definition 2.4.3. It is standard to conclude the uniqueness of a (Z, Z_ρ)-attractor, since any second candidate $\mathcal{A}_2$ would be attracted to $\mathcal{A}$ as well as attract $\mathcal{A}$ towards itself. Using the invariance and closedness we obtain

$$\mathrm{dist}_{Z_\rho}(\mathcal{A}, \mathcal{A}_2) = \mathrm{dist}_{Z_\rho}(\mathcal{A}_2, \mathcal{A}) = 0,$$

which implies $\mathcal{A} = \mathcal{A}_2$. We assume that the 'uniform' and the 'spatially localized' norm are connected as follows.

(A1) *The translations T_y are continuous w.r.t. the norm $\|\cdot\|_\rho$ and*

$$\|u\| = \sup\{\|T_y u\|_\rho : y \in \mathbb{R}\}.$$

For notational convenience we let $Z_u = (Z, \|\cdot\|)$ and $Z_\rho = (Z, \|\cdot\|_\rho)$ to indicate that Z is equipped with different norms. However, we mostly omit the subscript u to denote the topological space $Z = Z_u$. The uniform space Z_u is the original Banach space, whereas Z_ρ is only a normed space and is not necessarily complete. In fact, our interest lies precisely in those cases where the translations T_y are not uniformly bounded in the ρ-norm, which implies that Z_ρ is not complete.

We use the abbreviation $B_Z(r, u_0)$ for the closed ball of radius r in Z with center u_0, i.e., $B_Z(r, u_0) = \{u \in Z : \|u - u_0\| \leq r\}$. Moreover we let $B_Z(r) = B_Z(r, 0)$. As a direct consequence of (A1) we obtain the following result which will be needed in Theorem 8.3.22.

Lemma 8.3.19. *If $A \subset Z$ is contained in $B_Z(r)$ for some $r > 0$, then $\overline{A}^\rho = \mathrm{closure}_{Z_\rho}(A) \subset B_Z(r)$.*

Proof. Let $(u_n)_{n\in\mathbb{N}} \in B$ be a sequence with limit u in Z_ρ. Since $\|u_n\| < r$ and since T_y is continuous in Z_ρ we obtain for fixed y

$$\|T_y u\|_\rho \leq \|T_y u_n\|_\rho + \|T_y u_n - T_y u\|_\rho \leq r + \varepsilon_n$$

with $\varepsilon_n \to 0$ for $n \to \infty$. Thus, $\|u\| \leq r$ which is the desired result. □

Example 8.3.20. Consider $Z = C^0_{b,unif}$ and let Z_ρ be equipped with the norm

$$\|u\|_\rho = \sup_{x\in\mathbb{R}} |u(x)/(1 + x^2)|.$$

Convergence in Z_u is uniform convergence, whereas convergence in Z_ρ means uniform convergence on each compact interval. Consider $u : x \to \tanh(x)$ and let $A = \{T_y u : y \in \mathbb{R}\}$. Then A is closed in Z_u, but its closure in Z_ρ is $\overline{A}^\rho = A \cup \{v_{-1}, v_1\}$, where $v_{\pm 1} \equiv \pm 1$. ⌋

These two norms allow us to define two different distances between sets, dist_{Z_ρ} and dist_{Z_u}.

Definition 8.3.21. *We define*

$$\begin{aligned} \mathrm{dist}_{Z_\rho}(b, A) &= \inf_{a \in A} \|b - a\|_\rho, \\ \mathrm{dist}_{Z_u}(b, A) &= \inf_{a \in A} \|b - a\| = \inf_{a \in A} \sup_{y \in \mathbb{R}} \|T_y b - T_y a\|_\rho. \end{aligned}$$

For both distances we let $\mathrm{dist}(B, A) = \sup_{b \in B} \mathrm{dist}(b, A)$ *for* $B \subset Z$.

With these preparations we can show the existence of an (Z_u, Z_ρ)-attractor [**MS95**, Theorem 2.6].

Theorem 8.3.22. *Let Z_u, T_y, and Z_ρ be given as above such that (A1) holds. Moreover, let $\mathcal{S}_t$ be a C_0-semigroup on Z_u which is translational invariant ($T_y \mathcal{S}_t = \mathcal{S}_t T_y$) and has a nonempty, bounded, and positively invariant set $\mathcal{B} \subset Z_u$. Assume that the following additional assumptions hold:*

(A2) (localized continuity) *For each $t \geq 0$ the evolution operator $\mathcal{S}_t$ is continuous from Z_ρ into itself.*

(A3) (compactness) *For all subsets $B \subset \mathcal{B}$ there is a $t_0 > 0$ such that $\mathcal{S}_{t_0}(B)$ is pre-compact in Z_ρ.*

Then there exists an unique (Z_u, Z_ρ)-attractor $\mathcal{A}$ for $\mathcal{S}_t$ in $\mathcal{B}$. If, additionally, $\mathcal{B}$ is an absorbing set, then $\mathcal{A}$ is the global attractor. Moreover, $\mathcal{A}$ has the following properties:

(i) $\mathcal{A}$ is translationally invariant, i.e., $T_y \mathcal{A} = \mathcal{A}$ for all $y \in \mathbb{R}$.

(ii) Every $B \subset \mathcal{B}$ which is bounded in Z is attracted to $\mathcal{A}$ w.r.t. the distance $\mathrm{dist} Z_\rho$, *i.e.,*

$$\mathrm{dist}_{Z_\rho}(\mathcal{S}_t(B), \mathcal{A}) := \sup_{b \in B} \inf_{a \in \mathcal{A}} \|\mathcal{S}_t(b) - a\|_\rho \to 0 \quad \text{for } t \to \infty.$$

Proof. W.l.o.g. we can assume that the set $\mathcal{B}$ is translational invariant, i.e., $T_y \mathcal{B} = \mathcal{B}$ for all $y \in \mathbb{R}$. If not, take $B_1 = \cup_{y \in \mathbb{R}} T_y \mathcal{B}$ instead of $\mathcal{B}$. The attracting set is defined by

$$\mathcal{A} = \bigcap_{t \geq 0} A_t \quad \text{with } A_t = \overline{\mathcal{S}_t(\mathcal{B})}^\rho.$$

As $\mathcal{B}$ is positively invariant, the family $(A_t)_{t \geq 0}$ is a decreasing family, i.e., $A_{t_1} \subset A_{t_2}$ for $t_1 > t_2$. Hence, $\mathcal{A} \subset A_0$ and $A_0 = \overline{\mathcal{B}}^\rho$ is bounded by Lemma 8.3.19. Therefore, $\mathcal{A}$ is bounded in Z_u. Moreover, from (A3) the set A_{t_0}, and hence all A_t for $t \geq t_0$ are compact in Z_ρ. Thus, $(A_t)_{t \geq t_0}$ forms a decreasing family of compact and nonempty sets in Z_ρ, and $\mathcal{A} = \cap_{t\geq t_0} A_t$ is nonempty and compact in Z_ρ. As $\mathcal{A}$ is closed in Z_ρ it is also closed in Z_u. This proves part a) in the Definition 8.3.18 for a (Z_u, Z_ρ)-attractor.

As $T_y\mathcal{S}_t(\mathcal{B}) = \mathcal{S}_t(T_y\mathcal{B}) = \mathcal{S}_t(\mathcal{B})$ we find by taking the closure in Z_ρ and by using the boundedness of T_y the relation $T_yA_t = A_t$. This implies (i).

The more difficult part of the proof is to show that $\mathcal{A}$ is in fact an attractor. It remains to show the time invariance and the attractivity, i.e., part b) and c) in Definition 8.3.18.

1) (Time invariance) Let $v \in \mathcal{S}_t(\mathcal{A})$, i.e., there is a $u = \lim_{t_n\to\infty} \mathcal{S}_{t_n}(u_n)$ in Z_ρ with $u_n \in \mathcal{B}$ such that $v = \mathcal{S}_t(u)$. Because of (A2) (continuity of $\mathcal{S}_t$ in Z_ρ), we have $\mathcal{S}_{t_n}(\mathcal{S}_t(u_n)) = \mathcal{S}_t(\mathcal{S}_{t_n}(u_n)) \to \mathcal{S}_t(u) = v$ as $n \to \infty$ in Z_ρ. As $\mathcal{S}_t(u_n) \in \mathcal{B}$ we conclude $v \in \mathcal{A}$ and hence $\mathcal{S}_t(\mathcal{A}) \subset \mathcal{A}$.

For the opposite direction (and the attractivity discussed below) the compactness in Z_ρ plays a crucial role. Let $v \in \mathcal{A}$, then there exist $t_n \to \infty$ and $u_n \in \mathcal{B}$ with $t_n < t_{n+1}$ and $v = \lim_{t_n\to\infty} \mathcal{S}_{t_n}(u_n)$ in Z_ρ. For any $t > 0$ we wish to show $v \in \mathcal{S}_t(\mathcal{A})$. From (A3) the set $\{\mathcal{S}_{t_n-t}(u_n) : t_n - t \geq t_0(\mathcal{B})\} \subset \mathcal{S}_{t_0(\mathcal{B})}(\mathcal{B})$ is pre-compact in Z_ρ. Therefore, for a subsequence $w_i = \mathcal{S}_{t_{n_i}-t}(u_{n_i}) \to w$ in Z_ρ. Applying the continuous map $\mathcal{S}_t$ we find $v = \mathcal{S}_t(w)$. As $w_j \in A_{t_{n_i}-t}$ for $j \geq i$ and all A_t are closed, w lies in all A_t and hence in $\mathcal{A}$. Thus, $v = \mathcal{S}_t(w) \in \mathcal{S}_t(\mathcal{A})$ and $\mathcal{A} \subset \mathcal{S}_t(\mathcal{A})$ is proved.

2) (Attractivity in Z_ρ) We use the compactness to give a proof by contradiction. Let $B \subset \mathcal{B}$ be arbitrary. Assume that B is not attracted to $\mathcal{A}$, then there exist $C > 0$, sequences $t_n \to \infty$ and $u_n \in B$ such that $\mathrm{dist}_{Z_\rho}(\mathcal{S}_{t_n}(u_n), \mathcal{A}) > C > 0$ for all $n \in \mathbb{N}$. Because of compactness there is a subsequence such that $v_i = \mathcal{S}_{t_{n_i}}(u_{n_i})$ converges in Z_ρ to w. As w lies in $\mathcal{A}$ this is a contradiction. Hence, we have shown $\mathrm{dist}_{Z_\rho}(\mathcal{S}_t(B), \mathcal{A}) \to 0$ for $t \to \infty$, for all $B \subset \mathcal{B}$. □

Remark 8.3.23. We remark that due to the translation invariance of $\mathcal{A}$ we have convergence in every translated Z_ρ-norm, too, i.e.,

$$\mathrm{dist}^*_{Z_\rho}(\mathcal{S}_t(\mathcal{B}), \mathcal{A}) = \sup_{b\in B} \sup_{y\in\mathbb{R}} \inf_{a\in\mathcal{A}} \|T_y\mathcal{S}_t(b) - T_ya\|_\rho \to 0 \quad \text{for } t \to \infty.$$

We further remark that if Z_{ρ_1} and Z_{ρ_2} define equivalent norms in Z_u via (A1) then for the attractors constructed via Z_{ρ_i} we have $\mathcal{A}_1 = \mathcal{A}_2$. ⌋

Example 8.3.24. We return to Example 8.3.17 and consider the linear diffusion equation with Z, Z_ρ, and $\mathcal{A} = \{u \equiv \theta : \theta \in [-R, R]\}$. Hence, obviously all assumptions of Theorem 8.3.22 are met and we have attractivity in Z_ρ as well as in the intermediate distance $\mathrm{dist}^*_{Z_\rho}$. To substantiate the theory we consider solutions $u(t) = \mathcal{S}_t(u_0)$ with arbitrary initial conditions $u_0 \in \mathcal{B}$. Using the explicit solution formula we easily find $\|\partial_x u(t)\|_\infty \leq CR/\sqrt{t}$ for all $u_0 \in \mathcal{B}$. Choosing $\theta(y) = u(t,y)$, we obtain

$$\begin{aligned}\|T_y u(t) - \theta(y)\|_\rho &\le \sup_{x\in\mathbb{R}} \frac{1}{1+x^2}|u(t,x+y) - u(t,y)| \\ &\le \sup_{x\in\mathbb{R}} \frac{1}{1+x^2}\frac{C}{\sqrt{t}}|x|\|u_0\|_\infty \le C/\sqrt{t}.\end{aligned}$$

Thus, we conclude $\mathrm{dist}_{Z_\rho}(\mathcal{S}_t(\mathcal{B}),\mathcal{A}) = \mathrm{dist}^*_{Z_\rho}(\mathcal{S}_t(\mathcal{B}),\mathcal{A}) \le CR/\sqrt{t}$. Hence, for this linear case we obtain an explicit decay rate towards the attractor. ⌋

Due to the smoothing and global existence results for the GL equation (8.29) we then have

Theorem 8.3.25. *For every $\theta \ge 0$ the GL equation (8.29) has a global $(H^\theta_{\mathrm{ul}}, H^\theta_\rho)$-attractor $\mathcal{A}_{\mathcal{G}}$ which satisfies attractivity in $\mathrm{dist}^*_{H^\theta_\rho}$, is translational invariant, and invariant under the rotations $R_\phi : A \mapsto \mathrm{e}^{\mathrm{i}\phi}A$.*

8.3.5. GL rolls and their linearized stability. We now turn to spatially periodic solutions of the GL equation (8.29), so called rolls, and their the spectral stability. See also the exercises for further explicit solutions of the GL equation. The rolls can easily be found by the ansatz

$$u(x,t) = a\mathrm{e}^{\mathrm{i}(qx+\omega t)}, \quad a \in \mathbb{C}, \quad q \in \mathbb{R},$$

which yields $\mathrm{i}\omega a = \left[-(1+\mathrm{i}\alpha)q^2 + 1 - (1+\mathrm{i}\beta)|a|^2\right]a$. Separating real and imaginary part gives the conditions

$$|a|^2 = 1 - q^2 \quad \text{and} \quad \omega = -\alpha q^2 - \beta|a|^2. \tag{8.32}$$

Thus, we have a family of stationary solutions parameterized by the phase of $a \in \mathbb{C}$ and the wave number $q \in \mathbb{R}$ with $q^2 < 1$. We set $a = r\mathrm{e}^{\mathrm{i}\theta}$ with $r^2 = 1 - q^2$ and write $u_{q,\theta}$ for these rolls.

The spectral stability of a fixed $u_{q,\theta}$ is determined by the linearization around $u_{q,\theta}$ which reads

$$\partial_t\widetilde{v} = (1+\mathrm{i}\alpha)\partial_x^2\widetilde{v} + \widetilde{v} - (1+\mathrm{i}\beta)(2|u_{q,\theta}|^2\widetilde{v} + u_{q,\theta}^2\overline{\widetilde{v}}).$$

Due to the occurrence of $u_{q,\theta}^2$ this linear system has x-dependent coefficients. However, due to the S^1-symmetry of (8.29) the x-dependence can be removed by using the ansatz

$$\widetilde{v} = v\mathrm{e}^{\mathrm{i}(qx+\omega t)}.$$

We obtain

$$\partial_t v = (1+\mathrm{i}\alpha)(\partial_x^2 v + 2\mathrm{i}q\partial_x v) - (1+\mathrm{i}\beta)r^2(v+\overline{v}) \tag{8.33}$$

by using (8.32). The real and imaginary part of $v = v_r + \mathrm{i}v_i$ satisfy

$$\partial_t\begin{pmatrix} v_r \\ v_i\end{pmatrix} = \begin{pmatrix} 1 & -\alpha \\ \alpha & 1\end{pmatrix}\partial_x^2\begin{pmatrix} v_r \\ v_i\end{pmatrix} + 2q\begin{pmatrix} -\alpha & -1 \\ 1 & -\alpha\end{pmatrix}\begin{pmatrix} v_r \\ v_i\end{pmatrix} - 2r^2\begin{pmatrix} v_r \\ \beta v_r\end{pmatrix}. \tag{8.34}$$

This system is solved by $v(x,t) = e^{\lambda t + i\ell x} w$ with $w \in \mathbb{C}^2$, which yields the algebraic system $L(\lambda, q, \ell) w = 0$ with

$$L(\lambda, q, \ell) = \begin{pmatrix} -\ell^2 - 2q\alpha i\ell - 2r^2 - \lambda & \alpha\ell^2 - 2qi\ell \\ -\alpha\ell^2 + 2qi\ell - 2r^2\beta & -\ell^2 - 2q\alpha i\ell - \lambda \end{pmatrix}. \tag{8.35}$$

In order to find non-trivial solutions we need to solve $\det L(\lambda, q, \ell) = 0$. For fixed ℓ we have two eigenvalues $\lambda_{1,2}(\ell)$ which we order such that $\mathrm{Re}\lambda_1(\ell) > \mathrm{Re}\lambda_2(\ell)$.

We always have $\lambda_1(0) = 0$. Eckhaus [**Eck65**] recognized that depending on q an instability close to $\ell = 0$ can occur. As an example we start with $\alpha = \beta = 0$ where we have

$$\lambda_1(\ell) = -(r^2 + \ell^2) + \sqrt{r^4 - 4q^2\ell^2}.$$

To detect the instability close to $\ell = 0$ we expand $\lambda_1(\ell) = 0 + \lambda_1'(0)\ell + \frac{1}{2}\lambda_1''(0)\ell^2 + \mathcal{O}(\ell^3)$ and find

$$\lambda_1'(0) = 0 \quad \text{and} \quad \lambda_1''(0) = -2 + 4q^2/(1-q^2).$$

The condition for instability $\mathrm{Re}\lambda_1''(0) > 0$ then yields

$$q^2 > 1/3. \tag{8.36}$$

This instability is called sideband or Eckhaus instability [**Eck65**]. See Figure 8.6b).

For $\alpha, \beta \neq 0$ the calculations become somewhat more involved. Exercise 8.22 yields $\lambda_1'(0) = 2iq(\beta - \alpha)$ and

$$\lambda_1''(0) = -\frac{2}{1-q^2}(1 + \alpha\beta - q^2(3 + \alpha\beta + 2\beta^2)). \tag{8.37}$$

Thus, we find $\mathrm{Re}\lambda_1(\ell) > 0$ for small ℓ if the generalized Eckhaus criterion

$$q^2 > \frac{1 + \alpha\beta}{3 + \alpha\beta + 2\beta^2} \tag{8.38}$$

is satisfied. In particular, there are no stable rolls at all in the so called Benjamin-Feir unstable regime, see Figure 8.6 c),

$$\alpha\beta < -1. \tag{8.39}$$

8.3.6. Diffusive stability of Eckhaus-stable rolls. Even for spectrally stable rolls we have spectrum up to the imaginary axis and so the principle of linearized instability does not apply. However, it turns out that spectrally stable equilibria are stable in the nonlinear system w.r.t. small spatially localized perturbations. Here we give some heuristic arguments why this is true for the Eckhaus-stable solution $A = 1$ for the real GL equation, i.e., in case $\alpha = \beta = 0$. Later on in §14 this will be explained in more detail.

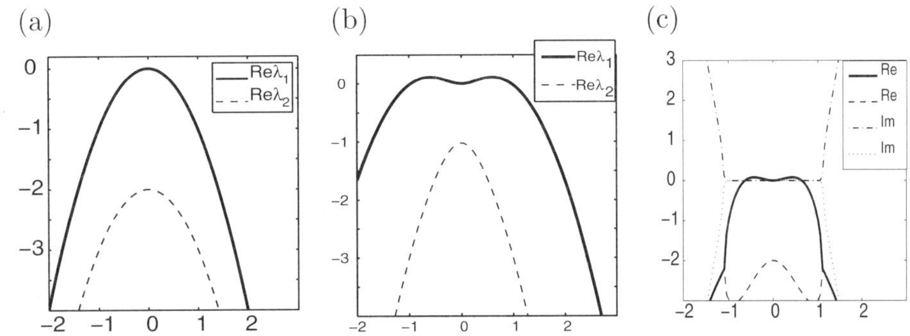

Figure 8.6. The curves of eigenvalues $\ell \mapsto \lambda_j(\ell)$. (a) $\alpha = \beta = 0$ and $q = 0$, stable case. (b) $\alpha = \beta = 0$ and $q = 0.7 > 1/\sqrt{3}$, Eckhaus instability. (c) $\alpha = -2, \beta = 1$ and $q = 0$, Benjamin-Feir instability.

We introduce polar coordinates $u(x,t) = r(x,t)\,\mathrm{e}^{\mathrm{i}\phi(x,t)}$ and find r and ϕ to satisfy

$$\partial_t \phi = \partial_x^2 \phi - 2\frac{\partial_x r}{r}\partial_x \phi, \tag{8.40}$$

$$\partial_t r = \partial_x^2 r + r\big(1-r^2-(\partial_x \phi)^2\big). \tag{8.41}$$

The deviation (ϕ, s) from the stable equilibrium $(1,0)$ defined by $r = 1 + s$ satisfies

$$\partial_t \phi = \partial_x^2 \phi - 2\frac{\partial_x s}{1+s}\partial_x \phi, \tag{8.42}$$

$$\partial_t s = \partial_x^2 s - 2s - 3s^2 - s^3 - (\partial_x \phi)^2 - s(\partial_x \phi)^2. \tag{8.43}$$

Therefore, on the linear level we have

$$\partial_t \phi = \partial_x^2 \phi \quad \text{and} \quad \partial_t s = \partial_x^2 s - 2s,$$

i.e., ϕ behaves diffusively, while s is exponentially damped. For spatially localized initial conditions we have that

$$\phi(x,t) = \frac{\phi^*}{\sqrt{t}}\mathrm{e}^{-x^2/4t} + \mathcal{O}(1/t)$$

for a constant $\phi^* \in \mathbb{R}$, such that $\phi \sim t^{-1/2}$, $\partial_t \phi \sim t^{-3/2}$, and $\partial_x^2 \phi \sim t^{-3/2}$ for $t \to \infty$. It turns out that s is slaved by ϕ and that in lowest order $s = -(\partial_x \phi)^2/2 \sim t^{-2}$ for $t \to \infty$. Therefore, (8.42) and (8.43) asymptotically behave like

$$\partial_t \phi = \partial_x^2 \phi + \mathcal{O}(t^{-7/2}), \tag{8.44}$$

$$\mathcal{O}(t^{-1}) = -2s - (\partial_x \phi)^2 + \mathcal{O}(t^{-1}) \tag{8.45}$$

for $t \to \infty$. Hence, all other terms vanish faster than those that we claimed to describe the asymptotic behavior. This argument can be made rigorous and the asymptotics

$$\phi(x,t) = \frac{\phi^*}{\sqrt{t}} e^{-x^2/4t} + \mathcal{O}(1/t) \qquad \text{and} \qquad s(x,t) = \mathcal{O}(1/t^2)$$

can be established. This has been done first in [**CEE92, BKL94**]. This kind of stability will be presented in detail in Chapter 14.

Further Reading. Besides its importance for applications, the NLS equation also plays a big role inside mathematics. Bourgain (in 1994) and Tao (in 2006) received Fields medals partly for work about the local and global existence and uniqueness of solutions in spaces of functions with low regularity, see, e.g., [**Bou99, Tao06**]. There are numerous books and review papers about the NLS equation, e.g. [**AS81, DJ89, SS99b, BK00, Fib15**]. A major result is that the NLS equation is a completely integrable Hamiltonian system. There exist so called inverse scattering schemes, namely the Ablowitz-Kaup-Newell-Segur (AKNS) scheme and the Zakharov-Shabat (ZS) scheme which allow to solve the NLS equation explicitly, cf. [**AS81, DJ89**]. The so called Birkhoff normal form for the NLS equation in case of periodic boundary conditions is discussed in [**GK14**], which also contains a modern account of the Hamiltonian formalism, of the Poisson bracket, and of the Lax pair formalism for the NLS equation in terms of the ZS operator.

Besides the motivation by the water wave problem, cf. §8.2.1, the KdV equation can be derived as an approximation to various other physical systems, see also [**Cri95**] and Chapter 12. The question occurs which phenomena of the KdV equation are robust under perturbations. Viewing the KdV equation as a completely integrable Hamiltonian system, this is analogous to the questions studied by the Kolmogorov-Arnold-Moser (KAM) theory, cf. §4.4, and has led to a development of KAM-like results for a number of different PDEs like the KdV equation. If one considers the KdV equation with periodic boundary conditions, temporally periodic or quasi-periodic solutions will persist under small perturbations, cf. [**KP03**]. The situation is more complicated and less well-understood for the equation on the whole line. The local and global existence of low regularity solutions of the KdV equation and related equations is a very active field of research. As already said, see, e.g., [**Tao06**, §4], [**LP09**, Ch. 7], or [**Koc15**, Ch. 6] for modern accounts.

The GL equation as a universal modulation equation occurs at the end of the 1960s in a number of papers, cf. [**NW69, Seg69, dES71**], cf. Chapter 10. There exist many further explicit solutions of (8.29), cf. [**AK02, Mie02**]. A famous example are the so called Bekki-Nozaki holes, see [**BN85, vSH92**]. There also exist stability results for such rather complicated solutions of the GL equation, cf. [**BNSZ14**]. Definition 8.3.18 about attractors on unbounded domains is a special case of that in [**BV90, BV92**]. The role of the GL equation in superconductivity, especially the occurrence of vortices, is discussed in [**BBH94, HT00, SS07, Ser15**].

Exercises

8.1. Find the coefficients β_1, β_2, and β_3 in terms of ν_1 and ν_2 for the transformations which bring (8.1) into the form (8.2).

8.2. Let $u = u_r + iu_i$ and write the NLS equation as a system of two real equations for (u_r, u_i). Let $u = re^{\mathrm{i}\phi}$ and find the equations for $(r, \phi)(x, t)$.

8.3. Show that, additionally to the usual translation invariance, the NLS equation is S^1-, Galilei-, and scaling-invariant, i.e., if $u = u(x, t)$ solves the NLS equation, then, for all $\phi, c, \eta \in \mathbb{R}$,

$$\begin{aligned} v(x,t;\phi) &= u(x,t)\mathrm{e}^{\mathrm{i}\phi}, && S^1\text{-invariance,} \\ v(x,t;c) &= u(x-ct,t)\mathrm{e}^{\mathrm{i}(c^2t-2cx)/4}, && \text{Galilei (or boost) -invariance,} \\ v(x,t;\eta) &= \eta u(\eta x, \eta^2 t), && \text{scaling-invariance,} \end{aligned}$$

solve the NLS equation, too.

8.4. Show that for arbitrary $\eta, c, \gamma, x_0 \in \mathbb{R}$ the pulse solutions

$$u(t,x;\eta,c,\gamma,x_0) = \sqrt{2}\eta\,\mathrm{sech}(\eta(x-x_0-ct))\mathrm{e}^{\mathrm{i}((c^2-4\eta^2)t-2cx+\gamma)/4}$$

are exact solutions of the focusing NLS equation.

8.5. a) Show that $u(x,t) = \dfrac{1-4(1+2\mathrm{i}t)}{1+2x^2+4t^2}\mathrm{e}^{\mathrm{i}t}$ and the Peregrine soliton $u_P(x,t) = \left(1 - \dfrac{4(1+2\mathrm{i}t)}{1+4x^2+4t^2}\right)\mathrm{e}^{\mathrm{i}t}$ are exact solutions of the focussing NLS.
b) Sketch the solutions (e.g., $|u|$) in the x–t plane. Does u have finite H? Does u_P?

8.6. Show that the curve of solutions $t \mapsto u(t, u_0)$ of the linear Schrödinger equation is continuous in $X = H^m(\mathbb{R}, \mathbb{C})$ with $m > 1/2$ if $u_0 \in X$.

8.7. Prove by direct calculation the conservation of the Hamiltonian (8.8) for the NLS equation, i.e., that $\frac{\mathrm{d}}{\mathrm{d}t}H(u(t)) = 0$.

8.8. Consider $\partial_t u = \mathrm{i}\partial_x^4 u$, $u = u(x,t) \in \mathbb{C}$. Show that:
a) This generates a C_0-semigroup $u(t) = \mathrm{e}^{\mathrm{i}t\partial_x^4}u_0$ in $L^2(\mathbb{R})$ with $\|u(t)\|_{L^2} = \|u_0\|_{L^2}$.
b) If $u_0 \in L^1(\mathbb{R})$, then $\|u(t)\|_{L^\infty} \leq Ct^{-1/4}\|u_0\|_{L^1}$.

8.9. Use Lemma 8.1.8 to establish a L^1-L^∞ estimate for the solutions of the discrete NLS equation $\mathrm{i}\partial_t u_n = -u_{n+1} + 2u_n - u_{n-1}$ with $n \in \mathbb{Z}$.
a) Solve the discrete NLS equation by the ansatz $u_n(t) = \mathrm{e}^{\mathrm{i}kn}\mathrm{e}^{\mathrm{i}\omega(k)t}$ with $k \in \mathbb{R}$.
b) Write down the solution in Fourier space as $\widehat{u}(k,t) = \widehat{G}(k,t)\widehat{u}(k,0)$.
c) Write down the solution in physical space as $u_n(t) = \sum_{m\in\mathbb{Z}} G_{n-m}(t)u_m(0)$, with $G_n(t) = \frac{1}{2\pi}\int_{-\pi}^{\pi} \mathrm{e}^{\mathrm{i}t\omega(k)}\mathrm{e}^{-\mathrm{i}kn}\,\mathrm{d}k$, and estimate $\sup_{n\in\mathbb{Z}}|G_n(t)|$ by separating the interval $[-\pi, \pi]$ in intervals where Lemma 8.1.8 can be applied.
d) Conclude that $\|(u_n(t))_{n\in\mathbb{Z}}\|_{\ell^\infty} \leq Ct^{-1/3}\|(u_n(0))_{n\in\mathbb{Z}}\|_{\ell^1}$.

8.10. Consider the traveling wave equation $-cv + v'' - 3v^2 = D$ for the KdV equation. Depending on c, D, discuss the existence of periodic travelling waves, and of solutions which have $\lim_{\xi\to\pm\infty} v(\xi) = D$ and $v(0) < D$.

8.11. Prove directly that the energy $\int_{\mathbb{R}} u^2(x,t)\,\mathrm{d}x$ is a conserved quantity for the KdV equation.

8.12. Find $\alpha \in \mathbb{R}$ such that $u(x,t) = \alpha c \operatorname{sech}^2\left(\sqrt{c}(x-4ct)\right)$, with arbitrary $c>0$ is an exact solution of KdV.

8.13. Let $L(t){=}L(t)^T{\in}\mathbb{R}^{d\times d}$ satisfy $L(0) = U(t)^T L(t) U(t)$ with $U(t)U(t)^T = I$. Show that $L = L(t)$ satisfies

$$\partial_t L(t){=}M(t)L(t){-}L(t)M(t)$$

with

$$M(t){=}-\left(\partial_t U(t)\right)U(t)^T{=}U(t)\partial_t U(t)^T{=}-M(t)^T,$$

and that as a consequence the eigenvalues of $L = L(t)$ are independent of t.

8.14. Consider the solution $u(x,t) = -12\dfrac{3+4\cosh(2\xi+24t)+\cosh(4\xi)}{(3\cosh(\xi-12t)+\cosh(3\xi+12t))^2}$ of the KdV equation, where $\xi = x-16t$. Prove that for $t\to\pm\infty$ this solution separates into two single waves, i.e., $u(x,t) \sim -8\operatorname{sech}^2\left(2\xi \mp \frac{1}{2}\log 3\right) - 2\operatorname{sech}^2\left(\eta \pm \frac{1}{2}\log 3\right)$ with $\eta = x-4t$, where $\sim$ means asymptotically equal.

8.15. Show that $\int xu + 3t^2u^2\,\mathrm{d}x$ is conserved for the KdV equation.

8.16. Show that the Kadomtsev-Petviashvili (KP) equation $\partial_x(\partial_t u{-}6u\partial_x u{+}\partial_x^3 u){+}3\partial_y^2 u = 0$ follows from the Lax pair $L = -\partial_x^2 + \partial_y + u$ and $M = -4\partial_x^3 + 6u\partial_x + 3u_x + 3\int^x \partial_y u\,\mathrm{d}\xi$.

8.17. Let $u = u_r + \mathrm{i}u_i$ and rewrite the GL equation (8.29) as a real system for (u_r, u_i).

8.18. Show that the GL equation $\partial_t u = \nu_2\partial_x^2 u + \nu_0 u - \nu_3|u|^2u$, with $\nu_j \in \mathbb{C}$, $\mathrm{Re}\nu_0, \mathrm{Re}\nu_2, \mathrm{Re}\nu_3 > 0$, and $x\in\mathbb{R}$, can be rescaled to the standard form

$$\partial_t u = (1+\mathrm{i}\alpha)\partial_x^2 u + u - (1+\mathrm{i}\beta)|u|^2u, \quad \alpha,\beta\in\mathbb{R}.$$

8.19. a) Show local existence of solutions of the GL equation (8.29) for $u_0 \in H^1(\mathbb{R})$. b) Derive a priori estimates to show global existence. Do solutions stay bounded in $H^1(\mathbb{R})$?

8.20. Show that the space of uniformly continuous functions can equivalently be characterized by $C^0_{b,unif}(\mathbb{R},\mathbb{R}) = \{\, u \in C^0_b(\mathbb{R},\mathbb{R}) \, : \, \|T_y u - u\|_{C^0_b} \to 0 \text{ as } y\to 0\,\}$.

8.21. Use the variation of constant formula and smoothing properties of the GL semigroup to establish a bound $\limsup_{t\to\infty}\|u(t)\|_{H^1_{\mathrm{ul}}} = \mathcal{O}(|\beta|)$ for $|\beta|\to\infty$ from $\limsup_{t\to\infty}\|u(t)\|_{L^2_{\mathrm{ul}}} \le \sqrt{4\pi\sqrt{2}}$ for the solutions of the GL equation.

8.22. Derive the expansion (8.37).

8.23. (Benjamin–Feir instability for the NLS equation) Discuss the stability of the solution $a(t) = a_0\mathrm{e}^{-\mathrm{i}a_0^2 t}$ of the NLS $\mathrm{i}\partial_t u - \partial_x^2 u - |u|^2u = 0$. For the linearization around a, make an ansatz $u(x,t) = a(t)(1+b(x,t))$ with $b(x,t) = b_1\mathrm{e}^{\mathrm{i}(\omega t + kx)} + b_2\mathrm{e}^{\mathrm{i}(\overline{\omega}t - kx)}$, and derive an algebraic system for the coefficients $b_{1,2}\in\mathbb{C}$. Show that this system possesses non-trivial solutions if $\omega^2 = 2k^2(a_0^2 - k^2)$, and from this derive that a is unstable w.r.t. perturbations with wave numbers k with $k^2 < a_0^2$.

Chapter 9

Reaction-Diffusion systems

In §7.1 we considered with the KPP and the Allen-Cahn equation the simplest examples of scalar nonlinear reaction-diffusion equations. In general, chemical reactions involve a number of different species, and so we consider in this section reaction-diffusion systems of the form

$$\partial_t u = D\Delta u + f(u), \tag{9.1}$$

with $t \geq 0$, $x \in \mathbb{R}^n$, $u(x,t) \in \mathbb{R}^N$, where $f : \mathbb{R}^N \to \mathbb{R}^N$ a smooth function, and $D \in \mathbb{R}^{N \times N}$ is a positive (semi-)definite diffusion matrix. Often D is diagonal with strictly positive diagonal elements $d_j > 0$, i.e., $D = \operatorname{diag}(d_1, \ldots, d_N)$.

Famous examples of two-component reaction-diffusion systems, where $a, b, d, \ldots$ denote parameters, are:
The Schnakenberg model [**Sch79**]

$$\partial_t u = \Delta u - u + u^2 v, \quad \partial_t v = d\Delta v + b - u^2 v. \tag{9.2}$$

The Brusselator [**PL68**]

$$\partial_t u = \Delta u + a - (b+1)u + u^2 v, \quad \partial_t v = d\Delta v + bu - u^2 v. \tag{9.3}$$

The Lengyel-Epstein model [**LE91**]

$$\partial_t u = \Delta u + a - u - \frac{4uv}{1+u^2}, \quad \partial_t v = d\Delta v + b(u - \frac{uv}{1+u^2}). \tag{9.4}$$

The Gray-Scott model [**GS83**]

$$\partial_t u = \Delta u - uv^2 + f(1-u), \quad \partial_t v = d\Delta v + uv^2 - (f+k)v. \tag{9.5}$$

The Gierer-Meinhard model [**GM72**],

$$(9.6) \qquad \partial_t u = \Delta u + a - u + u^2/v, \quad \partial_t v = d\Delta v + u^2 - v.$$

The FitzHugh-Nagumo model [**Fit69**]

$$(9.7) \qquad \partial_t u = \Delta u + u(1-u)(u-a) - v, \quad \partial_t v = d\Delta v + bu - cv.$$

While models (9.2)-(9.6) essentially are all from chemistry, (9.7) models the transmission of nerve impulses in some giant squid axon. Moreover, the Kolmogorov models from ecology, i.e., the 2-species interactions from Example 2.6.5, reappear as reaction-diffusion systems when combined with diffusion of one or both of the species.

Typical questions about reaction-diffusion systems are again the local and global existence of solutions in a suitable phase space, and the existence and stability of special solutions such as pulses and fronts. Concerning global existence, the crucial difference compared to scalar equations is that the maximum principle is no longer valid for systems. However, it can often be replaced by the method of invariant regions. Similarly, the existence of special solutions becomes more complicated due to the fact that the traveling wave system in general can no longer be written as a two-dimensional first order system, such that phase plane methods are no longer available but have to be replaced by more sophisticated methods, which often use so called trapping regions.

In §9.1 we start with modeling aspects and the local existence and uniqueness theory for reaction-diffusion systems, and discuss quadratic autocatalysis as an example of a two-dimensional system of the form (9.1). It turns out that when $d_1 = d_2$ such systems can be rescaled to the KPP equation from §7.1. We explain the concept of invariant regions which allows to prove global existence and uniqueness of solutions for many reaction-diffusion systems.

The FitzHugh-Nagumo system (9.7) and the Gray-Scott system (9.5) are used as examples in §9.2 to discuss, on a heuristic level, some typical phenomena of travelling pulses and pattern formation in reaction-diffusion systems. However, our main objective is the Turing instability of spatially homogeneous states, also called diffusion driven instability [**Tur52, Mur89**], which we consider in §9.3. This is a famous explanation for morphogenesis, which refers to pattern formation during growth in biological system. The bifurcating pattern very often can be described by the GL approximation, which will be studied in Part IV.

9.1. Modeling, and existence and uniqueness

Assume that a chemical system consists of d chemical species $\alpha_1, \ldots, \alpha_d$ which react according to the law of mass action,

$$\lambda_1\alpha_1 + \ldots + \lambda_d\alpha_d \to \mu_1\alpha_1 + \ldots + \mu_d\alpha_d.$$

The λ_i, μ_i are called stoichiometric coefficients. The reaction proceeds at rate

$$\dot{r} = k[\alpha_1]^{\lambda_1} \cdots [\alpha_d]^{\lambda_d}$$

where $u_i = [\alpha_i]$ is the dimensionless concentration of α_i, k is a constant which is often related to temperature T by Arrhenius law and which is proportional to $\exp(-E/(RT))$, where E is the so called activation energy of the reaction, and R is the universal gas constant. Conservation of mass yields

$$\dot{u}_i = (\mu_i - \lambda_i)\dot{r}. \tag{9.8}$$

An everyday example is $2H_2 + O_2 \to 2H_2O$. With $\alpha_1 = H_2, \alpha_2 = O_2, \alpha_3 = H_2O$ this yields $\lambda = (2, 1, 0), \mu = (0, 0, 2)$, and the reaction rate is proportional to $[H_2]^2[O_2]$. Thus, the ODE system for $u_1 = [H_2], u_2 = [O_2]$ and $u_3 = [H_2O]$ reads

$$\dot{u}_1 = -2ku_1^2u_2, \quad \dot{u}_2 = -ku_1^2u_2, \quad \dot{u}_3 = 2ku_1^2u_2.$$

Clearly, the last equation decouples from the first two and is usually dropped, since the product H_2O no longer takes part in the reaction, i.e., if u_1, u_2 are known, then $\dot{u}_3 = 2ku_1^2u_2$ can be solved by simple time integration. Alternatively, mass conservation can be used.

9.1.1. Local existence and uniquenss. As explained above, in spatially extended problems, i.e. $x \in \mathbb{R}^n$, this yields systems of the form (9.1). Before embarking on specific examples, we note the following local existence result, where as phase space we always choose

$$X = [C^0_{b,unif}(\mathbb{R})]^N, \quad \text{with norm} \quad \|u_0\|_X = \sup_{j=1,\ldots,N} \sup_{x\in\mathbb{R}} |u_j(x)|.$$

Moreover, we restrict to diagonal diffusion matrices $D = \mathrm{diag}(d_1, \ldots, d_N)$, with $d_i \geq 0$ the diffusion rate of the ith spezies, though the generalization to general positive (semi-)definite D is straightforward.

Theorem 9.1.1. *Let $u_0 \in X$. Then there exists a $t_0 > 0$ depending only on $\|u_0\|_X$ such that* (9.1) *has a unique (mild) solution $u \in C([0, t_0], X)$.*

Proof. The proof is very similar to the proof of Theorem 7.1.7. First we fix a $C_1 > 0$ and define the semigroup $\mathrm{e}^{tD\Delta}$ by

$$\mathrm{e}^{tD\Delta}u = \mathrm{diag}(\mathrm{e}^{td_1\partial_x^2}u_1, \ldots, \mathrm{e}^{td_N\partial_x^2}u_d),$$

where $e^{td_j\partial_x^2}u_j = \frac{1}{\sqrt{4\pi d_j t}}\int_{\mathbb{R}} e^{-\frac{x-y}{4d_jt}}u_j(y)\,dy$ if $d_j > 0$, cf. (7.6), and $e^{td_j\partial_x^2}u_j = u_j$ if $d_j = 0$. Then $e^{tD\Delta}$ is a C_0-semigroup in X and $\|e^{tD\Delta}u\|_X \le \|u\|_X$. Next we use the variation of constant formula and show that

$$F(u) = e^{tD\Delta}u_0 + \int_0^{t_0} e^{(t-s)D\Delta}f(u(s))\,ds$$

maps

$$M = C^0([0,t_0], \{u \in X : \|u - e^{tD\Delta}u_0\|_X \le C_1\}),$$

equipped with the norm

$$\|u\|_M = \sup_{0\le t\le t_0} \|u(t)\|_X,$$

into itself and is a contraction in M if $t_0 > 0$ is sufficiently small. □

Remark 9.1.2. If $d_j > 0$ for $j = 1, \dots, N$, then the solution $u(\cdot, t)$ becomes analytic for $t > 0$ and hence a classical solution. If one or more of the d_j are zero, then in general the solutions only stay as smooth as the initial conditions. ⌋

9.1.2. Quadratic autocatalysis. This section follows closely [**BK00**]. The reaction $A+B \to 2B$, with reaction rate $k[A][B]$, is called quadratic autocatalysis, since the greater the concentration of the catalyst B, the faster B is produced, i.e., B catalyses its own production, while it preys on A. With $a = [A]$ and $b = [B]$ this yields

$$\dot{a} = -kab, \quad \dot{b} = kab, \tag{9.9}$$

which, augmented by diffusion, will be studied below. By mass conservation, $\frac{d}{dt}(a+b) = 0$, hence $a+b = a_0+b_0$ and eliminating a yields $\dot{b} = kb(a_0+b_0-b)$. Separation of variables yields

$$a(t) = \frac{a_0(a_0+b_0)}{a_0 + b_0e^{k(a_0+b_0)t}}, \quad b(t) = \frac{b_0(a_0+b_0)e^{k(a_0+b_0)t}}{a_0 + b_0e^{k(a_0+b_0)t}},$$

such that $a \to 0$ and $b \to a_0 + b_0$ as $t \to \infty$. For all $b_0 > 0$, all of A is converted into B as $t \to \infty$. However, with diffusion the reduction to a scalar equation only works if the diffusion constants of a and b are equal, and thus the system form (9.9) must be kept. After suitable non-dimensionalization, the one-dimensional reaction-diffusion system associated to (9.9) reads

$$\partial_t a = \partial_x^2 a - ab, \quad \partial_t b = d\partial_x^2 b + ab, \tag{9.10}$$

where $d = D_B/D_A$ is the ratio of the diffusion constants. With $u = (a,b)$, (9.10) is of the form (9.1) with $D = \mathrm{diag}(1,d)$ and $f(u) = (-u_1u_2, u_1u_2)$. Local existence and uniqueness of solutions in $X = C_b^0 \times C_b^0$ follows from Theorem 9.1.1. To obtain global existence we need to bound $\|u(t)\|_X$. Unfortunately, the maximum principle is no longer available, and thus it must be substituted by different methods which we postpone to §9.1.3.

We are interested in traveling waves for (9.10), where first we assume that A and B diffuse with the same rate, such that $d = 1$. The ansatz $(a, b)(x, t) = (v_1(x - ct), v_2(x - ct))$ for travelling waves yields

$$\partial_\xi^2 v_1 + c\partial_\xi v_1 - v_1 v_2 = 0, \quad \partial_\xi^2 v_2 + c\partial_\xi v_2 + v_1 v_2 = 0, \tag{9.11}$$

hence, with $w = v_1 + v_2$, we find $\partial_\xi^2 w + c\partial_\xi w = 0$, which can be integrated to $w(\xi) = c_0 + c_1 \mathrm{e}^{-c\xi}$ where c_0, c_1 are constants of integration. Looking for a front with $w \to 1$ as $\xi \to \infty$ and w bounded for $\xi \to -\infty$ we obtain $c_0 = 1$ and $c_1 = 0$, hence $w \equiv 1$, and (9.11) reduces to the single equation

$$\partial_\xi v_2 + c\partial_\xi v_2 + v_2(1 - v_2) = 0$$

which is the traveling wave equation for the KPP equation and the analysis of §7.1 applies. Thus, there are fronts for (9.11) connecting $(v_1, v_2) = (1, 0)$ with $(v_1, v_2) = (0, 1)$ for all $c \geq 2$.

If the diffusion constants of A and B are not equal, then the travelling wave ansatz $(a, b)(x, t) = (v_1(x - ct), v_2(x - ct))$ yields

$$\partial_\xi^2 v_1 + c\partial_\xi v_1 - v_1 v_2 = 0, \quad d\partial_\xi^2 v_2 + c\partial_\xi v_2 + v_1 v_2 = 0. \tag{9.12}$$

As before we assume that $(a, b) \to (1, 0)$ as $\xi \to \infty$ (unreacted state far ahead of the wave) and $(a, b) \to (0, 1)$ as $\xi \to -\infty$ (fully reacted state far behind the wave). We may still add v_1 and v_2 and integrate once to obtain

$$\partial_\xi v_1 + d\partial_\xi v_2 + c(v_1 + v_2) = c,$$

but further integration and thus reduction to the KPP equation is no longer possible. Instead, with $v_3 = \partial_\xi v_2$ we obtain the first order system

$$\partial_\xi v_1 = c(1 - v_1 - v_2) - dv_3, \quad \partial_\xi v_2 = v_3, \quad \partial_\xi v_3 = -(v_1 v_2 + cv_3)/d. \tag{9.13}$$

Thus, we are no longer able to use phase plane analysis to discuss the existence of a heteroclinic orbit for (9.13) connecting $(1, 0, 0)$ with $(0, 1, 0)$. However, essentially the same method as for the KPP equation works. We first consider the linearization at the fixed points and find that at $(0, 1, 0)$ we have a one-dimensional unstable manifold $M_u^{(1)}$, and at $(1, 0, 0)$ a three-dimensional stable manifold $M_s^{(2)}$, such that we may expect that $M_u^{(1)}$ and $M_s^{(2)}$ intersect. Moreover, all eigenvalues at $(1, 0, 0)$ are real for $c \geq 2\sqrt{d}$ which hence yields the minimal front speed to obtain a monotonic front. Finally, in order to show that $M_u^{(1)}$ and $M_s^{(2)}$ intersect we consider in Exercise 9.2 the wedge shaped trapping region

$$R = \{(v_1, v_2, v_3) : 0 \leq v_1 \leq 1, 0 \leq v_2 \leq 1, -cv_2/(2d) \leq v_3 \leq 0\},$$

and conclude on the existence of fronts for (9.10).

Thus, given initial conditions for (9.10) of the form $a_0 \equiv 1$ (or some constant) and $b_0 \geq 0$ some localized perturbation of 0 we expect a pair of fronts with the minimal speed $c = 2\sqrt{d}$ to emerge, one propagating to the left, one propagating to the right. Numerical simulations of (9.10) show that this is indeed the case, see Figure 9.1. Note that the critical front speed only depends on the diffusion coefficient of the autocatalyst B, which moreover determines the steepness of the front. We remark that the shape of the front $v_{\text{het}}(\xi)$ can be obtained by considering asymptotic limits, in particular the limit $0 < d \ll 1$, but for this and more analysis we refer to the literature, e.g., [**BK00**].

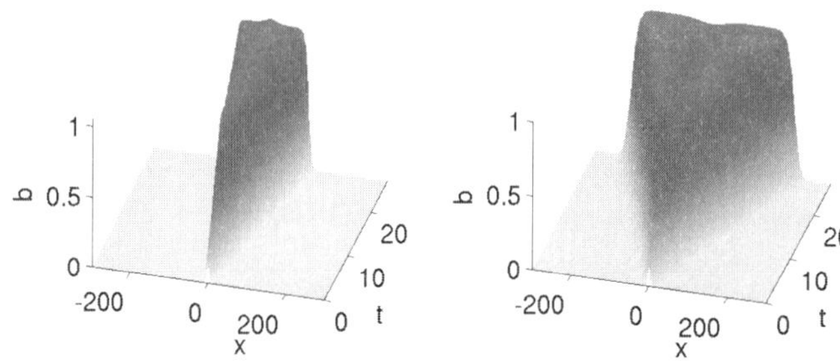

Figure 9.1. Fronts (b-component) for (9.10) with d=1 (left) and d=5 (right) for the initial condition $(u_0, v_0)(x) = (1, 0.2\,\text{sech}(x/2))$.

9.1.3. Invariant regions. For reaction-diffusion systems the maximum principle is no longer available. However, to show global existence in a similar way as for the KPP equation it can often be substituted by the method of contracting sets for the reaction term. This section follows [**Smo94**, §14], which treats the problem on general (bounded) spatial domains $\Omega \subset \mathbb{R}^m$, but since we restrict to the case $x \in \mathbb{R}$ our presentation is somewhat shorter.

Definition 9.1.3. *A closed bounded non-empty set $B \subset \mathbb{R}^N$ is called contracting for the vector field f if $f(u)$ points inward on ∂B, i.e., if $f(u) \cdot \nu(u) < 0$ for $u \in \partial B$, where ν is the outer normal to B.*

Contracting sets for the reaction f are often the first step to construct invariant regions for the reaction-diffusion system (9.1).

Definition 9.1.4. *A closed bounded non-empty set $\Sigma \subset \mathbb{R}^N$ is called positively invariant region for (9.1), if $u(x, 0) \in \Sigma$ for all $x \in \mathbb{R}$ implies $u(x, t) \in \Sigma$ for all $t > 0$ and $x \in \mathbb{R}$.*

The invariant regions we shall consider will be defined by functions $g_i \in C^2(\mathbb{R}^N, \mathbb{R})$, namely

$$\Sigma = \cap_{i=1}^m \{u \in \mathbb{R}^N : g_i(u) \leq 0\}. \tag{9.14}$$

Theorem 9.1.5. *For $u_0 \in \partial\Sigma$ with $g_i(u_0) = 0$ for some $i \in \{1, \ldots, N\}$, assume that the following holds.*

(1) $\partial_u g_i(u_0)$ *is a left eigenvector of* D*, i.e.,* $\partial_u g_i(u_0)D = \mu\partial_u g_i(u_0)$ *for some* $\mu \in \mathbb{R}$.

(2) If $\partial_u g_i(u_0)D = \mu\partial_u g_i(u_0)$ *with* $\mu \neq 0$*, then* g_i *is quasi-convex at* u_0*, i.e., if* $\partial_u g_i(u_0) \cdot \eta = 0$*, then* $\langle \partial_u^2 g_i(v_0)\eta, \eta\rangle \geq 0$.

(3) $\partial_u g_i(u_0) \cdot f(u_0) < 0$.

Then Σ *defined by* (9.14) *is invariant for* (9.1).

The **idea of the proof** is as for the maximum principle in §7.1.4. For a solution u coming to the boundary for (x_0, t_0) we have $\frac{\mathrm{d}}{\mathrm{d}t} g_i(u(x_0, t_0)) < 0$ such that the solution goes back to the interior of Σ. □

Remark 9.1.6. Since D is diagonal, any left eigenvector $\partial_u g$ of D is also a right eigenvector. If $D = \mathrm{diag}(d_1, \ldots, d_N)$ with $d_i \neq d_j$ for $i \neq j$, then the eigenvectors are obviously of the form $e_1 := (1, 0, \ldots, 0)$, $e_2 := (0, 1, 0, \ldots, 0)$, …, $e_N := (0, 0, \ldots, 0, 1)$, which shows that each g_i actually is a function of only one variable. Essentially, this means that Σ is a rectangle and that we consider the system in a "decoupled way". If, e.g., $d_i = d_j$ for exactly two $i \neq j$, then we have a two-dimensional eigenspace $\mathrm{span}\{e_i, e_j\}$ and hence also more freedom for g_i and g_j. ⌋

From Remark 9.1.6 we have the following corollary.

Corollary 9.1.7. *A rectangle*

$$\Sigma = \cap_{i=1}^N \{a_i \leq u_i \leq b_i\}$$

is invariant if $f(u)$ *points strictly inward on* $\partial\Sigma$*, i.e., if (3) in Theorem* 9.1.5 *holds.*

Proof. For notational simplicity we write $g_i = g$. We have, for $u = u(x, t)$,

$$\frac{\mathrm{d}}{\mathrm{d}t} g(u) = \partial_u g(u) \cdot \partial_t u = \partial_u g(u) \cdot (D\partial_x^2 u + f(u)).$$

By (1) we have, at $u = u(x_0, t_0)$, $\partial_u g(u)D = \mu\partial_u g(u)$, hence

$$\frac{\mathrm{d}}{\mathrm{d}t} g(u) = \partial_u g(u) \cdot (\mu\partial_x^2 u + f(u)). \tag{9.15}$$

Thus, let $\mu > 0$ and let $h(x) = g(u(x, t_0))$. Then $h(x_0) = h'(x_0) = 0$ and $h''(x_0) \leq 0$ by assumption on x_0. Therefore, $0 \geq h''(x_0) = \langle \partial_u^2 g \partial_x u, \partial_x u\rangle + \partial_u g \partial_x^2 u$. By (2) we have $\langle \partial_u^2 g \partial_x u, \partial_x u\rangle \geq 0$, thus $\partial_u g \partial_x^2 u \leq 0$, hence again $\frac{\mathrm{d}}{\mathrm{d}t} g(u) \leq \partial_u g(u) f(u) < 0$. □

Remark 9.1.8. Often, only rectangles Σ of the form (9.14) can be constructed such that f does not point outward on $\partial\Sigma$, i.e., f may vanish on $\partial\Sigma$ or may be tangential to $\partial\Sigma$. Then, instead of (3) in Theorem 9.1.5 we only have $\partial_u g_i(u_0) \cdot f(u_0) \leq 0$. However, under mild technical assumptions on f, which are fulfilled for smooth f, the assertion of Theorem 9.1.5 remains true, i.e., Σ is positively invariant, see [**Smo94**, Thm 14.11]. ⌋

9.2. Two classical examples

We discuss two specific examples of reaction-diffusion systems. The FitzHugh-Nagumo (FHN) system models transmission of nerve pulses in some giant squid axon. It is a simplification of the so called Hodgkin-Huxley model for the transmission of a nerve impulse though the axon of a giant squid's neuron, see [**BK00**] for the modeling. Therefore, additional to global existence, we shall in particular be interested in the existence of pulse solutions. The Gray-Scott model in §9.2.2 is a famous model for cubic autocatalysis which shows a multitude of different pattern. The details of the analysis of both, the FHN pulses and the Gray-Scott pattern are rather difficult. Therefore, here we only explain the basic phenomena using heuristic arguments and illustrations by numerical simulations.

9.2.1. The FitzHugh-Nagumo system. After suitable non-dimensionalization, the FHN system reads

$$\begin{aligned} \partial_t u &= \partial_x^2 u + f(u) - v, \\ \partial_t v &= d\partial_x^2 v + bu - cv, \end{aligned} \tag{9.16}$$

where $f(u) = u(1-u)(u-a)$, with parameters $a \in \mathbb{R}$, $b, c > 0$ and $d \geq 0$, where $u = u(x,t)$ is the voltage in the axon, and $v = v(x,t)$ is a so called recovery variable. From the modeling point of view the case $d = 0$ where only u diffuses makes sense. We also remark that here the modeling does not give the constraints $u, v \geq 0$.

The phase portrait for the ODE part

$$\dot{u} = f(u) - v, \quad \dot{v} = bu - cv, \tag{9.17}$$

of (9.16) is shown for some specific parameters in Figure 9.2. This is the

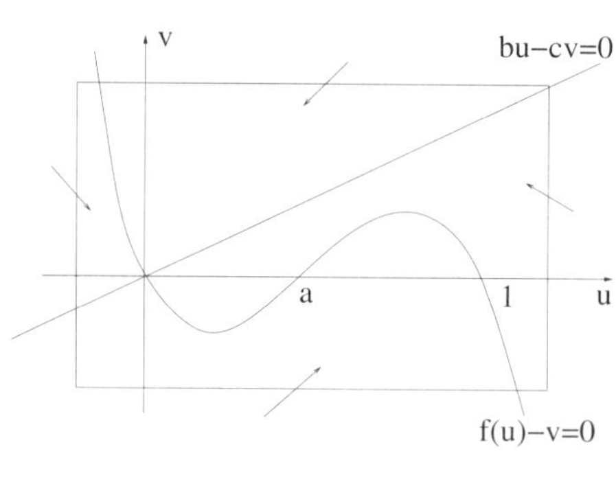

Figure 9.2. A contracting rectangle for (9.16)

so called excitable case with exactly one stable fixed point. In the so called oscillatory case the unique fixed point is unstable, while in the so called bistable case there are three fixed points, two stable and one unstable (see below).

In any case, there are large contracting sets, as illustrated by the rectangle in Figure 9.2. Thus, using Theorem 9.1.1, and Corollary 9.1.7 we conclude that for arbitrary $(u_0, v_0) \in X = [C^0_{b,unif}(\mathbb{R})]^2$ we have a global solution to (9.16).

From the modeling point of view, in suitable parameter regimes, (9.16) should possess pulse solutions, i.e., solutions of the form

$$(u, v)(x, t) = (\widetilde{u}, \widetilde{v})(x - ct)$$

with $(\widetilde{u}, \widetilde{v})(\xi) \to (\widetilde{u}, \widetilde{v})^*$ as $\xi \to \pm\infty$, where $(\widetilde{u}, \widetilde{v})^*$ is a fixed point of (9.17).

These pulses are found in the regime b, c and d small. Then (9.17) is often rescaled to the form

$$\varepsilon \dot{u} = f(u) - v, \quad \dot{v} = u - \gamma v, \tag{9.18}$$

where $0 < \varepsilon \ll 1$ is a small parameter. Since $\dot{u} = \frac{1}{\varepsilon}(f(u) - v)$ in (9.18) is large except if $f(u) - v = 0$, the u-dynamics is called fast in contrast to the slow dynamics in v.

Excitable dynamics. If $0 < a < 1/2$ and $\gamma < 4/(a-1)^2$, then $(u, v) = (0, 0)$ is the only fixed point of (9.18). It is asymptotically stable, with eigenvalues $\lambda_1 = -a/\varepsilon$ and $\lambda_2 = -1$. However, as is evident from the phase portrait in Figure 9.3, a finite perturbation with, e.g., $u > 0$ and $v < 0$ may lead to a large excursion with $u > 1$ before the orbit returns to $(0, 0)$ from the left. This is called excitable behavior. In particular, this happens for $\varepsilon > 0$ sufficiently small.

To understand this behavior we follow [**Kee88**, §12]. We first set $\varepsilon = 0$, which formally yields

$$v = f(u), \quad \dot{v} = u - \gamma v. \tag{9.19}$$

This is an implicit ODE for v. The algebraic equation $v = f(u)$ has three branches of solutions, namely

$$u = g_-(v), \quad u = g_0(v) \quad \text{and} \quad u = g_+(v),$$

see Figure 9.3. The lower/upper branches $g_-(v)$ and $g_+(v)$ exist only for $u < u^*_-$ and $u > u^*_+$, where $u^*_\pm = \frac{1}{2}(a + 1 \pm \sqrt{a^2 - a + 1})$, and the middle branch $g_0(v)$ exists for $u^*_- \le u \le u^*_+$. Equivalently, $g_-(v)$ and $g_+(v)$ exist for $v > v^*_- := f(u^*_-)$ and $v < v^*_+ := f(u^*_+)$, respectively. Thus, depending on the branch, (9.19) has three different meanings, namely $\dot{v} = g_\pm(v) - v$ or $\dot{v} = g_0(v) - v$. On the middle and upper branches, v increases monotonically, while $g_-(v) - v$ changes sign at $v = 0$.

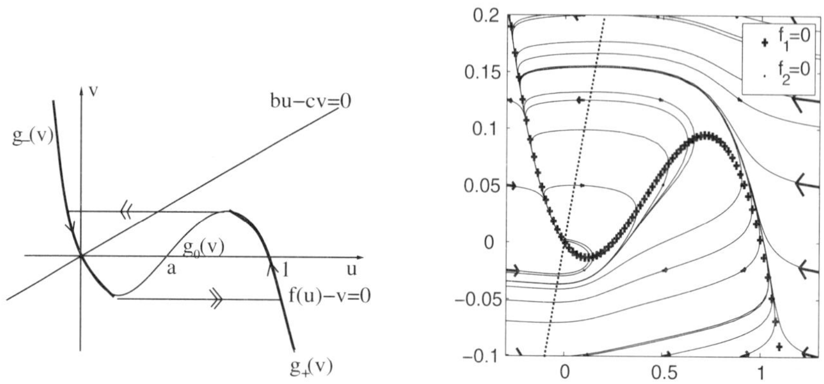

Figure 9.3. The mechanism of excitation for (9.16), explained in the text, and the phase portrait for (9.18) for $(a, \gamma, \varepsilon) = (0.25, 1, 0.01)$

On the other hand, in (9.18) we can make the change of variables $(U, V)(\tau) = (u, v)(\varepsilon\tau)$ to obtain

$$\dot{U} = f(U) - V, \quad \dot{V} = \varepsilon(U - V). \tag{9.20}$$

Now setting $\varepsilon = 0$ the slow dynamics in V vanish, i.e., $V = V_0$ some constant. Then $\dot{U} = f(U) - V_0$ and given $U_0 \neq g_0(V_0)$, $U(\tau)$ converges to a stable root of $f(U) - V_0$, i.e.,

$$\begin{aligned} U(\tau) &\to g_-(V_0), \quad \text{if} \quad U_0 < g_-(V_0), \\ U(\tau) &\to g_+(V_0), \quad \text{if} \quad U_0 > g_+(V_0). \end{aligned} \tag{9.21}$$

Given (u_0, v_0) we can now (heuristically) piece together the solution of (9.18) using the above two different scaling limits. Assume that $u > 0$ and $v < v_-^*$. Then v first stays essentially constant, and u rapidly relaxes to $u_+(v_0)$, according to (9.21). Then the slow dynamics in v kick in, i.e., v slowly increases on the upper branch. This branch ceases to exist when v reaches v_+^*, and the fast dynamics in u take over and rapidly take u to $u_-(v_+^*)$. After this, the solution slowly relaxes along the lower branch to the equilibrium $(u, v) = (0, 0)$.

This was a formal discussion. Nevertheless, it agrees very well with the phase portrait in Figure 9.4, and it can be made rigorous using so called singular perturbation theory. Due to the rapid excitation followed by a slow relaxation, this kind of dynamics is also called relaxation dynamics. If the initial perturbation of $(u, v) = (0, 0)$ is sufficiently small, e.g., $u < a$ and $v = 0$, then the solution directly relaxes back to $(u, v) = (0, 0)$. This is called sub-threshold dynamics.

Before we discuss how this is related to the existence of traveling pulses for the FHN PDE (9.16), we briefly explain two other regimes encountered in the FHN ODE (9.18).

Remark 9.2.1. Bistable and oscillatory dynamics. If $\gamma > 4/(a-1)^2$ in (9.18), then the nullclines $f(u) - v = 0$ and $u - \gamma v$ intersect in three points, namely $P = (0,0)$, and $Q = (u_-, u_-/\gamma)$ and $R = (u_+, u_+/\gamma)$ with $u_\pm = \frac{1+a}{2} \pm \sqrt{\frac{(1+a)^2}{4} - (a + 1/\gamma)}$. Now P and R are stable while Q is unstable, see Figure 9.5. The two new fixed points are created in a saddle (Q)-node (R) bifurcation at $\gamma = 4/(a-1)^2$. The phase portrait shows that for (u_0, v_0) a sufficiently large perturbation of P the solution converges to R, while for (u_0, v_0) a sufficiently large perturbation of R the solution converges to P.

If $a < 0$ and $\gamma > 4/(a-1)^2$, then we again have the unique fixed point $(u, v) = (0, 0)$, but now it lies on the unstable middle branch. The above (singular perturbation) analysis again applies and yields so called relaxation oscillations. ⌋

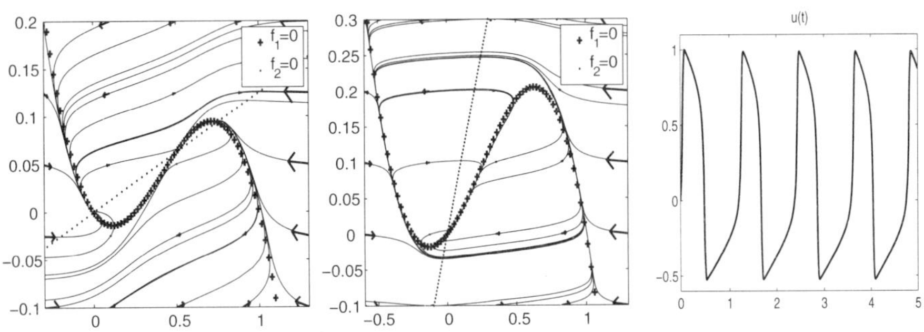

Figure 9.4. The bistable $(a, \gamma, \varepsilon) = (0.25, 8, 0.01)$ and the oscillatory $(a, \gamma, \varepsilon) = (-0.25, 1, 0.01)$ case: phase portrait and relaxation oscillations.

Travelling pulses for the FHN system. In the excitable regime we may expect pulse solutions for the FHN system. The heuristic reason is as follows. Assume that we start with a (suitable, sufficiently large) perturbation of $(u, v) \equiv (0, 0)$, localized near some x_0. Then, locally, the ODE dynamics wants to run the excitation. Meanwhile, diffusion pulls the solution beyond the threshold for excitation also in a neighborhood of the primary perturbation, and the process repeats. Thus, we may expect one or more (depending on the details of the IC) excitation pulses to emerge, see Figure 9.5 for two examples. Moreover, the pulses are asymptotically stable in the numerics, with a large basin of attraction as we do not need any fine tuning of the initial data.

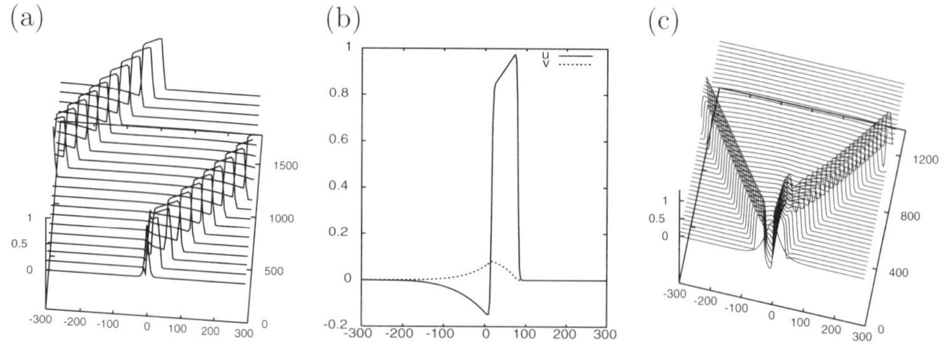

Figure 9.5. Numerical simulation of (9.16) on a large domain with periodic boundary conditions. (a) emergence of a pulse in the excitable regime, IC $(u_0, v_0)(x) = (\text{sech}(x/4), -\text{sech}(x/2)\sin(x/2))$, parameters $(a, b, c, d) = (1/4, 0.00075, 0.005, 0.1)$. (b) $(u, v)|_{t=1775}$. (c) for, e.g., $(u_0, v_0)(x) = (\text{sech}(x/10), -\text{sech}(x/5)\sin(x/5))$ we obtain two counter-propagating pulses, which annihilate upon collision. This behavior is typical but not the only possibility for pulses in reaction-diffusion systems.

The analysis for the existence and in particular for the stability of the FHN pulses is rather difficult, and we refer to the literature, see, e.g., [**Jon84, Kue15**].

9.2.2. The Gray Scott model. The Gray Scott model [**GS83**] is a famous model for cubic autocatalysis of two species according to $A+2B \to 3B$, rate k_1ab^2, where A is fed into the system at rate k_f and B decays with rate k_2b, and where the diffusion coefficients are D_A and D_B. After suitable nondimensionalization the system reads

$$\begin{aligned} \partial_t u &= \partial_x^2 u - uv^2 + f(1-u), \\ \partial_t v &= d\partial_x^2 v + uv^2 - (f+k)v, \end{aligned} \tag{9.22}$$

with typically $0 < d < 1$ in applications.

The nullclines of the ODE for (9.22) are given by

$$g_1 = 0: \quad u = f/(f+v^2), \qquad g_2 = 0: \quad u = (f+k)/v \quad \text{or} \quad v = 0.$$

We always have the stable fixed point $P = (1, 0)$. For $f > 4(f+k)^2$ there occurs a saddle-node bifurcation with saddle $Q = ((f+k)/v_-, v_-)$ and unstable node $R = ((f+k)/v_+, v_+)$ with

$$v_\pm = \frac{f}{2(f+k)} \pm \sqrt{\frac{f^2}{4(f+k)^2} - f}.$$

The radicand is positive for $f > 4(f+k)^2$, or equivalently $k < -f + \sqrt{f/4} =: f_{\text{sn}}(f)$. However, for $k < k_{\text{Hopf}}(f)$, R regains stability by a Hopf bifurcation such that R is surrounded by an unstable periodic orbit for $k_{\rm sn}(f) > k > k_{\rm Hopf}(f)$. At $(k, f) = (1/16, 1/16)$ the two curves $k_{\rm sn}(f)$ and $k_{\rm Hopf}(f)$ meet. This is called a Bogdanov-Takens or co-dimension 2-point, since in its neighborhood two parameters are needed to describe all possible bifurcations. See Figure 9.6 for some ODE phase portraits.

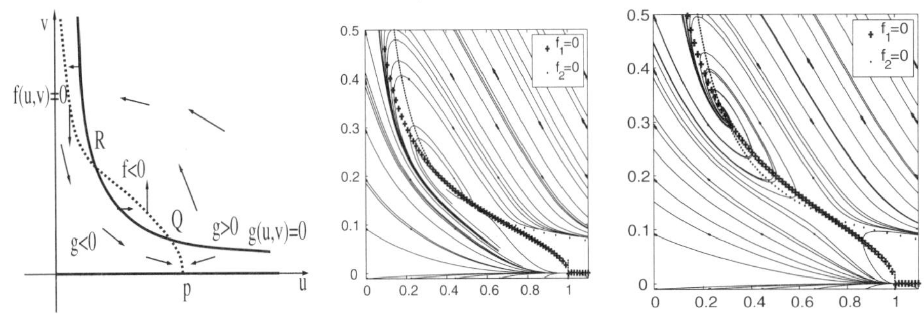

Figure 9.6. The schematic phase portrait for the Gray-Scott Sytem (9.22) right after the saddle (Q)-node (R) bifurcation in the first quadrant, and the phase portrait for $(f, k) = (0.023, 0.05)$ ($k_{\rm sn}(f) > k > k_{\rm Hopf}(f)$) and for $(f, k) = (0.04, 0.05)$ ($k_{\rm Hopf} > k$).

Thus, a (very rough) first characterization of the Gray-Scott system (9.22) is: (a) excitable for $k > k_{\rm Hopf}(f)$ and (b) bistable for $k < k_{\rm Hopf}(f)$. Thus, we again may expect pulses and fronts in cases (a) and (b), respectively. Again, the actual proof of existence of these traveling waves is beyond the scope of this book, not to mention their stability analysis. Therefore, we content ourselves with a small sample of numerical simulations, see Figure 9.7. Some very interesting numerical simulations of the Gray-Scott system leading to Sierpinsky-Gaskets of pulses can be found in [**HO00**], see also [**HO98**].

Remark 9.2.2. a) From the phase portrait we see that the Gray Scott model does not have contracting rectangles. Thus, we cannot obtain global existence with the methods from §9.1.3. Nevertheless, global existence holds and can be established using somewhat more involved (energy-type-) methods, see, e.g., [**FMW92**] for rather general results.
b) For specific parameter values in (9.22) some fronts and pulses can be obtained explicitly. For this see [**HPT00**]. ⌋

9.3. The Turing instability

The Turing instability [**Tur52, Mur89**] gives a mechanism for pattern formation in a reaction-diffusion models. Our starting point is a system of the form (9.1), where the reaction term $f = f_\mu$ depends smoothly on a parameter (vector) $\mu \in \mathbb{R}^p$, where w.l.o.g. we may restrict to $p = 1$. We

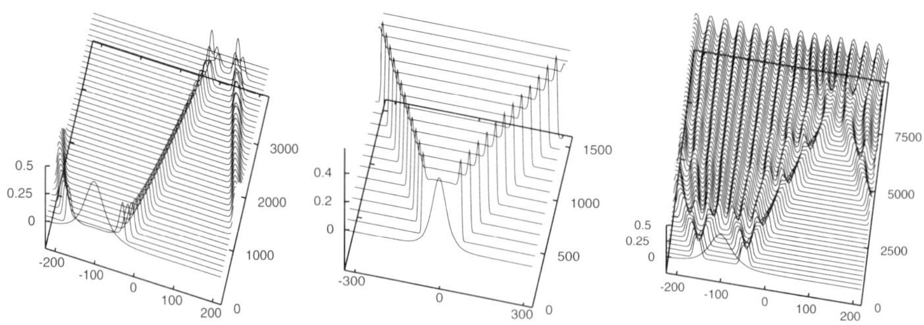

Figure 9.7. Numerical simulations of (9.22) on a large domain with periodic boundary conditions. (a) two counter-propagating pulses that annihilate upon collision ($d = 0.5$, $(f, k) = (0.023, 0.05)$). (b) Two fronts in the bistable regime $(f, k) = (0.04, 0.04)$. (c) self replicating pattern (SRP) ($d = 0.5$, $(f, k) = (0.038, 0.06)$).

assume that at some fixed μ the system has a stable spatially homogeneous equilibrium u^*, i.e., u^* is a stable fixed point of the ODE $\dot{u} = f(u)$.

Intuitively one would expect that diffusion cannot destabilize a homogenous state u^*. However, Turing [**Tur52**] recognized that this intuition is wrong if the diffusion matrix D is not a scalar multiple of the identity, and it turns out that u_0 may be unstable w.r.t. harmonic waves with some wavenumber $k_c \neq 0$. An attempt for an intuitive explanation using fire as the activator with fast diffusion and "sweating grasshoppers" as inhibitor with slow diffusion is made in [**Mur89**].

The general situation is as follows. The linearization

$$\partial_t v = D\Delta v + f'_\mu(u^*)v$$

of (9.1) around u^* has solutions of the form

$$v(x,t) = \mathrm{e}^{ikx+\lambda(k)t}\widehat{v}(k) + \text{c.c.},$$

where the eigenvalue $\lambda(k) \in \mathbb{C}$ and the vector $\widehat{v}(k) \in \mathbb{C}^N$ are determined from the $N \times N$ eigenvalue problem

$$(f'_\mu(u_0) - k^2 D)\widehat{v}(k) = \lambda(k)\widehat{v}(k),$$

i.e., there exist N curves $k \mapsto \lambda_j(k,\mu)$ with $j = 1, \dots, N$, which we order such that $\mathrm{Re}\lambda_j(k,\mu) \geq \mathrm{Re}\lambda_{j+1}(k,\mu)$.

The homogeneous state u^* is unstable if there exists a $k_0 \in \mathbb{R}$ with $\mathrm{Re}\lambda_1(k_0) > 0$. The instability appears as some parameter μ of the system is varied. For instance $\mathrm{Re}\lambda_1(k) < 0$ for all $k \in \mathbb{R}$ if $\mu < \mu_0$, while at $\mu = \mu_0$ we have $\mathrm{Re}\lambda_1(k_0) = 0$ for some $k_0 > 0$ and $\mathrm{Re}\lambda_1(k) < 0$ for $k \neq \pm k_0$. Finally, for $\mu > \mu_0$, $\mathrm{Re}\lambda_1(k) > 0$ for all k in two unstable bands, i.e., $\pm k \in (k_-(\varepsilon), k_+(\varepsilon))$, with $\varepsilon^2 = \mu - \mu_c$.

Thus, at criticality, i.e., at $\mu = \mu_c$, spatially periodic solutions may bifurcate, which is called pattern formation. It is called a) (proper) Turing bifurcation if $\mathrm{Im}(\lambda_1(k_c, \mu_c)) = 0$ and b) Turing-Hopf bifurcation if $\mathrm{Im}(\lambda_1(k_c, \mu_c)) \neq 0$. However, the instability may also appear at the wave number $k = 0$, i.e., the homogeneous solutions $u = u^*$ itself may become unstable as a fixed point of the reaction $\dot{u} = f_\mu(u)$ as the parameter μ is varied. It is called d) Hopf bifurcation if $\mathrm{Im}(\lambda_1(0, \mu_c)) \neq 0$.

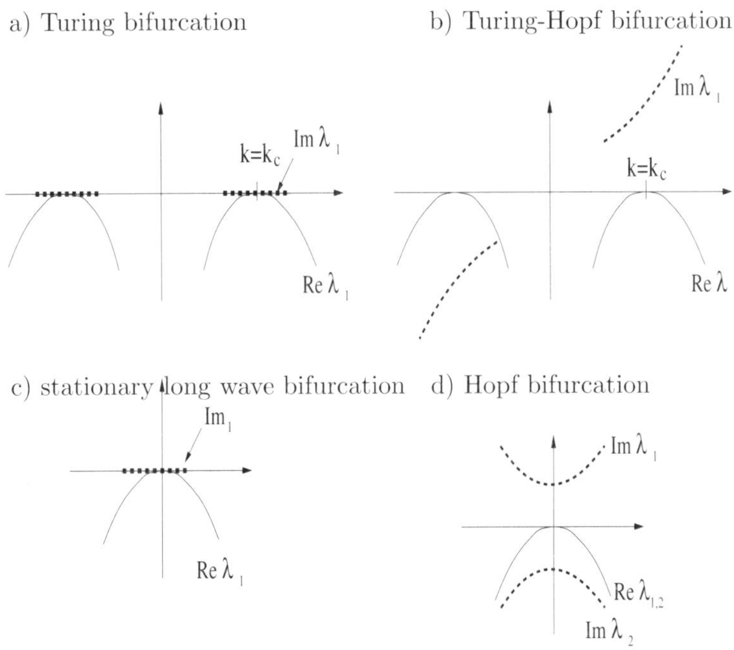

Figure 9.8. The four different instability scenarios

Example 9.3.1. We consider the so called Schnakenberg model [**Sch79, Mur89**]

$$\begin{aligned} \partial_t u &= \partial_x^2 u - u + u^2 v, \\ \partial_t v &= d\partial_x^2 v + b - u^2 v, \end{aligned} \tag{9.23}$$

with parameter $b > 0$ and the ratio $0 < d = d_{\mathcal{V}}/d_{\mathcal{U}}$ of the diffusion constants. Below we shall further assume that $d > 1$. To start the bifurcation analysis, we first need to determine the linearized stability of the unique homogeneous fixed point

$$(u^*, v^*) = (b, 1/b)$$

of (9.23). The new coordinates $(u, v) = (u^*, v^*) + (\widetilde{u}, \widetilde{v})$ yield

$$\partial_t \begin{pmatrix} \widetilde{u} \\ \widetilde{v} \end{pmatrix} = L(\partial_x) \begin{pmatrix} \widetilde{u} \\ \widetilde{v} \end{pmatrix} + g(\widetilde{u}, \widetilde{v}), \tag{9.24}$$

with

$$L(\partial_x)\begin{pmatrix}\widetilde{u}\\ \widetilde{v}\end{pmatrix} = \begin{pmatrix}\partial_x^2\widetilde{u}+\widetilde{u}+b^2\widetilde{v}\\ d\partial_x^2\widetilde{v}-2\widetilde{u}-b^2\widetilde{v}\end{pmatrix}, \quad g(\widetilde{u},\widetilde{v}) = (2u_0\widetilde{u}\widetilde{v}+\widetilde{u}^2 v_0+\widetilde{u}^2\widetilde{v})\begin{pmatrix}1\\ -1\end{pmatrix}.$$

Plugging the ansatz $(\widetilde{u},\widetilde{v})(x,t) = \mathrm{e}^{\mathrm{i}kx+\lambda(k)t}\varphi$ with $\varphi = \varphi(k) \in \mathbb{C}^2$ into the linearization $\partial_t(\widetilde{u},\widetilde{v}) = L(\partial_x)(\widetilde{u},\widetilde{v})$ yields the eigenvalue problem

$$\begin{pmatrix}-k^2+1 & b^2\\ -2 & -dk^2-b^2\end{pmatrix}\varphi = \lambda\varphi, \tag{9.25}$$

with wave number $k \in \mathbb{R}$ as parameter. For notational convenience in the following we drop the dependence on (b,d). The two curves of eigenvalues $\lambda_{1,2}(k)$ are determined by

$$\lambda^2 - \lambda[-(1+d)k^2+1-b^2] + [dk^4+(b^2-d)k^2+b^2] = 0. \tag{9.26}$$

The stability of the stationary point for the reaction system is determined by $\lambda_{1,2}(0)$. We obtain

$$\lambda_{1,2}(0) = (1-b^2)/2 \pm \sqrt{(1-b^2)^2/4-b^2}$$

such that $(b,1/b)$ is unstable for $b \in (0,1)$. We have $\mathrm{Im}(\lambda_{1,2}(0,b)) \neq 0$ for

$$b \in \left[\sqrt{(3-\sqrt{5})/2}, \sqrt{(3+\sqrt{5})/2}\right].$$

The homogeneous solution of (9.23) is asymptotically stable if $\mathrm{Re}(\lambda_{1,2}(k)) < 0$ for all $k \in \mathbb{R}$. Depending on the parameters, instability can set in in the two ways a) and d) described in Figure 9.8. See Figure 9.9 for details. ⌋

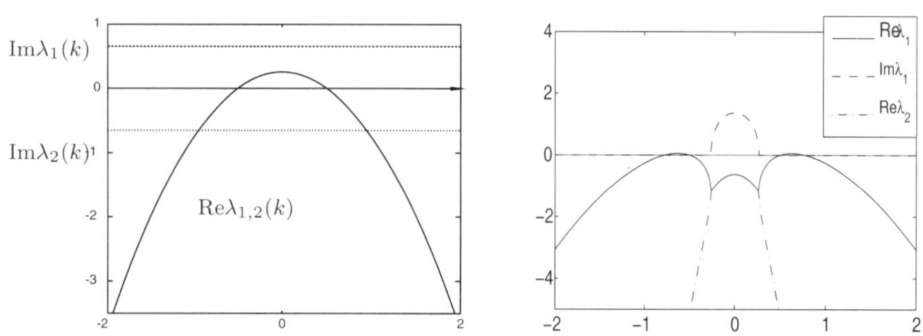

Figure 9.9. The curves of eigenvalues in the Hopf and in the Turing case. Here $(b,d) = (0.7,1)$ and $(b,d) = (3,60)$, respectively.

The general two-dimensional situation. Even for the simple model (9.23) the discussion of the dispersion relation (9.26) is rather unpleasant. Thus, we derive necessary and sufficient conditions for the occurrence of a Turing instability in a general two-component system

$$\partial_t u = \partial_x^2 u + f(u,v), \quad \partial_t v = d\partial_x^2 v + g(u,v). \tag{9.27}$$

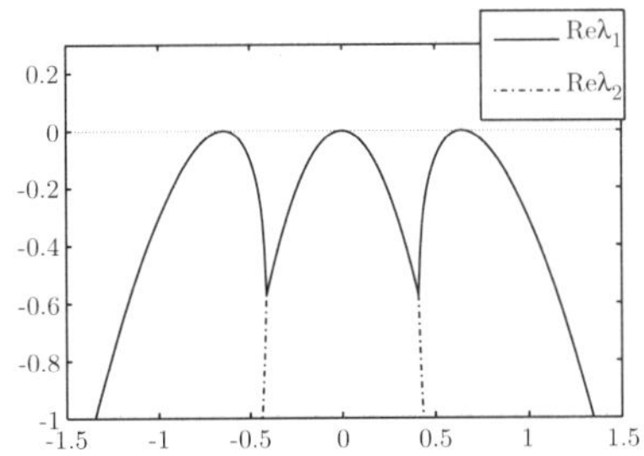

Figure 9.10. Transitional case: $b = 1, d = 5.83$.

We assume that the ODE

$$\dot{u} = f(u, v), \quad \dot{v} = g(u, v) \tag{9.28}$$

has some steady state (u^*, v^*). The eigenvalues λ of

$$A = \frac{\partial(f, g)}{\partial(u, v)} = \begin{pmatrix} f_u & f_v \\ g_u & g_v \end{pmatrix}_{(u^*, v^*)},$$

satisfy

$$\lambda^2 - (f_u + g_v)\lambda + (f_u g_v - f_v g_u) = 0.$$

Hence, the steady state (u^*, v^*) is stable for the ODE (9.28) if

$$\operatorname{tr} A = f_u + g_v < 0 \quad \text{and} \quad \det A = f_u g_v - f_v g_u > 0. \tag{9.29}$$

The eigenvalues λ of

$$\begin{pmatrix} \partial_x^2 & 0 \\ 0 & d\partial_x^2 \end{pmatrix} + A$$

satisfy

$$\lambda^2 + \lambda(k^2(1 + d) - (f_u + g_v)) + h(k^2) = 0,$$

with $h(k^2) = dk^4 - (df_u + g_v)k^2 + \det A$. Since already $f_u + g_v < 0$, a necessary condition for Turing instability of the steady state (u^*, v^*) is $h(k^2) < 0$ for some k, which requires $df_u + g_v > 0$, hence, by (9.29) $d \neq 1$ and $f_u g_v < 0$. Ultimately, we need $h_{\min} < 0$, where, by calculus,

$$h(k^2) = h_{\min} = \det A - (df_u + g_v)^2/(4d) \quad \text{at} \quad k^2 = k_m^2 = (df_u + g_v)/(2d).$$

At the bifurcation point the condition $h_{\min} = 0$ defines via $|A| = (df_u + g_v)^2/4d$ a critical diffusion ratio as a root of $d_c^2 f_u^2 + 2(2f_v g_u - f_u g_v)d_c + g_v^2 = 0$. The critical wave number k_c is given by

$$k_c^2 = (df_u + g_v)/(2d) = \sqrt{(f_u g_v - f_v g_u)/d_c}. \tag{9.30}$$

Lemma 9.3.2. *For a Turing-instability to occur in a system of the form* (9.27), *the conditions*

$$f_u + g_v < 0, \quad f_u g_v - f_v g_u > 0, \quad df_u + g_v > 0,$$
$$(df_u + g_v)^2 - 4d(f_u g_v - f_v g_u) > 0$$

are necessary and sufficient, with k_c given by (9.30).

Definition 9.3.3. For a reaction-diffusion system $\partial_t u = D\partial_x^2 u + f(u)$ with a spatially homogeneous equilibrium u^* which is asymptotically stable for the ODE $\dot{u} = f(u)$, the range of parameters for which there exists a $k \neq 0$ with $\text{Re}\lambda(k) > 0$ and $\text{Im}(\lambda(k))$=0 is called the Turing space for u^*.

Lemma 9.3.2 gives criteria for the onset of a Turing instability in a two-component reaction-diffusion system with diagonal diffusion. See Example 9.3.4 for the Schnakenberg model, and the exercises for further examples and generalizations. For $N \geq 3$-component reaction-diffusion systems, characterizations of Turing-instabilities as in Lemma 9.3.2 become rather complicated; see [**HNM14**] for some partial results. Moreover, in practice, non-diagonal diffusion matrices occur, for which it is often more convenient to find Turing instabilities numerically.

Example 9.3.4. For (9.23) we obtain $f_u = 1, f_v = b^2, g_u = -2, g_v = -b^2$ and $f_u g_v - f_v g_u = b^2$. Thus Lemma 9.3.2 yields $b > 1$, $b^2 > 0$, $db > b^3$ and $(db - b^3)^2 > 4db^4$. These are partly redundant, and reduce to $b > 1$ (for the stability of $(b, 1/b)$ at $k = 0$) and $d > (3 + \sqrt{8})b^2 \approx 5.83b^2$. For $d = d_c = (3+\sqrt{8})b^2$ we obtain $k_c = \sqrt{\frac{b^2}{d_c}}$. For instance, for $b = 1.5$ we obtain $d_c \approx 13.12$ and $k_c \approx 0.63$, cf. Figure 9.9. ⌋

For $x \in \mathbb{R}^n$, $n \geq 2$, the curves $\mathbb{R} \ni k \mapsto \lambda_j(k)$, $j = 1, \ldots, d$, become surfaces $\mathbb{R}^n \ni k \mapsto \lambda_j(k)$, which only depend on $|k|$ due to the rotational invariance of the Laplacian, e.g., for $n = 2$ we have a circle of critical wave-vectors $k \in \mathbb{R}^2$ with $|k| = k_c$.

Two-component systems with diagonal diffusion have played a major role in pattern formation for reaction-diffusion systems in the last decades. The examples at the beginning of Chapter 9 are very rich in "pattern formation" in particular for $x \in \mathbb{R}^2$, i.e., in suitable parameter regimes they show a wealth of spatially non-homogeneous stationary solutions, some of which show remarkable similarities to animal coat pattern (stripes on zebras, spots on leopards, stripes and spots on fishes). The mathematical analysis of these pattern starts with the investigation of the linear instabilities of the homogeneous states as described above, and then proceeds with a weakly nonlinear analysis via center manifold theory or the theory of modulation equations as considered in Part IV.

Further Reading. Besides [**Mur89, Smo94, BK00**], good introductory textbooks about reaction-diffusion systems, their mathematical analysis and their various types of solutions are [**Gri96, Nis02**]. Recent textbooks about reaction-diffusion systems and background in applications are [**GR12, MK15**]. The importance of reaction-diffusion systems in ecosystem modelling is stressed in [**Mer15**], which also includes a very accessible account on pattern formation and amplitude equations.

A relatively recent development are Turing(-like) bifurcations in infinite time horizon distributed optimal control problems for parabolic PDEs of reaction diffusion type. For these, the steady version of the so called canonical system, derived via Pontryagin's maximum principle, see [**GCF$^+$08**], is an elliptic system, which under certain condition has bifurcating spatially periodic solutions. In [**BX08, BX10**], this is called optimal diffusion induced instability. In [**Uec16, GU17**], the optimality of these patterned canonical steady states, and the associated optimal paths, have been studied numerically for some model problems. One result is that in certain systems a patterned harvesting in an otherwise spatially uniform system gives a higher yield; this may lead to many applications in the area of spatial resource economics, in particular w.r.t. to natural reserves, and also in the field of climate-economic models [**BEX14**].

A thorough treatment of (different versions of) the FitzHugh-Nagumo system, and many other systems of excitable or oscillatory type with multiple time scales is given in [**Kue15**], including many references. Much recent work on the existence and dynamics for pulses in reaction-diffusion systems focusses on singularly perturbed problems, and on heterogeneous media, see, e.g., [**NTYU07, DvHX16, DV15**] and the references therein. Another important class of reaction-diffusion systems arise in chemotaxis [**Hor03, Hor04**] and are of the form $\partial_t u = \text{div}(D(u)\nabla u) + f(u)$. For such systems, the diffusion coefficients depends on u which makes them *quasi*linear. The local existence theory for such systems is more complicated than for (9.1), requiring so called maximal regularity methods [**Lun95, Ama95**]. Blow-up of solutions is more frequent than in the semilinear case, see for instance [**SGKM95**]. See also [**BBTW15, TW15**] for coupling with fluid equations and related modeling issues, and for results on equations of mixed parabolic-elliptic type, that also frequently arise in applications.

Exercises

9.1. Show that any quadratic autocatalytic system (i.e., arbitrary $k, D_A, D_B > 0$)

$$\partial_t a = D_a \partial_x^2 a - kab, \quad \partial_t b = D_b \partial_x^2 b + kab$$

can be rescaled to the form $\partial_t u = \partial_x^2 u - uv$, $\partial_t v = d\partial_x^2 v + uv$.

9.2. Show that the vector field f defined by (9.13) points inward on ∂R. Use this to prove that there exists a heteroclinic orbit $v_{\text{het}}(\xi)$ connecting $(0,1,0)$ and $(1,0,0)$.

9.3. Consider the reaction–diffusion–systems

$$\partial_t u = \partial_x^2 u + uM(u,v), \qquad \partial_t v = d\partial_x^2 v + vN(u,v),$$

associated to the Kolmogorov form of the equations for 2–species interaction in mathematical ecology, see Example 2.6.5. For each type, (PP), (C) and (S), construct (weakly) invariant regions, either abstractly, or for the concrete examples given in Example 2.6.5.

9.4. Show that (9.18) is equivalent to (9.17) with $\varepsilon = b$, $\gamma = c/\varepsilon$ and $\tau = \varepsilon t$.

9.5. Let the linearization of a system (9.27) around $(u,v)^*$ be given by

$$w_t = Aw \text{ with } A = \begin{pmatrix} 1+\partial_x^2 & 4 \\ -1 & -3+d\partial_x^2 \end{pmatrix}.$$

Use Lemma 9.3.2 to find d_c such that $(u,v)^*$ is Turing unstable for $d>d_c$, and calulate k_c.

9.6. Sketch the Turing space for the unique fixed point of (simplified) Gierer–Meinhard system

$$\partial_t u = \partial_x^2 u + a - u + u^2/v, \quad \partial_t v = d\partial_x^2 v + u^2 - v, \quad a, d > 0.$$

9.7. In general, reaction diffusion systems may include so called cross-diffusion, where the diffusion of one species directly influences some other species. For simplicity we restrict to

$$\partial_t U = D\Delta U + F(U), \text{ where } U = \begin{pmatrix} u \\ v \end{pmatrix}, \quad D = \begin{pmatrix} 1 & 0 \\ d_3 & d_4 \end{pmatrix}. \tag{9.31}$$

Thus, if $d_3 > 0$, then "cross–diffusion" in $\partial_t U = D\Delta U$ yields a decrease/increase of v at the maxima/minima of u. Assume that (9.31) has a homogeneous steady state U_*. Show that U_* is Turing unstable if

$$f_u + g_v < 0, \quad f_u g_v - f_v g_u > 0, \quad d_4 f_u - d_3 f_v + g_v > 0,$$
$$(d_4 f_u - d_3 f_v + g_v)^2 - 4d_4(f_u g_v - f_v g_u) > 0,$$

with critical wave number k_c given by $k_c^2 = \dfrac{d_4 f_u - d_3 f_v + g_v}{2d_4}$.

Chapter 10

Dynamics of pattern and the GL equation

This is the first chapter of Part IV of this book. Part IV is about modulation theory which is applied to physically realistic systems over unbounded domains. Like for systems over bounded domains, which very often can be reduced to finite-dimensional ODEs, also for problems over unbounded domains often only a subset of the uncountable many degrees of freedom plays a role. These still uncountable many degrees of freedom are described by simpler PDEs, which in this book are called modulation equations. They can be derived via perturbation theory. In particular, the Korteveg-de Vries (KdV), the Nonlinear Schrödinger (NLS), and the Ginzburg-Landau (GL) equations from Chapter 8, play an important role as modulation equations, and one unifying theme of the remaining chapters of this book is what can (and what can not) be deduced about the original full system from these reduced PDEs.

Part IV consists of five chapters. The first three chapters are about the derivation and justification of such approximations. Here, in Chapter 10 we discuss the GL approximation, in Chapter 11 the NLS approximation, and in Chapter 12 the KdV approximation. The last two chapters handle aspects of the existence and stability theory of special solutions of nonlinear PDEs on unbounded domains which are related to modulation theory. We start in Chapter 13 with the construction of solutions via the spatial dynamics and invariant manifold approach. The final Chapter 14 is about the stability of solutions via the diffusive or dispersive stability approach.

The plan of this chapter is as follows. In §10.2 we introduce the GL approximation for the Swift-Hohenberg equation and justify it with a first approximation result. The universality of the GL approximation is explained in §10.3. We develop an abstract approximation theory in §10.4, which is subsequently applied to a number of examples, such as the Kuramoto-Shivashinsky equation, or in §10.5 to reaction-diffusion systems, where we also extend our theory to Hopf bifurcations at the wave number $k = 0$. In §10.6 we consider convection problems and extend our theory to planar systems and systems with a time-periodic forcing. In §10.7 we consider with the Couette-Taylor problem another classical hydrodynamical stability problem and extend our theory to Turing-Hopf bifurcations at wave numbers $k \neq 0$. In §10.8 we discuss the attractivity of the set of GL described solutions, and the comparison of the original pattern forming system attractor and the associated GL attractor. We close this chapter with an outlook on the Ginzburg-Landau approximation in stochastic PDEs, lattice differential equations, and delay equations.

10.1. Introduction

In §9.3 we already discussed, on a linear level, Alan Turing's remarkable observation [**Tur52**] that diffusion can destabilize ODE fixed points in the associated PDE, i.e., destabilize spatially homogeneous fixed points u^* of two-component reaction-diffusion systems of the form

$$\partial_t u = D\Delta u + f(u),$$

with $u(x,t) \in \mathbb{R}^2$, $x \in \mathbb{R}^d$, a reaction $f : \mathbb{R}^2 \to \mathbb{R}^2$, $\Delta = \sum_{j=1}^d \partial_{x_j}^2$, and $D \in \mathbb{R}^{2\times 2}$ a diagonal matrix with entries $d_1, d_2 > 0$. Until the work of Turing it was believed that whatever the reaction is, diffusion will homogenize the reactants in space. Turing found that this not true and that even simple reaction-diffusion systems can lead to a spatial structure of the solutions similar to the pattern on animals such as tigers, zebras, leopards and many others.

As already explained in §9.3 the idea behind this observation is surprisingly simple. The linearization

$$\partial_t v = D\Delta v + f'(u^*)v \tag{10.1}$$

around u^* is solved in case $x \in \mathbb{R}$ by

$$v(x,t) = \mathrm{e}^{ikx+\lambda t}\widehat{v}, \tag{10.2}$$

with $k \in \mathbb{R}$, $\lambda \in \mathbb{C}$, and $\widehat{v} \in \mathbb{C}^2$, where λ and $\widehat{v}$ are determined by the eigenvalue problem

$$-|k|^2 D\widehat{v} + f'(u^*)\widehat{v} = \lambda\widehat{v}.$$

For fixed $k \in \mathbb{R}$ this is a two-dimensional eigenvalue problem, and thus we obtain two curves of eigenvalues $k \mapsto \lambda_{1,2}(k)$ and two curves of eigenvectors $k \mapsto \widehat{v}_{1,2}(k)$. If $d_1 = d_2$, then obviously the maximum of the curves $\lambda_{1,2}$ is always obtained at the wave number $k = 0$. However, if $d_1 \neq d_2$, then the maximum can be obtained at a non-zero wave number $k = k_c \neq 0$. See Figure 9.8 a) and b), and Lemma 9.3.2.

If a control parameter is varied in such systems, then an instability can occur leading to linear growth of spatially periodic modes in some wave number range. Using $0 < \varepsilon^2 \ll 1$ for the distance to the bifurcation point in parameter space we may assume that this growth rate is of order $\mathcal{O}(\varepsilon^2)$, and generically, i.e., except for special cases where certain coefficients vanish, the growth of these modes will be saturated by the nonlinear terms at an $\mathcal{O}(\varepsilon)$ amplitude. Therefore, we have a so called weakly nonlinear regime.

The weakly nonlinear methods for the analysis of pattern forming systems differ fundamentally between finite and infinite domains. For finite domains, bifurcating pattern near the bifurcation point can be analyzed via the Lyapunov-Schmidt, cf. §3.1.3, or the center manifold reduction, cf. §3.2. After the reduction they can be described by the solutions of some ODEs. See also the subsequent §13.2. However, in many cases the wave length of the pattern is much smaller than the physical domain, and the description of the bifurcating pattern via Lyapunov-Schmidt reduction or center manifold theory becomes increasingly useless for larger and larger domains, as more and more modes become unstable. This can be seen as follows. If we have a large bounded domain with suitable boundary conditions, e.g., $x \in (0, l\pi)$ with large $l \in \mathbb{N}$ and homogeneous Neumann boundary conditions $\partial_n u = 0$, then we have discrete eigenvalues $\mathbb{Z}/l \ni k \mapsto \lambda_j(k)$ with $\lambda_j(k)$ from Lemma 9.3.2. For large l the discrete eigenvalues fill up the continuous curve $\mathbb{R} \ni k \mapsto \lambda_j(k)$, and Lemma 9.3.2 and formula (9.30) give good approximations for the occurence of the Turing instability and the critical wave number k_c. As a consequence the dimension of the center manifold and of the reduced equations grow. Since in parallel the size of the center-manifold gets smaller and smaller for large l these reduction methods become useless.

Thus, for the description of pattern forming systems on large domains close to the threshold of instability we need a new approach. Since the influence of far away boundaries is negligible in a certain sense for large domains, similar to the PDEs in Part III, these pattern forming systems are considered on unbounded domains to understand the intrinsic mechanism for the occurrence of the pattern. Hence, throughout the rest of this chapter all systems are considered on the whole real line or unbounded cylindrical domains. We remark that the influence of the far away boundaries on the dynamics have been estimated for instance in [**Mie97a**], see also [**SS00**].

The observation by Turing for (10.1) is nowadays accepted as the mechanism of pattern formation in animal furs. Moreover, this mechanism of pattern formation is not restricted to biological systems. The same kind of instability widely occurs in nature [**Man92, Pis06, Mar07, CG09**]. Examples are pattern in sand dunes, convection pattern in clouds or magma, desertification pattern in savannas, and many more. Additionally, this kind of instability occurs in laboratory experiments which have been created to understand the occurrence of pattern formation. The Rayleigh-Bénard problem and the Couette-Taylor problem will be discussed as examples in §10.6 and in §10.7.

In the second half of the 1960s a number of scientists started to describe the behavior of all these pattern forming systems close to the first instability by one universal amplitude or modulation equation, cf. [**dES71**]. In the simplest case, by a multiple scaling ansatz

$$v(x,t) = \varepsilon A(\varepsilon x, \varepsilon^2 t) e^{ik_c x} \widehat{v}_1(k_c) + c.c.$$

(using the notation following (10.1)), a GL equation

$$\partial_T A = \nu_2 \partial_X^2 A + \nu_0 A + \nu_3 A|A|^3, \tag{10.3}$$

with $T \geq 0$, $X \in \mathbb{R}$, coefficients ν_0, ν_2, $\nu_3 \in C$, and $A(X,T) \in \mathbb{C}$ has been derived. For many years, only the associated GL equations have been analyzed for the understanding of the pattern forming systems. In nonlinear science it is still widely accepted that the derivation of a GL equation implies that it is a good model for these systems, and that it is sufficient to analyze the GL equation instead of the original system. In the late 1980s mathematicians began to discuss to which extend this approximation is correct. The main purpose of this chapter is to give an introduction into the mathematical theory developed around this question.

The history of the approximation results starts with [**CE90**] where the (cubic) SH equation as original pattern forming system has been handled. The proof has been simplified in [**KSM92**]. The case of quadratic nonlinearities has been treated first in [**vH91**], and the proof has been simplified in [**Sch94c**] to the version presented here, and generalized to the Navier-Stokes equations in [**Sch94a**]. See also [**TBD$^+$96**]. There are attractivity results [**Eck93, Sch95a**], showing that every small solution can be described after a certain time by the GL equation. Combining the approximation and attractivity results allows to prove the upper semi-continuity of attractors [**MS95, Sch99c**], shadowing by pseudo-orbits, and global existence results for the pattern forming systems [**Sch94b, Sch99b**]. These results will be described in detail in §10.8.

We focus on the case of one unbounded space direction. The exceptions are §10.5.3, where Hopf bifurcations at the wave vector $k = 0$ are treated for $x \in \mathbb{R}^d$, and §10.6.3-§10.6.5, where we discuss amplitude equations for pattern forming systems with $x \in \mathbb{R}^2 \times \Sigma$, where Σ is a bounded cross section.

10.2. The Swift-Hohenberg equation

The simplest equation where the GL equation can be derived is the so called Swift-Hohenberg (SH) equation [**SH77**]

$$\partial_t u = -(1+\partial_x^2)^2 u + \alpha u - u^3, \tag{10.4}$$

with $u(x,t) \in \mathbb{R}$, $x \in \mathbb{R}$, $t \geq 0$, $\alpha \in \mathbb{R}$, and $-(1+\partial_x^2)^2 u = -\partial_x^4 u - 2\partial_x^2 u - u$. This problem has the spatially homogenous state $u^* \equiv 0$, which becomes unstable at $\alpha = 0$ via an instability at the non-zero wave number $k_c = 1$. The linearization $\partial_t v = -(1+\partial_x^2)^2 v + \alpha v$ of the SH equation at the trivial solution $u^* \equiv 0$ possesses the solutions $v(x,t) = \mathrm{e}^{ikx+\lambda(k)t}$ where

$$\lambda(k,\alpha) = -(1-k^2)^2 + \alpha. \tag{10.5}$$

It is clear that $u^* \equiv 0$ is asymptotically stable for $\alpha < 0$. The maxima of λ at $k = \pm 1$ are positive if $\alpha > 0$. For the SH equation the ansatz for the derivation of the GL equation is given by

$$\varepsilon\psi_{GL}(x,t) = \varepsilon A(X,T)\mathrm{e}^{\mathrm{i}x} + \varepsilon \overline{A}(X,T)\mathrm{e}^{-\mathrm{i}x} \tag{10.6}$$

where $0 < \alpha = \varepsilon^2 \ll 1$, $X = \varepsilon x$, and $T = \varepsilon^2 t$. The terms which do not cancel after inserting the GL approximation into the SH equation are collected in the residual

$$\mathrm{Res}(u) = -\partial_t u - (1+\partial_x^2)^2 u + \varepsilon^2 u - u^3.$$

If $\mathrm{Res}(u) = 0$, then u is an exact solution of the SH equation (10.4). With the abbreviation $\mathbf{E} = \mathrm{e}^{\mathrm{i}x}$ we find

$$\begin{aligned}
\mathrm{Res}(\varepsilon\psi_{GL}) = &- \mathbf{E}(\varepsilon^3\partial_T A) - \mathbf{E}(\varepsilon A) - 2\mathbf{E}\left(\varepsilon \mathrm{i}^2 A + 2\varepsilon^2 \mathrm{i}\partial_X A + \varepsilon^3 \partial_X^2 A\right) \\
&- \mathbf{E}\left(\varepsilon \mathrm{i}^4 A + 4\varepsilon^2 \mathrm{i}^3 \partial_X A + 6\varepsilon^3 \mathrm{i}^2 \partial_X^2 A + 4\varepsilon^4 \mathrm{i}\partial_X^3 A + \varepsilon^5 \partial_X^4 A\right) \\
&+ \mathbf{E}(\varepsilon^3 A) - \mathbf{E}(3\varepsilon^3 A^2 \overline{A}) - \mathbf{E}^3(\varepsilon^3 A^3) + \mathrm{c.c.} \\
= \quad & \varepsilon \mathbf{E}\cdot 0 + \varepsilon^2 \mathbf{E}\cdot 0 + \varepsilon^3 \mathbf{E}\left(-\partial_T A + 4\partial_X^2 A + A - 3A|A|^2\right) \\
&- \varepsilon^3 \mathbf{E}^3 A^3 - \varepsilon^4 \mathbf{E}(4\mathrm{i}\partial_X^3 A) - \varepsilon^5 \mathbf{E}(\partial_X^4 A) + \mathrm{c.c.}
\end{aligned}$$

The first line of the residual vanishes by choosing the envelope A of the bifurcating spatially periodic pattern $\mathrm{e}^{\mathrm{i}x}$ to satisfy the GL equation

$$\partial_T A = 4\partial_X^2 A + A - 3A|A|^3. \tag{10.7}$$

10.2.1. Special solutions. Before we explain to which extend the dynamics of the GL equation can be transferred to the original system, we look for special solutions of the GL equation and formally approximate with these GL solutions the dynamics of the SH equation. We consider three classes of GL solutions, namely space-independent solutions, i.e., $A = A(T)$, time-independent solutions, i.e., $A = A(X)$, and traveling wave solutions, i.e., $A = A(X - cT)$. We consider real-valued solutions which form an invariant subspace for the real GL equation (10.7).

Space-independent solutions. For space-independent solutions we get the one-dimensional complex-valued ODE

$$\frac{\mathrm{d}}{\mathrm{d}T}A = A - 3A|A|^2.$$

Real-valued solutions satisfy

$$\frac{\mathrm{d}}{\mathrm{d}T}A = A - 3A^3.$$

This last ODE possesses three equilibria, the unstable origin $A = 0$ and the two stable fixed points $A = \pm\sqrt{1/3}$. A solution $A = A(T)$ with initial condition $A(0) > 0$ satisfies $\lim_{T\to\infty} A(T) = \sqrt{1/3}$ with some exponential rate. Associated with this solution we find formally in the SH equation

$$\begin{aligned} u(x,t) =& \varepsilon A(T)\mathrm{e}^{\mathrm{i}x} + \varepsilon A(T)\mathrm{e}^{-\mathrm{i}x} + \text{h.o.t.} = 2\varepsilon A(T)\cos x + \text{h.o.t.} \\ \to& \ 2\varepsilon\sqrt{1/3}\cos x + \text{h.o.t.} \end{aligned} \tag{10.8}$$

for $T \to \infty$. Thus, we have formally found 2π-spatially periodic SH solutions converging towards a spatially periodic equilibrium for $T \to \infty$ as drawn in Figure 10.1.

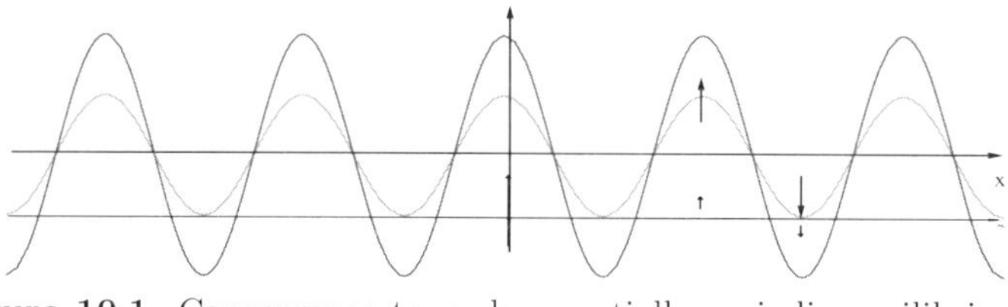

Figure 10.1. Convergence towards a spatially periodic equilibrium in the SH equation.

This can be made rigorous by center manifold theory, and it can be shown that in the subspace of 2π-spatially periodic SH solutions a supercritical pitchfork bifurcation occurs at $\alpha = 0$, cf. §13.2.2. Since the GL equation is S^1-invariant, with $T \mapsto A(T)$ also $T \mapsto A(T)\mathrm{e}^{\mathrm{i}\theta}$ for all $\theta \in \mathbb{R}$ is a solution. Associated to this solution for the SH equation we find

$$\begin{aligned} u(x,t) =& \varepsilon A(T)\mathrm{e}^{\mathrm{i}\theta}\mathrm{e}^{\mathrm{i}x} + \varepsilon A(T)\mathrm{e}^{-\mathrm{i}\theta}\mathrm{e}^{-\mathrm{i}x} + \text{h.o.t.} \\ =& 2\varepsilon A(\varepsilon^2 t)\cos(x+\theta) + \text{h.o.t.} \to 2\varepsilon\sqrt{1/3}\cos(x+\theta) + \text{h.o.t.} \end{aligned} \tag{10.9}$$

for $T \to \infty$. Hence, associated to the one-dimensional family of constant equilibria for the GL equation there is a one-dimensional family of 2π-spatially periodic equilibria for the SH equation in accordance with the translation invariance of the SH equation.

Time-independent solutions. Next we consider real-valued stationary solutions $A = A(X)$ of the GL equation. They can be found as solutions of the scalar second order ODE

$$0 = 4\partial_X^2 A + A - 3A^3.$$

In order to analyze this equation we consider the phase portrait of the equivalent first order system

$$\partial_X A = B, \quad \partial_X B = -A/4 + 3A^3/4, \tag{10.10}$$

cf. Exercise 2.10. We find the three equilibria from above, namely $(A, B) = (0, 0)$ and $(A, B) = (\pm 1/\sqrt{3}, 0)$. The last two equilibria are connected by two heteroclinic solutions $(A, B) = (A, B)_{\mathrm{het}}(\xi)$. Formally, these give approximations of equilibria of the SH equation of the form

$$u(x) = 2\varepsilon A_{\mathrm{het}}(\varepsilon x) \cos x + \text{h.o.t.}$$

with $u(x) = 2\varepsilon \cos x/\sqrt{3} + \mathcal{O}(\varepsilon^2)$ for $x \to \infty$ and with $u(x) = -2\varepsilon \cos x/\sqrt{3} + \mathcal{O}(\varepsilon^2) = 2\varepsilon \cos(x + \pi)/\sqrt{3} + \mathcal{O}(\varepsilon^2)$ for $x \to -\infty$. Therefore, these solutions connect two members of the same family of translates of the spatially periodic equilibrium $2\varepsilon \cos x/\sqrt{3}$ which differ by a phase shift of π. The transition from one equilibrium at $x \to -\infty$ to the other at $x \to \infty$ happens on a spatial scale of size $\mathcal{O}(1/\varepsilon)$. These solutions are sketched in Figure 10.2.

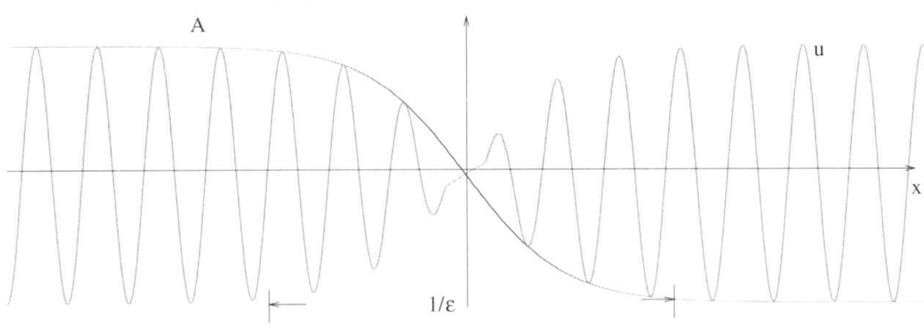

Figure 10.2. The spatially heteroclinic solution in the GL equation and for the SH equation. There is a phase difference between the asymptotically approached spatially periodic equilibrium for $x \to \infty$ and $x \to -\infty$.

In the phase portrait for (10.10) we also find periodic solutions $(A, B) = (A, B)_{\mathrm{per}}(X)$ surrounding the origin $(A, B) = (0, 0)$. For the SH equation we find formally the equilibria

$$u(x) = 2\varepsilon A_{\mathrm{per}}(\varepsilon x) \cos x + \text{h.o.t..}$$

The envelope $A = A_{\mathrm{per}}(X)$ is spatially periodic with some period of order $\mathcal{O}(1/\varepsilon)$ w.r.t. the fast spatial variable x and modulates the underlying pattern $\cos x$ which has some period of order $\mathcal{O}(1)$. These solutions are sketched in Figure 10.3.

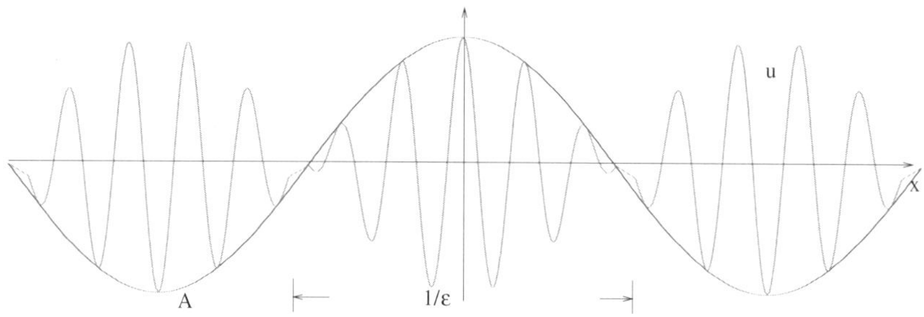

Figure 10.3. The spatially periodically modulated solution in the GL equation and for the SH equation.

Again with $X \mapsto A(X)$ also $X \mapsto A(X)e^{i\theta}$ for all $\theta \in \mathbb{R}$ is a solution of (10.7). Since the GL equation is translational invariant, with $X \mapsto A(X)$ also $X \mapsto A(X+X_0)$ for all $X_0 \in \mathbb{R}$ is a solution. Therefore, we find for the SH equation the two parameter families of approximate solutions

$$\begin{aligned} u(x) &= 2\varepsilon A_{\mathrm{het}}(X+X_0)\cos(x+\theta) + \text{h.o.t.}, \quad \text{and} \\ u(x) &= 2\varepsilon A_{\mathrm{per}}(X+X_0)\cos(x+\theta) + \text{h.o.t.}. \end{aligned} \tag{10.11}$$

Front solutions. Next we consider real-valued traveling waves $A = A(\zeta) = A(X - cT)$ of the GL equation. These satisfy

$$0 = 4\partial_\zeta^2 A + c\partial_\zeta A + A - 3A^3.$$

Again we consider the phase portrait of the equivalent first order system

$$\partial_\zeta A = B, \quad \partial_\zeta B = -cB/4 - A/4 + 3A^3/4, \tag{10.12}$$

cf. Exercise 2.10. For all values of $c \neq 0$ we find heteroclinic connections $(A,B) = (A,B)_{\mathrm{front}}(\zeta)$ between the fixed points $(A,B) = (\pm 1/\sqrt{3}, 0)$ and the origin $(A,B) = (0,0)$. Such solutions describe the spreading of the stable phases $A = \pm 1/\sqrt{3}$ into the unstable phase $A = 0$. The functions $\zeta \mapsto A_{\mathrm{front}}(\zeta)$ are monotonic for $|c| \geq 4$. For the SH equation the associated solutions

$$u(x,t) = 2\varepsilon A_{\mathrm{front}}(X - cT)\cos x + \text{h.o.t.} \tag{10.13}$$

are called modulating fronts. These solutions are sketched in Figure 10.4. With the arguments from above we find again a two parameter family of such approximate solutions, namely

$$u(x,t) = 2\varepsilon A_{\mathrm{front}}(X - cT + X_0)\cos(x+\theta) + \text{h.o.t.},$$

with $X_0, \theta \in \mathbb{R}$. Two GL fronts can be patched together to describe the growths of the solutions to small positive spatially localized initial conditions of $A \equiv 0$.

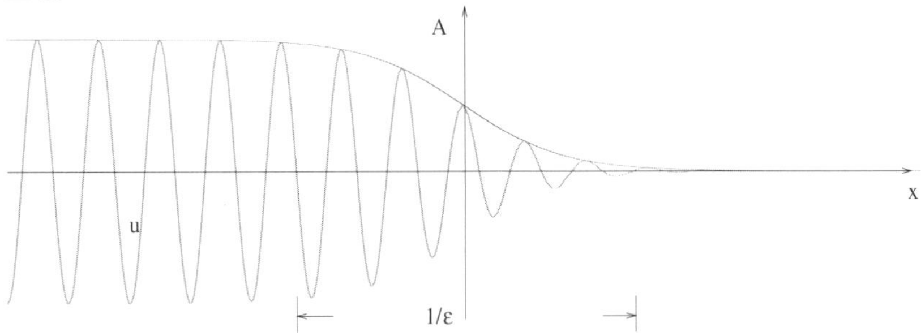

Figure 10.4. The front solution in the GL equation and the modulating front for the SH equation.

Figures 10.2-10.5 illustrate why the GL equation is called a modulation equation. There are a number of interesting mathematical questions about the approximation of solutions of the SH equation via the GL equation, e.g., how good is the formal approximation, and can we give existence and stability proofs for the formally constructed solutions? The first question is considered in §10.2.2, and the second for spatially periodic stationary solutions in §13.2.2. Their stability is discussed in §14.2.2. As a preview and an illustration of the quality of the approximation we give a numerical result in Figure 10.5.

Remark 10.2.1. For $\alpha > 0$ we have a continuum of wave numbers with $\lambda(k) > 0$ for $k \in (k_-, k_+)$, where $k_\pm = 1 \pm \sqrt{\alpha/2} + \mathcal{O}(\alpha)$. Thus, $2\pi/k$-periodic solutions of the SH equation occur for all wave numbers $k \in (k_-, k_+)$. Therefore, the two parameter families (10.9), (10.11) and (10.13) are embedded in larger families with the wave number as a third parameter. ⌋

10.2.2. Validity of the GL approximation cubic case. In this section we explain how to justify the GL equation in case of cubic nonlinearities in the original system. For expository reasons we restrict to the SH equation (10.4) as original system. Our main purpose is to prove that solutions of the SH equation (10.4) behave as predicted by the associated GL equation (10.7). In the following many different constants are denoted by the same symbol C if they can be chosen independently of the small perturbation parameter $0 \leq \varepsilon \ll 1$.

Formal smallness of the residual. As already said, the terms which do not cancel after inserting the GL approximation into the SH equation are collected in the residual

$$\mathrm{Res}(u) = -\partial_t u - (1 + \partial_x^2)^2 u + \varepsilon^2 u - u^3.$$

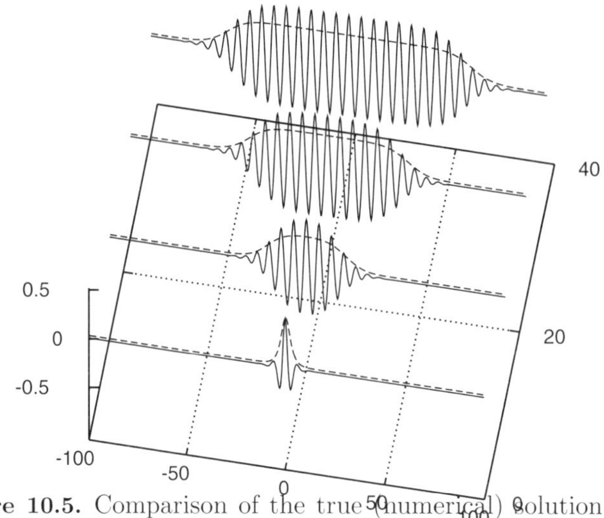

Figure 10.5. Comparison of the true (numerical) solution of the SH equation with ε=0.5 and initial condition $u_0(x)$=$A_0(\varepsilon x)\cos(x)$, $A_0(X) = 1/\cosh(X)$, with the (numerical) solution A (dashed line) of the GL equation with initial condition $A_0(X)$, indicating that $\psi_{GL}(x,t) = \varepsilon A(\varepsilon x, \varepsilon^2 t)\cos(x)$ gives a good approximation for all times considered.

The ansatz (10.6) and the abbreviation $\mathbf{E} = \mathrm{e}^{\mathrm{i}x}$ give

$$\begin{aligned}\mathrm{Res}(\varepsilon\psi_{GL}) =& \varepsilon^3\mathbf{E}\left(-\partial_T A + 4\partial_X^2 A + A - 3A|A|^2\right)\\ &- \varepsilon^3\mathbf{E}^3 A^3 - \varepsilon^4\mathbf{E}(4\mathrm{i}\partial_X^3 A) - \varepsilon^5\mathbf{E}(\partial_X^4 A) + \text{c.c.},\end{aligned}$$

and by choosing A to satisfy the GL equation (10.7) the first line of the residual vanishes. However, we still have terms of $\mathcal{O}(\varepsilon^3)$, which turn out to be too large for proving an approximation result.

By adding higher order terms to the approximation $\varepsilon\psi_{GL}$ the residual can be made arbitrarily small, i.e., for arbitrary but fixed $n \in \mathbb{N}$ there exists an approximation $\varepsilon\psi_n$ with formally $\varepsilon\psi_n - \varepsilon\psi_{GL} = \mathcal{O}(\varepsilon^3)$ and $\mathrm{Res}(\varepsilon\psi_n) = \mathcal{O}(\varepsilon^n)$. Since $\varepsilon\psi_n - \varepsilon\psi_{GL} = \mathcal{O}(\varepsilon^3)$ the approximation $\varepsilon\psi_n$ makes the same predictions as $\varepsilon\psi_{GL}$ about the behavior of the solutions u of the original system. We will show $\mathrm{Res}(\varepsilon\psi_n) = \mathcal{O}(\varepsilon^n)$ for $n = 4, 5$. With these examples the general situation will be clear.

In order to obtain

$$\mathrm{Res}(\varepsilon\psi_4) = \mathcal{O}(\varepsilon^4) \tag{10.14}$$

we define

$$\varepsilon\psi_4 = \varepsilon\psi_{GL} + \left(\varepsilon^3 A_3(\varepsilon x, \varepsilon^2 t)\mathbf{E}^3 + \text{c.c.}\right).$$

We then have

$$\mathrm{Res}(\varepsilon\psi_4) = \varepsilon^3\mathbf{E}^3(-A^3 - 64A_3) + \mathcal{O}(\varepsilon^4),$$

and by choosing $A_3 = -\frac{1}{64}A^3$ we obtain (10.14). In order to obtain

$$\text{Res}(\varepsilon\psi_5) = \mathcal{O}(\varepsilon^5) \tag{10.15}$$

we define

$$\varepsilon\psi_5 = \varepsilon\psi_4 + (\varepsilon^2 A_{12}(\varepsilon x, \varepsilon^2 t)\mathbf{E} + \varepsilon^4 A_{32}(\varepsilon x, \varepsilon^2 t)\mathbf{E}^3 + \text{c.c.})$$

where A_{12} and A_{32} are new functions which are chosen below. We find

$$\begin{aligned}\text{Res}(\varepsilon\psi_5) = \ & \varepsilon^4\mathbf{E}(-\partial_T A_{12} + 4\partial_X^2 A_{12} + A_{12} - 3A^2\overline{A_{12}} - 6|A|^2 A_{12} - 4\mathrm{i}\partial_X^3 A) \\ & +\varepsilon^4\mathbf{E}^3(-64A_{32} - 3A^2 A_{12} - 96\partial_X A_3) + \mathcal{O}(\varepsilon^5).\end{aligned}$$

By choosing A_{12} to satisfy the inhomogeneous linearized GL equation

$$\partial_T A_{12} = 4\partial_X^2 A_{12} + A_{12} - 3A^2\overline{A_{12}} - 6|A|^2 A_{12} - 4\mathrm{i}\partial_X^3 A, \tag{10.16}$$

and A_{32} to satisfy $-64A_{32} - 3A^2 A_{12} - 96\partial_X A_3 = 0$, we obtain (10.15). In order to achieve $\text{Res}(\varepsilon\psi_n) = \mathcal{O}(\varepsilon^n)$ we choose

$$\varepsilon\psi_n = \sum_{m=-N,\dots,N} \sum_{j=1}^{\widetilde{\alpha}(m)} \varepsilon^{\alpha(m)+j} A_{mj}(X,T)\mathbf{E}^m \tag{10.17}$$

with $N = n-1$, $X = \varepsilon x$, $T = \varepsilon^2 t$, and

m	0	1	2	3	$\cdots$	m	$\cdots$	N
$\alpha(m)$	2	1	2	3	$\cdots$	$\lvert\lvert m\rvert - 1\rvert + 1$	$\cdots$	N
$\widetilde{\alpha}(m)$	$N-1$	$N-2$	$N-1$	$N-2$	$\cdots$	$N+1-\alpha(m)-2\delta_{\lvert m\rvert 1}$	$\cdots$	1

where $\delta_{ij} = 1$ for $i = j$, and 0 else, see Figure 10.6.

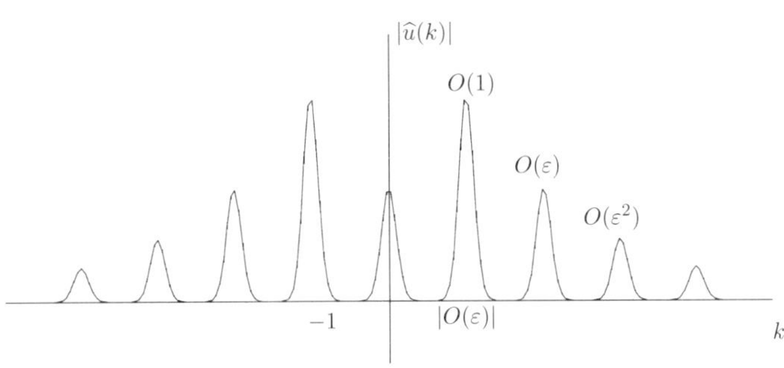

Figure 10.6. Fourier mode distribution of the GL approximation. Note that $\widehat{A}(\frac{k-1}{\varepsilon})$ is $\mathcal{O}(1)$ in L^∞, but $\mathcal{O}(\varepsilon)$ in L^1 and $\mathcal{O}(\varepsilon)$ in L^∞ in physical space.

As before A_{11} satisfies the GL equation. The evolution of the A_{1j} for $j \geq 2$ are determined by inhomogeneous linearized GL equations, and the A_{mj} for $m \neq \pm 1$ are determined by algebraic equations. The solutions A_{mj} of the resulting system exist as long as the solution A_{11} of the GL equation exists. In case of an odd nonlinearity such as for (10.4) we can set $A_{mj} = 0$ for m odd.

Estimates for the residual in C_b^0. The formal orders of the residual can be turned into estimates in norms. We find for instance

$$\| \operatorname{Res}(\varepsilon\psi_{\mathrm{GL}})\|_{C_b^0} \leq s_1 + s_2 + s_3,$$

where

$$\begin{aligned} s_1 &= \|3\varepsilon^3 E^3 A(\varepsilon\cdot)^3\|_{C_b^0} \leq 3\varepsilon^3 \|A\|_{C_b^0}^3 , \\ s_2 &= \|\varepsilon^4 \mathbf{E}(4\mathrm{i}\partial_X^3 A(\varepsilon\cdot))\|_{C_b^0} \leq 4\varepsilon^4 \|A\|_{C_b^3} , \\ s_3 &= \|\varepsilon^5 \mathbf{E}(\partial_X^4 A(\varepsilon\cdot))\|_{C_b^0} \leq \varepsilon^5 \|A\|_{C_b^4} , \end{aligned}$$

such that the following holds.

Lemma 10.2.2. *Let $A \in C([0,T_0], C_b^4)$ be a solution of the GL equation (10.7). Then for all $\varepsilon_0 \in (0,1]$ there exists a $C > 0$ such that for all $\varepsilon \in (0,\varepsilon_0)$ we have*

$$\sup_{t\in[0,T_0/\varepsilon^2]} \| \operatorname{Res}(\varepsilon\psi_{\mathrm{GL}}(t))\|_{C_b^0} \leq C\varepsilon^3.$$

Similarly, for every $n \in \mathbb{N}$ with $n \geq 4$ there exists an approximation $\varepsilon\psi_n$ such that the following holds. Let $A \in C([0,T_0], C_b^{\theta_A})$ with $\theta_A = 3(n-3)+1$ be a solution of the GL equation (10.7). Then for all $\varepsilon_0 \in (0,1]$ there exists a $C > 0$ such that for all $\varepsilon \in (0,\varepsilon_0)$ we have

$$\sup_{t\in[0,T_0/\varepsilon^2]} \| \operatorname{Res}(\varepsilon\psi_n(t))\|_{C_b^0} \leq C\varepsilon^n,$$

and

$$\sup_{t\in[0,T_0/\varepsilon^2]} \|\varepsilon\psi_{GL}(t) - \varepsilon\psi_n(t)\|_{C_b^0} \leq C\varepsilon^3.$$

Remark 10.2.3. Looking in more detail at the approximation $\varepsilon\psi_3$ we find that most regularity is lost in (10.16) for A_{12}. Since $\|\partial_X^4 A_{12}(\cdot,T)\|_{C_b^0}$ has to be estimated, we need $A_{12}(\cdot,T) \in C_b^4$, and if we do not use the smoothing properties of the linear part in (10.16), then we need $A(\cdot,T) \in C_b^7$. Since the structure of the approximation equations is the same for the next orders, we lose three derivatives in every step, such that the estimates are possible with $\theta_A = 3(n-3)+1$. However, by using the smoothing properties of the linear part in (10.16) and by constructing the approximation in a more clever way, θ_m can be chosen much smaller, see below. ⌋

Estimates for the residual in H^θ. For many pattern forming systems the C_b^θ spaces are no suitable phase spaces, because often it is hard to establish the existence of semigroups in these spaces. In this respect Sobolev spaces turn out to be a better choice. The estimates for the residual are less trivial than before due to the scaling properties of the L^2-norm. For $\varepsilon\psi_{GL}$ we find for $\theta > 1/2$ similar as before

$$\begin{aligned}&\|\operatorname{Res}(\varepsilon\psi_{GL})\|_{H^\theta}\\ &\qquad\leq C\left(\varepsilon^3\|A(\varepsilon\cdot)\|^2_{C_b^\theta}\|A(\varepsilon\cdot)\|_{H^\theta}+\varepsilon^4\|A(\varepsilon\cdot)\|_{H^{\theta+3}}+\varepsilon^5\|A(\varepsilon\cdot)\|_{H^{\theta+4}}\right).\end{aligned}$$

However, for $\varepsilon \in (0,1)$ we find

$$\begin{aligned}\|A(\varepsilon\cdot)\|_{L^2} &= \int_{\mathbb{R}}\left(|A(\varepsilon x)|^2\,\mathrm{d}x\right)^{1/2}\\ &= \varepsilon^{-1/2}\int_{\mathbb{R}}\left(|A(X)|^2\,\mathrm{d}X\right)^{1/2}\leq\varepsilon^{-1/2}\|A\|_{L^2},\end{aligned}\tag{10.18}$$

such that finally

$$\|\operatorname{Res}(\varepsilon\psi_{GL}(t))\|_{H^\theta}\leq C\varepsilon^{5/2}.$$

Nevertheless, for the proof of an approximation theorem this loss of $\varepsilon^{-1/2}$ is no problem, since as before the residual can be made arbitrarily small by adding higher order terms to the approximation.

Lemma 10.2.4. *For all $n \in \mathbb{N}$ with $n \geq 4$, and $\theta \geq 1$, there exists an approximation $\varepsilon\psi_n$ such that the following holds. Let $A \in C([0,T_0],H^{\theta_A})$ with $\theta_A = 3(n-3)+1+\theta$ be a solution of the GL equation* (10.7). *Then for all $\varepsilon_0 \in (0,1]$ there exists a $C > 0$ such that for all $\varepsilon \in (0,\varepsilon_0)$ we have*

$$\sup_{t\in[0,T_0/\varepsilon^2]}\|\operatorname{Res}(\varepsilon\psi_n(t))\|_{H^\theta}\leq C\varepsilon^{n-1/2}$$

and

$$\sup_{t\in[0,T_0/\varepsilon^2]}\|\varepsilon\psi_{GL}(t)-\varepsilon\psi_n(t)\|_{C_b^0}\leq C\varepsilon^3.$$

Estimates for the residual in H^θ_{ul}. In §8.3.1 we introduced the H^θ_{ul}-spaces to use the advantages of both spaces C_b^0 and H^θ, namely that C_b^0 contains non-decaying functions for $|x|\to\infty$ and the fact that Fourier transform is a suitable tool in H^θ. The residual is estimated in H^θ_{ul} as follows. For $\theta > 1/2$ we find

$$\|\operatorname{Res}(\varepsilon\psi_{GL})\|_{H^\theta_{\mathrm{ul}}}\leq C\left(\varepsilon^3\|A(\varepsilon\cdot)\|^3_{H^\theta_{\mathrm{ul}}}+\varepsilon^4\|A(\varepsilon\cdot)\|_{H^{\theta+3}_{\mathrm{ul}}}+\varepsilon^5\|A(\varepsilon\cdot)\|_{H^{\theta+4}_{\mathrm{ul}}}\right).$$

The loss of $\varepsilon^{-1/2}$ can be avoided by

$$\begin{aligned}\|A(\varepsilon\cdot)\|_{L^2_{\mathrm{ul}}} &= \sup_{y\in\mathbb{R}}\left(\int_y^{y+1}|A(\varepsilon x)|^2\,\mathrm{d}x\right)^{1/2}\\ &\leq \sup_{y\in\mathbb{R}}\left\{\left(\int_y^{y+1}1\,\mathrm{d}x\right)^{1/2}\sup_{x\in[y,y+1]}|A(\varepsilon x)|\right\}\leq\|A\|_{C_b^0}\leq C\|A\|_{H^1_{\mathrm{ul}}}\end{aligned}$$

such that we finally obtain

$$\| \operatorname{Res}(\varepsilon\psi_{GL}(t))\|_{H^\theta_{\mathrm{ul}}} \leq C\varepsilon^3.$$

Like before we have

Lemma 10.2.5. *For all $n \in \mathbb{N}$ with $n \geq 4$, and $\theta \geq 1$, there exists an approximation $\varepsilon\psi_n$ such that the following holds. Let $A \in C([0,T_0], H^{\theta_A}_{\mathrm{ul}})$ with $\theta_A = 3(n-3)+1+\theta$ be a solution of the GL equation (10.7). Then for all $\varepsilon_0 \in (0,1]$ there exists a $C>0$ such that for all $\varepsilon \in (0,\varepsilon_0)$ we have*

$$\sup_{t\in[0,T_0/\varepsilon^2]} \| \operatorname{Res}(\varepsilon\psi_n(t))\|_{H^\theta_{\mathrm{ul}}} \leq C\varepsilon^n,$$

and

$$\sup_{t\in[0,T_0/\varepsilon^2]} \|\varepsilon\psi_{GL}(t) - \varepsilon\psi_n(t)\|_{C^0_b} \leq C\varepsilon^3.$$

The equations for the error. Estimates for the residual, even in norms, are only a necessary condition for the GL equation to correctly predict the behavior of the original systems. Such estimates are not sufficient. The errors can accumulate in time and there are a number of counter-examples, cf. [**Sch95b, Sch05, SSZ15**], §10.6.3, and §11.5.3, showing that modulation equations, although derived in a formally correct way, make wrong predictions about the behavior of the original system.

The error

$$\varepsilon^\beta R = u - \varepsilon\psi,$$

i.e., the difference between the solution u and the approximation $\varepsilon\psi = \varepsilon\psi_n$, with β and n suitably chosen, is estimated with the help of Gronwall's inequality. It satisfies

$$\partial_t R = -(1+\partial_x^2)^2 R + \varepsilon^2 R - 3\varepsilon^2\psi^2 R - 3\varepsilon^{\beta+1}\psi R^2 - \varepsilon^{2\beta}R^3 + \varepsilon^{-\beta}\operatorname{Res}(\varepsilon\psi).$$

We use the variation of constant formula

$$R(t) = \mathrm{e}^{t\Lambda}R(0) + \varepsilon^2\int_0^t \mathrm{e}^{(t-\tau)\Lambda}F(\tau)\,\mathrm{d}\tau, \tag{10.19}$$

where $(\mathrm{e}^{t\Lambda})_{t\geq 0}$ is the semigroup generated by $\Lambda = -(1+\partial_x^2)^2$ and

$$\begin{aligned}\varepsilon^2 F(\tau) =& \varepsilon^2 R(\tau) - 3\varepsilon^2\psi(\tau)^2 R(\tau) - 3\varepsilon^{\beta+1}\psi(\tau)R(\tau)^2 \\ &- \varepsilon^{2\beta}R(\tau)^3 + \varepsilon^{-\beta}\operatorname{Res}(\varepsilon\psi(\tau)).\end{aligned}$$

We have to decide in which norm we want to estimate the error, and are not completely free in this. If A does not decay to zero for $x \to \pm\infty$, then the residual will be infinitely large in H^θ. There are essentially three ingredients which have to be estimated, namely first the residual terms which have already been estimated in the spaces C^0_b, H^θ, and H^θ_{ul} above. Secondly, we have to estimate the other terms in F, and thirdly the linear semigroup $(e^{\Lambda t})_{t\geq 0}$.

In order to make clear that estimates for F are easy to obtain independently of the chosen norm we will give these estimates in all three spaces. Therefore, we assume that $A \in C([0,T_0], C^{\theta_A}_{b,unif})$, or $A \in C([0,T_0], H^{\theta_A})$, or $A \in C([0,T_0], H^{\theta_A}_{\mathrm{ul}})$, is a solution of the GL equation with $\theta_A \geq 0$ sufficiently large. Then we obtain

Lemma 10.2.6. *For every $\theta \geq 1$ there is a $C > 0$ such that for all $\varepsilon \in (0,1]$ we have*

$$\|\varepsilon^2 F\|_{X^\theta} \leq C(\varepsilon^2 \|R\|_{X^\theta} + \varepsilon^{\beta+1}\|R\|^2_{X^\theta} + \varepsilon^{2\beta}\|R\|^3_{X^\theta} + \varepsilon^2),$$

where X^θ stands either for C^0_b, H^θ, or H^θ_{ul} depending on the chosen phase space for A.

Proof. All spaces X^θ are closed under multiplication if $\theta > 1/2$. Since the H^θ-norm of ψ is of order $\mathcal{O}(\varepsilon^{-1/2})$ we have to estimate ψ in the $C^{\widetilde{\theta}}_b$-norm, where $\widetilde{\theta} = 0$ in case $X^\theta = C^0_b$ and $\widetilde{\theta} = \theta$ in the other cases. Then the estimate follows from

$$\begin{aligned}
\|\psi^2 R\|_{X^\theta} &\leq C\|\psi\|^2_{C^{\widetilde{\theta}}_b}\|R\|_{X^\theta},\\
\|\psi R^2\|_{X^\theta} &\leq C\|\psi\|_{C^{\widetilde{\theta}}_b}\|R\|^2_{X^\theta},\\
\|R^3\|_{X^\theta} &\leq C\|R\|^3_{X^\theta},\\
\|\varepsilon^{-\beta}\operatorname{Res}(\varepsilon\psi)\|_{X^\theta} &\leq C\varepsilon^2,
\end{aligned}$$

where the last estimate is either a consequence of Lemma 10.2.2 or Lemma 10.2.4, or Lemma 10.2.5. Moreover, we have

$$\|\psi_{GL}\|_{C^{\widetilde{\theta}}_b} = \|x \mapsto (A(\varepsilon x)e^{ix} + c.c.)\|_{C^{\widetilde{\theta}}_b} \leq C\|A\|_{C^{\widetilde{\theta}}_b}$$

and similar estimates for the other terms in ψ. Sobolev's embedding theorem gives the additional bounds

$$\|A\|_{C^{\widetilde{\theta}}_b} \leq C\|A\|_{H^{\widetilde{\theta}+1}} \qquad \text{and} \qquad \|A\|_{C^{\widetilde{\theta}}_b} \leq C\|A\|_{H^{\widetilde{\theta}+1}_{\mathrm{ul}}}.$$

Therefore, we are done. □

The third step, the bound for the semigroup $(e^{t\Lambda})_{t\geq 0}$, depends in its complexity on the chosen space. Uniform bounds in the spaces C^0_b and H^θ_{ul} are hard to obtain. In order to keep this introductory section simple, we skip the proof in C^0_b and postpone the proof in H^θ_{ul} to §10.4.

Lemma 10.2.7. *For every $\theta \geq 0$ there exists a $C > 0$ such that for all $\varepsilon \in (0,1]$ the semigroup $(e^{t\Lambda})_{t\geq 0}$ generated by $\Lambda = -(1+\partial_x^2)^2$ satisfies*

$$\sup_{t\in[0,T_0/\varepsilon^2]} \|e^{t\Lambda}\|_{H^\theta \to H^\theta} \leq C.$$

Proof. We use that Fourier transform is an isomorphism from H^θ to H^0_θ, cf. Lemma 7.3.31, and obtain

$$\|e^{t\Lambda}u\|_{H^\theta} \leq C\|e^{-(1-\cdot k^2)^2 t}\widehat{u}\|_{H^0_\theta} \leq C\sup_{k\in\mathbb{R}}|e^{-(1-k^2)^2 t}|\|\widehat{u}\|_{H^0_\theta} \leq C\|u\|_{H^\theta}.$$

□

The associated estimate in $C^0_{b,unif}$ is less trivial and can be found in [**KSM92**, Lemma 2.3]. Our favorite choice is the space H^θ_{ul}.

Lemma 10.2.8. *For every $\theta \geq 0$ there exists a $C > 0$ such that for all $\varepsilon \in (0,1]$ the semigroup $(e^{t\Lambda})_{t\geq 0}$ generated by $\Lambda = -(1+\partial_x^2)^2$ satisfies*

$$\sup_{t\in[0,T_0/\varepsilon^2]} \|e^{t\Lambda}\|_{H^\theta_{\mathrm{ul}}\to H^\theta_{\mathrm{ul}}} \leq C.$$

Proof. See page 343. □

Now we have all estimates which are necessary to prove the following approximation result. In the following Theorem 10.2.9 and Remark 10.2.11 the space X^θ stands either for H^θ or H^θ_{ul}. In case $X^\theta = H^\theta$ we set $\beta = n-5/2$ and in the other cases $\beta = n-2$.

Theorem 10.2.9. *Let $n \in \mathbb{N}$, $\theta \geq 1$, and $A \in C([0,T_0], X^{\theta_A})$ with $\theta_A = 3(n-3)+1+\theta$ be a solution of the GL equation* (10.7). *Then there exist an $\varepsilon_0 > 0$ and a $C > 0$ such that for all $\varepsilon \in (0,\varepsilon_0)$ there are solutions u of the SH equation* (10.4) *which can be approximated by $\varepsilon\psi_n$ such that*

$$\sup_{t\in[0,T_0/\varepsilon^2]} \|u(t) - \varepsilon\psi_n(t)\|_{X^\theta} \leq C\varepsilon^\beta, \tag{10.20}$$

and thus

$$\sup_{t\in[0,T_0/\varepsilon^2]} \|u(t) - \varepsilon\psi_{GL}(t)\|_{X^\theta} \leq C\varepsilon^{3/2}, \tag{10.21}$$

where $\varepsilon\psi_n$ is the full ansatz (10.17), *and $\varepsilon\psi_{GL}$ is the lowest order approximation* (10.6).

Proof. We use Lemma 10.2.6 and Lemma 10.2.7, respectively Lemma 10.2.8, to bound the mild solutions of the equations for the error in the variation of constant formula (10.19). By Theorem 5.2.22 we have local existence and uniqueness of the solutions in C^0_b, H^θ, or H^θ_{ul} for $\theta > 1/2$. The solutions exist until the norm of the solutions becomes infinite, and thus we only have to derive a priori estimates to guarantee that the solutions exist and stay bounded for $t \in [0, T_0/\varepsilon^2]$.

$$
\begin{aligned}
\|R(t)\|_{X^\theta} &\le C\varepsilon^2 \int_0^t \Big(\|R(\tau)\|_{X^\theta} + \varepsilon^{\beta-1}\|R(\tau)\|_{X^\theta}^2 + \varepsilon^{2\beta-2}\|R(\tau)\|_{X^\theta}^3 + 1 \Big)\, \mathrm{d}\tau \\
&\le C\varepsilon^2 \int_0^t (\|R(\tau)\|_{X^\theta} + 2)\, \mathrm{d}\tau \\
&\le 2CT_0 + C\varepsilon^2 \int_0^t \|R(\tau)\|_{X^\theta}\, \mathrm{d}\tau,
\end{aligned}
$$

which holds as long as

$$
\varepsilon^{\beta-1}\|R(\tau)\|_{X^\theta}^2 + \varepsilon^{2\beta-2}\|R(\tau)\|_{X^\theta}^3 \le 1. \tag{10.22}
$$

Gronwall's inequality then yields

$$
\|R(t)\|_{X^\theta} \le 2CT_0 \mathrm{e}^{C\varepsilon^2 t} \le 2CT_0\mathrm{e}^{CT_0} =: M
$$

for all $t \in [0, T_0/\varepsilon^2]$. Choosing $\varepsilon_0 > 0$ such that $\varepsilon_0^{\beta-1}M^2 + \varepsilon_0^{2\beta-2}M^3 \le 1$ we have satisfied (10.22). This yields (10.20). The triangle inequality then yields (10.21), i.e.,

$$
\begin{aligned}
&\sup_{t\in[0,T_0/\varepsilon^2]} \|u(t) - \varepsilon\psi_{GL}(t)\|_{X^\theta} \\
&\qquad \le \sup_{t\in[0,T_0/\varepsilon^2]} \|u(t) - \varepsilon\psi_n(t)\|_{X^\theta} + \sup_{t\in[0,T_0/\varepsilon^2]} \|\varepsilon\psi_n(t) - \varepsilon\psi_{GL}(t)\|_{X^\theta} \\
&\qquad \le C\varepsilon^\beta + C\varepsilon^{3/2}
\end{aligned}
$$

□

Remark 10.2.10. For clarity we formulate the last step of Theorem 10.2.9 in a different way. We define $\widetilde{t} = \sup\{t : \|R(t)\|_{X^\theta} \le M\}$, with M defined as above. We are done, if we show $\widetilde{t} \ge T_0/\varepsilon^2$. Choosing $\varepsilon_0 > 0$ in the above way guarantees the validity of (10.22) for all $t \le \widetilde{t}$ and so the validity of

$$
\|R(t)\|_{X^\theta} \le 2CT_0 + C\varepsilon^2 \int_0^t \|R(\tau)\|_{X^\theta}\, \mathrm{d}\tau
$$

for all $t \le \widetilde{t}$. As above, Gronwall's inequality implies $\sup_{t\in[0,T_0\varepsilon^2]} \|R(t)\|_{X^\theta} \le M$, and so $\widetilde{t} \ge T_0/\varepsilon^2$. ⌋

Remark 10.2.11. The estimate (10.21) shows that already the original ansatz yields a suitable approximation, i.e., for $\varepsilon > 0$ sufficiently small we see the dynamics predicted by the GL equation in the original system. For times $t \in [0, T_0/\varepsilon^2]$ the error of order $\mathcal{O}(\varepsilon^{\min(3/2,\beta)})$ in X^θ and by Sobolev's embedding theorem also in C_b^0 is much smaller than the solution and approximation which are both of order $\mathcal{O}(\varepsilon)$ in C_b^0. ⌋

10.3. The universality of the GL equation

Above we already claimed that the GL equation

$$\partial_T A = \nu_2 \partial_X^2 A + \nu_0 A + \nu_3 A|A|^2, \tag{10.23}$$

with coefficients $\nu_0, \nu_2, \nu_3 \in \mathbb{C}$, appears as a universal modulation equation for many different systems. The underlying original system will condense in the values of the coefficients ν_0, ν_2, and ν_3 in (10.23). Here is another simple example.

Example 10.3.1. The Kuramoto-Shivashinsky (KS) equation

$$\partial_t u = -(1+\partial_x^2)^2 u + \alpha u + \partial_x(u^2), \tag{10.24}$$

with $u(x,t) \in \mathbb{R}$, $x \in \mathbb{R}$, $t \geq 0$, and $\alpha \in \mathbb{R}$ appears as a model for flame propagation [**KY76, Siv77, Kur80**]. The lowest order approximation

$$\varepsilon \Psi_{\mathrm{GL}} = \varepsilon A_1(X,T)\mathrm{e}^{\mathrm{i}x} + \varepsilon A_{-1}(X,T)\mathrm{e}^{-\mathrm{i}x}, \tag{10.25}$$

where $0 < \alpha = \varepsilon^2 \ll 1$ is a small perturbation parameter, is no longer sufficient for the derivation of the GL equation. The ansatz is now given by

$$\varepsilon \Psi = \varepsilon \Psi_{\mathrm{GL}} + \varepsilon^2 \Psi_h \tag{10.26}$$

where

$$\varepsilon \Psi_h = +\varepsilon^2 A_2(X,T)\mathrm{e}^{2\mathrm{i}x} + \varepsilon^2 A_{-2}(X,T)\mathrm{e}^{-2\mathrm{i}x} + \varepsilon^2 A_0(X,T),$$

and $A_{-j}(X,T) = \overline{A_j(X,T)}$. We obtain

$$\begin{aligned} \partial_T A_1 &= 4\partial_X^2 A_1 + A_1 + 2i(A_0 A_1 + A_2 A_{-1}), \\ 0 &= -9A_2 + 2\mathrm{i} A_1 A_1, \\ 0 &= -A_0. \end{aligned}$$

Eliminating A_0 and A_2 via the algebraic relations from the second and third line in the equation for A_1 shows that the envelope A_1 has to satisfy in lowest order the GL equation

$$\partial_T A = 4\partial_X^2 A + A - \frac{4}{9} A|A|^3. \tag{10.27}$$

⌋

In order to explain why the GL equation plays such an important role and occurs as universal amplitude or modulation equation describing the evolution of many pattern forming systems close to the first instability, we first review the derivation of the GL equation from the SH equation (10.4) from a different point of view. The following computations are of limited practical use, i.e., the are not useful for obtaining the nonlinear coefficient ν_3 of the GL equation, see also Remark 10.3.2 below, but they give the essential theoretical framework for understanding the universality of the GL equation.

The Fourier transformed SH equation. It turns out that Fourier transform is the key for understanding the universality. In Fourier space, the linearized SH equation is given by

$$\partial_t \widehat{u}(k,t) = \lambda(k,\alpha)\widehat{u}(k,t), \quad \text{with} \quad \lambda(k,\alpha) = -(1-k^2)^2 + \alpha.$$

For $\alpha = \varepsilon^2 > 0$ there are two intervals centered around $k = \pm 1$ on which $\lambda(k,\varepsilon^2) > 0$, namely $I_-(\varepsilon) \cup I_+(\varepsilon) = (-k_+(\varepsilon), -k_-(\varepsilon)) \cup (k_-(\varepsilon), k_+(\varepsilon))$ with

$$k_\pm(\varepsilon) = \sqrt{1 \pm \varepsilon} = 1 \pm \frac{1}{2}\varepsilon + \mathcal{O}(\varepsilon^2).$$

A wave $e^{ikx+\lambda(k,\varepsilon^2)t}$ associated with a $k \in I_+(\varepsilon)$ (or $k \in I_-(\varepsilon)$) will grow exponentially, whereas a wave $e^{ikx+\lambda(k,\varepsilon^2)t}$ associated with $k \notin \overline{I_\pm(\varepsilon)}$ will be damped with some exponential rate. In order to find a modulation equation, which describes the evolution of these unstable modes in an ε-neighborhood of $k = \pm k_c = \pm 1$, and which is independent of the small perturbation parameter ε we introduce K by $k = 1 + \varepsilon K$, such that

$$k \in I_+(\varepsilon) \Leftrightarrow K \in \left(-\frac{1}{2} + \mathcal{O}(\varepsilon), \frac{1}{2} + \mathcal{O}(\varepsilon)\right). \tag{10.28}$$

With

$$\begin{aligned} \lambda(1+\varepsilon K, \varepsilon^2) =& \lambda(1,0) + \partial_1\lambda(1,0)\varepsilon K + \frac{1}{2}\partial_1^2\lambda(1,0)\varepsilon^2 K^2 \\ & + \partial_2\lambda(1,0)\varepsilon^2 + \mathcal{O}(\varepsilon^3) \\ =& \varepsilon^2(1-4K^2) + \mathcal{O}(\varepsilon^3) \end{aligned} \tag{10.29}$$

we can now write a family of linearly unstable spatially periodic perturbations of $u^* = 0$ as

$$e^{i(1+\varepsilon K)x + \varepsilon^2(1-4K^2+\mathcal{O}(\varepsilon))t} = e^{iK\varepsilon x + (1-4K^2+\mathcal{O}(\varepsilon))(\varepsilon^2 t)} e^{ix}. \tag{10.30}$$

This has the form of a "basic pattern" e^{ix} modulated by a slowly varying amplitude

$$A_{\text{lin}}(X,T) = e^{iKX+(1-4K^2)T} + \mathcal{O}(\varepsilon), \text{ with } X = \varepsilon x \text{ and } T = \varepsilon^2 t. \tag{10.31}$$

Then, ignoring the $\mathcal{O}(\varepsilon)$ terms, we find

$$\partial_T A_{\text{lin}} = 4\partial_X^2 A_{\text{lin}} + A_{\text{lin}} \tag{10.32}$$

which is the linear part of the GL equation. It is natural to expect that the nonlinear problem (10.4) also has solutions which are small amplitude modulations of e^{ix}. The nonlinear terms determine the scaling of the amplitude of the bifurcating solutions and so, as above, we insert the ansatz

$$\varepsilon\psi_{GL}(x,t) = \varepsilon A(X,T)e^{ix} + \varepsilon\overline{A}(X,T)e^{-ix} \tag{10.33}$$

into the SH equation (10.4). As illustrated by Example 10.3.1, even if we have quadratic terms in the original system, the nonlinear term in the GL equation must always be of the form $\nu_3 A|A|^2$. This is due to the fact that the quadratic interaction of the unstable modes $\mathrm{e}^{\pm ix}$ gives modes concentrated at the wave numbers $k = 0, \pm 2$ and not at the unstable modes at $k = \pm 1$ in contrast to $\nu_3 A|A|^2$. A term proportional to $A|A|^4$ is also concentrated at $k = \pm 1$, but is of higher order.

The instability mechanism. With this idea in mind we consider a general dissipative PDE in a cylindrical domain $\mathbb{R} \times \Sigma$ with coordinates $x \in \mathbb{R}$ and $y \in \Sigma$, where $\Sigma \subset \mathbb{R}^d$ is an bounded domain. We assume that the system does not depend explicitly on $x \in \mathbb{R}$, and that there exists a trivial solution U^* which depends on $y \in \Sigma$ only. W.l.o.g. for our purposes we can assume that $U^* = 0$. Hence, we consider

$$\partial_t U = LU + N_2(U,U) + N_3(U,U,U) + \ldots$$

for an unknown function U. The operator L is linear and the terms N_j are j-multilinear and w.l.o.g. symmetric in their arguments. They can depend on $y \in \Sigma$ and on a control parameter $\alpha \in \mathbb{R}$, but they do not depend explicitly on $x \in \mathbb{R}$. If for instance N_2 is not symmetric, then define $\widetilde{N}_2(U,V) = \dfrac{1}{2}(N_2(U,V) + N_2(V,U))$, which has the properties that $\widetilde{N}_2(U,V) = \widetilde{N}_2(V,U)$ and $\widetilde{N}_2(U,U) = N_2(U,U)$.

Due to the translation invariance in x, the linearization $\partial_t U = LU$ around $U^* = 0$ possesses solutions of the form

$$U(x,y,t) = \varphi_n(k,y)\mathrm{e}^{ikx+\lambda_n(k)t},$$

with $k \in \mathbb{R}$. Under suitable assumptions on L and the boundary conditions on $\mathbb{R} \times \Sigma$, for fixed k the operator $-L_k\cdot = -\mathrm{e}^{-ikx}L(\mathrm{e}^{ikx}\cdot)$, acting on functions which live on the bounded cross section Σ, is elliptic. Therefore, L_k possesses a countable number of eigenvalues $\lambda_n(k)$ with associated eigenfunctions $\varphi_n(k,\cdot)$. We order the eigenvalues for fixed k in such a way that $\mathrm{Re}\ \lambda_j \geq \mathrm{Re}\ \lambda_{j+1}$.

We assume that $U^* = 0$ is stable for $\alpha < \alpha_c$, i.e., for $\alpha < \alpha_c$ all curves $k \mapsto \lambda_n(k)$ are in the left complex half-plane, respectively, all curves $k \mapsto \mathrm{Re}\lambda_n(k)$ are below the k-axis. We assume further that that for $\alpha = \alpha_c$ one curve λ_1 or a pair of complex conjugate curves touch the imaginary axis for a wave number $k = k_c \neq 0$. Since we consider a real-valued problem the same happens at $k = -k_c$. Generically, all other curves are strictly bounded away from the imaginary axis. See Figure 10.7.

For a simplification of the presentation we assume for the rest of this subsection that the linear operator L can be diagonalized, that only the curve $\lambda_1 = \lambda_1(k, \varepsilon^2)$ touches the k-axis at $(k, \alpha) = (k_c, 0)$, and that this

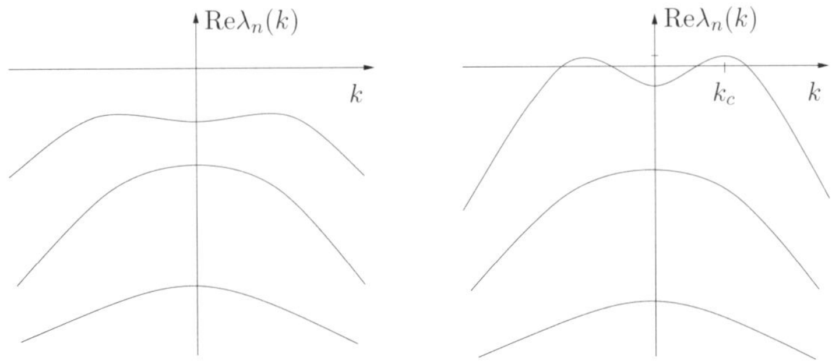

Figure 10.7. The curves of eigenvalues $k \mapsto \mathrm{Re}\lambda_n(k)$ in the stable case are shown in the left panel. In the right panel instability occurs due to one curve taking positive values at the wave numbers $\pm k_c$.

curve is real-valued like for the SH equation. Thus, we have $\lambda_1(k_c, 0) = 0$, $\partial_k \lambda_1(k_c, 0) = 0$, and $\partial_k^2 \lambda_1(k_c, 0) < 0$.

Derivation of the GL equation in general systems. After an expansion

$$U(x, y, t) = \int_{-\infty}^{\infty} \sum_{n \in \mathbb{N}} c_n(k, t) \varphi_n(k, y) \mathrm{e}^{ikx} \, \mathrm{d}k$$

in eigenfunctions the coefficients $c_n(k, t)$ satisfy

$$\begin{aligned} \partial_t c_n(k, t) &= \lambda_n(k) c_n(k, t) + \sum_{n_1, n_2 \in \mathbb{N}} \int_{-\infty}^{\infty} s_{2,n,n_1,n_2}(k, k-l, l) c_{n_1}(k-l) c_{n_2}(l) \, \mathrm{d}l \\ &+ \sum_{n_1, n_2 \in \mathbb{N}} \int_{-\infty}^{\infty} s_{3,n,n_1,n_2,n_3}(k, k-l, l-m, m) c_{n_1}(k-l) c_{n_2}(l-m) c_{n_3}(m) \, \mathrm{d}l \, \mathrm{d}m \\ &+ \ldots, \end{aligned}$$

with complex-valued kernels $s_{2,n,n_1,n_2}(k, k-l, l)$, $s_{3,n,n_1,n_2,n_3}(k, k-l, l-m, m)$, etc. Like the eigenvalues and eigenvectors, the kernels additionally depend on ε^2, e.g. $s_{2,n,n_1,n_2}(k, k-l, l; \varepsilon^2)$. We have explicit representations such as

$$s_{2,n,n_1,n_2}(k, k-l, l) = \left\langle \varphi_n^*(k), \mathrm{e}^{-ikx} N_2(\varphi_{n_1}(k-l)\mathrm{e}^{\mathrm{i}(k-l)x}, \varphi_{n_2}(l)\mathrm{e}^{\mathrm{i}lx}) \right\rangle_{L^2(\Sigma)}$$

where $\varphi_n^*(k)$ is the associated adjoint eigenfunction w.r.t. the scalar product $\langle \cdot, \cdot \rangle_{L^2(\Sigma)}$, with $\langle \varphi_1^*, \varphi_1 \rangle_{L^2(\Sigma)} = 1$.

The modes $\widehat{u}$ of the Fourier transformed SH equation solely correspond to the modes c_1. For general systems additionally to these modes we have infinitely many modes $c_2, c_3, \ldots$ which are damped with some exponential rate. Since the Fourier transform of $x \mapsto \varepsilon A(\varepsilon x, \varepsilon^2 t)\mathrm{e}^{ik_c x}$ is given by $k \mapsto$

$\widehat{A}(\varepsilon^{-1}(k-k_c),\varepsilon^2 t)$ we make the ansatz

$$\begin{aligned} c_1(k,t) =&\widehat{A}_{1,-1}(\varepsilon^{-1}(k+k_c),\varepsilon^2 t)+\widehat{A}_{1,1}(\varepsilon^{-1}(k-k_c),\varepsilon^2 t)\\ &+\varepsilon\widehat{A}_{1,-2}(\varepsilon^{-1}(k+2k_c),\varepsilon^2 t)+\varepsilon\widehat{A}_{1,0}(\varepsilon^{-1}k,\varepsilon^2 t)\\ &+\varepsilon\widehat{A}_{1,2}(\varepsilon^{-1}(k-2k_c),\varepsilon^2 t),\\ c_n(k,t) =&\varepsilon\widehat{A}_{n,-2}(\varepsilon^{-1}(k+2k_c),\varepsilon^2 t)+\varepsilon\widehat{A}_{n,0}(\varepsilon^{-1}k,\varepsilon^2 t)\\ &+\varepsilon\widehat{A}_{n,2}(\varepsilon^{-1}(k-2k_c),\varepsilon^2 t), \end{aligned}$$

for $n\geq 2$, with complex valued functions $\widehat{A}_{n,j}$, where $A_{n,-j}=\overline{A_{n,j}}$ in physical space. With $k-k_c=\varepsilon K$, and for instance

$$s_{211n}(k_c+\varepsilon K,k_c+\varepsilon(K-L),\varepsilon L;\varepsilon^2)-s_{211n}(k_c,k_c,0;0)=\mathcal{O}(\varepsilon(|K|+|L|)),$$

we find at ε^2 for the modes concentrated at k_c that

$$\begin{aligned} \partial_T\widehat{A}_{1,1}(K,T) =&\partial_\alpha\lambda_1(k_c,0)\widehat{A}_{1,1}(K,T)+\partial_k^2\lambda_1(k_c,0)K^2\widehat{A}_{1,1}(K,T)/2\\ &+2\sum_{n=1}^{\infty}s_{211n}(k_c,k_c,0;0)\int_{-\infty}^{\infty}\widehat{A}_{1,1}(K-\kappa,T)\widehat{A}_{n,0}(\kappa,T)\,\mathrm{d}\kappa\\ &+2\sum_{n=1}^{\infty}s_{211n}(k_c,-k_c,2k_c;0)\int_{-\infty}^{\infty}\widehat{A}_{1,-1}(K-\kappa,T)\widehat{A}_{n,2}(\kappa,T)\,\mathrm{d}\kappa\\ &+3s_{31111}(k_c,k_c,k_c,-k_c;0)\int_{-\infty}^{\infty}\int_{-\infty}^{\infty}\widehat{A}_{1,1}(K-\kappa_1,T)\\ &\qquad\times\widehat{A}_{1,1}(\kappa_1-\kappa_2,T)\widehat{A}_{1,-1}(\kappa_2,T)\,\mathrm{d}\kappa_2\,\mathrm{d}\kappa_1, \end{aligned}\tag{10.34}$$

where we used the symmetry of the bilinear terms in their arguments. With $k-jk_c=\varepsilon K$ we find at ε in the n-th equation for the modes concentrated at jk_c for $j=0,2$ and $n\geq 1$ that

$$\begin{aligned} 0 =&\lambda_n(0,0)\widehat{A}_{n,0}(K,T)\\ &+2s_{2n1-1}(0,-k_c,k_c;0)\int_{-\infty}^{\infty}\widehat{A}_{1,-1}(K-\kappa,T)\widehat{A}_{1,1}(\kappa,T)\,\mathrm{d}\kappa,\\ 0 =&\lambda_n(2k_c,0)\widehat{A}_{n,2}(K,T)\\ &+s_{2n11}(2k_c,k_c,k_c;0)\int_{-\infty}^{\infty}\widehat{A}_{1,1}(K-\kappa,T)\widehat{A}_{1,1}(\kappa,T)\,\mathrm{d}\kappa. \end{aligned}$$

Since $\mathrm{Re}\lambda_n(0,0)<0$ and $\mathrm{Re}\lambda_n(2k_c,0)<0$ these algebraic relations determine $\widehat{A}_{n,0}$ and $\widehat{A}_{n,2}$ in terms of $\widehat{A}_{1,1}$ and $\widehat{A}_{1,-1}$ such that (10.34) becomes the GL equation in Fourier space, namely

$$\begin{aligned} \partial_T\widehat{A}_{1,1}(K,T) =&\nu_0\widehat{A}_{1,1}(K,T)-\nu_2K^2\widehat{A}_{1,1}(K,T)\\ &+\nu_3\int_{-\infty}^{\infty}\int_{-\infty}^{\infty}\widehat{A}_{1,1}(K-\kappa,T)\widehat{A}_{1,1}(\kappa-\widetilde{\kappa},T)\widehat{A}_{1,-1}(\widetilde{\kappa},T)\,\mathrm{d}\widetilde{\kappa}\,\mathrm{d}\kappa, \end{aligned}\tag{10.35}$$

with coefficients

$$\nu_2 = \frac{1}{2}\partial_k^2 \lambda_1(k_c, 0), \qquad \nu_0 = \partial_\alpha \lambda_1(k_c, 0), \tag{10.36}$$

and

$$\begin{aligned} \nu_3 = & -4 \sum_{n=1}^{\infty} s_{211n}(k_c, k_c, 0; 0) s_{2n1-1}(0, -k_c, k_c; 0)/\lambda_n(0,0) \\ & -2 \sum_{n=1}^{\infty} s_{211n}(k_c, -k_c, 2k_c; 0) s_{2n11}(2k_c, k_c, k_c; 0)/\lambda_n(2k_c, 0) \\ & +3 s_{31111}(k_c, k_c, k_c, -k_c; 0). \end{aligned} \tag{10.37}$$

This derivation of the Fourier transformed GL equation (10.35) explains the universality of the GL equation (10.3) and the formulas (10.36) for ν_2 and ν_0. The universal behavior can also be seen when computing the higher order approximations. We find linearized GL equations at c_1 and $k = k_c$ and algebraic equations for all other modes, cf. Exercise 10.2. We leave the case of a complex critical curve of eigenvalues λ_1 to Exercise 10.3.

Remark 10.3.2. The formula (10.37), using the complete expansion in eigenfunctions, is conceptually useful but of little practical use. Instead, the exponentially damped modes can be collected in one big vector, and the value of ν_3 can be found by restricting to $2\pi/k_c$-spatially periodic functions, see for instance §10.5.1, and also Remark 10.5.2. This often allows to obtain this coefficient from the literature about the associated center manifold reduction. ⌋

10.4. An abstract approximation result

We provide an abstract approximation theorem which will allow us subsequently to establish the validity of the GL approximation for general pattern forming systems, such as reaction-diffusion systems or classical hydrodynamical stability problems. In order to make things more concrete and for the motivation of the following steps we in parallel consider the KS equation

$$\partial_t u = L_\varepsilon(\partial_x) u + \partial_x(u^2) \qquad \text{with} \qquad L_\varepsilon(\partial_x) = -(1+\partial_x^2)^2 + \varepsilon^2 \tag{10.38}$$

as an example, cf. Example 10.3.1.

Outline of the idea. In order to establish an approximation theorem like the one for the SH equation, for a given ansatz $\varepsilon\Psi$, we introduce the error $\varepsilon^\beta R = u - \varepsilon\Psi$ which satisfies

$$\partial_t R = L_\varepsilon(\partial_x) R + 2\varepsilon \partial_x(\Psi R) + \varepsilon^\beta \partial_x(R^2) + \varepsilon^{-\beta} \mathrm{Res}(\varepsilon\Psi),$$

where the residual $\mathrm{Res}(\varepsilon\Psi)$, defined by $\mathrm{Res}(u) = -\partial_t u + L_\varepsilon(\partial_x) u + \partial_x(u^2)$, contains all terms which do not cancel after inserting the ansatz $\varepsilon\Psi$ into

(10.38). As before, to any given n we can find an approximation $\varepsilon\Psi$ such that

$$\text{(10.39)} \qquad \operatorname{Res}(\varepsilon\Psi) = \mathcal{O}(\varepsilon^n) \qquad \text{and} \qquad \varepsilon\Psi_{\mathrm{GL}} - \varepsilon\Psi = \mathcal{O}(\varepsilon^2).$$

Thus, β and n can be chosen so large that no difficulty will occur in getting bounds for the terms $\varepsilon^\beta\partial_x^2(R^2) + \varepsilon^{-\beta}\operatorname{Res}(\varepsilon\Psi)$ for all $t \in [0, T_0/\varepsilon^2]$.

However, a direct estimate for all $t \in [0, T_0/\varepsilon^2]$ with the help of Gronwall's inequality is not possible due to the term $2\varepsilon\partial_x(\Psi R)$, which could lead to a growth rate $\mathcal{O}(\mathrm{e}^{r\varepsilon t})$ for an $r > 0$. This becomes unbounded for $t = T_0/\varepsilon^2$ and $\varepsilon \to 0$.

In order to overcome this difficulty with the term $2\varepsilon\partial_x^2(\Psi R)$ we introduce so called mode filters E_c and E_s. They are defined as multiplication operators in Fourier space by

$$E_j u = \mathcal{F}^{-1}(\chi_j \widehat{u}), \quad j = \pm c, s,$$

where $\chi_j : \mathbb{R} \to [0,1]$ is C_0^∞, with $\chi_{\pm c}(k) = 1 \Leftrightarrow |k \mp 1| \le 1/30$, $\chi_{\pm c}(k) = 0 \Leftrightarrow |k \mp 1| \ge 1/15$ and $\chi_s = 1 - \chi_c - \chi_{-c}$.

Remark 10.4.1. Since we would like to apply Lemma 8.3.7 about multipliers in H^θ_{ul}-spaces we have chosen $\chi_j \in C_0^\infty$. In H^θ-spaces we can take a simple cut-off function $\chi_j : \mathbb{R} \to \{0,1\}$, with $\chi_{\pm c}(k) = 1 \Leftrightarrow |k \mp 1| \le 1/30$, and $\chi_s = 1 - \chi_c - \chi_{-c}$. ⌋

Clearly we have $\varepsilon E_c\Psi = \mathcal{O}(\varepsilon)$ and $\varepsilon E_s\Psi = \mathcal{O}(\varepsilon^2)$. The idea is to split the error into a critical and a non-critical part which are scaled differently, i.e.,

$$\varepsilon^\beta R = \varepsilon^\beta R_c + \varepsilon^\beta R_{-c} + \varepsilon^{\beta+1} R_s,$$

where, depending on the chosen space, the support of $R_{\pm c}$ in Fourier space is contained in a set equal to or slightly larger than the support of $E_{\pm c}$, and where the support of R_s in Fourier space is contained in a set equal to or slightly larger than the support of E_s. The parts of the error satisfy a system of the form

$$\begin{aligned}
\partial_t R_c &= L_\varepsilon(\partial_x)R_c + \mathcal{O}(\varepsilon^2|R_c| + \varepsilon^2|R_s|) + \mathcal{O}(\varepsilon^2),\\
\partial_t R_s &= L_\varepsilon(\partial_x)R_s + \mathcal{O}(|R_c| + \varepsilon|R_s|) + \mathcal{O}(1).
\end{aligned}$$

This is because

E_c applied to the quadratic interaction of critical modes vanishes,

i.e., $E_c(\varepsilon\partial_x^2((E_c\Psi)R_c)) = 0$. Since R_s is exponentially damped by the semigroup generated by $L_\varepsilon(\partial_x)$, it is easy to obtain an estimate $R_s = \mathcal{O}(|R_c|) + \mathcal{O}(1)$. Inserting this into the first equation yields

$$\partial_t R_c = L_\varepsilon(\partial_x)R_c + \mathcal{O}(\varepsilon^2|R_c|) + \mathcal{O}(\varepsilon^2).$$

A direct estimate for all $t \in [0, T_0/\varepsilon^2]$ with the help of Gronwall's inequality is now possible and so we obtain $R_c = \mathcal{O}(1)$ and $R_s = \mathcal{O}(1)$ for all $t \in [0, T_0/\varepsilon^2]$.

In the following we will extract assumptions which will allow us to handle not only the KS equation, but very general pattern forming systems.

10.4.1. The abstract pattern forming system. Our starting point is a pattern forming system

$$\partial_t u = \Lambda u + N(u) \tag{10.40}$$

which is written with the help of the variation constant formula as

$$u(t) = \mathrm{e}^{t\Lambda} u(0) + \int_0^t \mathrm{e}^{(t-\tau)\Lambda} N(u(\tau))\,\mathrm{d}\tau. \tag{10.41}$$

We assume the existence of a formal approximation $\varepsilon\Psi$ of u which is given for all $t \in [0, T_0/\varepsilon^2]$, where $T_0 > 0$ is a constant independent of $0 < \varepsilon \ll 1$. The error $\varepsilon^\beta R = u - \varepsilon\Psi$, with $\beta \geq 3/2$, made by the approximation $\varepsilon\Psi$ satisfies

$$\begin{aligned} R(t) =& \mathrm{e}^{t\Lambda} R(0) + \varepsilon^{-\beta} \int_0^t \mathrm{e}^{(t-\tau)\Lambda} (N(\varepsilon\Psi(\tau) + \varepsilon^\beta R(\tau)) - N(\varepsilon\Psi(\tau)))\,\mathrm{d}\tau \\ &+ \varepsilon^{-\beta} \int_0^t \mathrm{e}^{(t-\tau)\Lambda} \mathrm{Res}(\varepsilon\Psi(\tau))\,\mathrm{d}\tau, \end{aligned}$$

where the residual is defined by

$$\mathrm{Res}(u) = -\partial_t u + \Lambda u + N(u).$$

In order to prove that $\varepsilon\Psi$ is a suitable approximation of u we need a number of assumptions. The first one is quite natural:

(A0) *$\mathcal{X}$ and $\mathcal{Y}$ are Banach spaces.*

Then we introduce mode filters E_c and E_s for extracting the critical and stable modes. For these we assume

(A1) *There exist linear operators E_c and E_s which are continuous both in $\mathcal{X}$ and $\mathcal{Y}$. Moreover, they commute with the semigroup $(\mathrm{e}^{t\Lambda})_{t\geq 0}$, i.e.,*

$$\mathrm{e}^{t\Lambda} E_c = E_c \mathrm{e}^{t\Lambda} \qquad \text{and} \qquad \mathrm{e}^{t\Lambda} E_s = E_s \mathrm{e}^{t\Lambda}.$$

Using these mode filters we split the error R in two parts which we scale differently, i.e., we set $R = R_c + \varepsilon R_s$ and define R_c and R_s to be the solutions of

$$
\begin{aligned}
R_c(t) = & \mathrm{e}^{t\Lambda} R_c(0) \\
& + \varepsilon^{-\beta} \int_0^t \mathrm{e}^{(t-\tau)\Lambda} E_c(N(\varepsilon\Psi(\tau) + \varepsilon^{\beta} R(\tau)) - N(\varepsilon\Psi(\tau)))\,\mathrm{d}\tau \qquad (10.42) \\
& + \varepsilon^{-\beta} \int_0^t \mathrm{e}^{(t-\tau)\Lambda} E_c \mathrm{Res}(\varepsilon\Psi(\tau))\,\mathrm{d}\tau, \\
R_s(t) = & \mathrm{e}^{\Lambda t} R_s(0) \\
& + \varepsilon^{-(\beta+1)} \int_0^t \mathrm{e}^{(t-\tau)\Lambda} E_s(N(\varepsilon\Psi(\tau) + \varepsilon^{\beta} R(\tau)) - N(\varepsilon\Psi(\tau)))\,\mathrm{d}\tau \qquad (10.43) \\
& + \varepsilon^{-(\beta+1)} \int_0^t \mathrm{e}^{(t-\tau)\Lambda} E_s \mathrm{Res}(\varepsilon\Psi(\tau))\,\mathrm{d}\tau.
\end{aligned}
$$

It turns out that this system is a good interface for coupling the systems which we have in mind to the abstract set-up. For the formulation of the other assumptions we need additional mode filters E_c^h and E_s^h. This is due to the fact that for functional analytic reasons E_c and E_s cannot be chosen as projections in H_{ul}^{θ}. We assume

(A2) *There exist linear operators E_c^h and E_s^h which are continuous both in $\mathcal{X}$ and $\mathcal{Y}$ and satisfy $E_c^h E_c = E_c E_c^h = E_c$ and $E_s^h E_s = E_s E_s^h = E_s$. The mode filters applied to the semigroup give the following estimates. There exist C_Λ, $\sigma_c \geq 0$, $\alpha \in [0,1)$, and a $\sigma_s > 0$ such that for all $t \geq 0$ and $\varepsilon \geq 0$ we have*

$$
\begin{aligned}
\|\mathrm{e}^{t\Lambda} E_c^h\|_{\mathcal{Y}\to\mathcal{Y}} \leq & C_\Lambda \mathrm{e}^{\sigma_c \varepsilon^2 t}, \\
\|\mathrm{e}^{t\Lambda} E_c^h\|_{\mathcal{X}\to\mathcal{Y}} \leq & C_\Lambda \mathrm{e}^{\sigma_c \varepsilon^2 t}, \\
\|\mathrm{e}^{t\Lambda} E_s^h\|_{\mathcal{Y}\to\mathcal{Y}} \leq & C_\Lambda \mathrm{e}^{-\sigma_s t}, \\
\|\mathrm{e}^{t\Lambda} E_s^h\|_{\mathcal{X}\to\mathcal{Y}} \leq & C_\Lambda \max(1, t^{-\alpha}) \mathrm{e}^{-\sigma_s t}.
\end{aligned}
$$

(A3) *The nonlinear terms obey the following estimates. There exist constants $C_{1,c}$ and $C_{1,s}$, monotonically growing functions $C_{2,c}(M_c, M_s)$ and $C_{2,s}(M_c, M_s)$ and an $\varepsilon_0 > 0$ such that for all $\varepsilon \in (0, \varepsilon_0)$ we have*

$$
\begin{aligned}
& \|\varepsilon^{-\beta} E_c(N(\varepsilon\Psi(\tau) + \varepsilon^{\beta} R(\tau)) - N(\varepsilon\Psi(\tau)))\|_{\mathcal{X}} \\
& \quad \leq C_{1,c}\varepsilon^2(\|R_c\|_{\mathcal{Y}} + \|R_s\|_{\mathcal{Y}}) + C_{2,c}(M_c, M_s)\min(\varepsilon^3, \varepsilon^{\beta})(\|R_c\|_{\mathcal{Y}} + \|R_s\|_{\mathcal{Y}})^2, \\
& \|\varepsilon^{-(\beta+1)} E_s(N(\varepsilon\Psi(\tau) + \varepsilon^{\beta} R(\tau)) - N(\varepsilon\Psi(\tau)))\|_{\mathcal{X}} \\
& \quad \leq C_{1,s}\|R_c\|_{\mathcal{Y}} + C_{1,s}\varepsilon\|R_s\|_{\mathcal{Y}} + C_{2,s}(M_c, M_s)\min(\varepsilon, \varepsilon^{\beta-1})(\|R_c\|_{\mathcal{Y}} + \|R_s\|_{\mathcal{Y}})^2,
\end{aligned}
$$

as long as $\|R_c\|_{\mathcal{Y}} \leq M_c$ and $\|R_s\|_{\mathcal{Y}} \leq M_s$.

(A4) *The residual terms obey the following estimates. There exist constants* C_{res} *and* $\varepsilon_0 > 0$ *such that for all* $\varepsilon \in (0, \varepsilon_0)$ *we have*

$$\sup_{\tau\in[0,T_0/\varepsilon^2]} \|E_c \mathrm{Res}(\varepsilon\Psi(\tau))\|_{\mathcal{Y}} \leq C_{res}\varepsilon^{\beta+2},$$

$$\sup_{\tau\in[0,T_0/\varepsilon^2]} \|E_s \mathrm{Res}(\varepsilon\Psi(\tau))\|_{\mathcal{Y}} \leq C_{res}\varepsilon^{\beta+1}.$$

Remark 10.4.2. As already said, the choice in **(A0)** depends on what kind of solutions we want to study, e.g., localized or not, and for which $\mathcal{X}$ and $\mathcal{Y}$ we can check **(A1)-(A4)**, and with what effort. Not for all systems considered below the assumptions will be checked in detail, e.g., for some hydrodynamical stability problems we will refer to the literature. In any case, for checking the assumptions **(A0)-(A4)** one usually does not have to start from scratch. Before mathematicians discussed the validity of the GL approximation, they justified the Landau approximation in case of spatially periodic boundary conditions with the help of the center manifold theorem. In order to apply the center manifold theorem one has to check a number of assumptions, which are relatively close to our assumptions **(A0)-(A4)**, see for instance the proof of Theorem 10.7.5. Moreover, the cubic coefficient in the Ginzburg-Landau equation and in the reduced ODE on the associated center manifold in case of spatially periodic boundary conditions coincide, cf. §13.2.2. ⌋

The following abstract approximation theorem will be applied to various systems below, where we use the triangle inequality as in (10.21) to formulate the results for the associated lowest order approximations $\varepsilon\Psi_{\mathrm{GL}}$, where $\|\varepsilon\Psi_{\mathrm{GL}} - \varepsilon\Psi\|_{\mathcal{Y}} \leq C\varepsilon^{\gamma}$, with $\gamma \geq 3/2$ depending on the chosen space $\mathcal{Y}$.

Theorem 10.4.3. *Assume the validity of* **(A0)-(A2)**, *and let* $\varepsilon\Psi$ *be an approximation such that* **(A3)-(A4)** *hold. Then there exist a* $C > 0$ *and an* $\varepsilon_0 > 0$ *such that for all* $\varepsilon \in (0, \varepsilon_0)$ *we have solutions* u *of the pattern forming system* (10.40) *with*

$$\sup_{t\in[0,T_0/\varepsilon^2]} \|u - \varepsilon\Psi\|_{\mathcal{Y}} \leq C\varepsilon^{\beta}.$$

Proof. Our starting point is (10.42)-(10.43). Taking the $\mathcal{Y}$-norm of both sides yields

$$\begin{aligned}\|R_c(t)\|_{\mathcal{Y}} &\leq \|e^{t\Lambda}E_c^h\|_{\mathcal{Y}\to\mathcal{Y}}\|R_c(0)\|_{\mathcal{Y}} \\ &+ \int_0^t \|e^{(t-\tau)\Lambda}E_c^h\|_{\mathcal{X}\to\mathcal{Y}}\|\varepsilon^{-\beta}E_c(N(\varepsilon\Psi(\tau)+\varepsilon^{\beta}R(\tau)) - N(\varepsilon\Psi(\tau)))\|_{\mathcal{X}}\,d\tau \\ &+ \varepsilon^{-\beta}\int_0^t \|e^{(t-\tau)\Lambda}E_c^h\|_{\mathcal{Y}\to\mathcal{Y}}\|E_c\mathrm{Res}(\varepsilon\Psi(\tau))\|_{\mathcal{Y}}\,d\tau,\end{aligned}$$

$$
\begin{aligned}
&\leq C_\Lambda e^{\sigma_c \varepsilon^2 t} \|R_c(0)\|_{\mathcal{Y}} \\
&\quad + \int_0^t C_\Lambda e^{\sigma_c \varepsilon^2 (t-\tau)} (C_{1,c}\varepsilon^2 (\|R_c(\tau)\|_{\mathcal{Y}} + \|R_s(\tau)\|_{\mathcal{Y}}) \\
&\qquad + C_{2,c}(M_c, M_s) \min(\varepsilon^3, \varepsilon^\beta)(\|R_c(\tau)\|_{\mathcal{Y}} + \|R_s(\tau)\|_{\mathcal{Y}})^2) \, \mathrm{d}\tau \\
&\quad + \varepsilon^{-\beta} \int_0^t C_\Lambda e^{\sigma_c \varepsilon^2 (t-\tau)} C_{res} \varepsilon^{\beta+2} \, \mathrm{d}\tau
\end{aligned}
$$

and

$$
\begin{aligned}
&\|R_s(t)\|_{\mathcal{Y}} \leq \|e^{t\Lambda} E_s^h\|_{\mathcal{Y}\to\mathcal{Y}} \|R_s(0)\|_{\mathcal{Y}} \\
&\quad + \int_0^t \|e^{(t-\tau)\Lambda} E_s^h\|_{\mathcal{X}\to\mathcal{Y}} \|\varepsilon^{-(\beta+1)} E_s(N(\varepsilon\Psi(\tau) + \varepsilon^\beta R(\tau)) - N(\varepsilon\Psi(\tau)))\|_{\mathcal{X}} \, \mathrm{d}\tau \\
&\quad + \varepsilon^{-(\beta+1)} \int_0^t \|e^{(t-\tau)\Lambda} E_s^h\|_{\mathcal{Y}\to\mathcal{Y}} \|E_s \mathrm{Res}(\varepsilon\Psi(\tau))\|_{\mathcal{Y}} \, \mathrm{d}\tau \\
&\leq C_\Lambda e^{-\sigma_s t} \|R_s(0)\|_{\mathcal{Y}} \\
&\quad + \int_0^t C_\Lambda \max(1, (t-\tau)^{-\alpha}) e^{-\sigma_s (t-\tau)} (C_{1,s} \|R_c(\tau)\|_{\mathcal{Y}} + C_{1,s}\varepsilon \|R_s(\tau)\|_{\mathcal{Y}} \\
&\qquad + C_{2,c}(M_c, M_s) \min(\varepsilon, \varepsilon^{\beta-1})(\|R_c(\tau)\|_{\mathcal{Y}} + \|R_s(\tau)\|_{\mathcal{Y}})^2) \, \mathrm{d}\tau \\
&\quad + \varepsilon^{-(\beta+1)} \int_0^t C_\Lambda e^{-\sigma_s (t-\tau)} C_{res} \varepsilon^{\beta+1} \, \mathrm{d}\tau,
\end{aligned}
$$

as long as $\|R_c\|_{\mathcal{Y}} \leq M_c$ and $\|R_s\|_{\mathcal{Y}} \leq M_s$. Introducing

$$
S_c(t) = \sup_{\tau \in [0,t]} \|R_c(\tau)\|_{\mathcal{Y}} \qquad \text{and} \qquad S_s(t) = \sup_{\tau \in [0,t]} \|R_c(\tau)\|_{\mathcal{Y}}
$$

allows for all $t \in [0, T_0/\varepsilon^2]$ to estimate that

$$
\begin{aligned}
\|R_c(t)\|_{\mathcal{Y}} \leq & C_\Lambda e^{\sigma_c T_0} S_c(0) + \int_0^t C_\Lambda e^{\sigma_c T_0} (C_{1,c}\varepsilon^2 (S_c(\tau) + S_s(\tau)) \\
& \qquad + C_{2,c}(M_c, M_s) \min(\varepsilon^3, \varepsilon^\beta)(S_c(\tau) + S_s(\tau))^2) \, \mathrm{d}\tau \\
& + T_0 C_\Lambda e^{\sigma_c T_0} C_{res}, \\
\|R_s(t)\|_{\mathcal{Y}} \leq & C_\Lambda S_s(0) + C_\Lambda C_\sigma (C_{1,s} S_c(t) + C_{1,s}\varepsilon S_s(t) \\
& \qquad + C_{2,s}(M_c, M_s) \min(\varepsilon, \varepsilon^{\beta-1})(S_c(t) + S_s(t))^2) \\
& + C_\Lambda C_\sigma C_{res},
\end{aligned}
$$

where $C_\sigma = \sup_{t\in[0,T_0/\varepsilon^2]} \int_0^t \max(1,(t-\tau)^{-\alpha}) e^{-\sigma_s(t-\tau)} d\tau < \infty$ is independent of $0 \leq \varepsilon \ll 1$. Since the right-hand sides of the last estimates increase monotonically w.r.t. t we can replace $\|R_c(t)\|_{\mathcal{Y}}$ and $\|R_s(t)\|_{\mathcal{Y}}$ on the left-hand side by $S_c(t)$ and $S_s(t)$, respectively. From the second inequality we then obtain: For all M_c and M_s there exists an $\varepsilon_1 > 0$ such that for all

$\varepsilon \in (0, \varepsilon_1)$ we have

$$S_s(t) \le (C_\Lambda S_s(0) + C_\Lambda C_\sigma C_{res} + 1) + 2C_\Lambda C_\sigma C_{1,s} S_c(t), \tag{10.44}$$

as long as $S_c(t) \le M_c$ and $S_s(t) \le M_s$, if $0 < \varepsilon_1 \ll 1$ is chosen so small that

$$C_\Lambda C_\sigma (C_{1,s} \varepsilon M_s + C_{2,s}(M_c, M_s) \min(\varepsilon, \varepsilon^{\beta-1})(M_c + M_s)^2) \le 1 \tag{10.45}$$

for all $\varepsilon \in (0, \varepsilon_1)$. Inserting (10.44) into the inequality for $S_c(t)$ yields

$$S_c(t) \le \beta_0 + \varepsilon^2 \int_0^t \beta_1 S_c(\tau) \, \mathrm{d}\tau,$$

where

$$\begin{aligned} \beta_0 =& C_\Lambda \mathrm{e}^{\sigma_c T_0} S_c(0) \\ &+ T_0 C_\Lambda \mathrm{e}^{\sigma_c T_0} C_{res} + T_0 C_\Lambda \mathrm{e}^{\sigma_c T_0} C_{1,c} (C_\Lambda S_s(0) + C_\Lambda C_\sigma C_{res} + 1) + 1, \\ \beta_1 =& C_\Lambda \mathrm{e}^{\sigma_c T_0} C_{1,c} (1 + 2 C_\Lambda C_\sigma C_{1,s}), \end{aligned}$$

if we choose $0 < \varepsilon_2 \ll 1$ so small that for given M_c and M_s we have

$$C_\Lambda \mathrm{e}^{\sigma_c T_0} T_0 C_{2,c}(M_c, M_s) \min(\varepsilon^3, \varepsilon^\beta)(M_c + M_2)^2) \le \varepsilon^2 \tag{10.46}$$

for all $\varepsilon \in (0, \varepsilon_2)$. Using Gronwall's inequality immediately yields

$$S_c(t) \le \beta_0 \mathrm{e}^{\beta_1 T_0} =: M_c,$$

and due to (10.44) we define

$$M_s := (C_\Lambda S_s(0) + C_\Lambda C_\sigma C_{res} + 1) + 2 C_\Lambda C_\sigma C_{1,s} M_c.$$

To this M_c and M_s we define $\varepsilon_0 = \min(\varepsilon_1, \varepsilon_2)$ where $\varepsilon_1 > 0$ is chosen so small that for all $\varepsilon \in (0, \varepsilon_1)$ condition (10.45) is satisfied and $\varepsilon_2 > 0$ so small that for all $\varepsilon \in (0, \varepsilon_2)$ condition (10.46) is satisfied. □

With the help of the mode filters we are now able to give the

Proof of Lemma 10.2.8. For the proof of the estimate we will use Lemma 8.3.7. Unfortunately, a direct estimate of $\|k \mapsto \mathrm{e}^{-(1-k^2)^2 t}\|_{C_b^2}$ would lead to some unwanted growth rates in time. Hence, we introduce two mode filters E_c and E_s satisfying $E_c + E_s = 1$ where E_c is defined by $\widehat{E}_c = \chi_1 + \chi_{-1}$, with $\chi \in C_0^\infty(\mathbb{R}, [0, 1])$ an even cut-off function with $\chi(k) = 1$ for $|k| \le 1/10$ and $\chi(k) = 0$ for $|k| \ge 1/5$, and where $\chi_j(k) = \chi(k - j)$. Then we find

$$\|\mathrm{e}^{t\Lambda}\|_{H^\theta_{\mathrm{ul}} \to H^\theta_{\mathrm{ul}}} \le \|E_c \mathrm{e}^{t\Lambda}\|_{H^\theta_{\mathrm{ul}} \to H^\theta_{\mathrm{ul}}} + \|E_s \mathrm{e}^{t\Lambda}\|_{H^\theta_{\mathrm{ul}} \to H^\theta_{\mathrm{ul}}}$$

and immediately

$$\|E_s \mathrm{e}^{t\Lambda}\|_{H^\theta_{\mathrm{ul}} \to H^\theta_{\mathrm{ul}}} \le C \|k \mapsto \widehat{E}_s(k) \mathrm{e}^{-(1-k^2)^2 t}\|_{C_b^2} \le C \mathrm{e}^{-\sigma t}$$

for a $\sigma > 0$ since the semigroup decays with some exponential rate for the wave numbers in the support of $\widehat{E}_s$. The estimate for the E_c part uses that for $\theta > 1/2$ we have

$$\begin{aligned}\|E_c u\|_{H^\theta_{\rm ul}} &\leq C\|E_c u\|_{L^2_{\rm ul}} \leq C\|\widehat{\chi}_1 u\|_{L^2_{\rm ul}} \leq C\|\widehat{\chi}_1 u\|_{L^\infty} \leq C\|\widehat{\chi}(u\mathrm{e}^{-\mathrm{i}k\cdot x})\|_{L^\infty} \\ &\leq C\|S_\varepsilon(\widehat{\chi}(u\mathrm{e}^{-\mathrm{i}k\cdot x}))\|_{L^\infty} \leq C\|S_\varepsilon(\widehat{\chi}(u\mathrm{e}^{-\mathrm{i}k\cdot x}))\|_{H^\theta_{\rm ul}} \\ &\leq C\|\widehat{\chi}(u\mathrm{e}^{-\mathrm{i}k\cdot x})\|_{H^\theta_{\rm ul}} \leq C\|E_c u\|_{H^\theta_{\rm ul}} \leq C\|E_c u\|_{L^2_{\rm ul}}\end{aligned}$$

where the scaling operator S_ε is defined by $(S_\varepsilon u)(x) = u(\varepsilon x)$. Hence, for functions with a fixed compact support in Fourier space we have the equivalence of the norms $\|E_c u\|_{H^\theta_{\rm ul}}$, respectively $\|E_c u\|_{L^2_{\rm ul}}$, and $\|S_\varepsilon(\widehat{\chi}(u\mathrm{e}^{-\mathrm{i}k\cdot x}))\|_{H^\theta_{\rm ul}}$. W.r.t. the last norm Lemma 8.3.7 requires us to estimate $\|k \mapsto \mathrm{e}^{\lambda_\varepsilon(k)t}\|_{C^2_b}$ with

$$\lambda_\varepsilon(k) = -(1-(\varepsilon k - 1))^2)^2.$$

Since $\lambda_\varepsilon(k) \sim -(\varepsilon k)^2$ for small k the term $\mathrm{e}^{\lambda_\varepsilon(k)t}$ behaves as $\mathrm{e}^{-k^2 T}$ with $T = \varepsilon^2 t$ and so its C^2_b-norm is uniformly bounded for t on the $\mathcal{O}(1/\varepsilon^2)$ time scale. □

10.4.2. The Kuramoto-Shivashinsky equation. For simplicity we will restrict ourselves to the case of Sobolev spaces H^θ. The case of uniformly local Sobolev spaces $H^\theta_{\rm ul}$ is given as Exercise 10.4; see also [**Sch94c**].

Since the nonlinear terms are given by $N(u) = \partial_x(u^2)$ we make the choice

(A0) *We set* $\mathcal{X} = H^{\theta-1}$ *and* $\mathcal{Y} = H^\theta$.

The semigroup $(\mathrm{e}^{t\Lambda})_{t\geq 0}$ is defined as a multiplier in Fourier space via $\mathrm{e}^{t\Lambda} = \mathcal{F}^{-1}\widehat{\mathrm{e}^{t\Lambda}}\mathcal{F}$ where $\widehat{\mathrm{e}^{t\Lambda}}(k) = \mathrm{e}^{(-(1-k^2)^2+\varepsilon^2)t}$. Therefore, we also introduce the mode filters E_c and E_s as multiplication operators in Fourier space, namely

$$E_j u = \mathcal{F}^{-1}\chi_j\mathcal{F}, \quad j = \pm c, s,$$

where $\chi_j : \mathbb{R} \to \{0,1\}$, with $\chi_{\pm c}(k) = 1 \Leftrightarrow |k \mp 1| \leq 1/30$, and $\chi_s = 1 - \chi_c - \chi_{-c}$. See Remark 10.4.1.

(A1) *We have the continuity of the* E_j *in* H^θ *due to*

$$\|E_j u\|_{H^\theta} \leq C_1\|\chi_j\widehat{u}\|_{L^2_\theta} \leq C_1\|\chi_j\|_{L^\infty}\|\widehat{u}\|_{L^2_\theta} \leq C_1\|\widehat{u}\|_{L^2_\theta} \leq C_2\|u\|_{H^\theta}.$$

Moreover, the E_j *commute with the semigroup* $(\mathrm{e}^{t\Lambda})_{t\geq 0}$ *due to*

$$\begin{aligned}\mathrm{e}^{t\Lambda}E_j &= \mathcal{F}^{-1}\widehat{\mathrm{e}^{t\Lambda}}\mathcal{F}\mathcal{F}^{-1}\chi_j\mathcal{F} = \mathcal{F}^{-1}\widehat{\mathrm{e}^{t\Lambda}}\chi_j\mathcal{F} = \mathcal{F}^{-1}\chi_j\widehat{\mathrm{e}^{t\Lambda}}\mathcal{F} \\ &= \mathcal{F}^{-1}\chi_j\mathcal{F}\mathcal{F}^{-1}\widehat{\mathrm{e}^{t\Lambda}}\mathcal{F} = E_j\mathrm{e}^{t\Lambda}\end{aligned}$$

and

$$(\widehat{\mathrm{e}^{t\Lambda}}\chi_j)(k) = \mathrm{e}^{(-(1-k^2)^2+\varepsilon^2)t}\chi_j(k) = \chi_j(k)\mathrm{e}^{(-(1-k^2)^2+\varepsilon^2)t} = (\chi_j\widehat{\mathrm{e}^{t\Lambda}})(k).$$

With this setup it is not difficult to verify Assumption **(A2)**.

(A2) *We choose* $E_c^h = E_c$ *and* $E_s^h = E_s$ *since* $E_c^2 = E_c$ *and* $E_s^2 = E_s$. *Then there exist positive constants* C_Λ, σ_c, *and* $\alpha \in [0,1)$, *and a strictly positive constant* σ_s *such that for all* $t \geq 0$ *and* $\varepsilon \geq 0$ *we have*

$$\begin{aligned}
\|e^{t\Lambda} E_c u\|_{H^\theta} \leq & C_1 \|\widehat{e^{t\Lambda}} \chi_c \widehat{u}\|_{L^2_\theta} \leq C_1 \sup_{k\in\mathbb{R}} |e^{(-(1-k^2)^2+\varepsilon^2)t} \chi_c(k)| \|\widehat{u}\|_{L^2_\theta} \\
\leq & C_\Lambda e^{\sigma_c \varepsilon^2 t} \|u\|_{H^\theta}, \\
\|e^{t\Lambda} E_c u\|_{H^\theta} \leq & C_1 \|\widehat{e^{t\Lambda}} \chi_c \widehat{u}\|_{L^2_\theta} \leq C_1 \sup_{k\in\mathbb{R}} |e^{(-(1-k^2)^2+\varepsilon^2)t} \chi_c(k)(1+k^2)^{1/2}| \|\widehat{u}\|_{L^2_{\theta-1}} \\
\leq & C_\Lambda e^{\sigma_c \varepsilon^2 t} \|u\|_{H^{\theta-1}}, \\
\|e^{t\Lambda} E_s u\|_{H^\theta} \leq & C_1 \|\widehat{e^{t\Lambda}} \chi_s \widehat{u}\|_{L^2_\theta} \leq C_1 \sup_{k\in\mathbb{R}} |e^{(-(1-k^2)^2+\varepsilon^2)t} \chi_s(k)| \|\widehat{u}\|_{L^2_\theta} \\
\leq & C_\Lambda e^{-\sigma_s t} \|u\|_{H^\theta}, \\
\|e^{t\Lambda} E_s u\|_{H^\theta} \leq & C_1 \|\widehat{e^{t\Lambda}} \chi_s \widehat{u}\|_{L^2_\theta} \leq C_1 \sup_{k\in\mathbb{R}} |e^{(-(1-k^2)^2+\varepsilon^2)t} \chi_s(k)(1+k^2)^{1/2}| \|\widehat{u}\|_{L^2_{\theta-1}} \\
\leq & C_\Lambda \max(1, t^{-1/4}) e^{-\sigma_s t} \|u\|_{H^{\theta-1}}.
\end{aligned}$$

Before we come to the assumptions **(A3)** and **(A4)** we must have a more detailed look at the mode filters applied to the approximation. We have

Lemma 10.4.4. *Let* $\theta_A \geq \max(\theta+1, 3)$ *and let* $A \in C([0,T_0], H^{\theta_A})$ *be a solution of the associated GL equation* (10.27). *There exist* C_ψ *and* $\varepsilon_0 > 0$ *such that for all* $\varepsilon \in (0, \varepsilon_0)$ *we have that* $\varepsilon\Psi_c = E_c(\varepsilon\Psi)$ *and* $\varepsilon^2\Psi_s = E_s(\varepsilon\Psi)$ *satisfy*

$$\sup_{t\in[0,T_0/\varepsilon^2]} \|\varepsilon\Psi_c\|_{C_b^\theta} \leq C_\psi \varepsilon \qquad \text{and} \qquad \sup_{t\in[0,T_0/\varepsilon^2]} \|\varepsilon^2\Psi_s\|_{C_b^\theta} \leq C_\psi \varepsilon^2.$$

Proof. The GL approximation for the solutions of the KS equation is of the form

$$\varepsilon\Psi(x,t) = \varepsilon A(\varepsilon x, \varepsilon^2 t) e^{ix} + \text{c.c.} + \varepsilon^2 \Psi_h(x,t)$$

where $\sup_{t\in[0,T_0/\varepsilon^2]} \|\varepsilon^2\Psi_h\|_{C_b^\theta} \leq C_\psi \varepsilon^2$. With

$$\begin{aligned}
\|E_s(A(\varepsilon\cdot)\mathbf{E})\|_{C_b^\theta} \leq & C \|E_s\widehat{(A(\varepsilon\cdot)\mathbf{E})}\|_{L^1_\theta} \leq C \|(1-\chi_1)\widehat{A}(\varepsilon^{-1}(\cdot - 1))\|_{L^1_\theta} \\
\leq & C \sup_{k\in\mathbb{R}} |(1-\chi_1(k))(1+k^2)^{\theta/2}(1+((k-1)/\varepsilon)^2)^{-s/2}| \|\widehat{A}\|_{L^1_\theta} \\
\leq & C\varepsilon^\theta \|\widehat{A}\|_{L^2_{\theta+1}} \leq C\varepsilon^s \|A\|_{H^{\theta+1}},
\end{aligned}$$

where $\mathbf{E} = e^{ix}$, we are done. □

The fact that $E_c(\Psi_c R_c) = E_c(R_c^2) = 0$ due to disjoint supports in Fourier space of E_c with $\Psi_c R_c$ and R_c^2, and the boundedness of E_c and E_s, yield the following estimates for the nonlinear terms

$$\begin{aligned}&N(\varepsilon\Psi(\tau)+\varepsilon^\beta R(\tau))-N(\varepsilon\Psi(\tau))\\&=\partial_x((\varepsilon\Psi_c+\varepsilon^2\Psi_s+\varepsilon^\beta R_c+\varepsilon^{\beta+1}R_s)^2)-\partial_x((\varepsilon\Psi_c+\varepsilon^2\Psi_s)^2).\end{aligned}$$

(A3) *We have*

$$\begin{aligned}&\varepsilon^{-\beta}\|E_c(N(\varepsilon\Psi(\tau)+\varepsilon^\beta R(\tau))-N(\varepsilon\Psi(\tau)))\|_{H^{\theta-1}}\\&\leq C(\varepsilon^2\|\Psi_c\|_{C_b^\theta}\|R_s\|_{H^\theta}+\varepsilon^2\|\Psi_s\|_{C_b^\theta}\|R_c\|_{H^\theta}+\varepsilon^3\|\Psi_s\|_{C_b^\theta}\|R_s\|_{H^\theta}\\&\qquad+\varepsilon^{\beta+1}\|R_c\|_{H^\theta}\|R_s\|_{H^\theta}+\varepsilon^{\beta+2}\|R_s\|_{H^\theta}^2)\end{aligned}$$

and

$$\begin{aligned}&\varepsilon^{-\beta-1}\|E_s(N(\varepsilon\Psi(\tau)+\varepsilon^\beta R(\tau))-N(\varepsilon\Psi(\tau)))\|_{H^{\theta-1}}\\&\leq C(\|\Psi_c\|_{C_b^\theta}\|R_s\|_{H^\theta}+\varepsilon\|\Psi_c\|_{C_b^\theta}\|R_s\|_{H^\theta}+\varepsilon\|\Psi_s\|_{C_b^\theta}\|R_c\|_{H^\theta}+\varepsilon^2\|\Psi_s\|_{C_b^\theta}\|R_s\|_{H^\theta}\\&\qquad+\varepsilon^{\beta-1}\|R_c\|_{H^\theta}^2+\varepsilon^\beta\|R_c\|_{H^\theta}\|R_s\|_{H^\theta}+\varepsilon^{\beta+1}\|R_s\|_{H^\theta}^2).\end{aligned}$$

As for the SH equation, due to $\|\Psi_c\|_{H^\theta} = \mathcal{O}(\varepsilon^{-1/2})$ and $\|\Psi_s\|_{H^\theta} = \mathcal{O}(\varepsilon^{-1/2})$ it is essential at this point to estimate Ψ_c and Ψ_s in C_b^θ and not in H^θ.

The residual terms are of the form

$$\operatorname{Res}(\varepsilon\Psi) = \sum_{j=-N}^{N} \varepsilon^{\beta_j} s_j(\varepsilon x)e^{ijx},$$

with coefficients β_j satisfying $\beta_{\pm 1} = 4$ and $\beta_j \geq 3$ for all other $j \in \mathbb{Z}\backslash\{-1,1\}$. Hence, it remains to prove

$$\|E_c\varepsilon^{\beta_j}s_j(\varepsilon\cdot_x)e^{ij\cdot_x}\|_{H^\theta} \leq C\varepsilon^{7/2},$$

but this follows exactly as the estimate for $\|E_s(A(\varepsilon\cdot)\mathbf{E})\|_{C_b^\theta}$ in the proof of Lemma 10.4.4. With this remark it is an easy exercise to check that the approximation from Example 10.3.1 satisfies the Assumption **(A4)** with $\beta = 3/2$ and $\varepsilon\Psi$ from (10.26).

(A4) *The residual terms obey the following estimates. There exist constants C_{res} and $\varepsilon_0 > 0$ such that for all $\varepsilon \in (0, \varepsilon_0)$ we have*

$$\begin{aligned}\sup_{\tau\in[0,T_0/\varepsilon^2]}\|E_c\operatorname{Res}(\varepsilon\Psi(\tau))\|_{H^\theta}&\leq C_{res}\varepsilon^{7/2},\\ \sup_{\tau\in[0,T_0/\varepsilon^2]}\|E_s\operatorname{Res}(\varepsilon\Psi(\tau))\|_{H^\theta}&\leq C_{res}\varepsilon^{5/2}.\end{aligned}$$

We delegate the missing details to the exercises. Here, we only remark that due to the fact that we have to expand the curves of eigenvalues at the wave number $k = k_c$, cf. (10.29) and Exercise 10.6, we need that the solutions of the GL equation have to be three spatial derivatives more regular than the solutions of the underlying pattern forming system.

Hence, the KS equation with the previous choices for the mode filters and H^θ as phase space satisfies the assumptions **(A0)-(A4)** and so we have the following approximation theorem.

Theorem 10.4.5. *Fix $\theta \geq 1$. Let $A \in C([0,T_0], H^{\theta+3})$ be a solution of the GL equation* (10.27)*. Then there exist a $C > 0$ and an $\varepsilon_0 > 0$ such that for all $\varepsilon \in (0, \varepsilon_0)$ we have solutions u of the KS equation* (10.38) *with*

$$\sup_{t \in [0, T_0/\varepsilon^2]} \|u - \varepsilon \Psi_{\mathrm{GL}}\|_{H^\theta} \leq C\varepsilon^{3/2}$$

where $\varepsilon\Psi_{\mathrm{GL}}$ is the lowest order approximation defined in (10.25).

10.5. Reaction-Diffusion systems

We apply the above theory to reaction-diffusion systems (9.1), i.e.,

$$\partial_t u = D_\mu \partial_x^2 u + f_\mu(u), \tag{10.47}$$

that show a Turing instability as discussed in §9.3. In (10.47) we explicitly denote the dependence on the bifurcation parameter μ. We show that the assumptions **(A0)-(A4)** are satisfied if a simple spectral assumption and some genericity conditions are valid. Moreover, we give the derivation of the GL equation from a more applied point of view, compared to the derivation in §10.3. Finally, we extend our theory to systems which show some Hopf-bifurcation at the wave number $k = 0$, with the Brusselator as prominent example.

10.5.1. Derivation of the GL equation. We consider a spatially homogeneous equilibrium u^* of the reaction-diffusion system (10.47) which undergoes a Turing instability. By a change of coordinates $u = u^* + v$ we find

$$\partial_t v = L_\mu(\partial_x) v + g_\mu(v), \tag{10.48}$$

where

$$\begin{aligned} L_\mu(\partial_x) =& D_\mu \partial_x^2 + \partial_u f_\mu(u^*), \\ g_\mu(v) =& f_\mu(u^* + v) - f_\mu(u^*) - \partial_u f_\mu(u^*) v = \mathcal{O}(v^2). \end{aligned}$$

We assume the occurrence of a Turing instability for (10.48) at the fixed point $v = 0$, i.e., for the eigenvalues $\lambda_j(k)$ for $j = 1, \ldots, d$ of $\widehat{L}_\mu(k) : \mathbb{C}^d \to \mathbb{C}^d$ defined by

$$L_\mu(\partial_x)(\mathrm{e}^{\mathrm{i}kx} \cdot) = \mathrm{e}^{\mathrm{i}kx} \widehat{L}_\mu(k) \cdot$$

we assume:

(Spec$_{\mathbf{RD}}$) *For $\mu < \mu_c$ the homogeneous state u^* is stable, i.e., we have $\mathrm{Re}\lambda_j(k,\mu) < 0$ for all $k \in \mathbb{R}$ and $j = 1,\dots,d$. At $\mu = \mu_c$ we have $\mathrm{Re}\lambda_1(k_c,\mu_c) = 0$ for some $k_c > 0$ and $\mathrm{Re}\lambda_j(k,\mu_c) < 0$ for $k \neq \pm k_c$ if $j = 1$ or for all $k \in \mathbb{R}$ if $j \neq 1$. Finally, for $\varepsilon^2 = \mu - \mu_c$, we assume that*

$$\mathrm{Re}\lambda_1(k,\mu) = \nu_0\varepsilon^2 - \nu_2(k-k_c)^2 + \mathcal{O}(\varepsilon^3 + |k-k_c|^3)$$

for all k in some neighborhood of k_c, with $\nu_0, \nu_2 > 0$.

Hence, $\mathrm{Re}\lambda_j(k,\mu) < -\sigma < 0$ for all eigenvalues and all $k \in \mathbb{R}$ except of a small neighborhood of k_c, and we have $\mathrm{Re}\lambda_1(k_c,\mu) = \mathcal{O}(\varepsilon^2) > 0$.

Remark 10.5.1. Since (10.48) is real-valued we have $\lambda_j(-k) = \overline{\lambda_j}(k)$, and due to the symmetry $x \mapsto -x$ we have $\lambda_j(-k) = \lambda_j(k)$, hence simple eigenvalues are real. Thus, in particular $\lambda_1(k) \in \mathbb{R}$ for k near k_c. However, we continue to use complex notations since the calculus stays the same if we consider a system of the form (10.48) without the symmetry $x \mapsto -x$. ⌋

As before the curve λ_1 near k_c determines the linear coefficients in the GL equation. In order to compute the coefficient in front of the cubic terms we write the nonlinear terms as

$$g(v) = B(v,v) + C(v,v,v) + \mathcal{O}(v^4),$$

with bilinear and symmetric $B(u,v)$ and trilinear and symmetric $C(u,v,w)$. In order to derive the GL equation we make the ansatz

$$v(x,t) = \varepsilon\Psi(x,t) = \varepsilon\Psi_{\mathrm{GL}}(x,t) + \varepsilon^2\Psi_h(x,t), \tag{10.49}$$

with

$$\begin{aligned}\varepsilon\Psi_{\mathrm{GL}}(x,t) &= \varepsilon A(X,T)\mathrm{e}^{\mathrm{i}k_c x}\phi_1 + \mathrm{c.c.},\\ \varepsilon^2\Psi_h(x,t) &= \frac{\varepsilon^2}{2}\Phi_0(X,T) + \varepsilon^2\mathrm{e}^{2\mathrm{i}k_c x}\Phi_2(X,T) + \mathrm{c.c.},\end{aligned}$$

where $X = \varepsilon x$, $T = \varepsilon^2 t$, $A(X,T) \in \mathbb{C}$, $\Phi_0(X,T), \Phi_2(X,T) \in \mathbb{C}^d$, and where ϕ_1 is the critical eigenvector, i.e.,

$$\widehat{L}_{\mu_c}(k_c)\phi_1 = \lambda_1(k_c,\mu_c)\phi_1 = 0. \tag{10.50}$$

Plugging (10.49) into (10.48) and sorting the terms w.r.t. coefficients of $\varepsilon^m\mathbf{E}^n$, where $\mathbf{E}^n = \mathrm{e}^{\mathrm{i}nk_c x}$, gives

$$\begin{aligned}\varepsilon\mathbf{E}: &\quad 0 = 0,\\ \varepsilon^2\mathbf{E}^2: &\quad 0 = \widehat{L}_{\mu_c}(2k_c)\Phi_2 + A^2B(\phi_1,\phi_1),\\ \varepsilon^2\mathbf{E}^0: &\quad 0 = \widehat{L}_{\mu_c}(0)\Phi_0 + 2|A|^2B(\phi_1,\overline{\phi_1}),\end{aligned}$$

and at $\varepsilon^3\mathbf{E}^1$ we find

$$\begin{aligned}\partial_T A = &\left[\partial_\mu\lambda_1(k_c,\mu_c) - \tfrac{1}{2}\partial_k^2\lambda_1(k_c,\mu_c)\partial_X^2\right] A \\ &+ \left\langle 2B(A\phi_1,\Phi_0) + 2B(\overline{A\phi_1},\Phi_2) + 3C(A\phi_1,A\phi_1,\overline{A\phi_1}),\phi_1^*\right\rangle .\end{aligned} \tag{10.51}$$

For obtaining (10.51) we applied the $\widehat{L}_{\mu_c}(k_c)$ invariant projection onto ϕ_1, defined via ϕ_1^*, the adjoint eigenvector of $\widehat{L}_{\mu_c}(k_c)$ to the eigenvalue $\lambda_1(k_c,\mu_c)$ satisfying $\widehat{L}_{\mu_c}(k_c)^*\phi_1^* = \overline{\lambda_1}(k_c,\mu_c)\phi_1^*$, and normalized as $\langle\phi_1^*,\phi_1\rangle = 1$. See Remark 10.5.2.

Due to our spectral assumptions $\widehat{L}_{\mu_c}(0)$ and $\widehat{L}_{\mu_c}(2k_c)$ are invertible such that we obtain

$$\begin{aligned}\Phi_2 &= -A^2\widehat{L}_{\mu_c}(2k_c)^{-1}B(\phi_1,\phi_1),\\ \Phi_0 &= -2|A|^2\widehat{L}_{\mu_c}(0)^{-1}B(\phi_1,\overline{\phi_1}).\end{aligned} \tag{10.52}$$

Then plugging (10.52) into (10.51) and setting

$$\phi_2 = \widehat{L}_{\mu_c}(2k_c)^{-1}B(\phi_1,\phi_1) \qquad \text{and} \qquad \phi_0 = \widehat{L}_{\mu_c}(0)^{-1}B(\phi_1,\overline{\phi_1}),$$

we obtain

$$\partial_T A = \nu_2\partial_X^2 A + \nu_0 A + \nu_3|A|^2 A, \tag{10.53}$$

where

$$\nu_3 = \left\langle -2B(\phi_1,\phi_0) - 2B(\phi_1,\phi_2) + 3C(\phi_1,\phi_1,\overline{\phi_1}),\phi_1^*\right\rangle . \tag{10.54}$$

Since $\lambda_1(k_c,\mu_c)$ and $\widehat{L}_{\mu_c}(k_c)$ are real-valued we also have $\phi_1 \in \mathbb{R}^d$, and $\widehat{L}(k_c)^* = \widehat{L}(k_c)^T$. In particular, the coefficients ν_2,ν_0,ν_3 in (10.53) are all real-valued.

Thus, under rather generic assumptions on (10.48) we find that the envelope A of the critical modes satisfy a GL equation. For a given problem, of course it remains to determine the coefficients ν_2,ν_0,ν_3.

Remark 10.5.2. In the literature, e.g., [**Man92, CH93b, Mie02, Pis06, Mer15**] the following derivation of the GL equation can be found. Plugging the extended ansatz

$$u(x,t) = \varepsilon\Psi(x,t) + \varepsilon^3\Phi_3(\varepsilon x,\varepsilon t)\mathrm{e}^{\mathrm{i}k_c x} + \text{c.c.}, \tag{10.55}$$

into (10.48) yields at $\varepsilon^3\mathbf{E}_1$

$$\begin{aligned}\widehat{L}(k_c)\Phi_3 = &-\partial_T A\phi_1 + \left[\nu_0 + \nu_2\partial_X^2\right] A\phi_1 \\ &+ 2B(A\phi_1,\Phi_0) + 2B(\overline{A\phi_1},\Phi_2) + 3C(A\phi_1,A\phi_1,\overline{A\phi_1}).\end{aligned} \tag{10.56}$$

The GL equation (10.53) is obtained as the solvability condition if we want to remove the terms at $\varepsilon^3\mathbf{E}^1$ from the residual. Since $\widehat{L}(k_c)$ is not invertible there exists a solution Φ_3 to (10.56) if and only if the right-hand side is orthogonal to the zero eigenvector of the adjoint of $\widehat{L}(k_c)$ which is precisely (10.53). ⌋

Example 10.5.3. (GL formalism for the Schnakenberg model) Figure 10.8 shows some numerical simulations for the Schnakenberg model (9.23). It shows that the GL approximation describes the qualitative behavior pretty well even in case of non-small ε. In detail, we consider (9.23) with $(b,d) = (1.5, 15)$, which is close to a Turing instability, cf. Figure 9.9. We have $k_c \approx 0.63$, $\lambda_1(k_c) \approx 0.063 =: \varepsilon^2$, hence $\varepsilon \approx 0.25$, and $\phi_1(k_c) \approx \begin{pmatrix} 0.97 \\ -0.24 \end{pmatrix}$. Moreover,

$$B(\phi_1,\phi_1) = B(\phi_1,\overline{\phi_1}) = q\begin{pmatrix} 1 \\ -1 \end{pmatrix}, \text{ with } q = 2b\phi_{11}\phi_{12} + \frac{1}{b}\phi_{11}^2 \approx -0.0563,$$

$$\Phi_2 = q\begin{pmatrix} -1.2 \\ 0.13 \end{pmatrix}, \ \Phi_0 = q\begin{pmatrix} 0 \\ 1 \end{pmatrix}, \ C(\phi_1,\phi_1,\phi_1) = \phi_1^2\phi_2\begin{pmatrix} 1 \\ -1 \end{pmatrix}, \ \phi_1^* \approx \begin{pmatrix} 1.03 \\ 0.28 \end{pmatrix},$$

and thus altogether $\nu_3 \approx -0.42$. The spatially independent equilibria of the GL equation are therefore $A_e \equiv \pm\sqrt{-1/\nu_3} \approx 1.55$, from which we get the approximate solutions

$$\begin{aligned} \begin{pmatrix} u \\ v \end{pmatrix} &= \begin{pmatrix} b \\ 1/b \end{pmatrix} + 2\varepsilon A_e \cos(k_c x + \vartheta)\phi_1 + \mathcal{O}(\varepsilon^2) \\ &\approx \begin{pmatrix} 1.5 + 0.75\cos(k_c x + \vartheta) \\ 0.67 - 0.18\cos(k_c x + \vartheta) \end{pmatrix} + \mathcal{O}(\varepsilon^2), \end{aligned}$$

which fit well with the numerical simulations for the Schnakenberg model (9.23) shown in Figure 10.8. ⌋

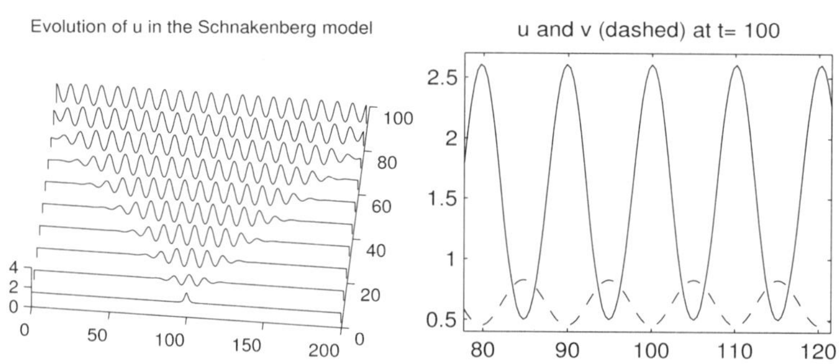

Figure 10.8. Pattern formation for the Schnakenberg model (9.23), $(b,d) = (1.5, 15)$, initial data $(u_0, v_0) = (b+\text{sech}(4x), 1/b+0.01\sec(4*x))$. Left: evolution of u. Right, $u(x,100)$ and $v(x,100)$ (dashed).

10.5.2. Validity of the GL approximation. The excellent agreement between numerical solutions and predictions of the GL equation is not restricted to the Schnakenberg model and this particular solution. It is the goal of the rest of this section to check that for the important class of reaction-diffusion systems the assumptions **(A0)**-**(A4)** of our abstract approximation theorem are satisfied, and that thus the GL approximation makes correct predictions if the simple spectral assumption **(Spec$_{\mathbf{RD}}$)** is valid. Since the nonlinear terms are given by $N(v) = g_\mu(v)$ we make the choice:

(A0) *We set* $X = Y = H^\theta_{\mathrm{ul}}(\mathbb{R}, \mathbb{R}^d)$.

The semigroup $(\mathrm{e}^{t\Lambda})_{t\geq 0}$ is defined by $\mathrm{e}^{t\Lambda} = \mathcal{F}^{-1}\widehat{\mathrm{e}^{t\Lambda}}\mathcal{F}$ where $\widehat{\mathrm{e}^{t\Lambda}}(k) = \mathrm{e}^{t\widehat{L}_\mu(k)}$. In order to extract the critical modes we define for all $k \in [k_c - \delta, k_c + \delta]$, with a $\delta > 0$ sufficiently small, a projection $P_c(k)$ on the critical eigenfunction $f_1(k)$ by

$$P_c(k) = \langle f_1^*(k), \cdot \rangle_{\mathbb{R}^d} f_1(k),$$

where $f_1^*(k)$ is the associated eigenfunction of the adjoint operator $\widehat{L}_\mu(k)^*$. We introduce the mode filter E_c as multiplication operators in Fourier space by

$$E_j = \mathcal{F}^{-1}\chi_j P_c \mathcal{F}, \quad j = \pm c$$

where $\chi_j : \mathbb{R} \to [0,1]$ is C_0^∞, with $\chi_{\pm c}(k) = 1 \Leftrightarrow |k \mp k_c| \leq \delta/4$, $\chi_{\pm c}(k) = 0 \Leftrightarrow |k \mp k_c| \geq \delta/2$, and $E_s = 1 - E_c - E_{-c}$. Next we choose

$$E_j^h = \mathcal{F}^{-1}\chi_j^h P_c \mathcal{F}, \quad j = \pm c,$$

where $\chi_j^h : \mathbb{R} \to [0,1]$ is C_0^∞, with $\chi_{\pm c}^h(k) = 1 \Leftrightarrow |k \mp k_c| \leq \delta/2$, $\chi_{\pm c}^h(k) = 0 \Leftrightarrow |k \mp k_c| \geq \delta$. Finally, we set

$$E_s^h = I - \mathcal{F}^{-1}\chi_s^h P_c \mathcal{F},$$

where $\chi_j^h : \mathbb{R} \to [0,1]$ is C_0^∞, with $\chi_{\pm c}^h(k) = 1 \Leftrightarrow |k \mp k_c| \leq \delta/8$, $\chi_{\pm c}^h(k) = 0 \Leftrightarrow |k \mp k_c| \geq \delta/4$. With this choice the validity of the assumptions **(A1)** and **(A2)** follow almost line for line as for the KS equation. See Lemma 10.2.8 for such estimates in H^θ_{ul}-spaces.

Since H^θ_{ul} is an algebra for $\theta > 1/2$ and since

$$E_c B(\Psi_c, R_c) = 0, \qquad E_c B(R_c, R_c) = 0$$

for $\Psi_c = E_c^h \Psi_c$ and $R_c = E_c^h R_c$ the Assumption **(A3)** follows again almost line for line as for the KS equation.

The residual terms are of the form

$$\mathrm{Res}(\varepsilon\Psi) = \sum_{j=-N}^{N} \varepsilon^{\beta_j} s_j(\varepsilon x, \varepsilon^2 t) \mathrm{e}^{\mathrm{i}jx},$$

with coefficients β_j satisfying $\beta_j \geq 3$ for all $j \in \{-N, N\}$. Hence, it remains to prove

$$\sup_{t\in[0,T_0/\varepsilon^2]} \|E_c \varepsilon^{\beta_j} s_j(\varepsilon x, \varepsilon^2 t) \mathrm{e}^{\mathrm{i}j\cdot x}\|_{H^\theta_{\mathrm{ul}}} \leq C\varepsilon^4,$$

but this follows exactly as the estimate for the KS equation (10.38). Therefore, **(A4)** is also valid for reaction-diffusion systems.

Theorem 10.5.4. *Fix $\theta \geq 1$, assume that the reaction-diffusion system* (10.48) *satisfies* **(Spec$_{\mathbf{RD}}$)**, *and let $A \in C([0,T_0], H^{\theta+3}_{\mathrm{ul}})$ be a solution of the GL equation* (10.53). *Then there exist a $C > 0$ and an $\varepsilon_0 > 0$ such that for all $\varepsilon \in (0, \varepsilon_0)$ we have solutions v of the reaction-diffusion system* (10.48) *with*

$$\sup_{t\in[0,T_0/\varepsilon^2]} \|v - \varepsilon\Psi_{\mathrm{GL}}\|_{H^\theta_{\mathrm{ul}}} \leq C\varepsilon^2,$$

where $\varepsilon\Psi_{\mathrm{GL}}$ is the associated GL approximation defined in (10.49).

10.5.3. The Hopf bifurcation case. The GL equation can also be derived in case when the bifurcating pattern is oscillatory in time without oscillations in space, i.e., when a Hopf bifurcation occurs at the wave number $k = 0$, see Figure 9.8 d). In this case, the number of unbounded spatial directions plays no role, and thus we consider $x \in \mathbb{R}^d$. The most famous example is the so called Brusselator, cf. (9.3) and [**Kur84**],

$$\partial_t v_1 = d_1 \Delta v_1 + a - (b+1)v_1 + v_1^2 v_2, \qquad \partial_t v_2 = d_2 \Delta v_2 + b v_1 - v_1^2 v_2,$$

where a, b, d_1, d_2 are non-negative constants, and v_1, v_2 are functions of $t \geq 0$ and $x \in \mathbb{R}^d$. For this reaction-diffusion system there exists a unique uniform steady state $(v_1^*, v_2^*) = (a, b/a)$. Introducing new coordinates $(v_1, v_2) = (v_1^* + u_1, v_2^* + u_2)$ gives the system

$$\begin{aligned} \partial_t u_1 &= d_1 \Delta u_1 + (b-1)u_1 + a^2 u_2 + f(u_1,u_2), \\ \partial_t u_2 &= d_2 \Delta u_2 - b u_1 - a^2 u_2 - f(u_1,u_2), \end{aligned} \tag{10.57}$$

where

$$f(u_1,u_2) = (b/a)u_1^2 + 2a u_1 u_2 + u_1^2 u_2.$$

We fix a and take b as a control parameter. The stability of $u_1 = u_2 = 0$ is determined by the linearization of (10.57). We find the eigenfunctions $(u_1(k), u_2(k))\mathrm{e}^{\mathrm{i}k\cdot x}$, where the associated eigenvalues λ satisfy

$$\lambda^2 + \alpha(|k|)\lambda + \beta(|k|) = 0,$$

with $|k|^2 = \sum_{j=1}^d k_j^2$ and $k = (k_1, \ldots, k_d) \in \mathbb{R}^d$. The constants are given by

$$\begin{aligned} \alpha(q) &= 1 + a^2 - b + (d_1 + d_2)q^2, \\ \beta(q) &= a^2 + (a^2 d_1 + (1-b)d_2)q^2 + d_1 d_2 q^4. \end{aligned}$$

The trivial solution becomes unstable via an oscillatory instability at the critical wave vector $k = 0$ for a critical value $b_c = 1 + a^2$ of the control parameter if $\sqrt{d_1/d_2} > a^{-1}(\sqrt{1+a^2} - 1)$. Drawing the surfaces of eigenvalues $\lambda_\pm$ we obtain near $k = 0$ the rotational symmetric Figure 9.8 d).

We are interested in the bifurcation scenario near the threshold of instability. Therefore, we introduce the small bifurcation parameter $0 < \varepsilon^2 \ll 1$ where $\varepsilon^2 = (b - b_c)/b_c$. In order to derive the GL equation from (10.57) we make the ansatz

$$u = \varepsilon\Psi_{\mathrm{GL}} + \varepsilon^2\Psi_h \tag{10.58}$$

where

$$\varepsilon\Psi_{\mathrm{GL}} = A(\varepsilon x, \varepsilon^2 t)e^{\mathrm{i}(\mathrm{Im}\lambda_+|_{k=0,\varepsilon=0})t}\begin{pmatrix} 1 \\ -1 + \mathrm{i}a^{-1} \end{pmatrix} + c.c., \quad \varepsilon^2\Psi_h = \mathcal{O}(\varepsilon^2),$$

where the vector is the eigenvector corresponding to the eigenvalue $\lambda_+ = \mathrm{i}a$ for $k = 0$ and $\varepsilon = 0$. Following the computations of [**Kur84**] shows that the amplitude A in lowest order satisfies a GL equation

$$\partial_T A = \nu_2 \sum_{j=1}^{p} \partial_{X_j}^2 A + \nu_0 A + \nu_3 |A|^2 A, \tag{10.59}$$

with $T = \varepsilon^2 t$, $X_j = \varepsilon x_j$, $A(X,T) \in \mathbb{C}$, and the coefficients

$$\begin{aligned}
\nu_0 &= (1 + a^2)/2, \\
\nu_2 &= (d_1 + d_2 - \mathrm{i}a(d_1 - d_2))/2, \\
\nu_3 &= -\frac{1}{2}\left(\frac{2 + a^2}{a^2} + \mathrm{i}\frac{4 - 7a^2 + 4a^4}{3a^3}\right).
\end{aligned}$$

The solutions of (10.57) which are described by the GL equation are slow modulations in time and in space of the spatially homogeneous temporally oscillating state $u = (1, -1 + \mathrm{i}a^{-1})e^{\mathrm{i}at}$.

The abstract set-up. In the following we consider an abstract reaction-diffusion system

$$\partial_t u = D\partial_x^2 u + f_\mu(u), \tag{10.60}$$

with diagonal diffusion matrix $D = \mathrm{diag}(d_1\Delta, \ldots, d_d\Delta)$, where $d_j > 0$. For this system we assume the existence of a stationary solution u^*, w.l.o.g. $u^* = 0$, such that the linearization

$$L_\mu(\partial_x) = D\Delta + \partial_u f_\mu(0)$$

around u^* satisfies the following spectral assumption.

(Spec$_\mathbf{H}$) *There exist $\rho_0, \varepsilon_0 > 0$, such that for $0 < \varepsilon < \varepsilon_0$ and $k \in U_{\rho_0}(0) = \{k \in \mathbb{R}^d : |k| < \rho_0\}$ the following holds: The matrices*

$$-\mathrm{diag}(d_1|k|^2, \ldots, d_d|k|^2) + \partial_u f_\mu(0)$$

possess two surfaces of eigenvalues $\widetilde{\lambda}_\pm \in C^\infty(U_{\rho_0}, \mathbb{C})$, *with associated eigenfunctions* $\widetilde{\varphi}_\pm \in C^\infty(U_{\rho_0}, \mathbb{C}^q)$, *such that*

$$\widetilde{\lambda}_\pm(k, \varepsilon^2) = \pm i\omega + \varepsilon^2(\lambda_0 \pm i\nu_0) + |k|^2(\lambda_2 \pm i\nu_2) + \mathcal{O}(\varepsilon^3 + |k|^3),$$

with constants $\lambda_0 > 0$, $\lambda_2 < 0$, *and* $\omega > 0$. *Denote by* $\Sigma^-(k, \varepsilon^2)$ *the set of all other eigenvalues of* $\widetilde{\Lambda}(k, \varepsilon^2)$. *Then, there exists an* ε*-independent constant* $\widetilde{\sigma}_0 > 0$ *such that*

$$\sup \mathrm{Re} \left(\bigcup_k \Sigma^-(k, \varepsilon^2) \right) < -\widetilde{\sigma}_0.$$

In general the abstract theorem 10.4.3 cannot be applied in this situation, even when restricting to $x \in \mathbb{R}$, due to the fact that Assumption **(A3)**, with the crucial property that $E_c N_2(\Psi_c, R_c) = 0$, is not satisfied in general. The quadratic interaction of critical modes again gives critical modes in case of a Hopf bifurcation at the wave vector $k = 0$. What helps is the oscillatory character of these quadratic terms. It will allow us to remove them by a normal form transformation similar to §3.3. This has been carried out in detail in [**Sch98b**]. We come back to normal form transformations in §11.4. After the transformation we have $E_c N_2(\Psi_c, R_c) = 0$ and so an analogue of Theorem 10.4.3 can be applied.

Theorem 10.5.5. *Fix* $s > d/2$ *and assume that the reaction-diffusion system* (10.60) *satisfies the spectral assumption* **(Spec$_\mathbf{H}$)**. *Let*

$$A \in C([0, T_0], H^{\theta+3}_{\mathrm{ul}})$$

be a solution of the associated GL equation. Then there exist a $C > 0$ *and an* $\varepsilon_0 > 0$ *such that for all* $\varepsilon \in (0, \varepsilon_0)$ *we have solutions* u *of the reaction-diffusion system* (10.60) *with*

$$\sup_{t \in [0, T_0/\varepsilon^2]} \|u - \varepsilon \Psi_{\mathrm{GL}}\|_{H^\theta_{\mathrm{ul}}} \leq C\varepsilon^2,$$

where $\varepsilon\Psi_{\mathrm{GL}}$ *is the associated GL approximation defined in* (10.58).

10.6. Convection problems

Hydrodynamic stability problems have always inspired new methods in dynamical systems theory. All classical hydrodynamic stability problems are experiments or Gedanken-experiments, which have been designed to understand and to extract special phenomena in more complicated situations. Examples considered in this chapter are convection problems and the Couette-Taylor problem. Further hydrodynamic stability problems such as Poiseuille flow, or the flow down an inclined plane are discussed in Chapter 14 about diffusive stability.

A convection cell is a characteristic fluid flow pattern which occurs for instance when a fluid is heated from below and cooled from above. Because of the heat loss at the top surface and the fact that the fluid is more dense at low temperature, a rising fluid becomes denser than the fluid underneath. Since it cannot descend through the rising fluid, it moves sidewards. At some distance the gravitational force overcomes the rising force and the fluid begins to descend. As it descends, it warms again and the cycle repeats itself. This mechanism plays a big role in geophysics. Examples are atmospheric or oceanic circulations, but also the movement of the tectonic plates on earth is induced by convection in the mantle of the earth. The forces leading to convection do not necessarily come from a heat-dependent density. Vibration induced convection occurs in powders and granular materials in vibrating containers, Marangoni convection is induced by a heat-dependent surface tension, and electro convection is observed in nematic crystals.

In this section we apply our previous theory to convection in porous media and the classical Bénard problem. Moreover, we comment on the case of planar systems and systems with a time-periodic forcing.

10.6.1. Convection in porous media. For so called black smokers on the ocean floor, convection takes place in a porous medium, namely in the highly fractured rock between the air or the sea at the top and the magma chambers at the bottom. Compared to classical hydrodynamical stability problems the associated system of PDEs for convection in porous media is easier since the velocity field of the fluid is determined by a constitutive law, namely Darcy's law, and does not have to be computed as a solution of the Navier-Stokes equations.

In detail, we are interested in the velocity field $u = (u_1, u_2)$ and the temperature field T of a fluid in a strip $\Omega = \mathbb{R} \times (0,1)$ of a porous medium, heated from below. If we denote the coordinates in the strip by $(x, y) \in \mathbb{R} \times (0,1)$, then we have to solve

$$\nabla \cdot u = 0, \tag{10.61}$$

$$u = -\nabla p + RTe_2, \tag{10.62}$$

$$\partial_t T + u \cdot \nabla T = \Delta T, \tag{10.63}$$

under the boundary conditions $T = 1$, $u_2 = 0$ at $y = 0$ and $T = 0$, $u_2 = 0$ at $y = 1$. Here, $\nabla = (\partial_x, \partial_y)^T$, $\Delta = \partial_x^2 + \partial_y^2$, $e_2 = (0,1)^T$, p denotes a pressure field, and the so called Rayleigh number R is a dimensionless parameter which is proportional for instance to the distance of the plates and the temperature difference between the upper and lower plate.

For a detailed derivation of (10.61)-(10.63) see for instance [**Fow97**, §14]. Equation (10.61) describes conservation of mass for an incompressible fluid, while (10.62) is based on the balance of forces. It is a constitutive law which relates the velocity field u, the pressure p and the temperature T. It comes from Darcy's law which relates the velocity field u, the pressure p and the density ρ. Here the relation between the temperature T and the density ρ is given by an affine law. Otherwise the small temperature induced density differences are neglected and ρ is considered to be constant for all $(x, y) \in \mathbb{R} \times (0, 1)$. This is called the Boussinesq approximation. Equation (10.63) describes the diffusion and transport of heat.

The trivial state and its stability. The trivial solution of (10.61)-(10.63) is given by the purely conducting state

$$u = 0, \quad T = 1 - y, \quad p = -\frac{R}{2}(1 - y)^2. \tag{10.64}$$

Since (10.61)-(10.63) is supposed to be a model for convection we expect that for large R, e.g., for large temperature difference between the upper and lower plate, convection sets in, resulting in some pattern of convection rolls. In the following we explain that this is indeed the case, and that it can be described via a GL approximation.

Our first step is to find the dispersion relation for the linearized system. We eliminate the pressure p by introducing the stream function ψ such that

$$u_1 = \partial_y \psi \quad \text{and} \quad u_2 = -\partial_x \psi,$$

and introduce the deviation Θ from the linear temperature profile by $T = 1 - y + \Theta$. This yields

$$\Delta\psi = -R\partial_x\Theta \quad \text{and} \quad \partial_t\Theta + \partial_x\psi + (\partial_y\psi)\partial_x\Theta - (\partial_x\psi)\partial_y\Theta = \Delta\Theta. \tag{10.65}$$

The linearized system is given by

$$\Delta\psi = -R\partial_x\Theta, \qquad \partial_t\Theta + \partial_x\psi = \Delta\Theta,$$

together with the boundary conditions $\Theta = \psi = 0$ at $y = 0, 1$. Due to the boundary conditions we make the ansatz

$$\psi = f \sin(n\pi y)e^{ikx+\lambda t}, \qquad \Theta = g \sin(n\pi y)e^{ikx+\lambda t},$$

with $n \in \mathbb{N}$, $k \in \mathbb{R}$, and complex-valued coefficients f and g. This gives the system of linear equations

$$-(\pi^2 n^2 + k^2)f = -ikRg, \qquad -(\pi^2 n^2 + k^2)g = ikf + \lambda g.$$

We find

$$f = \frac{ikR}{n^2\pi^2 + k^2} g \quad \text{and} \quad \lambda = \frac{Rk^2}{n^2\pi^2 + k^2} - (n^2\pi^2 + k^2), \tag{10.66}$$

i.e., we have a family of curves $k \mapsto \lambda_n(k) \in \mathbb{R}$ of eigenvalues with $n \in \mathbb{N}$ and $k \in \mathbb{R}$. It is easy to see that $\lambda_{n+1}(k) \leq \lambda_n(k) \in \mathbb{R}$ for each fixed $k \in \mathbb{R}$.

Moreover, $\lambda_n(k) \to -\infty$, for $|k| \to \infty$ and fixed n, or for $n \to \infty$ and fixed k.

The trivial solution $(\Theta, \psi) = (0,0)$ is stable if $\lambda_1(k) < 0$ for all $k \in \mathbb{R}$. Instability occurs when the curve λ_1 touches the axis $\lambda = 0$ at a wave number $k = k_c \in \mathbb{R}$ for a parameter value $R = R_c$. This leads to the conditions

$$\lambda_1 = \frac{Rk^2}{\pi^2 + k^2} - (\pi^2 + k^2) \stackrel{!}{=} 0$$

and

$$\partial_{k^2}\lambda_1 = \frac{R}{\pi^2 + k^2} - \frac{Rk^2}{(\pi^2 + k^2)^2} - 1 = \frac{\pi^2 R}{(\pi^2 + k^2)^2} - 1 \stackrel{!}{=} 0.$$

From this we find $\pi R^{1/2} = \pi^2 + k^2$ and $\lambda_1 = R - 2\pi R^{1/2}$. This shows that $\lambda_1 = 0$ for $R = R_c = 4\pi^2 \approx 39.48$ at the critical wave number $k = k_c = \pi$, see Figure 10.9.

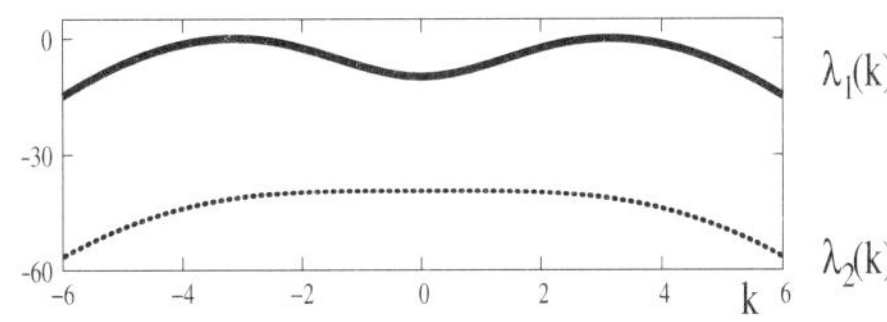

Figure 10.9. The curves of eigenvalues $k \mapsto \lambda_n(k)$ for $n = 1, 2$.

Remark 10.6.1. The problem can be extended periodically into the y-direction and can be solved in the subspace of odd functions w.r.t. y. Hence for a function

$$u(x,y) = \sum_{n\in\mathbb{N}} \int_{\mathbb{R}} \widehat{u}(k,n)e^{ikx}\,dk \sin(n\pi y)$$

the H^θ-norm is equivalent to the norm

$$\left(\sum_{n\in\mathbb{N}} \int_{\mathbb{R}} |\widehat{u}(k,n)|^2(1 + |k|^2 + |n|^2)^\theta\right)^{1/2}.$$

$\rfloor$

Remark 10.6.2. The value R_c for spectral instability coincides with the critical value $R_{c,e}$ for which energetic stability of the trivial solution can be shown. For this we multiply (10.65) with ψ and integrate over Ω to obtain

$$-\int_\Omega |\nabla\psi|^2\,dx\,dy = -R\int_\Omega \Theta\partial_x\psi\,dx\,dy,$$

hence $\|\nabla\psi\|_{L^2} \le R\|\Theta\|_{L^2}$. This, Poincaré's inequality $\|\Theta\|_{L^2} \le \frac{1}{2\pi}\|\nabla\Theta\|_{L^2}$, and the fact that the last two terms in the first line of the following formula cancel after integration by parts, gives

$$\frac{1}{2}\partial_t\|\Theta\|_{L^2}^2 = -\|\nabla\Theta\|_{L^2}^2 + \int_\Omega -(\partial_x\psi)\Theta - (\partial_y\psi)(\partial_x\Theta)\Theta + (\partial_x\psi)(\partial_y\Theta)\Theta \,\mathrm{d}x\,\mathrm{d}y$$
$$\leq -\|\nabla\Theta\|_{L^2}^2 + \|\partial_x\psi\|_{L^2}\|\Theta\|_{L^2} \leq (R-4\pi^2)\|\Theta\|_{L^2}^2.$$

Thus, Gronwall's inequality yields

$$\|\Theta(\cdot,t)\|_{L^2}^2 \leq \mathrm{e}^{t(R^2-4\pi^2)}\|\Theta(\cdot,0)\|_{L^2}^2. \tag{10.67}$$

Hence, for $R < R_{c,e} = 4\pi^2$ solutions decay exponentially to zero in L^2, while for $R > R_{c,e} = 4\pi^2$ the L^2-norm can grow exponentially. ⌋

The dynamical systems formulation. We have the spectral situation necessary for the derivation of the GL equation. Therefore, we would like to come back to the arguments of §10.3 where we explained the universality of the occurrence of the GL equation as an amplitude or modulation equation. In order to do so we first formulate (10.65) as a dynamical system, i.e., we have to get rid of the problem that (10.65) contains variables and equations without time derivatives.

(10.65) is a parabolic equation for Θ coupled to an elliptic equation for ψ. The evolution for (10.65) is therefore determined by specifying initial conditions $\Theta(x,y,0) = \Theta_0(x,y)$, while ψ is determined by the compatibilty condition $\Delta\psi = -R\partial_x\Theta$, $\psi|_{y=0,1} = 0$. Using (10.66), and

$$\Theta(x,y,t) = \sum_{n\in\mathbb{N}}\int_{\mathbb{R}} \widehat{\Theta}(k,n,t)\mathrm{e}^{\mathrm{i}kx}\,\mathrm{d}k\sin(n\pi y),$$

we find

$$\psi(x,y) = R\sum_{n\in\mathbb{N}}\int_{\mathbb{R}} \frac{\mathrm{i}k}{k^2+n^2\pi^2}\widehat{\Theta}(k,n,t)\mathrm{e}^{\mathrm{i}kx}\,\mathrm{d}k\sin(n\pi y),$$

which defines a linear operator M via

$$\psi = RM\Theta = R\mathcal{F}^{-1}\widehat{M}\mathcal{F}\Theta \quad \text{where} \quad \widehat{M}(k,n) = \frac{\mathrm{i}k}{k^2+n^2\pi^2}.$$

Thus, M is bounded from H^θ to $H^{\theta+1}$, cf. Exercise 10.7. The introduction of M allows to rewrite (10.65) as a single scalar parabolic equation

$$\partial_t\Theta = L\Theta + N(\Theta), \tag{10.68}$$

with

$$L = \Delta - RM\partial_x\Theta \quad \text{and} \quad N(\Theta) = -R(\partial_y(M\Theta))\partial_x\Theta + R(\partial_x(M\Theta))\partial_y\Theta.$$

However, there is no maximum principle for (10.68) as for, e.g., the KPP equation or the Burgers equation, because (10.68) is nonlocal in physical space.

Establishing local existence and uniqueness of solutions is straightforward in Sobolev spaces H^θ. The linear operator L generates a C_0-semigroup $(\mathrm{e}^{tL})_{t\geq 0}$ in H^θ, with the smoothing estimate

$$\begin{aligned}\left\|\mathrm{e}^{tL}\Theta_0\right\|_{H^{\theta+1}} \leq & C_1 \left\|\mathrm{e}^{t\widehat{L}}\widehat{\Theta}_0\right\|_{L^2_{\theta+1}} \leq C_1 \sup_{k\in\mathbb{R},m\in\mathbb{N}} |\mathrm{e}^{t\widehat{L}(k,m)}(1+k^2+m^2)^{1/2}| \left\|\widehat{\Theta}_0\right\|_{L^2_\theta} \\ \leq & C(1+t^{-1/2})\mathrm{e}^{\beta t} \left\|\Theta_0\right\|_{H^\theta}\end{aligned}$$

for a $\beta \in \mathbb{R}$ since $\widehat{L}(k,m) \leq \beta - k^2 - m^2$. Since M is smoothing we have estimates such as

$$\begin{aligned}\|\partial_y(M\Theta)\partial_x\Theta\|_{L^2} & \leq C\|\partial_y(M\Theta))\|_{L^\infty}\|\partial_x\Theta\|_{L^2} \leq C\|\partial_y(M\Theta))\|_{H^\theta}\|\partial_x\Theta\|_{L^2} \\ & \leq C\|\Theta\|_{H^\theta}\|\Theta\|_{H^1}\end{aligned}$$

for every $\theta > 1$ by the Sobolev embedding $H^\theta \hookrightarrow L^\infty$ for $\theta > d/2$. Hence, N is locally Lipschitz-continuous from H^θ to $H^{\theta-1}$ for $\theta > 1$ and using the method from §5.2.3 we can construct local mild solutions via the variation of constant formula

$$\Theta(t) = \mathrm{e}^{tL}\Theta_0 + \int_0^t \mathrm{e}^{(t-\tau)L} N(\Theta(\tau))\,\mathrm{d}\tau.$$

Therefore, we have the following local existence and uniqueness theorem.

Theorem 10.6.3. *Let $\theta > 1$. For every $\Theta_0 \in H^\theta$ there exists a $t_0 = t_0(\|\Theta_0\|_{H^\theta}) > 0$ and a unique local mild solution $\Theta \in C([0,t_0], H^\theta)$ of (10.68) with $\Theta|_{t=0} = \Theta_0$.*

As a consequence we additionally have $\psi = RM\Theta \in C([0,t_0], H^{\theta+1})$.

Derivation of the GL equation. Our starting point is System (10.68) which satisfies the spectral situation necessary for the derivation of the GL equation. We introduce the small perturbation parameter $0 < \varepsilon^2 \ll 1$ by $R = R_c + \varepsilon^2$. The solutions of the system can be expanded in eigenfunctions, i.e., we set

$$\Theta(x,y,t) = \sum_{n\in\mathbb{N}} \int_{\mathbb{R}} c_n(k,t)\mathrm{e}^{\mathrm{i}kx}\,\mathrm{d}k \sin(n\pi y).$$

In contrast to the reaction-diffusion systems considered in §10.5 for fixed k we have an infinite-dimensional problem. Nevertheless, the present system is special in the sense that the eigenfunctions $f_n(k,y) = \sin(n\pi y)$ do not depend on $k \in \mathbb{R}$. Moreover, due to the fact that the problem can be extended periodically in the y-direction we also have some convolution structure in the y-direction which allows us to compute an exact expression for the coefficient in front of the cubic terms of the GL equation. We find

$$\partial_t c_n(k,t) = \lambda_n(k)c_n(k,t) + \sum_{n_1,n_2\in\mathbb{N}} \int_{-\infty}^{\infty} s_{2,n,n_1,n_2}(k,k-l,l)c_{n_1}(k-l)c_{n_2}(l)\,\mathrm{d}l$$

with eigenvalues

$$\lambda_n(k,\varepsilon^2) = \frac{Rk^2}{n^2\pi^2+k^2} - (n^2\pi^2+k^2)$$

and symmetrized complex-valued kernels

$$\begin{aligned} s_{2,n,n_1,n_2}(k,k-l,l) =& \frac{1}{8}R\pi\frac{\mathrm{i}(k-l)}{(k-l)^2+n_1^2\pi^2}(-n_1\mathrm{i}l+\mathrm{i}(k-l)n_2)\delta_{n-n_1-n_2} \\ &+ \frac{1}{8}R\pi\frac{\mathrm{i}l}{l^2+n_2^2\pi^2}(-n_2\mathrm{i}(k-l)+\mathrm{i}ln_1)\delta_{n-n_1-n_2} \end{aligned}$$

since $\int_0^1 \sin(n\pi y)\cos(n_1\pi y)\sin(n_2\pi y)dy = \delta_{n-n_1-n_2}/4$. Following the calculations of §10.5 we have to compute

$$\begin{aligned} \nu_2 =& \partial_k^2\lambda_1(\pi;0)/2 = 2, \\ \nu_0 =& \partial_{\varepsilon^2}\lambda_1(\pi;0) = \frac{1}{2}, \\ \nu_3 =& -4\sum_{n=1}^{\infty} s_{211n}(k_c,k_c,0;0)s_{2n1-1}(0,-k_c,k_c;0)/\lambda_n(0;0) \\ & -2\sum_{n=1}^{\infty} s_{211n}(k_c,-k_c,2k_c;0)s_{2n11}(2k_c,k_c,k_c;0)/\lambda_n(2k_c;0) \\ =& -4s_{2112}(k_c,k_c,0;0)s_{221-1}(0,-k_c,k_c;0)/\lambda_2(0;0) \\ & -2s_{2112}(k_c,-k_c,2k_c;0)s_{2211}(2k_c,k_c,k_c;0)/\lambda_2(2k_c;0) = -4\pi^2. \end{aligned}$$

Thus, we finally find the GL equation

$$\partial_T A = 2\partial_X^2 A + \frac{1}{2}A - 4\pi^4|A|^2 A, \tag{10.69}$$

cf. Exercise 10.8. Via the solutions of the GL equation we describe slow spatial and temporal modulations of small amplitude of the underlying convection pattern via the approximation

$$\begin{aligned} \begin{pmatrix}\psi\\ \Theta\end{pmatrix}(x,y,t) &= \varepsilon\Psi_{\mathrm{GL}}(x,y,t) + \mathcal{O}(\varepsilon^2) \\ &= \varepsilon A(X,T)\begin{pmatrix}2\pi\mathrm{i}\\ 1\end{pmatrix}e^{\mathrm{i}\pi x}\sin\pi y + \mathrm{c.c.} + \mathcal{O}(\varepsilon^2), \end{aligned} \tag{10.70}$$

where $X = \varepsilon x$ and $T = \varepsilon^2 t$ are the long spatial and long temporal scale. An immediate observation is that the GL equation (10.69) possesses stable steady states $A = \frac{1}{2\sqrt{2}\pi^2}e^{\mathrm{i}\phi}$ which are constant in space, where $\phi \in [0, 2\pi)$ is a free phase. For $\phi = 0$ this formally yields the steady convection rolls

$$\begin{pmatrix}u_1\\ u_2\\ T\end{pmatrix} = \begin{pmatrix}0\\ 0\\ 1-y\end{pmatrix} + \frac{\varepsilon}{2\sqrt{2}\pi^2}\begin{pmatrix}4\pi^2\cos\pi y\sin\pi x\\ -4\pi^2\sin\pi y\cos\pi x\\ \sin\pi y\cos\pi x\end{pmatrix} + \mathcal{O}(\varepsilon^2). \tag{10.71}$$

As usual, these formal calculations do not guarantee that (10.71) is a good approximation of steady convection rolls for (10.61)-(10.63), nor that steady rolls exist at all. A rigorous existence proof for the stationary convection rolls follows with center manifold theory or the Lyapunov-Schmidt method. See §13 or Exercise 10.9.

The approximation result. The error estimates for the GL approximation follow immediately in H^θ-spaces, where all assumptions of our abstract approximation theorem 10.4.3 are easy to check. Therefore, we have

Theorem 10.6.4. *Fix $\theta > 1$ and let $A \in C([0,T_0], H^{\theta+3})$ be a solution of the GL equation* (10.69). *Then there exist $\varepsilon_0, C > 0$ such that for all $\varepsilon \in (0,\varepsilon_0)$ there are solutions of* (10.61)-(10.63) *with*

$$\sup_{t\in[0,T_0/\varepsilon^2]} \|(\psi,\Theta)(\cdot,t) - \varepsilon\Psi_{\mathrm{GL}}(\cdot,t)\|_{(H^\theta)^2} \leq C\varepsilon^{3/2}. \tag{10.72}$$

We refrain from working with H^θ_{ul}-spaces for this introductory example.

10.6.2. Bénard's problem. Next, we treat the two-dimensional Rayleigh-Bénard problem between two fixed plates, i.e., in the strip $\Omega = \mathbb{R}\times(0,\pi)$. The problem is very similar to the one of the last section except that no longer a porous medium is considered. Therefore, the Navier-Stokes equations replace Darcy's law. The so called Boussinesq approximation is used, i.e., the density is considered to be a constant except in the buoyancy term. The problem is supplemented with the mean flux condition $\int_0^\pi u_1\, dy = 0$ and the boundary conditions

$$\partial_y u_1|_{y=0} = u_2|_{y=0} = \partial_y u_1|_{y=\pi} = u_2|_{y=\pi} = 0, \quad T|_{y=0} = T_0, \quad T|_{y=\pi} = T_1,$$

where $T_0 \geq T_1$. These boundary conditions allow for a number of explicit calculations. For the physically more realistic boundary condition, $u_1|_{y=0,\pi} = 0$ instead of $\partial_y u_1|_{y=0,\pi} = 0$, the subsequent analysis can be carried out, too.

There is a trivial solution, the purely conducting state with an affine temperature profile and no motion of the fluid, namely

$$u = 0, \qquad T = T_0 + y(T_1 - T_0)/\pi.$$

This state is stable if the temperature difference between the lower and the upper plate is sufficiently small. If the temperature difference becomes large, then it loses stability and convection rolls appear. For even larger temperature differences the motion becomes more complicated and eventually turbulent. With $\Theta = T - T_0 - y(T_1 - T_0)/\pi$ we introduce the deviation from the linear heat profile and consider

$$\partial_t u = \Delta u - \nabla p - \rho\Theta\vec{e}_2 - (u\cdot\nabla)u, \tag{10.73}$$

$$\partial_t \Theta = \kappa\Delta\Theta + u_2 - (u\cdot\nabla)\Theta, \tag{10.74}$$

$$0 = \nabla\cdot u. \tag{10.75}$$

The equations contain two dimensionless parameters, namely the Rayleigh number $\rho = \beta_0(T_0 - T_1)h^3/(\pi^3\nu^2)$ and $\kappa = \delta/\nu$, where δ stands for the heat conductivity, ν for viscosity, β_0 for the buoyancy parameter, and h for the physical height of the fluid.

Linear stability analysis. In a first step we consider the linearization around the pure heat conducting state. We find

$$\begin{aligned} \partial_t u =& \Delta u - \nabla p - \rho\Theta\vec{e}_2, \\ \partial_t \Theta =& \kappa\Delta\Theta + u_2, \\ 0 =& \nabla \cdot u, \end{aligned}$$

which is solved by

$$\begin{pmatrix} u_1 \\ u_2 \\ \Theta \\ p \end{pmatrix}(x,y,t) = e^{\lambda t}e^{ikx}\begin{pmatrix} \widehat{u}_{1,m}\cos(my) \\ \widehat{u}_{2,m}\sin(my) \\ \widehat{\Theta}_m\sin(my) \\ \widehat{p}_m\cos(my) \end{pmatrix}$$

for $m \in \{1,2,3,\ldots\}$ and where

$$2\lambda_{m,\pm}(k) = -(\kappa+1)s \pm \sqrt{(\kappa+1)^2 s^2 + 4\rho k^2 s^{-1}},$$

and $s = k^2 + m^2$. The associated eigenfunctions in (u_1, u_2, Θ) are denoted in the following by $\varphi_{m,\pm}(k)$.

We are interested in the problem close to the threshold of instability, fix κ and vary the control parameter ρ. Due to the vanishing mean flux condition we have $m \geq 1$. It turns out that the trivial solution is unstable for $\rho > \rho_c = 27\kappa/4$. For $\rho = \rho_c + \varepsilon^2$ the curve of eigenvalues $\lambda_{1,+}(k)$ is positive in an ε-neighborhood around $\pm k_c = \pm 1/\sqrt{2}$. See Exercise 10.10.

The GL approximation. Defining the small bifurcation parameter ε in this way, allows us to define the GL approximation

$$(u,\Theta) = \varepsilon\Psi_{\mathrm{GL}}(x,y,t) + \mathcal{O}(\varepsilon^2) = \varepsilon A(X,T)e^{ik_c x}\varphi_{1,+}(k_c) + c.c. + \mathcal{O}(\varepsilon^2),$$

with $X = \varepsilon x$ and $T = \varepsilon^2 t$. Inserting this into (10.73)-(10.75) we find that the complex amplitude A, which modulates the spatially periodic pattern $e^{ik_c x}\varphi_{1,+}(k_c)$ slowly in time and in space, has to satisfy the GL equation

$$\partial_T A = \frac{4\kappa}{\kappa+1}\partial_X^2 A + \frac{2}{9(\kappa+1)}A - \frac{1}{8(\kappa+1)}A|A|^2. \tag{10.76}$$

For a derivation see Exercise 10.11. Since again the problem can be extended periodically into the y-direction, for the calculation of the cubic coefficient only finitely many modes play a role, and so this coefficient can be computed explicitly. For Bénard's problem our approximation theorem specializes to

Theorem 10.6.5. *Fix $\theta \geq 2$ and let $A \in C([0,T_0], H_{\mathrm{ul}}^{\theta+3}(\mathbb{R},\mathbb{C}))$ be a solution of the GL equation (10.76). Then there exists $\varepsilon_0, C > 0$ such that for all*

$\varepsilon \in (0, \varepsilon_0)$ *there are solutions* (u, Θ) *of Bénard's problem* (10.73)-(10.75) *with*

$$\sup_{t\in[0,T_0/\varepsilon^2]} \|(u,\Theta)(\cdot,\cdot,t) - \varepsilon\Psi_{\mathrm{GL}}(\cdot,\cdot,t)\|_{H^\theta_{\mathrm{ul}}} \leq C\varepsilon^2.$$

The proof follows by applying the abstract approximation theorem stated in Theorem 10.4.3. How to check the assumptions **(A0)-(A4)** can be found in [**Sch94a**] where a detailed proof of Theorem 10.6.5 has been given.

10.6.3. Remarks on planar pattern. In this section we consider pattern forming systems with more than one unbounded space directions. There, pattern such as rolls, hexagons, and squares occur. As a toy problem we consider

$$\partial_t u = -(1+\Delta)^2 u + \alpha\partial_y^2 u + \varepsilon^2 u + \beta|\nabla u|^2 + \gamma u^3, \tag{10.77}$$

where $\Delta = \partial_x^2 + \partial_y^2$, $(x, y) \in \mathbb{R}^2$, $t \geq 0$, and $\alpha, \beta \in \mathbb{R}$. The small bifurcation parameter is denoted by $0 < \varepsilon^2 \ll 1$. The linearized problem is solved by $v(x, y, t) = \mathrm{e}^{ikx+ily+\lambda t}$ with

$$\lambda = \lambda(k, l, \varepsilon^2) = -(1 - k^2 - l^2)^2 - \alpha l^2 + \varepsilon^2.$$

The parameter α in our model allows to break the rotational symmetry.

The rotational symmetric situation. For $\alpha = 0$ the surface of spectral values is rotational symmetric. Like for the Rayleigh-Bénard problem the Fourier modes belonging to the wave vectors in a ring, here $|k^2 + l^2 - 1| = \mathcal{O}(\varepsilon)$, have positive growth rates. Amplitude equations can be derived for the various pattern. The case $\beta = 0$ is a phenomenological model for Rayleigh-Bénard problem with a rigid top boundary and the case $\beta \neq 0$ for the problem with a free top boundary. In the following we first restrict to the case $\beta = 0$ and $\gamma = -1$.

i) For squares, the Fourier modes are concentrated in $\mathcal{O}(\varepsilon)$-neighborhoods of four orthogonal wave vectors. W.l.o.g. we choose $(1, 0)$, $(0, 1)$, $(-1, 0)$, and $(0, -1)$. Modulations of square pattern are described via the ansatz

$$u(x, y, t) = \varepsilon A_{10}(\varepsilon x, \varepsilon y)\mathrm{e}^{\mathrm{i}x} + \varepsilon A_{01}(\varepsilon x, \varepsilon y)\mathrm{e}^{\mathrm{i}y} + \mathrm{c.c.} + \mathcal{O}(\varepsilon^2).$$

This ansatz leads to the system, see Exercise 10.12,

$$\begin{aligned}\partial_T A_{10} &= 4\partial_X^2 A_{10} + A_{10} - 3A_{10}|A_{10}|^2 - 6A_{10}|A_{01}|^2,\\ \partial_T A_{01} &= 4\partial_Y^2 A_{01} + A_{01} - 3A_{01}|A_{01}|^2 - 6A_{01}|A_{10}|^2,\end{aligned}$$

which is degenerated in the sense that in the A_{10} equation there is no diffusion in Y-direction and similar for the A_{01} equation. This reflects the fact that a big part of the ring of unstable modes is not covered by our ansatz.

ii) For hexagons, the Fourier modes are concentrated in $\mathcal{O}(\varepsilon)$-neighborhoods of six wave vectors with angle $\pi/3$ in between. W.l.o.g. we choose

$(\pm 1, 0)$ and $(\pm 1/2, \pm\sqrt{3}/2)$. Modulations of hexagons are described via the ansatz

$$u(x,y,t) = \varepsilon A_1(\varepsilon x, \varepsilon y)e^{ix} + \varepsilon A_2(\varepsilon x, \varepsilon y)e^{i(\frac{1}{2}x + \frac{\sqrt{3}}{2}y)} + \varepsilon A_3(\varepsilon x, \varepsilon y)e^{i(\frac{1}{2}x - \frac{\sqrt{3}}{2}y)} + \text{c.c.} + \mathcal{O}(\varepsilon^2)$$

leading to the system

$$\begin{aligned}
\partial_T A_1 &= 4\partial_X^2 A_1 + A_1 - 3A_1|A_1|^2 - 6A_1(|A_2|^2 + |A_3|^2),\\
\partial_T A_2 &= 4(\frac{\sqrt{3}}{2}\partial_X + \frac{1}{2}\partial_Y)^2 A_2 + A_2 - 3A_2|A_2|^2 - 6A_2(|A_1|^2 + |A_3|^2),\\
\partial_T A_3 &= 4(\frac{\sqrt{3}}{2}\partial_X - \frac{1}{2}\partial_Y)^2 A_3 + A_3 - 3A_3|A_3|^2 - 6A_3(|A_1|^2 + |A_2|^2).
\end{aligned}$$

The degeneracy w.r.t. diffusion again reflects the fact that a big part of the ring of unstable modes is not covered by the ansatz.

iii) For rolls there are two possibilities. If the Fourier modes are concentrated in $\mathcal{O}(\varepsilon)$-neighborhoods of the two wave vectors $(\pm 1, 0)$, then the above system for the squares, but with $A_{01} = 0$, is obtained. The second possibility is an $\mathcal{O}(\varepsilon)$-concentration in k-direction and an $\mathcal{O}(\varepsilon^{1/2})$-concentration in l-direction near $(\pm 1, 0)$. The ansatz is then given

$$u(x,y,t) = \varepsilon A(\varepsilon x, \varepsilon^{1/2} y)e^{ix} + \overline{\varepsilon A(\varepsilon x, \varepsilon^{1/2} y)}e^{-ix} + \mathcal{O}(\varepsilon^2),$$

leading to the so called Newell-Whitehead equation [**NW69**]

$$\partial_T A = -4(i\partial_X - \partial_Y^2/2)^2 A + A - 3A|A|^2. \tag{10.78}$$

This anisotropic scaling allows a better approximation of the ring of unstable modes at the wave vector $(1, 0)$, see Exercise 10.12. Nevertheless, the Newell-Whitehead equation is a degenerated parabolic equation.

For $\beta = 0$ the validity of these approximations follows line for line as for the SH equation in §10.2. In case $\beta \neq 0$ quadratic terms are present in the problem and the essential assumption of the analysis in §10.4, namely that the quadratic interaction of critical modes gives non-critical modes, is no longer valid. In fact, it can be shown rigorously [**Sch95b**] that in case $\beta \neq 0$ the Newell-Whitehead equation makes wrong predictions about the dynamics of (10.77). See §10.6.4.

Although quadratic terms are present in Bénard's problem, in case of rigid boundaries the approximation property holds since the quadratic interaction of critical modes gives non-critical modes, due to reflection symmetry of the problem in the bounded direction. We refer to [**GS85**] for a discussion of the bifurcation scenarios for the Rayleigh-Bénard problem with rigid and free top boundaries. However, the validity proof is only straightforward in

H^s-spaces. In H^s_{ul}-spaces new functional difficulties occur due to the projection on the divergence-free vector fields in case of more than one unbounded space directions, cf. [**CZ15**]. For other approaches to describe planar pattern via amplitude equations see also [**Mel00, EINP03**], and the further reading at the end of this Chapter.

The anisotropic situation. Next we consider the case $\alpha > 0$, i.e., we break the rotational symmetry in (10.77). Then unstable modes occur in $\mathcal{O}(\varepsilon)$-neighborhoods of the wave vectors $(\pm 1, 0)$. With the ansatz

$$u(x,y,t) = \varepsilon A(\varepsilon x, \varepsilon y)e^{\mathrm{i}x} + \varepsilon \overline{A(\varepsilon x, \varepsilon y)}e^{-\mathrm{i}x} + \mathcal{O}(\varepsilon^2)$$

a two-dimensional GL equation can be derived

$$\partial_T A = 4\partial_X^2 A + \alpha \partial_Y^2 A + A - 3A|A|^2. \tag{10.79}$$

Our abstract approximation result, Theorem 10.4.3, applies in this situation since now the quadratic interaction of the critical modes gives non-critical modes.

The prototype of an anisotropic pattern forming system is electro-convection in nematic liquid crystals [**OD08**]. In this problem the pattern formation is due to some external time-periodic forcing. It can be brought back with Floquet theory to a problem with an autonomous linear part. The relevant nonlinear terms can be made autonomous with the help of normal form transformations. Hence, the above theory applies to time-periodic systems as well [**SU07**]. See §10.6.5.

10.6.4. A counter-example.

We recall the example from [**Sch95b**], which shows that in case $\alpha = \gamma = 0$ and $\beta = 1$ the predictions made by the Newell-Whitehead equation

$$\partial_T A = -4(\mathrm{i}\partial_X - \partial_Y^2/2)^2 A + A - 4A|A|^2/9. \tag{10.80}$$

for (10.77) in general are wrong, although the Newell-Whitehead equation is derived in a formally correct way. Although modulation equations are not primarily derived for the handling of the spatially periodic situation, they should be valid for this special case, too. We consider the Newell-Whitehead equation (10.80) with the initial condition

$$A(X,Y,0) = \frac{3}{2} + B(0)\mathrm{e}^{\mathrm{i}(\frac{\sqrt{3}y}{2\sqrt{\varepsilon}} - \frac{x}{2\varepsilon})} + B(0)\mathrm{e}^{\mathrm{i} - \frac{\sqrt{3}y}{2\sqrt{\varepsilon}} - \frac{x}{2\varepsilon})}, \tag{10.81}$$

with $B(0) = \mathcal{O}(\varepsilon) \in \mathbb{R}$. The solutions of (10.80) are then spatially periodic with periods $\frac{2\pi\sqrt{\varepsilon}}{\sqrt{3}}$ and $2\pi\varepsilon$. In this space, (10.80) possesses a two-dimensional center manifold tangential to the complex space spanned by $\mathrm{e}^{\mathrm{i}0}$. The reduced system on the center manifold is given by

$$\partial_T \widetilde{A} = \widetilde{A} - 4/9\widetilde{A}|\widetilde{A}|^2 + \text{h.o.t.}$$

Therefore, the Newell-Whitehead equation predicts on an $\mathcal{O}(1/\varepsilon^2)$-time scale w.r.t. t, that solutions u of (10.77) behave as $u(x,y,t) = \frac{3}{2}e^{ix} + c.c. + \mathcal{O}(\varepsilon^{3/2})$.

It turns out that this prediction is only correct on a much shorter time scale. The initial condition (10.81) for (10.80) leads to a hexagonal pattern for (10.77). Therefore, the solutions of (10.77) can be written as Fourier series

$$u(x,y,t) = \sum_{j,k\in\mathbb{Z}} u_{jk}(t)e^{ijx+ik(x/2+\sqrt{3}y/2)}.$$

Equation (10.77) possesses a six-dimensional attracting center manifold with coordinates u_{10}, u_{01}, u_{-11}, u_{-10}, u_{0-1} and u_{1-1}. The initial condition for (10.80) gives solutions for (10.77) with $a = u_{10} = u_{-10} \in \mathbb{R}$ and $b = u_{01} = u_{-11} = u_{0-1} = u_{1-1} \in \mathbb{R}$. In this two-dimensional invariant subspace on the center manifold the reduced equations are given by

$$\begin{aligned} \partial_t a &= \varepsilon^2 a + b^2 - 4/9a^3 - 2/3b^2 a + h.o.t., \\ \partial_t b &= \varepsilon^2 b + ba - 1/3a^2 b - 7/9b^3 + h.o.t.. \end{aligned}$$

The perfect roll solutions described by (10.80) are given by the stationary solution at $(a,b) \approx (3\varepsilon/2, 0)$. The linearization at this fixed point is given by $\partial_t\widetilde{a} = -2\varepsilon^2\widetilde{a} + h.o.t.$, $\partial_t\widetilde{b} = 3\varepsilon\widetilde{b}/2 + h.o.t.$.

There is a one-to-one relation between the regularity of the initial condition $A|_{T=0}$ of (10.80) and the order of $B(0)$. If $A|_{T=0}$ is chosen analytic in a strip in the complex plane, then there there is an $r > 0$ such that $b(0) = \varepsilon B(0) = \mathcal{O}(\varepsilon e^{-r/\varepsilon})$, cf. §6.2.2. Hence, although there is an exponential growth $e^{\frac{3\varepsilon}{2}t}$ in the b-equation, b stays less than $\mathcal{O}(\varepsilon^2)$ on an $\mathcal{O}(1/\varepsilon^2)$ time scale since $\varepsilon e^{-r/\varepsilon}e^{\frac{3\varepsilon}{2}t} \ll \varepsilon^2$ for $t \leq r/(3\varepsilon^2)$. However, if only n-times differentiability of $A|_{T=0}$ is assumed, then $b(0) = \mathcal{O}(\varepsilon^{n+1})$. The Newell-Whitehead equation makes wrong predictions, if b is of order $\mathcal{O}(\varepsilon)$ on a time scale shorter than $\mathcal{O}(1/\varepsilon^2)$, since then the error is of the same order as the solution. In a neighborhood of the stationary solution $(a,b) \sim (3\varepsilon/2, 0)$. we have $b(t) \approx e^{3\varepsilon t/2}b(0)$ and so $b(t) = \mathcal{O}(\varepsilon)$ for $t \approx (n+1)|\ln\varepsilon|/\varepsilon \ll 1/\varepsilon^2$. Hence the Newell-Whitehead approximation in general makes wrong predictions for initial conditions which are only n-times differentiable. By looking at the phase portrait of the (a,b)-system we see that the same behavior occurs in the nonlinear system, too.

10.6.5. Time-periodic systems. Here we explain how to transfer the ideas from the autonomous to the time-periodic case. Prototypes of pattern-forming systems with an external time-periodic forcing are electro-convection in nematic liquid crystals and the so called Faraday problem. In the first problem a time-periodic external electric field is applied since a time-independent external field would destroy the experiment by electrolysis, and in the second problem a container is shaken periodically. We restrict ourselves to the underlying ideas. The precise conditions on the underlying systems are similar to the previous ones and can be found in [**SU07**].

We consider a translationally invariant system

$$\partial_t V = M(t)V + N(t, V) \tag{10.82}$$

which depend on some parameter $\alpha \in \mathbb{R}$, on a cylindrical domain $\mathbb{R} \times \Sigma$, where $M(t)V$ stands for the linear and $N(t,V)$ for the nonlinear terms satisfying $M(t) = M(t + 2\pi/\omega_0)$ and $N(t,V) = N(t + 2\pi/\omega_0, V)$ for an $\omega_0 > 0$. As explained above, electro-convection is an anisotropic problem. For such problems the subsequent ideas also apply in domains $\mathbb{R}^2 \times \Sigma$, and yield systems of more complicated amplitude equations.

In order to analyze the stability of the trivial solution $V{=}0$ in (10.82), we consider the linearized system

$$\partial_t V = M(t)V. \tag{10.83}$$

Due to the translational invariance of the problem the solutions are given by Floquet-Fourier modes

$$V = \varphi_m(k, z, t)\mathrm{e}^{\mathrm{i}kx}\mathrm{e}^{\lambda_m(k)t}, \tag{10.84}$$

with $k \in \mathbb{R}$, $m \in \mathbb{N}$, $z \in \Sigma$, and φ_m periodic in t, i.e.

$$\varphi_m(\cdot, \cdot, t) = \varphi_m(\cdot, \cdot, t + 2\pi/\omega_0).$$

If $V = 0$ is asymptotically stable, then we have that $\operatorname{Re}\lambda_m(k) < 0$ for all $m \in \mathbb{N}$ and $k \in \mathbb{R}$. We assume that $V = 0$ becomes unstable at $\alpha = \alpha_c$, i.e., at $\alpha = \alpha_c$ there exists one curve of Floquet exponents λ_1 satisfying $\operatorname{Re}\lambda_1(k_c) = 0$ for some $k_c > 0$. We set $\varepsilon^2 = \alpha - \alpha_c > 0$, write $\lambda_m = \lambda_m(k, \varepsilon^2)$, and assume that all Floquet exponents possess a real part strictly less than $-\sigma_0$ for a $\sigma_0 > 0$, except of $\lambda_1(k)$ for k in small neighborhoods of $\pm k_c$. Hence we have the same assumptions as in the autonomous case, except that the curve of critical eigenvalues has to be replaced by a curve of critical Floquet exponents.

Like for the autonomous case, it turns out that for the mathematical analysis, it is essential to consider the Fourier transformed system w.r.t. the unbounded spatial variable. In Fourier space (10.82) is given by

$$\partial_t \widehat{V}(k,t) = \widehat{M}(k,t)\widehat{V}(k,t) + \widehat{N}(\widehat{V})(k,t), \tag{10.85}$$

with $k \in \mathbb{R}$ and $\widehat{V}(k,t)$ a vector-valued function of z. We now explain how (10.85) can be brought in a form similar to the one of §10.3 about the universality of the GL equation in the autonomous case.

Derivation of the GL equation. For the subsequent analysis it is sufficient that the critical Floquet exponents λ_1 of $\widehat{M}(k,t)$ are simple near k_c. However, for simplicity we assume that the linear operator $\widehat{M}(k,t)$ possesses for every $k \in \mathbb{R}$ and $t \in [0, 2\pi/\omega_0)$ a Floquet Schauder basis $(\varphi_j(k,t))_{j\in\mathbb{N}}$ of $L^2(\Sigma)$, with $\varphi_j(k,t) = \varphi_j(k,t+2\pi/\omega_0)$ satisfying

$$\partial_t \varphi_j(k,t) = \widehat{M}(k,t)\varphi_j(k,t) - \lambda_j(k)\varphi_j(k,t).$$

The functions φ_j are normalized by $\|\varphi_j(k,0)\|_{L^2}=1$. For defining projections onto the $\varphi_j(k,t)$ we consider the adjoint problem $-\partial_t \widehat{V}(k,t) = \widehat{M}^*(k,t)\widehat{V}(k,t)$. Consequently, also this problem has for every $k \in \mathbb{R}$ a Floquet Schauder basis $(\varphi_j^*(k,t))_{j\in\mathbb{N}}$ of $L^2(\Sigma)$, with $\varphi_j^*(k,t) = \varphi_j^*(k,t+2\pi/\omega_0)$, solving

$$-\partial_t \varphi_j^*(k,t) = \widehat{M}^*(k,t)\varphi_j^*(k,t) - \overline{\lambda_j(k)}\varphi_j^*(k,t),$$

and satisfying the orthogonality

$$\langle \varphi_i^*, \varphi_j \rangle_{L^2(\Sigma)} = \delta_{ij}. \tag{10.86}$$

A solution $\widehat{V}(k,t)$ of (10.85) is expanded in terms of the Floquet functions $\varphi_j(k,t)$, i.e.

$$\widehat{V}(k,t) = \sum_{j\in\mathbb{N}} \widehat{a}_j(k,t)\varphi_j(k,t), \qquad \text{with} \quad \widehat{a}_j(k,t) \in \mathbb{C}, \tag{10.87}$$

such that

$$\begin{aligned}\partial_t \left(\sum_{j\in\mathbb{N}} \widehat{a}_j(k,t)\varphi_j(k,t) \right) &= \sum_{j\in\mathbb{N}} ((\partial_t \widehat{a}_j(k,t))\varphi_j(k,t) + \widehat{a}_j(k,t)\partial_t\varphi_j(k,t)) \\ &= \sum_{j\in\mathbb{N}} \widehat{a}_j(k,t)\widehat{M}(k,t)\varphi_j(k,t) + \widehat{N}(\widehat{V})(k,t).\end{aligned}$$

In order to find the equations for the coefficient functions $\widehat{a}_j(k,t)$ we apply the adjoint eigenfunction $\langle \varphi_j^*(k,t), \cdot \rangle_{L^2(\Sigma)}$ and find

$$\partial_t \widehat{a}_j(k,t) = \widehat{\lambda}_j(k)\widehat{a}_j(k,t) + \langle \varphi_j^*(k,t), \widehat{N}(k,t)\rangle \tag{10.88}$$

for $j \in \mathbb{N}$. We used (10.86) and

$$\begin{aligned}-\langle \varphi_j^*(k,t), \partial_t\varphi_i(k,t)\rangle + &\langle \varphi_j^*(k,t), \widehat{M}(k,t)\varphi_i(k,t)\rangle \\ &= \langle \varphi_j^*(k,t), \widehat{\lambda}_j(k)\varphi_i(k,t)\rangle = \widehat{\lambda}_j(k)\delta_{ij}.\end{aligned}$$

Our derivation of the GL equation is now based on (10.88). We make the ansatz

$$\begin{aligned}a_1(x,t) =& \varepsilon A_1(X,T)e^{ik_cx} + \varepsilon^2 A_{2,1}(X,T)e^{2ik_cx} + \varepsilon^2 A_{0,1}(X,T) + c.c., \\ a_j(x,t) =& \varepsilon^2 A_{2,j}(X,T)e^{2ik_cx} + \varepsilon^2 A_{0,j}(X,T) + c.c.,\end{aligned}$$

where $j \in \mathbb{N}\backslash\{1\}$, $X = \varepsilon x$, and $T = \varepsilon^2 t$. With this ansatz we derive formally a GL equation with time-periodic coefficients. We write the nonlinearity of (10.82) in the form

$$N(V) = B(t,V,V) + C(t,V,V,V) + \mathcal{O}(V^4), \tag{10.89}$$

with bilinear and trilinear symmetric terms B and C and introduce the abbreviations

$$\begin{aligned}
&\widehat{B}_j(t,k,k-m,m) = \mathrm{e}^{-ikx}B(t,\varphi_1(k-m,t)\mathrm{e}^{\mathrm{i}(k-m)x},\varphi_j(m,t)\mathrm{e}^{\mathrm{i}mx}),\\
&\widehat{C}(t,k,k-l_1,l_1-l_2,l_2)\\
&\quad = \mathrm{e}^{-ikx}C(t,\varphi_1(k-l_1,t)\mathrm{e}^{\mathrm{i}(k-l_1)x},\varphi_1(l_1-l_2,t)\mathrm{e}^{\mathrm{i}(l_1-l_2)x},\varphi_1(l_2,t)\mathrm{e}^{\mathrm{i}l_2x}).
\end{aligned}$$

For $\varepsilon^2\mathrm{e}^{0ix}$ in the jth equation we obtain

$$\lambda_j(0)A_{0,j} = -2\langle\varphi_j^*,\widehat{B}_1(t,0,k_c,-k_c)\rangle|A_1|^2, \tag{10.90}$$

and for $\varepsilon^2\mathrm{e}^{2ik_cx}$ in the jth equation

$$\lambda_j(2k_c)A_{2,j} = -\langle\varphi_j^*,\widehat{B}_1(t,2k_c,k_c,k_c)\rangle A_1^2. \tag{10.91}$$

For $\varepsilon^3\mathrm{e}^{\mathrm{i}k_cx}$ in the equation for $j=1$ we obtain

$$\begin{aligned}
\partial_T A_1 =& \nu_0A_1+\nu_2\partial_X^2A_1\\
&+2\langle\varphi_1^*,\sum_{j\in\mathbb{N}\setminus\{1\}}\widehat{B}_j(t,k_c,k_c,0)\rangle A_1A_{0,j}\\
&+2\langle\varphi_1^*,\sum_{j\in\mathbb{N}\setminus\{1\}}\widehat{B}_j(t,k_c,-k_c,2k_c)\rangle A_{-1}A_{2,j}\\
&+3\langle\varphi_1^*,\widehat{C}(t,k_c,k_c,k_c,-k_c)\rangle A_1|A_1|^2,
\end{aligned} \tag{10.92}$$

with $\nu_0=\partial_{\varepsilon^2}\lambda_1(k_c,0)$ and $\nu_2=-\frac{1}{2}\partial_k^2\lambda_1(k_c,0)$. In (10.92) we replace $A_{0,j}$ by (10.90), $A_{2,j}$ by (10.91), and obtain the GL equation

$$\partial_T A_1 = \nu_0A_1+\nu_2\partial_X^2A_1+\nu_3(t,\varepsilon)A_1|A_1|^2, \tag{10.93}$$

with a time-periodic coefficient $\nu_3(t,\varepsilon)$, which depends smoothly on ε^2, and hence has a limit $\nu_3(t,0)$. Then averaging over the highly oscillating cubic coefficient $\nu_3(T/\varepsilon^2,0)$ shows that in lowest order it can replaced by its mean value ν_3, and thus we finally have the autonomous GL equation

$$\partial_T A_1 = \nu_0A_1+\nu_2\partial_X^2A_1+\nu_3A_1|A_1|^2. \tag{10.94}$$

Justification of the GL approximation. For the justification of this approximation we only remark, that, since the estimates for the linear and nonlinear terms are exactly the same as in the autonomous case, the proof of the approximation result goes along the lines of the autonomous case in §10.4. This has been worked out in [**SU07**].

10.7. The Couette-Taylor problem

This problem consists in finding the flow of a viscous incompressible fluid filling the domain $\Omega = \mathbb{R} \times \Sigma$ between two concentric cylinders of infinite length with radii r_1, r_2 and which rotate with angular velocities ω_1, ω_2. Our presentation follows the textbook [**CI94**]. We denote the bounded cross section by Σ. The flow is governed by the Navier-Stokes equations in Ω with no-slip boundary conditions. In this section we apply our previous theory to this classical hydrodynamical stability problem and extend our theory to Turing-Hopf bifurcations.

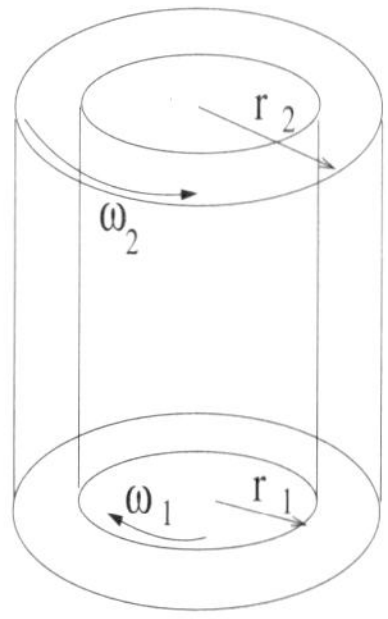

Figure 10.10. The Couette-Taylor problem.

10.7.1. The symmetries.

Before we start with the linear stability analysis we discuss the symmetries of the underlying PDEs. In cylindrical coordinates (r, ϕ, z) the velocity field $v = (v_r, v_\phi, v_z)$ and the pressure p satisfy

$$\begin{aligned} \partial_t v_r + v_r \partial_r v_r + \frac{v_\phi}{r} \partial_\phi v_r + v_z \partial_z v_r - \frac{v_\phi^2}{r} &= -\frac{1}{\rho} \partial_r p + \nu(\Delta v_r - \frac{2}{r^2} \partial_\phi v_\phi - \frac{v_r}{r^2}), \\ \partial_t v_\phi + v_r \partial_r v_\phi + \frac{v_\phi}{r} \partial_\phi v_\phi + v_z \partial_z v_\phi - \frac{v_\phi v_r}{r} &= -\frac{1}{\rho r} \partial_\phi p + \nu(\Delta v_\phi + \frac{2}{r^2} \partial_\phi v_r - \frac{v_\phi}{r^2}), \\ \partial_t v_z + v_r \partial_r v_z + \frac{v_\phi}{r} \partial_\phi v_z + v_z \partial_z v_z &= -\frac{1}{\rho} \partial_z p + \nu \Delta v_z, \end{aligned}$$

and

$$\partial_r v_r + \frac{1}{r} \partial_\phi v_\phi + \partial_z v_z + \frac{v_r}{r} = 0,$$

where $\Delta = \partial_r^2 + \frac{1}{r} \partial_r + \frac{1}{r^2} \partial_\phi^2 + \partial_z^2$. The equations are invariant under the transformations of the velocity field defined by

$$(\tau_a v)(r, \phi, z) = (v_r, v_\phi, v_z)(r, \phi, z + a), \tag{10.95}$$

$$(S v)(r, \phi, z) = (v_r, v_\phi, -v_z)(r, \phi, -z), \tag{10.96}$$

$$(R_\varphi v)(r, \phi, z) = (v_r, v_\phi, v_z)(r, \phi + \varphi, z). \tag{10.97}$$

The associated transformations for the pressure p are computed in Exercise 10.13. After non-dimensionalization the following three dimensionless parameters appear:

$$\omega = \omega_2/\omega_1, \quad \eta = r_1/r_2, \quad \mathcal{R} = r_1\omega_1(r_2 - r_1)/\nu,$$

with ν the kinetic viscosity, and $\mathcal{R}$ is called the Reynolds number. The non-dimensionalized equations which describe this physical system possess an exact solution, which is called the Couette flow. It is a pure azimuthal flow whose streamlines are concentric circles. In the cylindrical coordinates it is given by

$$\begin{aligned} U_{\mathrm{Cou}} &= (0, (U_{\mathrm{Cou}})_{(\phi)}(r), 0), \quad (U_{\mathrm{Cou}})_{(\phi)}(r) = Ar + B/r, \\ A &= \frac{\omega - \eta^2}{\eta(1+\eta)}, \quad B = \frac{\eta(1-\omega)}{(1-\eta)(1-\eta^2)}, \end{aligned} \tag{10.98}$$

with associated pressure $p_{\mathrm{Cou}} = \mathcal{R} \int ((U_{\mathrm{Cou}})_{(\phi)}(r))^2/r\,\mathrm{d}r$. The Couette flow U_{Cou} satisfies the symmetries (10.95)-(10.97).

Remark 10.7.1. The Couette-Taylor problem, written as an abstract dynamical system $\frac{\mathrm{d}}{\mathrm{d}t}x = f(x)$ is an example of an equivariant system. The transformations τ_a, S, and R_φ define a group Γ acting on a suitable chosen phase space M. The system is said to be Γ-equivariant if $f(\gamma x) = \gamma f(x)$ for all $\gamma \in \Gamma$ and $x \in M$. An important consequence is that if $t \mapsto x(t)$ solves the dynamical system, then so does $t \mapsto \gamma x(t)$ for all $\gamma \in \Gamma$. Works that discuss equivariant dynamical systems include [**GSS88, CK91, CL00, GS02, Hoy06**]. ⌋

We use the equivariance of the system in the analysis of the bifurcation scenario in §13.2.2, and in the derivation of the associated amplitude equations here. In the following we denote the coordinate in the unbounded axial direction by the variable x. The variable y is used for the coordinates in the bounded cross section Σ. Since we are interested in solutions which bifurcate from the Couette flow we choose this solution as the new origin of the Navier-Stokes equations. The deviation (U, q) from the Couette-flow $(U_{\mathrm{Cou}}, p_{\mathrm{Cou}})$ satisfies in Cartesian coordinates the system

$$\begin{aligned} \partial_t U &= \Delta U - \mathcal{R}[(U_{\mathrm{Cou}} \cdot \nabla)U + (U \cdot \nabla)U_{\mathrm{Cou}}] - \nabla q \\ &\quad -\mathcal{R}(U \cdot \nabla)U, \\ \nabla \cdot U &= 0, \end{aligned} \tag{10.99}$$

with Dirichlet-boundary conditions

$$U = 0 \quad \text{at } r = \eta/(1-\eta),\ r = 1/(1-\eta).$$

We choose the vanishing mean flux condition

$$[U_{(x)}]_\Sigma = \frac{1}{|\Sigma|} \int_\Sigma U_{(x)}(x, y)\,\mathrm{d}y = 0$$

for $x \in \mathbb{R}$ in order to solve this problem uniquely. Since $\partial_x [U_{(x)}]_\Sigma = 0$ due to $\nabla \cdot U = 0$, this can be achieved by adding to q an unbounded term αx. Herein, $U_{(x)}$ denotes the velocity component along the cylinder.

10.7.2. Linear stability analysis. The Couette flow, $U \equiv 0$ in (10.99), is linearly stable if all eigenvalues have a negative real part, and it is linearly unstable if at least one eigenvalue has a positive real part. A value $\lambda \in \mathbb{C}$ is an eigenvalue if we find non-trivial solutions of

$$\begin{gathered} \Delta U - \mathcal{R}[(U_{\text{Cou}} \cdot \nabla)U + (U \cdot \nabla)U_{\text{Cou}}] - \nabla q = \lambda U, \\ \nabla \cdot U = 0, \quad U|_{\mathbb{R}\times\partial\Sigma} = 0, \quad [U_{(x)}]_\Sigma = 0. \end{gathered} \tag{10.100}$$

Since U is translational invariant in the unbounded x-direction we can use Fourier series

$$U = \mathrm{e}^{\mathrm{i}kx} \widehat{U}_k^n(y), \quad \text{with } k \in \mathbb{R}, y \in \Sigma, n \in \mathbb{N}. \tag{10.101}$$

Since U is 2π-periodic in ϕ, the eigenfunctions $\widehat{U}_k^n$ are of the form $\widehat{U}_k^n(y) = \mathrm{e}^{im\phi} U_{k,m}^n(r)$ with $m = m(k,n) \in \mathbb{Z}$. For fixed k we obtain discrete eigenvalues with $\operatorname{Re}\lambda_n \geq \operatorname{Re}\lambda_{n+1}$ depending on the three parameters ω, η, and $\mathcal{R}$. Since the geometry of the apparatus can not be changed in the experiment we assume that $\eta \in (0,1)$ is fixed. According to [**CI94**] the following assertions hold for instance for $\eta \sim 0.883$.

In the $(\omega, \mathcal{R})$-plane there are two essentially different parameter regions called PR I and PR II in the following, see Figure 10.11. For $\mathcal{R} < \mathcal{R}_c(\omega)$ the largest eigenvalue satisfies $\operatorname{Re}\lambda_1(k, \mathcal{R}, \eta, \omega) < 0$.

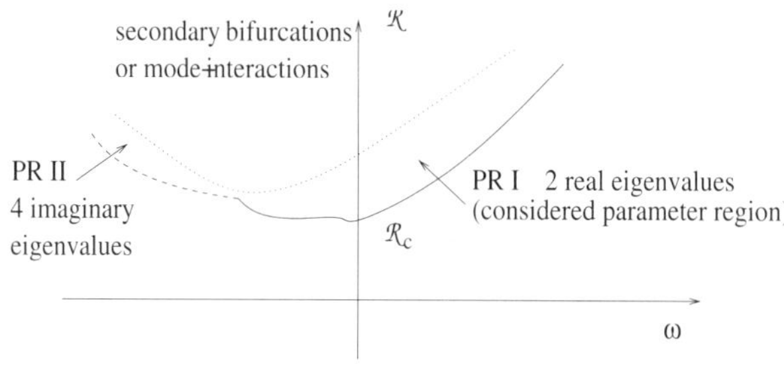

Figure 10.11. The bifurcation scenario for $\eta \sim 0.8$.

PR I: Numerical computations show that when $\omega = \omega_2/\omega_1 \geq 0$ or for ω slightly negative, the instability at $\mathcal{R} = \mathcal{R}_c(\omega)$ occurs via a curve of real eigenvalues crossing the k-axis at a wave number $k_c \neq 0$. See Figure 10.12.

PR II: For strongly counter-rotating cylinders, i.e., ω sufficiently negative, the instability at $\mathcal{R} = \mathcal{R}_c(\omega)$ comes via a pair of pure imaginary eigenvalues at a wave number $k_c \neq 0$. All eigenvalues which occur for the wave number k must also occur for the wave number $-k$ due to the reflection symmetry (10.96) of the problem. See Figure 10.13.

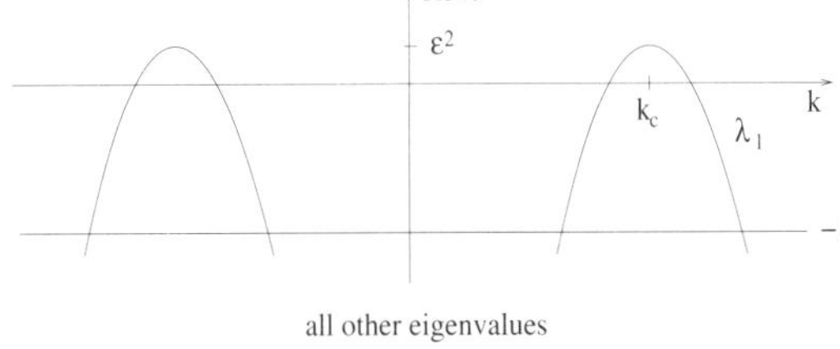

Figure 10.12. PR I: The real-valued curve λ_1 close to the wave numbers $\pm k_c$ and the rest of spectrum.

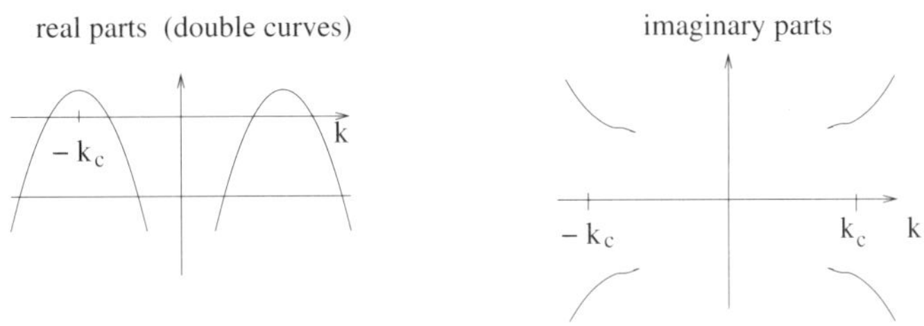

Figure 10.13. PR II:The complex-valued curves λ_1, λ_2 close to the wave numbers $\pm k_c$ and the rest of spectrum.

10.7.3. The GL formalism in PR I. The starting point for the GL formalism for the solution U of (10.99) in PR I is as usual the ansatz

$$U(x,y,t) = \varepsilon \Psi_{\mathrm{GL}}(x,y,t) + \mathcal{O}(\varepsilon^2), \tag{10.102}$$

with

$$\varepsilon \Psi_{\mathrm{GL}}(x,y,t) = \varepsilon A(X,T) \widehat{U}^1_{k_c,0}(y) \mathrm{e}^{\mathrm{i}k_c x} + \text{c.c.},$$

$T = \varepsilon^2 t$ and $X = \varepsilon x$. As above the idea is to approximately describe slow modulations in time and in space of the linearly most unstable pattern $\widehat{U}^1_{k_c,0}(y)\mathrm{e}^{\mathrm{i}k_c x}$ in the Couette-Taylor problem for $0 < \varepsilon^2 = \mathcal{R} - \mathcal{R}_c \ll 1$. Formally equating the coefficients of the $\varepsilon^{j_1}\mathrm{e}^{\mathrm{i}j_2 k_c x}$ for $j_1 \in \mathbb{N}, j_2 \in \mathbb{Z}$, to zero shows that the complex valued amplitude $A = A(X,T) \in \mathbb{C}$ should satisfy a GL equation

$$\partial_T A = \nu_2 \partial_X^2 A + \nu_0 A + \nu_3 |A|^2 A \tag{10.103}$$

for which the complex-valued coefficients satisfy $\mathrm{Re}\nu_2, \mathrm{Re}\nu_0, -\mathrm{Re}\nu_3 > 0$. For $\nu_2 = 0$ this equation is the lowest order approximation of the reduced system of §13.2.2 on the center manifold in case of $2\pi/k_c$-spatial periodicity. Therefore, the coefficients ν_0 and ν_3 are already known from the spatially periodic case. From the reflection symmetry (10.96) it follows that the coefficients ν_2, ν_0, and ν_3 have to be real-valued. The detailed argument is as follows. We find

$$S(\widehat{U}^1_{k_c,0}(y)\mathrm{e}^{\mathrm{i}k_c x}) = \overline{\widehat{U}^1_{k_c,0}(y)}\mathrm{e}^{-\mathrm{i}k_c x}$$

such that

$$\begin{aligned}S(\varepsilon\psi) =&\varepsilon A(-X,T)\ S(\widehat{U}^1_{k_c,0}(y)\mathrm{e}^{\mathrm{i}k_cx})\ + \text{c.c.} + \mathcal{O}(\varepsilon^2)\\ =&\varepsilon A(-X,T)\overline{\widehat{U}^1_{k_c,0}(y)\mathrm{e}^{-\mathrm{i}k_cx}}\ + \text{c.c.} + \mathcal{O}(\varepsilon^2).\end{aligned}$$

Therefore on the level of the amplitude function A the map S possesses a representation $S : A(X,T) \mapsto \overline{A(-X,T)}$. Since the Couette-Taylor problem is invariant under S also the associated GL equation has to be invariant under S. If we apply S we find

$$\partial_T\overline{A} = \nu_2\partial_X^2\overline{A} + \nu_0\overline{A} + \nu_3|\overline{A}|^2\overline{A}.$$

Since $\overline{A}$ satisfies the complex conjugate GL equation the coefficients necessarily have to satisfy $\nu_j = \overline{\nu_j} \in \mathbb{R}$.

In order to write (10.99) as a dynamical system in an infinite dimensional phase space we introduce a projection on the divergence free vector fields. The folowing results are from [**Sch99b**].

Lemma 10.7.2. *There exists a continuous projection* $\Pi_0 : (L^2_{\mathrm{ul}}(\Omega))^3 \to H$, *where*

$$H = \{U \in (L^2_{\mathrm{ul}}(\Omega))^3 : \nabla \cdot U = 0,\ U \cdot n|_{\mathbb{R}\times\partial\Sigma} = 0,\ [U_{(x)}]_\Sigma = 0\}$$

is equipped with the norm of $(L^2_{\mathrm{ul}}(\Omega))^3$. *This projection additionally satisfies* $\Pi_0\nabla q = 0$.

With the help of the projection Π_0 we define the unbounded linear operator $\Lambda_\mathcal{R}$ and the nonlinearity $\mathcal{N}(\mathcal{R},\cdot)$, depending smoothly on the parameters $\mathcal{R},\eta$, and ω, by

$$\begin{aligned}\Lambda_\mathcal{R}U =&\Pi_0(\Delta U - \mathcal{R}[(U_{\mathrm{Cou}}\cdot\nabla)U + (U\cdot\nabla)U_{\mathrm{Cou}}]),\\ \mathcal{N}(\mathcal{R},U) =& -\mathcal{R}\Pi_0(U\cdot\nabla)U,\end{aligned}$$

such that (10.99) writes as

$$\partial_t U = \Lambda_\mathcal{R}U + \mathcal{N}(\mathcal{R},U). \tag{10.104}$$

For the analysis of (10.104) we need the two spaces

$$Z = \{U \in H : U \in (H^2_{\mathrm{ul}}(\Omega))^3, U|_{\mathbb{R}\times\partial\Sigma} = 0\}$$

and

$$Z^* = \{U \in H : U \in (H^1_{\mathrm{ul}}(\Omega))^3\}.$$

It turns out that

Lemma 10.7.3. *The operator* $\Lambda_\mathcal{R}$ *is the generator of an analytic semigroup* $\mathrm{e}^{\Lambda_\mathcal{R}t}$ *in* H *with domain of definition* Z *satisfying*

$$\|\mathrm{e}^{\Lambda_\mathcal{R}t}\|_{Z^*\to Z} \le Ct^{-3/4}$$

for $t \in (0,1]$.

Lemma 10.7.4. *The nonlinearity $\mathcal{N}(\mathcal{R}, U)$ is locally Lipschitz-continuous from Z to Z^*.*

With these preparations we can now formulate the approximation theorem.

Theorem 10.7.5. *Let $\theta_A \geq 5$. Let $A \in C([0,T_0], H_{\mathrm{ul}}^{\theta_A}(\mathbb{R},\mathbb{C}))$ be a solution of the GL equation* (10.103)*. Then there exists $\varepsilon_0, C > 0$ such that for all $\varepsilon \in (0, \varepsilon_0)$ there are solutions U of the Couette-Taylor problem* (10.104) *with*

$$\sup_{t\in[0,T_0/\varepsilon^2]} \|U(t) - \varepsilon\Psi_{\mathrm{GL}}(t)\|_Z \leq C\varepsilon^2.$$

A slightly modified approximation theorem has been combined with the attractivity of the set of functions in GL form to show a number of other results such as the global existence of solutions in H_{ul}^{θ} for the weakly unstable Couette-Taylor problem [**Sch99b**] or the upper semicontinuity of attractors [**Sch99c**]. We will come back to this in §10.8.

The proof of Theorem 10.7.5 follows by applying the abstract approximation theorem stated in Theorem 10.4.3. How to check the assumptions **(A0)-(A4)** can be found in [**Sch99b**] where a detailed proof of Lemma 10.7.2, Lemma 10.7.3, Lemma 10.7.4, and Theorem 10.7.5 has been given. Following Remark 10.4.2, this essentially requires modifications of the results from [**CI94**], where $2\pi/k_0$-spatially periodic boundary conditions are considered and for instance estimates for the projection Π_0 on the divergence free vector fields and resolvent estimates are given in the spaces

$$\begin{aligned} H_{k_0} &= \{U \in (L^2_{per}(\Omega_{2\pi/k_0}))^3 : \nabla\cdot U = 0,\ U\cdot n|_{\mathbb{R}\times\partial\Sigma} = 0,\ [U_{(x)}]_\Sigma = 0\}, \\ Z_{k_0} &= \{U \in H_{k_0} : U \in (H^2_{per}(\Omega_{2\pi/k_0}))^3, U|_{\mathbb{R}\times\partial\Sigma} = 0\}, \end{aligned}$$

and

$$Z^*_{k_0} = \{U \in H_{k_0} : U \in (H^1_{per}(\Omega_{2\pi/k_0}))^3\},$$

where $\Omega_{2\pi/k_0} = \mathbb{R}/((2\pi/k_0)\mathbb{Z}) \times \Sigma$. Specializing such estimates to functions $e^{ikx}\widehat{U}(y)$ gives estimates which allow to apply the Multiplier Lemma 8.3.7 and to obtain the associated estimates in H_{ul}^{θ}-spaces. Again, see [**Sch94a**] or [**Sch99b**] for details.

10.7.4. PR II: The Turing-Hopf scenario. Panel b) is the last one in Figure 9.8 where a GL equation can be derived. It appears in reflection symmetric ($x \mapsto -x$) evolution problems when a trivial ground state loses stability and bifurcates into a temporally oscillating, spatially periodic pattern. A famous example of such a system is the Couette-Taylor problem in PR II, i.e., in case of strongly counter-rotating cylinders. In PR II the amplitude or modulation equations are given by a set of coupled singularly perturbed GL equations. Again the symmetries (10.95)-(10.97) can be used to show that in the associated amplitude equation only four and not eight different coefficients occur. See Exercise 10.14.

A simple example of an original system with these properties is a system of coupled KS equations

$$
\begin{aligned}
\partial_t u &= -(1+\partial_x^2)^2 u - \partial_x u + \varepsilon^2 u + \partial_x(u^2 + uv + v^2),\\
\partial_t v &= -(1+\partial_x^2)^2 v + \partial_x v + \varepsilon^2 v + \partial_x(u^2 + uv + v^2),
\end{aligned} \tag{10.105}
$$

where $t \geq 0$, $x \in \mathbb{R}$, and $(u(x,t), v(x,t)) \in \mathbb{R}^2$. As before $0 < \varepsilon^2 \ll 1$ is used as a small bifurcation parameter. The system is invariant under the transformation $(x, u, v) \mapsto (-x, -v, -u)$.

The linearization at $(u, v) \equiv (0, 0)$ possesses eigenfunctions of the form $(u, v)(x, t) = (a_k e^{ikx}, b_k e^{ikx})$, with coefficients $a_k, b_k \in \mathbb{C}$, wave number $k \in \mathbb{R}$, and associated eigenvalues

$$
\widehat{\lambda}_{1,2}(k, \varepsilon^2) = -(1-k^2)^2 \mp \mathrm{i}k + \varepsilon^2.
$$

We observe that $\mathrm{Re}\widehat{\lambda}_{1,2}(k, \varepsilon^2)$ is positive for wave numbers k close to the values $\pm k_c = \pm 1$. As above we expect that the bifurcating solutions are slow modulations in time and in space of the bifurcating pattern $e^{\mathrm{i}x}$ due to the form of $\widehat{\lambda}_{1,2}$. Therefore, we make the ansatz

$$
\begin{aligned}
u &= \varepsilon A(\varepsilon(x - c_g t), \varepsilon^2 t) e^{\mathrm{i}(x - \omega_0 t)} + \varepsilon^2 A_2(\varepsilon(x - c_g t), \varepsilon^2 t) e^{2\mathrm{i}(x - \omega_0 t)}\\
&\quad + \frac{1}{2}\varepsilon^2\, A_0^0(\varepsilon(x - c_g t), \varepsilon^2 t) + c.c.,\\
v &= \varepsilon B(\varepsilon(x + c_g t), \varepsilon^2 t) e^{\mathrm{i}(x + \omega_0 t)} + \varepsilon^2 B_2(\varepsilon(x + c_g t), \varepsilon^2 t) e^{2\mathrm{i}(x + \omega_0 t)}\\
&\quad + \frac{1}{2}\varepsilon^2\, B_0^0(\varepsilon(x + c_g t), \varepsilon^2 t) + c.c.,
\end{aligned}
$$

with $c_g, \omega_0 \in \mathbb{R}$. Inserting the ansatz into the original system and equating the coefficients in front of $\varepsilon^n e^{imx} e^{ij\omega_0 t}$ to zero gives $c_g = -1$ and $\omega_0 = 1$. Setting $X_1 = \varepsilon(x - c_g t)$, $X_2 = \varepsilon(x + c_g t)$, $T = \varepsilon^2 t$ and eliminating the algebraic equations for $m = 0, \pm 2$ we finally obtain

$$
\begin{aligned}
\partial_T A(X_1, T) &= 4\partial_{X_1}^2 A(X_1, T) + A(X_1, T)\\
&\qquad - \tfrac{6}{9}(A(X_1, T)|A(X_1, T)|^2 + a_1 + a_2),\\
\partial_T B(X_2, T) &= 4\partial_{X_2}^2 B(X_2, T) + B(X_2, T)\\
&\qquad - \tfrac{6}{9}(B(X_2, T)|B(X_2, T)|^2 + b_1 + b_2),
\end{aligned} \tag{10.106}
$$

where

$$
\begin{aligned}
a_1 =& A(X_1, T)\overline{B}(X_2, T)B(X_2, T),\\
a_2 =& A(X_1, T)\left[\overline{B}(X_2, T)A(X_1, T)e^{-2\mathrm{i}\alpha T/\varepsilon^2} + \overline{A}(X_1, T)B(X_2, T)e^{2\mathrm{i}\alpha T/\varepsilon^2}\right]\\
&+ B(X_2, T)\left[\overline{A}(X_1, T)B(X_2, T)e^{4\mathrm{i}\alpha T/\varepsilon^2} + \overline{B}(X_2, T)B(X_2, T)e^{2\mathrm{i}\alpha T/\varepsilon^2}\right],
\end{aligned}
$$

$$\begin{aligned}
b_1 =& B(X_2,T)\overline{A}(X_1,T)A(X_1,T),\\
b_2 =& A(X_1,T)\left[\overline{B}(X_2,T)B(X_2,T)\mathrm{e}^{-2\mathrm{i}\alpha T/\varepsilon^2}+A(X_1,T)\overline{B}(X_2,T)\mathrm{e}^{-4\mathrm{i}\alpha T/\varepsilon^2}\right]\\
&+\overline{A}(X_1,T)\left[B(X_2,T)B(X_2,T)\mathrm{e}^{2\mathrm{i}\alpha T/\varepsilon^2}+A(X_1,T)A(X_1,T)\mathrm{e}^{-2\mathrm{i}\alpha T/\varepsilon^2}\right].
\end{aligned}$$

A classical GL equation is only obtained if either $A = 0$ or $B = 0$. If this is not the case the coupled system of GL equations depends singularly on the small bifurcation parameter $0 < \varepsilon \ll 1$. Since a_2 and b_2 are highly oscillatory w.r.t. time they can be eliminated by a normal form transform, respectively averaging, cf. Exercise 10.5. Hence, a_2 and b_2 can be neglected in lowest order. The terms collected in a_1 and b_1 also depend singularly on the small bifurcation parameter $0 < \varepsilon \ll 1$ via X_1 and X_2. Since a_1 and b_1 cannot be removed in general the modulation equations in the Turing-Hopf case still depend on the small bifurcation parameter $0 < \varepsilon \ll 1$ and are given by

$$\begin{aligned}
\partial_T A(X_1,T) =& \nu_2\partial_{X_1}^2 A(X_1,T) + \nu_0 A(X_1,T)\\
&+ \nu_3 A(X_1,T)|A(X_1,T)|^2 + \nu_4 A(X_1,T)|B(X_2,T)|^2,\\
\partial_T B(X_2,T) =& \nu_2\partial_{X_2} B(X_2,T) + \nu_0 B(X_2,T)\\
&+ \nu_3 B(X_2,T)|B(X_2,T)|^2 + \nu_4 B(X_2,T)|A(X_1,T)|^2,
\end{aligned}$$

with coefficients $\nu_j \in \mathbb{C}$ for $j = 1,\ldots,4$. An approximation theorem for this system can be found for (10.105) in [**Sch97**], and for the Couette-Taylor problem in [**Sch99b**].

Remark 10.7.6. In the equation for A which depends on the space variable $X_1 = \varepsilon(x - \alpha t)$ the function B depends on the fast variable $X_1 + 2\alpha T/\varepsilon$ and vice versa in the equation for B. The idea is that due to the high velocity of order $\mathcal{O}(1/\varepsilon)$ of the counter-propagating waves they only influence each other by their mean value. Therefore, in the literature, cf. [**KDL90, Kno92, DFKM96**], a so called mean-field coupled system of GL equations is considered

$$\begin{aligned}
\partial_T A(X_1,T) &= 4\partial_{X_1}^2 A(X_1,T) + A(X_1,T)\\
&\quad -\tfrac{6}{9}(A(X_1,T)|A(X_1,T)|^2 + A(X_1,T)|\langle B\rangle_L(X_2,T)|^2),\\
\partial_T B(X_2,T) &= 4\partial_{X_2}^2 B(X_2,T) + B(X_2,T)\\
&\quad -\tfrac{6}{9}(B(X_2,T)|B(X_2,T)|^2 + B(X_2,T)|\langle A\rangle_L(X_1,T)|^2),
\end{aligned} \tag{10.107}$$

where

$$\langle F\rangle_L(Y,T) = \frac{1}{2L}\int_{-L}^{L} F(Y+X,T)\,\mathrm{d}X$$

for a fixed $L > 0$. In [**Sch97**, Theorem 3.1] it has been shown that the non-averaged system (10.106) can be approximated by the averaged system (10.107) up to an error $\mathcal{O}(\varepsilon^{\min(3/2,2-\nu)})$ w.r.t. the H^{θ}_{ul}-norm if $L=\mathcal{O}(1/\varepsilon^{\nu})$

with $0 \le \nu < 1$. As a consequence, the equations decouple in two classical GL equations for spatially localized solutions, cf. [**PW96**] in case of NLS approximations for dispersive wave systems, see also §11.7.2 and §12.1. ⌋

10.8. Attractors for pattern forming systems

Besides the approximation theorems showing that the GL equation makes correct predictions about the dynamics of the associated original systems, in this book we also show similar results for dispersive systems, which can be described by the NLS equation and the KdV equation. We explain that all three equations are universal modulation equations appearing for various original systems under rather weak assumptions, and that they predict the behavior of solutions starting in special sets of initial conditions for the associated original systems. In contrast to the cases described by the NLS equation and the KdV equation, an additional feature for dissipative systems is that the set of initial conditions for which the GL equation makes correct predictions is absorbing.

We now explain this attractivity property and some of its consequences, namely the global existence and uniqueness of solutions in a neighborhood of the unstable origin of the original system, the upper-semicontinuity of the rescaled original system attractor towards the associated GL attractor, and that every solution in the original system can be shadowed for all times by some pseudo-orbit for the GL equation.

This has been carried out in detail for the KS equation in [**Sch94b, MS95**] and for the Couette-Taylor problem in in [**Sch99b, Sch99c**]. For instance, for $0 < \varepsilon \ll 1$ there exists a $\mathcal{O}(1)$-neighborhood of the unstable Couette flow in H^2_{ul} in which the global existence and uniqueness of solutions of the 3D Couette-Taylor problem holds. We refrain from rewriting these papers and restrict ourselves to the explanation of the underlying ideas by considering the Swift-Hohenberg (SH) equation as the simplest dissipative system showing a pattern forming instability.

The plan of this section is as follows. In §10.8.1 we prove the existence of the SH attractor. In order to do so we use weighted energy estimates to establish the existence of an absorbing set. However, weighted energy estimates are not optimal for small values of the bifurcation parameter $0 < \varepsilon \ll 1$, since they only yield $\mathcal{O}(\varepsilon^{1/2})$ estimates for the diameter of the attractor in H^1_{ul}. In §10.8.2 we introduce pseudo-orbits of GL approximations, and use these in §10.8.4 to show a much stronger result, namely that the diameter of the attractor in H^1_{ul} is $\mathcal{O}(\varepsilon)$ for $\varepsilon \to 0$. In §10.8.5 we prove the upper-semicontinuity of the rescaled original system attractor towards the associated GL attractor.

Notation. As in §8.3 we choose the weight function ρ as

$$\rho(x) = 1/\cosh(x) \quad \text{or} \quad \rho(x) = 2/(2+x^2),$$

such that $|\rho''(x)| \leq \rho(x)$ for all $x \in \mathbb{R}$. We recall from Definition 8.3.21 that

$$\begin{aligned} \text{dist}_{Z_\rho}(b, A) &= \inf_{a\in A} \|b-a\|_\rho, \\ \text{dist}_{Z_u}(b, A) &= \inf_{a\in A} \|b-a\| = \inf_{a\in A}\sup_{y\in\mathbb{R}} \|T_y b - T_y a\|_\rho, \end{aligned}$$

where $(T_y a)(x) = a(x-y)$ and $\|a\|_\rho$ is the localized norm, for instance $\|a\|_{H^1_\rho} = \|a\rho\|_{H^1}$. Throughout this section we use

$$\begin{aligned} &Z = H^1_{\text{ul}} \text{ for the phase space of the SH equation, and} \\ &Y = H^1_{\text{ul}} \text{ for the phase space of the GL equation.} \end{aligned} \tag{10.108}$$

Thus, this distinction is somewhat artificial here, but it helps to identify SH variables vs. GL variables, and in other applications these spaces are really different. For instance, for the Navier-Stokes equations as an original system we may choose the space $Z = H^2_{\text{ul}}(\mathbb{R}\times\Sigma)$ for some bounded cross-section Σ, cf. [**Sch94a**]. Moreover, we use the notations Z_ρ and Y_ρ for the respective localized spaces.

10.8.1. Existence of the Swift-Hohenberg attractor. We establish the existence of a $(H^1_{\text{ul}}, H^1_\rho)$-attractor in the sense of Definition 8.3.18 for the SH equation

$$\partial_t u = -(1+\partial_x^2)^2 u + \alpha u - u^3 = Lu + \alpha u - u^3. \tag{10.109}$$

The presentation follows in many parts the lines of [**MS95**]. Since no maximum principle is available for the SH equation we use weighted energy estimates to establish the existence of an absorbing set. For $b \in (0,1]$ set $\rho_b(x) = \rho(bx)$. If we differentiate the weighted energy $\int \rho_b u^2\,dx$ w.r.t. time we obtain integrals $\int \rho_b u^4\,dx$ and $\int \rho_b u(1+\partial_x^2)^2 u\,dx$. The second integral can be estimated as follows.

Lemma 10.8.1. *a) For all $u \in H^4_{\text{ul}}$, the following estimate holds:*

$$-\int_{\mathbb{R}} \rho_b u(1+\partial_x^2)^2 u\,dx \leq (2b^2 + \tfrac{3}{2}b^4)\int_{\mathbb{R}} \rho_b u^2 dx. \tag{10.110}$$

b) For each $d > 0$ there is a $D > 0$ such that

$$-\int_{\mathbb{R}} \rho u(1+\partial_x^2)^2 u\,dx \leq -d\int_{\mathbb{R}} \rho u'^2 dx + D\int_{\mathbb{R}} \rho u^2 dx.$$

Proof. a) By several integrations by part we obtain

$$-\int_{\mathbb{R}} \rho_b u(1+\partial_x^2)^2 u\,dx = -\int_{\mathbb{R}} \{\rho_b(u^2 + 2uu'' + u''^2) - \rho_b'' u'^2 + \rho_b'' uu''\} dx.$$

Using $|\rho_b''| \le b^2\rho_b$ and

$$\begin{aligned} \textstyle\int \rho_b u'^2\,dx &= -\textstyle\int (\rho_b u')'u\,dx = -\textstyle\int \rho_b'(\frac{1}{2}u^2)'\,dx - \textstyle\int \rho_b u''u\,dx \\ &= \textstyle\int \frac{1}{2}\rho_b'' u^2 dx - \textstyle\int \rho_b uu'' dx \le \frac{b^2}{2}\textstyle\int \rho_b u^2\,dx + \textstyle\int \rho_b |uu''|\,dx \end{aligned}$$

we conclude

$$-\int_{\mathbb{R}} \rho_b u(1+\partial_x^2)^2 u\,dx \le \int_{\mathbb{R}} \rho_b \left\{(b^4/2-1)u^2 + 2(1+b^2)|uu''| - u''^2\right\} dx,$$

which yields the desired result after maximizing the integrand w.r.t. u''.

b) Applying the estimates of part a) with $b=1$ we find

$$\int \rho\left(-u(1+\partial_x^2)^2u + du'^2\right) dx \le \int \rho\left(\frac{d-1}{2}u^2 + (4+d)|uu''| - u''^2\right) dx.$$

Maximizing w.r.t. u'' gives the result with $D = (2d-2+(4+d)^2)/4$. □

With these estimates we can construct a global semiflow and an absorbing ball for (10.109).

Theorem 10.8.2. *The SH equation* (10.109) *defines a global semiflow* $u(t) = \mathcal{S}_t(u_0)$ *where for each* $t>0$ *the nonlinear map* $\mathcal{S}_t$ *maps bounded sets in* $Z = H^1_{\mathrm{ul}}$ *into bounded sets in* H^2_{ul}*. Moreover, for all* $\alpha \in \mathbb{R}$ *there is a constant* $\Delta_1(\alpha)$ *such that*

$$\limsup_{t\to\infty} \|u(t)\|_{H^1_{\mathrm{ul}}} \le \Delta_1(\alpha).$$

More precisely, there is a constant C *independent of* $\alpha \in [0,1]$ *such that* $\Delta_1(\alpha) \le C\alpha^{1/4}$.

Proof. The local existence and uniqueness of solutions and continuous dependence on the initial condition in H^1_{ul} follow as in §10.2.2. We consider the variation of constant formula

$$u(t) = \mathrm{e}^{tL}u(0) + \int_0^t \mathrm{e}^{(t-s)L}F(u(s))\,ds, \quad \text{where } F(u) = \alpha u - u^3.$$

F is a smooth map from H^1_{ul} into H^1_{ul}, and using Lemma 8.3.7 we have

$$\|\mathrm{e}^{tL}u\|_{H^{1+j}_{\mathrm{ul}}} \le C\max(1,t^{-j/4})\|u\|_{H^1_{\mathrm{ul}}} \tag{10.111}$$

for all $t>0$, with $C>0$ independent of α.

To show global existence it is sufficient to bound the H^1_{ul} norm. The L^2_{ul} estimate follows from standard weighted energy estimates using part a) of Lemma 10.8.1, namely

$$\begin{aligned} &\frac{1}{2}\frac{\mathrm{d}}{\mathrm{d}t}\int_{\mathbb{R}} \rho_b u^2\,dx = \int_{\mathbb{R}} \rho_b\{-u(1+\partial_x^2)^2u + \alpha u^2 - u^4\}dx \\ &\le 4b^2\|u\|^2_{\rho_b} + \int_{\mathbb{R}} \{\rho_b u^2(\alpha - u^2)\}dx \end{aligned}$$

$$\leq (4b^2+\alpha)\|u\|_{\rho_b}^2 - \int_{\mathbb{R}} \rho_b u^4\,dx \leq c_1 - c_2\|u\|_{\rho_b}^2,$$

with $c_1 = \frac{d^2}{2}\int_{\mathbb{R}} \rho_b dx$ and $c_2 = d - 4b^2 - \alpha$, and where $b \leq 1$ was used and $d \geq 0$ is arbitrary. Assuming $c_2 > 0$ and applying Gronwall's Lemma to the differential inequality yields

$$\|u(t)\|_{\rho_b}^2 \leq e^{-2c_2 t}\|u(0)\|_{\rho_b}^2 + \frac{c_1}{c_2}(1 - e^{-2c_2 t}).$$

Letting $t \to \infty$ and choosing the optimal d gives

$$\limsup_{t\to\infty} \|u(t)\|_{\rho_b}^2 \leq C\big(\alpha + b^2\big)\int_{\mathbb{R}} \rho_b\,dx = \Delta_0(\alpha, b).$$

With $\int_{\mathbb{R}} \rho_b dx = C/b$ we obtain the desired estimate for $\alpha \in [0,1]$ when optimizing w.r.t. $b = \mathcal{O}(\alpha^{1/2}) \in (0,1]$.

The estimate in H^1_{ul} is derived via the variation of constant formula

$$u(t) = e^{(t-\tau)L}u(\tau) + \int_\tau^t e^{(t-r)L}(\alpha u(r) - u(r)^3)dr.$$

Using the smoothing properties (10.111) and $\|u\|_{L^\infty}^2 \leq C\|u\|_{L^2_{\mathrm{ul}}}\|u\|_{H^1_{\mathrm{ul}}}$ the nonlinear terms can be estimated as

$$\begin{aligned}\|e^{rL}u^3\|_{H^1_{\mathrm{ul}}} \leq& \|e^{rL}\|_{L^2_{\mathrm{ul}}\to H^1_{\mathrm{ul}}}\|u^3\|_{L^2_{\mathrm{ul}}}\\ \leq& Cr^{-1/4}\|u\|_{L^2_{\mathrm{ul}}}\|u\|_{L^\infty}^2 \leq Cr^{-1/4}\|u\|_{L^2_{\mathrm{ul}}}^2\|u\|_{H^1_{\mathrm{ul}}}.\end{aligned}$$

For $s \in \{0,1\}$ we set $e_s(t) = \|u(t)\|_{H^s_{\mathrm{ul}}}$ and find the integral inequality

$$\begin{aligned}e_1(t) \leq& C\Big\{(t-\tau)^{-1/4}e_0(\tau) + \alpha\int_\tau^t (t-r)^{-1/4}e_0(r)\,dr\\ &+ \int_\tau^t (t-r)^{-1/4}e_0(r)^2 e_1(r)\,dr\Big\},\end{aligned} \tag{10.112}$$

where $\tau \in [t-1, t]$. We assume $e_0(r) \leq E$ on $[t-1,t]$ and with Lemma 7.3.19 we obtain

$$\left(\int_\tau^{\tau+\delta}\left|\int_\tau^t (t-r)^{-\kappa}e_1(r)\,dr\right|^p dt\right)^{1/p} \leq \int_\tau^{\tau+\delta}|t^{-\kappa}|dt\left(\int_\tau^{\tau+\delta}|e_1(t)|^p dt\right)^{1/p}.$$

We choose $p = 3$, and estimate $I_\delta = (\int_\tau^{\tau+\delta} e_1^3 dt)^{1/3}$ using (10.112) by

$$I_\delta \leq C\left\{E\delta^{3/4} + \alpha E\delta^{3/4} + E^2\delta^{3/4}I_\delta\right\}.$$

Thus, for sufficiently small $\delta \in (0,1]$ the integral I_δ can be estimated in terms of α, E, and δ, namely for

$$\delta = \min\{1, (4CE^2)^{-4/3}\}$$

we have $I_\delta \le 2CE(1+\alpha)\delta^{3/4}$. Inserting this result into (10.112) yields

$$e_1(\tau+\delta) \le C\Big\{E\delta^{-1/4} + \alpha E\delta^{3/4} + E^2\delta^{5/8}I_\delta\Big\}.$$

This shows that $u(t)$ is bounded in H^1_{ul} and that the bound only depends on the parameters and the L^2_{ul} bound E. For small $E>0$ we have $e_1 \le CE$. Hence, solutions exist globally and for $t\to\infty$ the bound E can be replaced by Δ_0. □

Theorem 10.8.3. *The SH equation possesses an absorbing set, and a global $(H^1_{\mathrm{ul}}, H^1_\rho)$-attractor $\mathcal{A}^\varepsilon$ in the sense of Definition* 8.3.18.

Proof. In order to apply Theorem 8.3.22 it remains to prove the continuity of $\mathcal{S}_t$ in the localized space H^1_ρ. This proof has been given in [**MS95**, Theorem 3.9]. □

10.8.2. Shadowing by GL pseudo-orbits. The estimates from Theorem 10.8.2 are rather bad for $0<\alpha=\varepsilon^2\ll 1$. We obtain $\Delta_1=\mathcal{O}(\varepsilon^{1/2})$ and not $\mathcal{O}(\varepsilon)$ which is the size of the bifurcating stationary solutions, and the size of the solutions, which can be approximated by the GL approximation. In this section we show that the result for small $0<\varepsilon^2\ll 1$ can be improved to $\Delta_1=\mathcal{O}(\varepsilon)$ by using a sequence of GL approximations. In order to do so we have to apply the GL approximation to solutions larger than $\mathcal{O}(\varepsilon)$. Therefore, we modify our approximation ansatz to

$$u(x,t)=\widetilde{\psi}_\delta(A)(x,t) := \delta(A(X,T)\mathrm{e}^{ix}+\overline{A}(X,T)\mathrm{e}^{-ix}) \tag{10.113}$$

with $T=\delta^2 t$, $X=\delta x$, where $\delta>0$ is a small number, but independent of $0<\varepsilon\ll 1$. Then the amplitude A is a solution of the associated GL equation

$$\partial_T A = 4\partial_X^2 A + \widetilde{\alpha}A - 3|A|^2A, \tag{10.114}$$

where $\widetilde{\alpha}=\varepsilon^2/\delta^2$. This does not lead to any conceptual differences, but has the advantage that we are able to consider solutions of size larger than ε. For the precise formulation of the results and the proofs we need a number of preparations.

Preparations. The solution operator of (10.109) in the space $Z=H^1_{\mathrm{ul}}$ is denoted with $\mathcal{S}^\varepsilon_t$ and for (10.114) in $Y=H^1_{\mathrm{ul}}$ with $\mathcal{G}_T$. To deal with the slow spatial scale $X=\delta x$ and the mode concentration at multiples of $k=1$ we need scaling operators

$$S_\delta : Z\to Y,\quad (S_\delta u)(X)=u(X/\delta),$$

and the multiplication operator

$$\theta : Z\to Z,\quad (\theta u)(x)=\mathrm{e}^{-\mathrm{i}x}u(x),$$

which is a translation operator in Fourier space.

We define mode filters which extract the critical modes close to $k = k_{\rm crit} = 1$. In order to do so we define an even cut-off function $\widehat{\phi}_0 \in C_0^\infty(\mathbb{R}, [0,1])$ with $\widehat{\phi}_0(k) = 1$ for $k \in [-1/6, 1/6]$ and $\widehat{\phi}_0(k) = 0$ for $k \notin [-1/3, 1/3]$. To extract the modes close to $k = -1, 0, 1$ we define

$$\widehat{e}_j(k) = \widehat{\phi}_0(k-j) \text{ for } j \in \{-1,0,1\}, \quad \widehat{e}_c = \widehat{e}_1 + \widehat{e}_{-1}, \quad \text{and } \widehat{e}_s = 1 - \widehat{e}_c.$$

By Lemma 8.3.7 we associate to $\widehat{e}_\sigma$, $\sigma \in \{s, c, 1, 0, -1\}$, an operator $E_\sigma : Z \to Z$ which extracts the Fourier modes belonging to wave numbers in the relevant intervals; $E_c u \in Z$ contains the critical modes of $u \in Z$, and $E_s u \in Z$ the stable modes of u.

The approximation ansatz (10.113), i.e., the map $\widetilde{\psi}_\delta : Y \to Z$, can be written as

$$\widetilde{\psi}_\delta(A) = \delta(\theta^{-1} S_\delta^{-1} A + \overline{\theta^{-1} S_\delta^{-1} A}).$$

and will be interpreted as a lift from the GL problem into the original problem (10.109). For technical reasons we introduce a modified lift $\psi_\delta : Y \to Z$ defined as

$$\psi_\delta(A) = \delta(\theta^{-1} E_0 S_\delta^{-1} A + \overline{\theta^{-1} E_0 S_\delta^{-1} A}), \tag{10.115}$$

i.e., via a cut-off of $S_\delta^{-1} A$ in Fourier space.

To map $u \in Z$ to an associated $A \in Y$ we have to extract the modes near $k = 1$, and subsequently rescale, and thus define

$$\Phi_\delta : Z \to Y, \quad u \mapsto \frac{1}{\delta} S_\delta \theta E_1 u. \tag{10.116}$$

Obviously $\Phi_\delta(u)$ contains all the information of $u_c = E_c u$, since $u_c = E_1 u + E_{-1} u$ with $E_{-1} u = \overline{E_1 u}$. Thus, if we control $\Phi_\delta(u)$ and u_s, then we control all of u.

With these connections between the underlying spaces we want to compare the dynamics of (10.109) and (10.114). This is done by lifting up the semigroup $\mathcal{G}_T$ from Y into Z, or by extracting the GL modes from $\mathcal{S}_t$.

Approximation, attractivity, and shadowing. We state two results which together yield a number of consequences for the dynamics of the original system and the comparison of attractors. The first result is an approximation theorem similar to, e.g., Theorem 10.2.9, or the abstract Approximation Theorem 10.4.3 with its applications Theorem 10.5.4 or Theorem 10.6.5, but with two improvements. It states that if

$$A_0 \in Y = H^1_{\rm ul}, \text{ with } \|u_0 - \psi_\delta(A_0)\|_Z \text{ sufficiently small,}$$

then $\psi_\delta[\mathcal{G}_{\delta^2 t}(A_0)] \in Z$ is a good approximation of $\mathcal{S}_t^\varepsilon(u_0)$, where $0 \le t \le \delta^{-2}$.

Theorem 10.8.4. (Improved Approximation) *For all $R_1, T_1, d > 0$ there exists $C, \delta_0 > 0$ such that for all $0 < \varepsilon \le \delta \le \delta_0$ the following holds. If*

$A_0 \in B_Y(R_1)$ and $u_0 \in Z$ with $\|u_0 - \psi_\delta(A_0)\|_Z \le d\delta^{5/4}$, then

$$\sup_{0\le t\le T_1/\delta^2} \|\mathcal{S}_t^\varepsilon(u_0) - \psi_\delta(\mathcal{G}_{\delta^2 t}(A_0))\|_Z \le C\delta^{5/4} \tag{10.117}$$

$$\|\Phi_\delta(\mathcal{S}^\varepsilon_{T_1/\delta^2}(u_0)) - G_{T_1}(A_0)\|_Y \le C\delta^{1/4}. \tag{10.118}$$

Thus, we relax the regularity requirements for A_0, and we allow for an initial error $u_0 - \psi_\delta(A_0)$. These two improvements allow to connect the approximation result with the second result, which is the attractivity of the GL set, i.e., of the set of all functions $u \in Z$ having in leading order the form of $\psi_\delta(A)$ for some appropriate $A \in Y$.

Theorem 10.8.5. (Attractivity) *For each $r_0 > 0$ there exists constants $C, T_0, R_1, \delta_0 > 0$ such that for all $0 < \varepsilon \le \delta \le \delta_0$ we have*

$$\mathrm{dist}_Z(\mathcal{S}^\varepsilon_{T_0/\delta^2}(B_Z(\delta r_0)), \psi_\delta(B_Y(R_1))) \le C\delta^{5/4}, \tag{10.119}$$

$$\mathrm{dist}_Y(\Phi_\delta(\mathcal{S}^\varepsilon_{T_0/\delta^2}(B_Z(\delta r_0))), B_Y(R_1)) \le C\delta^{1/4}, \tag{10.120}$$

where $B_Z(r) = \{u \in Z : \|u\|_Z \le r\}$ and $\mathrm{dist}_Z(A, B) = \sup_{a\in A} \inf_{b\in B} \|a - b\|_Z$.

Both theorems are slight generalizations (due to the additional scaling parameter $\delta \ge \varepsilon$) of [**Sch94b**, Lemmas 10-12] and have first been formulated in [**MS95**, Theorem 4.3 and Theorem 4.2]. We review their proofs in §10.8.3. As a first consequence of the two results and of the existence of the globally attracting set for the GL equation, cf. Corollary 8.3.25, we have the global existence and uniqueness of solutions in a neighborhood of the unstable origin of the pattern forming system.

Theorem 10.8.6. (Global existence and uniqueness) *There exist T_0, T_1, $\delta_0 > 0$ such that for all $0 \le \varepsilon \le \delta \le \delta_0$ and R_0 sufficiently large we have $\mathcal{S}_{(T_0+T_1)/\delta^2}(B_Z(\delta R_0)) \subset B_Z(\delta R_0)$. Therefore, solutions $\mathcal{S}_t(u_0)$ with initial conditions $u_0 \in B_Z(\delta R_0)$ stay bounded and exist globally in time.*

Remark 10.8.7. For the SH equation this is not a new result. However, for instance for the 3D Couette-Taylor problem, where no weighted a priori estimates are available, Theorem 10.8.6 gives a nontrivial global existence and uniqueness result, cf. [**Sch99b**]. For pattern forming systems global existence in the sense of Theorem 10.8.6 can be shown, whenever analogs to Theorem 10.8.4, Theorem 10.8.5, and Corollary 8.3.25 can be established. See Figure 10.14. ⌋

Moreover, it is possible to show that all solutions $u(t) = \mathcal{S}_t^\varepsilon(u_0)$ can be shadowed by the lift of a pseudo-orbit in the GL equation. Here the notion of pseudo-orbits is similar to but slightly different from the one in §2.4.3. A (T_1, κ)-pseudo-orbit is pieced together from true orbits of time span T_1 with jumps of maximal size κ in between:

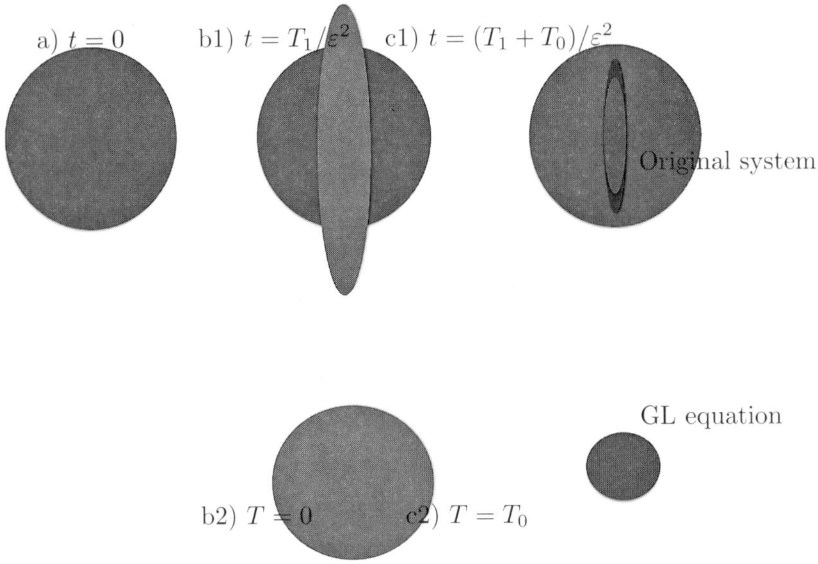

Figure 10.14. The GL formalism can be used to obtain a priori estimates for the solutions of the pattern forming system. A neighborhood (balls in a), b1), c1)) of the origin of the pattern forming system is mapped by the attractivity, cf. Theorem 10.8.5, into a set (ellipse in b1)) which can be described by the GL equation. The GL equation possesses an exponentially attracting absorbing ball, and hence the big ball in b2) is mapped into the smaller ball in c2). Therefore, the original neighborhood (balls in a), b1), c1)) of the pattern forming system is mapped after a time of order $\mathcal{O}(1/\varepsilon^2)$ into itself (ellipse in c1)) due to the approximation property, cf. Theorem 10.8.4. These a priori estimates combined with the local existence and uniqueness gives the global existence and uniqueness of solutions of the pattern forming system in a neighborhood of the weakly unstable origin.

Definition 10.8.8. *Let $T_1 > 0$ and $\kappa > 0$. We call a function $A = A(T)$ a* (T_1, κ)-pseudo-orbit *in Y for the GL equation* (10.114) *if for all $n \in \mathbb{N}$ we have*

$$A(nT_1 + \tau) = \mathcal{G}_\tau(A(nT)) \text{ for all } \tau \in [0, T_1), \text{ and}$$
$$\|A((n+1)T_1 - 0) - \mathcal{G}_{T_1}(A(nT_1))\|_Y \leq \kappa,$$

where $A(T-0) = \lim_{\tau \nearrow T} A(\tau)$.

Theorem 10.8.9. (Shadowing by pseudo-orbits) *For all $T_1 > 0$ there exist $\delta_0, C, T_0 > 0$ such that for all $\delta \in (0, \delta_0]$ the following is true. For all $\varepsilon \in (0, \delta]$ and all initial conditions u_0 with $\|u_0\|_Z \leq \delta$ the solution $u(t) = \mathcal{S}_t(u_0)$ exists for all time, and there is a $(T_1, C\delta^{1/4})$-pseudo-orbit A for* (10.114) *which satisfies $\|A(0)\|_Y \leq C$ and approximates $u(t)$ as*

$$\|u(t) - \psi_\delta(A(\delta^2 t - T_0))\|_Z \leq C\delta^{5/4} \quad \text{for all } t \geq T_0/\delta^2.$$

As already said, this will allow us to prove that the diameter of the attractor of the SH equation in H^1_{ul} is not only of size $\mathcal{O}(\varepsilon^{1/2})$, but of size $\mathcal{O}(\varepsilon)$ for $\varepsilon \to 0$. Moreover, it allows to show the upper-semicontinuity of the rescaled SH attractor towards the GL attractor, cf. §10.8.5.

In the following Examples 10.8.10 and 10.8.11 we illustrate the idea of pseudo-orbits (with $\delta = \varepsilon$) numerically for our two canonical examples, namely the SH equation (10.109), and the KS equation (10.24), for which the above results can be shown as well [**Sch94b**]. However, these numerics also illustrate some quantitative difference between cubic and quadratic systems: For the (cubic) SH equation, the next order correction to the ansatz $\psi_\varepsilon(A)$ is $\mathcal{O}(\varepsilon^3)$ and indeed "very small", also for "moderate" $\varepsilon > 0$. Therefore, pseudo-orbits with $\mathcal{O}(\varepsilon^2)$ jumps (in $u - \psi_\varepsilon(A)$) can be considered. In contrast, for the (quadratic) KS equation, the next order correction to the ansatz $\psi_\varepsilon(A)$ is $\mathcal{O}(\varepsilon^2)$.

Example 10.8.10. Pseudo-orbits for the SH equation. We consider (10.109), i.e.,

$$\partial_t u = -(1+\partial_x^2)^2 u + \varepsilon^2 u - u^3 \tag{10.121}$$

with associated GL equation

$$\partial_T A = 4\partial_X^2 A + A - 3|A|^2 A. \tag{10.122}$$

Numerically we need to work on a finite domain, and we choose $x \in (0, l)$, $l = 100\pi$ with periodic boundary conditions, and similarly $X = \varepsilon x \in (0, L)$, $L = \varepsilon l$, with periodic boundary conditions. We choose $\varepsilon = 0.5$,

$$A_0(X) = \mathrm{e}^{-\frac{(X-L/2)^2}{4}}, \quad u_0 = \varepsilon A_0 \mathrm{e}^{\mathrm{i}x} + \text{c.c.}, \tag{10.123}$$

and solve (10.121) and (10.122) with a Fourier spectral method, see [**Uec09**]. From the solution $u(\cdot, t)$ we extract $\Phi_\varepsilon(u)$ as in (10.116), i.e., with $\delta = \varepsilon$. According to Theorem 10.8.4, u and $\psi_\varepsilon(A)$ will stay $\mathcal{O}(\varepsilon^{5/4})$ close on a $1/\varepsilon^2$ time-scale. In fact, for the SH equation we can do better, and in the simulations we track $e_{u,\infty}(t) := \|u(t) - \psi_\varepsilon(A)(t)\|_\infty$ and reset A to $\Phi_\varepsilon(u)$ if $e_{u,\infty}(t) > \varepsilon^2$. Thus, we use a slightly different notion of pseudo-orbit than in Theorem 10.8.9, where the jump-times T_j and sizes $C\varepsilon^{1/4}$ are fixed on the GL level. Here we only jump "when needed", based on the L^∞-norm. However, clearly a $(T_1, C\varepsilon^{1/4})$ pseudo-orbit in the sense of Theorem 10.8.9 can be constructed just as easily. The results are given in Figure 10.15. ⌋

Example 10.8.11. Pseudo-orbits for the KS equation. We now repeat Example 10.8.10 for (10.24), i.e.,

$$\partial_t u = -(1+\partial_x^2)^2 u + \varepsilon^2 u + \partial_x (u)^2, \tag{10.124}$$

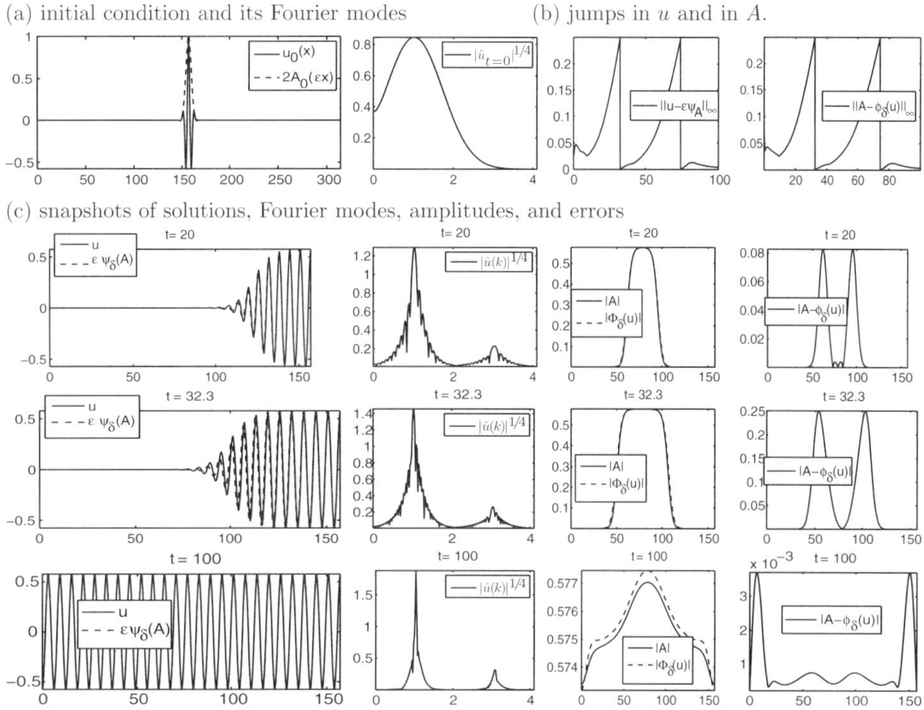

Figure 10.15. A pseudo-orbit for the SH equation, $\varepsilon = \delta = 0.5$, see Example 10.8.10. (a) shows the initial condition and its Fourier transform, where for clarity we do not plot $|\widehat{u}(k)|$ but $|\widehat{u}(k)|^{1/4}$. Since $\varepsilon = 0.5$ is relatively large, $\widehat{u}(k)$ is not very strongly localized around $k = \pm 1$. (b) shows the L^∞ errors and jumps in $u - \psi_\varepsilon(A)$ and $A - \Phi_\varepsilon(u)$. Since the $\mathcal{O}(\varepsilon^3)$ modes are quite small, it appears that $\|u - \psi_\varepsilon(A)\| = 0$ after resetting A, but this is in fact not the case; see Example 10.8.11 for clearer illustration. Finally, (c) shows, at selected time steps, the solutions (on half the domain for u and $\psi_\varepsilon(A)$), the Fourier transform, the extracted critical modes $\Phi_\varepsilon(u)$ and their comparison with the GL evolution. The rather narrow initial wave packet widens, and the solution develops the expected mode distribution; the error in $A - \Phi_\varepsilon(u)$ is essentially localized at the edges of the wave packet, and the error in $u - \psi_\varepsilon(A)$ can hardly be seen up to $t = 20$, say. The first resetting occurs at $t \approx 32.3$, and is triggered by the edges of the wave packet. This continues, with a second reset at $t \approx 74$, until the wave packet has filled the domain and the solution u converges to the stationary roll u_ε given in lowest order by $\frac{2\varepsilon}{\sqrt{3}} \cos(x)$, which corresponds to $A \equiv 1/\sqrt{3}$.

with associated GL equation

$$\partial_T A = 4\partial_X^2 A + A - \frac{4}{9}|A|^2 A. \tag{10.125}$$

Again we work on $x \in (0, l)$, $l = 100\pi$, $X = \varepsilon x \in (0, L)$, $L = \varepsilon l$, and choose the initial condition (10.123), but now we choose the smaller $\varepsilon = 0.25$. Moreover, we reset A to $\Phi_\varepsilon(u)$ if $e_{u,\infty}(t) > 2\varepsilon^{5/4}$. The results are given in Figure 10.16. ⌋

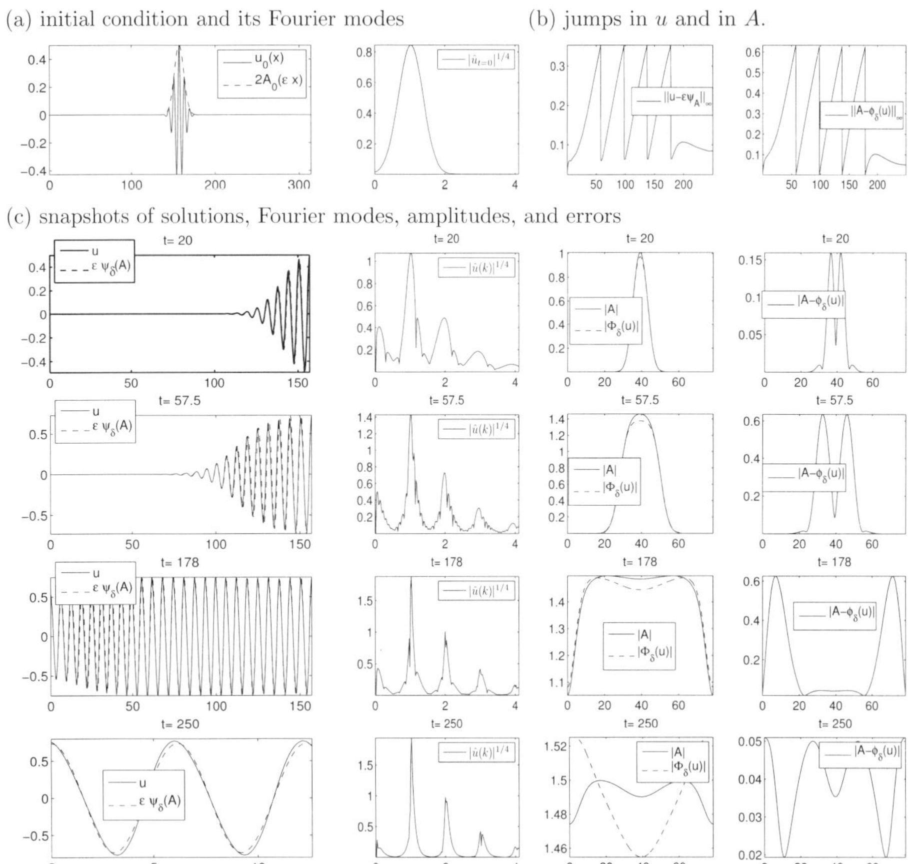

Figure 10.16. A pseudo-orbit for the KS equation, $\varepsilon = \delta = 0.25$, see Example 10.8.11. We mainly point out the differences to Figure 10.15. First of all, (b) clearly shows that after reset $0 \neq u - \psi_\varepsilon(A) = \mathcal{O}(\varepsilon^2)$ from the modes at $k = 0$ and $k = 2$, generated by the quadratic terms, see the Fourier plots in (c). Moreover, even though the criterion now is $e_{u,\infty}(t) > 2\varepsilon^{5/4}$ (as opposed to $e_{u,\infty}(t) > \varepsilon^2$ in our SH example), more jumps are necessary than in the SH equation; however, the jumps still occur after $\mathcal{O}(1/\varepsilon^2)$ t-intervals, which can be checked by varying ε. Finally, as in the SH example, the solution u converges to a stationary roll, here $u_\varepsilon = \frac{3\varepsilon}{2}\cos(x) + \mathcal{O}(\varepsilon^2)$, and $A \to \frac{3}{2}$. Again, even after convergence ($t > 250$, say), these $\mathcal{O}(\varepsilon^2)$ terms can be seen more clearly in (b) than the $\mathcal{O}(\varepsilon^3)$ terms in the SH example in Figure 10.15(b).

10.8.3. Ideas of the proofs. We want to prove that the set of initial conditions for which the GL equation makes correct predictions is absorbing in the sense of Theorem 10.8.5. In other words, given an arbitrary initial conditions $u_0 \in B_Z(\delta r_0)$, we want to show that the associated solution $t \mapsto u(t)$ of the SH equation (10.109) near the Turing instability develops in such a way that there exist a time $\widetilde{t}$ and functions $B = B(X) \in H^1_{\text{ul}}$ and $R = R(x) \in H^1_{\text{ul}}$ such that

$$u(x, \widetilde{t}) = \varepsilon B(\varepsilon x) + c.c. + \varepsilon^3 R(x).$$

Then we can use B as initial condition for the solution $A \in C([0, T_0], H^1_{\text{ul}})$ of the associated GL equation (10.7), and apply the Approximation Theorem 10.8.4 to predict the dynamics of the solutions $t \mapsto u(t)$ of (10.109) near the Turing instability.

For convenience, in the following reformulation and proof of the attractivity result we return to the case $\delta = \varepsilon$.

Theorem 10.8.12. *Let u_0 be an initial condition for (10.109) with $\|u_0\|_{H^1_{\text{ul}}} \leq R_0\varepsilon$. Then there exist $T_1 > 0$, $\varepsilon_0 > 0$ such that for all $\varepsilon \in (0, \varepsilon_0)$ the following holds. For $t = T_1/\varepsilon^2$ the solution $u(t, u_0)$ of (10.109) can be written as*

$$u(T_1/\varepsilon^2, u_0) = \varepsilon w_1 + \varepsilon w_{-1} + \varepsilon^3 w_s$$

with $w_1 = E^h_1 w_c$, $w_{-1} = \overline{w_1}$, and $w_s = E^h_c w_s$. There exists constants C_1, C_2 only depending on R_0 such that $\|S_{1/\varepsilon}\theta w_c\|_{H^1_{\text{ul}}} \leq C_1$ and $\|w_s\|_{H^1_{\text{ul}}} \leq C_2$.

Proof. The variation of constant formula applied to (10.109) yields

$$u(t) = \mathrm{e}^{t\Lambda} u_0 - \int_0^t \mathrm{e}^{(t-\tau)\Lambda} u^3(\tau) d\tau.$$

Similar to the proof of Theorem 10.8.2, by a simple contraction argument we have the existence of $\varepsilon_0 > 0$, $T_1 = T_1(R_0)$, $C_0 = C_0(R_0)$, such that for all $\varepsilon \in (0, \varepsilon_0)$ we have a unique solution $u \in C([0, T_1/\varepsilon^2], H^1_{\text{ul}})$ with $u(0) = u_0$ satisfying $\sup_{t\in[0,T_1/\varepsilon^2]} \|u(t)\|_{H^1_{\text{ul}}} \leq C_0\varepsilon$.

Then we define $w_s = \varepsilon^{-3} E_s u(T_1/\varepsilon^2)$ and use the exponential decay $\left\|\mathrm{e}^{t\Lambda} E^h_s u\right\|_{H^1_{\text{ul}}} \leq \mathrm{e}^{-\sigma_s t}\|u\|_{H^1_{\text{ul}}}$ with a $\sigma_s > 0$ independent of $0 < \varepsilon \ll 1$ such that

$$\begin{aligned}
\|w_s\|_{H^1_{\text{ul}}} =& \|\varepsilon^{-3} E_s \mathrm{e}^{T_1\varepsilon^{-2}\Lambda} u_0\|_{H^1_{\text{ul}}} + \|\varepsilon^{-3} E_s \int_0^{T_1/\varepsilon^2} \mathrm{e}^{(T_1\varepsilon^{-2}-\tau)\Lambda} u^3(\tau) d\tau\|_{H^1_{\text{ul}}} \\
\leq& \varepsilon^{-3}\mathrm{e}^{-\sigma_s T_1/\varepsilon^2} R_0\varepsilon + \varepsilon^{-3}\sigma_s^{-1}(1 - \mathrm{e}^{-\sigma_s T_1/\varepsilon^2}) C_0^3\varepsilon^3 \leq C_2 = \mathcal{O}(1)
\end{aligned}$$

for $\varepsilon \to 0$. Accordingly, we define $w_c = \varepsilon^{-1} E_c u(T_1/\varepsilon^2) = s_1 + s_2$ with

$$s_1 = \varepsilon^{-1} E_c \mathrm{e}^{T_1\varepsilon^{-2}\Lambda} u_0, \quad s_2 = -\varepsilon^{-1} E_c \int_0^{T_1/\varepsilon^2} \mathrm{e}^{(T_1/\varepsilon^{-2}-\tau)\Lambda} u^3(\tau) d\tau.$$

Using the estimate

$$\left\| \mathcal{S}_{1/\varepsilon}\theta \mathrm{e}^{t\Lambda} E_1^h u \right\|_{H^1_{\mathrm{ul}}} \leq C \max(1, (\varepsilon t)^{-1/2}) \| E_1^h u \|_{H^1_{\mathrm{ul}}} \tag{10.126}$$

we find

$$\|\mathcal{S}_{1/\varepsilon}\theta s_1\|_{H^1_{\mathrm{ul}}} \leq C \|\varepsilon^{-1} E_1^h u_0\|_{H^1_{\mathrm{ul}}} \leq C R_0 = \mathcal{O}(1)$$

and similarly

$$\begin{aligned} \|\mathcal{S}_{1/\varepsilon}\theta s_2\|_{H^1_{\mathrm{ul}}} \leq & \int_0^{T_1/\varepsilon^2} \|\mathcal{S}_{1/\varepsilon}\theta E_1^h \mathrm{e}^{(T_1/\varepsilon^2-\tau)\Lambda}\|_{H^1_{\mathrm{ul}} \to H^1_{\mathrm{ul}}} \,\mathrm{d}\tau \\ & \times \sup_{t\in[0,T_1/\varepsilon^2]} \|\varepsilon^{-1} E_1^h (u^3(\tau))\|_{H^1_{\mathrm{ul}}} \\ \leq & \varepsilon^{-1} T_1^{1/2} C_0^3 \varepsilon^3 = \mathcal{O}(\varepsilon^2) \end{aligned}$$

for $\varepsilon \to 0$. □

Remark 10.8.13. It can be proved that what is indicated in Theorem 10.8.12 with $\mathcal{O}(\varepsilon^2)$ has more structure, namely that for every $N \in \mathbb{N}$ the function $u|_{t=\mathcal{O}(1/\varepsilon^2)}$ possesses a representation

$$u|_{t=\mathcal{O}(1/\varepsilon^2)} = \sum_{m=0}^{N} \varepsilon^{\beta(m)} A_m(\varepsilon x) e^{imx} + \mathcal{O}(\varepsilon^{N+1}) \tag{10.127}$$

where the function $\beta : \mathbb{Z} \to \mathbb{N}$ is defined by $\beta(0) = 2$, $\beta(\pm 1) = 1$, and $\beta(m) = |m|$ for $|m| \geq 2$. The functions $A_m \in C_b^N$ satisfy $\|A_m\|_{C_b^N} \leq C_0$ for a constant C_0 independent of ε. This is exactly the structure of a higher order GL approximation. Its Fourier mode distribution is plotted in Figure 10.6. Such an attractivity property has first been stated in [**dES71**], and first been proved in [**Eck93**] by checking the occurrence of the Fourier mode distribution plotted in Figure 10.6 for $\widehat{u}|_{t=\mathcal{O}(1/\varepsilon^2)}$ in case of initial conditions $\widehat{u}|_{t=0} = \mathcal{O}(\varepsilon) \in L^1$. The result has been improved in [**Sch95a**]. ⌋

It remains to prove the improved approximation result Theorem 10.8.4 with low regularity $A_0 \in H^1_{\mathrm{ul}}$, compared to $A_0 \in H^5_{\mathrm{ul}}$ in Theorem 10.2.9. For simplicity we again set $\delta = \varepsilon$ and only explain the main idea, see [**Sch94b**] for more details. We reconsider the rather simple proof of Theorem 10.2.9 in §10, where we made the ansatz $u = \varepsilon\psi + \varepsilon^{3/2} R$ with

$$\varepsilon\psi = \varepsilon A(\varepsilon x, \varepsilon^2 t)\mathrm{e}^{ix} + \frac{1}{64}\varepsilon^3 A^3(\varepsilon x, \varepsilon^2 t)\mathrm{e}^{3ix} + c.c. \,.$$

With A a solution of the GL equation (10.7), we found the error R to satisfy

$$\partial_t R = \Lambda R - 3\varepsilon^2\psi^2 R - 3\varepsilon^{5/2}\psi R^2 - \varepsilon^3 R^3 - \varepsilon^{-3/2}\mathrm{Res}(\varepsilon\psi).$$

We are interested in mild solutions and consider the variation of constant formula

$$R(t) = e^{t\Lambda}R(0) + \int_0^t e^{(t-\tau)\Lambda}(-3\varepsilon^2\psi^2 R - 3\varepsilon^{5/2}\psi R^2 - \varepsilon^3 R^3)(\tau)d\tau + s_{\text{res}}(t),$$

with $s_{\text{res}}(t) = \varepsilon^{-3/2}\int_0^t e^{(t-\tau)\Lambda}\text{Res}(\varepsilon\psi)(\tau)d\tau$. In §10 we used that $\sup_{\tau\in[0,T_0/\varepsilon^2]}$ $\|\text{Res}(\varepsilon\psi)(\tau)\|_{H^1_{\text{ul}}}\leq C\varepsilon^{7/2}$ if $A\in C([0,T_0),H^5_{\text{ul}})$, but this can be improved. By using the smoothing properties $\|G(T)A_0\|_{H^{m+j}_{\text{ul}}} \leq C\max(1,T^{-j/2})\|A_0\|_{H^m_{\text{ul}}}$ of the linear semigroup for the GL equation, i.e., $G(T) = e^{4T\partial_X^2}$, and smoothing estimates and concentration properties for the linear semigroup $e^{t\Lambda}$ of the original system, such as (10.126), we obtain that for $A_0 \in H^1_{\text{ul}}$ we have

$$\sup_{\tau\in[0,T_0/\varepsilon^2]} \|s_{\text{res}}(t)\|_{H^1_{\text{ul}}} \leq C = \mathcal{O}(1),$$

and this turns out to be sufficient to prove (10.117). With similar estimates we can prove

$$\|\mathcal{S}_{1/\varepsilon}\theta E_1^h s_{\text{res}}(T_0/\varepsilon^2)\|_{H^1_{\text{ul}}} \leq C = \mathcal{O}(1),$$

which in the end implies (10.118).

10.8.4. Improved estimates. In this section we demonstrate how the pseudo-orbit method can be used to improve the estimate on the size of the attractor from $\mathcal{O}(\varepsilon^{1/2})$ to $\mathcal{O}(\varepsilon)$. First, extending the analysis of §8.3.3 we provide sharp bounds on the decay rates for the real (i.e., $c_2, c_3 \in \mathbb{R}$) GL equation $\partial_T A = c_2\partial_X^2 A + \widetilde{\alpha}A - c_3|A|^2 A$. The main tool is the maximum principle which is applicable for the equation for the radial part if polar coordinates are introduced.

Lemma 10.8.14. *Let $A(T) = \mathcal{G}_T(A_0)$ be a solution of the real GL equation with $A_0 \in H^1_{\text{ul}}$. For every $\nu > 0$ and $T > 0$ we have*

$$\|A(T)\|_{L^\infty} \leq \frac{2c_3}{\nu}\left(\frac{\widetilde{\alpha}+\nu}{3c_3}\right)^{3/2}(1-e^{-\nu T}) + e^{-\nu T}\|A_0\|_{L^\infty}. \tag{10.128}$$

Moreover, for each $\tau \in [0,T)$ we have

$$\|A(T)\|_{H^1_{\text{ul}}} \leq C\left(1 + \frac{1}{\sqrt{T-\tau}} + Me^{M(T-\tau)}\sqrt{T-\tau}\right)\|A(\tau)\|_{L^\infty}, \tag{10.129}$$

where $M = \widetilde{\alpha} + 3c_3\max\{\sqrt{\widetilde{\alpha}/c_3}, \|A(\tau)\|_{L^\infty}\}$ and C is a universal constant independent of $\widetilde{\alpha}$, c_3, and A.

Proof. Let $A(T,X) = r(T,X)e^{i\phi(T,X)}$, which for r yields the equation

$$\partial_T r = c_2\partial_X^2 r - (\partial_X\phi)^2 r + \widetilde{\alpha}r - c_3 r^3.$$

Using the maximum principle it is possible to compare r with the solution of $\partial_T a = \widetilde{\alpha}a - c_3 a^3$. Hence, if $\|A_0\|_{L^\infty} = a(0)$, then $\|A(T)\|_{L^\infty} \leq a(T)$ for all

$T > 0$. From $\widetilde{\alpha}a - c_3a^3 \leq 2c_3((\widetilde{\alpha}+\nu)/(3c_3))^{3/2} - \nu a$ the desired L^∞ estimate follows. In particular, we have $\|A(T)\|_{L^\infty} \leq \max\{\sqrt{\widetilde{\alpha}/c_3}, \|A(0)\|_{L^\infty}\}$.

The H^1_{ul} estimate is derived by using the variation of constant formula for $V = \partial_X A$. From

$$\partial_T V = \widetilde{\alpha}V + c_2\partial_X^2 V - c_3(2|A|^2V + A^2\overline{V})$$

we find

$$V(T) = G(T-\tau)\partial_X A(\tau) + \int_\tau^T G(T-r)\Big(\widetilde{\alpha}V(r) - c_3(\ldots)\Big)\,dr$$

and hence

$$\|V(T)\|_{L^2_{l,u}} \leq \frac{C}{\sqrt{T-\tau}}\|A(\tau)\|_{L^2_{l,u}} + \int_\tau^T C(\widetilde{\alpha} + 3c_3\|A(r)\|^2_{L^\infty})\|V(r)\|_{L^2_{l,u}}\,dr.$$

Applying Gronwall's inequality the result follows. □

With these preparations we estimate $\limsup_{T\to\infty}\|A(T)\|_Y$ for pseudo-orbits.

Lemma 10.8.15. *There exists a constant C such that for all $\widetilde{\alpha}, \kappa \in (0,1]$, all $T_1 \geq 1$, and all (T_1,κ)-pseudo-orbit $A = A(T)$ of (10.114) we have*

$$\limsup_{T\to\infty}\|A(T)\|_Y \leq C\Pi, \qquad \text{with } \Pi = \sqrt{\widetilde{\alpha}} + (\kappa/T_1)^{1/3} + \kappa. \tag{10.130}$$

Proof. We first estimate the L^∞-norm of $g_n = \|A(nT_1)\|_{L^\infty}$. Then (10.128) and the jump condition give

$$g_{n+1} \leq \frac{2c_3}{\nu}\left(\frac{\widetilde{\alpha}+\nu}{3c_3}\right)^{3/2}(1 - \mathrm{e}^{-\nu T_1}) + \mathrm{e}^{-\nu T_1}g_n + \kappa.$$

Hence, we find

$$\limsup_{n\to\infty} g_n \leq \frac{2c_3}{\nu}\left(\frac{\widetilde{\alpha}+\nu}{3c_3}\right)^{3/2} + \frac{\kappa}{1-\mathrm{e}^{-\nu T_1}}.$$

Using the estimate $1/(1-\mathrm{e}^{-\nu T_1}) \leq 1 + 1/(\nu T_1)$ and optimizing w.r.t. ν yields $\limsup_{T\to\infty}\|A(T)\|_{L^\infty} \leq C\Pi$. With (10.129) (where now $M \leq C$ and $\tau = T-1$) we conclude $\limsup_{n\to\infty}\|A(nT_1+s)\|_Y \leq C\Pi$ for all $s \in [1, T_1)$. The missing intervals can be estimated by the uniform continuity of the semigroup $\mathcal{G}_T$ for $T \in [0,1]$. □

Now Theorem 10.8.9 is applied iteratively with a decreasing sequence of δ_j such that in each step the size of $u(t) = \psi_{\delta_j}(A_j(\delta_j^2 t - \tau_j)) + \mathcal{O}(\delta_j^{5/4})$ is decreased.

Theorem 10.8.16. *There exist positive constants ε_0, δ_1, and C such that for all $\varepsilon \in (0,\varepsilon_0]$ and every solution $u(t) = \mathcal{S}_t(u_0)$ of (10.109) with $\|u_0\|_Z \leq \delta_1$ we have*

$$\limsup_{t\to\infty}\|u(t)\|_Z \leq C\varepsilon.$$

Proof. We take $\varepsilon_0 \le \delta_1 \le \delta_0$ where in the following δ_0, T_0, and T_1 are the same values as in Theorem 10.8.9. By induction over j we construct a monotonically decreasing sequence $(\delta_j)_{j=1,\ldots,J}$ where $\delta_J = C\varepsilon$ with a $C \ge 1$ independent of $0 \le \varepsilon \ll 1$.

Assume that we know $\|u(t_j)\|_Z \le \delta_j$, which is true for $j = 1$ with $t_1 = 0$. We apply Theorem 10.8.9 with $\delta = \delta_j \le \delta_0$ to obtain a $(T_1, C\delta_j^{1/4})$-pseudo-orbit A_j such that

$$\|u(t - t_j) - \psi_{\delta_j}(A_j(\delta_j^2 t - T_0))\|_Z \le C\delta_j^{5/4}.$$

Note that A_j satisfies (10.114) with $\widetilde{\alpha}_j = \varepsilon^2/\delta_j^2 \le 1$. Using the subsequent estimate (10.132) and the estimate (10.130) we find a time t_{j+1} with

$$\begin{aligned} \|u(t_{j+1})\|_Z \le & C\delta_j \|A_j(\delta_j^2(t_{j+1} + t_j) - T_0)\|_Y + C\delta_j^{5/4} \\ \le & C\delta_j \left(\varepsilon/\delta_j + \left(\delta_j^{1/4}\right)^{1/3}\right) + C\delta_j^{5/4} \le C_1\left(\varepsilon + \delta_j^{13/12}\right). \end{aligned} \tag{10.131}$$

Thus, we are finished, as we have shown that it is possible to define δ_j such that $\delta_{j+1} \le C_1(\varepsilon + \delta_j^{13/12})$. By decreasing δ_1 and ε_0, if necessary, such that $C_1\delta_1^{1/12} \le 1/3$ and $C_1\varepsilon_0 \le \min\{1, \delta_1/3\}$ we find $\delta_2 \le 2\delta_1/3$ for all $\varepsilon \in (0, \varepsilon_0]$.

We let $h(\varepsilon, \delta) = C_1(\varepsilon + \delta^{13/12})$ and define $\delta_{j+1} = h(\varepsilon, \delta_j)$. Then $\delta = h(\varepsilon, \delta)$ has a unique fixed point δ_ε^* in the interval $(0, \delta_1)$. Obviously, $\delta_\varepsilon^* \in (C_1\varepsilon, 2C_1\varepsilon)$ and $\delta > h(\varepsilon, \delta)$ for $\delta \in (\delta_\varepsilon^*, \delta_1]$. Thus, the sequence δ_j decays monotonically and has the limit δ_ε^*. Hence, there is a finite J such that $\delta_J \le 2C_1\varepsilon$, and we conclude that $\limsup_{t\to\infty} \|u(t)\|_Z \le \delta_{J+1} \le 2C_1\varepsilon$. □

Remark 10.8.17. a) From §10.8.1 we know that δ_1 in Theorem 10.8.16 can be taken as $C\varepsilon^{1/2}$. Using the construction in the proof we see that it is possible to assume $\delta_j = C\varepsilon^{\alpha_j}$ with $\alpha_{j+1} = \min\{\frac{13}{12}\alpha_j, 1\}$. Hence, it is sufficient to do $J = 10$ iteration steps. Therefore, the overall time needed for a solution starting in $B_Z(\delta)$ to reach the ball $B_Z(C\varepsilon)$ is of order $1/\varepsilon^2$.

b) More generally, if for some pattern forming system a small invariant set of size $\delta = \mathcal{O}(1)$ can be established (e.g., Theorem 10.8.6 for the case of the SH equation), then the pseudo-orbit approach allows to prove that in this invariant set there is an attractor of diameter $\mathcal{O}(\varepsilon)$, cf. [**MS95, Sch99b**]. ⌋

It remains to prove the following estimate.

Lemma 10.8.18. *The linear operator $\psi_\delta : Y \to Z$ satisfies for all $\delta \in (0, 1]$ the estimate*

$$\|E_s\psi_\delta(A)\|_Z \le C\delta^{3/2}\|A\|_Y \quad \textit{for all } A \in Y. \tag{10.132}$$

Proof. For $v = E_s\theta^{-1}E_0S_\delta^{-1}A$ we have $u = E_s\psi_\delta(A) = \delta(v + \overline{v})$, and it is sufficient to show that $\|v\|_Z \le C\delta^{1/2}\|A\|_Y$. We introduce the spaces

$Y^0 = L^2_{\rm ul}$ and $Z^0 = L^2_{\rm ul}$. As multiplier operators commute we find $v = \theta^{-1}FE_0S_\delta^{-1}A = \theta^{-1}E_0FS_\delta^{-1}A$ where $F : Z \to Z$ is associated with the multiplier $\widehat{f}(k) = \widehat{e}_s(k+1) = 1 - \widehat{\phi}_0(k) - \widehat{\phi}_0(k+2)$. The scaled operator $F_\delta = S_\delta F S_\delta^{-1} : Y \to Y$ is again a multiplier operator with kernel $\widehat{f_\delta}(K) = \widehat{f}(\delta K)$. We now have $v = \theta^{-1}E_0S_\delta^{-1}F_\delta A$ and estimate

$$\|v\|_Z \leq \|\theta^{-1}\|_{Z\to Z}\|E_0\|_{Z^0\to Z}\|S_\delta^{-1}\|_{Y^0\to Z^0}\|F_\delta\|_{Y\to Y^0}\|A\|_Y.$$

Obviously, $\|\theta^{-1}\| \leq C$ and $\|E_0\|_{Z^0\to Z} \leq C$ by Lemma 8.3.7 since $(1+k^2)^{1/2}\widehat{\phi}_0(k) \in C_b^2(\mathbb{R},\mathbb{R})$. Moreover, we have $\|S_\delta^{-1}\|_{Y^0\to Z^0} \leq C\delta^{-1/2}$ and

$$\|F_\delta\|_{Y\to Y^0} \leq C\|(1+K^2)^{-1/2}\widehat{f_\delta}(K)\|_{C_b^2(\mathbb{R},\mathbb{R})} \leq C\delta,$$

since $\widehat{f_\delta}(K) = 0$ for $|K| \leq 1/(6\delta)$. These estimates give the desired result. □

10.8.5. Upper semicontinuity of the rescaled SH attractor. In Theorems 8.3.25 and 10.8.3 we have proved the existence of the (global) attractors $\mathcal{A}_\mathcal{G} \subset Y$ for $\mathcal{G}_T$ and $\mathcal{A}^\varepsilon \subset Z$ for $\mathcal{S}_t^\varepsilon$, respectively. Our aim is to compare these attractors and the dynamics on them. A natural question is whether the distance of $\mathcal{A}^\varepsilon$ from $\psi_\varepsilon(\mathcal{A}_\mathcal{G})$ in Z tends to zero faster than ε, which is the diameter of the two sets. The two principles, attractivity and approximation property, can be used to establish the upper-semicontinuity of the rescaled SH attractor $\Phi_\varepsilon\mathcal{A}^\varepsilon$, i.e., it converges to the GL attractor $\mathcal{A}_\mathcal{G}$, as the bifurcation parameter goes to zero. Like for other problems of this kind, lower semicontinuity cannot be expected in general, cf. Example 2.4.9 or [**HR90**].

Theorem 10.8.19. *For every $\sigma > 0$ there exist $C, \varepsilon_0 > 0$ such that for all $\varepsilon \in (0, \varepsilon_0]$ we have the estimates*

$$\mathrm{dist}_Y(\Phi_\varepsilon\mathcal{A}^\varepsilon, \mathcal{A}_\mathcal{G}) \leq \sigma \quad \textit{and} \quad \mathrm{dist}_Z(E_s\mathcal{A}^\varepsilon, \{0\}) \leq C\varepsilon^{5/4}.$$

Proof. From Theorem 10.8.16 we know that $\mathcal{A}^\varepsilon$ is contained in $B_Z(\varepsilon r_0)$ for some $r_0 > 0$. Let $v \in \mathcal{A}^\varepsilon$. Since $\mathcal{A}^\varepsilon$ is invariant under the flow $\mathcal{S}_t^\varepsilon$, there is a $u_0 \in \mathcal{A}^\varepsilon$ such that $v = \mathcal{S}_{T_0/\varepsilon^2}(u_0)$, where T_0 is chosen according to Theorem 10.8.5. Hence,

$$\begin{aligned}\|E_sv\|_Z = \|E_s\mathcal{S}^\varepsilon_{T_0/\varepsilon^2}(u_0)\|_Z &\leq \|E_s(\mathcal{S}^\varepsilon_{T_0/\varepsilon^2}(u_0) - \psi_\varepsilon(A_0))\|_Z + \|E_s\psi_\varepsilon(A_0)\|_Z \\ &\leq C\varepsilon^{5/4},\end{aligned}$$

where (10.119) and (10.132) are used. This shows the second estimate.

From (10.120) we find $R_1 > 0$ such that $\mathrm{dist}_Y(\Phi_\varepsilon\mathcal{A}^\varepsilon, B_Y(R_1)) \leq C\varepsilon^{1/4}$. Since $\mathcal{A}_\mathcal{G}$ is an attractor, there exists, for given σ, a time $T_2 > 0$ such that $\mathrm{dist}_Y(\mathcal{G}_{T_2}B_Y(R_1), \mathcal{A}_\mathcal{G}) \leq \sigma/2$. Now let $v \in \mathcal{A}^\varepsilon$ be arbitrary. By the invariance of $\mathcal{A}^\varepsilon$ there is a $u_0 \in \mathcal{A}^\varepsilon$ with $v = \mathcal{S}^\varepsilon_{T_2/\varepsilon^2}(u_0)$. Applying the approximation result (10.118) with $T_1 = T_2$ and (10.113) we find

$$\mathrm{dist}_Y(\Phi_\varepsilon v, \mathcal{G}_{T_2}(B_Y(R_1))) = \mathrm{dist}_Y(\Phi_\varepsilon \mathcal{S}^\varepsilon_{T_2/\varepsilon^2}(u_0), \mathcal{G}_{T_2}(B_Y(R_1))) \leq C\varepsilon^{1/4}.$$

Choosing $\varepsilon_0 > 0$ so small that $C\varepsilon_0^{1/4} < \sigma/2$, we complete the proof by

$$\begin{aligned}\mathrm{dist}_Y(\Phi_\varepsilon \mathcal{A}^\varepsilon, \mathcal{A}_\mathcal{G}) \leq& \mathrm{dist}_Y(\Phi_\varepsilon \mathcal{A}^\varepsilon, \mathcal{G}_{T_2}(B_Y(R_1))) \\ &+ \mathrm{dist}_Y(\mathcal{G}_{T_2}(B_Y(R_1)), \mathcal{A}_\mathcal{G}) \leq \sigma.\end{aligned}$$

10.9. Further remarks

We close this chapter with a few remarks about other approaches which have been used to justify the GL approximation, justification results for other dissipative systems, and so called beyond-all-order asymptotics.

Remark 10.9.1. One direction has been to establish the GL reduction as an exact reduction method with the help of a Lyapunov-Schmidt reduction, cf. [**Mie92, Mel98, Mel00**]. Unfortunately, the reduction is only exact for so called essential solutions. These are solutions which exist globally forward and backward in time. ⌋

Remark 10.9.2. An attempt to generalize the center manifold theorem, cf. §3.2 and §13.1, to the case of unbounded domains has been made in [**Sch96c**]. The Fourier mode distribution of the attractive solutions shows an exponential concentration around integer multiples of the critical wave number k_c. This allows to modify the original system in such a way that by an exponentially small change w.r.t. the small perturbation parameter $0 < \varepsilon \ll 1$ a system can be created which possesses an $\mathcal{O}(1)$ spectral gap. An exact center manifold reduction is possible for the modified system which allows to approximate the original system on an $\mathcal{O}(1/\varepsilon^3)$ time scale which is much longer than the natural GL time scale $\mathcal{O}(1/\varepsilon^2)$. However, in all these attempts from this and the previous remark the reduced systems are only valuable from a theoretical point of view, but no practical simplification. ⌋

Remark 10.9.3. A number of approximation theorems have been proven in slightly modified situations. The degenerated situation that the coefficient in front of the cubic terms of the GL equation vanishes at the bifurcation point has been discussed in [**She97, BS07**]. Problems with a weak spatial periodicity have been considered in, e.g., [**SD98, SU01**]. The time-periodic situation has been treated in [**SU07**].

Convection problems, or more general pattern forming systems with a free surface possess an additional curve of eigenvalues touching the $\mathrm{Re}\,\lambda = 0$ line at the wave number $k = 0$ for all values of the bifurcation parameter. The spectrum for such systems is of the form sketched in Figure 10.17.

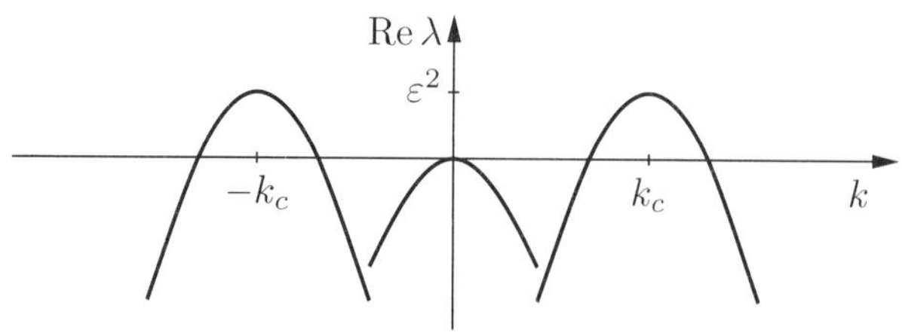

Figure 10.17. Spectrum for the derivation of the GL equation in case of some additional conservation law leading to a zero eigenvalue at $k = 0$. The rest of the spectrum lies strictly below the line $\mathrm{Re}\lambda = 0$. The magnitude of the instability defines the small perturbation parameter $0 < \varepsilon \ll 1$.

If such systems become unstable via an instability at a non-zero wave number k_c, then a GL like system can be derived as the amplitude equation of the system. The situation, occurs for instance in the Bénard-Marangoni problem and for the flow down an inclined plane. In the classical case, cf. Figure 9.8, the proof of the approximation theorem is based on the fact that the quadratic interaction of the critical modes, i.e., the modes with positive or zero growth rates, gives only non-critical modes, i.e., modes which are damped with some exponential rates. This is no longer true for the Bénard-Marangoni problem and the flow down an inclined plane. Nevertheless, approximation results have been established in [**HSZ11**] in case of different group velocities in $k = 0$ and $k = k_c$ and in [**SZ13, Zim16**] in case of vanishing group velocities. See [**DKSZ16**] for an attractivity result.

Such systems can also become unstable via a long wave instability at the wave number $k = 0$. Approximation results for the Cahn-Hilliard equation, which is one of the modulation equations in this case, can be found in [**Sch99a**]. See also [**DHV04**]. ⌋

Remark 10.9.4. A lattice differential pattern forming system is given by a discrete SH equation

$$\partial_t u_n = -u_{n-2} + 2u_{n-1} - 3u_n + 2u_{n+1} - u_{n+2} + \varepsilon^2 u_n - u_n^3.$$

The linearization at $u = 0$ is solved by $u_n = e^{ikn+\lambda t}$ with

$$\lambda = -e^{-2ik} + 2e^{-ik} - 3 + 2e^{ik} - e^{2ik} + \varepsilon^2 = -2\cos(2k) + 4\cos(k) - 3 + \varepsilon^2.$$

By solving $\lambda' = 0$ it is easy to see that the critical wave numbers are given by $\pm k_c = \pm\pi/3$. Again by looking at the Fourier transformed system we obtain the abstract pattern forming system from §10.3. The approximation proofs from §10 can transferred almost line for line from the PDE situation to the lattice differential equation situation. We leave this as an exercise to the reader. We only remark that one has to be careful when computing higher order approximations not to forget terms as A^5 which break the S^1-symmetry of the modulation equation. ⌋

Remark 10.9.5. The GL equation can also be derived for delay-differential equations

$$\frac{\mathrm{d}}{\mathrm{d}t}u(t) = g(u(t), u(t-\tau))$$

with large delay $\tau = 1/\varepsilon$ and $0 < \varepsilon \ll 1$. The linearization at $u = 0$ is given by

$$\frac{\mathrm{d}}{\mathrm{d}t}u(t) = au(t) + bu(t-\tau),$$

with coefficients $a, b \in \mathbb{R}$. It is solved by $u = e^{\lambda t}$ with λ satisfying

$$\lambda = a + be^{-\lambda\tau}.$$

By adjusting a and b, an instability can be created and a GL equation can be derived. This formal analysis has been justified via some approximation result in [**YLWM13**]. ⌋

Remark 10.9.6. One of the major open problems in the theory of modulation equations is the justification of the stochastic GL equation,

$$\partial_T A = \nu_1 \partial_X^2 A + \nu_2 A + \nu_3 A|A|^2 + \xi_{GL},$$

with coefficients $\nu_j \in \mathbb{C}$, and ξ_{GL} white noise in time and space. It occurs for pattern forming systems with thermal fluctuations in the background such as the stochastic SH equation

$$\partial_t u = -(1+\partial_x^2)^2 u + \alpha\varepsilon^2 u - u^3 + \varepsilon^3 \xi_{SH},$$

with coefficient $\alpha \in \mathbb{R}$, small parameter $0 < \varepsilon \ll 1$, and ξ_{SH} white noise in time and space. Various results already have been established. A first approximation result has been shown in [**BMS01**] in case of 2π-spatially periodic boundary conditions. In [**BHP05**] such a result has been established in case of large domains of size $\mathcal{O}(1/\varepsilon)$ with spatially periodic boundary conditions. Other kinds of stochastic forcing and stochastic convergence have been considered in a number of papers, cf. [**BM13, MBK14**]. A promising approach to tackle the full problem is given in [**BR13**]. ⌋

Remark 10.9.7. There are a number of results, e.g., [**CK09, DMCK11, KC13**], that go beyond the asymptotic expansion of solutions of pattern forming systems in powers of ε. For instance, in [**DMCK11**] an expansion $u = \sum_{n=1}^{N} \varepsilon^n A_n + r_N$ for stationary solutions of the cubic quintic SH equation

$$\partial_t u = \varepsilon^2 u - (1+\partial_x^2)^2 u + \varepsilon s u^3 - u^5$$

is considered, where the A_n fulfill GL type amplitude equations as above, and it is shown that by optimally choosing N the error r_N can be made *exponentially* small, i.e.,

$$\|r_N\|_{L^2} \leq C_1 \varepsilon^{-3} \mathrm{e}^{C_2/\varepsilon}.$$

In particular, this can be used to find stationary fronts between periodic solutions and the zero solution, cf. Figure 10.4, and to find localized pattern, i.e., solutions that are homoclinic to the origin but are close to periodic solutions in an (arbitrary) large interval. These fronts and homoclinics, which come in branches snaking around the so called Maxwell point, have already been predicted in [**Pom86**] due to a locking between the GL envelope front and the phase of the underlying periodic pattern. However, to see this phase locking one needs precisely the "beyond all (algebraic) order" asymptotics. See also [**LSAC08, BKL**$^+$**09, ALB**$^+$**10, UW14**] for related results (partly numerical), including localized pattern and snaking in 2D. ⌋

Further Reading. There is abundant physical, chemical and biological literature describing and/or applying the amplitude equation (over bounded domains) or modulation equation (over unbounded domains) formalism to various pattern forming systems. As already noted, for reaction diffusion problems a main textbook reference is [**Mur89**]. For hydrodynamical problems we recommend [**Man92, CI94, Str04**], and a small selection of physically oriented general pattern formation and amplitude and modulation equations textbooks and reviews is [**CH93b, DK98, Pis06**].

Convection in porous media is discussed in [**Fow97**, Chapter 14] and [**Str08**]. See also [**Str04, AGGP12**] for the handling of various convection problems, [**Lap10**] for an comprehensive work on thermal convection, and [**Rad13**] for so called double-diffusive convection; for all these systems GL type modulation equations can be derived.

For reviews of and outlooks on the active field of the derivation and analysis of modulation equations for planar pattern, here only touched upon in §10.6.3, see for instance [**Pis06**, Chapter 4], [**CG09**, Chapter 7], or [**Mer15**, Chapter 6]. Additionally, see [**GS02, Hoy06**] for the crucial symmetry considerations in this field, and [**CK99, CK01**] for steps to take this to the three-dimensional case. Most of the above books and reviews also treat formal consequences of the derived modulation equations, for instance the various instabilities of planar patterns, such as Eckhaus, zig-zag, cross-roll, and some more, that may be discussed on the level of amplitude equations; as a starting point we again recommend [**Hoy06**, Chapters 8 and 9].

Exercises

10.1. Consider the Kuramoto-Shivashinsky-KdV equation

$$\partial_t u = -(1+\partial_x^2)^2 u + \partial_x u + \partial_x^3 u + \varepsilon^2 u + \partial_x(u^2),$$

with $x \in \mathbb{R}$, $t \geq 0$, $0 < \varepsilon \ll 1$, and $u(x,t) \in \mathbb{R}$. Make an ansatz

$$\begin{aligned} u(x,t) =& \varepsilon A_1(\varepsilon(x-ct), \varepsilon^2 t) e^{i(x-\omega t)} \\ &+ \varepsilon^2 A_2(\varepsilon(x-ct), \varepsilon^2 t) e^{2i(x-\omega t)} + \frac{\varepsilon^2}{2} A_0(\varepsilon(x-ct), \varepsilon^2 t) + \text{c.c.} \end{aligned}$$

and derive equations for A_0, A_1, and A_2. Eliminate A_0 and A_2 to derive a GL equation for A_1.

10.2. Compute the higher order GL approximations for the SH equations. How do the higher order approximations for the abstract system in §10.3 look like?

10.3. Replace the assumption on page 334 on the critical curve of eigenvalues λ_1 by the assumptions that $\lambda_1(k_c,0) = i\omega_0 \in i\mathbb{R}$, $\partial_k \lambda_1(k_c,0) = -ic_g$, and $\partial_k^2 \text{Re}\lambda_1(k_c,0) < 0$, for a wave number $k = k_c > 0$. Derive the GL equation for this situation.

10.4. Apply Theorem 10.4.3 to the KS equation by checking the assumptions (A0)-(A4) in H^θ_{ul}-spaces.

10.5. Prove that a_2 and b_2 in (10.106) can be eliminated from the terms of order $\mathcal{O}(1)$ by a normal form transform, respectively averaging.

10.6. Let $\theta, \theta_0 \geq 0$ and let $g(k)$ satisfy $|g(k)| \leq C|k-k_0|^{\theta_0}$. Prove that

$$\|g(\cdot)\varepsilon^{-1}\widehat{A}(\varepsilon^{-1}(\cdot - k_0))\|_{L^2_\theta} \leq C\varepsilon^{\theta_0 - 1/2}\|\widehat{A}\|_{L^2_{\theta+\theta_0}}.$$

Prove the associated estimate in H^θ- and in H^θ_{ul}-spaces.

10.7. Show that the linear operator M, cf. page 358, defined via

$$M\Theta = \mathcal{F}^{-1}\widehat{M}\mathcal{F}\Theta, \quad \text{where} \quad \widehat{M}(k,n) = \frac{ik}{k^2 + n^2\pi^2}, \qquad k \in \mathbb{R}, n \in \mathbb{N},$$

is bounded from H^s to H^{s+1}.

10.8. Compute the GL equation (10.69) for the system (10.61)-(10.63) describing convection in porous media. Hint: Use a classical perturbation ansatz.

10.9. Use the Lyapunov-Schmidt method to prove the existence of stationary spatially periodic convection rolls for the system (10.61)-(10.63) describing convection in porous media.

10.10. Derive the curves of eigenvalues

$$2\lambda_{m,\pm}(k) = -(\kappa+1)s \pm \sqrt{(\kappa+1)^2 s^2 + 4\rho k^2 s^{-1}}.$$

and the associated eigenfunctions for Bénard's problem from §10.6.2. Show that the trivial solution becomes unstable for $\rho_c = 27\kappa/4$ at the wave number $\pm k_c = \pm 1/\sqrt{2}$.

10.11. Derive the GL equation (10.76) for Bénard's problem from §10.6.2.

10.12. Expand $\lambda(k,l) = -(1-k^2-l^2)^2$ at the wave vector $(1,0)$, first with $k = 1 + \varepsilon K$ and $l = \varepsilon L$ w.r.t. ε. Do the same with $k = 1 + \varepsilon K$ and $l = \varepsilon^{1/2} L$.

10.13. Compute for (10.95)-(10.97) the associated transformations for the pressure.

10.14. Use the symmetries (10.95)-(10.97) to show that in the associated amplitude equation (10.106) in PR II only four and not eight different coefficients occur. In order to do so make a GL ansatz in PR II and compute representations of the transformations τ_a, S, and R_φ on the level of the amplitudes.

Chapter 11

Wave packets and the NLS equation

The transport of information through glass fibers by light is one of today's key technologies. Information is encoded digitally by ones and zeroes, in one approach by sending a light pulse through the optical fiber or not. From a physical point of view such a light pulse consists of an underlying electromagnetic carrier wave moving with phase velocity c_p and of a pulse-like envelope moving with group velocity c_g.

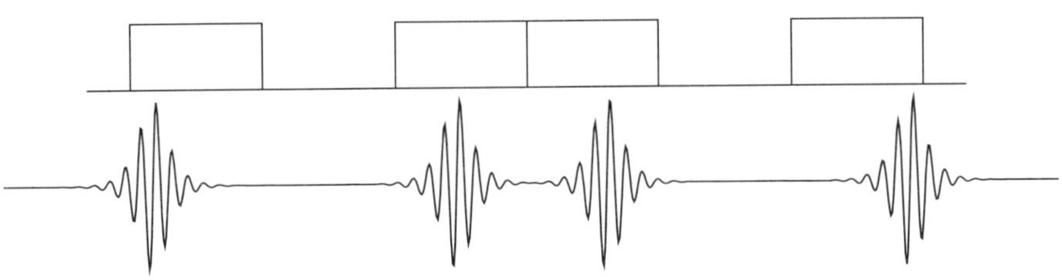

Figure 11.1. 0s and 1s are encoded physically by sending a light pulse or not; thus, for instance, the above series of electromagnetic waves encodes the sequence 101101.

The analysis of the evolution of the associated physical system is a non-trivial task. It shows linear and nonlinear dispersion and (weak) dissipation. As a result the pulses are smeared out which eventually causes an unwanted loss of information. A numerical investigation of the problem leads to a very large system due to the multiple scaling character of the problem. The wave length of the underlying carrier wave is around 10^{-7} m. Resolving such small oscillations in a fiber of 100 km $=10^5$ m gives in a one-dimensional uniform and not very accurate spatial discretization 10^{12} points, even if we ignore the transverse directions and the temporal discretization. Hence, a direct simulation of Maxwell's equations which describe these electromagnetic waves is very expensive, if not impossible. Therefore, before making any numerical investigation, the system has to be analyzed and simpler, numerically more suitable, models have to be derived.

It turns out that the multiple scaling character of the problem is not only a curse, but also a blessing, since it allows to separate the dynamics of the envelope from the dynamics of the carrier wave, such that by multiple scaling analysis the NLS equation can be derived for the description of the slow modulations in time and space of the envelope of the spatially and temporarily oscillating wave packet. Due to the immense reduction of the dimension of the discretized problem by this procedure the NLS equation turned out to be a very successful model. Even though arguably its most important application is in nonlinear optics, e.g., [**Agr01**], the NLS equation has also been derived for water waves [**Zak68, Osb10**], for waves in DNA [**SH94b**] and other discrete chains, for Bose-Einstein condensates [**Pel11**], in plasma physics [**Deb05**, Chapter 10], and in many other fields as a universal envelope or modulation equation, cf. also [**Mil06**, Chapter 10]. In this chapter we explain its justification by approximation theorems for model problems. We explain its universal character, give an overview about approximation results, and explain some applications.

11.1. Introduction

The Nonlinear Schrödinger (NLS) equation

$$\partial_T A = \mathrm{i}\nu_1 \partial_X^2 A + \mathrm{i}\nu_2 A\,|A|^2\,, \tag{11.1}$$

with $T \in \mathbb{R}$, $X \in \mathbb{R}$, $\nu_1, \nu_2 \in \mathbb{R}$, and $A(X,T) \in \mathbb{C}$ is a universal modulation equation which can be derived via multiple scaling analysis in order to describe slow modulations in time and space of the envelope of a spatially and temporarily oscillating wave packet. For instance, for the nonlinear wave equation

$$\partial_t^2 u = \partial_x^2 u - u - u^3, \qquad (x \in \mathbb{R},\ t \in \mathbb{R},\ u(x,t) \in \mathbb{R}), \tag{11.2}$$

also called the cubic Klein-Gordon equation, the ansatz for the derivation of the NLS equation is

$$\varepsilon\psi_{\mathrm{NLS}} = \varepsilon A\left(\varepsilon(x - c_g t), \varepsilon^2 t\right) \mathrm{e}^{\mathrm{i}(k_0 x + \omega_0 t)} + \mathrm{c.c.}, \tag{11.3}$$

where $0 < \varepsilon \ll 1$ is a small perturbation parameter, where c_g is the group velocity, and where the basic temporal and basic spatial wave number ω_0 and k_0 are related by the linear dispersion relation $\omega_0^2 = k_0^2 + 1$. We obtain that the envelope A of the underlying carrier wave $e^{i(k_0 x+\omega_0 t)}$ has to satisfy in lowest order the NLS equation

$$2i\omega_0 \partial_T A = (1-c_g^2)\partial_X^2 A - 3A\,|A|^2 . \tag{11.4}$$

The dynamics of the NLS equation has been discussed in §8.1. The pulse solutions found in §8.1.1 correspond to modulating pulse solutions in the original system, cf. Figure 11.2.

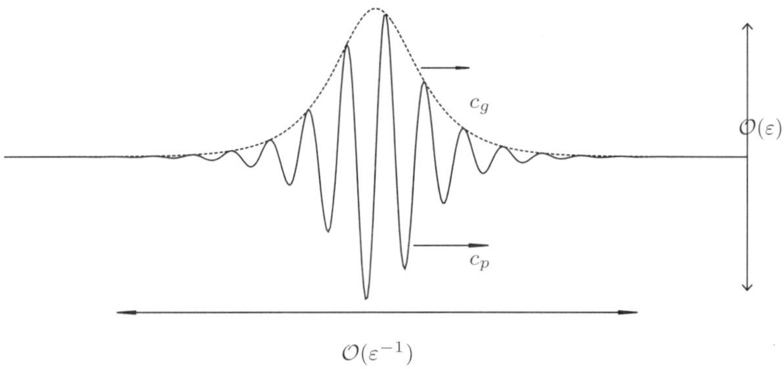

Figure 11.2. A modulating pulse described by the NLS equation. The envelope advancing with group velocity c_g in the laboratory frame modulates the underlying carrier wave $e^{i(k_0 x+\omega_0 t)}$ advancing with group velocity c_p. The envelope evolves approximately as a solution of the NLS equation.

Here we explain mathematical results which justify this formal approximation and show that the NLS equation makes correct predictions about the behavior of the solutions in the original system. In case of no quadratic terms in the original system the proof of error estimates turns out to be rather easy. The estimates follow by a simple application of Gronwall's inequality. This is the situation as it occurs in nonlinear optics due to symmetries. A complete proof of the estimates in this situation is given in §11.2. The proof of the approximation property in case of quadratic terms is presented in §11.4. In this case there are serious difficulties due to the fact that solutions of order $\mathcal{O}(\varepsilon)$ have to be bounded on the long $\mathcal{O}(1/\varepsilon^2)$ time scale w.r.t. t, which corresponds to an $\mathcal{O}(1)$ time scale w.r.t. $T=\varepsilon^2 t$. However, if a non-resonance condition is satisfied, then by averaging or normal form techniques the quadratic terms can be eliminated and this case can be brought back to the situation discussed in §11.2. In §11.5 we explain how the theory from the previous sections can be extended to the situation of additional resonances and to quasilinear systems. In §11.6 we transfer the theory to problems with spatially periodic coefficients and explain why standing light pulses can theoretically occur in photonic crystals. In this chapter we concentrate on nonlinear wave equations as original systems.

Nevertheless, the NLS equation is a universal modulation equation. This is explained in §11.3 and an in §11.7 the connection to nonlinear optics is given.

It turns out that the formal derivation of the NLS equation, its universality, and for cubic nonlinearities also its justification, has many similarities with the derivation, justification and universality of the GL equation for the simple model problems in §10.2 and §10.3. Nevertheless, we repeat most details to better be able to explain the crucial differences, which mainly lie in different suitable phase spaces, in the role of so called non-resonance conditions, and in the lack of attractivity properties of the set of modulated waves as in §10.8.

11.2. Justification in case of cubic nonlinearities

In this section we explain how to justify the NLS equation in case of cubic nonlinearities in the original system. For expository reasons we restrict ourselves to the cubic Klein-Gordon equation (11.2) as original system. Our main purpose is to prove that its solutions behave as predicted by the associated NLS equation (11.4). The NLS approximation (11.3) is formally a good approximation if the terms which do not cancel after inserting $\varepsilon\psi_{\mathrm{NLS}}$ into (11.2) are small. They are collected in the residual

$$\mathrm{Res}(u) = -\partial_t^2 u + \partial_x^2 u - u - u^3. \tag{11.5}$$

If $\mathrm{Res}(u) = 0$, then u is an exact solution of (11.2). With the abbreviation $\mathbf{E} = \mathrm{e}^{\mathrm{i}(k_0 x + \omega_0 t)}$ we find

$$\begin{aligned}\mathrm{Res}(\varepsilon\psi_{\mathrm{NLS}}) =& \varepsilon\mathbf{E}\left((\omega_0^2 - k_0^2 - 1)A\right)\\ &+ \varepsilon^2\mathbf{E}\left((2\mathrm{i}k_0 - 2\mathrm{i}c_g\omega_0)\partial_X A\right)\\ &+ \varepsilon^3\mathbf{E}\left((-2\mathrm{i}\omega_0\partial_T A + (1-c_g^2)\partial_X^2 A - 3A\left|A\right|^2\right)\\ &+ \varepsilon^3\mathbf{E}^3(-A^3)\\ &+ \varepsilon^4\mathbf{E}(2c_g\partial_X\partial_T A)\\ &+ \varepsilon^5\mathbf{E}(-\partial_T^2 A) + \mathrm{c.c.}.\end{aligned}$$

By choosing $\omega = \omega_0$ and $k = k_0$ to satisfy the linear dispersion relation

$$\omega^2 = k^2 + 1,$$

by choosing c_g to be the linear group velocity

$$c_g = \frac{\mathrm{d}}{\mathrm{d}k}\omega\Big|_{k=k_0,\omega=\omega_0} = \frac{k_0}{\omega_0},$$

and by choosing A to satisfy the NLS equation

$$2\mathrm{i}\omega_0\partial_T A = (1-c_g^2)\partial_X^2 A - 3A\left|A\right|^2, \tag{11.6}$$

the first three lines in the residual cancel. However, we still have $\text{Res}(\varepsilon\psi_{\text{NLS}}) = \mathcal{O}(\varepsilon^3)$.

Formal smallness of the residual. It turns out that by adding higher order terms to the approximation $\varepsilon\psi_{\text{NLS}}$ the residual can be made arbitrarily small, i.e., for arbitrary, but fixed $n \in \mathbb{N}$ with $n \geq 3$ there exists an approximation $\varepsilon\psi_n$ with $\varepsilon\psi_n - \varepsilon\psi_{\text{NLS}} = \mathcal{O}(\varepsilon^3)$ and $\text{Res}(\varepsilon\psi_n) = \mathcal{O}(\varepsilon^n)$. Since $\varepsilon\psi_n - \varepsilon\psi_{\text{NLS}} = \mathcal{O}(\varepsilon^3)$ the approximation $\varepsilon\psi_n$ makes the same predictions as $\varepsilon\psi_{\text{NLS}}$ about the behavior of the solutions u of the original system. We will show $\text{Res}(\varepsilon\psi_n) = \mathcal{O}(\varepsilon^n)$ for $n = 4, 5$. With these two examples the general situation can be understood.

In order to obtain

$$\text{Res}(\varepsilon\psi_4) = \mathcal{O}(\varepsilon^4) \tag{11.7}$$

we define

$$\varepsilon\psi_4 = \varepsilon\psi_{\text{NLS}} + \left(\varepsilon^3 A_3\left(\varepsilon(x - c_g t), \varepsilon^2 t\right)\mathbf{E}^3 + \text{c.c.}\right).$$

We find

$$\text{Res}(\varepsilon\psi_4) = \varepsilon^3\mathbf{E}^3\left(-A^3 - (9\omega_0^2 - 9k_0^2 - 1)A_3\right) + \mathcal{O}(\varepsilon^4).$$

Due to the non-resonance $9\omega_0^2 - 9k_0^2 - 1 = 9(k_0^2 + 1) - 9k_0^2 - 1 = 8 \neq 0$ we can choose $A_3 = -(9\omega_0^2 - 9k_0^2 - 1)^{-1}A^3$ in order to achieve (11.7). In order to achieve

$$\text{Res}(\varepsilon\psi_5) = \mathcal{O}(\varepsilon^5) \tag{11.8}$$

we define

$$\varepsilon\psi_5 = \varepsilon\psi_4 + \left(\varepsilon^2 A_{12}(\varepsilon(x - c_g t), \varepsilon^2 t)\mathbf{E} + \varepsilon^4 A_{32}(\varepsilon(x - c_g t), \varepsilon^2 t)\mathbf{E}^3 + \text{c.c.}\right)$$

where A_{12} and A_{32} are new functions to be chosen below. We find

$$\begin{aligned}\text{Res}(\varepsilon\psi_5) =& \varepsilon^4\mathbf{E}\left(-2\mathrm{i}\omega_0\partial_T A_{12} + (1 - c_g^2)\partial_X^2 A_{12}\right)\\ &+ \varepsilon^4\mathbf{E}\left(-3A^2\overline{A_{12}} - 6|A|^2 A_{12} - 2c_g\partial_X\partial_T A\right)\\ &+ \varepsilon^4\mathbf{E}^3((9\omega_0^2 - 9k_0^2 - 1)A_{32} - 3A^2A_{12}) + \mathcal{O}(\varepsilon^5) + \text{c.c.}.\end{aligned}$$

By choosing A_{12} to satisfy the linearized NLS equation

$$-2\mathrm{i}\omega_0\partial_T A_{12} + (1 - c_g^2)\partial_X^2 A_{12} - 3A^2\overline{A_{12}} - 6|A|^2 A_{12} - 2c_g\partial_X\partial_T A = 0 \tag{11.9}$$

and A_{32} to satisfy

$$(9\omega_0^2 - 9k_0^2 - 1)A_{32} - 3A^2A_{12} = 0$$

we achieve (11.8). In order to achieve $\text{Res}(\varepsilon\psi_n) = \mathcal{O}(\varepsilon^n)$ we choose

$$\varepsilon\psi_n = \sum_{m=-N,\ldots,N}\ \sum_{j=1}^{\widetilde{\alpha}(m)} \varepsilon^{\alpha(m)+(j-1)} A_{mj}(X, T)\mathbf{E}^m$$

with $N = n-1$, $X = \varepsilon(x - c_g t)$, $T = \varepsilon^2 t$, and $\alpha, \widetilde{\alpha}$ chosen according to

m	0	1	2	3	$\cdots$	m	$\cdots$	N
$\alpha(m)$	2	1	2	3	$\cdots$	$\lvert\lvert m\rvert - 1\rvert + 1$	$\cdots$	N
$\widetilde{\alpha}(m)$	$N-1$	$N-2$	$N-1$	$N-2$	$\cdots$	$N + 1 - \alpha(m) - 2\delta_{\lvert m\rvert 1}$	$\cdots$	1

The mode distribution of the NLS approximation is similar to one for the GL approximation which is sketched in Figure 10.6. As before A_{11} satisfies the NLS equation, the A_{1j} for $j \geq 2$ linearized NLS equations, and the A_{mj} for $m \neq \pm 1$ algebraic equations, which are linear in A_{mj} and can be solved w.r.t. the A_{mj} due to the validity of the non-resonance conditions

$$(m\omega_0)^2 - (mk_0)^2 - 1 = (m\omega(k_0))^2 - \omega(mk_0)^2 \neq 0, \tag{11.10}$$

for $m = -N, \ldots, N$, where $\omega(k) = \sqrt{1+k^2}$. In case of an odd nonlinearity such as for (11.2) we can set $A_{mj} = 0$ for m even.

Estimates for the residual. The formal orders of the residual can be improved to estimates in norms. We find for instance

$$\|\operatorname{Res}(\varepsilon\psi_{\mathrm{NLS}})\|_{C_b^0} \leq s_1 + s_2 + s_3$$

where

$$\begin{aligned} s_1 &= 2\|\varepsilon^3 E^3 A^3\|_{C_b^0} \leq 2\varepsilon^3 \|A\|_{C_b^0}^3, \\ s_2 &= 4\|\varepsilon^4 E c_g \partial_X \partial_T A\|_{C_b^0} \leq 4\varepsilon^4 c_g \|\partial_T A\|_{C_b^1}, \\ s_3 &= 2\|\varepsilon^5 E \partial_T^2 A\|_{C_b^0} \leq 2\varepsilon^5 \|\partial_T^2 A\|_{C_b^0}. \end{aligned}$$

We can use the right-hand side of the NLS equation to estimate $\|\partial_T A\|_{C_b^1}$ and $\|\partial_T^2 A\|_{C_b^0}$. For instance we have

$$\|\partial_T A\|_{C_b^1} \leq \frac{1}{2\omega_0}\left((1 - c_g^2)\|\partial_X^2 A\|_{C_b^1} + 3\|A\|_{C_b^1}^3\right) < \infty$$

if $A \in C_b^3$. Similarly, we find $\|\partial_T^2 A\|_{C_b^0} < \infty$ if $A \in C_b^4$. This is completely analogous to the GL case in §10.2.

Lemma 11.2.1. *Let $A \in C([0, T_0], C_b^4)$ be a solution of the NLS equation and $\varepsilon_0 \in (0, 1]$. There exists a $C > 0$ such that for all $\varepsilon \in (0, \varepsilon_0)$ we have*

$$\sup_{t \in [0, T_0/\varepsilon^2]} \|\operatorname{Res}(\varepsilon\psi_{\mathrm{NLS}}(t))\|_{C_b^0} \leq C\varepsilon^3.$$

Similarly, for every $n \in \mathbb{N}$ with $n \geq 4$ there exists an approximation $\varepsilon\psi_n$ such that the following holds. Let $A \in C([0, T_0], C_b^{\theta_A})$ with $\theta_A = 3(n-3)+1$ be a solution of the NLS equation. Then for all $\varepsilon_0 \in (0, 1]$ there exists a $C > 0$ such that for all $\varepsilon \in (0, \varepsilon_0)$:

$$\sup_{t \in [0, T_0/\varepsilon^2]} \|\operatorname{Res}(\varepsilon\psi_n(t))\|_{C_b^0} \leq C\varepsilon^n$$

and

$$\sup_{t\in[0,T_0/\varepsilon^2]} \|\varepsilon\psi_{NLS}(t) - \varepsilon\psi_n(t)\|_{C_b^0} \le C\varepsilon^3.$$

For (11.2) we have $\sup_{t\in[0,T_0/\varepsilon^2]} \|\varepsilon\psi_{NLS}(t)-\varepsilon\psi_\beta(t)\|_{C_b^0} \le C\varepsilon^3$. If we look in more detail at the approximation $\varepsilon\psi_5$ we recognize that most regularity is lost in (11.9) for A_{12}. Since $\|\partial_T^2 A_{12}(\cdot,T)\|_{C_b^0}$ has to be estimated, we need $A_{12}(\cdot,T) \in C_b^4$, and so we need $\partial_X\partial_T A(\cdot,T) \in C_b^4$, respectively $A(\cdot,T) \in C_b^7$. Since the structure of the approximation equations is the same for the next orders we lose three derivatives in each step such that the estimates are possible with $\theta_A = 3(n-3)+1$. In fact θ_A can be chosen much smaller by a number of simple tricks, cf. §11.5.1.

The equations for the error. Estimates for the residual, even in norms, are only a necessary condition for showing that the NLS equation makes correct predictions about the behavior of the original systems. By no means they are sufficient. The errors can sum up in time and there are a number of counter-examples, cf. [**Sch95b, SSZ15**], showing that formally derived modulation equations make wrong predictions about the behavior of the original system.

The error $\varepsilon^\beta R = u - \varepsilon\psi$, the difference between the solution u and the approximation $\varepsilon\psi = \varepsilon\psi_n$, with β and n suitably chosen, is estimated with the help of Gronwall's inequality. It satisfies

$$\partial_t^2 R = \partial_x^2 R - R - 3\varepsilon^2\psi^2 R - 3\varepsilon^{\beta+1}\psi R^2 - \varepsilon^{2\beta}R^3 - \varepsilon^{-\beta}\operatorname{Res}(\varepsilon\psi). \tag{11.11}$$

Although there is local existence and uniqueness for (11.11) in C_b^m-spaces by the method of characteristics, here some crucial differences to Chapter 10 arise, since these spaces and this method are not suitable for obtaining estimates on the long time scale $\mathcal{O}(1/\varepsilon^2)$.

Estimates for the residual in Sobolev spaces. Sobolev spaces turn out to be more suitable for (11.11). Hence, we assume that

$A \in C([0,T_0],H^{\theta_A})$ is a solution of the NLS equation with $\theta_A \ge 0$ sufficiently large.

As a first step we have to re-estimate the residual in Sobolev spaces, taking into account the scaling properties of the L^2-norm. As in §10.2.2 we find

$$\begin{aligned}\|\operatorname{Res}(\varepsilon\psi_{\mathrm{NLS}})\|_{H^\theta} \le & C\Big(\varepsilon^3\|A(\varepsilon\cdot)\|_{C_b^\theta}^2\|A(\varepsilon\cdot)\|_{H^\theta} \\ & + \varepsilon^4\|\partial_X\partial_T A(\varepsilon\cdot)\|_{H^\theta} + \varepsilon^5\|\partial_T^2 A(\varepsilon\cdot)\|_{H^\theta}\Big) \\ = & \mathcal{O}\left(\varepsilon^3\|A(\varepsilon\cdot)\|_{H^{\theta+4}}\right).\end{aligned}$$

However,

$$\|A(\varepsilon\cdot)\|_{L^2} = \left(\int_{\mathbb{R}} |A(\varepsilon x)|^2 \mathrm{d}x\right)^{1/2} = \varepsilon^{-1/2}\left(\int_{\mathbb{R}} |A(X)|^2 \mathrm{d}X\right)^{1/2} = \varepsilon^{-1/2}\|A\|_{L^2}$$

such that finally

$$\| \operatorname{Res}(\varepsilon\psi_{\mathrm{NLS}}(t))\|_{H^\theta} = \mathcal{O}\left(\varepsilon^{5/2}\|A\|_{H^{\theta+4}}\right).$$

It is essential that we estimate A^3 with $\|A(\varepsilon\cdot)\|^2_{C_b^\theta}\|A(\varepsilon\cdot)\|_{H^\theta}$ and not with $\|A(\varepsilon\cdot)\|^3_{H^\theta}$.

Nevertheless, ultimately this loss of $\varepsilon^{-1/2}$ is no problem, since as before, the residual can be made arbitrarily small by adding higher order terms to the approximation.

Lemma 11.2.2. *For all $n \in \mathbb{N}$ with $n \geq 4$ and $\theta \geq 1$ the following holds. Let $A \in C([0,T_0], H^{\theta_A})$ with $\theta_A = 3(n-3)+1+\theta$ be a solution of the NLS equation. Then for all $\varepsilon_0 \in (0,1]$ there exists a $C > 0$ such that for all $\varepsilon \in (0,\varepsilon_0)$ there is an approximation $\varepsilon\psi_n$ with*

$$\sup_{t\in[0,T_0/\varepsilon^2]} \| \operatorname{Res}(\varepsilon\psi_n(t))\|_{H^\theta} \leq C\varepsilon^{n-1/2}$$

and

$$\sup_{t\in[0,T_0/\varepsilon^2]} \|\varepsilon\psi_{NLS}(t) - \varepsilon\psi_n(t)\|_{H^\theta} \leq C\varepsilon^{3/2}.$$

The equations for the error in Fourier space. For (11.2) simple energy estimates are possible, cf. Exercise 11.1. However, in order to have a method which also works for more general systems we use semigroup theory and write the equations for the error as first order system in Fourier space. We set $\beta = n - 5/2$, choose $\varepsilon\psi = \varepsilon\psi_n$ and find

$$\partial_t^2 \widehat{R} = -\omega^2\widehat{R} - 3\varepsilon^2\widehat{\psi}^{*2} * \widehat{R} - 3\varepsilon^{\beta+1}\widehat{\psi} * \widehat{R}^{*2} - \varepsilon^{2\beta}\widehat{R}^{*3} + \varepsilon^{-\beta}\widehat{\operatorname{Res}(\varepsilon\psi)},$$

where $\omega(k) = \sqrt{k^2+1}$. This is conveniently written as a first order system

$$\partial_t \widehat{R}_1 = \mathrm{i}\omega\widehat{R}_2,$$
$$\partial_t \widehat{R}_2 = \mathrm{i}\omega\widehat{R}_1 + \varepsilon^2\widehat{f},$$

where

$$\widehat{f} = \frac{1}{\mathrm{i}\omega}\left(-3\widehat{\psi}^{*2} * \widehat{R} - 3\varepsilon^{\beta-1}\widehat{\psi} * \widehat{R}^{*2} - \varepsilon^{2\beta-2}\widehat{R}^{*3} + \varepsilon^{-\beta-2}\widehat{\operatorname{Res}(\varepsilon\psi)}\right).$$

This system is abbreviated in the following as

$$\partial_t\widehat{\mathcal{R}}(k,t) = \Lambda(k)\widehat{\mathcal{R}}(k,t) + \varepsilon^2\widehat{F}(k,t),$$

with

$$\Lambda(k) = \begin{pmatrix} 0 & \mathrm{i}\omega(k) \\ \mathrm{i}\omega(k) & 0 \end{pmatrix}, \qquad \widehat{F}(k,t) = \begin{pmatrix} 0 \\ \widehat{f}(k,t) \end{pmatrix}.$$

We use the variation of constant formula

$$\widehat{\mathcal{R}}(k,t) = \mathrm{e}^{t\Lambda(k)}\widehat{\mathcal{R}}(k,0) + \varepsilon^2 \int_0^t \mathrm{e}^{(t-\tau)\Lambda(k)} \widehat{F}(k,\tau)\,\mathrm{d}\tau$$

in order to estimate the solutions of this system.

Lemma 11.2.3. *The semigroup $(\mathrm{e}^{t\Lambda(k)})_{t\geq 0}$ is uniformly bounded in every H^0_θ, cf. Definition 7.3.30 and Lemma 7.3.31, i.e., there exists a $C > 0$ such that we have $\sup_{t\in\mathbb{R}} \|\mathrm{e}^{t\Lambda}\|_{H^0_\theta\to H^0_\theta} \leq C$.*

Proof. We have $\Lambda(k) = SD(k)S^{-1}$ where

$$S = \begin{pmatrix} 1 & 1 \\ 1 & -1 \end{pmatrix} \qquad \text{and} \qquad D(k) = \begin{pmatrix} \mathrm{i}\omega(k) & 0 \\ 0 & -\mathrm{i}\omega(k) \end{pmatrix}$$

such that $\mathrm{e}^{t\Lambda(k)} = S\mathrm{e}^{tD(k)}S^{-1}$. Hence,

$$\|\mathrm{e}^{t\Lambda}\widehat{u}\|_{H^0_\theta} \leq \sup_{k\in\mathbb{R}} \|\mathrm{e}^{t\Lambda(k)}\|_{\mathbb{C}^2\to\mathbb{C}^2} \|\widehat{u}\|_{H^0_\theta}$$

and further

$$\begin{aligned} \sup_{k\in\mathbb{R}} \|\mathrm{e}^{t\Lambda(k)}\|_{\mathbb{C}^2\to\mathbb{C}^2} &\leq \|S\|_{\mathbb{C}^2\to\mathbb{C}^2} \sup_{k\in\mathbb{R}} \|\mathrm{e}^{tD(k)}\|_{\mathbb{C}^2\to\mathbb{C}^2} \cdot \|S^{-1}\|_{\mathbb{C}^2\to\mathbb{C}^2} \\ &\leq \|S\|_{\mathbb{C}^2\to\mathbb{C}^2} \|S^{-1}\|_{\mathbb{C}^2\to\mathbb{C}^2} < \infty. \end{aligned}$$

□

Lemma 11.2.4. *For every $\theta \geq 1$ there is a $C > 0$ such that for all $\varepsilon \in (0, 1]$ we have*

$$\|\widehat{F}\|_{H^0_\theta} \leq C\left(\|\widehat{\mathcal{R}}\|_{H^0_\theta} + \varepsilon^{\beta-1}\|\widehat{\mathcal{R}}\|^2_{H^0_\theta} + \varepsilon^{2\beta-2}\|\widehat{\mathcal{R}}\|^3_{H^0_\theta} + 1\right).$$

Proof. The estimate follows from

$$\begin{aligned} &\left\|\frac{1}{\omega}\widehat{u}\right\|_{H^0_\theta} \leq \|\widehat{u}\|_{H^0_\theta}, \quad \|\varepsilon^{-\beta}\widehat{\mathrm{Res}(\varepsilon\psi)}\|_{H^0_\theta} \leq C, \quad \|\widehat{R}^{*3}\|_{H^0_\theta} \leq C\|\widehat{R}\|^3_{H^0_\theta}, \\ &\left\|\widehat{\psi} * \widehat{\psi} * \widehat{R}\right\|_{H^0_\theta} \leq C\|\widehat{\psi}\|^2_{L^1_\theta}\|\widehat{R}\|_{H^0_\theta}, \quad \|\widehat{\psi} * \widehat{R} * \widehat{R}\|_{H^0_\theta} \leq C\|\widehat{\psi}\|_{L^1_\theta}\|\widehat{R}\|^2_{H^0_\theta}, \end{aligned}$$

and

$$\begin{aligned} \|\widehat{\psi}\|_{L^1_\theta} &= 2\left\|\frac{1}{\varepsilon}A\Big(\frac{\cdot - k_0}{\varepsilon}\Big) + \text{h.o.t.}\right\|_{L^1_\theta} \leq C\left\|\frac{1}{\varepsilon}A\Big(\frac{\cdot}{\varepsilon}\Big)\right\|_{L^1_\theta} + \text{h.o.t.} \\ &\leq C\|A\|_{L^1_\theta} + \text{h.o.t.} \leq C\|A\|_{H^0_{\theta+1}} + \text{h.o.t.} \end{aligned}$$

□

Remark 11.2.5. Note that $\|\widehat{\psi}\|_{H^0_\theta} = \mathcal{O}(\varepsilon^{-1/2})$ such that $\widehat{\psi}$ has to be estimated in the L^1_θ-norm, respectively ψ in the C^θ_b-norm in order to get ε^2 for the most dangerous term $\varepsilon^2\widehat{\psi} * \widehat{\psi} * \widehat{R}$. This power is necessary to obtain estimates on the natural time scale $\mathcal{O}(1/\varepsilon^2)$ w.r.t. t. ⌟

Using the previous lemmas shows that

$$\begin{aligned}\|\widehat{\mathcal{R}}(t)\|_{H^0_\theta} &\leq C\varepsilon^2 \int_0^t \left(\|\widehat{\mathcal{R}}(\tau)\|_{H^0_\theta} + \varepsilon^{\beta-1}\|\widehat{\mathcal{R}}(\tau)\|^2_{H^0_\theta} + \varepsilon^{2\beta-2}\|\widehat{\mathcal{R}}(\tau)\|^3_{H^0_\theta} + 1\right) \mathrm{d}\tau \\ &\leq C\varepsilon^2 \int_0^t \left(\|\widehat{\mathcal{R}}(\tau)\|_{H^0_\theta} + 2\right) \mathrm{d}\tau \leq 2CT_0 + C\varepsilon^2 \int_0^t \|\widehat{\mathcal{R}}(\tau)\|_{H^0_\theta} \mathrm{d}\tau,\end{aligned}$$

which holds as long as

$$\varepsilon^{\beta-1}\|\widehat{\mathcal{R}}(\tau)\|^2_{H^0_\theta} + \varepsilon^{2\beta-2}\|\widehat{\mathcal{R}}(\tau)\|^3_{H^0_\theta} \leq 1. \tag{11.12}$$

Gronwall's inequality then yields

$$\|\widehat{\mathcal{R}}(t)\|_{H^0_\theta} \leq 2CT_0 \mathrm{e}^{C\varepsilon^2 t} \leq 2CT_0 \mathrm{e}^{CT_0} = M$$

for all $t \in [0, T_0/\varepsilon^2]$. Choosing $\varepsilon_0 > 0$ such that $\varepsilon_0^{\beta-1}M^2 + \varepsilon_0^{2\beta-2}M^3 \leq 1$ we have satisfied the condition (11.12) and so proved the following approximation result.

Theorem 11.2.6. *For all $n \in \mathbb{N}$ with $n \geq 4$ and $\theta \geq 1$ the following holds: Let $A \in C([0,T_0], H^{\theta_A})$ with $\theta_A = 3(n-3)+1+\theta$ be a solution of the NLS equation* (11.6). *Then there exists an $\varepsilon_0 > 0$ and a $C > 0$ such that for all $\varepsilon \in (0, \varepsilon_0)$ there are solutions u of the original system* (11.2) *which can be approximated by $\varepsilon\psi_n$ with*

$$\sup_{t\in[0,T_0/\varepsilon^2]} \|u(t) - \varepsilon\psi_n(t)\|_{H^\theta} < C\varepsilon^{n-5/2}$$

and as a consequence

$$\sup_{t\in[0,T_0/\varepsilon^2]} \|u(t) - \varepsilon\psi_{NLS}(t)\|_{H^\theta} < C\varepsilon^{3/2}.$$

Remark 11.2.7. We have local existence and uniqueness of the solutions of the nonlinear wave equation (11.2), respectively the equations for the error (11.11), in the spaces where we proved the error estimates: Fix $\theta \geq 1$ and let $(u_0, u_1) \in H^{\theta+1} \times H^\theta$. Then there exists a $t_0 > 0$ such that (11.2) possesses a unique solution $u \in C([-t_0, t_0], H^{\theta+1})$ with $u|_{t=0} = u_0$ and $\partial_t u|_{t=0} = u_1$.

In order to construct solutions of (11.2) we use the formula

$$u(x,t) = \frac{1}{2}(u_0(x+t)+u_0(x-t))+\frac{1}{2}\int_{x-t}^{x+t} u_1(\xi)\,\mathrm{d}\xi+\int_0^t \int_{x-(t-s)}^{x+(t-s)} f(y,s)\,\mathrm{d}y\,\mathrm{d}s$$

with $f(x,t) = -u(x,t) - u(x,t)^3$ which is based on the solution formula for the inhomogeneous wave equation. For $t_0 > 0$ sufficiently small the right-hand side $F(u)$ is a contraction in the space $C([-t_0, t_0], H^{\theta+1})$. Thus, there exists a unique fixed point $u^* = F(u^*)$ which is a classical solution of (11.2) if $m \geq 2$.

The solutions exist as long as the norm of the solutions stay bounded. By using the error estimates as a priori estimates we can guarantee that the solutions stay bounded for $t \in [0, T_0/\varepsilon^2]$ and so we can apply the local existence and uniqueness again and again to guarantee the existence and uniqueness of the solutions of the error equations which are obtained from (11.2) by a smooth change of variables. ⌋

11.3. The universality of the NLS equation

As already said, the NLS approximation can be derived in various systems. In order to explain why this is the case, and why the NLS equation plays such an important role, we review the derivation of the NLS equation from (11.2) from a different point of view. This derivation will explain why the NLS equation occurs as a universal modulation equation describing the evolution of modulated wave packets. The underlying system will condense in the values of the coefficients ν_1 and ν_2 in (11.1).

The Fourier transformed system. As for the GL approximation in §10.3 it turns out that Fourier transform is the key for the understanding of the universality. Hence, we consider (11.2) in Fourier space. The Fourier transform $\widehat{u}$ satisfies

$$\partial_t^2 \widehat{u}(k,t) = -\omega^2(k)\widehat{u}(k,t) - \widehat{u}^{*3}(k,t), \tag{11.13}$$

where $\omega(k) = \sqrt{k^2+1}$. By introducing $\widehat{w}(k) = (\widehat{u}(k), \frac{1}{\mathrm{i}\omega(k)}\partial_t \widehat{u}(k))$ we rewrite (11.13) into the first order system

$$\partial_t \widehat{w}(k,t) = \widehat{M}(k)\widehat{w}(k,t) + \widehat{N}(\widehat{w})(k,t), \tag{11.14}$$

where

$$\widehat{M}(k) = \begin{pmatrix} 0 & \mathrm{i}\omega(k) \\ \mathrm{i}\omega(k) & 0 \end{pmatrix}, \quad \widehat{N}(\widehat{w})(k,t) = \begin{pmatrix} 0 \\ \frac{-1}{\mathrm{i}\omega(k)}\widehat{u}^{*3}(k,t) \end{pmatrix}.$$

This system is diagonalized for fixed wave number k. For (11.14) the associated transformation $\widehat{S} = \frac{1}{\sqrt{2}}\begin{pmatrix} 1 & 1 \\ 1 & -1 \end{pmatrix}$ is independent of k and unitary, i.e., $\widehat{S}^{-1} = \widehat{S}^*$. The transformed variable $\widehat{z} = \widehat{S}^*\widehat{w}$ satisfies the diagonalized system

$$\partial_t \widehat{z} = \widehat{\Lambda}\widehat{z} + \widehat{S}^*\widehat{N}(\widehat{S}\widehat{z}), \tag{11.15}$$

with $\widehat{\Lambda}(k) = \mathrm{diag}(i\omega(k), -i\omega(k))$. It turns out that the NLS equation can be derived whenever the original system can be transformed in a system of this form.

General dispersive wave systems. The nonlinear wave equation (11.2) is an example of a dispersive wave system

$$\partial_t U = LU + N_2(U,U) + N_3(U,U,U) + \ldots$$

for an unknown function $U = U(x,t)$ with values in $\mathbb{C}^N$ and $x, t \in \mathbb{R}$. The operator L is linear and skew symmetric, and the terms N_j are j-multilinear and w.l.o.g. symmetric in their arguments, and do not depend explicitly on $x \in \mathbb{R}$. Due to this translation invariance w.r.t. $x \in \mathbb{R}$ the linearization $\partial_t U = LU$ around $U^* = 0$ possesses solutions of the form

$$U(x,t) = f_n(k) \mathrm{e}^{\mathrm{i}kx} \mathrm{e}^{\mathrm{i}\omega_n(k)t},$$

with $k \in \mathbb{R}$ and $n \in I$, where I is some finite index set and $f_n(k) \in \mathbb{C}^N$. The case of $U(\cdot,t) : \Omega \times \mathbb{R} \to \mathbb{R}^N$, where $\Omega = \mathbb{R} \times \Sigma$, $\Sigma \subset \mathbb{R}^m$ a bounded cross section with $m \geq 1$ leads to a countable index set $I = \mathbb{N}$ or $I = \mathbb{Z}$ and to $f_n(k) \in L^2(\Sigma, \mathbb{C}^N)$. We sort the eigenvalues for fixed k as $\omega_n \leq \omega_{n+1}$. For real-valued systems we assume $\omega_n = -\omega_{-n}$. After an expansion

$$U(x,t) = \int_{-\infty}^{\infty} \sum_{n \in I} c_n(k,t) f_n(k) \mathrm{e}^{\mathrm{i}kx} \, \mathrm{d}k$$

in eigenfunctions, the coefficients $c_n(k,t)$ satisfy

$$\begin{aligned}
&\partial_t c_n(k,t) \\
&= \mathrm{i}\omega_n(k) c_n(k,t) + \sum_{n_1,n_2 \in I} \int_{-\infty}^{\infty} s_{2,n,n_1,n_2}(k, k-l, l) c_{n_1}(k-l) c_{n_2}(l) \, \mathrm{d}l \\
&\quad + \sum_{n_1,n_2 \in I} \int_{-\infty}^{\infty} s_{3,n,n_1,n_2,n_3}(k, k-l, l-m, m) c_{n_1}(k-l) c_{n_2}(l-m) c_{n_3}(m) \, \mathrm{d}l \, \mathrm{d}m + \ldots.
\end{aligned}$$

with complex-valued kernels $s_{2,n,n_1,n_2}(k, k-l, l)$, $s_{3,n,n_1,n_2,n_3}(k, k-l, l-m, m)$, etc. We have for instance

$$s_{2,n,n_1,n_2}(k, k-l, l) = \langle f_n^*(k), \mathrm{e}^{-\mathrm{i}kx} N_2[f_{n_1}(k-l)\mathrm{e}^{\mathrm{i}(k-l)x}, f_{n_2}(l)\mathrm{e}^{\mathrm{i}lx}] \rangle_{L^2}$$

where $f_n^*(k)$ is the associated adjoint eigenfunction w.r.t. the scalar product $\langle \cdot, \cdot \rangle_{L^2}$.

Derivation of the NLS equation for (11.15). Taking the Fourier transform of the ansatz in physical space leads to the ansatz

$$\begin{aligned}
\widehat{z}(k,t) &= \varepsilon\varepsilon^{-1} \widehat{A}_1 \left(\frac{k-k_0}{\varepsilon}, \varepsilon^2 t \right) \mathrm{e}^{\mathrm{i}\omega(k_0)t} \mathrm{e}^{\mathrm{i}c_g(k-k_0)t} \vec{e}_1 \\
&\quad + \varepsilon\varepsilon^{-1} \widehat{A}_{-1} \left(\frac{k+k_0}{\varepsilon}, \varepsilon^2 t \right) \mathrm{e}^{-i\omega(k_0)t} \mathrm{e}^{\mathrm{i}c_g(k+k_0)t} \vec{e}_2
\end{aligned} \tag{11.16}$$

for (11.15), where

$$\vec{e}_1 = \begin{pmatrix} 1 \\ 0 \end{pmatrix} \quad \text{and} \quad \vec{e}_2 = \begin{pmatrix} 0 \\ 1 \end{pmatrix}.$$

The notation $\varepsilon\varepsilon^{-1}$ refers to the amplitude scaling ε and the wave number scaling ε^{-1}, but in the following we shorten this to $\varepsilon\varepsilon^{-1} = 1$. Since the Fourier modes of the wave packet are concentrated in an $\mathcal{O}(\varepsilon)$ neighborhood of the basic wave numbers $\pm k_0$ the evolution of the wave packet will be strongly determined by the curves $\pm\omega$ at $\pm k_0$. At $\mathrm{e}^{\mathrm{i}\omega(k_0)t}\mathrm{e}^{\mathrm{i}c_{\mathrm{g}}(k-k_0)t}\vec{e}_1$ we find

$$\begin{aligned}
&\mathrm{i}\omega(k_0)\widehat{A}_1 + \mathrm{i}\varepsilon c_g K\widehat{A}_1 + \varepsilon^2\partial_T\widehat{A}_1 \\
&\quad=\mathrm{i}\omega(k_0)\widehat{A}_1 + \mathrm{i}\varepsilon\partial_k\omega(k_0)K\widehat{A}_1 + \frac{\mathrm{i}}{2}\varepsilon^2\partial_k^2\omega(k_0)K^2\widehat{A}_1 \\
&\quad\quad+\varepsilon^2\frac{3\mathrm{i}}{4\omega(k_0)}\widehat{A}_1 * \widehat{A}_1 * \widehat{A}_{-1} + \mathcal{O}(\varepsilon^3),
\end{aligned}$$

where $k = k_0 + \varepsilon K$, $\widehat{A}_1 = \widehat{A}_1(K,T)$, and where we used

$$\begin{aligned}
&\int_{\mathbb{R}} \widehat{A}\left(\frac{k-l-k_0}{\varepsilon}, T\right)\mathrm{e}^{\mathrm{i}\omega(k_0)t}\mathrm{e}^{\mathrm{i}c_{\mathrm{g}}(k-l-k_0)t}\widehat{A}\left(\frac{l-k_0}{\varepsilon}, T\right)\mathrm{e}^{\mathrm{i}\omega(k_0)t}\mathrm{e}^{\mathrm{i}c_{\mathrm{g}}(l-k_0)t}\,\mathrm{d}l \\
&\quad= \varepsilon\int_{\mathbb{R}} \widehat{A}\left(\frac{k-2k_0}{\varepsilon} - m, T\right)\widehat{A}(m,T)\mathrm{e}^{2\mathrm{i}\omega(k_0)t}\mathrm{e}^{\mathrm{i}c_{\mathrm{g}}(k-2k_0)t}\,\mathrm{d}m\,.
\end{aligned}$$

At ε^0 and ε^1 we obtain the linear dispersion relation and the linear group velocity. At ε^2 we obtain a NLS equation. Undoing the transformation $w = Sz$ and $u = w_1$ gives a multiple of the original approximation

$$\begin{aligned}
\widehat{u}(k,t) =& \frac{1}{\sqrt{2}}\left(\widehat{A}_1\left(\frac{k-k_0}{\varepsilon}, \varepsilon^2 t\right)\mathrm{e}^{\mathrm{i}\omega(k_0)t}\mathrm{e}^{\mathrm{i}c_{\mathrm{g}}(k-k_0)t}\right. \\
&\left.+\widehat{A}_{-1}\left(\frac{k+k_0}{\varepsilon}, \varepsilon^2 t\right)\mathrm{e}^{-\mathrm{i}\omega(k_0)t}\mathrm{e}^{\mathrm{i}c_{\mathrm{g}}(k+k_0)t}\right).
\end{aligned}$$

Derivation of the NLS equation in the general situation. By this procedure it is clear that the NLS equation occurs as a modulation equation for dispersive wave systems whenever the Fourier transform of the initial condition is strongly concentrated at a wave number $k_0 \neq 0$ and when the concentration and the amplitude are of the correct order. It also occurs in case $k_0 = 0$ if $\omega(0) \neq 0$, cf. Exercise 11.1.

W.l.o.g. let us assume that we derive the NLS equation for the curve ω_1 and that we have a real-valued system. We make the ansatz

$$\begin{aligned}
c_1(k,t) =& \widehat{A}_{1,1}(\varepsilon^{-1}(k-k_0), \varepsilon^2 t)\mathrm{e}^{\mathrm{i}\omega(k_0)t}\mathrm{e}^{\mathrm{i}c_{\mathrm{g}}(k-k_0)t} \\
&+\varepsilon\widehat{A}_{1,-2}(\varepsilon^{-1}(k+2k_0), \varepsilon^2 t)\mathrm{e}^{-2\mathrm{i}\omega(k_0)t}\mathrm{e}^{\mathrm{i}c_{\mathrm{g}}(k+2k_0)t} \\
&+\varepsilon\widehat{A}_{1,0}(\varepsilon^{-1}k, \varepsilon^2 t)\mathrm{e}^{\mathrm{i}c_{\mathrm{g}}kt} \\
&+\varepsilon\widehat{A}_{1,2}(\varepsilon^{-1}(k-2k_0), \varepsilon^2 t)\mathrm{e}^{2\mathrm{i}\omega(k_0)t}\mathrm{e}^{\mathrm{i}c_{\mathrm{g}}(k-2k_0)t},
\end{aligned}$$

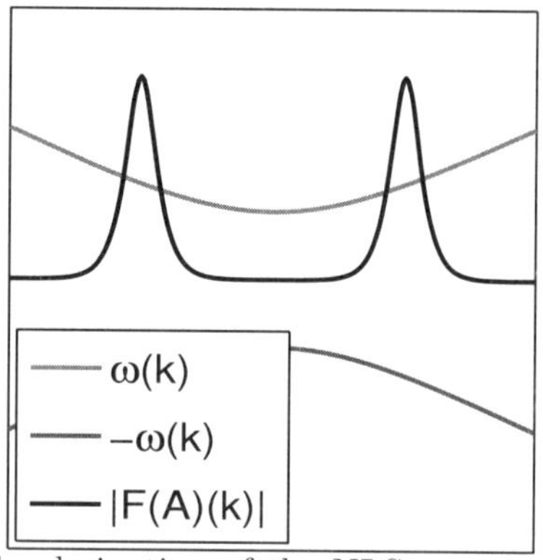

Figure 11.3. The derivation of the NLS equation is based on the concentration of the Fourier modes at a certain wave number. The figure shows the curves of eigenvalues and the concentration of the Fourier modes. Hence, for the evolution of these modes only the curves of eigenvalues close to these wave numbers play a role.

$$\begin{aligned}
c_{-1}(k,t) =&\widehat{A}_{-1,-1}(\varepsilon^{-1}(k+k_0),\varepsilon^2 t)\mathrm{e}^{-\mathrm{i}\omega(k_0)t}\mathrm{e}^{\mathrm{i}c_g(k+k_0)t})\\
&+\varepsilon\widehat{A}_{-1,-2}(\varepsilon^{-1}(k+2k_0),\varepsilon^2 t)\mathrm{e}^{-2\mathrm{i}\omega(k_0)t}\mathrm{e}^{\mathrm{i}c_g(k+2k_0)t}\\
&+\varepsilon\widehat{A}_{-1,0}(\varepsilon^{-1}k,\varepsilon^2 t)\mathrm{e}^{\mathrm{i}c_g kt}\\
&+\varepsilon\widehat{A}_{-1,2}(\varepsilon^{-1}(k-2k_0),\varepsilon^2 t)\mathrm{e}^{2\mathrm{i}\omega(k_0)t}\mathrm{e}^{\mathrm{i}c_g(k-2k_0)t},\\
c_n(k,t) =&\varepsilon\widehat{A}_{n,-2}(\varepsilon^{-1}(k+2k_0),\varepsilon^2 t)\mathrm{e}^{-2\mathrm{i}\omega(k_0)t}\mathrm{e}^{\mathrm{i}c_g(k+2k_0)t}\\
&+\varepsilon\widehat{A}_{n,0}(\varepsilon^{-1}k,\varepsilon^2 t)\mathrm{e}^{\mathrm{i}c_g kt}\\
&+\varepsilon\widehat{A}_{n,2}(\varepsilon^{-1}(k-2k_0),\varepsilon^2 t)\mathrm{e}^{2\mathrm{i}\omega(k_0)t}\mathrm{e}^{\mathrm{i}c_g(k-2k_0)t},
\end{aligned}$$

for $|n| \geq 2$ and complex valued functions $\widehat{A}_{n,j}$, where $A_{n,-j} = \overline{A_{n,j}}$ in physical space. With $k - k_0 = \varepsilon K$ we find at ε^2 for the modes concentrated at k_0 that

$$\begin{aligned}
&\partial_T\widehat{A}_{1,1}(K,T)\\
&=i\partial_k^2\omega_1(k_0)K^2\widehat{A}_{1,1}(K,T)/2\\
&\quad+2\sum_{n\in\mathbb{Z}} s_{2,1,1,n}(k_0,k_0,0)\int_{-\infty}^{\infty}\widehat{A}_{1,1}(K-\kappa,T)\widehat{A}_{n,0}(\kappa,T)\,\mathrm{d}\kappa\\
&\quad+2\sum_{n\in\mathbb{Z}} s_{2,1,1,n}(k_0,-k_0,2k_0)\int_{-\infty}^{\infty}\widehat{A}_{1,-1}(K-\kappa,T)\widehat{A}_{n,2}(\kappa,T)\,\mathrm{d}\kappa\\
&\quad+3s_{3,1,1,1,1}(k_0,k_0,k_0,-k_0)\int_{-\infty}^{\infty}\int_{-\infty}^{\infty}\widehat{A}_{1,1}(K-\kappa_1,T)\\
&\qquad\qquad\times\widehat{A}_{1,1}(\kappa_1-\kappa_2,T)\widehat{A}_{1,-1}(\kappa_2,T)\,\mathrm{d}\kappa_2\,\mathrm{d}\kappa_1,
\end{aligned}$$

where we used the symmetry of the multilinear terms in their arguments, cf. (11.21). With $k - jk_0 = \varepsilon K$ we find at ε in the n-th equation for the modes concentrated at jk_c for $j = 0, 2$ that

$$\begin{aligned}
0 =& \mathrm{i}\omega_n(0)A_{n,0}(K,T) \\
&+ 2s_{2n1-1}(0,-k_c,k_c)\int_{-\infty}^{\infty}\widehat{A}_{1,-1}(K-\kappa,T)\widehat{A}_{1,1}(\kappa,T)\,\mathrm{d}\kappa, \\
0 =& \mathrm{i}(\omega_n(2k_0) - 2\omega_1(k_0))A_{n,2}(K,T) \\
&+ s_{2n11}(2k_c,k_c,k_c)\int_{-\infty}^{\infty}\widehat{A}_{1,1}(K-\kappa,T)\widehat{A}_{1,1}(\kappa,T)\,\mathrm{d}\kappa,
\end{aligned}$$

see (11.22)-(11.23) for an example. Under the non-resonance conditions

$$\omega_n(0) \neq 0, \qquad \omega_n(2k_0) - 2\omega_1(k_0) \neq 0,$$

these two algebraic relations determine $A_{n,0}$ and $A_{n,2}$ in terms of $A_{1,1}$ and $A_{1,-1}$ such that $A_{1,1}$ finally satisfies the NLS equation in Fourier space, namely

$$\begin{aligned}
\partial_T\widehat{A}_{1,1}(K,T) =& -\mathrm{i}\nu_1 K^2\widehat{A}_{1,1}(K,T) \\
&+ \mathrm{i}\nu_2\int_{-\infty}^{\infty}\int_{-\infty}^{\infty}\widehat{A}_{1,1}(K-\kappa,T)\widehat{A}_{1,1}(\kappa-\widetilde{\kappa},T)\widehat{A}_{1,-1}(\widetilde{\kappa},T)\,\mathrm{d}\widetilde{\kappa}\,\mathrm{d}\kappa,
\end{aligned}$$

with coefficients

$$\begin{aligned}
\nu_1 =& -\partial_k^2\omega_1(k_0)/2, \\
\nu_2 =& 4\sum_{n\in\mathbb{Z}} s_{2,1,1,n}(k_0,k_0,0)s_{2,n,1,-1}(0,-k_c,k_c)/\omega_n(0) \\
&+ 2\sum_{n\in\mathbb{Z}} s_{2,1,1,n}(k_0,-k_0,2k_0)s_{2,n,1,1}(2k_0,k_0,k_0)/\omega_n(2k_0) \\
&- 3s_{3,1,1,1,1}(k_0,k_0,k_0,-k_0).
\end{aligned}$$

Therefore, the NLS equation is the universal modulation equation describing slow modulations in time and space of a propagating wave packet in dispersive systems

$$\partial_t\widehat{c}_j(k,t) = \mathrm{i}\omega_j(k)\widehat{c}_j(k,t) + \text{ nonlinear terms,} \tag{11.17}$$

where j is in some index set and where the nonlinear terms have some convolution structure.

Remark 11.3.1. We have seen that for (11.2) the valdity of the non-resonance condition (11.10) is necessary for the construction of higher order approximations. For general dispersive wave systems (11.17) for the validity of a counterpart to Lemma 11.2.1 the validity of the non-resonance condition

$$(m\omega_1(k_0))^2 - \omega_j(mk_0)^2 \neq 0, \tag{11.18}$$

is necessary for $m{=}-N,\ldots,N$, where $N{=}n-1$, and j in some index set. ⌋

11.4. Quadratic nonlinearities

In this section we explain how to justify the NLS approximation in case of quadratic nonlinearities. For expository reasons we restrict ourselves to

$$\partial_t^2 u = \partial_x^2 u - u + u^2, \tag{11.19}$$

with $x \in \mathbb{R}$, $t \in \mathbb{R}$, and $u(x,t) \in \mathbb{R}$ as original system. The ansatz for the derivation of the NLS equation is then given by

$$\begin{aligned} \varepsilon\psi_{\text{NLS}} =& \varepsilon A_1\left(\varepsilon(x-c_g t), \varepsilon^2 t\right) \mathrm{e}^{\mathrm{i}(k_0 x + \omega_0 t)} + \text{c.c.} \\ & + \varepsilon^2 A_2\left(\varepsilon(x-c_g t), \varepsilon^2 t\right) \mathrm{e}^{2\mathrm{i}(k_0 x + \omega_0 t)} + \text{c.c.} \\ & + \varepsilon^2 A_0\left(\varepsilon(x-c_g t), \varepsilon^2 t\right). \end{aligned} \tag{11.20}$$

We find as before at εE the linear dispersion relation and at $\varepsilon^2 E$ the condition for the linear group velocity c_g. At $\varepsilon^3 E$ we find

$$2\mathrm{i}\omega_0 \partial_T A_1 = (1-c_g^2)\partial_X^2 A_1 + 2A_1 A_0 + 2A_2 A_{-1}. \tag{11.21}$$

The algebraic relations which are found at $\varepsilon^2 E^0$ and $\varepsilon^2 E^2$

$$\varepsilon^2 E^0: \quad 0 = -A_0 + 2A_1 A_{-1}, \tag{11.22}$$

$$\varepsilon^2 E^2: \quad 0 = -(-4\omega_0^2 + 4k_0^2 + 1)A_2 + A_1^2, \tag{11.23}$$

can be solved w.r.t. A_0 and A_2 since

$$-4\omega_0^2 + 4k_0^2 + 1 = -(2\omega(k_0))^2 + \omega(2k_0)^2 \neq 0.$$

Inserting the solution for A_0 and A_2 into the equation for A_1 finally yields the NLS equation

$$2\mathrm{i}\omega_0 \partial_T A_1 = (1-c_g^2)\partial_X^2 A_1 + \gamma A_1 |A_1|^2, \tag{11.24}$$

with

$$\gamma = 4 + \frac{2}{-4\omega_0^2 + 4k_0^2 + 1}.$$

Like in case of cubic nonlinearities the residual

$$\text{Res}(u) = -\partial_t^2 u + \partial_x^2 u - u + u^2$$

can be made arbitrarily small by adding higher order terms, i.e., we have

Lemma 11.4.1. *For all $n \in \mathbb{N}$ with $n \geq 4$ and $\theta \geq 1$ the following holds. For $A \in C([0,T_0], H^{\theta_A})$ with $\theta_A = 3(n-3)+1+\theta$ and $\varepsilon_0 \in (0,1]$ there exists a $C > 0$ such that for all $\varepsilon \in (0,\varepsilon_0)$ there is an approximation $\varepsilon\psi_n$ with*

$$\sup_{t\in[0,T_0/\varepsilon^2]} \|\text{Res}(\varepsilon\psi_n(t))\|_{H^\theta} \leq C\varepsilon^{n-1/2}$$

and

$$\sup_{t\in[0,T_0/\varepsilon^2]} \|\varepsilon\psi_{\text{NLS}}(t) - \varepsilon\psi_n(t)\|_{C_b^0} \leq C\varepsilon^2.$$

In order to prove that the solution A_1 of the NLS equation (11.24) predicts the behavior of the solutions u of the original system correctly we estimate as before the difference $\varepsilon^\beta R = u - \varepsilon\psi$ between the correct solution u and its approximation, and as before we choose $\varepsilon\psi = \varepsilon\psi_n$ with $\beta = n - 5/2$. This difference satisfies

$$\partial_t^2 R = \partial_x^2 R - R + 2\varepsilon\psi R + \varepsilon^\beta R^2 + \varepsilon^{-\beta}\,\mathrm{Res}(\varepsilon\psi).$$

As a first order system in Fourier space this yields

$$\begin{aligned} \partial_t \widehat{R}_1 &= \mathrm{i}\omega \widehat{R}_2, \\ \partial_t \widehat{R}_2 &= \mathrm{i}\omega \widehat{R}_1 + \tfrac{1}{\mathrm{i}\omega}\left(2\varepsilon\widehat{\psi} * \widehat{R}_1 + \varepsilon^\beta \widehat{R}_1^{*2} + \varepsilon^{-\beta}\widehat{\mathrm{Res}(\varepsilon\psi)}\right), \end{aligned} \tag{11.25}$$

where again $\omega(k) = \sqrt{k^2+1}$. The simple argument of the last section no longer works because of the new $\mathcal{O}(\varepsilon)$-term $2\varepsilon\widehat{\psi} * \widehat{R}_1$. In principle this term can give some exponential growth of order $\mathcal{O}(\exp(\varepsilon t))$ which is not $\mathcal{O}(1)$-bounded on the time scale of order $\mathcal{O}(1/\varepsilon^2)$. However, this term is oscillatory in time and can be eliminated by averaging or a normal form transformation such that it finally has an $\mathcal{O}(1)$-influence on the size of the solutions. This observation goes back to [**Kal88**].

Remark 11.4.2. Normal form transformations have already been considered in Part I of this book. For PDEs the idea is very similar. For the abstract evolutionary system

$$\partial_t u = Au + N_Q(u) + N_c(u),$$

where N_Q stands for quadratic and N_c for the higher order terms, we seek a near identity change of coordinates $v = u - K(u)$ to eliminate the quadratic terms $N_Q(u)$ and to transfer them into higher order terms. We find

$$\begin{aligned} \partial_t v =& \partial_t u - K'(u)\partial_t u \\ =& Au + N_Q(u) + N_c(u) - K'(u)Au - K'(u)N_Q(u) - K'(u)N_c(u) \\ =& Av + AK(u) - K'(u)Au + N_Q(u) + N_c(u) - K'(u)N_Q(u) - K'(u)N_c(u). \end{aligned}$$

In order to eliminate the quadratic terms we choose K to satisfy

$$AK(u) - K'(u)Au + N_Q(u) = 0$$

such that after the transformation

$$\partial_t v = Av + N_c(u) - K'(u)N_Q(u) - K'(u)N_c(u).$$

As an instructive example [**Str89**, Page 38] we consider

$$i\partial_t u = \Delta u - \nabla u \cdot \nabla u \tag{11.26}$$

and choose $v = u - \frac{1}{2}u^2$, i.e., $K(u) = \frac{1}{2}u^2$, cf. Example 11.4.3 for an explanation how to find this transformation. We have

$$AK(u) - K'(u)Au + N_Q(u) = -\mathrm{i}\Delta(\frac{1}{2}u^2) - u(-\mathrm{i}\Delta u) + \mathrm{i}\nabla u \cdot \nabla u = 0$$

such that

$$\partial_t v = Av - K'(u)N_Q(u) = -\mathrm{i}\Delta u - u\mathrm{i}\nabla u \cdot \nabla u$$

In order to obtain an evolutionary problem for v we have to invert the transformation $v = u - K(u)$. For small v this is possible by applying the implicit function theorem. ⌋

We apply this idea to the equation for the error and eliminate the $\mathcal{O}(\varepsilon)$-term $2\varepsilon\widehat{\psi} * \widehat{R}_1$ with a normal form transformation. In order to do so we first diagonalize (11.25) by introducing

$$\mathcal{R} = S^{-1}\begin{pmatrix} R_1 \\ R_2 \end{pmatrix}, \quad \widehat{\Psi} = S^{-1}\begin{pmatrix} \widehat{\psi} \\ \frac{1}{\mathrm{i}\omega(k)}\partial_t\widehat{\psi}(k)) \end{pmatrix}, \text{ where } S = \frac{1}{\sqrt{2}}\begin{pmatrix} 1 & 1 \\ 1 & -1 \end{pmatrix}.$$

We find

$$\partial_t \mathcal{R} = \Lambda\mathcal{R} + 2\varepsilon B(\Psi,\mathcal{R}) + \varepsilon^{\beta}B(\mathcal{R},\mathcal{R}) + \varepsilon^{-\beta}\mathrm{RES}(\varepsilon\Psi), \tag{11.27}$$

with Λ a symmetric linear map and $B(\cdot,\cdot)$ a bilinear map, which are given in Fourier space by

$$\begin{aligned} \widehat{\Lambda} &= \begin{pmatrix} \mathrm{i}\omega & 0 \\ 0 & -\mathrm{i}\omega \end{pmatrix}, & \mathrm{RES}(\varepsilon\Psi) &= S^{-1}\begin{pmatrix} 0 \\ \mathrm{Res}(\varepsilon\psi) \end{pmatrix}, \\ \widehat{B}(\widehat{U},\widehat{U}) &= \tfrac{1}{\mathrm{i}\omega}S^{-1}\widetilde{B}(S\widehat{U},S\widehat{U}), & \widetilde{B}(\widehat{U},\widehat{V}) &= \begin{pmatrix} 0 \\ \widehat{U}_1 * \widehat{V}_1 \end{pmatrix}, \end{aligned}$$

where $\widehat{U} = (\widehat{U}_1, \widehat{U}_2)$. Then we make a near identity change of variables

$$w = \mathcal{R} + \varepsilon Q(\Psi,\mathcal{R}) \tag{11.28}$$

with Q an autonomous bilinear map. This gives

$$\begin{aligned} \partial_t w =& \partial_t\mathcal{R} + \varepsilon Q(\partial_t\Psi,\mathcal{R}) + \varepsilon Q(\Psi,\partial_t\mathcal{R}) \\ =& \Lambda\mathcal{R} + 2\varepsilon B(\Psi,\mathcal{R}) + \varepsilon^{\beta}B(\mathcal{R},\mathcal{R}) + \varepsilon^{-\beta}\mathrm{RES}(\varepsilon\Psi) + \varepsilon Q(\partial_t\Psi,\mathcal{R}) \\ &+ \varepsilon Q(\Psi, \Lambda\mathcal{R} + 2\varepsilon B(\Psi,\mathcal{R}) + \varepsilon^{\beta}B(\mathcal{R},\mathcal{R}) + \varepsilon^{-\beta}\mathrm{RES}(\varepsilon\Psi)) \\ =& \Lambda w - \varepsilon\Lambda Q(\Psi,\mathcal{R}) + 2\varepsilon B(\Psi,\mathcal{R}) + \varepsilon^{\beta}B(\mathcal{R},\mathcal{R}) + \varepsilon^{-\beta}\mathrm{RES}(\varepsilon\Psi) \\ &+ \varepsilon Q(\partial_t\Psi,\mathcal{R}) + \varepsilon Q(\Psi, \Lambda\mathcal{R} + 2\varepsilon B(\Psi,\mathcal{R}) + \varepsilon^{\beta}B(\mathcal{R},\mathcal{R}) + \varepsilon^{-\beta}\mathrm{RES}(\varepsilon\Psi)), \end{aligned}$$

and so

$$\begin{aligned} \partial_t w &= \Lambda w + \varepsilon\,(-\Lambda Q(\Psi,\mathcal{R}) + Q(\partial_t\Psi,\mathcal{R}) \\ &\qquad + Q(\Psi,\Lambda\mathcal{R}) + 2B(\Psi,\mathcal{R})) + \mathcal{O}(\varepsilon^2). \end{aligned} \tag{11.29}$$

In order to eliminate the dangerous term $2\varepsilon B(\Psi, w)$ we have to find a bilinear Q such that

$$-\Lambda Q(\Psi,\mathcal{R}) + Q(\partial_t\Psi,\mathcal{R}) + Q(\Psi,\Lambda w) + 2B(\Psi,\mathcal{R}) = 0. \tag{11.30}$$

In this form this equation is hard to analyze. For its simplification we first use that

$$\begin{aligned}\varepsilon\partial_t\Psi =&\sqrt{2}\partial_t\begin{pmatrix}\varepsilon A_1\left(\varepsilon(x-c_gt),\varepsilon^2t\right)\mathrm{e}^{\mathrm{i}(k_0x+\omega_0t)}+\mathcal{O}(\varepsilon^2)\\ \varepsilon A_{-1}\left(\varepsilon(x-c_gt),\varepsilon^2t\right)\mathrm{e}^{-\mathrm{i}(k_0x+\omega_0t)}+\mathcal{O}(\varepsilon^2)\end{pmatrix}\\ =&\sqrt{2}\begin{pmatrix}\mathrm{i}\omega_0\varepsilon A_1\left(\varepsilon(x-c_gt),\varepsilon^2t\right)\mathrm{e}^{\mathrm{i}(k_0x+\omega_0t)}+\varepsilon^2G_1\\ -\mathrm{i}\omega_0\varepsilon A_{-1}\left(\varepsilon(x-c_gt),\varepsilon^2t\right)\mathrm{e}^{-\mathrm{i}(k_0x+\omega_0t)}+\varepsilon^2G_{-1}\end{pmatrix}\end{aligned}$$

with $\|\widehat{G}\|_{L^1_\theta}=\mathcal{O}(1)$. Hence, (11.30) is given in lowest order by

$$-\Lambda Q(\Psi,\mathcal{R})+Q(\Upsilon,\mathcal{R})+Q(\Psi,\Lambda w)+2B(\Psi,\mathcal{R})=0, \tag{11.31}$$

where $\Upsilon=\begin{pmatrix}(\Lambda(k_0)\Psi)_1\\(\Lambda(-k_0)\Psi)_{-1}\end{pmatrix}$. In Fourier space we have for the j-th component of B that

$$(\widehat{B}(\widehat{\psi},\widehat{\mathcal{R}}))_j=\sum_{m,n=1,2}\int_{\mathbb{R}}\widehat{b}^j_{mn}(k,k-l,l)\widehat{\psi}_m(k-l)\widehat{\mathcal{R}}_n(l)\,\mathrm{d}l,$$

with $\widehat{b}^j_{mn}=\widehat{b}^j_{mn}(k,k-l,l)$ some smooth kernel. Thus, we make the same ansatz for the j-th component of Q, namely

$$(\widehat{Q}(\widehat{\psi},\widehat{\mathcal{R}}))_j=\sum_{m,n=1,2}\int_{\mathbb{R}}\widehat{q}^j_{mn}(k,k-l,l)\widehat{\psi}_m(k-l)\widehat{\mathcal{R}}_n(l)\,\mathrm{d}l,$$

with $\widehat{q}^j_{mn}=\widehat{q}^j_{mn}(k,k-l,l)$ some kernel which we have to compute. Inserting these representations in (11.31) yields the relations

$$\mathrm{i}(\omega_j(k)-\omega_1(k_0)-\omega_n(l))\widehat{q}^j_{1n}(k,k-l,l)=2\widehat{b}^j_{1n}(k,k-l,l), \tag{11.32}$$

and

$$\mathrm{i}(\omega_j(k)-\omega_2(-k_0)-\omega_n(l))\widehat{q}^j_{2n}(k,k-l,l)=2\widehat{b}^j_{2n}(k,k-l,l), \tag{11.33}$$

with $\omega_{1,2}(k)=\pm\omega(k)$.

Example 11.4.3. For system (11.26), where the approximation is replaced by a general function and where only one curve of eigenvalues is involved, the associated relation would be

$$\mathrm{i}(\omega(k)-\omega(k-l)-\omega(l))\widehat{q}(k,k-l,l)=\widehat{b}(k,k-l,l),$$

with $\omega(k)=-k^2$ and $\widehat{b}(k,k-l,l)=\mathrm{i}(k-l)l$. We find

$$\omega(k)-\omega(k-l)-\omega(l)=-k^2+(k-l)^2+l^2=-2(k-l)l$$

such that $\widehat{q}(k,k-l,l)=-\frac{1}{2}$. ⌋

In Fourier space, the approximation $\varepsilon\psi$ has order one amplitude only close to the wave numbers k_0 in the first component and close to the wave numbers $-k_0$ in the second component. Therefore, only wave numbers $|k-l-k_0|\leq\delta$ for (11.32) and only wave numbers $|k-l+k_0|\leq\delta$ for (11.33) have to be taken into account for a small $\delta > 0$ independent of $0 < \varepsilon \ll 1$. For the other wave numbers we set $\widehat{q}^j_{mn} = 0$. Hence, in order to solve (11.32) w.r.t. the $\widehat{q}^j_{mn}$ we need the non-resonance conditions

$$(11.34)\qquad \inf_{j,n\in\{1,2\}} \inf_{k,l\in\mathbb{R},|k-l|\leq\delta} |(\omega_j(k) - \omega_1(k_0) - \omega_n(l))| \geq C > 0$$

and

$$(11.35)\qquad \inf_{j,n\in\{1,2\}} \inf_{k,l\in\mathbb{R},|k-l|\leq\delta} |(\omega_j(k) - \omega_2(-k_0) - \omega_n(l))| \geq C > 0$$

for this $\delta > 0$ fixed. The validity of the non-resonance conditions can be checked graphically by looking for intersections of the curves $k \mapsto \pm\omega(k)$ and $k \mapsto \omega(k_0) \pm \omega(k - k_0)$, see Figure 11.4. Since the asymptotes $k \mapsto \pm k$ and $k \mapsto \omega(k_0) \pm (k - k_0)$ to these curves are separated, for no value of k_0 a quadratic resonance occurs.

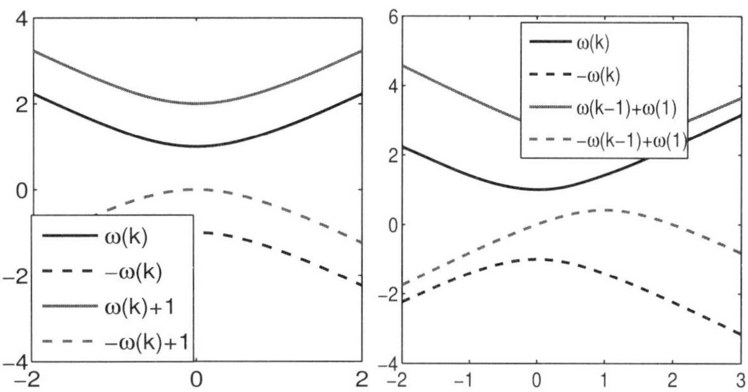

Figure 11.4. The curves $k \mapsto \pm\omega(k)$ and $k \mapsto \omega(k_0) \pm \omega(k - k_0)$ in case $k_0 = 0$ and $k_0 = 1$. There is no intersection and so no quadratic resonance.

Since

$$\sup_{j,m,n\in\{1,2\}} \sup_{k,l\in\mathbb{R},|k-l|\leq\delta} |\widehat{b}^j_{mn}(k, k-l, l)| \leq C < \infty,$$

(11.34) immediately implies

$$\sup_{j,m,n\in\{1,2\}} \sup_{k,l\in\mathbb{R},|k-l|\leq\delta} |\widehat{q}^j_{mn}(k, k-l, l)| \leq C < \infty.$$

As a consequence we obtain

$$\|(Q(\psi, \mathcal{R}))\|_{H^\theta} \leq C\|\psi\|_{C^\theta_b}\|\mathcal{R}\|_{H^\theta} \leq C\|\widehat{\psi}\|_{L^1_\theta}\|\widehat{\mathcal{R}}\|_{L^2_\theta}.$$

Thus, the transformation (11.28) can be inverted with the help of Neumann's series for $\varepsilon > 0$ sufficiently small. We denote the inverse by $\mathcal{R} = \mathcal{R}_\varepsilon(w) = \mathcal{O}(1)$. Therefore, (11.29) transforms into

$$(11.36)\qquad \partial_t w = \Lambda w + \varepsilon^2 F$$

with

$$\begin{aligned}\varepsilon^2 F =& \varepsilon^{\beta} B(\mathcal{R}_\varepsilon(w), \mathcal{R}_\varepsilon(w)) + \varepsilon^{-\beta}\mathrm{RES}(\varepsilon\Psi) + \varepsilon^2 Q(G, \mathcal{R}_\varepsilon(w)) \\ &+ \varepsilon Q(\Psi, 2\varepsilon B(\Psi, \mathcal{R}_\varepsilon(w)) + \varepsilon^{\beta} B(\mathcal{R}_\varepsilon(w), \mathcal{R}_\varepsilon(w)) + \varepsilon^{-\beta}\mathrm{RES}(\varepsilon\Psi)).\end{aligned}$$

Since $\varepsilon^2 F$ obeys the same estimates as the one in Lemma 11.2.4 the rest of the proof of the approximation property from §11.2 applies line for line to (11.36).

11.5. Extension of the theory

There are various physically relevant systems, especially the water wave problem, where the previous non-resonance conditions (11.38) and (11.39) are not satisfied. These are systems with trivial resonances at the wave number $k = 0$, additional non-trivial resonances, and quasilinear nonlinearities. Here we give a short overview about the strategies which have been developed in last years to overcome these difficulties. We start this section with some technical improvements of the previous analysis.

11.5.1. Some technical improvements. By two simple technical changes in the previous proofs the non-resonance conditions can be simplified and less regularity for the initial conditions of the NLS equation is needed.

Less regularity is needed. Since in Fourier space the approximation is strongly concentrated at integer multiples of the basic wave number k_0 we only make a small error if we cut-off the approximation outside small neighborhoods of $k_0\mathbb{Z}$.

The formal expansion of the curve of eigenvalues and of the kernels in the multilinear maps can be estimated with the aid of the following lemma.

Lemma 11.5.1. *Let $\theta, \theta_0 \geq 0$, $\theta_\infty \in \mathbb{R}$, and let $g(k)$ satisfy*

$$|g(k)| \leq C \min(|k - k_0|^{\theta_0}, (1 + |k|)^{\theta_\infty}).$$

Then

$$\|g(\cdot)\varepsilon^{-1}\widehat{A}(\varepsilon^{-1}(\cdot - k_0))\|_{L^2_\theta} \leq C\varepsilon^{\theta_0 - 1/2}\|\widehat{A}\|_{L^2_{\max(\theta+\theta_\infty, \theta_0)}}.$$

Proof. This follows immediately from the fact that the left-hand side of this inequality can be estimated by

$$\begin{aligned}&\leq \sup_{k\in\mathbb{R}} \left| g(k)\Big(1+\frac{\varepsilon}{|k-k_0|}\Big)^{-\max(\theta+\theta_\infty,\theta_0)}(1+|k-k_0|)^{\theta}\right| \; \|\varepsilon^{-1}\widehat{A}(\varepsilon^{-1}\cdot)\|_{L^2_{\max(\theta+\theta_\infty,\theta_0)}} \\ &\leq \sup_{K\in\mathbb{R}} \left| C\min\Big(\frac{|\varepsilon K|^{\theta_0}}{(1+|K|)^{\theta_0}}, \frac{(1+|k_0+\varepsilon K|)^{\theta+\theta_\infty}}{(1+|K|)^{\theta+\theta_\infty}}\Big)\right| \; \varepsilon^{-1/2}\|\widehat{A}(\cdot)\|_{L^2_{\max(\theta+\theta_\infty,\theta_0)}}\end{aligned}$$

where the loss of $\varepsilon^{-1/2}$ is due to the scaling properties of the L^2-norm. □

This lemma can be applied for instance to

$$g(k) = \omega(k) - \omega_0 - c_g k - \frac{1}{2}\omega''(k_0)k^2 = \mathcal{O}(|k|^3)$$

such that the residual terms can also be estimated for the diagonalized system in Fourier space.

Example 11.5.2. With the simple approach from §11.2 we need $\theta_A \geq \theta + 4$ in order to estimate the error in H^θ if we take the approximation ψ_2. The reason is as follows. The highest loss of regularity comes from $\|\partial_T^2 A_3\|_{H^\theta}$ which can be estimated by $\|\partial_T^2 A\|_{H^\theta}$. This can be estimated further via the right-hand side of the NLS equation by $\|A\|_{H^{\theta+4}}$. For the associated diagonalized system we have to expand ω at k_0 up to order three, i.e., we use

$$\begin{aligned}|\omega(k) - (\omega(k_0) + \omega'(k_0)(k-k_0) + \frac{1}{2}\omega''(k_0)(k-k_0)^2)\chi(|k-k_0|)| \\ \leq C\min(|k-k_0|^3, |k|)\end{aligned}$$

where χ is a C_0^∞ function with $\chi(k) \in [0,1]$, $\chi(k) = 0$ for $|k| \geq 2$ and $\chi(k) = 1$ for $|k| \leq 1$. Hence, the application of Lemma 11.5.1 allows to reduce the above value of θ_A to a value satisfying $\theta_A \geq \max(\theta + 1, 3)$ since for all other terms less derivatives are needed. ⌋

Sometimes it is possible to reduce the value for θ_A even further by using a cut-off function in Fourier space to mollify the original approximation. Let again $\chi \in C_0^\infty$ with $0 \leq \chi(k) \leq 1$, $\chi(k) = 0$ for $|k| \geq 2$, and $\chi(k) = 1$ for $|k| \leq 1$. Instead of $\varepsilon^{-1}\widehat{A}(\varepsilon^{-1}\cdot)$ we work with $\chi(\cdot)\varepsilon^{-1}\widehat{A}(\varepsilon^{-1}\cdot)$. Then we have

$$\|g(\cdot)\varepsilon^{-1}\widehat{A}(\varepsilon^{-1}(\cdot))\|_{L^2_\theta} \leq C\varepsilon^{\theta-1/2}\|\widehat{A}\|_{L^2_\theta} \tag{11.37}$$

for $g(k) = 1 - \chi(k)$. Hence, the modified approximation and the original approximation are $\mathcal{O}(\varepsilon^{\theta-1/2})$ close to each other in H^θ if the ansatz functions A are in H^θ. Moreover, we have

Lemma 11.5.3. *Let $\theta, \theta_0 \geq 0$, $\theta_\infty \in \mathbb{R}$, and let $g(k)$ satisfy*

$$|g_1(k)| \leq C\min(|k-k_0|^{\theta_0}, (1+|k|)^{\theta_\infty}).$$

Then

$$\|g_1(\cdot)\chi(\cdot)\varepsilon^{-1}\widehat{A}(\varepsilon^{-1}(\cdot - k_0))\|_{L^2_\theta} \leq C\varepsilon^{\theta_0-1/2}\|\widehat{A}\|_{L^2_{\theta_0}}.$$

Proof. This follows by applying Lemma 11.5.1 to $g(k) = g_1(k)\chi(k)$. □

Therefore, less regularity θ_A is needed for estimating the residual terms in the error equations. However, the original approximation and the cut-off approximation have to be shown to be $\mathcal{O}(\varepsilon^{3/2})$ close. Therefore, in order to have such an estimate in H^θ we need at least $A \in H^{\theta+1}$, cf. (11.37). For an approximation result optimized with this respect see [**MU16**].

Weakening of the non-resonance condition. Checking the non-resonance conditions (11.34) and (11.35) is non-trivial due to the fact that a two-dimensional function has to be bounded away from zero. The following version of Lemma 11.5.1 which avoids the loss of $\varepsilon^{-1/2}$ due to the scaling properties of the L^1-norm allows to reduce the checking of the non-resonance condition to a one-dimensional problem

Lemma 11.5.4. *Let $\theta, \theta_0 \geq 0$ and let $g(k)$ satisfy $|g(k)| \leq C|k - k_0|^{\theta_0}$. Then*

$$\|g(\cdot)\varepsilon^{-1}\widehat{A}(\varepsilon^{-1}(\cdot - k_0))\|_{L^1_\theta} \leq C\varepsilon^{\theta_0}\|\widehat{A}\|_{L^1_{\theta+\theta_0}}.$$

In order to remove l from (11.34) and (11.35) we use the concentration of the approximation at the wave numbers $k = \pm k_0$ and estimate with the help of Lemma 11.5.4

$$\begin{aligned} &\left\| \int_{\mathbb{R}} (\omega_n(l) - \omega_n(\cdot - k_0))\widehat{\psi}(\cdot - l)\widehat{w}_n(l)\,\mathrm{d}l \right\|_{L^2_\theta} \\ &= \left\| \int_{\mathbb{R}} (\omega_n(\cdot - (\cdot - l)) - \omega_n(\cdot - k_0))\widehat{\psi}(\cdot - l)\widehat{w}_n(l)\,\mathrm{d}l \right\|_{L^2_\theta} \\ &\leq \left\| \sup_{k\in\mathbb{R}} |\omega_n(k - (\cdot - l)) - \omega_n(k - k_0)||\widehat{\psi}(\cdot - l)| \right\|_{L^1_\theta} \|w_n\|_{L^2_\theta} \leq C\varepsilon\|w_n\|_{L^2_\theta}. \end{aligned}$$

Hence, if we replace $\omega_n(l)$ by $\omega_n(k - k_0)$ in (11.29) we produce additional terms of order $\mathcal{O}(\varepsilon^2)$ which can be included into the terms which do not make trouble in obtaining error estimates on the $\mathcal{O}(1/\varepsilon^2)$ time scale. Hence, the non-resonance condition (11.34) can be weakened to

$$\inf_{j,n\in\{1,2\}} \inf_{k\in\mathbb{R}} |\omega_j(k) - \omega_1(k_0) - \omega_n(k - k_0)| \geq C > 0 \tag{11.38}$$

and (11.35) can be weakened to

$$\inf_{j,n\in\{1,2\}} \inf_{k\in\mathbb{R}} |\omega_j(k) - \omega_2(-k_0) - \omega_n(k - k_0)| \geq C > 0. \tag{11.39}$$

11.5.2. Systems with a trivial resonance at $k = 0$. In physical systems with conserved quantities very often the eigenvalue zero occurs at the wave number $k = 0$, i.e., $\omega_j(0) = 0$ for a j in some index set. The most prominent example is the so called water wave problem, cf. §12.2.1, for which the NLS equation has been derived first [**Zak68**]. The fact $\omega_j(0) = 0$ will violate the non-resonance condition (11.34) and will always lead to quadratic resonances which at a first view will not allow to remove the quadratic terms. However, due to the fact that the eigenvalue zero is created by a conserved quantity also the nonlinear terms vanish at the wave number $k = 0$. The simplest example for such a system is the so called Boussinesq equation

$$\partial_t^2 u = \partial_x^2 u + \partial_x^2\partial_t^2 u + \partial_x^2(u^2), \tag{11.40}$$

with $x \in \mathbb{R}$, $t \in \mathbb{R}$, and $u(x,t) \in \mathbb{R}$. This model occurs as a long wave limit for the water wave problem, cf. Chapter 12. The linearized problem possesses solutions $e^{ikx+i\omega t}$ with dispersion relation

$$\omega^2 = \frac{k^2}{1+k^2}. \tag{11.41}$$

Hence, the non-resonance condition (11.38) is not satisfied as can be seen in Figure 11.5. The resonance at the wave number $k=0$ is trivial since the nonlinear terms vanish at the wave number $k=0$, too. However, the resonance at the wave number $k=k_0$ is non-trivial. One possibility to get rid of this second resonance is a k-dependent scaling of the error function.

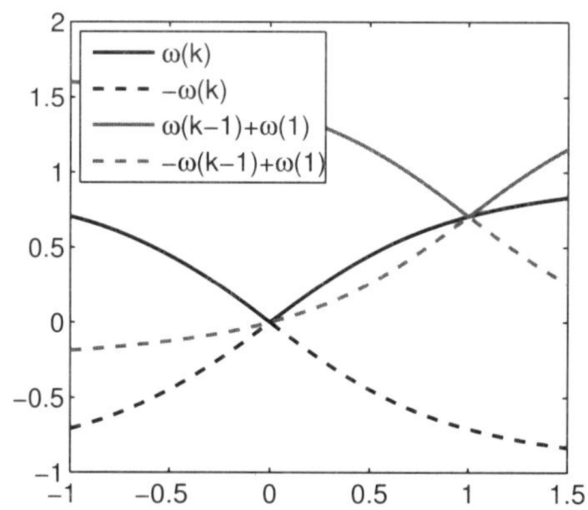

Figure 11.5. Intersection of the curves $k \mapsto \pm\omega(k)$ and $k \mapsto \omega(k_0) \pm \omega(k-k_0)$, here with $k_0 = 1$. There are two intersections at $k=0$ and $k=k_0$.

Before we explain this in more detail we derive the NLS equation for the Boussinesq model (11.40). As before we make the ansatz

$$\begin{aligned} \varepsilon\psi_{\text{NLSbouss}} =& \varepsilon A_1\left(\varepsilon(x-c_gt),\varepsilon^2 t\right) e^{i(k_0x+\omega_0 t)} + \text{c.c.} \\ &+ \varepsilon^2 A_2\left(\varepsilon(x-c_gt),\varepsilon^2 t\right) e^{2i(k_0x+\omega_0 t)} + \text{c.c.} \\ &+ \varepsilon^2 A_0\left(\varepsilon(x-c_gt),\varepsilon^2 t\right). \end{aligned} \tag{11.42}$$

We find as before at $\varepsilon\mathbf{E}$ the linear dispersion relation (11.41) and at $\varepsilon^2\mathbf{E}$ the condition for the linear group velocity c_g. At $\varepsilon^3\mathbf{E}$ we find

$$2i\omega_0(1+k_0^2)\partial_T A_1 = (1-c_g^2-\omega_0^2)\partial_X^2 A_1 - 2k_0^2(A_1A_0 + A_2A_{-1}).$$

The algebraic relations which are found at $\varepsilon^4\mathbf{E}^0$ and $\varepsilon^2\mathbf{E}^2$

$$\begin{aligned} \varepsilon^4\mathbf{E}^0: &\quad 0 = (1-c_g^2)^2\partial_X^2 A_0 + 2\partial_X^2 A_1A_{-1}, \\ \varepsilon^2\mathbf{E}^2: &\quad 0 = -(-4\omega_0^2 + 4k_0^2 + 16\omega_0^2k_0^2)A_2 - 4k_0^2A_1^2, \end{aligned}$$

can be solved w.r.t. A_0 since $(1-c_g^2)^2 \neq 0$ and w.r.t. A_2 since $-4\omega_0^2+4k_0^2+16\omega_0^2k_0^2 \neq 0$. Inserting the solutions for A_0 and A_2 into the equation for A_1 finally yields the NLS equation

$$2\mathrm{i}\omega_0(1+k_0^2)\partial_T A_1 = (1-c_g^2-\omega_0^2)\partial_X^2 A_1 + \gamma A_1 |A_1|^2, \tag{11.43}$$

with

$$\gamma = \frac{4k_0^2}{1-c_g^2} + \frac{k_0^4}{-\omega_0^2+k_0^2+4\omega_0^2k_0^2}.$$

The error $\varepsilon^\beta R = u - \varepsilon\Psi$ satisfies

$$\partial_t^2 R = \partial_x^2 R + \partial_x^2\partial_t^2 R + 2\varepsilon\partial_x^2(\Psi R) + \mathcal{O}(\varepsilon^2).$$

Writing this as a first order system gives two equations of the form

$$\partial_t \widehat{R}_j(k,t) = \mathrm{i}\omega_j(k)\widehat{R}_j(k,t) + \varepsilon\rho_j(k)\int_{\mathbb{R}} \widehat{\psi}(k-m,t)\widehat{R}_j(m,t)\,dm + \mathcal{O}(\varepsilon^2),$$

with $\omega_j(0) = \rho_j(0) = 0$, but non-vanishing $\omega_j'(0)$ and $\rho_j'(0)$. The approximation $\varepsilon\psi$ is concentrated at the wave numbers $\pm k_0$. Hence, a subsystem is given by

$$\begin{aligned}
\partial_t \widehat{R}_j(k_0,t) =& \mathrm{i}\omega_j(k_0)\widehat{R}_j(k_0,t) + \varepsilon\rho_j(k_0)\ \widehat{\psi}(k_0,t)\widehat{R}_j(0,t) + h.o.t.,\\
\partial_t \widehat{R}_j(0,t) =& \mathrm{i}\omega_j(0)\widehat{R}_j(0,t) + \varepsilon\rho_j(0)\widehat{\psi}(-k_0,t)\widehat{R}_j(k_0,t) + h.o.t..
\end{aligned}$$

For the second equation we have

$$\omega_j(0) - \omega_j(-k_0) - \omega_j(k_0) = 0 + \omega_j(k_0) - \omega_j(k_0) = 0$$

such that the non-resonance condition is not satisfied for $k=0$. However, we have $\rho_j(0) = 0$ such that the nonlinear terms vanish for this resonant wave number, too. Hence, this resonance is called trivial. Since

$$\omega_j(k) - \omega_j(-k_0) - \omega_j(k+k_0) = \mathcal{O}(|k|) \qquad \text{and} \qquad \rho_j(k) = \mathcal{O}(|k|),$$

with the denominator in the normal form transform also the nominator vanishes and therefore the quadratic nonlinear terms close to $k=0$ can be eliminated. For the first of these equations we have

$$\omega_j(-k_0) - \omega_j(k_0) - \omega_j(0) = \omega_j(k_0) - 0 - \omega_j(k_0) = 0$$

such that the non-resonance condition is not satisfied for $k = -k_0$ and similarly for $k = k_0$. However, we have $\rho_j(k_0) \neq 0$ such that the nonlinear terms do not vanish for this resonant wave number. Hence, this resonance is called non-trivial. One way to get rid of this difficulty is to scale $R_j(0)$ with $\varepsilon^{\beta+1}$ instead of ε^β. Doing so we obtain

$$\begin{aligned}
\partial_t \widehat{R}_j(k_0,t) =& \mathrm{i}\omega_j(k_0)\widehat{R}_j(k_0,t) + \varepsilon^2\rho_j(k_0)\widehat{\psi}(k_0,t)\widehat{R}_j(0,t) + h.o.t.,\\
\partial_t \widehat{R}_j(0,t) =& \mathrm{i}\omega_j(0)\widehat{R}_j(0,t) + \rho_j(0)\widehat{\psi}(-k_0,t)\widehat{R}_j(k_0,t) + h.o.t..
\end{aligned}$$

Hence, in the first equation the nonlinear terms do not make problems anymore. However, in the second equation we have now terms of order $\mathcal{O}(1)$ which can be eliminated in the full system with the argument from above. The terms of order $\mathcal{O}(\varepsilon)$ resulting from this transformation in the second equation can either be eliminated by another transformation or are of long wave form, i.e., of a similar form as $\varepsilon\partial_x(B(\varepsilon x)R(x))$ in physical space. Such terms can be estimated by energy estimates to have an $\mathcal{O}(\varepsilon^2)$ influence on the dynamics, cf. Chapter 12.

This can be made rigorous for the full system by making the ansatz

$$\varepsilon^{\beta}\vartheta(k)\widehat{R} = u - \varepsilon\Psi$$

for the error function R, where $\vartheta(k) = \min(\varepsilon+|k|/\delta, 1)$ with $\delta > 0$ sufficiently small, but independent of $0 < \varepsilon \ll 1$. This has been carried out in [**Sch98a**] with a correction explained in [**DS06**].

The above idea has been transferred to the water wave problem without surface tension in case of finite depth in [**DSW16**]. However, the water wave problem is a quasilinear problem. Below we will explain the additional difficulties occurring for such systems. But also the so called FPU system falls into this class, cf. Exercise 11.4. In [**Sch10**] it has been explained that the proofs of the approximation theorems given for the PDE systems can be transferred almost line for line to the FPU system by looking at the Fourier transformed FPU system.

11.5.3. Stable and unstable non-trivial resonances. For the water wave problem with small surface tension, i.e., for surface tension parameter $\sigma \in (0, 1/3)$ additional resonances are present in the system, i.e., there exist spatial wave numbers k_1, k_2, k_3 and associated temporal wave numbers ω_1, ω_2, ω_3 satisfying

$$k_1 + k_2 + k_3 = 0 \qquad \text{and} \qquad \omega_1 + \omega_2 + \omega_3 = 0,$$

cf. Exercise 11.6 and §12.2.1. The same happens for dispersive wave systems with spatially periodic coefficients, or the poly-atomic FPU model. There are at least two different approaches to get rid of these resonances.

Solutions of the NLS equation which are analytic in a strip in the complex plane decay with some exponential rate in Fourier space. By nonlinear interaction, the solutions of the original system have a Fourier mode distribution which is strongly localized at integer multiples of the basic wave number k_0. In between, the solutions will be exponentially small, i.e., the original system can be solved in a weighted L^1-space equipped with the norm

$$\|\widehat{u}\|_{L^1_w} = \int_{\mathbb{R}} |\widehat{u}(k)|w(k)\,dk$$

with

$$1/w(k) = \sup_{m\in\mathbb{Z}} |e^{-\alpha\varepsilon^{-1}|k-mk_0|}|. \tag{11.44}$$

Hence, the Fourier modes associated to the resonant wave numbers are exponentially small initially, i.e., of order $\mathcal{O}(\exp(-r\varepsilon^{-1}))$ for an $r > 0$, independent of $0 < \varepsilon \ll 1$. The quadratic resonances will lead to growth rates $\mathcal{O}(\exp(\varepsilon t))$ for solutions of order $\mathcal{O}(\varepsilon)$. Hence, it takes a time of order $\mathcal{O}(1/\varepsilon^2)$ to have $\mathcal{O}(\exp(-r\varepsilon^{-1}))\mathcal{O}(\exp(\varepsilon t)) = \mathcal{O}(1)$, i.e., it takes the time scale of the NLS equation for the error to grow to the order of the NLS approximation. This idea can be used to prove error estimates on an $\mathcal{O}(1/\varepsilon^2)$ time scale for the validity of the NLS approximation also in case of additional non-trivial resonances, if the solutions of the NLS equation are analytic in a strip in the complex plane, and if the set of wave numbers resonant to k_0 is separated from the set of integer multiples of the basic wave number k_0. This idea can be made rigorous by making the coefficient α time-dependent, i.e., by choosing $\alpha(t) = \alpha_0 - \check{\beta}\varepsilon^2 t$ in (11.44). This idea has been explained in [**Sch98c**] and carried out in [**DHSZ16**].

The second approach is based on a more detailed analysis of the resonances. Consider a basic wave number $k_1 = k_0$, resonant to wave numbers k_2 and k_3. The ansatz

$$u(x,t) = \varepsilon A_1(\varepsilon t)e^{i(k_1x+\omega_1 t)}\varphi_1 + \varepsilon A_2(\varepsilon t)e^{i(k_2x+\omega_2 t)}\varphi_2 + \varepsilon A_3(\varepsilon t)e^{i(k_3x+\omega_3 t)}\varphi_3,$$

with vectors φ_j, then yields a three wave interaction (TWI) system

$$\partial_T A_1 = i\gamma_1\overline{A_2A_3}, \qquad \partial_T A_2 = i\gamma_2\overline{A_1A_3}, \qquad \partial_T A_3 = i\gamma_3\overline{A_1A_2},$$

with coefficients $\gamma_j \in \mathbb{R}$, associated to the resonances, cf. [**Sch05**, §3.3]. In [**Sch05**, Theorem 3.8] a NLS approximation theorem has been shown in case that the subspace $\{A_2 = A_3 = 0\}$ associated to the wave number $k_1 = k_0$ is stable in the TWI system, cf. Figure 11.6. The proof is based on a mixture of normal form transforms for the non-resonant wave numbers and energy estimates for the resonant wave numbers. In [**DS06**] the ideas of [**Sch98a**] and [**Sch05**] are brought together to handle Boussinesq equations which model the water wave problem in case of small positive surface tension.

There is also a counter-example [**Sch05**, §4.1] showing that the NLS equation fails to approximate solutions in the original system in case of an unstable k_0-subspace in the associated TWI system and periodic boundary conditions in the original system. This idea has been carried out in [**SSZ15**] for the water wave problem with surface tension and periodic boundary conditions showing that there exists a continuum of wave numbers and values of surface tension where the NLS approximation does not make correct predictions.

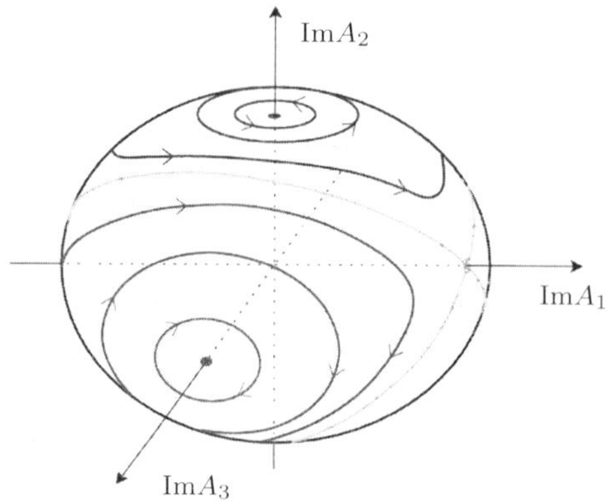

Figure 11.6. The phase portrait of the TWI system in the invariant subspace $\operatorname{Re} A_1 = \operatorname{Re} A_2 = \operatorname{Re} A_3 = 0$. The energy surface is an ellipsoid since due to conservation of energy not all j have the same sign. The axes are invariant subspaces associated to the wave numbers k_j. There are one unstable and two stable subspaces.

The situation on the whole real line for an unstable resonance is still open. In this case the different group velocities $\omega'(k_j)$ at the resonant wave numbers k_j no longer can be neglected. For a thorough discussion see [**Sch05**, §4.2]. A recent attempt to understand this situation can be found in [**MN13**].

11.5.4. Quasilinear quadratic nonlinearities. The first and very general NLS approximation theorem was proved in [**Kal88**] for general quasilinear dispersive wave systems. However, the occurrence of quasilinear quadratic terms has been excluded explicitly. A typical example is given by the quasilinear wave equation

$$\partial_t^2 u = \partial_x^2 u - u + \frac{1}{2}\partial_x^2(u^2) + u^3, \tag{11.45}$$

with $x \in \mathbb{R}$, $t \in \mathbb{R}$, and $u(x,t) \in \mathbb{R}$. With the NLS ansatz (11.42) we obtain the linear dispersion relation $\omega_0^2 = k_0^2 + 1$, the group velocity $c_g = k_0/\omega_0$, and the NLS equation,

$$i\nu_1 \partial_T A + \nu_2 \partial_X^2 A + \nu_3 A|A|^2 = 0, \tag{11.46}$$

with

$$\nu_1 = 2\omega_0, \qquad \nu_2 = (1 - c^2), \qquad \nu_3 = \frac{9\alpha_c - 2k_0^4}{3}, \tag{11.47}$$

for the description of small spatio-temporal modulations of the underlying carrier wave $e^{i(k_0 x + \omega_0 t)}$. As above, the idea is to use a normal form transform to eliminate the quadratic terms. For (11.45) the quasilinear quadratic nonlinearity causes the required normal form transform to lose regularity such that the original quasilinear system can no longer be handled after applying the transformation (see §11.4). This can be seen as follows. Writing the error equations as first order system gives nonlinear terms with growth rates proportional to $|k|$ for $|k| \to \infty$. This results in the normal form transform in a nominator proportional to $|k|$ and in a denominator proportional to

$$\omega(k) - \omega(k_0) - \omega(k - k_0) = \mathcal{O}(1)$$

for $|k| \to \infty$. Thus, the normal form transform is of the form identity plus a term which is small but loses one derivative. Hence, the normal form transform can no longer be inverted with Neumann's series.

So far there are only a few quasilinear systems where approximation results for the NLS approximation could have been established. One example is where the right-hand side of the quasilinear dispersive wave system only loses half a derivative, i.e., a factor of $\sqrt{k}$ in Fourier space, as a result of the normal form transformation. In this case the elimination of the quadratic terms is still possible and the transformed system can be handled with the Cauchy-Kowalevskaya theorem [**SW11**]. The Lagrangian formulation of the water wave problem in case of finite depth and zero surface tension falls into this class, cf. [**DSW16**]. In case of zero surface tension and infinite depth such a result has been established in [**TW12, Tot15**] by finding a special transformation which allows to eliminate all quadratic terms for this particular system without loss of regularity. Another example is in the context of the KdV equation where the result can be obtained by applying the Miura transformation [**Sch11**], cf. Exercise 11.5. In [**CS13**] numerical evidence is given that the NLS approximation is valid for (11.45). Very recently it turned out that the solutions of the transformed quasilinear system can be estimated by more clever energy estimates [**Dül16, CW16, DH16**].

Although relevant systems can be handled with the last approach a validity theory for general dispersive wave systems with quasilinear quadratic terms is still an open problem.

11.6. Pulse dynamics in photonic crystals

One of the major goals of photonics is the construction of 'electronic' devices where the electrons are completely replaced by photons. Photonic crystals turned out to be a suitable tool for the construction of such devices. They consist of a dielectric material such as glass with a periodic structure with a period comparable to the wave length of light. Due to the periodic structure the linearized problem is no longer solved by Fourier modes, but by so called Bloch modes. The curves of eigenvalues plotted as a function over the Bloch wave numbers can now possess horizontal tangencies, i.e., vanishing group velocities. Thus, in principle, standing light pulses are possible. At the horizontal tangencies very often spectral gaps occur, i.e., there are temporal wave numbers for which the associated wave cannot travel through the photonic crystal. The first fact allows to use photonic crystals as optical storage the second fact to use them for filtering. We will explain the possibility of standing light pulses in photonic crystals by using the NLS approximation.

For simplicity we restrict ourselves again to a nonlinear wave equation

$$\partial_t^2 u = \chi_1 \partial_x^2 u - \chi_2 u + \chi_3 u^3, \tag{11.48}$$

but now with spatially periodic coefficients $\chi_j(x) = \chi_j(x+L)$ for a $L > 0$, $x \in \mathbb{R}$, $t \in \mathbb{R}$, and $u(x,t) \in \mathbb{R}$. Moreover, we assume $\chi_1(x) \geq \gamma_1 > 0$ and $\chi_2(x) \geq \gamma_2 > 0$.

11.6.1. The eigenvalue problem for photonic crystals.

We start by reviewing a number of well known results, cf. [**Eas73, Sca99**]. For notational simplicity assume here $L = 2\pi$. The linear problem

$$\partial_t^2 u = \chi_1 \partial_x^2 u - \chi_2 u$$

possesses solutions $u(x,t) = \mathrm{e}^{\mathrm{i}\omega t} v(x)$ where v satisfies

$$-\omega^2 v = \chi_1 \partial_x^2 v - \chi_2 v.$$

Uniformly bounded solutions are given by Bloch waves

$$v(x) = w(x)\mathrm{e}^{\mathrm{i}\ell x},$$

with $w(x) = w(x+2\pi)$ satisfying the eigenvalue problem

$$L(\ell)w = -\chi_1(\partial_x + \mathrm{i}\ell)^2 w + \chi_2 w = \omega^2 w.$$

With $w(x) = w(x+2\pi)$ also $w(x)\mathrm{e}^{\mathrm{i}nx}$ for $n \in \mathbb{Z}$ is 2π-periodic. Hence we can restrict ourselves to $\ell \in [-\frac{1}{2}, \frac{1}{2})$, the so called Brillouin zone.

Lemma 11.6.1. *The properties of $L(\ell)$ are as follows:*

a) $L(\ell)$ is self-adjoint in $L^2_{\mathrm{per}}(\chi_1^{-1}\,\mathrm{d}x)$.

b) $L(\ell)$ is positive definite.

c) $L(\ell)$ has discrete spectrum with ∞ the only accumulation point. All eigenvalues are real, semi-simple, and non-negative.

Proof. a) follows from

$$\begin{aligned}
(L(\ell)w_1, w_2)_{L^2(\chi_1^{-1}\,\mathrm{d}x)} &= \int_0^{2\pi} -\chi_1((\partial_x + \mathrm{i}\ell)^2 w_1)\overline{w}_2 \frac{1}{\chi_1}\,\mathrm{d}x + \int_0^{2\pi} (\chi_2 w_1)\overline{w}_2 \frac{1}{\chi_1}\,\mathrm{d}x \\
&= \int_0^{2\pi} ((\partial_x + \mathrm{i}\ell)w_1)\overline{(\partial_x + \mathrm{i}\ell)w_2}\,\mathrm{d}x + \int_0^{2\pi} \chi_2 w_1 \overline{w}_2 \frac{1}{\chi_1}\,\mathrm{d}x \\
&= (w_1, L(\ell)w_2)_{L^2(\chi_1^{-1}\,\mathrm{d}x)}.
\end{aligned}$$

b) follows from

$$(L(\ell)w, w)_{L^2(\chi_1^{-1}\,\mathrm{d}x)} = \int_0^{2\pi} |(\partial_x + \mathrm{i}\ell)w|^2 + \frac{\chi_2}{\chi_1}|w|^2 \,\mathrm{d}x > 0, \qquad (w \neq 0).$$

c) follows from the fact that for fixed ℓ the operator $L(\ell) : H^2_{\mathrm{per}}(\mathbb{R}/2\pi\mathbb{Z}) \to L^2_{\mathrm{per}}(\mathbb{R}/2\pi\mathbb{Z})$ is elliptic, i.e., $L(\ell)$ has a compact resolvent and so $L(\ell)$ has discrete spectrum with ∞ the only accumulation point. Due to the self-adjointness all eigenvalues are real and semi-simple. Due to the positive definiteness all eigenvalues are non-negative. □

Lemma 11.6.2. *Except of intersection points the curves of eigenvalues $\ell \mapsto \omega_n(\ell)$ of $L(\ell)$ are smooth.*

Proof. This follows by a simple perturbation argument, cf. [**Kat95**]. □

11.6.2. The computation of the eigenvalues. Here and the subsequent example we assume $L = 1$, hence $\ell \in [-\pi, \pi)$. The eigenvalue problem

$$\chi_1 \partial_x^2 u - \chi_2 u = -\lambda u \tag{11.49}$$

is a second order scalar ODE with spatially periodic coefficients. We write (11.49) as first order system

$$\begin{array}{rcl} \partial_x u(x) &=& v, \\ \partial_x v(x) &=& -s(x)\lambda u(x) + q(x)u(x), \end{array} \tag{11.50}$$

where $s(x) = 1/\chi_1(x)$ and $q(x) = \chi_2(x)/\chi_1(x)$. The fundamental matrix of (11.50) is denoted by $\Phi_\lambda = \Phi_\lambda(x, x_0)$ where $\Phi_\lambda(x_0, x_0) = I$. Floquet's theorem 2.1.17 shows that

$$\Phi_\lambda(x, x_0) = P_\lambda(x, x_0)\mathrm{e}^{(x-x_0)M_\lambda},$$

with $P_\lambda(x, x_0) = P_\lambda(x + 1, x_0)$ and a matrix M_λ, independent of x and x_0. Note that M_λ is not unique since $\mathrm{e}^{2\pi i n} = 1$ for $n \in \mathbb{Z}$. The eigenvalues of M_λ are the Floquet exponents. The two eigenvalues ρ_- and ρ_+ of the monodromy matrix $C_\lambda = \mathrm{e}^{M_\lambda}$ are the Floquet multipliers. Since the trace of the linear vector field on the right-hand side of (11.50) vanishes we have the conservation of the phase volume for (11.50) leading to $\rho_+\rho_- = 1$, cf. the proof of Theorem 4.1.3. Hence, the Floquet multipliers can be computed via the characteristic polynomial

$$\rho^2 - D(\lambda)\rho + 1 = 0$$

and are given by

$$\rho_\pm(\lambda) = \frac{1}{2}D(\lambda) \pm \frac{1}{2}\sqrt{(D(\lambda))^2 - 4},$$

where the trace of the monodromy matrix, $D(\lambda) = \mathrm{trace}\, C_\lambda$, is called the discriminant. We find that

a) if $|D(\lambda)| > 2$ then the Floquet multipliers $\rho_\pm(\lambda)$ are real. As a consequence the solutions have exponential growth or decay w.r.t. x.

b) if $|D(\lambda)| < 2$ then the Floquet multipliers $\rho_\pm(\lambda)$ are on the complex unit circle. As a consequence the solutions are uniformly bounded w.r.t. x.

c) if $|D(\lambda)| = 2$ then the Floquet multipliers $\rho_\pm(\lambda)$ are 1 or -1. In this case we have at most linear growth

Example 11.6.3. We consider the eigenvalue problem (11.49) with

$$s(x) = \chi_{[0,6/13]} + 16\chi_{(6/13,7/13)} + \chi_{[7/13,1]}(x \bmod 1)$$

and $q(x) = \mu \in \mathbb{R}$ from [**BCBLS11**]. For this choice the ODE can be solved and the monodromy matrix can be computed explicitly. We find for the discriminant

$$D(\lambda) = \frac{25}{8}\cos\left(\frac{16}{13}\sqrt{\lambda+\mu}\right) - \frac{9}{8}\cos\left(\frac{8}{13}\sqrt{\lambda+\mu}\right). \tag{11.51}$$

See Exercise 11.8. The graph $\lambda \mapsto D(\lambda)$ of the discriminant and the associated dispersion relation can be found in Figure 11.7 for $\mu = 0$. ⌋

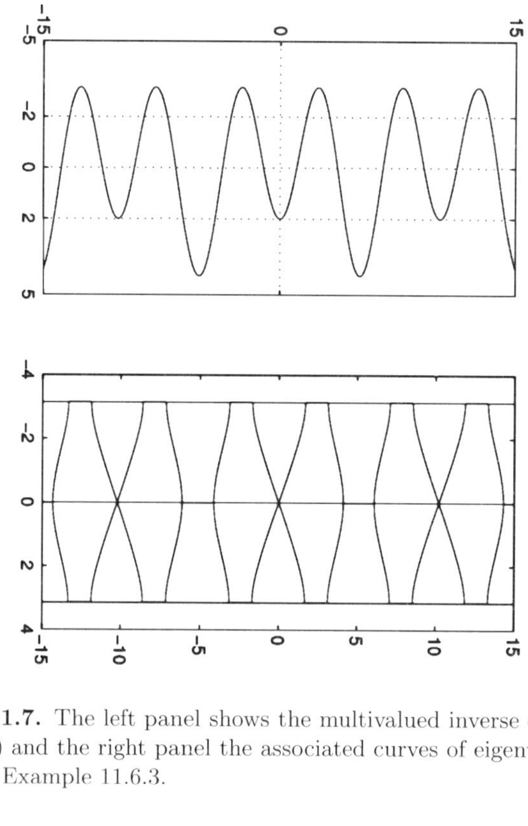

Figure 11.7. The left panel shows the multivalued inverse of the map $\lambda \mapsto D(\lambda)$ and the right panel the associated curves of eigenvalues $k \mapsto \omega^2(k)$ for Example 11.6.3.

For frequencies ω with $\omega^2 = \lambda$ which fall into a spectral gap the incoming wave is damped in the photonic crystal with some exponential rate w.r.t. the depth of penetration x. Hence, photonic crystals can be used as filters. This is one of the reasons why the wings of a butterfly show their colorful appearance.

Remark 11.6.4. There is a relation between the regularity of the coefficients χ_1 and χ_2 on one side and the size of the spectral gaps in

$$\{\omega_n(l) : l \in (-1/2, 1/2],\ n \in \mathbb{Z} \setminus \{0\}\} \subset \mathbb{R}$$

on the other side. For continuous or even smoother χ_1 and χ_2 the spectral gaps become smaller as n increases. According to [**Eas73, Nti76**], the size of the gaps which are found at $\omega \sim n$ decays at least with $1/n^{\theta+1}$ for $n \to \infty$ if $\chi_1 \in C_b^\theta$ and $\chi_2 \in C_b^{\theta-2}$, i.e., the more regular χ_1 and χ_2 are, the faster the gaps close.

Remark 11.6.5. Spectral gaps can be obtained from a spatially homogeneous situation by adding small spatially periodic perturbations to the coefficients. As an example consider the eigenvalue problem

$$(\partial_x + \mathrm{i}\ell)^2 w(x) + \omega^2(1 + 2\varepsilon\cos(2x))w(x) = 0,$$

with $w(x) = w(x + 2\pi)$, $\ell \in [-\frac{1}{2}, \frac{1}{2})$, and $0 \leq \varepsilon \ll 1$. For $\varepsilon = 0$ the problem is given by

$$(\partial_x + \mathrm{i}\ell)^2 w + \omega^2 w = 0,$$

and can be solved by $w(x) = \mathrm{e}^{\mathrm{i}nx}$ with $n \in \mathbb{Z}$ and associated eigenvalues $\omega_n^2(\ell) = (n + \ell)^2$. Hence, at $(\ell, \omega) = (0, 1)$ there is a crossing of the curves of eigenvalues. All single eigenvalues ω vary smoothly w.r.t. small ε since $2\varepsilon\cos(2x)\cdot$ is a small perturbation of the operator $(\partial_x + \mathrm{i}\ell)^2$. Hence, the smooth curves $\ell \mapsto \omega_n(\ell)$ will only vary slightly w.r.t. ε. However, at the crossing points the curves can split. As an example we consider the point $(\ell, \omega) = (0, 1)$. We use $2\cos 2x = \mathrm{e}^{2\mathrm{i}x} + \mathrm{e}^{-2\mathrm{i}x}$, make the ansatz $w(x) = \sum_{n \in 2\mathbb{Z}+1} c_n \mathrm{e}^{\mathrm{i}nx}$, and set $\omega^2 = 1 + \widetilde{\omega}^2$. We obtain

$$-(1 + \ell)^2 c_1 + (1 + \widetilde{\omega}^2)[c_1 + \varepsilon(c_3 + c_{-1})] = 0, \tag{11.52}$$

$$-(-1 + \ell)^2 c_{-1} + (1 + \widetilde{\omega}^2)[c_{-1} + \varepsilon(c_1 + c_{-3})] = 0, \tag{11.53}$$

$$-(3 + \ell)^2 c_3 + (1 + \widetilde{\omega}^2)[c_3 + \varepsilon(c_1 + c_5)] = 0, \tag{11.54}$$

$$\vdots$$

The equations (11.54)-... can be solved w.r.t. $c_3, c_{-3}, c_5, \ldots$, if $\ell, \widetilde{\omega}^2, \varepsilon$ are small, i.e., there exist functions

$$c_j = c_j(\ell, \widetilde{\omega}^2, \varepsilon, c_{-1}, c_1) = \mathcal{O}(|\ell| + |\widetilde{\omega}^2| + |\varepsilon|)\mathcal{O}(|c_1| + |c_{-1}|)$$

which are linear w.r.t. c_1 and c_{-1}. Inserting this into (11.52)-(11.53) gives

$$0 = (\mp 2\ell - \ell^2 + \widetilde{\omega}^2)c_{\pm 1} + \varepsilon(1+\widetilde{\omega}^2)c_{\mp 1} + \varepsilon\mathcal{O}(|\ell| + |\widetilde{\omega}|^2 + |\varepsilon|)\mathcal{O}(|c_1| + |c_{-1}|),$$

or equivalently,

$$\begin{pmatrix} -2\ell - \ell^2 + \widetilde{\omega}^2 + \text{h.o.t.} & (1+\widetilde{\omega}^2)\varepsilon + \text{h.o.t.} \\ (1+\widetilde{\omega}^2)\varepsilon + \text{h.o.t.} & 2\ell - \ell^2 + \widetilde{\omega}^2 + \text{h.o.t.} \end{pmatrix} \begin{pmatrix} c_1 \\ c_{-1} \end{pmatrix} = \begin{pmatrix} 0 \\ 0 \end{pmatrix}.$$

In order to have non-trivial solutions we need a vanishing determinant. We find

$$(\widetilde{\omega}^2)^2 - \varepsilon^2 + \text{h.o.t.} = 0,$$

i.e., $(\omega^2)_{1/2} = \pm\varepsilon$+h.o.t., i.e., there is a splitting of the eigenvalues. In higher space dimensions spectral gaps cannot be obtained by a small perturbation of the spatially homogeneous situation. For examples of spectral gaps in 2D and 3D see [**Kuc93, BFL$^+$07**]. ⌋

11.6.3. Bloch transform. To derive and justify the NLS equation for (11.48) we follow [**BSTU06**] and adapt the Fourier space approach for the constant coefficient problem (11.2) from §11.2 to Bloch space. For Schwartz functions $u \in \mathcal{S}$, the Bloch transform is defined by

$$\widetilde{u}(\ell, x) = (\mathcal{T}u)(\ell, x) = \sum_{j\in\mathbb{Z}} e^{ijx}\widehat{u}(\ell + j), \tag{11.55}$$

and its inverse by

$$u(x) = (\mathcal{T}^{-1}\widetilde{u})(x) = \int_{-1/2}^{1/2} e^{i\ell x}\widetilde{u}(\ell, x)\, d\ell. \tag{11.56}$$

By construction we have

$$\widetilde{u}(\ell, x) = \widetilde{u}(\ell, x + 2\pi) \quad \text{and} \quad \widetilde{u}(\ell, x) = \widetilde{u}(\ell + 1, x)e^{ix}. \tag{11.57}$$

The Bloch transform turns out to be an isomorphism between $H^\theta(\mathbb{R}, \mathbb{C})$ and $L^2((-1/2, 1/2], H^\theta([0, 2\pi), \mathbb{C}))$, cf. [**RS75b, Sca99**], where

$$\|\widetilde{u}\|_{L^2((-1/2,1/2],H^\theta([0,2\pi),\mathbb{C}))} = \left(\int_{-1/2}^{1/2} \|\widetilde{u}(\ell,\cdot)\|^2_{H^\theta[0,2\pi]} d\ell\right)^{1/2}.$$

Multiplication $u(x)v(x)$ in physical space corresponds in Bloch space to the operation

$$(\widetilde{u} \star \widetilde{v})(\ell, x) = \int_{-1/2}^{1/2} \widetilde{u}(\ell - m, x)\widetilde{v}(m, x)\, dm, \tag{11.58}$$

where (11.57) has to be used for $|\ell - m| > 1/2$. However, if $\chi : \mathbb{R} \to \mathbb{R}$ is 2π-periodic w.r.t. x, then

$$\mathcal{T}(\chi u)(\ell, x) = \chi(x)(\mathcal{T}u)(\ell, x).$$

Applying the Bloch transform to (11.48) gives

$$\partial_t^2 \widetilde{u}(\ell,x) = -\widetilde{L}(\ell,\partial_x)\widetilde{u}(\ell,x) + \chi_3(x)\widetilde{u}^{\star 3}(\ell,x), \tag{11.59}$$

where the Bloch operators $\widetilde{L}(\ell,\partial_x) : H^2([0,2\pi)) \to L^2([0,2\pi))$ are given by

$$\widetilde{L}(\ell,\partial_x)\widetilde{u}(\ell,\cdot)(x) = -\chi_1(x)(\partial_x + \mathrm{i}\ell)^2\widetilde{u}(\ell,x) + \chi_2(x)\widetilde{u}(\ell,x).$$

According to Lemma 11.6.1 for fixed ℓ these operators are self-adjoint and positive definite in the space $L^2(\chi_1^{-1}\,\mathrm{d}x)$. The induced norm $\|\cdot\|_{L^2(\chi_1^{-1}\,\mathrm{d}x)}$ and the usual L^2-norm are equivalent since $\chi_1(x) \geq \gamma_1 > 0$ for a constant γ_1 independent of x by assumption. Thus, for each fixed ℓ there exists a Schauder basis $(f_j(\ell,\cdot))_{j\in\mathbb{N}}$ of $L^2([0,2\pi))$ of eigenfunctions of $\widetilde{L}(\ell,\partial_x)$ with strictly positive eigenvalues $\lambda_j(\ell) > 0$, i.e., $\widetilde{L}(\ell,\partial_x)f_j(\ell,\cdot) = \lambda_j(\ell)f_j(\ell,\cdot)$. We make the ansatz

$$\widetilde{u}(\ell,x,t) = \sum_{j\in\mathbb{N}} \widetilde{u}_j(\ell,t)f_j(\ell,x).$$

Since $\widetilde{L}(\ell,\partial_x)$ is self-adjoint in $L^2(\chi_1^{-1}\,\mathrm{d}x)$, the eigenfunctions $(f_j(\ell,\cdot))_{j\in\mathbb{N}}$ can be chosen to form an orthonormal basis of $L^2(\chi_1^{-1}\,\mathrm{d}x)$ for each fixed ℓ. Hence,

$$\widetilde{u}_j(\ell,t) = \langle f_j(\ell,\cdot), \widetilde{u}(\ell,\cdot,t)\rangle_{L^2(\chi_1^{-1}\,\mathrm{d}x)},$$

and therefore

$$\begin{aligned}\partial_t^2\widetilde{u}_j(\ell,t) &= -\lambda_j(\ell)\widetilde{u}_j(\ell,t) + \langle f_j(\ell,\cdot), \chi_3(\cdot)\widetilde{u}^{\star 3}(\ell,\cdot,t)\rangle_{L^2(\chi_1^{-1}\,\mathrm{d}x)} \\ &= -\lambda_j(\ell)\widetilde{u}_j(\ell,t) + \sum_{j_1,j_2,j_3\in\mathbb{N}} \int_{-1/2}^{1/2}\int_{-1/2}^{1/2} b^j_{j_1j_2j_3}(\ell,\ell-\ell_1,\ell_1-\ell_2,\ell_2) \\ &\qquad \times \widetilde{u}_{j_1}(\ell-\ell_1,t)\widetilde{u}_{j_2}(\ell_1-\ell_2,t)\widetilde{u}_{j_3}(\ell_2,t)\,\mathrm{d}\ell_2\,\mathrm{d}\ell_1,\end{aligned} \tag{11.60}$$

where

$$\begin{aligned}&b^j_{j_1j_2j_3}(\ell,\ell-\ell_1,\ell_1-\ell_2,\ell_2) \\ &\quad= \langle f_j(\ell,\cdot), \chi_3(\cdot)f_{j_1}(\ell-\ell_{j_1},\cdot)f_{j_2}(\ell_{j_1}-\ell_{j_2},\cdot)f_{j_3}(\ell_{j_2}-,\cdot)\rangle_{L^2(\chi_1^{-1}\,\mathrm{d}x)}.\end{aligned}$$

Since the nonlinear terms have some convolution structure we have a system as in §11.3. Thus, we can proceed exactly as in §11.3 in order to derive a NLS equation. However, due to the special structure of (11.60) we keep the second order system and proceed as in §11.2 and make the ansatz

$$\widetilde{u}_{n_0}(\ell,t) = \varepsilon\varepsilon^{-1}\widetilde{A}_1\left(\frac{\cdot-\ell_0}{\varepsilon},\varepsilon^2 t\right)\mathbf{E}^1 + \varepsilon\varepsilon^{-1}\widetilde{A}_{-1}\left(\frac{\cdot+\ell_0}{\varepsilon},\varepsilon^2 t\right)\mathbf{E}^{-1}$$

for a $n_0 \in \mathbb{N}$ where $\mathbf{E}^j = e^{j\mathrm{i}\omega_{n_0}(\ell_0)t}e^{\mathrm{i}\omega'_{n_0}(\ell_0)(\ell-j\ell_0)t}$. In physical space the ansatz corresponds to

$$u(x,t) = \varepsilon\psi_A(x,t) = \varepsilon A(\varepsilon(x+c_g t),\varepsilon^2 t)f_{n_0}(\ell_0,x)e^{\mathrm{i}\ell_0 x}e^{\mathrm{i}\omega_{n_0}(\ell_0)t} + \text{c.c.}, \tag{11.61}$$

where again $T = \varepsilon^2 t$, $X = \varepsilon(x + c_g t)$, and $A(X,T) \in \mathbb{C}$. We have a cancelation at $\varepsilon\varepsilon^{-1}\mathbf{E}$ and $\varepsilon^2\varepsilon^{-1}\mathbf{E}$, and at $\varepsilon^3\varepsilon^{-1}\mathbf{E}$ we find the NLS equation

$$\begin{aligned} 2\mathrm{i}\omega_{n_0}(\ell_0)\partial_T\widetilde{A}_1 &= -(\lambda''_{n_0}(\ell_0) - 2(\omega'_{n_0}(\ell_0))^2)\kappa^2\widetilde{A}_1/2 \\ &\quad +\gamma\widetilde{A}_1 * \widetilde{A}_1 * \widetilde{A}_{-1}, \end{aligned} \tag{11.62}$$

where $T = \varepsilon^2 t$, $\kappa = \varepsilon^{-1}(\ell - \ell_0)$, and $\gamma = 3b^{n_0}_{n_0 n_0 n_0}(\ell_0, 0, 2\ell_0, -\ell_0)$, i.e.,

$$\gamma = \frac{3}{2\omega_{n_0}(\ell_0)}\int_0^{2\pi}\frac{\chi_3(x)}{\chi_1(x)}|f_{n_0}(\ell_0, x)|^4\,\mathrm{d}x \in \mathbb{R}, \tag{11.63}$$

while $A_{-1} = \mathcal{F}^{-1}\widetilde{A}_{-1}$ satisfies the complex conjugate equation. In order to obtain (11.62), in the convolution term we use

$$\begin{aligned} &\int_{-1/(2\varepsilon)}^{1/(2\varepsilon)}\int_{-1/(2\varepsilon)}^{1/(2\varepsilon)} b^{n_0}_{n_0 n_0 n_0}(\ell_0 + \varepsilon\kappa, \varepsilon(\kappa - \kappa_1), 2\ell_0 + \varepsilon(\kappa - \kappa_1), -\ell_0 + \varepsilon\kappa_2)) \\ &\qquad\qquad \times \widetilde{A}_1(\kappa - \kappa_1)\widetilde{A}_1(\kappa_1 - \kappa_2)\widetilde{A}_{-1}(\kappa_2)\,\mathrm{d}\kappa_2\,\mathrm{d}\kappa_1 \\ &\to b^{n_0}_{n_0 n_0 n_0}(\ell_0, 0, 2\ell_0, -\ell_0)\int_{-\infty}^{\infty}\int_{-\infty}^{\infty}\widetilde{A}_1(\kappa - \kappa_1)\widetilde{A}_1(\kappa_1 - \kappa_2)\widetilde{A}_{-1}(\kappa_2)\,\mathrm{d}\kappa_2\,\mathrm{d}\kappa_1 \end{aligned}$$

and the symmetry of the kernel. The derivation from (11.62) is consistent with the derivation from the associated first order system since for instance

$$-(\lambda''_{n_0}(\ell_0) - 2(\omega'_{n_0}(\ell_0))^2)/(4\mathrm{i}\omega_{n_0}) = -\left((\omega_{n_0}(\ell_0)^2)'' - 2(\omega'_{n_0}(\ell_0))^2\right)/(4\mathrm{i}\omega_{n_0})$$

which equals $\mathrm{i}\omega''(\ell_0)/2$.

11.6.4. An approximation result. The derivation of the NLS equation, the fact that the NLS equation possesses standing pulses, and the possibility of vanishing group velocities, in principle gives the possibility of standing light pulses. In order to show that these standing light pulses exist at least on an $\mathcal{O}(1/\varepsilon^2)$ time interval we again prove an approximation result.

We consider the following situation. We assume that one of the curves of eigenvalues, namely λ_{n_0}, has a horizontal tangency and that there is no other curve in a neighborhood of this point in the (ℓ, λ)-plane. Hence, $\lambda' = 2\omega\omega' = 0$ in this point and so we have a vanishing group velocity for an associated modulated wave packet. For simplicity we assume that $\lambda_1'(0) = 0$ with $\lambda_1(0) > 0$.

In order to bring together the NLS equation whose solutions in Fourier space are given on the complete real line with the Bloch wave representation of the nonlinear wave equation we introduce a cut-off operator $\chi \in C_0^\infty$ with $\chi(\ell) \in [0, 1]$, $\chi(\ell) = 1$ for $\ell \in [-1/5, 1/5]$, and $\chi(\ell) = 0$ for $|\ell| \geq 2/5$, and an extension operator $\mathcal{P}$ which extends a function with length of support less than 1 to a function on the complete real axis with period 1. Then we make the modified ansatz

$$\widetilde{u}_1(\ell,t) = \varepsilon\varepsilon^{-1}\mathcal{P}(\chi(\cdot)\widetilde{A}_1\left(\frac{\cdot}{\varepsilon},\varepsilon^2 t\right)\mathbf{E}^1)(\ell) + \varepsilon\varepsilon^{-1}\mathcal{P}(\chi(\cdot)\widetilde{A}_{-1}\left(\frac{\cdot}{\varepsilon},\varepsilon^2 t\right)\mathbf{E}^{-1})(\ell)$$

and find the NLS equation

$$2\mathrm{i}\omega_1(0)\partial_T\widetilde{A}_1 = -(\lambda_1''(0) - 2(\omega_1'(0))^2)\kappa^2\widetilde{A}_1/2 + \gamma\widetilde{A}_1 * \widetilde{A}_1 * \widetilde{A}_{-1}, \tag{11.64}$$

where now $\gamma = 3\widetilde{b}^1_{111}(0,0,0,0) \in \mathbb{R}$, while $A_{-1} = \mathcal{F}^{-1}\widetilde{A}_{-1}$ satisfies again the complex conjugate equation. In case of a cubic nonlinearity for the proof of the approximation result it is not necessary to expand the problem in eigenfunctions, and so we set

$$\widetilde{u}(\ell,x,t) = \widetilde{u}_1(\ell,t)f_1(\ell,x) + \widetilde{u}^{\perp}(\ell,x,t),$$

with $\langle (f_1(\ell,\cdot),\widetilde{u}^{\perp}(\ell,\cdot,t)\rangle_{L^2(\chi_1^{-1}\,\mathrm{d}x)} = 0$. Moreover, the separation will be made in such a way that the support of $\widetilde{u}_1(\ell,t)$ is contained in $[-2/5,2/5]$. The two functions $\widetilde{u}_1(\ell,t)$ and $\widetilde{u}^{\perp}(\ell,x,t)$ are defined to satisfy

$$\begin{aligned}
\partial_t^2\widetilde{u}_1(\ell,t) &= -\lambda_1(\ell)\widetilde{u}_1(\ell,t) + E_c(\ell)\langle f_1(\ell,\cdot),\ \chi_3(\cdot)\widetilde{u}^{\star 3}(\ell,\cdot,t)\rangle_{L^2(\chi_1^{-1}\,\mathrm{d}x)},\\
\partial_t^2\widetilde{u}^{\perp}(\ell,x,t) &= -\widetilde{L}(\ell,\partial_x)\widetilde{u}^{\perp}(\ell,x,t) + \chi_3(x)\widetilde{u}^{\star 3}(\ell,x)\\
&\quad - E_c(\ell)\langle f_1(\ell,\cdot),\ \chi_3(\cdot)\widetilde{u}^{\star 3}(\ell,\cdot,t)\rangle_{L^2(\chi_1^{-1}\,\mathrm{d}x)}f_1(\ell,x),
\end{aligned}$$

where $E_c \in C_0^\infty$, with $E_c(\ell) \in [0,1]$, with $E_c(\ell) = 1$ for $\ell \in [-1/5,1/5]$, and $\chi(\ell) = 0$ for $|\ell| \geq 2/5$. Note that χ is defined on the Fourier wave numbers, where E_c is defined on the Bloch wave numbers.

We add higher order terms to the ansatz to make the residual smaller, i.e., we consider

$$\begin{aligned}
\widetilde{u}_1(\ell,t) =&\varepsilon\varepsilon^{-1}\mathcal{P}(\chi(\cdot)\widetilde{A}_1\left(\frac{\cdot}{\varepsilon},\varepsilon^2 t\right)\mathbf{E}^1)(\ell) + \varepsilon\varepsilon^{-1}\mathcal{P}(\chi(\cdot)\widetilde{A}_{-1}\left(\frac{\cdot}{\varepsilon},\varepsilon^2 t\right)\mathbf{E}^{-1})(\ell)\\
&+ \varepsilon^3\varepsilon^{-1}\mathcal{P}(\chi(\cdot)\widetilde{A}_3\left(\frac{\cdot}{\varepsilon},\varepsilon^2 t\right)\mathbf{E}^3)(\ell)\\
&+ \varepsilon^3\varepsilon^{-1}\mathcal{P}(\chi(\cdot)\widetilde{A}_{-3}\left(\frac{\cdot}{\varepsilon},\varepsilon^3 t\right)\mathbf{E}^{-3})(\ell),\\
\widetilde{u}^{\perp}(\ell,x) =&\varepsilon^3\varepsilon^{-1}\widetilde{u}_1^{\perp}(\frac{\ell}{\varepsilon},x,\varepsilon^2 t)\mathbf{E}^1 + \varepsilon^3\varepsilon^{-1}\widetilde{u}_{-1}^{\perp}(\frac{\ell}{\varepsilon},x,\varepsilon^2 t)\mathbf{E}^{-1}\\
&+ \varepsilon^3\varepsilon^{-1}\widetilde{u}_3^{\perp}(\frac{\ell}{\varepsilon},x,\varepsilon^2 t)\mathbf{E}^3 + \varepsilon^3\varepsilon^{-1}\widetilde{u}_{-3}^{\perp}(\frac{\ell}{\varepsilon},x,\varepsilon^2 t)\mathbf{E}^{-3}.
\end{aligned}$$

As before we find $\widetilde{A}_1$ as a solution of the NLS equation (11.64) and $\widetilde{A}_{-1}$ as a solution of the complex conjugate equation. Moreover we choose

$$\begin{aligned}
&-9\omega_1^2(0)\widetilde{A}_3(\kappa,T)\\
&\qquad= -\lambda_1(0)\widetilde{A}_3(\kappa,T) + \langle f_1(0,\cdot),\ \chi_3(\cdot)(f_1(0,\cdot))^3\rangle_{L^2(\chi_1^{-1}\,\mathrm{d}x)}\widetilde{A}_1^{*3}(\kappa,T),\\
&-\omega_1^2(0)\widetilde{u}_1^{\perp}(\ell,x,t)
\end{aligned}$$

$$
\begin{aligned}
&= - \widetilde{L}(\varepsilon\kappa,\partial_x)\widetilde{u}_1^{\perp}(\kappa,x,T) + 3\chi_3(x)(\widetilde{A}_1 f_1)^{*2} * \widetilde{A}_{-1} f_{-1})(\kappa,x) \\
&\quad - E_c(\ell)\langle f_1(\varepsilon\kappa,\cdot),\, 3\chi_3(\cdot)(\widetilde{A}_1 f_1)^{*2} * \widetilde{A}_{-1} f_{-1})(\kappa,\cdot,T)\rangle_{L^2(\chi_1^{-1}\,\mathrm{d}x)} f_1(\varepsilon\kappa,x),
\end{aligned}
$$

$$
\begin{aligned}
-9\omega_1^2(0)&\widetilde{u}_3^{\perp}(\ell,x,t) \\
&= - \widetilde{L}(\varepsilon\kappa,\partial_x)\widetilde{u}_3^{\perp}(\kappa,x,T) + \chi_3(x)(\widetilde{A}_1 f_1 * \widetilde{A}_1 f_1 * \widetilde{A}_1 f_1)(\kappa,x) \\
&\quad - E_c(\ell)\langle f_1(\varepsilon\kappa,\cdot),\, \chi_3(\cdot)(\widetilde{A}_1 f_1)^{*3}(\kappa,\cdot,T)\rangle_{L^2(\chi_1^{-1}\,\mathrm{d}x)} f_1(\varepsilon\kappa,x),
\end{aligned}
$$

and $\widetilde{A}_{-3}$, $\widetilde{u}_{-1}^{\perp}$, and $\widetilde{u}_{-3}^{\perp}$ as the solutions of the associated complex conjugate equations. By this choice we eliminate all nonlinear terms w.r.t. $\widetilde{A}_1$ and $\widetilde{A}_{-1}$ in the part of the residual belonging to the $u^{\perp}$ equation. This has the advantage that we do not need more information about the spectrum and the associated eigenfunctions, especially we avoid to estimate derivatives of the eigenfunctions w.r.t. ℓ.

In order to solve the equations for $\widetilde{A}_3$, $\widetilde{u}_1^{\perp}$, and $\widetilde{u}_3^{\perp}$ we need the non-resonance conditions

$$
\begin{aligned}
-9\omega_1^2(0) &\neq -\lambda_1(0), \\
-\omega_1^2(0) &\notin \mathrm{spec}(-\widetilde{L}(\varepsilon\kappa,\partial_x))|_{\{f_1(\varepsilon\kappa,\cdot)\}^{\perp}}, \\
-9\omega_1^2(0) &\notin \mathrm{spec}(-\widetilde{L}(\varepsilon\kappa,\partial_x))|_{\{f_1(\varepsilon\kappa,\cdot)\}^{\perp}}.
\end{aligned}
$$

If these conditions are not satisfied for $\ell = \varepsilon\kappa$ for which $E_c(\ell) = 1$ we can make the support of E_c smaller to satisfy the non-resonance conditions. This is possible if

$$
\begin{aligned}
-\omega_1^2(0) &\notin \mathrm{spec}(-\widetilde{L}(0,\partial_x))|_{\{f_1(0,\cdot)\}^{\perp}}, \\
-9\omega_1^2(0) &\notin \mathrm{spec}(-\widetilde{L}(0,\partial_x))|_{\{f_1(0,\cdot)\}^{\perp}}.
\end{aligned}
$$

These formal calculations can be made rigorous, as in §11.2 it can be shown that the residual is of order $\mathcal{O}(\varepsilon^{7/2})$ in H^1 in physical space.

The error estimates can be proved in physical space. Let $\varepsilon\Psi$ be the approximation in physical space satisfying

$$
\sup_{t\in[0,T_0/\varepsilon^2]} \|\mathrm{Res}(\varepsilon\Psi)\|_{H^1} \leq C\varepsilon^{7/2}
$$

where

$$
\mathrm{Res}(u) = -\partial_t^2 u + \chi_1\partial_x^2 u - \chi_2 u + \chi_3 u^3.
$$

The error $\varepsilon^{3/2}R = u - \varepsilon\psi$ satisfies

$$
\partial_t^2 R = \chi_1\partial_x^2 R - \chi_2 R + 3\varepsilon^2\chi_3\psi^2 R + 3\varepsilon^{5/2}\chi_3\psi R^2 + \varepsilon^3\chi_3 R^3 + \varepsilon^{-3/2}\mathrm{Res}(\varepsilon\Psi).
$$

Multiplying this equation with $\chi_1^{-1}\partial_t R$ and performing the integration $\int_{-\infty}^{\infty}\ldots dx$ yields

$$
\partial_t E \leq C_1\varepsilon^2 E + C_2\varepsilon^{5/2}E^{3/2} + C_3\varepsilon^3 E^2 + C_4\varepsilon^2,
$$

with ε-independent constants $C_1, \ldots, C_4$ and where

$$E = \int_{-\infty}^{\infty} \chi_1^{-1}(\partial_t R)^2 + (\partial_x R)^2 + \chi_1^{-1}\chi_2 R^2 dx.$$

A simple application of Gronwall's inequality yields as in §11.2 the following approximation theorem.

Theorem 11.6.6. *Let the Fourier transform of $A \in C([0, T_0], H^4)$ be a solution of the NLS equation* (11.64)*. Then there exists an $\varepsilon_0 > 0$ and a $C > 0$ such that for all $\varepsilon \in (0, \varepsilon_0)$ there are solutions u of the original system* (11.48) *which can be approximated by $\varepsilon\Psi$ w.r.t. the H^1-norm such that*

$$\sup_{t\in[0,T_0/\varepsilon^2]} \|u(t) - \varepsilon\Psi(t)\|_{H^1} < C\varepsilon^{3/2}.$$

If the NLS equation (11.64) is a focussing one, then it possesses standing time-periodic pulse solutions, see §8.1.1. This in combination with Theorem 11.6.6 shows the existence of approximate standing light pulses on an $\mathcal{O}(1/\varepsilon^2)$ time scale. Since we have a finite speed of propagation for (11.48) with the help of spatial dynamics and invariant manifold theory, cf. Chapter 13, the existence of the standing modulating pulses can even be shown on $\mathcal{O}(1/\varepsilon^n)$ time scales for fixed $n > 2$ if certain non-resonance conditions are satisfied [**LBCB$^+$09**]. If the coefficients χ_1 and χ_2 are chosen in a special way, then standing modulating pulses can be shown [**BCBLS11**] to exist for all $t \in \mathbb{R}$.

11.6.5. Gap solitons. Another popular class of models for light propagation in photonic crystals starts directly with a Nonlinear Schrödinger (or Gross–Pitaevsky) equation in the form

$$\text{(11.65)}\qquad \mathrm{i}E_t = -\Delta E + V(x)E + \sigma|E|^2E, \quad V : \mathbb{R}^2 \to \mathbb{R},\ x \in \mathbb{R}^2,\ t \in \mathbb{R},$$

with $E = E(x, t) \in \mathbb{C}$, $\sigma = \pm 1$, and where w.l.o.g. the potential V is 2π–periodic in each coordinate. Here E is the "out–of–plane" electric field, i.e., perpendicular to the 2D periodicity structure, and the amplitude of V is called the contrast of the material. An interesting problem then is to search for so called gap solitons $E(x, t) = \phi(x)\mathrm{e}^{-\mathrm{i}\omega t}$, where $\phi \in \mathbb{C}$ is a localized solution of the stationary problem

$$\text{(11.66)}\qquad (-\Delta + V(x) - \omega)\phi + \sigma|\phi|^2\phi = 0.$$

Localized means $|\phi(x)| \to 0$ exponentially as $|x| \to 0$, and this implies that ω has to lie in a gap of the essential spectrum of the operator $L := -\Delta + V(x)$, hence the name "gap soliton". These have been discussed the physics literature since the early 1990ties, see, e.g., [**Ace00**] for a review. In 1D, where gaps open for small contrasts as discussed above, they are typically described by so called coupled mode equations, cf. Exercise 11.9. In two and more space dimensions gaps are more difficult to open [**Kuc93**], and in particular one needs a finite contrast. Rigorous proofs and asymptotics for 2D gap solitons, based on the reduction of (11.65) to systems of homogeneous NLS equations, here also called coupled mode equations, can be found for instance in [**DPS09**] (for the case of a separable potential) and in [**DU09**] (for general V).

11.7. Nonlinear optics

It is the purpose of this section to explain how the previous analysis is related to the motivation given at the beginning of this Chapter, namely the transport of information by light pulses through glass fibers. Moreover, we explain why the rate of information transported through the fiber can be increased by multiplexing, i.e., by taking simultaneously pulses with different carrier waves.

11.7.1. Maxwell's equations in glass fibers. Electromagnetic waves are described by Maxwell's equations. For $x = (x_1, x_2, x_3) \in \mathbb{R}^3$ and $t \in \mathbb{R}$ they are given by

$$\nabla \cdot E = \rho, \tag{11.67}$$

$$\nabla \times E = -\partial_t B, \tag{11.68}$$

$$\nabla \times B = \partial_t E + j, \tag{11.69}$$

$$\nabla \cdot B = 0, \tag{11.70}$$

where $E = E(x,t) \in \mathbb{R}^3$ is the electric field, $B = B(x,t) \in \mathbb{R}^3$ the magnetic field, $j = j(x,t) \in \mathbb{R}^3$ an electric current density, and $\rho = \rho(x,t) \in \mathbb{R}$ an electric charge density. There is conservation of charges, i.e., additionally we have $\partial_t \rho + \nabla \cdot j = 0$. There are no magnetic monopoles, i.e., (11.67)-(11.70) are not symmetric, since there is no magnetic current density and no magnetic charge density. We put the vacuum velocity of light and the vacuum electric permittivity to one.

Remark 11.7.1. For some purposes, alternative formulations of Maxwell's equations are useful. For time-independent solutions the system decouples into electrostatics

$$\nabla \cdot E = \rho, \quad \nabla \times E = 0$$

and magnetostatics

$$\nabla \times B = j, \quad \nabla \cdot B = 0.$$

Hence, there exist a scalar potential ϕ with $E = \nabla\phi$ satisfying $\Delta\phi = \rho$ and a vector potential A with $B = \nabla \times A$ satisfying $\nabla \times \nabla \times A = j$.

The integral formulation of Maxwell's equations requires less regularity. For a volume V with surface ∂V we find

$$\int_{\partial V} E \cdot n = \int_V \rho \quad \text{and} \quad \int_{\partial V} B \cdot n = 0.$$

For a surface S with boundary ∂S we find

$$\int_{\partial S} \langle E, \mathrm{d}s \rangle = -\int_S \partial_t B \quad \text{and} \quad \int_{\partial S} \langle B, \mathrm{d}s \rangle = \int_S \partial_t E + j. \qquad \lrcorner$$

Introducing the polarization. For given initial conditions $E|_{t=0}$ and $B|_{t=0}$ and given external densities $\rho = \rho(x,t)$ and $j = j(x,t)$ the solutions $E = E(x,t)$ and $B = B(x,t)$ of Maxwell's equations can be computed. In a medium, the electric field E and the magnetic field B affect the motion of the charged particles in the medium leading to another electric and magnetic field, namely the polarization and the magnetization. Glass fibers and photonic crystals are insulators, i.e., there are no free charge carriers. Moreover, there is no magnetization. However, the applied electric field yields to a shift of the electron density, i.e., to a charge density distribution ρ_{pol} in the atoms. Due to the conservation of charges there is an induced current density j_{pol} satisfying

$$\partial_t \rho_{\text{pol}} + \nabla \cdot j_{\text{pol}} = 0.$$

These induced densities on the other hand induce via (11.67) and (11.69) an electric field $P = P(x,t)$, which is called the polarization such that

$$\rho_{\text{pol}} = -\nabla \cdot P \quad \text{and} \quad \nabla \cdot j_{\text{pol}} = -\partial_t \rho_{\text{pol}} = \partial_t \nabla \cdot P,$$

leading to $\partial_t P = j_{\text{pol}}$. Thus, Maxwell's equations in glass fibers and photonic crystals are given by

$$\nabla \cdot E = -\nabla \cdot P, \tag{11.71}$$

$$\nabla \times E = -\partial_t B, \tag{11.72}$$

$$\nabla \times B = \partial_t E + \partial_t P, \tag{11.73}$$

$$\nabla \cdot B = 0. \tag{11.74}$$

This system can be simplified to a single equation. Applying $\nabla \times$ to (11.72) yields

$$\nabla \times (\nabla \times E) = -\partial_t (\nabla \times B).$$

Using

$$\nabla \times (\nabla \times E) = \nabla(\nabla \cdot E) - \Delta E$$

and (11.73) yields

$$\nabla(\nabla \cdot E) - \Delta E = -\partial_t^2 P - \partial_t^2 E. \tag{11.75}$$

For linearly polarized light, e.g. $E = u(x_3, t)\vec{e}_{x_1}$ and $P = p(x_3, t)\vec{e}_{x_1}$ we find $\nabla \cdot P = 0$, $\nabla \cdot E = 0$, and so

$$\partial_{x_3}^2 u = \partial_t^2 u + \partial_t^2 p, \tag{11.76}$$

where $u(x_3, t) \in \mathbb{R}$ and $p(x_3, t) \in \mathbb{R}$.

Modeling the polarization. In order to have an evolutionary problem we have to close (11.75) or (11.76) by a constitutive law $P = P(E)$ or $p = p(u)$, respectively. There exist various models. Basically the law of motion of a particle with coordinates $x = x(t)$ of mass m and charge q is given by

$$m \frac{\mathrm{d}^2}{\mathrm{d}t^2} x = q(E + \frac{\mathrm{d}x}{\mathrm{d}t} \times B).$$

In the simplest model the polarization is modeled as an oscillator and the influence of the magnetic field B is neglected. Thus we suppose that for an atom placed in an electric field the center $x = x(t)$ of the electron density obeys the equation

$$m \left(\frac{\mathrm{d}^2}{\mathrm{d}t^2} x + \omega_0^2 x \right) = q_{\mathrm{e}} E,$$

with q_{e} the elementary charge, ω_0 some normalized temporal wave number, and m the mass of the electron. On the continuum level in some mean-field limit for the electron position we find for the polarization

$$\partial_t^2 P + \omega_0^2 P = dE, \tag{11.77}$$

with a constant $d \in \mathbb{R}$. In case of damping by thermalization we have

$$m \left(\frac{\mathrm{d}^2}{\mathrm{d}t^2} x + \gamma \frac{\mathrm{d}}{\mathrm{d}t} x + \omega_0^2 x \right) = q_{\mathrm{e}} E$$

for a $\gamma > 0$. On the continuum level we find

$$\partial_t^2 P + \gamma \partial_t P + \omega_0^2 P = dE. \tag{11.78}$$

For larger values of E the linear oscillator has to be replaced by a nonlinear oscillator. Due to the symmetries of the problem very often there are no quadratic terms. As a nonlinear example we consider the oscillator model

$$m \left(\frac{\mathrm{d}^2}{\mathrm{d}t^2} x + \gamma \frac{\mathrm{d}}{\mathrm{d}t} x + \omega_0^2 x + r|x|^2 x \right) = q_{\mathrm{e}} E$$

for an $r \in \mathbb{R}$. On the continuum level we find

$$\partial_t^2 P + \gamma \partial_t P + \omega_0^2 P + rP|P|^2 = dE. \tag{11.79}$$

In general, matter consists of different atoms which combine to various molecules. This is modeled by considering various kind of oscillators. Hence, we finally come to a system

$$\nabla(\nabla \cdot E) - \Delta E = -\, \partial_t^2 P - \partial_t^2 E, \tag{11.80}$$

$$P = \sum_{j=0}^{n} P_j, \tag{11.81}$$

$$\partial_t^2 P_j + \gamma_j \partial_t P_j + \omega_j^2 P_j + r_j |P_j|^2 P_j = d_j E, \tag{11.82}$$

where the $d_j > 0$ are constants taking into account different masses, different numbers of atoms, etc., of the various kinds of atoms. In order to solve this system uniquely we need initial conditions for $E|_{t=0}, \partial_t E|_{t=0}, P_j|_{t=0}$ and $\partial_t P_j|_{t=0}$ for $j = 0, \ldots, n$. We refer to [**SU03b**] for a mathematical analysis of light pulse propagation in glass fibers with this modeling in case of damping, i.e., $\gamma_j > 0$.

Remark 11.7.2. a) (11.80)-(11.82) models isotropic media. Anisotropic media can be modeled for instance by choosing γ_j, ω_j^2, d_j, and r_j as tensors.

b) In the linear case $r_j = 0$ in (11.82), the constitutive laws can be solved explicitly. For $E = E_0 \mathrm{e}^{\mathrm{i}\omega t}$ we find $P = P_0 \mathrm{e}^{\mathrm{i}\omega t} + \widetilde{P}(t)$, with $\widetilde{P}(t) \to 0$ with some exponential rate for $t \to \infty$ and

$$P_0 = \alpha(\omega) E_0$$

where

$$\alpha(\omega) = \sum_{j=0}^{n} \frac{d_j}{-\omega^2 + \mathrm{i}\gamma_j \omega + \omega_j^2}.$$

The numbers γ_j, d_j, and ω_j can be used to fit the constitutive law to experimental data. There is a remarkable good agreement with experimental observations.

c) More generally, in the linear case there always is a Green's function χ_1 such that P can be expressed in terms of E, i.e.,

$$P(t) = \int_{-\infty}^{t} \chi_1(t-\tau) E(\tau) \, \mathrm{d}\tau. \tag{11.83}$$

In isotropic materials χ_1 is scalar. In glass fibers the above constants and also χ_1 only depend on the transverse variables, in photonic crystals there is a periodic dependence on the spatial variables. Although there is no Green's function for the nonlinear system, similar to the linear situation the constitutive law for the polarization is very often modeled by

$$P(t) = \int_{-\infty}^{t} \chi_1(t-\tau) E(\tau) \, \mathrm{d}\tau + \int_{-\infty}^{t} \chi_3(t-\tau) |E(\tau)|^2 E(\tau) \, \mathrm{d}\tau, \tag{11.84}$$

i.e., again (11.81)-(11.82) is replaced by (11.84).

d) In nonlinear optics very often time and space are interchanged. Due to the finite size of the fibers and due to the experimental data which can be measured initial conditions are posed at one end of the fiber, namely at $x = 0$, and one is interested in the solution at the end of the fiber, namely at $x =$

x_e, i.e., x is considered as evolutionary variable, and t as unbounded variable. From a mathematical point of view there is no difference if no dissipation is considered. However, very often phenomenologically dissipation is added to the NLS equation. It has been explained in [**SU03b**] that then it is highly problematic if x and t is interchanged. ⌋

In a one-dimensional optical fiber without damping and one kind of oscillators in the material we have the so called Maxwell-Lorentz system

$$\partial_x^2 u = \partial_t^2 u + \partial_t^2 p, \tag{11.85}$$

$$\partial_t^2 p + \omega_0^2 p - r|p|^2 p = du, \tag{11.86}$$

with coefficients ω_0, r, and d. The linearized problem

$$\partial_x^2 u = \partial_t^2 u + \partial_t^2 p, \qquad \partial_t^2 p + \omega_0^2 p = du$$

possesses solutions

$$u(x,t) = u_k e^{i(kx+\omega t)}, \quad p(x,t) = p_k e^{i(kx+\omega t)}$$

which yields

$$\begin{pmatrix} -k^2+\omega^2 & \omega^2 \\ -d & \omega_0^2-\omega^2 \end{pmatrix} \begin{pmatrix} u_k \\ p_k \end{pmatrix} = \begin{pmatrix} 0 \\ 0 \end{pmatrix}.$$

We have non-trivial solutions if the determinant vanishes, i.e., if

$$(-k^2+\omega^2)(\omega_0^2-\omega^2) + d\omega^2 = 0.$$

We find four curves of solutions $\omega = \omega_{1,2,3,4}(k)$ which are sketched in Figure 11.8. The Fourier transformed system can be written as first order system and then diagonalized, leading to

$$\begin{aligned} \partial_t \widehat{c}_1(k,t) &= i\omega_1(k)\widehat{c}_1(k,t) + \text{ nonlinear terms }, \\ &\vdots \\ \partial_t \widehat{c}_4(k,t) &= i\omega_4(k)\widehat{c}_4(k,t) + \text{ nonlinear terms }, \end{aligned}$$

such that the Maxwell-Lorentz system falls in the abstract class of systems considered in §11.3 for which an NLS equation can be derived.

Remark 11.7.3. Other models for the polarization are in use. The choice $\partial_t^2 p = -u - u^3$ leads to the Klein-Gordon model (11.2) for which we justified the NLS approximation in §11.2. Hence, the pulse dynamics present in the NLS equation is present in the models used for the description of the propagation of light pulses in glass fibers, too. Experimental observations confirm this approximation and modeling of reality. ⌋

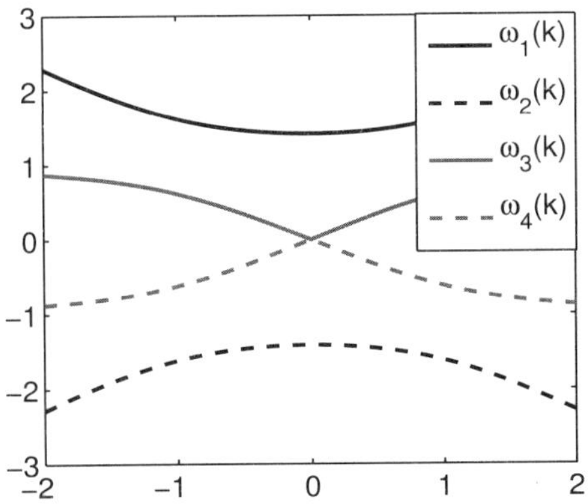

Figure 11.8. The curves of eigenvalues for the Maxwell-Lorentz system

11.7.2. Multiplexing. So far we considered pulses modulating one carrier wave with some basic wave number k_0. Now we use two or more carrier waves with different basic wave numbers $k_1, \ldots, k_N$. It turns out that pulses belonging to different carrier waves do not interact in lowest order, and thus the use of more than one carrier wave allows to increase the rate of information through the fiber. This concept is called multiplexing. In the following we explain the underlying ideas with the help of the NLS approximation.

Let us consider here again the nonlinear wave equation with cubic nonlinearity

$$\partial_t^2 u = \partial_x^2 u - u - u^3$$

as original system. For the situation of N different carrier waves we make the ansatz

$$\varepsilon\psi_{\text{multiNLS}} = \sum_{j=1}^{N} \varepsilon A_j \left(\varepsilon(x - c_j t), \varepsilon^2 t\right) \mathrm{e}^{\mathrm{i}(k_j x + \omega_j t)} + \text{c.c.}.$$

A system of coupled NLS equations is obtained, namely

$$2\mathrm{i}\omega_j \partial_T A_j = (1 - c_j^2)\partial_{X_j}^2 A_j - 3A_j |A_j|^2 - \text{coupling terms}_j$$

for $j = 1, \ldots, N$, where the jth coupling term is given by

$$6 \sum_{|n|=1,\ldots,N, n\neq j} A_j |A_n|^2 ,$$

where we assumed for simplicity that there are no resonant wave numbers $k_{j_1}, \ldots, k_{j_4}$ with $k_{j_1} + \ldots + k_{j_4} = 0$ and $\omega_{j_1} + \ldots + \omega_{j_4} = 0$. If this assumption is not satisfied there will be additional coupling terms $A_{j_2} A_{j_3} A_{j_4}$ in the equation for A_{-j_1} which however can be handled as explained below, too.

At a first view there seems to be a full coupling between all equations. However, looking more closely at the coupling terms shows that they have different arguments if $c_j \neq c_n$. We have for example

$$\begin{aligned} A_j |A_n|^2 =& A_j \left(\varepsilon(x - c_j t), \varepsilon^2 t\right) \left|A_n \left(\varepsilon(x - c_n t), \varepsilon^2 t\right)\right|^2 \\ =& A_j(X_j, T) \left|A_n \left(X_j - \varepsilon^{-1}(c_n - c_j)T, T\right)\right|^2 . \end{aligned}$$

Hence, two spatially localized functions interact only on an $\mathcal{O}(\varepsilon)$-time interval w.r.t. the T-time scale of the NLS equation if $c_j \neq c_n$. Thus, the influence of the coupling terms on the dynamics of the NLS equations is only $\mathcal{O}(\varepsilon)$. Hence, for spatially localized solutions the NLS equations decouple and the dynamics of the modulations of the carrier waves can be computed for each carrier wave individually by solving

$$2\mathrm{i}\omega_j \partial_T A_j = (1 - c_j^2)\partial_{X_j}^2 A_j - 3A_j |A_j|^2 .$$

The argument that spatially localized wave packets with different group velocities do not interact in lowest order has been made rigorous in case of the above NLS approximation first in [**PW96**]. The idea has been generalized in [**BF06**] for the interaction of wave packets in various original systems. It has also been used in case of the interaction of counter-propagating long waves [**Kal89, SW00b, SW02**], cf. §12.1. In case of dissipative systems mean-field coupled GL equations take the role of the NLS equation [**Sch97**].

The interaction of pulse solutions to different carrier waves can be described very precisely such that [**PW96**] and [**BF06**] can be improved strongly. The more detailed description is

$$\varepsilon\psi_{\text{multiNLS}} = \sum_{j=1}^{N} \varepsilon A_j \left(\varepsilon(x - c_j t - \varepsilon\psi_j(x,t)), \varepsilon^2 t\right) \mathrm{e}^{\mathrm{i}(k_j x + \omega_j t + \varepsilon\Omega_j(x,t))} + \text{c.c.},$$

where the A_j satisfy decoupled equations and there are explicit formulae [**CBSU07**] for the pulse shifts $\varepsilon\psi_j$ and the phase shifts $\varepsilon\Omega_j$. The internal dynamics of the wave packets (described via the A_j) and the interaction dynamics of the wave packets (described via ψ_j and Ω_j) can be separated up to very high order. An almost complete description of the interaction of general localized NLS described wave packets can be found for the nonlinear wave equation in [**CBCSU08**] and in [**CBS12**] for general dispersive wave systems. As a consequence the shift of the underlying carrier wave and the shift of the envelope both can be shown to be of order $\mathcal{O}(\varepsilon)$ instead of $\mathcal{O}(1)$ w.r.t. the original x-variable. See also [**SUW11**] for an application to oscillator chains.

Further Reading. The description of waves in nonlinear optics by modulation equations is an active field of research. Three main current topics are:

- ultra-short pulses [**SW04, CdR15, PS13, New16**], see also Exercise 11.10;

- the (numerical) analysis of 3D photonic crystals, and in particular their design to achieve favorable band structures [**BFL+07**, **STE+10**];
- gap solitons, cf. §11.6.5, and related phenomena. Besides gap solitons, models of type (11.65) support other solution families, for instance so called Nonlinear Bloch Modes and "Out–of-Gap" solitons, see, e.g., [**Yan10**] for a comprehensive overview, or [**DU16**] for recent rigorous results. See also [**DD13**] for related results for (simplified) nonlinear Maxwell's equations.

NLS equations also occur as singular limits of other dipersive PDE systems. Examples are the Zakharov system [**AA88**] or the Klein-Gordon-Zakharov system [**MN05**]. In [**UGS+15**], a so called shrinking limit of the NLS equation on fat graphs has been considered, and an analysis somewhat similar to 1D and 2D photonic crystals deals with wave propagation on infinite periodic metric graphs, used as models for nanotubes or graphen, see, e.g. [**GPS17**].

Exercises

11.1. Consider the nonlinear wave equation $\partial_t^2 u = \partial_x^2 u - u - u^3$, with $x, t, u(x,t) \in \mathbb{R}$. Make an ansatz $u(x,t) = \varepsilon A(\varepsilon x, \varepsilon^2 t)e^{it} + \text{c.c.}$ to derive a NLS equation for A. Estimate the residual and prove an approximation result by considering the energy

$$E = \int_{\mathbb{R}} u^2 + (\partial_x u)^2 + (\partial_t u)^2 \, dx.$$

11.2. Consider the nonlinear wave equation $\partial_t^2 u = \Delta u - u - u^3$, with $x, y, t, u(x,y,t) \in \mathbb{R}$. Make an ansatz

$$u(x,t) = \varepsilon A(\varepsilon(x - c_0 t), \varepsilon y, \varepsilon^2 t)e^{i(k_0 x + \omega_0 t)} + \text{c.c.}$$

to derive a 2D NLS equation for A.

11.3. Consider the lattice differential equation

$$\partial_t^2 u_n = u_{n+1} - 2u_n + u_{n-1} - 3u_n - u_n^2,$$

with $u_n = u_n(t) \in \mathbb{R}$ for $n \in \mathbb{Z}$.

a) Find solutions $u_n(t) = e^{i(kn - \omega t)}$ of the linearized problem. Compute the dispersion relation.

b) Check the validity of the non-resonance condition.

c) Make the ansatz $u_n(t) = \varepsilon A(\varepsilon n, \varepsilon^2 t)e^{i\omega_0 t} + c.c. + \mathcal{O}(\varepsilon^2)$ and derive a NLS equation for the amplitude $A = A(X,T)$.

11.4. Consider the FPU system $\partial_t^2 u_n = W'(u_{n+1}) - 2W'(u_n) + W'(u_{n-1})$, with W an analytic function, i.e., $W'(u) = au + bu^2 + \dots$. Derive an evolution equation for $u(k,t) = \sum_{n\in\mathbb{Z}} u_n(t)e^{ikn}$, cf. [**Mil06**]. Find differences and similarities between this evolution equation and the Fourier transform of a nonlinear dispersive PDE.

11.5. Consider the KdV equation $\partial_t u - 6u\partial_x u + \partial_x^3 u = 0$.

a) Apply the Miura transformation $u = v^2 + \partial_x v$ and derive the mKdV equation

$$\partial_t v - 6v^2 \partial_x v + \partial_x^3 v = 0.$$

b) Plug in the ansatz $\varepsilon\psi_v(x,t) = \varepsilon A(\varepsilon(x-ct), \varepsilon^2 t)\mathrm{e}^{\mathrm{i}(kx+\omega t)} + c.c.$ into the mKdV equation and derive a NLS equation for A.

c) Justify this approximation by energy estimates and conclude the validity of an approximation theorem for the approximation of the KdV equation by the NLS equation [**Sch11**].

11.6. The dispersion relation of the water wave problem with surface tension is given by

$$\omega^2 = (k + \sigma k^3)\tanh(k),$$

with surface tension parameter $\sigma \geq 0$. Show that for $\sigma \in (0, 1/3)$ beside $k = 0$ and $k = k_0$ there are two additional resonances k_1 and k_2 with $k_0 + k_1 + k_2 = 0$ and $\omega_0 + \omega_1 + \omega_2 = 0$.

11.7. Consider the system for the resonant three wave interaction

$$\dot{A}_1 = \mathrm{i}\gamma_1 \overline{A}_2 \overline{A}_3, \qquad \dot{A}_2 = \mathrm{i}\gamma_2 \overline{A}_1 \overline{A}_3, \qquad \dot{A}_3 = \mathrm{i}\gamma_3 \overline{A}_1 \overline{A}_2,$$

for $A_j(t) \in \mathbb{C}$ and coefficients $\gamma_j \in \mathbb{R}$.

a) Prove that this is a Hamiltonian system with

$$q_n = \mathrm{sgn}(\gamma_n)\frac{A_n}{\sqrt{|\gamma_n|}}, \qquad p_n = \frac{\overline{A}_n}{\sqrt{|\gamma_n|}}, \qquad H = \mathrm{i}(A_1A_2A_3 + \overline{A}_1\overline{A}_2\overline{A}_3).$$

b) Prove that the quantities

$$J_1 = \frac{A_1\overline{A}_1}{\gamma_1} - \frac{A_2\overline{A}_2}{\gamma_2}, \qquad J_2 = \frac{A_1\overline{A}_1}{\gamma_1} - \frac{A_3\overline{A}_3}{\gamma_3}, \qquad J_3 = \frac{A_1\overline{A}_2}{\gamma_2} - \frac{A_3\overline{A}_3}{\gamma_3}$$

are conserved under the flow of the system.

c) Prove for the Poisson brackets that $\{H, J_1\} = \{H, J_2\} = \{J_1, J_2\} = 0$. Use this to show the complete integrability of the Hamiltonian system.

11.8. Solve the eigenvalue problem $\partial_x u(x) = v$, $\partial_x v(x) = -s(x)\lambda u(x)$, with the 1-periodic function $s(x) = \chi_{[0,6/13]} + 16\chi_{(6/13,7/13)} + \chi_{[7/13,1]}(x \bmod 1)$. Compute the monodromy matrix C_λ and the discriminant $D(\lambda) = \mathrm{trace}\, C_\lambda$. Hint. On intervals where s is constant, the problem can be solved explicitly. We look for $u \in C_b^1$.

11.9. Consider the nonlinear wave equation $\partial_t^2 u = \partial_x^2 u + u + 2\varepsilon^2\cos(2x)u - u^3$, with spatially periodic perturbed coefficients, i.e., $0 \leq \varepsilon \ll 1$. For $\varepsilon > 0$ a spectral gap of order $\mathcal{O}(\varepsilon^2)$ occurs, which is too small to derive an NLS equation. In this case with the ansatz

$$u(x,t) = \varepsilon a(\varepsilon^2 x, \varepsilon^2 t)\mathrm{e}^{\mathrm{i}(x-\omega_0 t)} + \varepsilon b(\varepsilon^2 x, \varepsilon^2 t)\mathrm{e}^{-\mathrm{i}(x+\omega_0 t)} + \text{c.c.}$$

derive the coupled mode system, cf. [**SU01**]

$$\begin{aligned} -2\mathrm{i}\omega_0\partial_T a &= 2\mathrm{i}\partial_x a + ib - 3a|a|^2 - 6a|b|^2, \\ -2\mathrm{i}\omega_0\partial_T b &= -2\mathrm{i}\partial_x b + \mathrm{i}a - 3b|b|^2 - 6b|a|^2. \end{aligned}$$

11.10. If the pulses become very narrow, then the so called short pulse equation

$$\partial_\xi\partial_\tau A = A + \partial_\xi^2(A^3),$$

can be derived. Consider the quasilinear wave equation $\partial_t^2 u = \partial_x^2 u + u + \partial_x^2(u^3)$, and make the ansatz

$$u(t,x) = 2\varepsilon A(\tau, \xi), \quad \tau = \varepsilon t, \quad \xi = \frac{x-t}{2\varepsilon}$$

to derive the short pulse equation, cf. [**PS13**].

11.11. Show that the inhomogeneous Maxwell equations

$$\partial_t E = \nabla \times B - J, \quad \nabla \cdot E = \rho, \quad \partial_t B = -\nabla \times E, \quad \nabla \cdot B = 0,$$

can be transformed to the inhomogeneous wave equations

$$\partial_t^2 u - \Delta u = \rho, \quad \partial_t^2 A - \Delta A = J,$$

where A is a vector potential of B, i.e., $\nabla \times A = B$, and u is a scalar potential of E, i.e., $E + \partial_t A = -\nabla u$, see Remark 11.7.1.
Hint. Consider a suitable gauge transform, i.e., adding $\nabla\lambda$ to A and subtracting $-\partial_t \lambda$ from u.

Chapter 12

Long waves and their modulation equations

It is the purpose of this chapter to explain the role of the KdV equation as a long wave modulation equation for various dispersive wave systems. Such approximations have a long history in science, and they not only play a role in the description of the water wave problem but also for other nonlinear dispersive systems such as the equations of plasma physics or the FPU model.

One of the simplest systems where the KdV equation can be derived is the so called Boussinesq equation

$$\partial_t^2 u = \partial_x^2 u - \partial_x^4 u + \partial_x^2(u^2), \tag{12.1}$$

with $x \in \mathbb{R}$, $t \in \mathbb{R}$, and $u(x,t) \in \mathbb{R}$. The long wave ansatz

$$u(x,t) = \varepsilon^2 v(\xi, \tau) = \varepsilon^2 v(\varepsilon x, \varepsilon t),$$

where $0<\varepsilon\ll 1$ is a perturbation parameter, yields $\partial_\tau^2 v = \partial_\xi^2 v + \mathcal{O}(\varepsilon^2)$, hence

$$\partial_\tau^2 v = \partial_\xi^2 v \tag{12.2}$$

in lowest order. The linear wave equation (12.2) is solved by

$$v(\xi,\tau) = v_+(\xi - \tau) + v_-(\xi + \tau),$$

with functions v_+ and v_- satisfying

$$v(\xi, 0) = v_+(\xi) + v_-(\xi) \quad \text{and} \quad \partial_\tau v(\xi, 0) = -v_+'(\xi) + v_-'(\xi).$$

Hence, the formal long wave approximation predicts a wave moving to the left and a wave moving to the right. However, this approximation is formally only valid on a time scale of order $\mathcal{O}(1/\varepsilon)$. Nevertheless the right and left moving wave will be separated for spatially localized initial conditions. Afterwards the wave packets will evolve individually. For each of the wave packets a KdV equation can be derived, and these are formally valid on the much longer time scale of order $\mathcal{O}(1/\varepsilon^3)$. In detail, inserting the ansatz

$$u(x,t) = \varepsilon^2 A_+(\varepsilon(x-t), \varepsilon^3 t) + \varepsilon^2 A_-(\varepsilon(x+t), \varepsilon^3 t). \tag{12.3}$$

into the Boussinesq equation and equating the coefficient at ε^6 to zero yields the system of two decoupled KdV equations

$$2\partial_T A_+ - \partial_X^3 A_+ + \partial_X(A_+^2) = 0, \quad -2\partial_T A_- - \partial_X^3 A_- + \partial_X(A_-^2) = 0, \tag{12.4}$$

one for each of the wave packets. There are some coupling terms which however turn out to be of higher order for spatially localized initial conditions. Estimates that this formal KdV approximation and true solutions of the Boussinesq model stay close together over the natural KdV time scale are a non-trivial task since solutions of order $\mathcal{O}(\varepsilon^2)$ have to be estimated on an $\mathcal{O}(1/\varepsilon^3)$ time scale. This question will be discussed in §12.1.

It turns out that the set of KdV equations can be derived whenever the dispersion relation for harmonic waves $e^{i(kx+\omega(k)t)}$ gives two curves of eigenvalues $k \mapsto \omega_\pm(k)$ with $\omega_\pm(0) = 0$, and the nonlinear terms also vanish at the spatial wave number $k = 0$. This is the reason why the KdV equation occurs as a universal modulation equation, as discussed in §12.2. There, we introduce the water wave problem, the equations of plasma physics and the FPU system and show that they belong to the class of systems for which the KdV equation can be derived. There are many other modulation equations which can be derived in the long wave limit. Some of them are independent of the small perturbation parameter ε such as the inviscid Burgers equation or the Whitham system and some depend on the small perturbation parameter ε such as the regularized long wave equation or the Benjamin-Bona-Mahony system. We will comment on the use of these equations and on their validity in §12.3. In §12.4 we explain that in the so called long wave limit, where a big number of these modulation equations have been derived, the above splitting in two decoupled KdV equations is sufficient to describe the original system.

12.1. An approximation result

The goal of this section is to prove an approximation result for a single KdV equation. This is a non-trivial task due to the fact that solutions of order $\mathcal{O}(\varepsilon^2)$ have to be estimated on an $\mathcal{O}(1/\varepsilon^3)$ time scale. Hence, one power of ε is missing. There are at least two essentially different approaches.

In the first approach one works in a family of spaces of analytic functions where the strip of analyticity shrinks with time. This shrinking corresponds to some smoothing which allows to gain the missing power of ε by using the derivative in front of the KdV nonlinearity. This approach has been used in [**KN86**], but as pointed out in [**CSS92**] the error estimates for the KdV approximation there have only been shown on an $\mathcal{O}(1/\varepsilon)$ time scale instead of the $\mathcal{O}(1/\varepsilon^3)$ KdV time scale. However, as pointed out in [**Sch96b**] the approach of [**KN86**] can be extended to the correct $\mathcal{O}(1/\varepsilon^3)$ KdV time scale. This approach turns out to be very robust, but does not give optimal results, cf. [**CDS15**] for another example.

In this book we will present the second approach which in general gives better estimates and is conceptionally easier, but less robust. For the Boussinesq equation (12.1) the KdV approximation can be justified with a simple energy estimate. An energy is constructed in such a way that the influence of the terms of order $\mathcal{O}(\varepsilon^2)$ on the growth of the solutions can be estimated to be of order $\mathcal{O}(\varepsilon^3)$. The papers [**Cra85, SW00b, SW02, Dül12**] where the KdV approximation has been justified for the water wave problem work this way. The major difficulty in transferring this proof from the Boussinesq equation to the water wave problem or other applications is to bring together the energy estimates with the local existence and uniqueness theory for the various systems. For the Boussinesq equation our approximation result is as follows.

Theorem 12.1.1. *Fix $\theta > 1/2$ and let $A_+ = A \in C([0,T_0], H^{\theta+4})$ and $A_- = 0$ be solutions of the system of KdV equations* (12.4). *Then there exist $\varepsilon_0 > 0$ and $C > 0$ such that for all $\varepsilon \in (0,\varepsilon_0]$ we have solutions u of* (12.1) *with*

$$\sup_{t\in[0,T_0/\varepsilon^3]} \|u(\cdot,t) - \varepsilon^2 A(\varepsilon(\cdot - t),\varepsilon^3 t)\|_{H^\theta} \leq C\varepsilon^{7/2}.$$

Proof. For notational simplicity we only consider the case $\theta = 2$. The residual

$$\mathrm{Res}(u) = -\partial_t^2 u(x,t) + \partial_x^2 u(x,t) - \partial_x^4 u(x,t) + \partial_x^2(u(x,t)^2)$$

quantifies how much a function u fails to satisfy the Boussinesq model (12.1). For the approximation

$$\varepsilon^2\Psi(x,t,\varepsilon) = \varepsilon^2 A(\varepsilon(x-ct),\varepsilon^3 t)$$

we find

$$\begin{aligned}\mathrm{Res}(\varepsilon^2\Psi) =& -\varepsilon^4 c^2\partial_X^2 A + 2\varepsilon^6\partial_T\partial_X A - \varepsilon^8\partial_T^2 A\\ &+\varepsilon^4\partial_X^2 A - \varepsilon^6\partial_X^4 A + \varepsilon^6\partial_X^2(A^2)\\ =& -\varepsilon^8\partial_T^2 A\end{aligned}$$

if we choose $c^2 = 1$ and A to satisfy the first of the two KdV equations of (12.4). Therefore, we have

Lemma 12.1.2. *Fix* $\widetilde{\theta} \geq 0$ *and let* $A_+ = A \in C([0, T_0], H^{\widetilde{\theta}+6})$ *and* $A_- = 0$ *be solutions of the system of KdV equations* (12.4). *Then there exist* $\varepsilon_0 > 0$, C_{res} *such that for all* $\varepsilon \in (0, \varepsilon_0)$ *we have*

$$\sup_{t \in [0, T_0/\varepsilon^3]} \| \operatorname{Res}(\varepsilon^2 \Psi(\cdot, t, \varepsilon)) \|_{H^{\widetilde{\theta}}} \leq C_{\text{res}} \varepsilon^{15/2}$$

and

$$\sup_{t \in [0, T_0/\varepsilon^3]} \| \partial_x^{-1} \operatorname{Res}(\varepsilon^2 \Psi(\cdot, t, \varepsilon)) \|_{H^{\widetilde{\theta}}} \leq C_{\text{res}} \varepsilon^{13/2}.$$

Proof. Using the KdV equation allows us to write

$$\begin{aligned} 4\partial_T^2 A =& -2\partial_T(\partial_X^3 A - \partial_X(A^2)) = 2(\partial_X^3 \partial_T A - 2\partial_X(A \partial_T A)) \\ =& \partial_X^3(\partial_X^3 A - \partial_X(A^2)) - 2\partial_X(A(\partial_X^3 A - \partial_X(A^2))). \end{aligned}$$

This shows that $A(\cdot, T) \in H^{\widetilde{\theta}+6}$ is necessary to estimate the residual in $H^{\widetilde{\theta}}$. Moreover, we can apply $\partial_x^{-1} = \varepsilon^{-1} \partial_X^{-1}$ to the residual terms which however loses a factor ε^{-1}. The remaining formal error of order $\mathcal{O}(\varepsilon^7)$ is reduced further by a factor $\varepsilon^{-1/2}$ due to the scaling properties of the L^2-norm, cf. (10.18). □

The difference $\varepsilon^{7/2} R = u - \varepsilon^2 \Psi$ satisfies

$$\partial_t^2 R = \partial_x^2 R - \partial_x^4 R + 2\varepsilon^2 \partial_x^2(\Psi R) + \varepsilon^{7/2} \partial_x^2(R^2) + \varepsilon^{-7/2} \operatorname{Res}(\varepsilon^2 \Psi). \tag{12.5}$$

In order to estimate R we construct an energy E with $\frac{d}{dt} E = \mathcal{O}(\varepsilon^3)$. Thus, we multiply the error equation (12.5) with $\partial_t R$ and integrate it w.r.t. x. We find with integration by parts and $\tau = \varepsilon t$ that

$$\begin{aligned} \int_{\mathbb{R}} (\partial_t R) \partial_t^2 R \, dx =& \frac{1}{2} \frac{d}{dt} \int_{\mathbb{R}} (\partial_t R)^2 \, dx, \\ \int_{\mathbb{R}} (\partial_t R) \partial_x^2 R \, dx =& -\frac{1}{2} \frac{d}{dt} \int_{\mathbb{R}} (\partial_x R)^2 \, dx, \\ -\int_{\mathbb{R}} (\partial_t R) \partial_x^4 R \, dx =& -\frac{1}{2} \frac{d}{dt} \int_{\mathbb{R}} (\partial_x^2 R)^2 \, dx, \end{aligned}$$

$$\begin{aligned} \int_{\mathbb{R}} (\partial_t R) \partial_x^2 (\Psi R) \, dx =& -\int_{\mathbb{R}} (\partial_t \partial_x R) \Psi (\partial_x R) \, dx - \varepsilon \int_{\mathbb{R}} (\partial_t \partial_x R)(\partial_X \Psi) R \, dx \\ =& -\frac{d}{dt} \int_{\mathbb{R}} \Psi (\partial_x R)^2 \, dx/2 + \varepsilon \int_{\mathbb{R}} (\partial_\tau \Psi)(\partial_x R)^2 \, dx \\ &+ \varepsilon \int_{\mathbb{R}} (\partial_t R)(\partial_X \Psi) \partial_x R \, dx + \varepsilon^2 \int_{\mathbb{R}} (\partial_t R)(\partial_X^2 \Psi) R \, dx, \end{aligned}$$

$$\begin{aligned} \int_{\mathbb{R}} (\partial_t R) \partial_x^2 (R^2) \, dx =& -2 \int_{\mathbb{R}} (\partial_t \partial_x R) R (\partial_x R) \, dx \\ =& -\frac{d}{dt} \int_{\mathbb{R}} R (\partial_x R)^2 \, dx + \int_{\mathbb{R}} (\partial_x R)^2 (\partial_t R) \, dx. \end{aligned}$$

All terms which can be written as a time derivative are included in E. The others are estimated, namely

$$\begin{aligned}
\left|\int_{\mathbb{R}} (\partial_t R) \operatorname{Res}(\varepsilon^2\Psi)\,\mathrm{d}x\right| \leq & \|\partial_t R\|_{L^2} \|\operatorname{Res}(\varepsilon^2\Psi)\|_{L^2}, \\
\left|\int_{\mathbb{R}} (\partial_\tau \Psi)(\partial_x R)^2\,\mathrm{d}x\right| \leq & \|\partial_\tau \Psi\|_{L^\infty} \|\partial_x R\|_{L^2}^2, \\
\left|\int_{\mathbb{R}} (\partial_t R)(\partial_X \Psi)\partial_x R\,\mathrm{d}x\right| \leq & \|\partial_X \Psi\|_{L^\infty} \|\partial_t R\|_{L^2} \|\partial_x R\|_{L^2}, \\
\left|\int_{\mathbb{R}} (\partial_t R)(\partial_X^2 \Psi) R\,\mathrm{d}x\right| \leq & \|\partial_X^2 \Psi\|_{L^\infty} \|\partial_t R\|_{L^2} \|R\|_{L^2}, \\
\left|\int_{\mathbb{R}} (\partial_x R)^2(\partial_t R)\,\mathrm{d}x\right| \leq & \|\partial_x R\|_{L^\infty} \|\partial_t R\|_{L^2} \|\partial_x R\|_{L^2} \\
\leq & C\|\partial_x R\|_{H^1} \|\partial_t R\|_{L^2} \|\partial_x R\|_{L^2},
\end{aligned}$$

with C a constant coming from Sobolev's embedding theorem. Next we multiply the error equation with $-\partial_t\partial_x^{-2}R$ which is defined via its Fourier transform w.r.t. x, i.e. $\partial_x^{-1}R = \mathcal{F}^{-1}((ik)^{-1}\widehat{R})$ integrate it w.r.t. x, and find

$$\begin{aligned}
-\int_{\mathbb{R}} (\partial_t\partial_x^{-2}R)\partial_t^2 R\,\mathrm{d}x = & \frac{1}{2}\frac{\mathrm{d}}{\mathrm{d}t}\int_{\mathbb{R}} (\partial_t\partial_x^{-1}R)^2\,\mathrm{d}x, \\
-\int_{\mathbb{R}} (\partial_t\partial_x^{-2}R)\partial_x^2 R\,\mathrm{d}x = & -\frac{1}{2}\frac{\mathrm{d}}{\mathrm{d}t}\int_{\mathbb{R}} R^2\,\mathrm{d}x, \\
\int_{\mathbb{R}} (\partial_t\partial_x^{-2}R)\partial_x^4 R\,\mathrm{d}x = & -\frac{1}{2}\frac{\mathrm{d}}{\mathrm{d}t}\int_{\mathbb{R}} (\partial_x R)^2\,\mathrm{d}x, \\
-\int_{\mathbb{R}} (\partial_t\partial_x^{-2}R)\partial_x^2(\Psi R)\,\mathrm{d}x = & -\int_{\mathbb{R}} (\partial_t R)\Psi R\,\mathrm{d}x \\
= & -\frac{\mathrm{d}}{\mathrm{d}t}\int_{\mathbb{R}} \Psi R^2\,\mathrm{d}x/2 + \varepsilon\int_{\mathbb{R}} (\partial_\tau\Psi)R^2\,\mathrm{d}x, \\
-\int_{\mathbb{R}} (\partial_t\partial_x^{-2}R)\partial_x^2(R^2)\,\mathrm{d}x = & -2\int_{\mathbb{R}} (\partial_t R)R^2\,\mathrm{d}x = -\frac{2}{3}\frac{\mathrm{d}}{\mathrm{d}t}\int_{\mathbb{R}} R^3\,\mathrm{d}x, \\
-\int_{\mathbb{R}} (\partial_t\partial_x^{-2}R)\operatorname{Res}(\varepsilon^2\Psi)\,\mathrm{d}x = & \int_{\mathbb{R}} (\partial_t\partial_x^{-1}R)\partial_x^{-1}\operatorname{Res}(\varepsilon^2\Psi)\,\mathrm{d}x.
\end{aligned}$$

We can estimate

$$\begin{aligned}
\left|\int_{\mathbb{R}} (\partial_t\partial_x^{-1}R)\partial_x^{-1}\operatorname{Res}(\varepsilon^2\Psi)\,\mathrm{d}x\right| \leq & \|\partial_t\partial_x^{-1}R\|_{L^2} \|\partial_x^{-1}\operatorname{Res}(\varepsilon^2\Psi)\|_{L^2}, \\
\left|\int_{\mathbb{R}} (\partial_\tau\Psi)R^2\,\mathrm{d}x\right| \leq & \|\partial_\tau\Psi\|_{L^\infty} \|R\|_{L^2}^2.
\end{aligned}$$

We collect all total time derivatives in the energy

$$
\begin{aligned}
E = \int_{\mathbb{R}} & (\partial_t \partial_x^{-1} R)^2 + R^2 + (\partial_t R)^2 + 2(\partial_x R)^2 + (\partial_x^2 R)^2 \\
& + 2\varepsilon^2 \Psi R^2 + 2\varepsilon^2 \Psi (\partial_x R)^2 + 4\varepsilon^{7/2} R^3/3 + 2\varepsilon^{7/2} R (\partial_x R)^2 \, \mathrm{d}x. \qquad (12.6)
\end{aligned}
$$

Since for $\varepsilon > 0$ sufficiently small the second line of E can be estimated by the first line we find that for all $M > 0$ there exist $C_1, \varepsilon_1 > 0$ such that for all $\varepsilon \in (0, \varepsilon_1)$ we have

$$\|R\|_{H^2} \leq C_1 E^{1/2}$$

as long as $E \leq M$. Therefore, E satisfies the inequality

$$\frac{\mathrm{d}}{\mathrm{d}t} E \leq C\varepsilon^3 E + C\varepsilon^{7/2} E^{3/2} + C\varepsilon^3 E^{1/2} \leq 2C\varepsilon^3 E + C\varepsilon^{7/2} E^{3/2} + C\varepsilon^3, \qquad (12.7)$$

with a constant C independent of $\varepsilon \in (0, \varepsilon_1)$. Under the assumption that $C\varepsilon^{1/2} E^{1/2} \leq 1$ we obtain

$$\frac{\mathrm{d}}{\mathrm{d}t} E \leq (2C+1)\varepsilon^3 E + C\varepsilon^3.$$

Gronwall's inequality immediately gives the bound

$$\sup_{t \in [0, T_0/\varepsilon^3]} E(t) = C T_0 \mathrm{e}^{(2C+1)T_0} =: M = \mathcal{O}(1)$$

and so $\sup_{t \in [0, T_0/\varepsilon^3]} \|R(t)\|_{H^2} = \mathcal{O}(1)$. Finally choosing $\varepsilon_2 > 0$ so small that $C\varepsilon_2^{1/2} M^{1/2} \leq 1$ gives the required estimate for all $\varepsilon \in (0, \varepsilon_0)$ with $\varepsilon_0 = \min(\varepsilon_1, \varepsilon_2) > 0$. Since the residual has only to be estimated in L^2, we need $A \in H^6$ to estimate the error in H^2. For general θ we set $\widetilde{\theta} = \theta - 2$ such that $\theta + 4 = \widetilde{\theta} + 6$ in Lemma 12.1.2. $\square$

12.2. The universality of the KdV equation

As already said, the KdV approximation can be derived for various systems. In order to explain why this is the case and why the KdV equation plays such an important role, we review the derivation of the KdV equation from a different point of view. This derivation will explain why the KdV equation occurs as a universal modulation equation whenever the curves of eigenvalues $k \mapsto \omega_\pm(k)$ of the underlying dispersive wave system satisfy $\omega_\pm(0) = 0$, and the nonlinear terms also vanish at the wave number $k = 0$. In [**Bri13**] it has been shown that this is the case under very weak conditions. The underlying system will condense in the values of the coefficients of the associated KdV equation. We present with the water wave problem, with the FPU system, and with the equations of plasma physics three dispersive systems for which the KdV approximation plays a role.

As in §10.3 and §11.3 it turns out that Fourier transform is the key for the understanding of the universality. Hence, we consider the Boussinesq equation (12.1) in Fourier space. The Fourier transform $\widehat{u}$ satisfies

$$\partial_t^2 \widehat{u}(k,t) = -\omega^2(k)\widehat{u}(k,t) - \rho(k)\widehat{u}^{*2}(k,t), \tag{12.8}$$

where $\omega(k) = \sqrt{k^2+k^4}\ \mathrm{sign}(k)$ and $\rho(k) = k^2$. By introducing $\widehat{w}(k) = (\widehat{u}(k), \frac{1}{\omega(k)}\partial_t \widehat{u}(k))$ we rewrite (12.8) into the first order system

$$\partial_t \widehat{w}(k,t) = \widehat{M}(k)\widehat{w}(k,t) + \widehat{N}(\widehat{w})(k,t), \tag{12.9}$$

where

$$\widehat{M}(k) = \begin{pmatrix} 0 & i\omega(k) \\ i\omega(k) & 0 \end{pmatrix}, \quad \widehat{N}(\widehat{w})(k,t) = \begin{pmatrix} 0 \\ -\frac{\rho(k)}{i\omega(k)}\widehat{u}^{*2}(k,t) \end{pmatrix}.$$

This system is diagonalized for fixed wave number k. For (12.9) the associated transformation

$$\widehat{U} = \frac{1}{\sqrt{2}}\begin{pmatrix} 1 & 1 \\ 1 & -1 \end{pmatrix}$$

is independent of k and unitary, i.e., $\widehat{U}^{-1} = \widehat{U}^*$, cf. §11.1 where the same transformation has been used a number of times. The transformed variable $\widehat{z} = \widehat{U}^*\widehat{w}$ satisfies the diagonalized system

$$\partial_t \widehat{z} = \widehat{\Lambda}\widehat{z} + \widehat{U}^*\widehat{N}(\widehat{U}\widehat{z}), \tag{12.10}$$

with $\widehat{\Lambda}(k) = \mathrm{diag}(i\omega(k), -i\omega(k))$. In accordance with (12.3), for (12.10) we make the ansatz

$$\begin{aligned} \widehat{z}(k,t) &= \varepsilon^2\varepsilon^{-1}\widehat{A}_+\left(\frac{k}{\varepsilon}, \varepsilon^3 t\right)\mathrm{e}^{ickt}\vec{e}_1 \\ &\quad + \varepsilon^2\varepsilon^{-1}\widehat{A}_-\left(\frac{k}{\varepsilon}, \varepsilon^3 t\right)\mathrm{e}^{-ickt}\vec{e}_2, \end{aligned} \tag{12.11}$$

where

$$\vec{e}_1 = \begin{pmatrix} 1 \\ 0 \end{pmatrix} \quad \text{and} \quad \vec{e}_2 = \begin{pmatrix} 0 \\ 1 \end{pmatrix}.$$

Since the Fourier modes of the wave packets are concentrated in an $\mathcal{O}(\varepsilon)$ neighborhood of the wave number $k=0$, the evolution of the wave packets will strongly be determined by the curves $\pm\omega$ at $k=0$. At $\mathrm{e}^{ickt}\vec{e}_1$ we find

$$\begin{aligned} &\mathrm{i}\varepsilon^2 cK\widehat{A}_1 + \varepsilon^4\partial_T\widehat{A}_1 \\ &= \mathrm{i}\varepsilon^2\partial_k\omega(0)K\widehat{A}_1 + \frac{\mathrm{i}}{6}\varepsilon^4\partial_k^3\omega(0)K^3\widehat{A}_1 - \varepsilon^4\partial_k\frac{\rho(0)}{\omega(0)}\mathrm{i}K\widehat{A}_1 * \widehat{A}_1 + \mathcal{O}(\varepsilon^6), \end{aligned}$$

where $k = \varepsilon K$. At ε^2 we obtain the linear group velocity. At ε^4 we obtain a KdV equation.

By this procedure it is clear that the KdV equation occurs as modulation equation of dispersive systems

$$\partial_t \widehat{z}_j(k,t) = \mathrm{i}\omega_j(k)\widehat{z}_j(k,t) + \text{ nonlinear terms} \tag{12.12}$$

for j in some index set whenever the Fourier transform of the initial condition is strongly concentrated at $k=0$, when the linear ($\omega_1(0) = \omega_{-1}(0) = 0$) and nonlinear terms vanish at this wave number, and when the concentration and the amplitude are of the correct order.

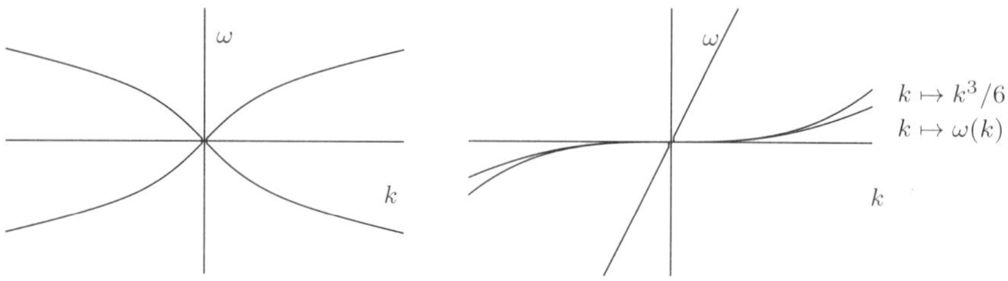

Figure 12.1. The derivation of the KdV equation is based on the concentration of the Fourier modes at the wave number $k = 0$. The left panel shows the curves of eigenvalues. Hence, for the evolution of these concentrated Fourier modes only the curves of eigenvalues close to the wave number $k = 0$ plays a role. The right panel therefore shows a blow-up of these curves in a frame moving with the group velocity $\omega'(0)$ and an expansion of one curve at the wave number $k = 0$.

12.2.1. The water wave problem. The most famous system where the KdV equation can be derived is the so called water wave problem. Following §12.1, there are no conceptional difficulties to justify this approximation, but the realization of this approach for the water wave problem is rather lengthy and involves a big number of estimates. Therefore, it is the only goal of this section to explain that the water wave problem falls into the class of systems for which the KdV equation can be derived. References to the KdV approximation for the full water wave problem can be found at the end of the chapter.

The water wave problem consists in finding the irrotational flow of an inviscid incompressible fluid in an infinitely long canal of fixed finite depth with an unknown free top surface subject to gravitational forces. The bottom is impermeable and for expository reasons surface tension is neglected in this subsection. The coordinates are denoted in horizontal direction by $x_1 \in \mathbb{R}$ and in the vertical bounded direction by x_2. The fluid fills the domain $\Omega(t)$ in between the bottom $\{(x_1, -h) : x_1 \in \mathbb{R}\}$ and the unknown free top surface $\Gamma(t)$. In the Eulerian formulation the free surface $\Gamma(t)$ is parameterized as function η over the bottom, i.e., we have

$$\Gamma(t) = \{(x_1, x_2) : x_2 = \eta(x_1, t)\}.$$

The velocity field $u = (u_1, u_2)(x,t) \in \mathbb{R}^2$ and the pressure field $p = p(x,t) \in \mathbb{R}$ in $\Omega(t)$ are governed by Euler's equations consisting of the balance of forces

$$\partial_t u + (u \cdot \nabla)u = -\nabla p + g(0, -1), \tag{12.13}$$

and the incompressibility condition

$$\partial_{x_1} u_1 + \partial_{x_2} u_2 = 0, \tag{12.14}$$

with the constant of gravitational acceleration $g \in \mathbb{R}$. W.l.o.g. we assume $h = 1$ in the following.

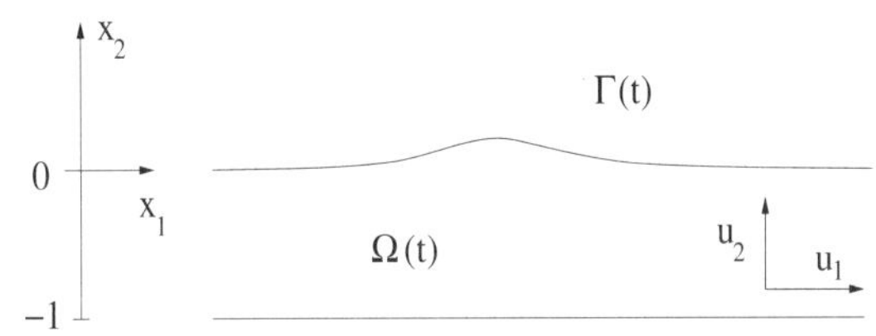

Figure 12.2. The water wave problem

It turns out that a solution with an irrotational initial condition stays irrotational under the evolution of Euler's equation, cf (6.5), and so we can assume

$$\partial_{x_2} u_1 - \partial_{x_1} u_2 = 0, \tag{12.15}$$

i.e., vanishing vorticity. Hence, we have a potential flow $u = \nabla\phi$ with potential $\phi : \Omega(t) \to \mathbb{R}$. The impermeability of the bottom yields the boundary condition $u_2|_{x_2=-1} = 0$. This system is completed with the boundary conditions for the free surface Γ. We assume that $(1, u_1, u_2)$ is tangential to $(t, \Gamma(t))$. Differentiating $x_2 = \eta(x_1, t)$ w.r.t. t, this is transferred into the dynamic boundary condition $u_2 = (\partial_{x_1}\eta)u_1 + \partial_t\eta$, respectively

$$\partial_t \eta = \partial_{x_2}\phi - (\partial_{x_1}\phi)\partial_{x_1}\eta. \tag{12.16}$$

We finally assume that for fixed time t the pressure p is constant along the free surface $\Gamma(t)$ such that after inserting $u = \nabla\phi$ into (12.13), this equation transforms into the Bernoulli equation

$$\partial_t \phi = \frac{1}{2}((\partial_{x_1}\phi)^2 + (\partial_{x_2}\phi)^2) - g\eta \tag{12.17}$$

on the free surface $\Gamma(t)$, cf. Exercise 12.5. If surface tension is included, then the last assumption has to be modified.

It turns out that the evolution of the problem is completely determined by the elevation $\eta = \eta(x_1, t)$ of the top surface and the horizontal velocity component $w = w(x_1, t) = \partial_1 \phi(x_1, \eta(x_1, t), t)$ at the top surface. Inserting $u = \nabla\phi$ into (12.14) yields $\Delta\phi = 0$ which can be solved uniquely up to a constant if the boundary conditions $u_2|_{x_2=-1} = \partial_{x_2}\phi|_{x_2=-1} = 0$ and $w = \partial_1\phi(x_1, \eta(x_1, t), t)$ are given. Hence all derivatives on the right-hand side of (12.16) and (12.17) can be computed if η and w are known, and the water wave problem can be written as an evolutionary system in η and w, alone.

The linear water wave problem. Next we explain that the water wave problem belongs to the class of systems for which the KdV equation can be derived and which has been described above. In order to do so we start with the linear problem

$$\partial_t \eta = \partial_{x_2}\phi|_{x_2=0}, \qquad \partial_t\phi|_{x_2=0} = -\eta, \tag{12.18}$$

where ϕ solves $\Delta\phi = 0$ in $\Omega_0 = \{(x_1, x_2) : x_1 \in \mathbb{R}, x_2 \in (-1, 0)\}$ under the boundary conditions $\partial_{x_2}\phi|_{x_2=-1} = 0$ and $\phi|_{x_2=0} = \Phi$.

Lemma 12.2.1. *The operator $\Phi \mapsto \partial_{x_2}\phi|_{x_2=0}$ is a multiplication operator in Fourier space defined by $\widehat{\partial_{x_2}\phi|_{x_2=0}}(k) = k(\tanh k)\widehat{\Phi}(k)$.*

Proof. The Fourier transform $\widehat{\phi}$ satisfies

$$-k^2\widehat{\phi}(k, x_2) + \partial_{x_2}^2\widehat{\phi}(k, x_2) = 0$$

with the boundary conditions $\partial_{x_2}\widehat{\phi}|_{x_2=-1} = 0$. We obtain solutions

$$\widehat{\phi}(k, x_2) = c_k \cosh(k(1 + x_2)),$$

and so $\widehat{\Phi}|_{x_2=0}(k) = c_k \cosh(k)$ and $\widehat{\partial_{x_2}\phi_{x_2=0}}(k) = kc_k \sinh(k)$. Therefore, we have

$$\widehat{\partial_{x_2}\phi_{x_2=0}}(k) = k(\tanh k)\widehat{\Phi}(k). \qquad \square$$

Remark 12.2.2. In a canal of finite depth h the factor $\tanh(hk)$ is obtained. In the limit of infinite depth we obtain the Heaviside function $\widehat{H}$, where $\widehat{H}(k) = 1$ for $k \geq 0$ and $\widehat{H}(k) = -1$ for $k < 0$. ⌋

Hence, (12.18) possesses solutions $(\Phi, \eta) = (\Phi_k, \eta_k)e^{i(kx+\omega t)}$ with the linear dispersion relation

$$\omega^2 = k\tanh(k), \tag{12.19}$$

and we have two curves of eigenvalues $i\omega_{1,2}$ satisfying $i\omega_{1,2}(0) = 0$. We explained above that this is a necessary condition for the derivation of the KdV equation.

In the equation for w all terms on the right hand side have a ∂_{x_1} in front by construction. Due to the conservation of mass, i.e. $\int \eta(x_1, t)\,dx_1 = const.$ also all terms on the right-hand side in the equation for η must vanish for

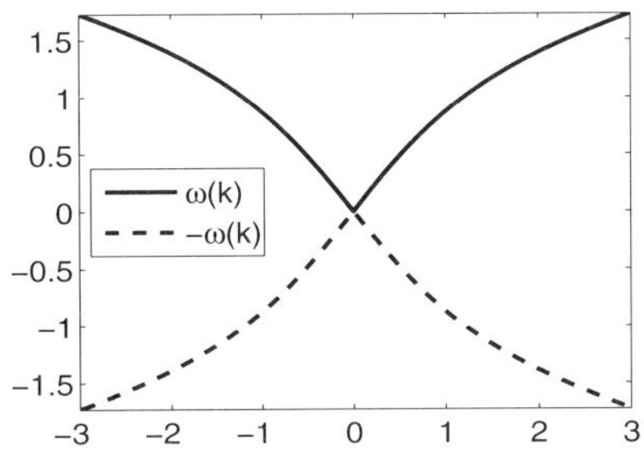

Figure 12.3. The curves of eigenvalues $k \mapsto \omega_{1,2}(k)$ in case of zero surface tension.

$k = 0$. Hence the water wave problem falls into the class of systems described in §12.2.

The KdV approximation. In case of positive surface tension (12.17) changes into

$$\partial_t \phi = \frac{1}{2}((\partial_{x_1}\phi)^2 + (\partial_{x_2}\phi)^2) - g\eta + \sigma\partial_{x_1}\left(\frac{\partial_{x_1}\eta}{\sqrt{1+(\partial_{x_1}\eta)^2}}\right), \tag{12.20}$$

where σ is the surface tension parameter. The linear dispersion relation then modifies into

$$\omega^2 = (k + \sigma k^3)\tanh(k). \tag{12.21}$$

There are essentially three situations, namely $\sigma \in [0, 1/3)$, $\sigma = 1/3$, and $\sigma > 1/3$. Expanding the curves of eigenvalues $k \mapsto \omega_{1,2}(k)$ at the wave number $k = 0$ gives

$$\omega_1(k) = k + \frac{1}{2}(\sigma - \frac{1}{3})k^3 + \mathcal{O}(k^5).$$

Hence, the sign in front of the cubic terms changes at $\sigma = 1/3$. At the same value the inflection points of the curves $k \mapsto \omega_{1,2}(k)$ disappear. See Figure 12.4.

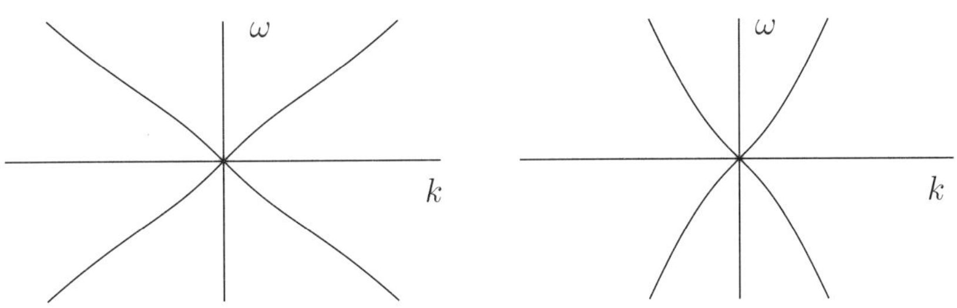

Figure 12.4. The curves of eigenvalues $k \mapsto \omega_{1,2}(k)$ of the water wave problem with surface tension with $\sigma \in (0, 1/3)$ in the left panel and $\sigma > 1/3$ in the right panel.

This is reflected in the associated KdV equations

$$\begin{aligned} 2\partial_T A_+ =& (\sigma - \frac{1}{3}) \partial_X^3 A_+ - \nu_2 \partial_X (A_+^2), \\ 2\partial_T A_- =& (\frac{1}{3} - \sigma) \partial_X^3 A_- + \nu_2 \partial_X (A_-^2), \end{aligned}$$

with coefficient $\nu_2 \in \mathbb{R}$, which are derived by the ansatz

$$\begin{pmatrix} \eta \\ w \end{pmatrix} (x,t) = \varepsilon^2 A_+(\varepsilon(x-t), \varepsilon^3 t) \begin{pmatrix} 1 \\ 1 \end{pmatrix} + \varepsilon^2 A_-(\varepsilon(x+t), \varepsilon^3 t) \begin{pmatrix} 1 \\ -1 \end{pmatrix}.$$

It turns out that $\nu_2 > 0$ and so the solitary waves change at $\sigma = 1/3$ from waves of elevation to waves of depression.

At $\sigma = 1/3$ the coefficient in front of k^3 vanishes. The regime $\sigma = 1/3 + \nu\varepsilon^2$, with $\nu \in \mathbb{R}$, independent of $0 < \varepsilon \ll 1$, is called the Kawahara regime. With the modified ansatz

$$\begin{pmatrix} \eta \\ w \end{pmatrix} (x,t) = \varepsilon^4 A_+(\varepsilon(x-t), \varepsilon^5 t) \begin{pmatrix} 1 \\ 1 \end{pmatrix} + \varepsilon^4 A_-(\varepsilon(x+t), \varepsilon^5 t) \begin{pmatrix} 1 \\ -1 \end{pmatrix}$$

a system of two decoupled Kawahara equations

$$\begin{aligned} -2\partial_T A_+ &= \nu \partial_X^3 A_+ - \nu_1 \partial_X^5 A_+ + \nu_2 \partial_X (A_+^2), \\ 2\partial_T A_- &= \nu \partial_X^3 A_- - \nu_1 \partial_X^5 A_- + \nu_2 \partial_X (A_-^2), \end{aligned} \tag{12.22}$$

with coefficients $\nu_j \in \mathbb{R}$ can be derived. The Kawahara equation has been derived first in [**Kaw72**], describes waves longer than the ones in the KdV scaling, and possesses rather complicated solutions such as multi-pulse solutions or solutions with a spatially chaotic behavior [**BGT96**]. Error estimates for the Kawahara approximation can be found in [**SW01**].

12.2.2. The FPU system. The most famous conservative lattice differential equation where the KdV and NLS (see Exercise 11.4) equation can be derived is the so called Fermi-Pasta-Ulam (FPU) system

$$\partial_t^2 q_j = W'(q_{j+1}(t) - q_j(t)) - W'(q_j(t) - q_{j-1}(t)) \ , \quad j \in \mathbb{Z} \ . \tag{12.23}$$

It was first studied numerically by Fermi, Pasta, and Ulam [**FPU55**] for a finite set of oscillators in order to see how energy was spread through the various modes of the system by the nonlinear coupling via the interparticle forces which are described by the potential function $W : \mathbb{R} \to \mathbb{R}$. They found that at low energy most trajectories did not "thermalize" as expected, but rather exhibited a regular motion. This observation has been explained in [**ZK65**], where Kruskal and Zabusky derived the KdV equation as a formal approximation to the FPU system and in studying the KdV equation numerically they found soliton dynamics. A rigorous proof that long waves in the FPU system can be approximated via the KdV equation has been given in [**SW00a**].

It is the goal of this section to explain how lattice equations can be brought into the abstract form from §11.3 and the beginning of §12.2 for the derivation of these equations. The FPU system (12.23) is rewritten in terms of the difference variables $u(j,t) = q_{j+1}(t) - q_j(t)$, so that (12.23) becomes

$$\partial_t^2 u(j,t) = W'(u(j+1,t)) - 2W'(u(j,t)) + W'(u(j-1,t)) \,, \quad j \in \mathbb{Z} \,. \tag{12.24}$$

With

$$W'(u) = a_1 u + a_2 u^2 + a_3 u^3 + \ldots, \qquad (a_1 > 0, a_j \in \mathbb{R} \text{ for } j \in \mathbb{N}), \tag{12.25}$$

the linear problem is given by

$$\partial_t^2 u(j,t) = a_1(u(j+1,t) - 2u(j,t) + u(j-1,t)). \tag{12.26}$$

It is solved by

$$u(j,t) = \mathrm{e}^{\mathrm{i}(kj+\omega t)} \tag{12.27}$$

for all $k \in \mathbb{R}$, and ω and k related through the linear dispersion relation

$$\omega^2 = -a_1(\mathrm{e}^{ik} - 2 + \mathrm{e}^{-ik}) = 2a_1(1 - \cos k). \tag{12.28}$$

In order to simplify the notation in the following we assume that

$$W'(u) = u + u^2, \tag{12.29}$$

i.e., we set $a_1 = a_2 = 1$, and except of the dangerous quadratic terms we neglect all higher order terms. In order to relate the FPU model to the abstract systems considered in §11.3 and at the beginning of §12.2 we consider the Fourier transformed FPU model. Since

$$\begin{aligned}
&\mathcal{F}(u(\cdot+1)^2 - 2u(\cdot)^2 + u(\cdot-1)^2)(k) \\
&\quad = \frac{1}{2\pi}\sum_{j\in\mathbb{Z}}(u(j+1)^2 - 2u(j)^2 + u(j-1)^2)\mathrm{e}^{-ikj} \\
&\quad = \frac{1}{2\pi}\sum_{j\in\mathbb{Z}}(\mathrm{e}^{ik} - 2 + \mathrm{e}^{-ik})u(j)^2\mathrm{e}^{-ikj} = \frac{1}{2\pi}\sum_{j\in\mathbb{Z}} -\omega(k)^2 u(j)^2\mathrm{e}^{-ikj} \\
&\quad = -\omega(k)^2\frac{1}{2\pi}\sum_{j\in\mathbb{Z}} u(j)^2\mathrm{e}^{-ikj} = -\omega(k)^2\int_{-\pi}^{\pi}\widehat{u}(k-l)\widehat{u}(l)\,\mathrm{d}l,
\end{aligned}$$

the FPU system in Fourier space is given by

$$\partial_t^2\widehat{u}(k,t) = -\omega^2(k)\widehat{u}(k,t) - \omega^2(k)\int_{-\pi}^{\pi}\widehat{u}(k-m,t)\widehat{u}(m,t)\,\mathrm{d}m \tag{12.30}$$

where ω^2 has been defined in (12.28) and where in the convolution integrals the 2π-periodicity of $\widehat{u}$ has to be used. Hence, for the derivation of the NLS and KdV equations and for the justification of the associated approximations we can proceed as in §11.1 and as above. A proof that the NLS approximation makes correct predictions for the FPU model can be found in [**Sch10**]. The KdV and NLS approximation has been justified for the poly-atomic FPU model in [**CCPS12**].

12.2.3. The equations of plasma physics. A plasma is modeled as an ionized fluid consisting of positively charged ions and negatively charged electrons which interact through the electro-magnetic field they generate. The determining equations for the densities, velocities, and pressures of the fluids and the electromagnetic field consist of the equations for the conservation of mass and momentum for the ion fluid and the electron fluid and Maxwell's equation. Under a number of assumptions these equations are simplified. The consideration of a 'cold' ion-fluid allows to neglect the ion pressure. Since the electrons are much lighter than the ions their inertia is neglected. An isothermal constitutive law between the electron pressure and density is chosen, and the time-dependent Maxwell's equations are replaced by electrostatics. In the one-dimensional situation the non-dimensionalized equations are then given by an Euler-Poisson system

$$\partial_t n + \partial_x (nu) = 0, \qquad \partial_t u + u\partial_x u + \partial_x \varphi = 0, \qquad -\partial_x^2 \varphi + \mathrm{e}^{\varphi} = n, \tag{12.31}$$

with φ the electric potential, n the ion density, and u the ion velocity. This system possesses a trivial solution $(n, u, \varphi) = (1, 0, 0)$. We introduce the density fluctuation $v = n - 1$ and find

$$\partial_t v + \partial_x ((1 + v)u) = 0, \qquad \partial_t u + u\partial_x u + \partial_x \varphi = 0, \tag{12.32}$$

and

$$-\partial_x^2 \varphi + \mathrm{e}^{\varphi} = 1 + v.$$

The linearized system is given by

$$\partial_t v + \partial_x u = 0, \qquad \partial_t u + \partial_x \varphi = 0, \qquad -\partial_x^2 \varphi + \varphi = v. \tag{12.33}$$

It is solved by the Fourier modes $(v, u, \varphi) = (\widehat{v}, \widehat{u}, \widehat{\varphi})\mathrm{e}^{\mathrm{i}kx + \mathrm{i}\omega t}$. We find with $\widehat{\varphi} = \widehat{v}/(1 + k^2)$ that

$$\omega^2 = \frac{k^2}{1 + k^2}.$$

For small v the third equation of (12.32) can be solved w.r.t. φ, i.e., we have

$$\varphi(v) = (-\partial_x^2 + 1)^{-1} v + S(v),$$

with $\|S(v)\|_{H^{s+2}} \leq C\|v\|_{H^s}^2$ for small v and $s > 1/2$. Hence, (12.32) can be written as evolutionary system in u and v alone, namely

$$\partial_t v + \partial_x ((1 + v)u) = 0, \qquad \partial_t u + \partial_x (u^2)/2 + \partial_x ((-\partial_x^2 + 1)^{-1} v + S(v)) = 0.$$

This representation shows that the right-hand side vanishes in Fourier space at the wave number $k = 0$ and so this system falls into the class of systems which has been described above and for which a KdV equation can be derived. With the ansatz

$$\begin{pmatrix} v \\ u \end{pmatrix} = \varepsilon^2 A(\varepsilon(x-t), \varepsilon^3 t) \begin{pmatrix} 1 \\ 1 \end{pmatrix}$$

we find the KdV equation

$$\partial_T A + \frac{1}{2}\partial_X^3 A + A\partial_X A = 0.$$

An approximation result for this KdV approximation has been proven in [**GP14**].

12.3. Whitham, Boussinesq, BBM, etc.

There are many other modulation equations which have been derived in the long wave limit. It is the purpose of this section to comment on the use of these equations and to discuss their validity. We start with modulation equations which are independent of ε and then come to the zoo of modulation equations which has been derived for the water wave problem by using lower order relations to modify the higher order terms.

Modulation equations which are independent of ε. For (12.1) we make the ansatz

$$u(x,t) = \varepsilon^{\alpha} A(X,T), \tag{12.34}$$

where $X = \varepsilon(x-t)$, $T = \varepsilon^{1+\alpha}t$, $A(X,T) \in \mathbb{R}$, and $\alpha > 0$. For $\alpha > 2$ the Airy equation $\partial_T A = \frac{1}{2}\partial_X^3 A$ occurs. The KdV equation is recovered for $\alpha = 2$, and for $\alpha \in (0,2)$ the inviscid Burgers equation

$$\partial_T A = -\frac{1}{2}\partial_X(A^2) \tag{12.35}$$

is obtained. There is another long wave limit which leads to an ε-independent non-trivial amplitude equation. With the ansatz

$$u(x,t) = U(X,T), \tag{12.36}$$

where $X = \varepsilon x$, $T = \varepsilon t$, $U(X,T) \in \mathbb{R}$, and $0 < \varepsilon \ll 1$ a small perturbation parameter, we obtain

$$\partial_T^2 U = \partial_X^2 U + \partial_X^2(U^2) \tag{12.37}$$

which can be written as a system of conservation laws

$$\partial_T U = \partial_X V, \qquad \partial_T V = \partial_X U + \partial_X(U^2). \tag{12.38}$$

This so called Whitham limit has been considered first in the description of slow modulations in time and space of a periodic traveling wave in a dispersive wave system [**Whi99, CS98**]. Since solutions of order $\mathcal{O}(1)$ are considered an additional smallness condition in the approximation theorem is necessary [**DS09, BDS16**]. The proof of approximation results for these modulation equations goes almost line for line as in §12.1.

The zoo of ε-dependent long wave modulation equations. Other modulation equations, such as the regularized long wave equation, the BBM system, etc., have been derived in the long wave limit, too. There have been many scientific discussions about the question which equation of this zoo of long wave modulation equations is the most suitable one for the description of the long wave limit. The Boussinesq equation (12.1) and modified versions of it can be derived in the long wave limit from the water wave problem, too. However, then it still contains the small perturbation parameter ε, i.e., it is of the form

$$\partial_\tau^2 v = \partial_\xi^2 v - \varepsilon^2 \partial_\xi^4 v + \varepsilon^2 \partial_\xi^2 (v^2), \tag{12.39}$$

with $u = \varepsilon^2 v$, $\tau = \varepsilon t$, and $\xi = \varepsilon x$. Very often the statement can be found that it only can be derived under the assumption of unidirectional waves. This is wrong. For instance in [**Cra85, SW00b**] it has been derived without using this assumption for the so called Lagrangian formulation of the water wave problem. For the Eulerian formulation of the water wave problem we have $\eta(x,t) \sim \varepsilon^2 v(\varepsilon x, \varepsilon t)$, while the expression for w is more complicated. Instead of the water wave problem we take (12.39) as starting point for the derivation of the above mentioned modulation equations. Using $\partial_\tau^2 v = \partial_\xi^2 v + \mathcal{O}(\varepsilon^2)$ yields the system

$$\partial_\tau^2 v = \partial_\xi^2 v - \varepsilon^2 \partial_\tau^2 \partial_\xi^2 v + \varepsilon^2 \partial_\xi^2 (v^2) + \mathcal{O}(\varepsilon^4) \tag{12.40}$$

which in the long wave limit is formally equivalent to (12.39) up to terms of order $\mathcal{O}(\varepsilon^4)$. Ignoring the terms of order $\mathcal{O}(\varepsilon^4)$ however gives an ill-posed system. In principle by this approach one gets a two parameter family of equations, namely

$$\partial_\tau^2 v = \partial_\xi^2 v - (1-\alpha)\varepsilon^2 \partial_\xi^4 v - \alpha \varepsilon^2 \partial_\tau^2 \partial_\xi^2 v + (1-\beta)\varepsilon^2 \partial_\xi^2 (v^2) + \beta \varepsilon^2 \partial_\tau^2 (v^2),$$

with arbitrary $\alpha, \beta \in \mathbb{R}$. If our starting Boussinesq system contains $+\partial_x^4 u$ instead of $-\partial_x^4 u$ then it is ill posed. However, the counterpart to (12.40) then is well-posed.

This is not the end of the game of creating a zoo of other long wave limit modulation equations. We write the Boussinesq equation (12.39) in long wave form as a first order system

$$\partial_\tau v = \partial_\xi w, \qquad \partial_\tau w = \partial_\xi v - \varepsilon^2 \partial_\xi^3 v + \varepsilon^2 \partial_\xi (v^2).$$

This system is next diagonalized in lowest order, i.e., we introduce $Y = v + w$ and $Z = v - w$, and find

$$\partial_\tau Y = \partial_\xi Y - \varepsilon^2 \partial_\xi^3 (Y+Z)/2 + \varepsilon^2 \partial_\xi ((Y+Z)^2)/4, \tag{12.41}$$

$$\partial_\tau Z = -\partial_\xi Z + \varepsilon^2 \partial_\xi^3 (Y+Z)/2 - \varepsilon^2 \partial_\xi ((Y+Z)^2)/4. \tag{12.42}$$

Now we can use

$$\partial_\tau Y = \partial_\xi Y + \mathcal{O}(\varepsilon^2), \quad \text{and} \quad \partial_\tau Z = -\partial_\xi Z + \mathcal{O}(\varepsilon^2) \tag{12.43}$$

to manipulate the right-hand side. We will mainly leave this to the reader, and only remark that with this procedure a large number of modulation equations can and have been obtained in the literature.

In order to derive the KdV equation from (12.41)-(12.42) we set $Z = 0$ and obtain in the Y equation

$$\partial_\tau Y = \partial_\xi Y - \varepsilon^2 \partial_\xi^3 Y/2 + \varepsilon^2 \partial_\xi (Y^2)/4 \tag{12.44}$$

By going into the moving frame $Y(\xi, \tau) = A(\xi - \tau, \varepsilon^2 \tau)$ we arrive at the standard form of the KdV equation

$$\partial_T A = \frac{1}{2} \partial_X^3 A + \frac{1}{4} \partial_X (A^2)$$

which is now independent of the small perturbation parameter $0 < \varepsilon \ll 1$. Starting from (12.44) by using again (12.43) allows to derive the so called regularized long wave equation

$$\partial_\tau Y = \partial_\xi Y - \frac{1}{2} \varepsilon^2 \partial_\tau \partial_\xi^2 Y + \frac{1}{4} \varepsilon^2 \partial_\xi (Y^2). \tag{12.45}$$

This equation is ill-posed, but for the ill-posed version of (12.1), namely $\partial_t^2 u = \partial_x^2 u + \partial_x^4 u + \partial_x^2 (u^2)$ one obtains a well-posed regularized long wave equation, namely

$$\partial_\tau Y = \partial_\xi Y + \frac{1}{2} \varepsilon^2 \partial_\tau \partial_\xi^2 Y + \frac{1}{4} \varepsilon^2 \partial_\xi (Y^2).$$

Since this equation can be written as

$$\partial_\tau Y = (1 - \varepsilon^2 \partial_\xi^2)^{-1} (\partial_\xi Y + \frac{1}{4} \varepsilon^2 \partial_\xi (Y^2)) \tag{12.46}$$

the local existence and uniqueness of solutions can be established with the Picard-Lindelöf theorem for every $\varepsilon > 0$. However, the equation still depends on $\varepsilon > 0$ and the advantage of a simpler local existence and uniqueness theory vanishes for $\varepsilon \to 0$ since the Lipschitz-constant on the right-hand side of (12.46) tends to infinity for $\varepsilon \to 0$.

All bi-directional models can be approximated for localized initial conditions by a system of two (decoupled) KdV equations and all uni-directional models can be approximated by a single KdV equation which in contrast to the other equations are independent of the small perturbation parameter $0 < \varepsilon \ll 1$. Even more is true. For the water wave problem the complete zoo of long wave modulation equations is derived in the same limit, namely for initial conditions of the form

$$\eta(x_1, 0) = \varepsilon^2 \Phi_0(\varepsilon x_1) \qquad \text{and} \qquad w(x_1, 0) = \varepsilon^2 \Phi_1(\varepsilon x_1).$$

We explain in the next subsection that at this point the analysis goes beyond the formal calculations and makes a decision which of the equations should be taken to approximate the long wave limit. The answer of [**SW00b, SW02**] is that two ε-independent decoupled KdV equations are sufficient to describe this limit. On the $\mathcal{O}(1/\varepsilon^3)$ time scale for small $\varepsilon > 0$ no other small amplitude dynamics can be found in the original system via the other ε-dependent modulation equations.

12.4. The long wave limit

We consider the Boussinesq equation (12.1) with initial conditions

$$u(x,0) = \varepsilon^2 \Phi_0(\varepsilon x), \qquad \partial_t(\partial_x^2 - \partial_x^4)^{-1/2} u(x,0) = \varepsilon^2 \Phi_1(\varepsilon x). \tag{12.47}$$

It is the goal of this section to prove that for small $\varepsilon > 0$ and for spatially decaying $\Phi_0, \Phi_1 \in H_2^\theta \cap H^{\theta+4}$ the dynamics of (12.1) can be described by two decoupled KdV equations describing wave packets moving to the left and to the right. In order to do so we set

$$A_+(X,0) = \frac{1}{\sqrt{2}\varepsilon^2}(\Phi_0(X) + \Phi_1(X)),$$
$$A_-(X,0) = \frac{1}{\sqrt{2}\varepsilon^2}(\Phi_0(X) - \Phi_1(X)),$$

and prove the following approximation theorem.

Theorem 12.4.1. *Fix $\theta > 1/2$, and let $A_+ \in C([0,T_0], H_2^{\theta+4})$ and $A_- \in C([0,T_0], H_2^{\theta+6})$ be solutions of the system of KdV equations* (12.4). *Then there exist $\varepsilon_0 > 0$ and $C > 0$ such that for all $\varepsilon \in (0, \varepsilon_0]$ we have*

$$\sup_{t \in [0,T_0/\varepsilon^3]} \|u(\cdot,t) - (\varepsilon^2 A_+(\varepsilon(\cdot - t), \varepsilon^3 t) + \varepsilon^2 A_-(\varepsilon(\cdot + t), \varepsilon^3 t))\|_{H^\theta} \leq C\varepsilon^{7/2}$$

for the solutions u of (12.1) *with initial conditions* (12.47).

Proof. For notational simplicity we only consider the case $\theta = 2$. For the approximation

$$\varepsilon^2 \Psi(x,t,\varepsilon) = \varepsilon^2 A_+(\varepsilon(x - ct), \varepsilon^3 t) + \varepsilon^2 A_-(\varepsilon(x + ct), \varepsilon^3 t)$$

we find using the previous calculations that

$$\mathrm{Res}(\varepsilon^2 \Psi) = -\varepsilon^8 \partial_T^2 A_+ - \varepsilon^8 \partial_T^2 A_- + 2\varepsilon^6 \partial_X^2 (A_+ A_-).$$

As before we have

$$4\partial_T^2 A_+ = \partial_X^3(\partial_X^3 A_+ - \partial_X(A_+^2)) - 2\partial_X(A_+(\partial_X^3 A_+ - \partial_X(A_+^2))).$$

and similarly for A_-. Next we have

$$
\begin{aligned}
&\|A_+(\cdot - c\varepsilon^{-2}T, T)A_-(\cdot + c\varepsilon^{-2}T, T)\|_{H^2} \\
\leq &\sup_{X\in\mathbb{R}} |\frac{1}{(1+(X+c\varepsilon^{-2}T)^2)(1+(X-c\varepsilon^{-2}T)^2)}| \|A_+(\cdot,T)\|_{H^2_2}\|A_-(\cdot,T)\|_{H^2_2} \\
\leq &\frac{C}{1+(\varepsilon^{-2}T)^2}\|A_+(\cdot,T)\|_{H^2_2}\|A_-(\cdot,T)\|_{H^2_2}.
\end{aligned}
$$

Therefore, we have with the previous arguments

Lemma 12.4.2. *Fix $\widetilde{\theta} \geq 0$, and let $A_+ \in C([0,T_0], H_2^{\widetilde{\theta}+6})$ and $A_- \in C([0,T_0], H_2^{\widetilde{\theta}+6})$ be solutions of the system of KdV equations (12.4). Then there exist $\varepsilon_0 > 0$ and $C_{\mathrm{res}} > 0$ such that for all $\varepsilon \in (0,\varepsilon_0)$ we have*

$$
\| \mathrm{Res}(\varepsilon^2\Psi(\cdot,t,\varepsilon))\|_{H^{\widetilde{\theta}}} \leq C_{\mathrm{res}}(\varepsilon^{15/2} + \varepsilon^{11/2}\frac{1}{1+(\varepsilon t)^2})
$$

and

$$
\|\partial_x^{-1}\mathrm{Res}(\varepsilon^2\Psi(\cdot,t,\varepsilon))\|_{H^{\widetilde{\theta}}} \leq C_{\mathrm{res}}(\varepsilon^{13/2} + \varepsilon^{9/2}\frac{1}{1+(\varepsilon t)^2})
$$

for all $t \in [0, T_0/\varepsilon^3]$.

As before the difference $\varepsilon^{7/2}R = u - \varepsilon^2\Psi$ satisfies (12.5) and as before we estimate R with the help of the energy E which has been defined in (12.6). Similarly, to the derivation of (12.7), we find now

$$
\frac{\mathrm{d}}{\mathrm{d}t}E \leq 2C\varepsilon^3 E + C\varepsilon^{7/2}E^{3/2} + C\varepsilon^3 + C\varepsilon\frac{1}{1+(\varepsilon t)^2}, \tag{12.48}
$$

with a constant C independent of $\varepsilon \in (0,\varepsilon_1)$. Under the assumption that $C\varepsilon^{1/2}E^{1/2} \leq 1$ we obtain

$$
\frac{\mathrm{d}}{\mathrm{d}t}E \leq (2C+1)\varepsilon^3 E + C\varepsilon^3 + C\varepsilon\frac{1}{1+(\varepsilon t)^2}.
$$

Using $\int_{-\infty}^{\infty}\varepsilon\frac{1}{1+(\varepsilon t)^2}\,dt = \pi$, Gronwall's inequality immediately gives the bound

$$
\sup_{t\in[0,T_0/\varepsilon^3]} E(t) = C(T_0+\pi)\mathrm{e}^{(2C+1)T_0} =: M = \mathcal{O}(1).
$$

Finally choosing $\varepsilon_2 > 0$ so small that $C\varepsilon_2^{1/2}M^{1/2} \leq 1$ gives the required estimate for all $\varepsilon \in (0,\varepsilon_0)$ with $\varepsilon_0 = \min(\varepsilon_1,\varepsilon_2) > 0$. □

Remark 12.4.3. a) We leave it to the reader to check that initial conditions in H^θ are sufficient, but then the relative error is $o(1)$ for $\varepsilon \to 0$.
b) The existence of solutions for the KdV equation in H_2^θ-spaces is guaranteed for initial conditions in $H_2^\theta \cap H^{\theta+4}$, cf. [**SW00b**, Lemma 6.4]. ⌋

Remark 12.4.4. The analogous result for the water wave problem gives an almost complete description of its long-time behavior in the long wave limit, cf. [**SW00b, SW02**]. On a time scale $\mathcal{O}(1/\varepsilon)$ the solutions split up into two wave packets, one moving to the right and one to the left. These wave packets evolve independently as solutions of the system of KdV equations. Their long-time behavior can be computed explicitly with the help of the inverse scattering transform, cf. Remark 8.2.10. Some solitons which are ordered w.r.t. their height evolve out of a dispersive remainder, see Figure 12.5, cf. [**EvH81**]. Approximation results such as Theorem 12.1.1 imply that the same behavior will be observed in (12.1) in the long wave limit. The approximation results from [**SW00b, SW02**] show the same behavior for the water wave problem. ⌋

Figure 12.5. The long time behavior for the water wave problem in the long wave limit for $t = T_0/\varepsilon^3$ with T_0 large. The solutions of the water wave problem in the long wave limit splits up into two wave packets, one moving to the right and one to the left, where each of these wave packets evolves independently as a solution of a KdV equation. For large T_0 solitons evolve out of a dispersive remainder.

Further Reading. There are various formulations of the water wave problem which differ in the chosen parametrization of the top surface. For the different formulations there are a number of local existence and uniqueness theorems. For the Eulerian formulation in §12.2.1, local existence and uniqueness results have been shown for instance in [**Shi76, KN79**] with the Cauchy-Kowalevslaya approach, and in [**Lan05**] with a Nash-Moser approach. For all other formulations the proofs are adaptions of the local existence and uniqueness theory for quasilinear hyperbolic equations [**Kat75**]. For the Lagrangian formulation such results have been shown for instance in [**Nal74, Yos82, Yos83, Cra85, Wu97, Wu99, SW00b, Igu01, SW02**]. For the arc-length description of $\Gamma(t)$ a local existence and uniqueness theorem has been shown in [**AM05**]. The theorems can be distinguished w.r.t. 2D or 3D, finite or infinite depth, with or without surface tension, and regularity of the initial conditions. See also [**ABZ14**]. Recently, a number of almost global and global existence and uniqueness results have been established [**AL08, Wu09, Wu11, GMS12, AD15**]. We refer to [**Lan13**] as recent textbook.

The Camassa-Holm equation has been derived [**CH93a**] as a modulation equation for the description of unidirectional surface water waves of an irrotational, inviscid fluid in an infinitely long canal of fixed constant depth, see also [**FL96, Joh02, DGH03**]. Similarly to the KdV equation, it is obtained in the limit of small amplitude and in the limit of long waves, but where now the limits are considered independently. The peakon equation is a special form of the Camassa-Holm equation and has attracted a lot of interest due to its complete integrability, and due to solitary wave solutions with a discontinuous derivative at the peak, which look like breaking water waves. See for instance [**CH93b, ACHM94, ACF$^+$99, ACF$^+$01**] and the references therein. The question whether solutions of the water wave problem can really be approximated by solutions of the Camassa-Holm equation has been addressed in a number of papers. In [**KS05**] an approximation theorem has been established which says that in the KdV-limit solutions of the Camassa-Holm equation remain close to the solutions of the associated KdV equation. Therefore, on the KdV time scale no intrinsic Camassa-Holm dynamics can be seen in the water wave problem since this dynamics happens on a much longer time scale than the KdV dynamics. In [**KS05**] also a discussion about the non-validity of the peakon equation in this limit can be found. A different asymptotic limit has been considered in [**CL09**] where approximation and blow up results are established. The blow up results are related to wave breaking in the full water wave problem. Higher order long wave limit approximations have been established in [**WW02, Wri06**].

The KdV limit of the Vlasov-Poisson system has been considered in [**Han13**]. Long wave limits in higher space dimensions are considered in many papers. In [**LLS13**] the Zakharov-Kusnetzov equation has been derived for the Euler-Poisson system. The validity and limitations of the KP equation has been discussed in [**GS01a**].

Exercises

12.1. Prove that solutions $u(x,t) = v(x+t, \varepsilon^2 t)$ of the regularized long wave equation

$$\partial_t u = \partial_x u + \varepsilon^2 \partial_x^3 u + \varepsilon^2 \partial_x (u^2)$$

can be approximated via the solutions v of the associated KdV equation on an $\mathcal{O}(1/\varepsilon^2)$ time scale.

12.2. Use simultanuosly

$$\varepsilon^2 \partial_\xi^3 Y = \varepsilon^2 \partial_\xi^2 \partial_\tau Y + \mathcal{O}(\varepsilon^4) \quad \text{and} \quad \varepsilon^2 \partial_\xi^3 Z = \varepsilon^2 \partial_\xi^2 \partial_\tau Z + \mathcal{O}(\varepsilon^4)$$

in (12.41)-(12.42) with different prefactors α_j, and bring these terms to the right-hand side. Then use

$$\partial_\tau Y - \alpha \varepsilon^2 \partial_\xi^2 \partial_\tau Y = (1 - \alpha \varepsilon^2 \partial_\xi^2) \partial_\tau Y$$

together with

$$\begin{pmatrix} 1-\alpha_1\varepsilon^2\partial_\xi^2 & -\alpha_2\varepsilon^2\partial_\xi^2 \\ -\alpha_3\varepsilon^2\partial_\xi^2 & 1-\alpha_4\varepsilon^2\partial_\xi^2 \end{pmatrix}^{-1} = \begin{pmatrix} 1 & 0 \\ 0 & 1 \end{pmatrix} + \varepsilon^2 \begin{pmatrix} \alpha_1\partial_\xi^2 & \alpha_2\partial_\xi^2 \\ \alpha_3\partial_\xi^2 & \alpha_4\partial_\xi^2 \end{pmatrix} + \mathcal{O}(\varepsilon^4).$$

to manipulate the coefficients on the right-hand side further and derive an approximation system which is no longer symmetric w.r.t. to the interchange of Y and Z.

12.3. Discuss the form of the smooth curves $\omega_{1,2} = \omega_{1,2}(k)$ defined by

$$\omega^2 = (k^2 + \mu k^4)/(1 + k^2)$$

for $|k| \to \infty$ and $|k| \to 0$. Compute $\partial_k^j \omega_1|_{k=0}$ for $j = 0, \ldots, 4$. Is there a change of sign?

12.4. Consider the two-dimensional Boussinesq model

$$\partial_t^2 u = \Delta u + \partial_t^2 \Delta u + \Delta(u^2)$$

with $u = u(x, y, t) \in \mathbb{R}$, x, y, $t \in \mathbb{R}$, and $\Delta = \partial_x^2 + \partial_y^2$ for modeling three-dimensional water waves. By making the ansatz $u = \varepsilon^2 A(\varepsilon(x-t), \varepsilon^2 y, \varepsilon^3 t)$ derive the Kadomtsev-Petviashvili (KP) equation

$$\partial_X \partial_T A = \partial_X(\nu_1 \partial_X^3 A + \nu_2 A \partial_X A) + \nu_3 \partial_Y^2 A,$$

with $X = \varepsilon(x-t)$, $Y = \varepsilon^2 y$. Compute the values of the coefficients ν_1, ν_2, and $\nu_3 \in \mathbb{R}$. *Remark.* The solutions of the KP equation describe unidirectional waves slowly modulated in the direction normal to the direction of propagation. For this model problem it has been pointed out in [**GS01b**] that the approximation property for the KP equation depends strongly on the chosen initial condition of the KP equation.

12.5. Derive the Bernoulli equation (12.17) from the Euler equations (12.13).

12.6. Derive a KdV equation for $\partial_t^2 u = \partial_x^2 u + \partial_x^2 \partial_t^2 u - \mu(\partial_x^4 u + \partial_x^4 \partial_t^2) u + \partial_x^2(u^2)$ by making the ansatz $u(x,t) = \varepsilon^2 A(\varepsilon(x - c_0 t), \varepsilon^3 t)$.

12.7. Derive a Kawahara equation $\partial_T A = c_1 \partial_X^3 A + c_2 \partial_X^5 A + c_3 \partial_X(A^2)$ with $T \in \mathbb{R}$, $X \in \mathbb{R}$, coefficients $c_j \in \mathbb{R}$, and amplitude $A(X,T) \in \mathbb{R}$ for

$$\partial_t^2 u = \partial_x^2 u + \partial_x^2 \partial_t^2 u - \mu(\partial_x^4 u + \partial_x^4 \partial_t^2 u) + \partial_x^2(u^2),$$

in case $1 + \mu = \varepsilon^2 \nu$, by making the ansatz $u(x,t) = \varepsilon^4 A(\varepsilon(x - c_0 t), \varepsilon^5 t)$.

12.8. Find solitary wave solutions of the Boussinesq equation

$$\partial_t^2 u - \partial_x^2 u + 3\partial_x^2(u^2) - \partial_x^4 u = 0$$

in the form $u(x,t) = a \operatorname{sech}^2(b(x - ct))$. Discuss the well-posedness of this Boussinesq equation.

Chapter 13

Center manifold reduction and spatial dynamics

In Chapters 10-12, our focus has been on the approximation of the dynamics of complicated PDEs on the real line or cylindrical domains by simple modulation equations. Often, by this approximation we described special solutions such as bifurcating spatially periodic pattern, solitary waves, modulating front solutions, or modulating pulse solutions. It is not clear that the corresponding exact solutions really exist in the original system, too. The last decades saw big efforts and successes in the construction of these special solutions using the Lyapunov-Schmidt reduction and the center manifold reduction. There are a number of overview articles [**Van89, VI92**] and textbooks [**Car81, HI11**] where especially center manifold theory and its applications are well explained. We concentrate on the aspects of the theory and its applications which have to do with modulation equations.

13.1. The center manifold theorem

The center manifold theorem has been introduced in §3.2 in the finite-dimensional situation. Here we explain how this approach can be generalized to the infinite-dimensional case. In the finite-dimensional case we considered

$$\begin{aligned} \dot{u}_c &= B_c u_c + \widetilde{g}_c(u_c, u_s, u_u), \\ \dot{u}_s &= B_s u_s + \widetilde{g}_s(u_c, u_s, u_u), \\ \dot{u}_u &= B_u u_u + \widetilde{g}_u(u_c, u_s, u_u), \end{aligned} \tag{13.1}$$

with u_c, u_s, u_u some finite-dimensional vectors, B_c a matrix with eigenvalues on the imaginary axis, B_s a matrix with eigenvalues with negative real part, B_u a matrix with eigenvalues with positive real part, and $\widetilde{g}_c, \widetilde{g}_s, \widetilde{g}_u : C^{r+1}$-maps without constant and linear terms. Then the center manifold theorem, stated in Theorem 3.2.1, guarantees the existence of a neighborhood U of $u_c = 0$, and of a C^r-map $h : u_c \mapsto h(u_c)$ defined in U, such that the manifold

$$W_c = \{u = u_c \oplus h(u_c) : u_c \in U, (u_s, u_u) = h(u_c)\}.$$

is invariant under the flow of (13.1). W_c is called a center manifold. The reduced flow on this in general non-unique center manifold is determined by

$$\dot{u}_c = B_c u_c + \widetilde{g}_c(u_c, h_s(u_c), h_u(u_c)), \tag{13.2}$$

The function $h = (h_s, h_u)$ contains no constant and no linear terms w.r.t. u_c.

In the infinite-dimensional case the situation is as above except that u_s and u_u are now allowed to be infinite-dimensional vectors. The variable u_c will still be of finite dimension such that (13.2) is a finite-dimensional ODE. Hence, the center manifold theorem allows to reduce an infinite-dimensional problem to a finite-dimensional one.

However, a number of things which are obvious in the finite-dimensional case are less clear in the infinite-dimensional case, and so we have to impose a number of additional assumptions. In order to formulate these assumptions we have to take a look at the existence proof. Usually the system

$$\begin{aligned} u_c(t) &= e^{tB_c} u_c(0) + \int_0^t e^{(t-\tau)B_c} \widetilde{g}_c((u_c, u_s, u_u)(\tau)) \, d\tau, \\ u_s(t) &= \int_{-\infty}^t e^{(t-\tau)B_s} \widetilde{g}_s((u_c, u_s, u_u)(\tau)) \, d\tau, \\ u_u(t) &= -\int_t^{\infty} e^{(t-\tau)B_u} \widetilde{g}_u((u_c, u_s, u_u)(\tau)) \, d\tau, \end{aligned} \tag{13.3}$$

which is obtained from the variation of constant formula, is used for the construction of the center manifold.

Remark 13.1.1. To simplify our notation we do not distinguish between $(u_s, u_c, u_u) = (0, u_c, 0)$ and u_c, and similarly for u_s and u_u, i.e., in our notation we can have $u = u_s \oplus u_c \oplus u_u$ and $u = u_s + u_c + u_u$. From the context the exact meaning will be clear. ⌋

For $u \in \ell_{2,\theta}$ sufficiently small the right-hand side of (13.3) turns out to be a contraction in a space of slowly exponentially growing functions. In the finite-dimensional case e^{tB_c}, e^{tB_s}, e^{tB_u} are well-defined and continuous from $\mathbb{R}^d$ to $\mathbb{R}^d$. In the infinite-dimensional case this has to be assumed. For our purposes it is sufficient to assume the existence of $\beta_+ > 0$ and $\beta_- < 0$ such that $(e^{tB_s})_{t\geq 0}$ and $(e^{tB_u})_{t\leq 0}$ define C_0-semigroups in some space $\ell_{2,\theta}$ which satisfy

$$\begin{aligned} i) \qquad & \|e^{tB_u}\|_{\ell_{2,\theta}\to\ell_{2,\theta}} \leq M e^{\beta_+ t}, \quad \forall\, t \leq 0, \\ ii) \qquad & \|e^{tB_s}\|_{\ell_{2,\theta}\to\ell_{2,\theta}} \leq M e^{\beta_- t}, \quad \forall\, t \geq 0. \end{aligned}$$

with a constant M. Moreover, the nonlinear terms $\widetilde{g}_c$, $\widetilde{g}_s$, and $\widetilde{g}_u$ should be C^{r+1} as functions from $\ell_{2,\theta}$ to $\ell_{2,\theta}$. According to §5.1.2 in case of diagonal matrices B_u and B_s the assumptions on the semigroups follow from the position of the spectral values, for instance by assuming that

$$\beta_+ = \inf\{\operatorname{Re}\lambda : \lambda \in \sigma(B_u)\} > 0, \quad \text{and} \quad \beta_- = \sup\{\operatorname{Re}\lambda : \lambda \in \sigma(B_s)\} < 0.$$

Since B_c is finite-dimensional we have for all $\varepsilon > 0$ the existence of a $M > 0$ such that

$$\|e^{tB_c}\|_{\ell_{2,\theta}\to\ell_{2,\theta}} \leq M e^{\varepsilon|t|}, \quad \forall\, t \in \mathbb{R}. \tag{13.4}$$

By assuming the existence of e^{tB_c}, with this last estimate we can construct a center manifold also in case that B_c is infinite-dimensional, cf. [**GS01b, GS05**]. Similar to §5.1.2 the assumptions on the semigroup and nonlinearity can be weakened.

The cut-off. Center manifolds in general only exist in a small neighborhood of the fixed point, here $u = 0$. In order to construct them, we modify the original system outside this neighborhood without changing the dynamics close to the fixed point. To do so, we modify the nonlinearity $\widetilde{g}$ outside a neighborhood of $u = 0$ by some cut-off function $\chi \in C^\infty([0,\infty),\mathbb{R})$ with the properties (i) $0 \leq \chi(r) \leq 1$ for all $r \in [0,\infty)$, (ii) $\chi(r) = 1$, if $r \in [0,1]$ and (iii) $\chi(r) = 0$, if $r \geq 2$. We define $\widetilde{g}_\rho(u) = \widetilde{g}(u)\chi(\rho^{-1}\|u\|_{\ell_{2,\theta}})$. Since $\widetilde{g}_\rho(u) = \widetilde{g}(u)$ for $u \in B_\rho = \{u \in \ell_{2,\theta} : \|u\|_{\ell_{2,\theta}} \leq \rho\}$ the vector fields of the modified system and of the original system (13.1) coincide for all $u \in B_\rho$. In particular we have

Lemma 13.1.2. *Let $\widetilde{g} \in C^k(\ell_{2,\theta},\ell_{2,\theta})$ for a $k \geq 1$ and $\widetilde{g}_\rho$ defined as above. Then $\widetilde{g}_\rho \in C^k(\ell_{2,\theta},\ell_{2,\theta})$ and*

$$\lim_{\rho\to 0} \sup_{u\in\ell_{2,\theta}} \|D\widetilde{g}_\rho(u)\|_{\ell_{2,\theta}\to\ell_{2,\theta}} = 0.$$

Proof. The proof is obvious, since $\widetilde{g}(u) = \mathcal{O}(\|u\|^2_{\ell_{2,\theta}})$ for $u \to 0$. □

Existence and invariance. Since we are only interested in the dynamics close to $u = 0$, we consider in the following (13.1) with $\widetilde{g}$ replaced by $g = \widetilde{g}_\rho$ with g satisfying $g \in C^k(\ell_{2,\theta},\ell_{2,\theta})$ for a $k \geq 1$, $g(0) = 0$, and $Dg(0) = 0$. The existence theorem for the center manifold is a follows.

Theorem 13.1.3. *Let $\eta \in (0,\beta)$ with $\beta = \min(-\beta_-,\beta_+)$. There exists a $\delta_0 > 0$ such that for g with $\sup_{u\in\ell_{2,\theta}} \|Dg(u)\|_{\ell_{2,\theta}} \leq \delta_0$ the center manifold,* *a smooth manifold in* $\ell_{2,\theta}$ *which is invariant under* (13.1) *and tangential to the center subspace, can be characterized as*

$$W_c = \{u_0 \in \ell_{2,\theta} : \sup_{t\in\mathbb{R}} \mathrm{e}^{-\eta|t|}\|u(t,u_0)\|_{\ell_{2,\theta}} < \infty\}$$

for a suitable chosen small $\eta > 0$. *This set is unique. In detail, there exists a neighborhood* $U \subset \ell_{2,\theta}$ *of* $u_c = 0$ *and a* $(k-1)$*-times differentiable* $h : u_c \mapsto (u_s, u_u)$ *such that* $W_c = \{u_c + h(u_c) : u_c \in U\}$.

Remark 13.1.4. The uniqueness formulated in this theorem is not contradicting the non-uniqueness shown in Example 3.2.4, since the uniquely constructed manifold depends on the choice of the cut-off function. ⌋

The full proof of Theorem 13.1.3 is rather lengthy. We restrict ourselves to a few points and refer to [**Van89**] and [**VI92**] for more details. We will discuss the invariance, the existence, and the Lipschitz-continuity of the center manifold, but leave out the proof of the $(k-1)$-times differentiability.

Proof of Theorem 13.1.3. i) The invariance of W_c follows directly from the definition. We have $u(\tau, u_0) \in W_c$, if $u_0 \in W_c$, since

$$\sup_{t\in\mathbb{R}} \mathrm{e}^{-\eta|t|}\|u(t,u(\tau,u_0))\|_{\ell_{2,\theta}} \leq \mathrm{e}^{\eta|\tau|} \sup_{t\in\mathbb{R}} \mathrm{e}^{-\eta|t+\tau|}\|u(t+\tau,u_0))\|_{\ell_{2,\theta}} < \infty.$$

ii) For the existence proof we use the variation of constant formula (13.3) which we abbreviate as

$$u = Su_c(0) + KG(u) \tag{13.5}$$

where

$$(Su_c(0))(t) = \mathrm{e}^{tB_c}u_c(0), \quad G(u)(t) = g(u(t)),$$

and

$$\begin{aligned}(Ky)(t) = &\int_0^t \mathrm{e}^{(t-\tau)B_c}y_c(\tau))\,\mathrm{d}\tau \\ &+ \int_{-\infty}^t \mathrm{e}^{(t-\tau)B_s}y_s(\tau))\,\mathrm{d}\tau - \int_t^\infty \mathrm{e}^{(t-\tau)B_u}y_u(\tau))\,\mathrm{d}\tau.\end{aligned}$$

We prove the existence of a fixed point of (13.5) by showing that the right-hand side is a contraction in the metric space

$$Y_\eta = \{u \in C^0(\mathbb{R}, \ell_{2,\theta}) : \|y\|_\eta = \sup_{t\in\mathbb{R}} \mathrm{e}^{-\eta|t|}\|y(t)\|_{\ell_{2,\theta}} < \infty\}.$$

In order to do so we use a number of estimates. From (13.4) we obtain for all $\varepsilon > 0$ a $M > 0$ such that

$$\begin{aligned}\|Su_c(0)\|_{Y_\eta} &= \sup_{t\in\mathbb{R}} \mathrm{e}^{-\eta|t|}\|\mathrm{e}^{tB_c}u_c(0)\|_{\ell_{2,\theta}} \leq \sup_{t\in\mathbb{R}} \mathrm{e}^{-\eta|t|}M(\varepsilon)\mathrm{e}^{\varepsilon|t|}\|u_c(0)\|_{\ell_{2,\theta}} \\ &\leq M(\varepsilon)\|u_c(0)\|_{\ell_{2,\theta}}.\end{aligned}$$

Lemma 13.1.5. *For $g \in C_b^0(\ell_{2,\theta}, \ell_{2,\theta})$ we have $G : Y_\eta \to Y_\eta$. If $g \in C_b^1(\ell_{2,\theta}, \ell_{2,\theta})$ then for all $\eta > 0$ we get*

$$\|G(y_1) - G(y_2)\|_{Y_\eta} \leq \|Dg\|_{C_b^0} \|y_1 - y_2\|_{Y_\eta}$$

for $y_1, y_2 \in Y_\eta$.

Proof. The first statement is obvious, since g is bounded. From the mean value theorem it follows that

$$\begin{aligned}\|G(y_1) - G(y_2)\|_{Y_\eta} &\leq \sup_{t\in\mathbb{R}} \mathrm{e}^{-\eta|t|} \|g(y_1(t)) - g(y_2(t))\|_{\ell_{2,\theta}} \\ &\leq \sup_{t\in\mathbb{R}} \mathrm{e}^{-\eta|t|} \|Dg\|_{C_b^0} \|y_1(t) - y_2(t)\|_{\ell_{2,\theta}} \leq \|Dg\|_{C_b^0} \|y_1 - y_2\|_{Y_\eta}\end{aligned}$$

for $y_1, y_2 \in Y_\eta$. □

Lemma 13.1.6. *For each $\eta \in (\varepsilon, \beta)$ the map K is a bounded linear operator from Y_η into Y_η, i.e., there is a function $\gamma : (\varepsilon, \beta) \to \mathbb{R}$ such that*

$$\|K\|_{Y_\eta \to Y_\eta} \leq \gamma(\eta).$$

Proof. We write the last two terms of K as $\int_\infty^\infty \kappa(t-\tau) y(\tau)\, \mathrm{d}\tau$. For $\eta \in (\varepsilon, \beta)$ and $y \in Y_\eta$ it follows

$$\begin{aligned}&\mathrm{e}^{-\eta|t|} \|(Ky)(t)\|_{\ell_{2,\theta}} \\ &\leq \|y\|_\eta \sup_{t\in\mathbb{R}} \mathrm{e}^{-\eta|t|} \left[\left| \int_0^t \|\mathrm{e}^{(t-\tau)B_c}\| \mathrm{e}^{\eta|\tau|}\, \mathrm{d}\tau \right| + \int_{-\infty}^\infty \|\kappa(t-\tau)\| \mathrm{e}^{\eta|\tau|}\, \mathrm{d}\tau \right] \\ &\leq \|y\|_\eta \sup_{t\in\mathbb{R}} \left[\left| \int_0^t \|\mathrm{e}^{(t-\tau)B_c}\| \mathrm{e}^{-\eta|t-\tau|}\, \mathrm{d}\tau \right| + \int_{-\infty}^\infty \|\kappa(t-\tau)\| \mathrm{e}^{\eta|t-\tau|}\, \mathrm{d}\tau \right] \\ &\leq \|y\|_\eta \left[\max\left(\int_0^\infty \|\mathrm{e}^{\tau B_c}\| \mathrm{e}^{-\eta\tau}\, \mathrm{d}\tau, \int_{-\infty}^0 \|\mathrm{e}^{\tau B_c}\| \mathrm{e}^{\eta\tau}\, \mathrm{d}\tau \right) \right. \\ &\qquad \left. + \int_{-\infty}^\infty \|\kappa(\tau)\| \mathrm{e}^{\eta|\tau|}\, \mathrm{d}\tau \right] \\ &\leq \|y\|_\eta M(\varepsilon) \left[(\eta - \varepsilon)^{-1} + 2(\beta - \eta)^{-1} \right]\end{aligned}$$

which implies the assertion. □

The last three estimates show that, for $\delta_0 > 0$ sufficiently small, the map

$$F(u) := Su_c(0) + KG(u)$$

is a contraction in Y_η for all $\eta \in (\varepsilon, \beta)$ and $g \in C_b^1(\mathbb{R}^d)$ with contraction constant $\|K\|_\eta \|Dg\|_{C_b^0} < 1$. Hence, F possesses a unique fixed point which is a solution of (13.5). We write this solution of (13.5) as $u = \Psi(Su_c(0))$.

iii) We define

$$\psi(u_c(0)) := \Psi(Su_c(0))(0) \text{ and set } h = (0, \psi_s, \psi_u).$$

The continuity of Ψ implies the continuity of ψ and h. Since for $g \in C_b^1$ the map Ψ is Lipschitz-continuous the same is true for ψ and h. Thus, we have proved the existence of a Lipschitz-continuous center manifold. □

Remark 13.1.7. a) To complete the proof of Theorem 13.1.3 it remains to prove that h is in C^{r-1}. This turns out to be rather complicated, because G is in general not differentiable from Y_η into Y_η. However, it turns out that G is k-times continuously differentiable from Y_{η_1} into Y_{η_2}, if $\eta_1 > k\eta_2$. For the heuristics of this consider $g(x) = x^k$ and $x(t) = \mathrm{e}^{\eta|t|}$.

b) Stable and unstable manifolds can be constructed similarly, cf. [**Van89**]. For instance, in case of a stable manifold a fixed point is constructed for the map

$$F(u) := Su_s(0) + K_s G(u)$$

in the space

$$Z_\eta^+ = \{u \in C^0(\mathbb{R}^+, \ell_{2,\theta}) : \|u\|_\eta = \sup_{t \geq 0} \mathrm{e}^{\eta t} \|u(t)\|_{\ell_{2,\theta}} < \infty\},$$

where $(Su_s(0))(t) = \mathrm{e}^{tB_s} u_s(0)$, $G(u)(t) = g(u(t))$, and

$$(K_s z)(t) = \int_0^t \mathrm{e}^{(t-\tau)B_s} z_s(\tau)\,\mathrm{d}\tau - \int_t^\infty \mathrm{e}^{(t-\tau)B_c} z_c(\tau)\,\mathrm{d}\tau - \int_t^\infty \mathrm{e}^{(t-\tau)B_u} z_u(\tau)\,\mathrm{d}\tau.$$

⌋

13.2. Local bifurcation theory on bounded domains

In this section we explain the classical application of the center manifold theorem to PDEs for two examples.

13.2.1. The Allen-Cahn equation. We reconsider the Allen-Cahn equation

$$\partial_t u = \partial_x^2 u + \alpha u - u^3, \tag{13.6}$$

with $\alpha \in \mathbb{R}$, $u = u(x,t) \in \mathbb{R}$, $t \geq 0$, and $x \in [0, \pi]$ under the boundary conditions

$$u(0,t) = u(\pi,t) = 0.$$

In §5.3 we already obtained the global bifurcation diagram sketched in the left panel of Figure 5.5 by using its gradient structure and by analyzing the set of stationary solutions with a simple phase plane analysis. Hence, we already know that there is a sequence of supercritical pitchfork bifurcations of fixed points for values $\alpha = m^2$ with $m \in \mathbb{N}$.

Nevertheless, we will use the Allen-Cahn equation as first example for the application of the center manifold theorem in the infinite-dimensional situation. We use Theorem 13.1.3 to establish the existence of the supercritical pitchfork bifurcations of fixed points with an alternative approach.

We start with the first bifurcation which occurs for $\alpha = 1$. We introduce a bifurcation parameter $\varepsilon^2 = \alpha - 1$. Making the ansatz

$$u(x,t) = \sum_{n=1}^{\infty} u_n(t)\sin(nx)$$

allows to rewrite the Allen-Cahn equation (13.6) as

$$\partial_t u_n = \lambda_n u_n + g_n(u_1, u_2, \ldots), \qquad (n \in \mathbb{N}),$$

with $\lambda_n = -n^2 + \alpha$ and

$$g_n = -\frac{2}{\pi}\int_0^{\pi} \sin(nx)u(x,t)^3 dx = \mathcal{O}(|u_1|^3 + |u_2|^3 + \ldots). \tag{13.7}$$

For $\alpha = 1$ we have one central eigenvalue, namely λ_1. The other eigenvalues are in the left half plane and so we have

$$u_c = (u_1, 0, 0, \ldots) \quad \text{and} \quad u_s = (0, u_2, u_3, \ldots).$$

We already know from §5.3.1 that

$$g_c = (g_1, 0, 0, \ldots) \quad \text{and} \quad g_s = (0, g_2, g_3, \ldots)$$

are smooth maps from $\ell_{2,\theta}$ to $\ell_{2,\theta}$ if $\theta \geq 1$. Hence, all assumptions for the application of the center manifold theorem as stated in Theorem 13.1.3 are satisfied. As in the finite-dimensional case, we add the equation $\dot{\varepsilon} = 0$ in order to apply the center manifold theorem not only for $\varepsilon = 0$, but also for small $\varepsilon \neq 0$. As a consequence there exist smooth functions h_j such that for the solutions on the center manifold

$$u_j = h_j(u_1, \varepsilon^2) = \mathcal{O}(u_1^3). \tag{13.8}$$

Therefore, the reduced system on the center manifold is given by

$$\partial_t u_1 = \lambda_1 u_1 + g_1(u_1, h_2(u_1, \varepsilon^2), \ldots).$$

Due to (13.8) we have

$$g_1(u_1, h_2(u_1, \varepsilon^2), \ldots) = g_1(u_1, 0, 0, , \ldots) + \mathcal{O}(u_1^5).$$

From the representation (13.7) we find

$$\partial_t u_1 = \lambda_1 u_1 - \gamma u_1^3 + \mathcal{O}(u_1^5),$$

with

$$\gamma = \frac{2}{\pi}\int_0^{\pi} \sin^4(x)\,\mathrm{d}x = \frac{3}{4}.$$

Hence, a supercritical pitchfork bifurcation occurs.

13.2.2. Rolls in the Swift-Hohenberg equation. In §10 we introduced the Swift-Hohenberg (SH) equation

$$\partial_t u = -(1+\partial_x^2)^2 u + \alpha u - u^3, \tag{13.9}$$

with $t \geq 0$, $x \in \mathbb{R}$, $u(x,t) \in \mathbb{R}$, and bifurcation parameter $\alpha \in \mathbb{R}$. In this section we apply the center manifold theorem, stated in Theorem 13.1.3, to prove rigorously the existence of spatially periodic equilibria for (13.9) which bifurcate from the trivial fixed point $u = 0$. In order to do so we impose periodic boundary conditions $u(x,t) = u(x+2\pi, t)$ on (13.9). This second example shows that for Turing instabilities the flow on the center manifold which is used for the rigorous construction of spatially periodic equilibria is related to the flow of the associated Ginzburg-Landau equation.

For the linearized system $\partial_t v = -(1+\partial_x^2)^2 v + \alpha v$ with $v(x,t) = v(x+2\pi, t)$ we find solutions $v(x,t) = \mathrm{e}^{\mathrm{i}kx+\lambda(k)t}$ with $\lambda_k = \lambda(k) = -(1-k^2)^2 + \alpha$, but now with $k \in \mathbb{Z}$ instead of $k \in \mathbb{R}$. We have $\lambda_k < 0$ for all $k \in \mathbb{Z}$ if $\alpha < 0$. For $\alpha = 0$ we have two zero eigenvalues $\lambda_1 = \lambda_{-1} = 0$. The rest of the spectrum satisfies $\lambda_k \leq -1$ for all $k \in \mathbb{Z} \setminus \{-1, 1\}$. Thus, we can apply the center manifold theorem to reduce the infinite-dimensional problem to a two-dimensional center-manifold. In order to apply the center manifold theorem also in case $\alpha \neq 0$ we extend (13.9) by the equation $\dot{\alpha} = 0$.

We introduce coordinates on the center manifold through

$$u(x,t) = c_1(t)\mathrm{e}^{\mathrm{i}x} + c_{-1}(t)\mathrm{e}^{-\mathrm{i}x} + h(c_1(t), c_{-1}(t), \alpha), \tag{13.10}$$

where h is the reduction function which is of the form

$$h(c_1, c_{-1}) = \sum_{k \in \mathbb{Z}\setminus\{-1,1\}} h_k(c_1, c_{-1}, \alpha)\mathrm{e}^{\mathrm{i}kx}.$$

The reduced system on the center manifold can be written as

$$\partial_t c_1 = \lambda_1 c_1 + f_1(c_1, c_{-1}, \alpha), \quad \partial_t c_{-1} = \lambda_{-1} c_{-1} + f_{-1}(c_1, c_{-1}, \alpha),$$

and we have $c_j(t) = \overline{c_{-j}(t)}$.

Remark 13.2.1. The ansatz (10.17) for the derivation of the GL equation for the SH equation and (13.10) for the computation of the center manifold reduction are very similar. By restricting the GL ansatz to solutions which are constant in space, i.e., $A_{ij}(X,T) = \underline{A}_{ij}(T)$, and setting $\alpha = \varepsilon^2$,

$$c_1(t) = \sum_{j=0}^{\infty} \varepsilon^{1+j} \underline{A}_{1j}(\varepsilon^2 t) \quad \text{and} \quad h_k(c_1, c_{-1}, \alpha) = \sum_{j=0}^{\infty} \varepsilon^{k+j} \underline{A}_{kj}(\varepsilon^2 t)$$

for $k \in (2\mathbb{Z}+1) \setminus \{-1, 1\}$, it is obvious that the coefficients in the center manifold reduction can be obtained by the coefficients from the GL approximation. ⌋

For the quadratic approximation of f_1 and f_{-1} in terms of c_1, c_{-1}, α in principle we have to compute 10 coefficients, for the cubic approximation 20 coefficients, etc.. Hence, it makes sense to use the symmetries of the problem in order to reduce the complexity of f_1 and f_{-1}. The SH equation is invariant under $u \mapsto -u$, which yields via (13.10) to $(c_1, c_{-1}) \mapsto -(c_1, c_{-1})$. It is translational invariant, i.e., $u(\cdot) \mapsto u(\cdot + y)$, which corresponds to the invariance of the reduced system under $(c_1, c_{-1}) \mapsto (c_1 e^{iy}, c_{-1} e^{-iy})$. Since $c_1 = \overline{c_{-1}}$ it is sufficient to consider the equation for c_1 alone, which must be of the form

$$\partial_t c_1 = \lambda_1 c_1 + c_1 g(|c_1|^2, \alpha)$$

due to these symmetries. The last symmetry is the reflection symmetry for the SH equation, namely $u(\cdot) \mapsto u(-\cdot)$, which corresponds to $(c_1, c_{-1}) \mapsto (c_{-1}, c_1)$. Hence, the function g must be real-valued. We introduce polar coordinates $c_1 = re^{i\phi}$ and find

$$\partial_t r = \lambda_1 r + rg(r^2, \alpha), \quad \partial_t \phi = 0.$$

Since the nonlinear term of the SH equation is given by $\alpha u - u^3$ we have

$$h_k(c_1, c_{-1}) = \mathcal{O}(|\alpha||c_{\pm 1}| + |c_{\pm 1}|^3),$$

and as a consequence

$$f_1(c_1, c_{-1}) = \alpha c_1 - 3c_1^2 c_{-1} + \mathcal{O}(|\alpha|^2 |c_{\pm 1}| + |\alpha||c_{\pm 1}|^3 + |c_{\pm 1}|^5).$$

Hence,

$$\partial_t r = \alpha r - 3r^3 + \mathcal{O}(\alpha^2 r + \alpha r^3 + r^5), \quad \partial_t \phi = 0.$$

The stationary solutions satisfy

$$\widetilde{G}(r, \alpha) = \alpha r - 3r^3 + \mathcal{O}(\alpha^2 r + \alpha r^3 + r^5) = 0$$

To balance the lowest order terms we consider the scaled version

$$G(A, \varepsilon) = \varepsilon^{-3} \widetilde{G}(\varepsilon A, \varepsilon^2) = A - 3A|A|^2 + \mathcal{O}(\varepsilon^2) = 0.$$

Since i) $G(\pm 1/\sqrt{3}, 0) = 0$ and ii) $\partial_A G(\pm 1/\sqrt{3}, 0) \neq 0$ we can apply the implicit function theorem to find solutions $A_\pm = A_\pm(\varepsilon) = \pm 1/\sqrt{3} + \mathcal{O}(\varepsilon^2)$ with $G(A_\pm(\varepsilon), \varepsilon) = 0$. Hence, we proved

Theorem 13.2.2. *At $\alpha = 0$ a supercritical pitchfork bifurcation of families of 2π-spatially periodic equilibria occurs for the SH equation. They are of the form*

$$u(x, t) = \pm 2\sqrt{\alpha} \cos(x + x_0) + \mathcal{O}(\alpha^{3/2})$$

for $\alpha > 0$ and arbitrary phase $x_0 \in \mathbb{R}$.

Obviously the same construction is possible for all spatial periods close to 2π.

Remark 13.2.3. Following Remark 13.2.1, the previous analysis applies more or less for all pattern forming systems where a GL equation can be derived. Examples which have been considered are reaction-diffusion systems in §10.5, convection problems in §10.6, and the Couette-Taylor problem in §10.7. The center manifold reduction for the Couette-Taylor problem has been carried out for instance in [**IA98**]. The bifurcating spatially periodic equilibria are called Taylor vortices and are of the form

$$U_{\mathrm{TV}}[q,\phi,\varepsilon](x,z) = \varepsilon A_0[q,\phi](X)\widehat{U}^1_{k_c,0}(z)\mathrm{e}^{\mathrm{i}k_c x} + \mathrm{c.c.} + h[q,\varepsilon](\varepsilon A_0, \varepsilon\bar{A}_0, \varepsilon)$$

where

$$A_0[q,\phi](X) = \sqrt{-\frac{\nu_0 - \nu_2 q^2}{\nu_3}}\mathrm{e}^{\mathrm{i}(qX+\phi)}$$

is an equilibrium of the associated GL equation (10.103). The critical wave number k_c of the most unstable pattern $\widehat{U}^1_{k_c,0}(z)\mathrm{e}^{\mathrm{i}k_c x}$ and the bifurcation parameter $0 \le \varepsilon^2 \ll 1$ have been introduced in §10.7. The function h from the center manifold reduction satisfies $h[q,\varepsilon](\varepsilon A_0, \varepsilon\bar{A}_0, \varepsilon) = \mathcal{O}(\varepsilon^2)$. Since our center manifold theorem, Theorem 13.1.3, has not been formulated to apply in this situation, more abstract versions like [**HI11**, Theorem 2.9] have to be used for the reduction. ⌋

13.3. Spatial dynamics for elliptic problems in a strip

For nonlinear problems the nowadays classical method of spatial dynamics combined with center manifold theory goes back to Kirchgässner [**Kir82**]. He studied elliptic problems in unbounded cylindrical domains by considering the unbounded space variable as evolutionary variable. As an example consider

$$\Delta u = \alpha u - u^3 \tag{13.11}$$

on the cylindrical domain $\mathbb{R} \times (0,\pi)$ with Dirichlet boundary conditions $u|_{y=0,\pi} = 0$. In order to find uniformly bounded solutions we write the problem as a dynamical system w.r.t. the unbounded space variable $x \in \mathbb{R}$. In order to apply our center manifold theorem 13.1.3 we make the ansatz

$$u(x,t) = \sum_{n=1}^{\infty} u_n(x)\sin(ny)$$

and find

$$(\partial_x^2 - n^2)u_n(x) = \alpha u_n + g_n(u_1, u_2, \ldots), \tag{13.12}$$

with

$$g_n = -\frac{2}{\pi}\int_0^{\pi} \sin(nx)u(x,t)^3\,\mathrm{d}x = \mathcal{O}(|u_1|^3 + |u_2|^3 + \ldots) \tag{13.13}$$

for $n \geq 1$. This can be written as a first order system

$$(\partial_x - n)u_n(x) = v_n(x), \qquad (\partial_x + n)v_n(x) = \alpha u_n + g_n(u_1, u_2, \ldots).$$

For $\alpha = -1$ we have two central eigenvalues for u_1. The other eigenvalues are then given by $\lambda_{\pm n} = \pm\sqrt{n^2 + \alpha}$ for $n \geq 2$, i.e., for all $\alpha \approx -1$ the other eigenvalues form two real sequences going to $\pm\infty$. See Figure 13.1.

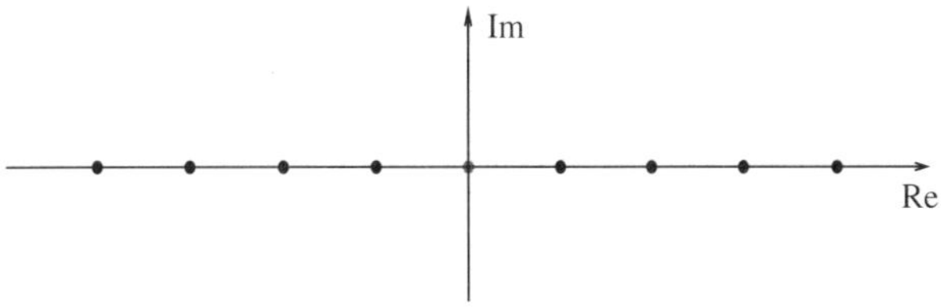

Figure 13.1. Spectrum of the linearized spatial dynamics formulation for (13.12). There are two central eigenvalues leading to a Jordan block, and infinitely many unstable and infinitely many stable eigenvalues up to $\pm\infty$.

Hence, the initial value problem for (13.12) is ill-posed in every $\ell_{2,\theta}$-space. Nevertheless, the center manifold theorem, Theorem 13.1.3, can be applied. The equations for $n \geq 2$ can be written as

$$u_n' = \lambda_n v_n, \qquad v_n' = \lambda_n u_n + \lambda_n^{-1} g_n(u_0, u_1, \ldots).$$

This system can be diagonalized by introducing $c_n = u_n + v_n$ and $c_{-n} = u_n - v_n$ which satisfy

$$c_n' = \lambda_n c_n + g_n, \qquad c_{-n}' = -\lambda_n c_{-n} - g_n.$$

Since functions satisfying Dirichlet boundary conditions on $(0, \pi)$ can be extended to odd 2π-periodic functions w.r.t. y, and since odd functions to the power three are odd again, we can consider (13.11) alternatively in the invariant subspace of odd 2π-periodic functions w.r.t. y. As a consequence the nonlinearity is a smooth map from $\ell_{2,\theta}$ to $\ell_{2,\theta}$ for every $\theta > 1/2$. Hence, all assumptions of Theorem 13.1.3 are satisfied and so there exist smooth functions $u_j = h_j(u_1, \alpha)$ for $j \geq 2$, such that on the center manifold the reduced system is given by

$$u_1'' = (\alpha + 1)u_1 + g_1(u_1, h_2(u_1, \alpha), \ldots) = \alpha u_1 - \frac{3}{4}u_1^3 + \mathcal{O}(u_1^5).$$

Ignoring the higher order terms yields

$$u_1' = v_1, \qquad v_1' = (\alpha + 1)u_1 - \frac{3}{4}u_1^3. \tag{13.14}$$

For $\alpha + 1 > 0$ small, this system possesses two homoclinic orbits at the origin which correspond to spatially localized solutions of the original elliptic problem (13.11).

Remark 13.3.1. It remains to prove the persistence of the homoclinic orbits if the higher order terms $\mathcal{O}(u_1^5)$ are not ignored. In general a homoclinic orbit would break up. However, the original system is reversible, i.e., invariant under $(x, u) \mapsto (-x, u)$. The cut-off function in the center manifold construction can be chosen in such a way that the reduction preserves this property, cf. [**HI11**, §2.3.3]. Hence, the reduced system is invariant under $(x, u_1, v_1) \mapsto (-x, u_1, -v_1)$ and so the phase portrait must be reflection symmetric at the u_1-axis which implies the persistence of the homoclinic orbit under higher order perturbations, cf. Figure 2.20. ⌋

13.4. Applications

The center manifold theorem in combination with spatial dynamics is nowadays a well established tool which has successfully been used in many applications. Examples are the construction of solitary surface water waves, of standing light pulses in photonic crystals, or of breather solutions in lattice differential equations. In this section we concentrate on the aspects which have to do with modulation equations when the center manifold theorem is applied, and explain how the spectrum of the linearized spatial dynamics formulation is related to the spectrum of the linearized time-dependent problem and how the reduced systems on the center manifold are related to the associated modulation equations. We will concentrate on the construction of traveling wave solutions, modulating front solutions, and breather solutions. We refrain from rewriting the literature and therefore skip all functional analytic difficulties which have to be overcome in applying the center manifold theorem and the discussion of the reduced system.

13.4.1. Solitary waves for the water problem. The center manifold theorem in combination with spatial dynamics has especially been used in the construction of solitary waves for the water wave problem as introduced in §12.2.1, cf. [**IK90, Ioo95**]. The water wave problem can be written as evolutionary system

$$\partial_t U = \Lambda(\partial_x, \partial_y)U + \mathcal{O}(\|U\|^2), \tag{13.15}$$

with $x \in \mathbb{R}$, and $y \in \Sigma$, with Σ a bounded fixed cross section. We are interested in waves of permanent form, i.e.,

$$U(x, y, t) = V(x - ct, y) = V(\xi, y).$$

Inserting this ansatz into the water wave problem yields

$$-c\partial_\xi V = \Lambda(\partial_\xi, \partial_y)V + \mathcal{O}(\|V\|^2). \tag{13.16}$$

Written as an evolutionary system w.r.t. ξ we obtain the spatial dynamics formulation

$$\partial_\xi W = \mathcal{M}(\partial_y)W + \mathcal{O}(\|W\|^2). \tag{13.17}$$

It is the purpose of the subsequent discussion to relate properties of the temporal evolutionary system (13.15) and of spatial dynamics formulation (13.17).

The spectrum. Solutions of the linearized spatial dynamics formulation $\partial_\xi w = Mw$ are of the form

$$W(y,\xi) = e^{\mu\xi}\widehat{W}(y).$$

The eigenvalues μ can be found from the eigenvalue problem

$$\mu\widehat{W} = \mathcal{M}(\partial_y)\widehat{W}.$$

Undoing the transformation which allowed to write (13.16) into (13.17) yields to solutions

$$V(y,\xi) = e^{\mu\xi}\widehat{V}(y)$$

and to the eigenvalue problem

$$-c\mu\widehat{V}(y) = \Lambda(\mu,\partial_y)\widehat{V}(y). \tag{13.18}$$

On the other hand the solutions of the linearized temporal dynamics formulation (13.15) are of the form

$$U(x,y,t) = e^{\lambda t}e^{ikx}\widehat{U}(y)$$

leading to the eigenvalue problem

$$\lambda\widehat{U}(y) = \Lambda(ik,\partial_y)\widehat{U}(y). \tag{13.19}$$

In general we have a countable set of curves of eigenvalues $k \mapsto \lambda_n(k)$ with $n \in \mathbb{N}$.

Comparing (13.18) with (13.19) shows that both coincide for the central eigenvalues $\mu = ik$ of the spatial dynamics formulation. Hence, these eigenvalues can be obtained by solving

$$\lambda_n(k) = -cik, \tag{13.20}$$

where the wave speed c is a free parameter. For the water wave problem we have

$$\lambda_\pm(k) = i\omega_\pm(k) = \pm i\text{sign}(k)\sqrt{\tanh(k)(k+\sigma k^3)}.$$

In Figure 13.2 we analyze (13.20) in case of zero surface tension, in Figure 13.3 in case of surface tension $\sigma > 1/3$, and in Figure 13.4 in case of surface tension $0 < \sigma < 1/3$. We always find an intersection of $k \mapsto -ck$ with ω_+ and ω_- at the wave number $k = 0$, but since the associated nonlinear terms at the wave number $k = 0$ vanish identically, cf. the derivation of the KdV equation, these two central eigenvalues play no role and can be eliminated.

i) The reduction in case $\sigma = 0$. For $|c| < 1$ there are two intersections of $k \mapsto \omega_\pm(k)$ and $k \mapsto -ck$ at wave numbers k with $|k| > 0$, which lead to two central eigenvalues. Since the spatial dynamics formulation (13.17) of the water wave problem can be written as an infinite-dimensional Hamiltonian system it can be expected that an infinite-dimensional version of Lyapunov's subcenter theorem can be applied. Therefore, the existence of spatially periodic traveling waves can be established, cf. [**Str26**]. If $|c|$ approaches 1 these two central eigenvalues collide in zero and leave the imaginary axis along the real axis. For $|c| > 1$ there exists a homoclinic orbit to the origin. This scenario is analogous to the following example.

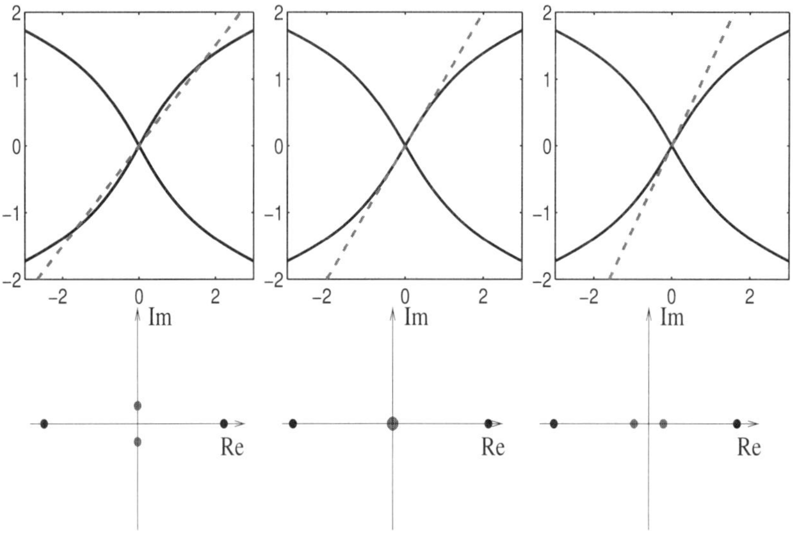

Figure 13.2. The curves of eigenvalues of the temporal dynamics formulation (13.15) intersected with the curve $k \mapsto -ck$ lead to the central eigenvalues of the spatial dynamics formulation (13.17) of the water wave problem without surface tension.

Example 13.4.1. We are interested in waves of permanent form for the Boussinesq model

$$\partial_t^2 u = \partial_x^2 u + \partial_x^2 \partial_t^2 u + \partial_x^2 (u^2).$$

We make the ansatz $u(x,t) = v(x-ct) = v(\xi)$ and find the fourth order ODE

$$c^2 \partial_\xi^2 v = \partial_\xi^2 v + c^2 \partial_\xi^4 v + \partial_\xi^2 (v^2). \tag{13.21}$$

The linearized problem is solved by $v(\xi) = e^{\lambda\xi}\widehat{v}$ where $0 = -c^2\lambda^2 + \lambda^2 + \lambda^4 = \lambda^2(\lambda^2 + 1 - c^2)$. Independently of the velocity c there are always two eigenvalues 0, but since the nonlinear terms at the wave number $k = 0$ also vanish identically, these two central eigenvalues play no role and can be eliminated. In fact, (13.21) can be integrated twice, i.e.,

$$c^2 v = v + c^2 \partial_\xi^2 v + v^2. \tag{13.22}$$

The constants of integration vanish since we are interested in solitary waves. We write (13.22) as the first order system

$$\partial_\xi v_0 = v_1, \qquad \partial_\xi v_1 = c^{-2}((c^2-1)v_0 - v_0^2). \tag{13.23}$$

We have two fixed points $(v_0, v_1) = (0,0)$ and $(v_0, v_1) = (c^{-2}(c^2-1), 0)$. For $c^2 < 1$ the origin is a center surrounded by a family of periodic solutions. For $c^2 > 1$ the origin is a saddle, and by looking at the phase portrait we find a homoclinic solution at the origin which corresponds to the solitary wave we are interested in and which exists for velocities $c^2 > 1$. ⌋

Remark 13.4.2. The ansatz

$$U(x,y,t) = \varepsilon^2 A(\varepsilon(x-t), \varepsilon^3 t)\varphi(y) + \mathcal{O}(\varepsilon^3).$$

for the derivation of the KdV equation for the water wave problem and the ansatz

$$V(\xi, y) = V_1(\xi)\varphi(y) + h(V_1, \partial_\xi V_1, c),$$

with $\xi = x - ct$, for the computation of the center manifold reduction are very similar. By restricting the KdV ansatz to traveling wave solutions, i.e., $A(X,T) = \underline{A}(X - \underline{c}T)$, and setting $c - 1 = \varepsilon^2$,

$$V_1(\xi) = \varepsilon^2 \underline{A}(\varepsilon\xi) + \mathcal{O}(\varepsilon^3) \quad \text{and} \quad h(V_1, \partial_\xi V_1, c) = \mathcal{O}(\varepsilon^3),$$

it is obvious that the coefficients in the center manifold reduction can be obtained by the coefficients from the KdV approximation. ⌋

From Remark 13.4.2 and the fact that the associated KdV equation possesses solitary waves A_{sol} and the reversibility of the water wave problem it follows that the reduced system on the center manifold possesses a homoclinic orbit for every c, with $|c| > 1$ not to big. By making the previous ideas rigorous the following result [**Ioo98**] can be shown.

Theorem 13.4.3. *There exists a $\delta_0 > 0$ such that for $\delta \in (0, \delta_0)$ the water wave problem without surface tension possesses a solitary wave solution $U(x,y,t) = V(x-ct, y)$ with $\lim_{|\xi|\to\infty} V(\xi, y) = 0$ satisfying*

$$\sup_{\xi\in\mathbb{R}, y\in[0,1]} |V(\xi,y) - \delta^2 A_{\text{sol}}(\delta\xi)\varphi(y)| \leq C\delta^3,$$

$c = 1 + \delta^2/\sqrt{2} + \mathcal{O}(\delta^4)$ or $c = -1 - \delta^2/\sqrt{2} + \mathcal{O}(\delta^4)$, where A_{sol} is the rescaled 1-soliton solution of the associated KdV equation.

ii) The reduction in case $\sigma > 1/3$. For $|c| > 1$ there are two intersection points of $k \mapsto \omega_{\pm}(k)$ and $k \mapsto -ck$ at wave numbers k with $|k| > 0$ which lead to two central eigenvalues. The existence of spatially periodic traveling waves can be established [**Zei71**] by applying an infinite-dimensional version of Lyapunov's subcenter theorem. If $|c|$ approaches 1 these two central eigenvalues collide in zero and leave the imaginary axis along the real axis. For $|c| < 1$ there exists a homoclinic orbit to the origin. This scenario is exactly as in the case $\sigma = 0$. However, as we already know from the KdV approximation the solitary wave for $\sigma > 1/3$ is a wave of depression and not a wave of elevation.

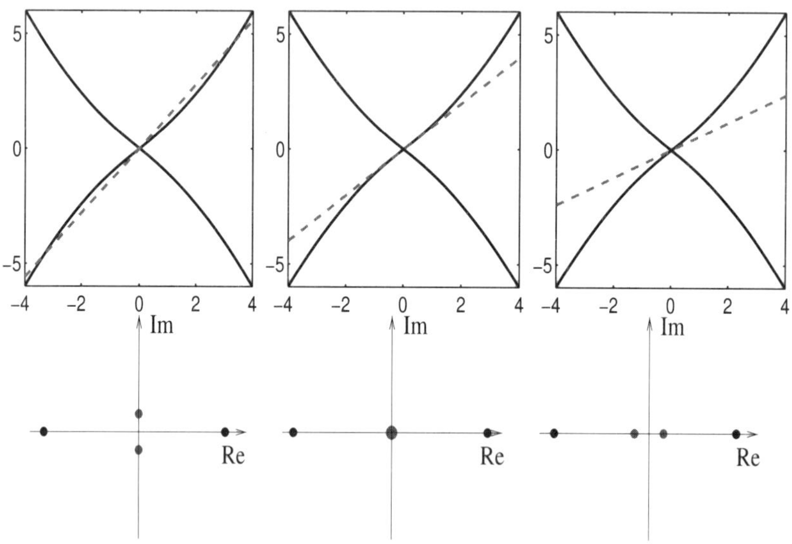

Figure 13.3. The curves of eigenvalues of the temporal dynamics formulation (13.15) intersected with the curve $k \mapsto -ck$ lead to the central eigenvalues of the spatial dynamics formulation (13.17) of the water wave problem with surface tension $\sigma > 1/3$.

iii) The reduction in case $0 < \sigma < 1/3$. For small values of $|c|$ except of the trivial intersection at $k = 0$ there are no further intersection points. For a value $|c| = c_{\min} < 1$ the curve $k \mapsto -ck$ is tangent to $k \mapsto \omega(k)$ for a $k \geq 0$ which implies that two pairs of eigenvalues collide on the imaginary axis. At this point the phase and group velocity of the associated NLS approximation coincide. Thus, the modulating pulse solutions described by the NLS approximation become traveling waves and can rigorously be established with the present center manifold approach. For $c_{\min} < |c| < 1$ there are four intersection points at wave numbers k with $|k| > 0$ which lead to four central eigenvalues. The existence of spatially periodic traveling waves can be established also in this case [**Zei71**]. As $|c|$ approaches 1, two of the four central eigenvalues collide in zero and leave the imaginary axis along the real axis. The solitary waves in this case can be obtained via the KdV approximation. However, the solitary waves do not decay to zero for $|\xi| \to \infty$, but have some small oscillatory tails. The reason for this fact is that it is very unlikely that a one-dimensional unstable manifold intersects with a one-dimensional stable manifold in a four-dimensional space. However, the three-dimensional center-unstable manifold and three-dimensional center-stable manifold intersect due to the reversibility of the water wave problem and so for $|\xi| \to \infty$ the solutions in the intersection converge towards the center manifold which is filled with periodic solutions.

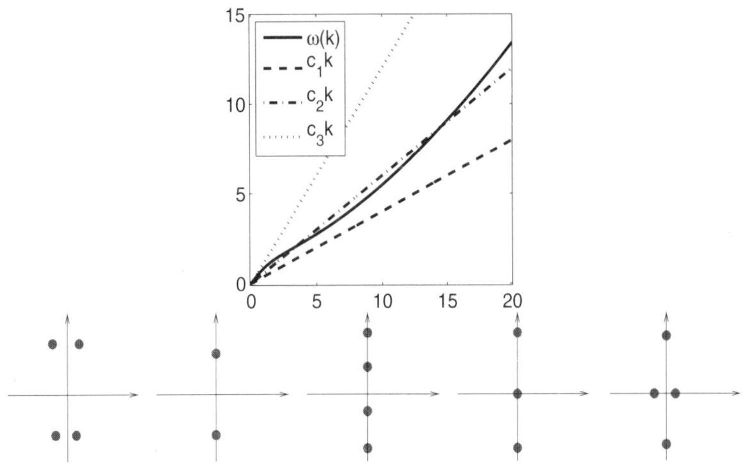

Figure 13.4. The curves of eigenvalues of the temporal dynamics formulation (13.15) intersected with the curves $k \mapsto -ck$ for different values of c lead to the central eigenvalues of the spatial dynamics formulation (13.17) of the water wave problem with surface tension $0 < \sigma < 1/3$.

Remark 13.4.4. In case $\sigma = 1/3$ and $c = 1$ the Kawahara equation (12.22) occurs as a modulation equation. The reduced system on the center manifold of the spatial dynamics formulation possesses a Jordan block of size 4. Near this point more complicated traveling wave solutions, such as multi-pulse solutions, can be found, cf. [**BGT96**]. ⌋

13.4.2. Modulating fronts for the Swift-Hohenberg equation. We are interested in the construction of modulating front solutions with the center manifold approach and spatial dynamics for the Swift-Hohenberg (SH) equation. Such solution describe the spreading of the stable spatially periodic equilibria u_ε constructed in §13.2.2 into the regions of the unstable trivial solution $u = 0$, i.e., we are interested in solutions

$$u(x, t) = v(x - ct, x)$$

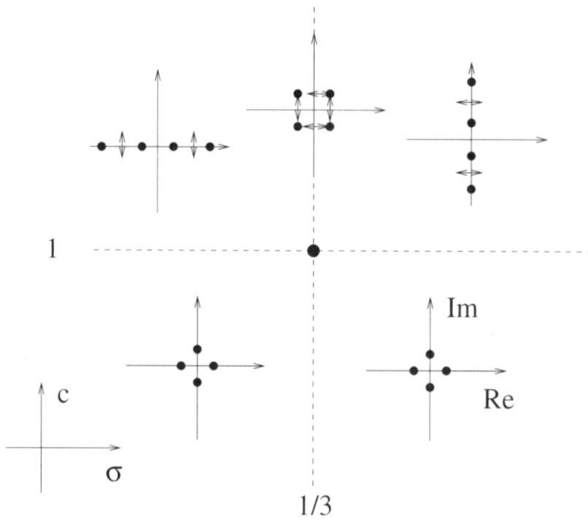

Figure 13.5. Spectrum of the linearized spatial dynamics formulation in the (σ, c) plane for $(\sigma, c) \approx (1/3, 1)$.

where v is 2π-periodic in its second argument and satisfies

$$\lim_{\xi \to \infty} v(\xi, p) = 0, \qquad \lim_{\xi \to -\infty} v(\xi, p) = u_\varepsilon(p).$$

Then, $v = v(\xi, p)$ satisfies $-c\partial_\xi v = -\left(1 + (\partial_\xi + \partial_p)^2\right)^2 v + \alpha v - v^3$ which written as a first order system yields

$$\begin{aligned} \partial_\xi v_0 &= -\, \partial_p v_0 + v_1, \\ \partial_\xi v_1 &= -\, \partial_p v_1 + v_2, \\ \partial_\xi v_2 &= -\, \partial_p v_2 + v_3, \\ \partial_\xi v_3 &= -\, \partial_p v_3 - 2v_2 - v_0 + c(-\partial_p v_0 + v_1) + \alpha v_0 - v_0^3. \end{aligned} \tag{13.24}$$

With $V = (v_0, v_1, v_2, v_3)$ this system is abbreviated as

$$\partial_\xi V = -\partial_p V + K_{c,\alpha}(\partial_p) V + N(V)$$

where $K_{c,\alpha}(\partial_p)V$ stands for the linear and $N(V)$ for the nonlinear terms. Since the modulating front solutions bifurcate out of the trivial solution we consider first the linearized problem

$$\partial_\xi V = -\partial_p V + K_{c,\alpha}(\partial_p) V.$$

Undoing the above transformations, this system is equivalent to

$$-c\partial_\xi v = -\left(1 + (\partial_\xi + \partial_p)^2\right)^2 v + \alpha v.$$

Since $v(\xi, p)$ is 2π-periodic w.r.t. p we can use Fourier series and look for solutions $v(\xi, p) = e^{\mu\xi + imp}$. This ansatz yields

$$-c\mu = -\left(1 + (\mu + im)^2\right)^2 + \alpha.$$

In order to use the center manifold theorem we have to compute the central eigenvalues $\mu = \mathrm{i}k \in \mathrm{i}\mathbb{R}$ which are solutions of

$$-c\mathrm{i}k = -\left(1 + (\mathrm{i}k + \mathrm{i}m)^2\right)^2 + \alpha = \lambda(k + m, \alpha).$$

For $c = 0$ and $\alpha = 0$ there are infinitely many intersection points. Since the eigenvalue curve λ possesses maxima in $k+m = \pm 1$, the solution $k = \pm 1 - m$ is double for each of these ms, which leads to a Jordan block of size two in the spatial dynamics formulation. Hence, for $c = 0$ and $\alpha = 0$ there are infinitely many eigenvalues on the imaginary axis. Thus, at a first view a reduction to a finite-dimensional center manifold does not seem possible. However, at a second view an interesting phenomenon occurs. In order to see the phenomenon take one of the front solutions

$$A(X, T) = A_f(X - \widetilde{c}T)$$

of the associated GL equation with

$$\lim_{\xi \to \infty} A_f(\xi) = 0 \quad \text{and} \quad \lim_{\xi \to -\infty} A_f(\xi) = 1/\sqrt{3}$$

which yields an approximation of the modulating front solution

$$\begin{aligned} v(\xi, p) =& v(x - ct, x) = 2\varepsilon A_f(X - \widetilde{c}T)\cos(x) + \mathcal{O}(\varepsilon^2) \\ =& 2\varepsilon A_f(\varepsilon x - \widetilde{c}\varepsilon^2 t)\cos(x) + \mathcal{O}(\varepsilon^2) = 2\varepsilon A_f(\varepsilon\xi)\cos(p) + \mathcal{O}(\varepsilon^2), \end{aligned}$$

with velocity $c = \varepsilon\widetilde{c} = \mathcal{O}(\varepsilon)$. Hence, we should simultaneously change $\alpha = \varepsilon^2$ and $c = \varepsilon\widetilde{c}$. Interestingly, by doing so the eigenvalues leave the imaginary axis with different velocities for $\varepsilon > 0$. The four eigenvalues from before stay in an $\mathcal{O}(\varepsilon)$-neighborhood of the imaginary axis, whereas the other eigenvalues are at least $\mathcal{O}(\varepsilon^{1/2})$-bounded away from the imaginary axis. Thus, we have a spectral gap of size $\mathcal{O}(\varepsilon^{1/2})$ which allows us to prove the existence of an invariant four-dimensional manifold of size $\mathcal{O}(\varepsilon^{1/2})$. Since the bifurcating modulating front solutions in which we are interested are of order $\mathcal{O}(\varepsilon)$, they are contained in this manifold. As above, in lowest order the flow on the manifold is given via the associated flow of the GL equation. The approximate heteroclinic solutions persist due to the transversal intersection of the four-dimensional unstable manifold of the origin and the two-dimensional stable manifold of $A = 1/\sqrt{3}$. The detailed analysis can be found in [**CE86, EW91**] for the SH equation and in [**HCS99**] for the Couette-Taylor problem.

13.4.3. Breathers and generalized breathers. We close this chapter with the construction of spatially localized time-periodic solutions, so called breather solutions. It turns out that the occurrence of such solutions with finite energy is rather rare. In general they only exist with small, but non-vanishing tails. Such solutions are called generalized breather solutions. Beside the non-persistence of breathers for perturbations of the sine-Gordon equation we also discuss the existence of breathers for nonlinear wave equations with spatially periodic coefficients.

The non-persistence of breathers for perturbations of the sine-Gordon equation. The sine-Gordon equation

$$\partial_t^2 u = \partial_x^2 u - \sin(u), \tag{13.25}$$

with $t \in \mathbb{R}$, $x \in \mathbb{R}$, and $u(x,t) \in \mathbb{R}$, originally came up in differential geometry in the form $\partial_\xi \partial_\eta u = \sin(u)$, describing surfaces with a constant negative curvature. It was found to govern the propagation of a dislocation in a crystal whose periodicity is represented by $\sin u$, it was posed as a tentative model of an elementary particle, and it was shown to be an equivalent form of the so called Thirring model, cf. [**DJ89**]. It turns out to be a completely integrable Hamiltonian system.

Remark 13.4.5. The name sine-Gordon equation is a pun referring to the Klein-Gordon equation

$$\partial_t^2 u = \partial_x^2 u - m^2 u, \tag{13.26}$$

which is just the linear wave equation with the additional term $-m^2 u$, where m is the rest mass of the particle. The Klein-Gordon equation is the linearized version of the sine-Gordon equation and of the nonlinear Klein-Gordon equation (11.2), which in physics is often called ϕ^4-model, and has been derived in 1928 as a relativistic version of the Schrödinger equation describing free particles, i.e., it is invariant under Lorentz transformations. Hence it attempts to unite quantum mechanics and special relativity, but it has some serious flaws; see [**Law90**] for a very basic introduction. ⌋

Traveling wave solutions $u(x,t) = v(x-ct) = v(\xi)$ of (13.25) satisfy

$$(c^2 - 1)\partial_\xi^2 v = -\sin(v).$$

For $|c| > 1$ this is the equation of the pendulum with the two heteroclinic connections. These traveling wave solutions are called kink and antikink for (13.25). For $|c| < 1$ the phase portrait is shifted by $v \mapsto v + \pi$. Hence, we have heteroclinic solutions no longer between $(-\pi, 0)$ and $(\pi, 0)$ but between $(0,0)$ and $(2\pi, 0)$.

There is a family of explicitly known spatially localized time-periodic solutions of the sine-Gordon equation, namely

$$u(x,t) = 4\arctan\left(\frac{(1-\lambda^2)^{1/2}}{\lambda}\frac{\sin(\lambda(t-t_0))}{\cosh((1-\lambda^2)^{1/2})(x-x_0))}\right) \tag{13.27}$$

for arbitrary x_0, t_0, and λ, with $0 < |\lambda| < 1$. Due to the dynamical behavior of these solutions they are called breather solutions. Surprisingly, it turns out that the sine-Gordon equation is the only nonlinear wave equation

$$\partial_t^2 u = \partial_x^2 u - u + g(u), \tag{13.28}$$

with $g : \mathbb{R} \to \mathbb{R}$ a smooth, odd function which satisfies $g(u) = \mathcal{O}(u^3)$ and $g'''(0) > 0$ for which such breather solutions exist [**Den93, BMW94**]. In the following we explain why this 'non-persistence of breathers' result holds. Moreover, we explain a number of positive results about the existence of generalized breather solutions.

Remark 13.4.6. With the ansatz

$$u(x,t) = \varepsilon A(\varepsilon x, \varepsilon^2 t)\mathrm{e}^{\mathrm{i}t} + c.c.$$

the NLS equation

$$2i\partial_T A = \partial_X^2 A - A|A|^2$$

can be derived which possesses localized time-periodic solutions. From the approximation results in §11.1 it is known that these solutions approximately exist in the nonlinear wave equations, too, on a time interval of length $\mathcal{O}(1/\varepsilon^2)$. However, the sine-Gordon equation up to rescaling is the only of the nonlinear wave equations for which these solutions exist as localized time-periodic solution for all times. In fact, taking the limit $\lambda = 1 - \varepsilon^2 \to 1$ in (13.27) shows that sine-Gordon breathers can be approximated by the localized time-periodic solutions of the associated NLS equation. ⌋

In order to construct time-periodic solutions for (13.28) we can use Fourier series w.r.t. the time variable t, i.e., we write

$$u(x,t) = \sum_{k\in\mathbb{Z}} \widehat{u}_k(x)\mathrm{e}^{\mathrm{i}\omega k t}.$$

The Fourier coefficients $\widehat{u}(x)$ satisfy

$$-\omega^2 k^2 \widehat{u}_k(x) = \partial_x^2 \widehat{u}_k(x) - \widehat{u}_k(x) + g_k(u)(x),$$

where

$$g_k(u)(x) = \frac{\omega}{2\pi}\int_0^{2\pi/\omega} \mathrm{e}^{-\mathrm{i}\omega k t} g(u(x,t))\,dt.$$

The linearization of this spatial dynamics formulation possesses solutions of the form $\widehat{u}_k(x) = \mathrm{e}^{\lambda x}$ with

$$\lambda^2 = -\omega^2 k^2 + 1.$$

From Remark 13.4.6 it follows that for small amplitude solutions we have approximately $\omega = 1$. Since we have an odd nonlinearity we only have to consider $k \in 2\mathbb{Z} + 1$. Hence, at the bifurcation point all eigenvalues are on the imaginary axis. For ω a little bit smaller than 1 there is one stable and one unstable eigenvalue and still infinitely many central eigenvalues. Breather solutions lie in the intersection of the one-dimensional stable and one-dimensional unstable manifold, which is very unlikely to exist, and the sine-Gordon equation up to rescaling turns out to be the only equation where this intersection occurs.

Remark 13.4.7. For all other nonlinear wave equations only generalized modulating pulse solutions can be expected. Such solutions do not decay to zero for $|x| \to \infty$, but have some small oscillatory tails of order $\mathcal{O}(\varepsilon^n)$, with $n \in \mathbb{N}$ arbitrary, but fixed, where $0 < \varepsilon \ll 1$ is the amplitude of the generalized modulating pulse solution. They can be found in the intersection of the infinite-dimensional center-unstable manifold and infinite-dimensional center-stable manifold. Both have co-dimension one and they intersect due to the reversibility of (13.28). For $|x| \to \infty$ the solutions in the intersection converge towards the center manifold. The boundedness for $|x| \to \infty$ is not at all obvious. In general the solutions on the center manifold can grow slowly and so the existence of generalized modulating pulse solutions can only be established for $|x| \leq 1/\varepsilon^n$. Since for (13.28) there is a finite speed of propagation such solutions also exist for all $|t| \leq 1/\varepsilon^n$ which is much larger than the $\mathcal{O}(1/\varepsilon^2)$ time scale of the NLS approximation. If the Hamiltonian restricted to the center manifold is positive definite, then the solutions stay bounded for all $x \in \mathbb{R}$ and hence for all $t \in \mathbb{R}$. See [**GS01b, GS05, GS08**] for detailed proofs. ⌋

Remark 13.4.8. Breathers for nonlinear wave equations

$$s(x)\partial_t^2 u(x,t) = \partial_x^2 u(x,t) - q(x)u(x,t) + r(x)u(x,t)^3, \tag{13.29}$$

with $x \in \mathbb{R}$, $t \in \mathbb{R}$, $u(x,t) \in \mathbb{R}$, and spatially periodic real-valued coefficient functions s, q, and r, i.e.,

$$s(x) = s(x+2\pi), \quad q(x) = q(x+2\pi), \quad \text{and} \quad r(x) = r(x+2\pi),$$

can be constructed with center manifold theory and spatial dynamics, too. The spatial dynamics formulation, with $x \in \mathbb{R}$ as new evolutionary variable, is solved in the space of $2\pi/\omega$-time-periodic functions. The spatial dynamics formulation is non-autonomous due to the x-dependence of the coefficients s, q, and r. Breather solutions are homoclinic solutions of the spatial dynamics formulation. We have seen that for spatially constant coefficients such solutions do not exist since for the linearization around the origin except of two eigenvalues all other eigenvalues are on the imaginary axis. However, by making the problem x-dependent and adjusting s and q in a proper way all the Floquet exponents can be moved off the imaginary axis and the existence of a two-dimensional center manifold with two homoclinic solutions can be established. The equations on the center manifold are related to a NLS equation which is the associated modulation equation for this problem, cf. §11.6. The details have been worked out in [**BCBLS11**]. ⌋

Further Reading. The previous theory is restricted to semilinear systems. Some theory for quasilinear systems can be found in [**Mie88, Ren92, Jam03, Sch15**]. The rigorous construction of stationary and time-periodic pattern in hydrodynamical stability problems in cylindrical domains can be found in [**DIM89, IM91**]. Three-dimensional water waves are constructed via center manifold reductions in [**GHS02, GH03**]. Recently, a lot of research has been done for non-irrotational water waves, cf. [**Con11**]. For the analysis of the reduced equations on the center manifold symmetries play a big role. The mathematical theory behind this approach can be found in [**GS85, GSS88, CL00**]. More generally, invariant manifold theory plays a big role in the construction of more and more complicated bifurcating solutions in pattern forming systems. Spiral waves have been constructed for instance in [**SSW99**], or dislocations of pattern for instance in [**HS12**].

Exercises

13.1. Consider the Allen-Cahn equation $\partial_t u = \partial_x^2 u + \alpha u - u^3$, $\alpha \in \mathbb{R}$, $u = u(x,t) \in \mathbb{R}$, $t \geq 0$, and $x \in [0,\pi]$ under the boundary conditions $u(0,t) = u(\pi,t) = 0$. Make a center manifold reduction for $\alpha = 4$ and use the reduced equation to prove the occurrence of a supercritical pitchfork bifurcation.

13.2. Consider $\Delta u = \alpha u - u^2$ for $u = u(x,y) \in \mathbb{R}$ on the cylindrical domain $(x,y) \in \mathbb{R} \times (0,\pi)$ with Neumann boundary conditions $\partial_y u|_{y=0,\pi} = 0$.
a) Write the elliptic problem as an evolutionary problem w.r.t. x as new time variable.
b) Compute the spectrum of the linearization around $u = 0$ of this spatial dynamics formulation for α close to zero.
c) Compute the reduction function h and the reduced ODE on the center manifold. (Hint: Consider y-independent solutions).
d) Discuss the flow on the center manifold for $\alpha < 0$, $\alpha = 0$, and $\alpha > 0$. Plot the solutions $x \mapsto u(x,y)$.

13.3. Discuss whether the solutions $\lambda_{1,2,3,4}$ of $1 - c^2 + (c^2 - \mu)\lambda^2 - \mu c^2 \lambda^4 = 0$ are on the imaginary or real axis. Sketch the result in the (μ, c)-plane.

13.4. (Explicit formula for the kink solution) Verify that for arbitrary C and λ, with $|\lambda| < 1$,

$$u(x,t) = 4 \arctan(C \exp((x - \lambda t)/(1 - \lambda^2)^{1/2}))$$

is a solution of the sine-Gordon equation $\partial_t^2 u = \partial_x^2 u - \sin u$. Sketch the solution.

13.5. Show that the sine-Gordon equation $\partial_t^2 u = \partial_x^2 u - \sin u$ is invariant under the transformation $\widetilde{x} = \gamma(x - vt), \widetilde{t} = \gamma(t - vx)$, where $\gamma = 1/(1 - v^2)^{1/2}$ and $|v| < 1$. Hence, conclude that

$$u(x,t) = 4 \arctan \left(\frac{(1 - \lambda^2)^{1/2}}{\lambda} \frac{\sin(\gamma \lambda (t - vx - t_0))}{\cosh(\gamma (1 - \lambda^2)^{1/2})(x - vt - x_0)} \right)$$

solves the sine-Gordon equation. Sketch the evolution of the solution.

Chapter 14

Diffusive stability

A basic concept in stability theory is the control of the nonlinear terms by the linearized problem if the linearization dissipates energy with an exponential rate. Such a behavior occurs when the spectrum lies in the left half plane strictly bounded away from the imaginary axis, cf. Theorem 2.3.4 for the ODE case, and Theorem 5.2.23 and Theorem 6.2.18 for examples in the PDE case. However, for PDEs posed on unbounded spatial domains there are various stability problems where we have essential spectrum up to the imaginary axis. Especially, the bifurcating Turing pattern discussed in Chapter 10 and 13 fall into this class. Therefore, it is the goal of this section to explain how nonlinear stability can be established in this case, too. We restrict ourselves to problems which are related to some modulation equation, as the real or complex GL equation.

For the problems we are interested in, the linear problem shows some diffusive behavior and polynomial decay rates. As a consequence, not all nonlinear terms can be controlled by the linearized problem. In order to see the difference to the case of exponential decay rates we consider the stability of $u = 0$ for the two toy problems

$$\frac{\mathrm{d}}{\mathrm{d}t} u = -u + u^p, \tag{14.1}$$

$$\partial_t u = \partial_x^2 u + u^p. \tag{14.2}$$

For (14.1) the linearization $\partial_t u = -u$ around $u = 0$ is solved by $u(t) = \mathrm{e}^{-t} u(0)$. For the terms occurring in (14.1) we therefore find $\frac{\mathrm{d}}{\mathrm{d}t} u \sim \mathrm{e}^{-t}$, $-u \sim \mathrm{e}^{-t}$, and $u^p \sim \mathrm{e}^{-pt}$. Hence for all $p > 1$ the nonlinear terms vanish much faster than the linear ones, and so it can be expected that the nonlinear problem behaves asymptotically as predicted by the linear one.

For the nonlinear diffusion equation (14.2), the linearized problem $\partial_t u = \partial_x^2 u$ in general shows no decay rate at all. However, for spatially localized initial conditions polynomial decay can be observed. It turns out that the solutions of the linearized problem with the slowest decay are given by $u(x,t) = ct^{-1/2}\mathrm{e}^{-x^2/(4t)}$. For the terms occurring in (14.2) we then find the asymptotic decay

$$\partial_t u \sim t^{-3/2}, \qquad \partial_x^2 u \sim t^{-3/2} \qquad \text{and} \qquad u^p \sim t^{-p/2},$$

cf. Exercise 14.1. Hence only for $p > 3$ the nonlinear terms vanish faster than the linear ones, and so only for $p > 3$ it can be expected that the nonlinear problem behaves asymptotically as predicted by the linear one. Nonlinearities which can be controlled with the help of the linearized problem in this way are called irrelevant w.r.t. diffusion or simply irrelevant. Nonlinearities exactly at the boundary, $p = 3$, are called critical. Surprisingly, it turns out that for many interesting and rather complicated problems from physics the nonlinear terms are irrelevant. It is has been observed at the beginning of the 1990s that the concept of diffusive behavior and irrelevant nonlinearities plays an important role in stability questions of pattern forming systems.

The plan of this chapter is as follows. In §14.1 we recall from §7.1 the most important properties of the linear diffusion equation from a slightly different point of view, and introduce the concept of irrelevant nonlinearities and a number of different methods to establish the irrelevance of nonlinearities. Then in §14.2.1 we show the occurrence of diffusive behavior and irrelevant nonlinearities in the real GL equation. In §14.2.2 we explain how this approach can be transferred to pattern forming systems in order to show the diffusive stability of spatially periodic equilibria in the SH equation, the Couette-Taylor problem or Bénard's problem. The rest of this section is devoted to critical nonlinearities. In §14.3.1 we discuss exponentially long transient behavior in unstable Poiseuille flow, and in §14.3.2 we consider the Burgers equation as limit equation. Examples are the inclined film problem or self-similar mixing of phases in pattern forming systems where the group velocity depends on the wave number. In §14.4 we introduce phase diffusion equations, which are modulation equations occurring in diffusive stability theory. Finally, in §14.5 we revisit similarities and differences between diffusive and dispersive dynamics.

14.1. Linear and nonlinear diffusive behavior

14.1.1. Linear diffusive behavior. Solutions of the linear diffusion equation

$$\partial_t u = \partial_x^2 u, \tag{14.3}$$

with $u|_{t=0} = u_0$, $x \in \mathbb{R}$, $t \geq 0$, and $u(x,t) \in \mathbb{R}$, can be written explicitly as

$$u(x,t) = (\mathrm{e}^{t\partial_x^2} u_0)(x) = \frac{1}{\sqrt{4\pi t}} \int_{-\infty}^{\infty} \mathrm{e}^{-\frac{(x-y)^2}{4t}} u_0(y)\,\mathrm{d}y. \tag{14.4}$$

Spatially constant functions stay constant in time, but for $u_0 \in L^q$ we obtain from Lemma 7.3.19 that, with some constant C independent of t,

$$\|u(t)\|_{L^p} \leq \|G(t)\|_{L^r} \|u_0\|_{L^q} \leq Ct^{-1/(2r)} \|u_0\|_{L^q}, \tag{14.5}$$

where $1/p = 1/q - 1/r$ for $1 \leq q \leq p \leq \infty$, and $G(x,t) = \dfrac{1}{\sqrt{4\pi t}} \mathrm{e}^{-\frac{x^2}{4t}}$. Thus, spatially localized initial conditions give rise to solutions with polynomial decay rates. Moreover, the solutions become flatter and flatter, since we have for instance

$$\|\partial_x^n u\|_{L^\infty} \leq \|\partial_x^n G(t)\|_{L^1} \|u_0\|_{L^\infty} \leq Ct^{-n/2} \|u_0\|_{L^\infty}.$$

The decay always happens in the same self-similar way which can be seen by looking at the Fourier transform of (14.3). The Fourier transformed diffusion equation $\partial_t \widehat{u} = -k^2 \widehat{u}$ is solved by $\widehat{u}(k,t) = \mathrm{e}^{-k^2 t} \widehat{u}_0(k)$. The self-similar behavior can be extracted by considering the renormalized solution

$$\widehat{u}(k/\sqrt{t}, t) = \mathrm{e}^{-k^2} \widehat{u}_0(k/\sqrt{t}) = \mathrm{e}^{-k^2} \left(\sum_{j=0}^{n} t^{-j/2} k^j \widehat{u}_0^{(j)}(0) + o(t^{-n/2}) \right),$$

if $\widehat{u}_0$ is n-times differentiable. Since smoothness in Fourier space corresponds to decay rates in physical space, solutions to spatially localized initial conditions vanish in a universal manner. Loosely speaking, if the initial conditions decay like $|x|^{-n}$, then

$$u(x,t) = \sum_{j=0}^{n-1} t^{-(j+1)/2} \widehat{u}_0^{(j)}(0) H_j(x/\sqrt{t}) + O(t^{-n/2})$$

for $t \to \infty$, where H_j is a multiple of the jth Hermite polynomial. See Exercise 14.2. Moreover, it is easy to see that

$$\|\widehat{u}(k/\sqrt{t}, t) - \mathrm{e}^{-k^2} \widehat{u}_0(0)\|_{H_2^\theta} \leq Ct^{-1/2}$$

for all $\theta \geq 2$. Using the fact that Fourier transform is an isomorphism from H_m^θ to H_θ^m, cf. Lemma 7.3.31, we obtain equivalently

$$\|\sqrt{t}\, u(x\sqrt{t}, t) - \sqrt{\pi}\, \widehat{u}_0(0) \mathrm{e}^{-x^2/4}\|_{H_\theta^2} \leq Ct^{-1/2}. \tag{14.6}$$

Most of the above ideas hold if we have a linear evolution operator $\mathrm{e}^{\lambda(k)t}$ with eigenvalues $\lambda(k) \sim -k^2$ for $k \to 0$. This is the reason why diffusive behavior can be observed in a big variety of problems, as we will see. Textbooks about self-similar solutions in various parabolic problems are [**Bar96, SR10**].

Remark 14.1.1. By introducing a logarithmic time scale the polynomial decay rates can be transferred into exponential decay rates. This allows to interpret the Hermite functions as eigenfunctions of a renormalized diffusion operator. We follow the lines of [**Way97**] and introduce the new variable w and the new coordinates ξ and τ by

$$u(x,t) = t^{-1/2} w(x/\sqrt{t}, \log t) = \mathrm{e}^{-\tau/2} w(\xi, \tau). \tag{14.7}$$

The transformed diffusion equation is then given by

$$\partial_\tau w = \partial_\xi^2 w + (\xi/2)\partial_\xi w + w/2. \tag{14.8}$$

The spectral problem in Fourier space, for instance in H_2^θ, reads

$$-k^2 \widehat{w} - (k/2)\partial_k \widehat{w} - \lambda \widehat{w} = \widehat{f}, \tag{14.9}$$

with $\widehat{f} \in H_2^\theta$. The eigenfunctions $\widehat{\psi}_s(k) = k^s \mathrm{e}^{-k^2}$ to the real eigenvalues $\lambda_s = -s/2$ are parameterized with $s \in \mathbb{R}$. Since $\partial_k^j \widehat{\psi}_s \in L^2$ is required for $j = 0, 1, \ldots, \theta$, this leads to a set of θ eigenvalues $\{\lambda_j = -j/2 : j = 0, 1, \ldots, \theta - 1\}$ and to essential spectrum $\{\lambda \in \mathbb{C} : \mathrm{Re}\lambda < -\theta/2 - 1/4\}$. In order to see this, one has to consider the singularity of $\partial_k^j \widehat{\psi}_s$ at $k = 0$ for large j. ⌋

14.1.2. Irrelevant nonlinearities. Following partly the lines of [**MSU01**] we now explain for the nonlinear diffusion equation

$$\partial_t u = \partial_x^2 u + c u^p, \qquad u|_{t=0} = u_0, \tag{14.10}$$

with $t \geq 0$, $x \in \mathbb{R}$, $p \in \mathbb{N}$, and $c \in \mathbb{R} \setminus \{0\}$, how the polynomial decay rates for the linearized problem can be used to control the nonlinear terms.

For initial conditions $u_0 > 0$ constant in space and $c > 0$ we always have blow-up. However, for spatially localized initial conditions the situation is more complicated. In case $p = 2$ and $c > 0$ we still have blow-up of solutions to arbitrary small non-negative initial conditions with non-vanishing integral, cf. [**Wei81**]. In case $p = 3$ the sign of c decides about stability ($c < 0$) and instability ($c > 0$), but for $p \geq 4$ the sign of c does not play a role and small spatially localized perturbations vanish for $t \to \infty$ with the same polynomial decay rate as in the linear case.

There are at least three methods to prove the last statement, namely L^1-L^∞ estimates, the construction of Lyapunov functions, and the discrete and continuous renormalization approach. We will present all three of them in order to make clear that when L^1-L^∞ estimates work also the renormalization approach works and vice versa, although the statements seem different. Moreover, we explain why the Lyapunov function approach in general gives weaker results.

14.1.3. L^1-L^∞ estimates. This method relies on the L^q-L^p estimate (14.5), the variation of constants formula, and suitable estimates for the nonlinearity.

Lemma 14.1.2. *Let $p > 3$. For all $C > 0$ there exists a $\delta > 0$ such that solutions u of (14.10) with $\|u_0\|_{L^1} + \|u_0\|_{L^\infty} \le \delta$ satisfy*

$$\|u(t)\|_{L^1} \le C \quad \textit{and} \quad \|u(t)\|_{L^\infty} \le C(1+t)^{-\frac{1}{2}}$$

for all $t \ge 0$.

Proof. We introduce

$$a(t) = \sup_{0\le s\le t} \|u(s)\|_{L^1}, \qquad b(t) = \sup_{0\le s\le t} \|(1+s)^{1/2}u(s)\|_{L^\infty}.$$

We use the variation of constant formula (2.6) estimate successively the L^1-norm and the L^∞-norm of the terms on right-hand side. We have, cf. Exercise 14.3,

$$\|e^{t\partial_x^2}u_0\|_{L^1} \le \|u_0\|_{L^1} \tag{14.11}$$

and

$$\|e^{t\partial_x^2}u_0\|_{L^\infty} \le \begin{cases} \|u_0\|_{L^\infty}, & t \le 1 \\ \frac{1}{\sqrt{4\pi t}}\|u_0\|_{L^1}, & t > 1 \end{cases} \tag{14.12}$$

such that

$$\|(1+t)^{1/2}e^{t\partial_x^2}u_0\|_{L^\infty} \le \|u_0\|_{L^1} + \|u_0\|_{L^\infty}.$$

For the nonlinear terms we use $\|u^p\|_{L^\infty} \le \|u\|_{L^\infty}^p$ and $\|u^p\|_{L^1} \le \|u\|_{L^\infty}^{p-1}\|u\|_{L^1}$. We obtain

$$\begin{aligned}
&(1+t)^{1/2}\left\|\int_0^t e^{(t-s)\partial_x^2}u^p(s)\,ds\right\|_{L^\infty} \le (1+t)^{1/2}\int_0^t \|e^{(t-s)\partial_x^2}\|_{L^1\to L^\infty}\|u^p\|_{L^1}\,ds \\
&\le \Big((1+t)^{1/2}\int_0^{t/2}(t-s)^{-1/2}(1+s)^{-(p-1)/2}\,ds \\
&\qquad +(1+t)^{1/2}\int_{t/2}^t(t-s)^{-1/2}(1+s)^{-(p-1)/2}\,ds\Big)\cdot b(t)^{p-1}a(t) \\
&\le \Big((1+t)^{1/2}\int_0^{t/2}(t-t/2)^{-1/2}(1+s)^{-(p-1)/2}\,ds \\
&\qquad +(1+t)^{1/2}\int_{t/2}^t(t-s)^{-1/2}(1+t/2)^{-(p-1)/2}\,ds\Big)\cdot b(t)^{p-1}a(t) \\
&\le C_1\, b(t)^{p-1}a(t),
\end{aligned}$$

with a constant C_1 independent of t for $p > 3$, and

$$\begin{aligned}
&\left\|\int_0^t e^{(t-s)\partial_x^2}u^p(s)\,ds\right\|_{L^1} \le \int_0^t \|e^{(t-s)\partial_x^2}\|_{L^1\to L^1}\|u^p\|_{L^1}\,ds \\
&\le \int_0^t(1+s)^{-(p-1)/2}\,ds\cdot b(t)^{p-1}a(t) \le C_2 b(t)^{p-1}a(t),
\end{aligned}$$

with a constant C_2 independent of t for $p > 3$. Thus,

$$a(t) \le a(0)+b(0)+|c|C_2 b(t)^{p-1}a(t), \qquad b(t) \le a(0)+b(0)+|c|C_1 b(t)^{p-1}a(t).$$

Introducing $y(t) = a(t) + b(t)$ and adding the two inequalities yields

$$y(t) \leq 2y(0) + (C_1 + C_2)y(t)^p. \tag{14.13}$$

For $y(0) < \delta$, with $\delta > 0$ sufficiently small, the two curves $y \mapsto y$ and $y \mapsto y(0) + (C_1 + C_2)y^p$ possess two intersection points y_1, y_2, with $y_1 < y_2$. Inequality (14.13) is valid for $y \leq y_1$ and $y \geq y_2$. Since $y(0) \leq y_1$ and since we have continuity of $t \mapsto y(t)$ we must have $y(t) \leq y_1$ for all $t \geq 0$. □

According to Remark 14.1.1 we cannot expect faster decay rates even if the initial conditions vanish much faster than a general L^1-function for $|x| \to \infty$, and for L^∞-initial conditions there is no decay. For functions with an intermediate spatial decay, e.g., L^2 initial conditions, we obtain the following. On a linear level we have via Lemma 7.3.19 that

$$\|u(t)\|_{L^\infty} = \|G(t) * u_0\|_{L^\infty} \leq \|G(t)\|_{L^2}\|u_0\|_{L^2},$$

with $G(x,t) = \frac{1}{\sqrt{4\pi t}} e^{-\frac{x^2}{4t}}$. We find

$$\|G(t)\|_{L^2} = \left(\int_{\mathbb{R}} \left(\frac{1}{\sqrt{4\pi t}} e^{-\frac{x^2}{4t}} \right)^2 dx \right)^{1/2} = \left(\int_{\mathbb{R}} \frac{1}{4\pi t} e^{-\frac{2x^2}{4}} dy\sqrt{t} \right)^{1/2} \sim t^{-1/4},$$

such that $\|u(t)\|_{L^\infty} \leq Ct^{-1/4}\|u_0\|_{L^2}$. By introducing

$$a(t) = \sup_{0 \leq s \leq t} \|u(s)\|_{L^2} \quad \text{and} \quad b(t) = \sup_{0 \leq s \leq t} \|(1+s)^{1/4} u(s)\|_{L^\infty},$$

the following result easily can be proved almost line for line as above.

Corollary 14.1.3. *Let $p > 5$. For all $C > 0$ there exists a $\delta > 0$ such that solutions u of (14.10) with $\|u_0\|_{L^2} + \|u_0\|_{L^\infty} \leq \delta$ satisfy*

$$\|u(t)\|_{L^2} \leq C \quad \text{and} \quad \|u(t)\|_{L^\infty} \leq C(1+t)^{-\frac{1}{4}}$$

for all $t \geq 0$. Similarly, for $u_0 \in L^q \cap L^\infty$ small, $p > 2q + 1$, and $q \in [1, \infty)$, we have

$$\|u(t)\|_{L^q} \leq C \quad \text{and} \quad \|u(t)\|_{L^\infty} \leq C(1+t)^{-\frac{1}{2q}}.$$

14.1.4. Lyapunov functions. The usage of Lyapunov functions is well established in nonlinear stability problems if the nonlinearity has the correct sign such as in $\partial_t u = \partial_x^2 u - u^3$. For small initial conditions Lyapunov functions can be used even if the nonlinearity has the wrong sign. For exponentially stable equilibria this is obvious, but also for diffusive stability problems the approach can be used, cf. [**EW94, GM98**]. We do not get the optimal power p for irrelevant nonlinearities, but the method is also applicable on unbounded domains $\Omega \subset \mathbb{R}^d$ with $\Omega \neq \mathbb{R}^d$, cf. [**ES98**].

As an example we consider again

$$\partial_t u = \partial_x^2 u + cu^p, \qquad u|_{t=0} = u_0,$$

in case $\Omega = \mathbb{R}$. We introduce the functionals $I(u) = \int_{\mathbb{R}} u^2 \, \mathrm{d}x$, $J(u) = \int_{\mathbb{R}} (\partial_x u)^2 \, \mathrm{d}x$ and $K(u) = \int_{\mathbb{R}} (\partial_x^2 u)^2 \, \mathrm{d}x$. For initial conditions with an L^2-decay for $|x| \to \infty$, due to the L^2-L^∞ estimate (14.5) we can only expect to handle nonlinearities for $p > 5$.

With $\|u\|_{L^\infty}^2 \leq I^{1/2} J^{1/2}$ and $J^2 \leq I\,K$ we obtain

$$\begin{aligned}
\frac{1}{2}\frac{\mathrm{d}}{\mathrm{d}t} I &= \int_{\mathbb{R}} u \partial_t u \, \mathrm{d}x = \int_{\mathbb{R}} u \partial_x^2 u + c u^{p+1} \, \mathrm{d}x \\
&= \int_{\mathbb{R}} -(\partial_x u)^2 + c u^{p+1} \, \mathrm{d}x \;\leq\; -J + |c| \|u(x)\|_{L^\infty}^{p-1} \int_{\mathbb{R}} u^2 \, \mathrm{d}x \\
&\leq -J + |c| I^{(p+3)/4} J^{(p-1)/4} = -J\,[1 - |c| I^{(p+3)/4} J^{(p-5)/4}],
\end{aligned}$$

and

$$\begin{aligned}
\frac{1}{2}\frac{\mathrm{d}}{\mathrm{d}t} J &= \int_{\mathbb{R}} (\partial_x u)(\partial_t \partial_x u) \, \mathrm{d}x \\
&= \int_{\mathbb{R}} (\partial_x u)(\partial_x^3 u) + c p u^{p-1} (\partial_x u)^2 \, \mathrm{d}x \;\leq\; -\int_{\mathbb{R}} (\partial_x^2 u)^2 \, \mathrm{d}x + |c| p \|u\|_{L^\infty}^{p-1} J \\
&\leq -K + |c| p I^{(p-1)/4} J^{(p+3)/4} \;\leq\; -K\,[1 - |c| p I^{(p+3)/4} J^{(p-5)/4}].
\end{aligned}$$

Hence, we have $\dot{I} \leq 0$ and $\dot{J} \leq 0$ if $I(u_0)^{(p+3)/4} J(u_0)^{(p-5)/4} > 0$ is sufficiently small. Thus, we have proved the following result.

Lemma 14.1.4. *Let $p \geq 5$. Then, there exists a $\delta_0 > 0$ such that for all $\delta \in (0, \delta_0)$ and all solutions u of (14.10) with $\|u_0\|_{H^1} \leq \delta$ we have $\sup_{t \geq 0} \|u(t)\|_{H^1} \leq \delta$.*

Note that the method also allows to handle the boundary case $p = 5$. In higher space dimensions the critical exponent obtained with this approach can be reduced to $p = 1 + 4/d$, cf. Exercise 14.4.

14.1.5. The continuous renormalization process. In addition to some stability result, the renormalization process gives the asymptotics of the decay towards zero. It relies on formula (14.6). By a fixed point argument we prove not only for the linear but also for the nonlinear system (14.10) that the renormalized solution $\sqrt{t}\, u(t, x\sqrt{t})$ converges towards a multiple of the Gaussian $\mathrm{e}^{-x^2/4}$. There are two approaches, a continuous and a discrete one. In the continuous approach the system satisfied by the renormalized solution is considered directly, where additionally a logarithmic time scale is used to transfer the polynomial into exponential decay rates. As in Remark 14.1.1 we introduce the new variable w and the new coordinates ξ and τ by

$$u(x,t) = t^{-1/2} w(x/\sqrt{t}, \log t) = \mathrm{e}^{-\tau/2} w(\xi, \tau). \tag{14.14}$$

The transformed equation is given by

$$\partial_\tau w = w/2 + (\xi/2)\partial_\xi w + \partial_\xi^2 w + \mathrm{e}^{(3-p)\tau/2} w^p, \tag{14.15}$$

where we choose $w|_{\tau=0} \in H_2^2$. Moreover, w.l.o.g. for our purposes we set $c = 1$. In H_2^2 we have a zero eigenvalue, and the rest of the spectrum is left of the line $\{\lambda \in \mathbb{C} : \mathrm{Re}\lambda = -1/2\}$, cf. Remark 14.1.1. Since the solutions of the linearized problem are uniformly bounded and since the nonlinear terms vanish with an exponential rate there exist $\delta, C > 0$ such that $\sup_{\tau\geq 0} \|w(\tau)\|_{H_2^2} < C$ for the solutions w of (14.15) if $\|w|_{\tau=0}\|_{H_2^2} < \delta$ and $p > 3$. If we denote by w_0 the part of w belonging to the eigenvalue 0 and with w_1 the rest of w, then by integration of the variation of constant formula w.r.t. time we can conclude that

$$w_0(\cdot,\tau) = w_{\lim}\mathrm{e}^{-(\cdot)^2/4} + \mathcal{O}(\mathrm{e}^{\frac{3-p}{2}\tau}) \quad \text{and} \quad w_1(\tau) = \mathcal{O}(\mathrm{e}^{\max\{-\frac{1}{2},\frac{3-p}{2}\}\tau})$$

for $\tau \to \infty$, where $w_{\lim} \in \mathbb{R}$ is a constant only depending on the initial conditions. This leads to the following convergence result, cf. [**Way97**].

Theorem 14.1.5. *Let $p > 3$. There exist $\delta, C > 0$ such that for all solutions u of (14.10) with $\|u|_{t=0}\|_{H_2^2} \leq \delta$, there exists a $w_{\lim} \in \mathbb{R}$ such that*

$$\|\sqrt{t}\,u(\cdot\sqrt{t},t) - w_{\lim}\mathrm{e}^{-(\cdot)^2/4}\|_{H_2^2} \leq C(1+t)^{-\max\{1/2,(3-p)/2\}} \quad \textit{for all } t \geq 0.$$

14.1.6. The discrete renormalization approach. In the discrete approach, cf. [**BK92, BKL94, Gal94**], a sequence of problems is considered which formally converges towards the linear diffusion equation. Define

$$u_n(x,\tau) = L^n u(L^n x, L^{2n}\tau), \tag{14.16}$$

with $L > 1$ fixed and $n \in \mathbb{N}$. Then u_n satisfies

$$\partial_\tau u_n = \partial_x^2 u_n + L^{n(3-p)} u_n^p \quad \text{for } \tau \in [L^{-2}, 1] \tag{14.17}$$

and

$$u_n(x, L^{-2}) = L u_{n-1}(Lx, 1). \tag{14.18}$$

Formally, for $p > 3$ the influence of the nonlinear terms vanishes geometrically for $n \to \infty$. In the limit we obtain the linear diffusion equation. By a fixed point argument we prove for the nonlinear problem the convergence of the sequence of functions $u_n|_{\tau=1} = L^n u(L^n\cdot, L^{2n})$ towards the limit of the linear problem, namely a multiple of the Gaussian $\mathrm{e}^{-x^2/4}$. We use this approach to give a

Proof of Theorem 14.1.5. In Fourier space (14.10) is given by

$$\partial_t \widehat{u} = -k^2\widehat{u} + \widehat{u}^{*p}, \tag{14.19}$$

where $\widehat{u}^{*p}$ denotes the p-times convolution of a function $\widehat{u}$, i.e., $\widehat{u}^{*p} = \widehat{u} * \ldots * \widehat{u}$. Corresponding to (14.16) we rescale $\widehat{u}$ to $\widehat{u}_n(k,\tau) = \widehat{u}(L^{-n}k, L^{2n}\tau)$ for a fixed $L > 1$. Note that due to the scaling properties of Fourier transform there is no L^n in front of $\widehat{u}$. We obtain

$$\partial_\tau \widehat{u}_n = -k^2\widehat{u}_n + L^{n(3-p)}\widehat{u}_n^{*p}, \tag{14.20}$$

where a factor $L^{n(1-p)}$ in front of the convolution terms comes from the substitution in the integrals. In order to simplify notation we denote all constants with the same symbol C if they can be chosen independently of L. We solve (14.19) via (14.20) by the following renormalization procedure.

$(\mathbf{RG})_{\mathbf{n}\to\mathbf{n+1}}$: Equation (14.20) is solved on the time interval $[1/L^2, 1]$. Then $\widehat{u}_{n+1}(k, 1/L^2) = \widehat{u}_n(k/L, 1)$ is taken as initial condition for $n+1$.

For solving (14.20) we use the variation of constant formula

$$\begin{aligned} \widehat{u}_n(k,\tau) &= \mathrm{e}^{-k^2(\tau-1/L^2)}\widehat{u}_{n-1}(k/L,1) \\ &\quad + L^{n(3-p)} \int_{1/L^2}^{\tau} \mathrm{e}^{-k^2(\tau-s)}\widehat{u}_n^{*p}(k,s)\,\mathrm{d}s. \end{aligned} \tag{14.21}$$

The size $R_n := \sup_{\tau\in[1/L^2,1]} \|\widehat{u}_n(\tau)\|_{H_2^2}$ of the solutions $\widehat{u}_n$ on the time interval $[1/L^2, 1]$ can be controlled in terms of the initial conditions.

Lemma 14.1.6. *There exist δ, C_1 and $L_0 > 0$ such that for all $\widehat{u}_{n-1}|_{\tau=1} \in H_2^2$ with $\|\widehat{u}_{n-1}|_{\tau=1}\|_{H_2^2} < \delta$ and all $L > L_0$ the solution $\widehat{u}_n \in C([1/L^2,1], H_2^2)$ exists and satisfies*

$$R_n \leq C_1 L^{5/2} \|\widehat{u}_{n-1}|_{\tau=1}\|_{H_2^2}.$$

Proof. Rescaling gives the estimate

$$\|v(\cdot/L)\|_{H_2^2} \leq C L^{5/2} \|v\|_{H_2^2}.$$

Thus, the linear terms in (14.21) can be estimated by

$$\begin{aligned} &\sup_{\tau\in[1/L^2,1]} \|\mathrm{e}^{-(\cdot)^2(\tau-1/L^2)}\widehat{u}_{n-1}(\cdot/L,1)\|_{H_2^2} \\ &\leq \sup_{\tau\in[1/L^2,1]} \|\mathrm{e}^{-(\cdot)^2(\tau-1/L^2)}\|_{C_b^2} \|\widehat{u}_{n-1}(\cdot/L,1)\|_{H_2^2} \leq C L^{5/2} \|\widehat{u}_{n-1}(\cdot,1)\|_{H_2^2}. \end{aligned}$$

This estimate for the semigroup and Sobolev's embedding theorem show for the nonlinear terms

$$\sup_{\tau\in[1/L^2,1]} \left\| L^{n(3-p)} \int_{1/L^2}^{\tau} \mathrm{e}^{-k^2(\tau-s)}\widehat{u}_n^{*p}(\cdot,s)\,\mathrm{d}s \right\|_{H_2^2} \leq C L^{n(3-p)} R_n^p.$$

Choosing $\delta > 0$ sufficiently small and L_0 sufficiently large it is easy to see that the right-hand side of (14.21) defines a contraction in a ball of radius $\mathcal{O}(L^{5/2}\|\widehat{u}_{n-1}|_{\tau=1}\|_{H_2^2})$ in the space $C([1/L^2,1], H_2^2)$. Therefore, the solution which is the fixed point of this contraction stays in H_2^2 in a ball of the same radius $\mathcal{O}(L^{5/2}\|\widehat{u}_{n-1}|_{T=1}\|_{H_2^2})$. □

In order to show the convergence of the points $\widehat{u}_n|_{\tau=1}$ towards $A_\infty\psi$ in H_2^2, with some $A_\infty \in \mathbb{R}$ and the Gaussian $\psi(k) = \mathrm{e}^{-k^2}$, let us define the projection $\Pi\widehat{u}_n = \widehat{u}_n|_{k=0}$. We are done, if we prove the convergence of

$$\widehat{\rho}_n = \widehat{u}_n|_{T=1} - A_n\psi,$$

with $A_n = \Pi \widehat{u}_n|_{\mathcal{T}=1}$ towards zero and the existence of the limit $\lim_{n\to\infty} A_n$. By definition we have $\widehat{\rho}_n|_{k=0} = 0$. The variables A_n and $\widehat{\rho}_n$ satisfy

$$\begin{aligned} A_{n+1} =& A_n + \Pi(L^{n(3-p)} \int_{1/L^2}^{1} \mathrm{e}^{-k^2(1-s)} \widehat{u}_{n+1}^{*p}(k,s)\,\mathrm{d}s), \\ \widehat{\rho}_{n+1}(k) =& \mathrm{e}^{-k^2(1-1/L^2)} \widehat{\rho}_n(k/L) + L^{n(3-p)} \int_{1/L^2}^{1} \mathrm{e}^{-k^2(1-s)} \widehat{u}_{n+1}^{*p}(k,s)\,\mathrm{d}s \\ &+ \mathrm{e}^{-k^2(1-1/L^2)} A_n \psi(k/L) - A_{n+1}\psi(k). \end{aligned}$$

For the terms on the right-hand side we get

$$\begin{aligned} &\|L^{n(3-p)} \int_{1/L^2}^{1} \mathrm{e}^{-(\cdot)^2(1-s)} \widehat{u}_{n+1}^{*p}(\cdot,s)\,\mathrm{d}s\|_{H_2^2} \leq C R_n^p L^{n(3-p)}, \\ &|\Pi u| \leq C\|u\|_{H_2^2}, \\ &\|\mathrm{e}^{-(\cdot)^2(1-1/L^2)} \widehat{\rho}_n(\cdot/L)\|_{H_2^2} \leq (C/L)\|\widehat{\rho}_n\|_{H_2^2}, \\ &\|\mathrm{e}^{-(\cdot)^2(1-1/L^2)} A_n\psi(\cdot/L) - A_{n+1}\psi(\cdot)\|_{H_2^2} \leq C|A_{n+1} - A_n|. \end{aligned}$$

Using these estimates we obtain

$$\begin{aligned} |A_{n+1} - A_n| \leq& C L^{n(3-p)} R_n^p, \\ \|\widehat{\rho}_{n+1}\|_{H_2^2} \leq& (C/L)\|\widehat{\rho}_n\|_{H_2^2} + C L^{n(3-p)} R_n^p. \end{aligned}$$

Using Lemma 14.1.6 and the abbreviations $r_n = \|\widehat{\rho}_n\|_{H_2^2}$ gives

$$\begin{aligned} |A_{n+1} - A_n| \leq& C L^{n(3-p)} (L^{5/2})^p (A_n + r_n)^p, \\ r_n \leq& (C/L) r_n + C L^{n(3-p)} (L^{5/2})^p (A_n + r_n)^p. \end{aligned}$$

We now choose $L_0 > 1$ and $n_0 > 0$ so big that for $n > n_0$ we have

$$\begin{aligned} |A_{n+1} - A_n| \leq& L^{(n-n_0)(p-3)} (A_n + r_n)^p, \\ r_{n+1} \leq& r_n/10 + L^{(n-n_0)(p-3)} (A_n + r_n)^p, \end{aligned}$$

if $\sup_{n\in\mathbb{N}}(|A_n| + r_n) < \delta$, which was needed in the derivation of this estimate since we used Lemma 14.1.6. If $\sup_{n\in\mathbb{N}}(|A_n| + r_n) < \delta$, then the sequence of A_n converges geometrically to a limit $A_\infty < \infty$ and we find that $\lim_{n\to\infty} r_n = 0$. Since the quantities $|A_n|$ and r_n increase only for at most n_0 steps we can stay smaller than $\delta > 0$ from Lemma 14.1.6 if we choose $|A_1|$ and r_1 of order $\mathcal{O}(L^{-m})$, for an $m > 0$ sufficiently large.

Since the convergence goes as $L^{-\min(1,p-3)n}$, since Fourier transform is an isomorphism in H_2^2, and since the convergence holds for all $L \in [L_0, L_0^2]$ we obtain Theorem 14.1.5. □

As a consequence we have

Corollary 14.1.7. *The fixed point* $u^* = 0$ *in* (14.10) *is* (H_2^2, L^∞)*-stable in the sense of Definition* 7.1.19.

Proof. Using the notation from the proof of Theorem 14.1.5 the statement follows from

$$\begin{aligned}\|u(\cdot,t)\|_{L^\infty} \leq & Ct^{-1/2}\|\mathrm{e}^{-(\cdot)^2/(4t)}\|_{L^\infty} + Ct^{-1}\|\rho(\cdot/\sqrt{t},t)\|_{L^\infty} \\ = & Ct^{-1/2} + Ct^{-1}\|\rho(\cdot,t)\|_{L^\infty} \\ \leq & Ct^{-1/2} + Ct^{-1}\|\rho(\cdot,t)\|_{H_2^2} = \mathcal{O}(t^{-1/2}).\end{aligned}$$

□

Remark 14.1.8. a) The above methods work for more general nonlinearities, too. If we call the exponent p of the nonlinear term u^p the degree of irrelevance, then $u^{p_1}(\partial_x u)^{p_2}(\partial_x^2 u)^{p_3}$ has the degree of irrelevance $p = p_1 + 2p_2 + 3p_3$. For $x \in \mathbb{R}$ and the diffusion equation there are only three relevant monomial nonlinearities, namely u^2, u^3, and $u\partial_x u$.

b) In higher space dimensions the solutions of the linear diffusion equation $\partial_t u = \Delta u$, with $x \in \mathbb{R}^d$, satisfy the estimate

$$\|u(t)\|_{L^\infty} \leq Ct^{-d/2}\|u\|_{L^1}.$$

With d increasing, the number of directions increases in which energy can diffuse. From the above analysis it is easy to see that in $\mathbb{R}^2$ the only relevant monomial is u^2. In $\mathbb{R}^d$ with $d \geq 3$ all smooth nonlinearities are irrelevant w.r.t. diffusion. See [**Uec99**] for an abstract theorem. ⌋

14.2. Diffusive stability of spatially periodic equilibria

Diffusive stability results for the GL equation have been shown first in [**CEE92, BK92**]. We already discussed this topic in §8.3.6 in an informal way. It is the purpose of this section to make the formal arguments of §8.3.6 rigorous. The previous arguments have to be extended in order to handle additional exponentially damped terms. The presented approach is a generalization of [**MSU01**] and is less involved than the ones given in [**CEE92, BK92**]. The scheme developed subsequently is very general and it will be explained afterwards how it can be used to prove for instance the diffusive stability of spatially periodic equilibria in pattern forming systems.

14.2.1. Diffusive stability in the GL equation. The real GL equation

$$\partial_T A = \partial_X^2 A + A - A|A|^2 \tag{14.22}$$

with $X \in \mathbb{R}$, $T \geq 0$, and $A(X,T) \in \mathbb{C}$ possesses a two-dimensional family of stationary solutions

$$A = A_{\mathrm{per}}[q,\phi_0](X) = \sqrt{1-q^2}\mathrm{e}^{iqX+i\phi_0}, \tag{14.23}$$

parameterized by $\phi_0, q \in \mathbb{R}$. In §8.3.6 we discussed the linearization around these equilibria and found diffusive modes and exponentially damped modes. They can be separated easily by introducing polar coordinates

$$A(X,T) = r(X,T)e^{i\phi(X,T)}$$

in the real GL equation, which then transforms into

$$\partial_T r = \partial_X^2 r + r - (\partial_X\phi)^2 r - r^3, \qquad \partial_T\phi = \partial_X^2\phi + \frac{2(\partial_X r)(\partial_X\phi)}{r}.$$

For simplicity we restrict ourselves to $(r,\phi) = (1,0)$ as the stationary pattern. We introduce the deviation (s,ϕ) from the equilibrium by setting $r = 1+s$ and find

$$\begin{aligned} \partial_T s &= \partial_X^2 s - 2s - (\partial_X\phi)^2 - (\partial_X\phi)^2 s - 3s^2 - s^3, \\ \partial_T\phi &= \partial_X^2\phi + \frac{2(\partial_X s)(\partial_X\phi)}{1+s}. \end{aligned} \tag{14.24}$$

As we already know from §8.3.6, the linearized system

$$\partial_T s = \partial_X^2 s - 2s, \qquad \partial_T\phi = \partial_X^2\phi,$$

possesses solutions $(s,\phi) = (s_k,\phi_k)e^{ikx+\mu_j(k)t}$, with $\mu_1(k) = -k^2$ and $\mu_2(k) = -k^2-2$. Hence, the component s which corresponds to μ_2 is exponentially damped, and so we expect s to be slaved to ϕ, i.e., we expect s to behave in lowest order as $(\partial_X\phi)^2$. Since ϕ behaves diffusively, we expect decay rates $\phi \sim t^{-1/2}$ and $s \sim t^{-2}$, and that ϕ formally satisfies

$$\partial_T\phi = \partial_X^2\phi - \partial_X((\partial_X\phi)^2))(\partial_X\phi) + h.o.t.. \tag{14.25}$$

In §8.3.6 we argued that with these asymptotics the nonlinear terms on the right-hand side have a degree of irrelevance of seven and hence are irrelevant w.r.t. linear diffusion. Thus, with a slight improvement of the previous methods the following L^1-L^∞ diffusive stability result can be established. The proof will be given below.

Theorem 14.2.1. *The stationary solution $A=1$ of (14.22) is diffusively stable in the following sense. For all $C>0$ there exists a $\delta>0$ such that solutions $A = 1+v$ of (14.22) with $\|\widehat{v}|_{T=0}\|_{L^1} + \|\widehat{v}|_{T=0}\|_{L^\infty} \leq \delta$ satisfy*

$$\|\widehat{v}(T)\|_{L^\infty} \leq C, \qquad \|\widehat{v}(T)\|_{L^1} \leq C(1+T)^{-1/2}$$

for all $T \geq 0$.

Since the Fourier transform is continuous from L^1 to L^∞ we have as a direct consequence the following result.

Corollary 14.2.2. *For all $C>0$ there exists a $\delta>0$ such that solutions $A = 1+v$ of (14.22) with $\|\widehat{v}|_{T=0}\|_{L^1} + \|v|_{T=0}\|_{L^1} \leq \delta$ satisfy*

$$\|v(T)\|_{L^\infty} \leq C(1+T)^{-1/2}$$

for all $T \geq 0$.

Remark 14.2.3. In order to obtain an irrelevant nonlinearity with a degree of irrelevance of seven we would have to handle (14.24) as a quasilinear system, for instance by choosing $s \in H^1$ and $\phi \in H^2$. By handling it as a semilinear one, for instance by choosing $s \in H^1$ and $\phi \in H^1$, we would come from seven to six. This procedure then would not allow us to handle systems with a degree of irrelevance of four due to the loss of one. In order to avoid the more complicated analytic tools used for quasilinear systems we choose $s \in H^1$ and $\phi \in H^{2-\vartheta}$ with $\vartheta \in (0,1)$ which finally leads only to a loss of ϑ in the degree of irrelevance. Since $\vartheta > 0$ can be chosen arbitrarily small, the scheme used in the proof of Theorem 14.2.1 will allow us to handle all systems with irrelevant nonlinearities. ⌋

Remark 14.2.4. Before we give a proof of Theorem 14.2.1 we consider time-periodic solutions for the complex GL equation

$$\partial_T A = (1+\mathrm{i}\alpha)\partial_X^2 A + A - (1+\mathrm{i}\beta)A|A|^2, \tag{14.26}$$

where $\alpha, \beta \in \mathbb{R}$, $X \in \mathbb{R}$, $T \geq 0$, and $A(X,T) \in \mathbb{C}$, in order to make clear that a more abstract consideration of the problem is advantageous. The complex GL equation possesses wave trains

$$A(X,T) = \sqrt{1-k^2}\mathrm{e}^{\mathrm{i}(kX-\omega(k)T)},$$

where $\omega(k) = \beta + (\alpha - \beta)k^2$. We introduce the amplitude and phase deviations (s, ϕ) of the wave train to $k = 0$ via

$$A(X,T) = (1+s(X,T))\mathrm{e}^{\mathrm{i}(-\beta t + \phi(X,T))},$$

which satisfy

$$\begin{aligned}
\partial_T s =& \partial_X^2 s - 2s - (\partial_X \phi)^2 - (\partial_X \phi)^2 s - 2\alpha(\partial_X s)(\partial_X \phi) \\
& - \alpha \partial_X^2 \phi - \alpha(\partial_X^2 \phi)s - 3s^2 - s^3, \\
\partial_T \phi =& \partial_X^2 \phi + \alpha \frac{\partial_X^2 s}{1+s} - \alpha(\partial_X \phi)^2 + \frac{2(\partial_X s)(\partial_X \phi)}{1+s} - 2\beta s - \beta s^2.
\end{aligned}$$

With the choice $s \in H^2$ and $\phi \in H^2$ this system is a quasilinear system although in the original coordinates the nonlinear terms contain no derivative at all. Note that there is no choice of H^s spaces which makes it a semilinear system. Therefore, already for this simple problem it will be advantageous to choose a more abstract approach as we do below. ⌋

Proof of Theorem 14.2.1. We go back to the GL equation (14.22). The deviation $v = A - 1$ from the equilibrium $A = 1$ satisfies

$$\partial_T v = Lv + N(v) \tag{14.27}$$

where

$$Lv = \partial_X^2 v + v - (2v + \overline{v}), \qquad N(v) = -(v^2 + 2v\overline{v} + v^2\overline{v}).$$

The eigenvalue problem $Lv = \mu v$ is solved by Fourier modes $v(x,t) = \widehat{\varphi}_{c,s}(k)\mathrm{e}^{ikx}$ with eigenvalues $\mu_{c,s}(k)$. In the Eckhaus stable domain we have constants $\sigma_c > 0$ and $\sigma_s > 0$ such that

$$\mathrm{Re}\mu_c(k) \leq -\sigma_c k^2 \qquad \text{and} \qquad \mathrm{Re}\mu_s(k) \leq -\sigma_s$$

for all $k \in \mathbb{R}$, cf. §8.3.5. At this point it turns out to be advantageous to work in Fourier space. Hence we consider

$$\partial_T \widehat{v} = \widehat{L}\widehat{v} + \widehat{N}(\widehat{v}), \tag{14.28}$$

where $\widehat{v} = \mathcal{F}v$, $\widehat{L} = \mathcal{F}L\mathcal{F}^{-1}$ and $\widehat{N}(\widehat{v}) = \mathcal{F}(N(\mathcal{F}^{-1}\widehat{v}))$. There exists a $k_1 > 0$ such that for all $|k| \leq k_1$ the two curves of eigenvalues are separated, and so we define

$$\widehat{p}_c(k)\widehat{v} = \chi(k)\langle \widehat{\varphi}_c^*(k), \widehat{v}\rangle,$$

where $\chi(k) = 1$ for $|k| \leq k_1/2$, and $\chi(k) = 0$ for $|k| > k_1/2$, and where $\widehat{\varphi}_c^*(k)$ is the eigenvector associated to the adjoint eigenvalue problem normalized by $\langle \widehat{\varphi}_c^*(k), \widehat{\varphi}_c(k)\rangle = 1$. Moreover, define $P_s(k)\widehat{v} = \widehat{v} - (\widehat{p}_c(k)\widehat{v})\widehat{\varphi}_c(k)$. We write $\widehat{v} = \widehat{v}_c\widehat{\varphi}_c + \widehat{v}_s$ and use the projections to separate (14.28) into two parts, namely

$$\partial_T \widehat{v}_c = \mu_c \widehat{v}_c + p_c\widehat{N}(\widehat{v}), \qquad \partial_T \widehat{v}_s = \widehat{L}P_s\widehat{v}_s + P_s\widehat{N}(\widehat{v}).$$

For the semigroups generated by μ_c and $\widehat{L}P_s$ we have the estimates

$$\begin{aligned}
\|\mathrm{e}^{\mu_c T}\widehat{v}_c\|_{L^1} &\leq \|\mathrm{e}^{\mu_c T}\|_{L^\infty}\|\widehat{v}_c\|_{L^1} \leq \|\widehat{v}_c\|_{L^1},\\
\|\mathrm{e}^{\mu_c T}\widehat{v}_c\|_{L^\infty} &\leq \|\mathrm{e}^{\mu_c T}\|_{L^\infty}\|\widehat{v}_c\|_{L^\infty} \leq \|\widehat{v}_c\|_{L^\infty},\\
\|\mathrm{e}^{\mu_c T}\widehat{v}_c\|_{L^1} &\leq \|\mathrm{e}^{\mu_c T}\|_{L^1}\|\widehat{v}_c\|_{L^\infty} \leq CT^{-1/2}\|\widehat{v}_c\|_{L^\infty},\\
\|\mathrm{e}^{T\widehat{L}P_s}\widehat{v}_s\|_{L^1} &\leq C\mathrm{e}^{-\sigma_s T/2}\|\widehat{v}_s\|_{L^1},\\
\|\mathrm{e}^{T\widehat{L}P_s}\widehat{v}_s\|_{L^\infty} &\leq C\mathrm{e}^{-\sigma_s T/2}\|\widehat{v}_s\|_{L^\infty}.
\end{aligned}$$

After showing the linear diffusive behavior we have to establish the irrelevance of the nonlinearity w.r.t. to this diffusive behavior. We stay a little bit more abstract than necessary at this point and expand the nonlinear terms into

$$\begin{aligned}
p_c\widehat{N}(\widehat{v}) &= B_{2,1}(\widehat{v}_c, \widehat{v}_c) + B_{2,2}(\widehat{v}_c, \widehat{v}_s) + B_3(\widehat{v}_c, \widehat{v}_c, \widehat{v}_c) + g_c(\widehat{v}_c, \widehat{v}_s),\\
P_s\widehat{N}(\widehat{v}) &= B_{2,3}(\widehat{v}_c, \widehat{v}_c) + g_s(\widehat{v}_c, \widehat{v}_s),
\end{aligned}$$

where the $B_{2,j}$ are symmetric bilinear maps, where B_3 is a symmetric trilinear map, and where g_c and g_s stand for the remaining terms. The splitting is motivated as follows. If $\widehat{v}_c$ decays like $t^{-1/2}$, then $\widehat{v}_s$, which is slaved to $\widehat{v}_c$, decays at least like t^{-1}. Then $g_c = \mathcal{O}(|\widehat{v}_c|^4 + |\widehat{v}_c|^2|\widehat{v}_s| + |\widehat{v}_s|^2)$ decays like t^{-2} and $g_s = \mathcal{O}(|\widehat{v}_c|^3 + |\widehat{v}_c||\widehat{v}_s| + |\widehat{v}_s|^2)$ decays like $t^{-3/2}$. They are therefore both irrelevant w.r.t. diffusive behavior.

In order to prove the irrelevance of the other terms we make a change of coordinates which removes $\widehat{v}_s$ from the relevant terms. We set

$$\widehat{v}_c = \widehat{w}_c, \qquad \widehat{v}_s = -\widehat{L}^{-1} E_s B_{2,3}(\widehat{v}_c, \widehat{v}_c) + \widehat{w}_s \tag{14.29}$$

Note that $\widehat{L}^{-1}E_s$ exists for the considered values of k and that $\widehat{w}_s$ decays like $t^{-3/2}$. We obtain

$$\begin{aligned} \partial_T \widehat{w}_c &= \mu_c \widehat{w}_c + N_1(\widehat{w}_c) + N_2(\widehat{w}_c, \widehat{w}_s), \\ \partial_T \widehat{w}_s &= \widehat{L}_s P_s \widehat{w}_s + N_3(\widehat{w}_c, \widehat{w}_s), \end{aligned} \tag{14.30}$$

where

$$\begin{aligned} N_1(\widehat{w}_c) =& B_{2,1}(\widehat{w}_c, \widehat{w}_c) + B_{2,2}(\widehat{w}_c, -\widehat{L}^{-1} E_s B_{2,3}(\widehat{w}_c, \widehat{w}_c)) \\ &+ B_3(\widehat{w}_c, \widehat{w}_c, \widehat{w}_c), \\ N_2(\widehat{w}_c, \widehat{w}_s) =& g_c(\widehat{w}_c, -\widehat{L}^{-1} E_s B_{2,3}(\widehat{w}_c, \widehat{w}_c) + \widehat{w}_s) + B_{2,2}(\widehat{w}_c, \widehat{w}_s), \\ N_3(\widehat{w}_c, \widehat{w}_s) =& g_s(\widehat{w}_c, -\widehat{L}^{-1} E_s B_{2,3}(\widehat{v}_c, \widehat{v}_c) + \widehat{w}_s) + \partial_t(\widehat{L}^{-1} E_s B_{2,3}(\widehat{w}_c, \widehat{w}_c)). \end{aligned}$$

By this transformation N_2 decays like t^{-2} and N_3 decays like $t^{-3/2}$ such that they are both irrelevant w.r.t. diffusive behavior. In order to prove the irrelevance of $N_1(\widehat{w}_c)$ we write it as

$$\begin{aligned} N_1(\widehat{w}_c) =& \int_{\mathbb{R}} K_2(k, k-l, l) \widehat{w}_c(k-l) \widehat{w}_c(l)\, \mathrm{d}l \\ &+ \int_{\mathbb{R}} \int_{\mathbb{R}} K_3(k, k-l, l-m, m) \widehat{w}_c(k-l) \widehat{w}_c(l-m) \widehat{w}_c(m)\, \mathrm{d}m\, \mathrm{d}l, \end{aligned}$$

with smooth kernels $K_2(k, k-l, l)$ and $K_3(k, k-l, l-m, m)$. A derivative corresponds in Fourier space to a multiplication with a wave number k, and hence terms are more irrelevant, if they are multiplied with a kernel vanishing at $k = l = 0$. The faster the kernel vanishes, the more irrelevant are the associated terms. In §8.3.6 we already computed formally that the nonlinear terms are irrelevant w.r.t. diffusion. Since after the transformation (14.29) the $\widehat{w}_c$-equation of the abstract system (14.30) coincides with (14.25) up to terms with asymptotic behavior of $\mathcal{O}(t^{-3/2})$, we must have that the quadratic terms have a decay of order $\mathcal{O}(t^{-2})$. Therefore, K_2 has to vanish quadratically and K_3 at least linearly, i.e., we have the estimates

$$|K_1(\ell, \ell - m, m)| \leq C|\ell^2 + (\ell - m)^2 + m^2|, \tag{14.31}$$

$$|K_2(\ell, \ell - m, m-k, k)| \leq C(|\ell| + |\ell - m| + |m-k| + |k|). \tag{14.32}$$

For the special choice of the stationary solution, $A = 1$, we have no quadratic terms at all. However, in general (14.31) occurs, and so in the following we will use the weaker estimate (14.31). See also §14.4.1 and §14.4.2.

We proceed as in §14.1.3 and consider the variation of constant formula

$$
\begin{aligned}
\widehat{w}_c(T) =& \mathrm{e}^{\mu_c T}\widehat{w}_c(0) + \int_0^T \mathrm{e}^{\mu_c(T-\tau)}(N_1(\widehat{w}_c) + N_2(\widehat{w}_c,\widehat{w}_s))(\tau)\,\mathrm{d}\tau,\\
\widehat{w}_s(T) =& \mathrm{e}^{T\widehat{L}_s P_s}\widehat{w}_s(0) + \int_0^T \mathrm{e}^{(T-\tau)\widehat{L}_s P_s} N_3(\widehat{w}_c,\widehat{w}_s)(\tau)\,\mathrm{d}\tau
\end{aligned}
\tag{14.33}
$$

for (14.27). In the following we use the abbreviations

$$
\begin{aligned}
a_c(T) &= \sup_{0\le\tau\le T} \|\widehat{w}_c(\tau)\|_{L^\infty},\\
b_c(T) &= \sup_{0\le\tau\le T} \|(1+\tau)^{1/2}\widehat{w}_c(\tau)\|_{L^1},\\
c_{c,\nu}(T) &= \sup_{0\le\tau\le T} \|(1+\tau)^{\nu/2}|k|^\nu\widehat{w}_c(\tau)\|_{L^\infty},\\
d_{c,\nu}(T) &= \sup_{0\le\tau\le T} \|(1+\tau)^{(\nu+1)/2}|k|^\nu\widehat{w}_c(\tau)\|_{L^1}.
\end{aligned}
$$

with $c_c = c_{c,1}$, $d_c = d_{c,1}$, and

$$
\begin{aligned}
a_s(T) &= \sup_{0\le\tau\le T} \|(1+\tau)^{\nu/2}\widehat{w}_s(\tau)\|_{L^\infty},\\
b_s(T) &= \sup_{0\le\tau\le T} \|(1+\tau)^{(\nu+1)/2}\widehat{w}_s(\tau)\|_{L^1},
\end{aligned}
$$

with $\nu < 2$ fixed arbitrarily close to 2, and denote many different constants with the same symbol C if they can be chosen independently of $a_c(T), \ldots, b_s(T)$, and T.

From the previous representations and Lemma 7.3.19 we find

$$
\begin{aligned}
\|N_1(\widehat{w}_c)\|_{L^1} \le& C(\||k|\widehat{w}_c\|_{L^1}^2 + \|\widehat{w}_c\|_{L^1}\||k|^2\widehat{w}_c\|_{L^1} + \|\widehat{w}_c\|_{L^1}^2\||k|\widehat{w}_c\|_{L^1}),\\
\|N_2(\widehat{w}_c,\widehat{w}_s)\|_{L^1} \le& C(\|\widehat{w}_c\|_{L^1}^4 + \|\widehat{w}_c\|_{L^1}\|\widehat{w}_s\|_{L^1} + \|\widehat{w}_s\|_{L^1}^2),\\
\|N_3(\widehat{w}_c,\widehat{w}_s)\|_{L^1} \le& C(\|\widehat{w}_c\|_{L^1}^3 + \|\widehat{w}_c\|_{L^1}\|\widehat{w}_s\|_{L^1} + \|\widehat{w}_s\|_{L^1}^2 + \|\widehat{w}_c\|_{L^1}\|\partial_T\widehat{w}_c\|_{L^1}),
\end{aligned}
$$

and

$$
\begin{aligned}
\|N_1(\widehat{w}_c)\|_{L^\infty} \le& C(\||k|\widehat{w}_c\|_{L^1}\||k|\widehat{w}_c\|_{L^\infty} + \|\widehat{w}_c\|_{L^1}\||k|^2\widehat{w}_c\|_{L^\infty}\\
&+ \|\widehat{w}_c\|_{L^1}^2\||k|\widehat{w}_c\|_{L^\infty}),\\
\|N_2(\widehat{w}_c,\widehat{w}_s)\|_{L^\infty} \le& C(\|\widehat{w}_c\|_{L^1}^3\|\widehat{w}_c\|_{L^\infty} + \|\widehat{w}_c\|_{L^\infty}\|\widehat{w}_s\|_{L^1} + \|\widehat{w}_s\|_{L^1}\|\widehat{w}_s\|_{L^\infty}),\\
\|N_3(\widehat{w}_c,\widehat{w}_s)\|_{L^\infty} \le& C(\|\widehat{w}_c\|_{L^1}^2\|\widehat{w}_c\|_{L^\infty} + \|\widehat{w}_c\|_{L^\infty}\|\widehat{w}_s\|_{L^1} + \|\widehat{w}_s\|_{L^1}\|\widehat{w}_s\|_{L^\infty}\\
&+ \|\widehat{w}_c\|_{L^\infty}\|\partial_T\widehat{w}_c\|_{L^1}),
\end{aligned}
$$

where $\|\partial_T\widehat{w}_c\|_{L^1}$ on the right-hand side will be replaced via

$$
\|\partial_T\widehat{w}_c\|_{L^1} \le C(\||k|^2\widehat{w}_c\|_{L^1} + \|N_1(\widehat{w}_c)\|_{L^1} + \|N_2(\widehat{w}_c,\widehat{w}_s)\|_{L^1}).
$$

Since $\widehat{w}_c$ has compact support in Fourier space, $|k|^2\widehat{w}_c$ can be estimated in terms of $|k|^\nu\widehat{w}_c$ for every $\nu \in [0,2)$.

The proof of Theorem 14.2.1 proceeds in three steps, first in A) estimates for the terms in the $\widehat{w}_c$ equation, in B) estimates for the terms in the $\widehat{w}_s$ equation, which are brought together in C).

A1) We estimate

$$\begin{aligned}
&\left\| \int_0^T e^{\mu_c(T-\tau)} N_1(\widehat{w}_c)(\tau)\,d\tau \right\|_{L^\infty} \leq \int_0^T \|e^{\mu_c(T-\tau)}\|_{L^\infty\to L^\infty} \|N_1(\widehat{w}_c)(\tau)\|_{L^\infty}\,d\tau \\
\leq & C \int_0^T (1+\tau)^{-3/2}\,d\tau \cdot (c_c(T)d_c(T) + c_{c,\mu}(T)b_c(T) + b_c^2(T)c_c(T)) \\
\leq & C(c_c(T)d_c(T) + c_{c,\mu}(T)b_c(T) + b_c^2(T)c_c(T)).
\end{aligned}$$

Similarly we find

$$\left\| \int_0^T e^{\mu_c(T-\tau)} N_2(\widehat{w}_c, \widehat{w}_s)(\tau)\,d\tau \right\|_{L^\infty} \leq C(b_c^3(T)a_c(T) + a_c(T)b_s(T) + a_s(T)b_s(T)).$$

A2) Next we estimate

$$\begin{aligned}
&(1+T)^{1/2} \left\| \int_0^T e^{\mu_c(T-\tau)} N_1(\widehat{w}_c)(\tau)\,d\tau \right\|_{L^1} \\
\leq & (1+T)^{1/2} \int_0^T \|e^{\mu_c(T-\tau)}\|_{L^\infty\to L^1} \|N_1(\widehat{w}_c)(\tau)\|_{L^\infty}\,d\tau \\
\leq & C(1+T)^{1/2} \int_0^T (T-\tau)^{-1/2}(1+\tau)^{-3/2}\,d\tau \\
& \qquad \times (c_c(T)d_c(T) + c_{c,\mu}(T)b_c(T) + b_c^2(T)c_c(T)) \\
\leq & C(1+T)^{1/2} \int_0^{T/2} (T/2)^{-1/2}(1+\tau)^{-3/2}\,d\tau \\
& \qquad \times (c_c(T)d_c(T) + c_{c,\mu}(T)b_c(T) + b_c^2(T)c_c(T)) \\
& + C(1+T)^{1/2} \int_{T/2}^T (T-\tau)^{-1/2}(1+T/2)^{-3/2}\,d\tau \\
& \qquad \times (c_c(T)d_c(T) + c_{c,\mu}(T)b_c(T) + b_c^2(T)c_c(T)) \\
\leq & C(c_c(T)d_c(T) + c_{c,\mu}(T)b_c(T) + b_c^2(T)c_c(T)).
\end{aligned}$$

Similarly we find

$$\begin{aligned}
&\left\| (1+T)^{1/2} \int_0^T e^{\mu_c(T-\tau)} N_2(\widehat{w}_c, \widehat{w}_s)(\tau)\,d\tau \right\|_{L^1} \\
& \qquad \leq C(b_c^3(T)a_c(T) + a_c(T)b_s(T) + a_s(T)b_s(T)).
\end{aligned}$$

A3) We estimate

$$
\begin{aligned}
&(1+T)^{\nu/2}\left\|\int_0^T e^{\mu_c(T-\tau)}|k|^\nu N_1(\widehat{w}_c)(\tau)\,d\tau\right\|_{L^\infty}\\
\le&(1+T)^{\nu/2}\int_0^T \|e^{\mu_c(T-\tau)}|k|^\nu\|_{L^\infty\to L^\infty}\|N_1(\widehat{w}_c)(\tau)\|_{L^\infty}\,d\tau\\
\le&(1+T)^{\nu/2}C\int_0^T (T-\tau)^{-\nu/2}(1+\tau)^{-3/2}\,d\tau\cdot(c_c(T)d_c(T)+b_c^2(T)c_c(T))\\
\le&C(c_c(T)d_c(T)+b_c^2(T)c_c(T)).
\end{aligned}
$$

again by separating $\int_0^T = \int_0^{T/2} + \int_{T/2}^T$. Similarly we find

$$
\begin{aligned}
(1+T)^{\nu/2}&\left\|\int_0^T e^{\mu_c(T-\tau)}|k|^\nu N_2(\widehat{w}_c,\widehat{w}_s)(\tau)\,d\tau\right\|_{L^\infty}\\
&\le C(b_c^3(T)a_c(T)+a_c(T)b_s(T)+a_s(T)b_s(T)).
\end{aligned}
$$

A4) The last estimate for the diffusive part is

$$
\begin{aligned}
&(1+T)^{(\nu+1)/2}\left\|\int_0^T e^{\mu_c(T-\tau)}|k|^\nu N_1(\widehat{w}_c)(\tau)\,d\tau\right\|_{L^1}\\
\le&(1+T)^{(\nu+1)/2}\int_0^{T-1} \|e^{\mu_c(T-\tau)}|k|^\nu\|_{L^\infty\to L^1}\|N_1(\widehat{w}_c)(\tau)\|_{L^\infty}\,d\tau\\
&+(1+T)^{(\nu+1)/2}\int_{T-1}^T \|e^{\mu_c(T-\tau)}|k|^\nu\|_{L^1\to L^1}\|N_1(\widehat{w}_c)(\tau)\|_{L^1}\,d\tau\\
\le&(1+T)^{(\nu+1)/2}C\int_0^{T-1} (T-\tau)^{-(\nu+1)/2}(1+\tau)^{-3/2}\,d\tau\\
&\qquad\qquad\times(c_c(T)d_c(T)+b_c^2(T)c_c(T))\\
&+(1+T)^{(\nu+1)/2}C\int_{T-1}^T (T-\tau)^{-\nu/2}(1+\tau)^{-2}\,d\tau\cdot(d_c^2(T)+b_c^2(T)d_c(T))\\
\le&s_1+C(c_c(T)d_c(T)+b_c^2(T)c_c(T)).
\end{aligned}
$$

We split $\int_0^{T-1}\dots = \int_0^{T/2}\dots + \int_{T/2}^{T-1}\dots$, respectively $s_1 = s_2 + s_3$, and find

$$
s_2 \le(1+T)^{(\nu+1)/2}\int_0^{T/2}(T/2)^{-(\nu+1)/2}(1+\tau)^{-3/2}\,d\tau\cdot(d_c^2(T)+b_c^2(T)d_c(T)).
$$

Moreover,

$$
\begin{aligned}
s_3 \le&(1+T)^{(\nu+1)/2}\int_{T/2}^{T-1}(T-\tau)^{-(\nu+1)/2}(1+T/2)^{-3/2}\,d\tau\\
&\qquad\qquad\times(d_c^2(T)+b_c^2(T)d_c(T))
\end{aligned}
$$

such that finally

$$s_1 \leq C(d_c^2(T) + b_c^2(T)d_c(T)).$$

Similarly we find

$$(1+T)^{(\nu+1)/2}\left\|\int_0^T e^{\mu_c(T-\tau)}|k|^\nu N_2(\widehat{w}_c, \widehat{w}_s)(\tau)\,d\tau\right\|_{L^1} \leq C(d_c^2(T) + b_c^2(T)d_c(T) + b_c^4(T) + b_c(T)b_s(T) + b_s^2(T)).$$

B1) For the linearly exponentially damped part we first estimate

$$\begin{aligned}
&(1+T)^{\nu/2}\left\|\int_0^T e^{(T-\tau)\widehat{L}_s P_s} N_3(\widehat{w}_c, \widehat{w}_s)(\tau)\,d\tau\right\|_{L^\infty} \\
\leq &(1+T)^{\nu/2}\int_0^T \|e^{(T-\tau)\widehat{L}_s P_s}\|_{L^\infty\to L^\infty}\|N_3(\widehat{w}_c, \widehat{w}_s)(\tau)\|_{L^\infty}\,d\tau \\
\leq &(1+T)^{\nu/2}\int_0^T e^{-\sigma_s(T-\tau)}(1+\tau)^{-\nu/2}\,d\tau \\
&\quad\times C(b_c^2(T)a_c(T)+a_c(T)b_s(T)+a_s(T)b_s(T) \\
&\qquad +a_c(T)(d_{c,\nu}+d_c^2(T)+b_c^2(T)d_c(T)+b_c^4(T)+b_c(T)b_s(T)+b_s^2(T))) \\
\leq &C(b_c^2(T)a_c(T)+a_c(T)b_s(T)+a_s(T)b_s(T) \\
&\qquad +a_c(T)(d_{c,\nu}(T)+d_c^2(T)+b_c^2(T)d_c(T)+b_c^4(T)+b_c(T)b_s(T)+b_s^2(T)))
\end{aligned}$$

due to the uniform boundedness of

$$\begin{aligned}
(1+T)^{\nu/2}&\int_0^T e^{-\sigma_s(T-\tau)}(1+\tau)^{-\nu/2}\,d\tau \\
&\leq (1+T)^{\nu/2}\int_0^{T/2} e^{-\sigma_s T}(1+\tau)^{-\nu/2}\,d\tau \\
&\quad +(1+T)^{\nu/2}\int_{T/2}^T e^{-\sigma_s(T-\tau)}(1+T/2)^{-\nu/2}\,d\tau.
\end{aligned}$$

B2) Secondly, we estimate

$$\begin{aligned}
&(1+T)^{(\nu+1)/2}\left\|\int_0^T e^{(T-\tau)\widehat{L}_s P_s} N_3(\widehat{w}_c, \widehat{w}_s)(\tau)\,d\tau\right\|_{L^1} \\
\leq &(1+T)^{(\nu+1)/2}\int_0^T \|e^{(T-\tau)\widehat{L}_s P_s}\|_{L^1\to L^1}\|N_3(\widehat{w}_c, \widehat{w}_s)(\tau)\|_{L^1}\,d\tau \\
\leq &(1+T)^{(\nu+1)/2}\int_0^T e^{-\sigma_s(T-\tau)}(1+\tau)^{-(\nu+1)/2}\,d\tau \\
&\quad\times C(b_c^3(T)+b_c(T)b_s(T)+b_s^2(T) \\
&\qquad +b_c(T)(d_{c,\nu}+d_c^2(T)+b_c^2(T)d_c(T)+b_c^4(T)+b_c(T)b_s(T)+b_s^2(T))) \\
\leq &C(b_c^3(T)+b_c(T)b_s(T)+b_s^2(T) \\
&\qquad +b_c(T)(d_{c,\nu}+d_c^2(T)+b_c^2(T)d_c(T)+b_c^4(T)+b_c(T)b_s(T)+b_s^2(T)))
\end{aligned}$$

due to the uniform boundedness of

$$\begin{aligned}(1+T)^{(\nu+1)/2} \int_0^T & e^{-\sigma_s(T-\tau)}(1+\tau)^{-(\nu+1)/2}\, d\tau \\ \leq & (1+T)^{(\nu+1)/2} \int_0^{T/2} e^{-\sigma_s T}(1+\tau)^{-(\nu+1)/2}\, d\tau \\ & + (1+T)^{(\nu+1)/2} \int_{T/2}^{T} e^{-\sigma_s(T-\tau)}(1+T/2)^{-(\nu+1)/2}\, d\tau.\end{aligned}$$

C) We set

$r(T) = a_c(T) + b_c(T) + c_c(T) + d_c(T) + c_{c,\nu}(T) + d_{c,\nu}(T) + a_s(T) + b_s(T).$

Summing up all estimates yields an inequality

$$r(T) \leq r(0) + f(r(T))$$

where f is at least quadratic in its argument. Comparing the curves $r \mapsto r$ and $r \mapsto \delta + r^2$ it is easy to see that r cannot go beyond 2δ if $\delta > 0$ is sufficiently small. Hence, if $r(0) < \delta$, with $\delta > 0$ sufficiently small, we have the existence of a $C > 0$ such that $r(T) \leq C$ for all $T \geq 0$. □

We will apply the scheme of the proof a second time, namely for showing the diffusive stability of the time-periodic solution $A(X,T) = e^{-i\beta t}$ of the complex GL equation (14.26). In order to do so we introduce $B(X,T) = A(X,T)e^{i\beta t}$ which satisfies

$$\partial_T B = (1 + i\alpha)\partial_X^2 B + (1 + i\beta)B - (1 + i\beta)B|B|^2.$$

This equation possesses the equilibrium $B = 1$ for which we would like to prove diffusive stability. The deviation $v = B - 1$ satisfies

$$\partial_T v = Lv + N(v) \tag{14.34}$$

where

$$\begin{aligned} Lv =& (1 + i\alpha)\partial_X^2 v + (1 + i\beta)v - (1 + i\beta)(2v + \overline{v}), \\ N(v) =& -(1 + i\beta)(v^2 + 2v\overline{v} + v^2\overline{v}). \end{aligned}$$

The eigenvalue problem $Lv = \mu v$ is solved by Fourier modes $v(x,t) = \widehat{v}_{1,2}(k)e^{ikx}$ with eigenvalues $\mu_{1,2}(k)$. In the Eckhaus stable domain we have constants $\sigma_c > 0$ and $\sigma_s > 0$ such that

$$\mathrm{Re}\mu_1(k) \leq -\sigma_c k^2 \qquad \text{and} \qquad \mathrm{Re}\mu_2(k) \leq -\sigma_s$$

for all $k \in \mathbb{R}$. System (14.34) has exactly the same properties as System (14.27). Therefore, we can follow the proof of Theorem 14.2.1 starting at (14.28) line for line. The formal calculations for the irrelevance of the nonlinear terms can be found in the subsequent §14.4.2. With the same arguments as above it can be concluded that the kernels K_2 and K_3 satisfy (14.31)-(14.32).

Theorem 14.2.5. *The time-periodic solution* $A = \mathrm{e}^{-\mathrm{i}\beta t}$ *of* (14.26) *is diffusively stable in the following sense. For all* $C > 0$ *there exists a* $\delta > 0$ *such that for solutions* $A = \mathrm{e}^{-\mathrm{i}\beta t} + v$ *of* (14.22) *with* $\|\widehat{v}|_{T=0}\|_{L^1} + \|\widehat{v}|_{T=0}\|_{L^\infty} \leq \delta$ *we have*

$$\|\widehat{v}(T)\|_{L^\infty} \leq C, \qquad \|\widehat{v}(T)\|_{L^1} \leq C(1+T)^{-1/2}$$

for all $T \geq 0$.

Remark 14.2.6. Obviously, with the presented method the diffusive stability of all Eckhaus-stable equilibria of (14.22) and of all Eckhaus-stable time-periodic solutions of (14.26) can be established. We already explained that whenever the L^1-L^∞ approach works the renormalization approach works, too. See [**BK92**] for a detailed proof for the real GL equation. The renormalization approach gives more informations than the L^1-L^∞ approach, but needs more localization of the initial conditions. ⌋

Theorem 14.2.7. *There exists a* $\delta > 0$ *such that for all* $p \in (0, 1/2)$ *there exists a* $C_1 > 0$ *such that for all* $(\phi_0, s_0) \in H_2^2 \times H_2^2$ *with* $\|\phi\|_{H_2^2} + \|s\|_{H_2^2} \leq \delta$ *there exists a unique global solution* $(\phi, s) \in C([1, \infty), H_2^2 \times H_2^2)$ *of* (14.24) *to initial conditions* (ϕ_0, s_0), *and a* $\phi_{\lim} \in \mathbb{R}$ *such that*

(14.35)
$$\|T^{1/2}\phi(T^{1/2}\cdot, T) - \phi_{\lim}\mathrm{e}^{-\cdot^2/4}\|_{H_2^2} + \|T^{1/2}s(T^{1/2}\cdot, T)\|_{H_2^2} \leq C_1 T^{-1/2+p},$$

and thus in particular

$$\|\phi(\cdot, T) - \frac{\phi_{\lim}}{\sqrt{T}}\mathrm{e}^{-\cdot^2/4T}\|_\infty + \|s(\cdot, T)\|_\infty \leq C_1 T^{-(1-p)}. \tag{14.36}$$

14.2.2. Diffusive stability of rolls in pattern forming systems. In the last section we have seen the diffusive stability of $A = 1$ in the real and of $A = \mathrm{e}^{-i\beta t}$ in the complex GL equation. In Chapter 10 we proved that the GL equation makes correct predictions on an $\mathcal{O}(1/\varepsilon^2)$-time scale about the dynamics of various pattern forming systems such as the SH equation or the Couette-Taylor problem. However, since we only have an exponentially growing error bound for this approximation the polynomial decay rates observed for the GL equation are insufficient to conclude diffusive stability of the spatially periodic pattern in the original system. Thus, a direct proof of diffusive stability is necessary. It will be based on a linear stability analysis via Bloch waves and the computation of the effective nonlinearity such that finally the approach of §14.2.1 can be transferred.

We refrain from greatest generality and restrict ourselves to the SH equation (10.4). The subsequent analysis is by no means restricted to bifurcating solutions or equilibria. It works for all spectrally stable spatially periodic equilibria or traveling waves.

In order to keep the notation on a reasonable level we consider the stability of the spatially periodic equilibrium

$$u_{per,\varepsilon}[x] = \varepsilon\sqrt{\frac{2}{3}}\cos(x) + \mathcal{O}(\varepsilon^2)$$

for the SH equation, cf. §13.2.2. It corresponds to the diffusively stable equilibrium $A = 1/\sqrt{3}$ in the associated GL equation

$$\partial_T A = 4\partial_X^2 A + A - 3A|A|^2,$$

cf. §10.2. The linearization of (14.37) around a roll $u_{per,\varepsilon}$ yields a linear operator with periodic coefficents. Thus, the Fourier transform has to be replaced by the Bloch transform and the splitting into a diffusive part and an exponentially damped remainder becomes more complicated. Finally, it turns out that also the nonlinear terms are best studied in Bloch space where the effective nonlinearity can be computed. The presentation is not optimized w.r.t. to the small perturbation parameter $0 < \varepsilon \ll 1$. In fact, the set of initial conditions for which the associated solutions converge towards the equilibrium is of the same size as the bifurcating equilibrium w.r.t. $0 < \varepsilon \ll 1$. See [**Sch96a, Sch98d**].

In the SH equation we take $u_{per,\varepsilon}$ as new origin, i.e., we set $u = u_{per,\varepsilon} + v$ and obtain

$$\partial_t v = Lv + N(v) \tag{14.37}$$

with

$$Lv = -(1+\partial_x^2)^2 v + \varepsilon^2 v - 3u_{per,\varepsilon}^2 v, \qquad N(v) = -3u_{per,\varepsilon}v^2 - v^3.$$

Linear diffusive behavior. In a first step we analyze the linearized problem $\partial_t v = Lv$. In addition to the spectral properties we gain a useful representation of the semigroup which allows us to analyze the asymptotic behavior of the perturbations. The linearized problem is solved by Bloch waves, cf. §11.6.3,

$$v(x,t) = e^{\mu t} e^{i\ell x}\widetilde{v}(\ell,x), \quad \text{with} \quad \widetilde{v}(\ell,x) = \widetilde{v}(\ell,x+2\pi)$$

satisfying

$$L_\ell \widetilde{v}(\ell,x) = \mu\widetilde{v}(\ell,x) \tag{14.38}$$

where

$$L_\ell \widetilde{v}(\ell,x) := -(1+(i\ell+\partial_x)^2)^2\widetilde{v}(\ell,x) + \varepsilon^2\widetilde{v}(\ell,x) - 3u_{per,\varepsilon}(x)^2\widetilde{v}(\ell,x).$$

For fixed $\ell \in [-1/2, 1/2)$ the operator L_ℓ is self-adjoint in $L^2(\mathbb{T})$. Therefore, all eigenvalues are real-valued. The operator $L_\ell : H^4(\mathbb{T}) \to L^2(\mathbb{T})$ has a compact resolvent $(L_\ell - \mu I)^{-1}$ for some $\mu \in \mathbb{C}$. From this, for fixed ℓ, it follows the existence of a discrete set of eigenvalues

$$\{\mu_j(\ell) \in \mathbb{C} : j \in \mathbb{N},\ \mu_j \geq \mu_{j+1} \to -\infty \ \text{ for } \ j \to \infty\}$$

and a corresponding set of eigenfunctions $\{f_j(\ell) : j \in \mathbb{N}\}$. We normalize f_j such that $\|f_j(\ell, \cdot)\|_{L^2(\mathbb{T})} = 1$.

For $\varepsilon = 0$ we have $u_{per,\varepsilon} = 0$ and so the eigenvalue problem can be solved explicitly. The well known curve $\lambda(k) = -(1-k^2)^2$ is cut into pieces of length one. These pieces are plotted via $k = m+\ell$, with $m \in \mathbb{Z}$ and $\ell \in [-1/2, 1/2)$, as functions w.r.t. ℓ, i.e., we consider the spatial homogenous situation artificially as spatially periodic. Spectral perturbation theory for fixed ℓ, cf. [**Kat95**] or Lemma 11.6.2, yields the smoothness of the curves of eigenvalues with the intersection points as exception. Hence, it is easy to see that for fixed $\ell_1 = \mathcal{O}(1) > 0$ for $\varepsilon \to 0$ there exist $\varepsilon_0 > 0$ and $\sigma_0(\ell_1) = \mathcal{O}(1) > 0$ for $\varepsilon \to 0$, such that for all $\varepsilon \in (0, \varepsilon_0)$ and all $|\ell| > \ell_1$ all eigenvalues satisfy $\mu_j(\ell) < -\sigma_0$. Except of the two curves μ_1, μ_2 touching zero this is also true for all other eigenvalues, i.e., $\sup_{j\geq 3} \sup_{\ell\in[-1/2,1/2)} \mu_j(\ell) < -\sigma_0$.

In order to understand what happens with the two curves μ_1, μ_2 for $\varepsilon > 0$ we proceed as follows. The eigenvalue problem

$$\mathcal{G}(\ell, \varepsilon, \mu)\widetilde{v} = L_\ell \widetilde{v} - \mu\widetilde{v} = 0$$

always has the trivial solution $\widetilde{v} = 0$ for all ℓ, μ, and ε. Solutions can bifurcate from this trivial branch if $(\partial_{\widetilde{v}}\mathcal{G}(\ell, \varepsilon, 0))^{-1}$ does not exist. We know already that $\partial_w \mathcal{G}(0, 0, 0) = L_0|_{\varepsilon=0}$ is not invertible and has a two-dimensional kernel spanned by $U_1 = \sin x$ and $U_2 = \cos x$. Thus, we apply the Lyapunov-Schmidt reduction method to compute the bifurcating solutions. Let P be the orthogonal projection on this kernel and let $\widetilde{v} = aU_1 + bU_2 + \mathcal{V}$ with $(1-P)\mathcal{V} = \mathcal{V}$. The solution of the hyperbolic part $(1-P)\mathcal{G}(\ell, \varepsilon, \mu, a, b, \mathcal{V}) = 0$ is denoted by $\mathcal{V} = \mathcal{V}(\ell, \varepsilon, \mu, a, b)$. Inserting this into $P\mathcal{G}(\ell, \varepsilon, \mu, a, b, \mathcal{V}) = 0$ gives the bifurcation equation

$$\mathcal{G}_1 \begin{pmatrix} a \\ b \end{pmatrix} = 0,$$

where

$$\mathcal{G}_1 = \begin{pmatrix} \rho - \mu & -\mathrm{i}\delta \\ \mathrm{i}\delta & \rho + c(\varepsilon) - \mu \end{pmatrix} + \mathcal{O}(\varepsilon^4) \begin{pmatrix} \mathcal{O}(\ell^2 + |\mu|) & \mathcal{O}(|\ell| + |\mu|) \\ \mathcal{O}(|\ell| + |\mu|) & \mathcal{O}(\ell^2 + |\mu|) \end{pmatrix},$$

with

$$\rho = -4\ell^2 - \ell^4, \quad \delta = -4\ell^3, \quad c(\varepsilon) = -2\varepsilon^2 + \mathcal{O}(\varepsilon^4).$$

Computing $\mu_{1,2}(\ell, \varepsilon)$ in such a way that the determinant of $\mathcal{G}_1$ vanishes, gives

Lemma 14.2.8. *The curves of eigenvalues* $\mu_{1,2}$ *have the expansions*

$$\mu_1(\ell) = -c_1(\varepsilon)\ell^2 + \mathcal{O}(\ell^4), \quad \mu_2(\ell) = -2\varepsilon^2 - c_1(\varepsilon)\ell^2 + \mathcal{O}(\ell^4 + \varepsilon^4),$$

where $c_1(\varepsilon) = 4 + \mathcal{O}(\varepsilon^2)$.

Irrelevance of the nonlinear terms. In order to prove the irrelevance of the nonlinear terms we proceed as in §14.2.1 except that Fourier transform is replaced by Bloch transform. Hence we consider

$$\partial_t \widetilde{v} = \widetilde{L}\widetilde{v} + \widetilde{N}(\widetilde{v}), \tag{14.39}$$

where $\widetilde{v} = \mathcal{B}v$, $\widetilde{L} = \mathcal{B}L\mathcal{B}^{-1}$, and $\widetilde{N}(\widetilde{v}) = \mathcal{B}(N(\mathcal{B}^{-1}\widetilde{v}))$. There exists an $\ell_1 > 0$ such that for all $|\ell| \le \ell_1$ the first curve of eigenvalues is separated from the rest. For fixed ℓ the operator L_ℓ is self-adjoint in $L^2(\mathbb{T})$ and therefore

$$p_c(\ell)\widetilde{v}(\ell) = \chi(\ell)\,\langle \widetilde{v}(\ell,\cdot), f_1(\ell,\cdot)\rangle_{L^2(\mathbb{T})}\, f_1(\ell,\cdot),$$

where $\chi(\ell) = 1$ for $|\ell| \le \ell_1/2$ and $\chi(\ell) = 0$ for $|\ell| > \ell_1/2$, allows to separate the diffusive modes from the exponentially damped modes. Moreover, define $P_s(\ell)\widetilde{v} = \widetilde{v} - (\widetilde{p}_c(\ell)\widetilde{v})\widetilde{f}_c(\ell)$.

We write $\widetilde{v} = \widetilde{v}_c\widetilde{f}_c + \widetilde{v}_s$ and use the projections to separate (14.39) into two parts, namely

$$\partial_t \widetilde{v}_c = \mu_c \widetilde{v}_c + p_c\widetilde{N}(\widetilde{v}), \qquad \partial_t \widetilde{v}_s = \widetilde{L}P_s\widetilde{v}_s + P_s\widetilde{N}(\widetilde{v}).$$

For the semigroups generated by μ_c and $\widetilde{L}P_s$ we have the estimates

$$\begin{aligned}
\|e^{\mu_c t}\widetilde{v}_c\|_{L^1} &\le \|e^{\mu_c t}\|_{L^\infty}\|\widetilde{v}_c\|_{L^1} \le \|\widetilde{v}_c\|_{L^1},\\
\|e^{\mu_c t}\widetilde{v}_c\|_{L^\infty} &\le \|e^{\mu_c t}\|_{L^\infty}\|\widetilde{v}_c\|_{L^\infty} \le \|\widetilde{v}_c\|_{L^\infty},\\
\|e^{\mu_c t}\widetilde{v}_c\|_{L^1} &\le \|e^{\mu_c t}\|_{L^1}\|\widetilde{v}_c\|_{L^\infty} \le Ct^{-1/2}\|\widetilde{v}_c\|_{L^\infty},\\
\|e^{\widetilde{L}P_s t}\widetilde{v}_s\|_{L^1} &\le Ce^{-\sigma_s t/2}\|\widetilde{v}_s\|_{L^1},\\
\|e^{\widetilde{L}P_s t}\widetilde{v}_s\|_{L^\infty} &\le Ce^{-\sigma_s t/2}\|\widetilde{v}_s\|_{L^\infty},
\end{aligned}$$

with a constant $\sigma_s > 0$, where we used the abbreviation

$$L^1 = L^1([-1/2,1/2), H^1(\mathbb{T})) \qquad \text{and} \qquad L^\infty = L^\infty([-1/2,1/2), H^1(\mathbb{T})).$$

By the choice $H^1(\mathbb{T})$ the nonlinear terms are well-defined subsequently. After showing the linear diffusive behavior we have to establish the irrelevance of the nonlinearity w.r.t. to this diffusive behavior. With the same motivation as before we expand the nonlinear terms into

$$\begin{aligned}
p_c\widetilde{N}(\widetilde{v}) &= B_{2,1}(\widetilde{v}_c,\widetilde{v}_c) + B_{2,2}(\widetilde{v}_c,\widetilde{v}_s) + B_3(\widetilde{v}_c,\widetilde{v}_c,\widetilde{v}_c) + g_c(\widetilde{v}_c,\widetilde{v}_s),\\
P_s\widetilde{N}(\widetilde{v}) &= B_{2,3}(\widetilde{v}_c,\widetilde{v}_c) + g_s(\widetilde{v}_c,\widetilde{v}_s)
\end{aligned}$$

where the $B_{2,j}$ are symmetric bilinear maps, B_3 symmetric trilinear maps, and where g_c and g_s stand for the remaining terms. In order to prove the irrelevance all these terms we make a change of coordinates. We set

$$\widetilde{v}_c = \widetilde{w}_c, \qquad \widetilde{v}_s = -\widetilde{L}^{-1} E_s B_{2,3}(\widetilde{v}_c, \widetilde{v}_c) + \widetilde{w}_s \tag{14.40}$$

and obtain

$$\begin{array}{rcl} \partial_t \widetilde{w}_c & = & \mu_c \widetilde{w}_c + N_1(\widetilde{w}_c) + N_2(\widetilde{w}_c, \widetilde{w}_s), \\ \partial_t \widetilde{w}_s & = & \widetilde{L}_s P_s \widetilde{w}_s + N_3(\widetilde{w}_c, \widetilde{w}_s), \end{array} \tag{14.41}$$

where

$$\begin{aligned} N_1(\widetilde{w}_c) =& B_{2,1}(\widetilde{w}_c, \widetilde{w}_c) + B_{2,2}(\widetilde{w}_c, -\widetilde{L}^{-1} E_s B_{2,3}(\widetilde{w}_c, \widetilde{w}_c)) + B_3(\widetilde{w}_c, \widetilde{w}_c, \widetilde{w}_c), \\ N_2(\widetilde{w}_c, \widetilde{w}_s) =& g_c(\widetilde{w}_c, -\widetilde{L}^{-1} E_s B_{2,3}(\widetilde{w}_c, \widetilde{w}_c) + \widetilde{w}_s) + B_{2,2}(\widetilde{w}_c, \widetilde{w}_s), \\ N_3(\widetilde{w}_c, \widetilde{w}_s) =& g_s(\widetilde{w}_c, -\widetilde{L}^{-1} E_s B_{2,3}(\widetilde{v}_c, \widetilde{v}_c) + \widetilde{w}_s) + \partial_t(\widetilde{L}^{-1} E_s B_{2,3}(\widetilde{w}_c, \widetilde{w}_c)). \end{aligned}$$

By this transformation N_2 decays like t^{-2} and N_3 decays like $t^{-3/2}$ such that they are both irrelevant w.r.t. diffusive behavior.

In order to prove the irrelevance of $N_1(\widetilde{w}_c)$ we write it as

$$\begin{aligned} N_1(\widetilde{w}_c) =& \int_{\mathbb{R}} K_2(\ell, \ell - \ell_1, \ell_1) \widetilde{v}_c(\ell - \ell_1) \widetilde{v}_c(\ell_1) \, \mathrm{d}\ell_1 \\ &+ \int_{\mathbb{R}} \int_{\mathbb{R}} K_3(\ell, \ell - \ell_1, \ell_1 - \ell_2, \ell_2) \widetilde{v}_c(\ell - \ell_1) \widetilde{v}_c(\ell_1 - \ell_2) \widetilde{v}_c(\ell_2) \, \mathrm{d}\ell_2 \, \mathrm{d}\ell_1, \end{aligned}$$

with smooth kernels $K_2(\ell, \ell - \ell_1, \ell_1)$ and $K_3(\ell, \ell - \ell_1, \ell_1 - \ell_2, \ell_2)$. As before, the faster the kernels vanish near $(k_1, \ldots, k_j) = (0, \ldots, 0)$, the more irrelevant are the associated terms. We have

$$K_2(0,0,0) = 0, \qquad K_3(0,0,0,0) = 0 \tag{14.42}$$

by considering the SH equation in the space of 2π-spatially periodic functions. There exists a one-dimensional center manifold associated to family of equilibria $\{u_{per,\varepsilon}(\cdot + x_0) : x_0 \in \mathbb{R}\}$, cf. §13.2.2. The flow on the center manifold is given by

$$\partial_t \widetilde{w}_c(0) = \mu_1(0) \widetilde{w}_c(0) + \beta_2 \widetilde{w}_c(0)^2 + \beta_3 \widetilde{w}_c(0)^3 + \ldots, \tag{14.43}$$

with coefficients $\beta_j \in \mathbb{R}$. Since the flow on the center manifold is trivial, we must have $\partial_t \widetilde{w}_c(0, t) = 0$ and so $\beta_2 = \beta_3 = 0$. Since after the transform (14.40) the right-hand side of (14.41) in the space of 2π-spatially periodic functions for $\widetilde{w}_s = 0$ coincides with the right-hand side of (14.43) up to terms of order $\mathcal{O}(|\widetilde{w}_c|^4)$ we must have

$$\beta_2 = K_2(0,0,0), \qquad \beta_3 = K_3(0,0,0,0)$$

and so (14.42). So it remains to prove that the first derivatives of K_2 vanish at $(0,0,0)$, too. This property can be computed explicitly for the SH equation using that the curve of eigenfunctions $f_1(\ell)$ has the following useful property.

Lemma 14.2.9. *The eigenfunction f_1 can be expanded as $f_1(\ell) = g_0 + \mathrm{i}\ell g_1 + \mathcal{O}(\ell^2)$, where $g_0, g_1 \in H^1(\mathbb{T})$ are real-valued functions.*

Proof. We have $g_0 = \partial_x u_{per,\varepsilon}/\|\partial_x u_{per,\varepsilon}(x)\|_{L^2(0,\pi)}$. The operator L_ℓ is of the form $L_\ell = A + \mathrm{i}\ell B + \mathcal{O}(\ell^2)$, where A and B are real ℓ-independent operators. The eigenvalues μ_1 satisfy $\mu_1(\ell) = \mathcal{O}(\ell^2)$. Let $h_1 \in L^2(\mathbb{T})$ also be real-valued. Inserting the ansatz $f_1(\ell) = g_0 + \mathrm{i}\ell g_1 + \ell h_1 + \mathcal{O}(\ell^2)$ into $L_\ell f_1(\ell) = \mu_1(\ell) f_1(\ell)$ gives for the terms of $\mathcal{O}(\ell)$ the condition $\mathrm{i}Ag_1 + Ah_1 + \mathrm{i}Bg_0 = 0$. Thus, $h_1 \in \ker A$, i.e., $h_1 \parallel g_0 = \beta \partial_x u_{per,\varepsilon}$ for a $\beta \in \mathbb{R}$. Since we have normalized f_1 we have to choose $h_1 = 0$. □

Using the above representation of the projection $P_1(\ell)$ shows

$$K_2(\ell, \ell - m, m) = 3 \int_{\mathbb{R}} \overline{f_1(\ell, x)} f_1(\ell - m, x) f_1(m, x) \widetilde{u}_0(x)\, dx.$$

Using Lemma 14.2.9 shows that

$$\begin{aligned} K_2(\ell, \ell{-}m, m) = 3 \int_{\mathbb{R}} \Big[& g_0(x) g_0(x) g_0(x) \widetilde{u}_0(x) \\ & - \mathrm{i} l g_1(x) g_0(x) g_0(x) \widetilde{u}_0(x) + g_0(x) \mathrm{i}(\ell{-}m) g_1(x) g_0(x) \widetilde{u}_0(x) \\ & + g_0(x) g_0(x) \mathrm{i} m g_1(x) \widetilde{u}_0(x) + \mathcal{O}(l^2 + (\ell{-}m)^2 + m^2) \Big]\, dx. \end{aligned}$$

Note that $\widetilde{u}_0(x)$ is an even function, and so $g_0(x) = \partial_x u_{per}/\|\partial_x u_{per}\|_{L^2(0,2\pi)}$ is an odd function. Thus, the integral over the zero order terms goes over an odd function and vanishes. Since additionally the first order terms cancel, we have shown

$$\begin{aligned} |K_2(\ell, \ell{-}m, m)| &\leq C|\ell^2 + (\ell{-}m)^2 + m^2|, \\ |K_3(\ell, \ell - m, m - k, k)| &\leq C(|\ell| + |\ell - m| + |m - k| + |k|). \end{aligned}$$

Therefore, we have a system with exactly the same properties as for the GL equation, and so we get

Theorem 14.2.10. *The stationary solution $u = u_{per,\varepsilon}$ of (10.4) is diffusively stable in the following sense. For all $C > 0$ there exists a $\delta > 0$ such that for solutions $u = u_{per,\varepsilon} + v$ of (14.22) with $\|\widetilde{v}|_{t=0}\|_{L^1} + \|\widetilde{v}|_{t=0}\|_{L^\infty} \leq \delta$ we have*

$$\|\widetilde{v}(t)\|_{L^\infty} \leq C, \qquad \|\widetilde{v}(t)\|_{L^1} \leq C(1+t)^{-1/2}$$

for all $t \geq 0$.

Since the Bloch transform is continuous from L^1 to L^∞ and due to $H^1(\mathbb{T}) \subset L^\infty(\mathbb{T})$ we have

Corollary 14.2.11. *For all* $C > 0$ *there exists a* $\delta > 0$ *such that solutions* $u = u_{per,\varepsilon} + v$ *of* (14.22) *with* $\|\widetilde{v}|_{t=0}\|_{L^1} + \|v|_{t=0}\|_{L^1} \leq \delta$ *satisfy*

$$\|v(t)\|_{L^\infty(\mathbb{R},\mathbb{R})} \leq C(1+t)^{-1/2}$$

for all $t \geq 0$.

Some remarks. We already explained that whenever the L^1-L^∞ approach works, then the renormalization approach also works. With the discrete renormalization approach introduced in §14.1.6, in [**Sch96a**] the following diffusive stability result has been shown.

Theorem 14.2.12. *There exists an* $\varepsilon_0 > 0$ *such that the following holds. For all* $\varepsilon \in (0, \varepsilon_0)$ *there exists a* $C_1(\varepsilon) > 0$, *with* $C_1(\varepsilon) = \mathcal{O}(1)$ *as* $\varepsilon \to 0$, *such that for initial conditions* $u_{per,\varepsilon} + v|_{t=0}$ *of the SH equation* (10.4), *with* $v|_{t=0}$ *sufficiently small in* H^2_2, *there exists an* $A_\infty \in \mathbb{R}$ *and for all* $p \in (0, 1/2)$ *a* $C > 0$ *such that*

$$\|v(x,t) - \frac{1}{\sqrt{t}} A_\infty \mathrm{e}^{-\frac{x^2}{4C_1(\varepsilon)t}} \partial_x u_{per,\varepsilon}[x]\|_{L^\infty} \leq Ct^{p-1} \quad \text{for} \quad t \to \infty.$$

The associated nonlinear diffusive stability result for spectrally stable Taylor vortices $U_{\mathrm{TV}}[q, \phi, \varepsilon]$ can be found in [**Sch98d**].

In [**SS99a**] the authors consider the situation where the homogeneous steady state at infinity for traveling pulses in reaction-diffusion systems becomes unstable and bifurcates to a spatially periodic Turing pattern. The spectral situation for the bifurcating modulating pulse is as for the spatially periodic equilibria except of some zero eigenvalue coming from the translational invariance of the original traveling wave. In [**GSU04**] it has been shown that the previous diffusive dynamics persists under the perturbation by the mode associated with the zero eigenvalue, and that therefore such modulating traveling waves are nonlinearly stable w.r.t. spatially localized perturbations.

The main difficulties for the two-dimensional case appear on the linear level, see, e.g., [**Mie97b, Mie97c**]. Once linear diffusive stability is established, the nonlinear problem becomes simpler in higher space dimensions due to the faster decay rate t^{-1}, see, e.g., [**Uec99**].

14.3. The critical case

After the consideration of problems with nonlinearities which are irrelevant w.r.t. diffusion we consider now problems with nonlinearities which are critical w.r.t. diffusion. For systems with an effective $+u^3$ nonlinearity it takes an $\mathcal{O}(e^{-1/\delta})$ time scale before the $+u^3$ wins against diffusion, with δ the effective size of the initial condition. Systems with an effective $u\partial_x u$

nonlinearity show the same decay rates as diffusion but the limiting profile is no longer a Gaussian.

14.3.1. Exponentially long transients in unstable Poiseuille flow. There exists a number of ODEs and PDEs where an instability manifests only after an exponentially long time in terms of the norm of the perturbation. Famous examples are the theorem of Nekhorosov, cf. Theorem 4.4.8, the long time existence of solutions on a time interval of length of order $\mathcal{O}(\exp(1/\delta))$ for nonlinear wave equations [**JK84**] with initial data of order $\mathcal{O}(\delta)$, and the exponentially slow evolution of interfaces in bistable nonlinear diffusion equations with δ being the order of the diffusion coefficient [**CP89, CP90**]. For nonlinear diffusion problems with an effective $+u^3$ nonlinearity the same phenomenon occurs. Poiseuille flow at criticality falls into this class. There is an exponentially long transient decay before positive growth rates occur [**SU03a**].

The starting point of our explanations is again the nonlinear diffusion equation

$$(14.44)\quad \partial_t u = \partial_x^2 u + u^p, \quad u|_{t=0} = u_0, \quad t \geq 0, \quad x \in \mathbb{R}, \quad u = u(x,t) \in \mathbb{R},$$

but now at the threshold of irrelevant nonlinearities, namely $p = 3$. The result is as follows. Take an initial condition in $L^1 \cap L^\infty$ of norm less than δ with $\delta > 0$ sufficiently small. Then the associated solution will be less than 4δ for $t \in [0, \exp(q/\delta^2)]$ with a constant $q > 0$ independent of δ. This means that in terms of the norm of the initial conditions it takes an exponentially long time for the solution to leave a small neighborhood of the origin. This instability is almost not observable since on this very long time interval the solution decays with a rate $\mathcal{O}(t^{-1/2})$ in L^∞, i.e., for $t = \exp(q/\delta^2)$ in L^∞ the solution has a norm of order $\mathcal{O}(\exp(-q/\delta^2))$ in L^∞. Before the explosion takes place the solution becomes flatter and flatter until the mass, respectively the L^1-norm, is sufficiently big to start the explosion. The phenomenon is robust under adding higher order terms, i.e., it also holds for

$$\partial_t u = \partial_x^2 u + u^3 + H(u, \partial_x u, \partial_x^2 u),$$

where $|H(u, \partial_x u, \partial_x^2 u)| \leq C(|u|^4 + |\partial_x u|^2 + |u||\partial_x^2 u|)$, with a constant C, i.e., H is an irrelevant nonlinearity w.r.t. linear diffusion. Moreover, adding the critical term $u\partial_x u$, will not affect the transient decay rates and the above phenomenon, cf. §14.3.2, Thus, whenever the lowest order nonlinear terms are critical in the above sense we can expect the above long time stability, where after a time $\mathcal{O}(\exp(1/\delta^\gamma))$, for a $\gamma > 0$, an explosion may occur.

We now give the precise statement and the proof for the nonlinear diffusion equation at criticality.

Theorem 14.3.1. *Consider (14.44) with $p = 3$. There exist $q > 0$ and $\delta_0 > 0$ such that for all $\delta \in (0, \delta_0)$ the following holds. For an initial condition u_0 with $\|u_0\|_{L^1} + \|u_0\|_{L^\infty} \leq \delta$ the associated solution $u = u(t)$ of (14.44) with $u(0) = u_0$ exists for all $t \in [0, \exp(q/\delta^2)]$ and satisfies*

$$\sup_{t\in[0,\exp(q/\delta^2)]} \|u(t)\|_{L^1\cap L^\infty} \leq 4\delta.$$

Proof. We use the explicit solution $u(x,t) = \frac{1}{\sqrt{2\pi t}} \int_{\mathbb{R}} \mathrm{e}^{-(x-y)^2/(4t)} u_0(y)\,\mathrm{d}y$ of the linear diffusion equation, abbreviated as $u(t) = \mathrm{e}^{t\partial_x^2} u_0$, and recall that for $p > q$ we obtain

$$\|u(t)\|_{L^p} \leq Ct^{-1/(2r)} \|u_0\|_{L^q}, \quad \text{where} \quad 1/p = 1/q - 1/r,\ 1 \leq p \leq q \leq \infty,$$

for some constant C independent of time. Next we consider the variation of constant formula

$$u(t) = \mathrm{e}^{t\partial_x^2} u_0 + \int_0^t \mathrm{e}^{(t-s)\partial_x^2} u^3(s)\,\mathrm{d}s$$

for (14.44). With $\|u^3\|_{L^\infty} \leq \|u\|_{L^\infty}^3$ and $\|u\|_{L^1}^3 \leq \|u\|_{L^\infty}^2 \|u\|_{L^1}$, the abbreviations

$$a(t) = \sup_{0\leq s\leq t} \|u(s)\|_{L^1} \quad \text{and} \quad b(t) = \sup_{0\leq s\leq t} \|(1+s)^{1/2} u(s)\|_{L^\infty}$$

and the estimate (14.5) we obtain similar to before

$$\begin{aligned}
&(1+t)^{1/2} \left\| \int_0^t \mathrm{e}^{(t-s)\partial_x^2} u^3(s)\,\mathrm{d}s \right\|_{L^\infty} \\
&\leq (1+t)^{1/2} \int_0^t \|\mathrm{e}^{(t-s)\partial_x^2}\|_{L^1\to L^\infty} \|u^3(s)\|_{L^1}\,\mathrm{d}s \\
&\leq (1+t)^{1/2} \int_0^t (t-s)^{-1/2}(1+s)^{-1}\,\mathrm{d}s \cdot b(t)^2 a(t) \leq C_1 \ln(1+t) b(t)^2 a(t),
\end{aligned}$$

with a constant C_1 independent of t. Furthermore, we have

$$\begin{aligned}
\left\| \int_0^t \mathrm{e}^{(t-s)\partial_x^2} u^3(s)\,\mathrm{d}s \right\|_{L^1} &\leq \int_0^t \|\mathrm{e}^{(t-s)\partial_x^2}\|_{L^1\to L^1} \|u^3(s)\|_{L^1}\,\mathrm{d}s \\
\leq C_2 \int_0^t (1+s)^{-1}\,\mathrm{d}s \cdot b(t)^2 a(t) &\leq C_2 \ln(1+t) b(t)^2 a(t),
\end{aligned}$$

with a constant C_2 independent of t such that finally

$$\begin{aligned}
a(t) &\leq a(0) + b(0) + C_1 \ln(1+t) b(t)^2 a(t), \\
b(t) &\leq a(0) + b(0) + C_2 \ln(1+t) b(t)^2 a(t).
\end{aligned}$$

We introduce $\widetilde{a}(t)$ and $\widetilde{b}(t)$ by $a(t) = \delta\widetilde{a}(t)$ and $b(t) = \delta\widetilde{b}(t)$ which satisfy

$$\begin{aligned}
\widetilde{a}(t) &\leq \widetilde{a}(0) + C_1\delta^2 \ln(1+t)\widetilde{b}(t)^2 \widetilde{a}(t), \\
\widetilde{b}(t) &\leq \widetilde{b}(0) + \widetilde{a}(0) + C_2\delta^2 \ln(1+t)\widetilde{b}(t)^2 \widetilde{a}(t).
\end{aligned}$$

If $\widetilde{a}(0)+\widetilde{b}(0)\leq 1$ we have $\widetilde{a}(t)+\widetilde{b}(t)\leq 4$ as long as for instance

$$\max\{C_1,C_2\}\delta^2\ln(1+t)\leq 1/64,$$

i.e., for all $t\in[0,\exp(q/\delta^2)]$ for a positive constant q. □

We have additionally proved

Proposition 14.3.2. *Under the assumptions of Theorem* 14.3.1 *we have a constant $C>0$ independent of $\delta\in(0,\delta_0)$ such that*

$$\|u(\cdot,t)\|_{L^\infty}\leq C\delta(1+t)^{-1/2}$$

for all $t\in[0,\exp(q/\delta^2)]$, i.e.,

$$\|u(\cdot,\exp(q/\delta^2))\|_{L^\infty}\leq \widetilde{C}\delta\exp(-q/(2\delta^2)),$$

with another constant $\widetilde{C}>0$ independent of $\delta\in(0,\delta_0)$.

The analogue to Theorem 14.3.1 and Proposition 14.3.2 has been proved in [**SU03a**] for Poiseuille flow. This makes the measurement of the critical Reynolds number $\mathcal{R}_c$ a delicate experiment. Initial perturbations may be amplified on a short transient time scale due to the non-normality of the linearization around the laminar flow, see, e.g., [**SH94a**]. After the short time transient growth the solutions seem to decay for a very long time, i.e., at criticality for an exponentially long time. The better the experiment is performed the more and more this observation plays a role and the longer it takes to observe the final growth of localized perturbations. For more details see Figure 14.1.

14.3.2. The Burgers equation as a limit. For diffusion in 1D there are three critical nonlinear monomials, namely $-u^3$ which leads to diffusive stability, $+u^3$ which leads to instability and which we discussed in detail in §14.3.1, and $\partial_x(u^2)$ which is the topic of this section. In contrast to $\pm u^3$ which yields a logarithmic correction to the growth rate, $\partial_x(u^2)$ will only change the limit profile. In §7.4 we already considered the Burgers equation, given by

$$\partial_t u=\partial_x^2 u-\frac{1}{2}\partial_x(u^2),\quad t\geq 0,\ x\in\mathbb{R},\ u=u(x,t)\in\mathbb{R}.$$

By the Cole-Hopf transformation we found that solutions u to spatially localized initial conditions in the Burgers equation satisfy

$$\sqrt{t}u(x\sqrt{t},t)\to-\frac{2z}{\sqrt{4\pi}}\frac{e^{-x^2/4}}{1+z\operatorname{erf}(x)}=:f_z^*(x),$$

for $t\to\infty$ with a rate $\mathcal{O}(1/\sqrt{t})$, and thus the renormalized solutions converge towards a non-Gaussian limit. In the following we explain why we see

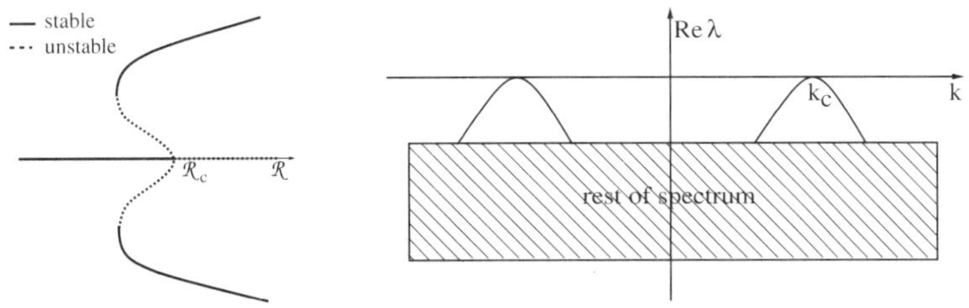

Figure 14.1. The left panel shows a simplified schematic bifurcation diagram for Poiseuille flow. A direct transition to turbulent flow with large amplitudes is observed experimentally based on the fact that, in contrast to the first bifurcation in the other hydrodynamical stability problems which we considered so far, the basic laminar flow becomes unstable at a critical Reynolds number $\mathcal{R}_c$ via a subcritical bifurcation. The right panel shows the curve of critical eigenvalues and the rest of the spectrum for Poiseuille flow at the threshold of instability. From the form of $k \mapsto \mathrm{Re}\lambda_1(k)$ close to wave number k_c, namely $\mathrm{Re}\lambda_1(k) = -C_1(k-k_c)^2 + \mathcal{O}((k-k_c)^3)$ for a constant $C_1 > 0$, it is clear that the linearized system exhibits some diffusive behavior. However, the nonlinear terms are critical, but not irrelevant w.r.t. diffusion. This can be seen easily by making a GL ansatz which shows that at criticality, i.e., $\mathcal{R} = \mathcal{R}_c$, the complex-valued amplitude $A(X,T)$ satisfies in lowest order a GL equation $\partial_T A = (C_1 + \mathrm{i}C_2)\partial_X^2 A + \gamma A|A|^2$, but now with $\mathrm{Re}\gamma > 0$ due to the subcritical bifurcation.

the same asymptotic decay rates in a perturbed Burgers equation

$$\partial_t u = \partial_x^2 u - \frac{1}{2}\partial_x(u^2) + H(u, \partial_x u), \qquad u|_{t=0} = u_0, \tag{14.45}$$

where H is a smooth function $\mathbb{R}^2 \to \mathbb{R}$ only containing irrelevant nonlinearities.

Lemma 14.3.3. *For all $C > 0$ there exists a $\delta > 0$ such that solutions u of (14.45) with $\|u_0\|_{L^1} + \|u_0\|_{L^\infty} \leq \delta$ satisfy*

$$\|u(t)\|_{L^1} \leq C \quad \text{and} \quad \|u(t)\|_{L^\infty} \leq C(1+t)^{-1/2}$$

for all $t \geq 0$.

Proof. For notational simplicity we restrict ourselves to the example $H(u, \partial_x u) = u^p$ with $p > 3$. We consider the variation of constant formula

$$u(t) = \mathrm{e}^{t\partial_x^2}u_0 - \frac{1}{2}\int_0^t \mathrm{e}^{(t-s)\partial_x^2}\partial_x(u^2(s))\,\mathrm{d}s + \int_0^t \mathrm{e}^{(t-s)\partial_x^2}u^p(s)\,\mathrm{d}s \tag{14.46}$$

for (14.45). We follow Lemma 14.1.2 where estimates of $\int_0^t \mathrm{e}^{(t-s)\partial_x^2}u^p(s)\,\mathrm{d}s$ in terms of

$$a(t) = \sup_{0\leq s\leq t}\|u(s)\|_{L^1}, \qquad b(t) = \sup_{0\leq s\leq t}\|(1+s)^{1/2}u(s)\|_{L^\infty}$$

can be found. The second term on the right-hand side of (14.46) is estimated as follows. We have

$$
\begin{aligned}
&(1+t)^{1/2}\left\|\int_0^t e^{(t-s)\partial_x^2}\partial_x(u^2(s))\,ds\right\|_{L^\infty} \\
&\le (1+t)^{1/2}\int_0^{t/2} \|e^{(t-s)\partial_x^2}\partial_x\|_{L^1\to L^\infty}\|u^2\|_{L^1}\,ds \\
&\quad + (1+t)^{1/2}\int_{t/2}^{t} \|e^{(t-s)\partial_x^2}\partial_x\|_{L^\infty\to L^\infty}\|u^2\|_{L^\infty}\,ds \\
&\le ((1+t)^{1/2}\int_0^{t/2} (t-s)^{-1}(1+s)^{-1/2}\,ds\cdot b(t)a(t) \\
&\quad +(1+t)^{1/2}\int_{t/2}^{t} (t-s)^{-1/2}(1+s)^{-(p-1)/2}\,ds)\cdot b(t)^2 \\
&\le ((1+t)^{1/2}\int_0^{t/2} (t-t/2)^{-1}(1+s)^{-1/2}\,ds\cdot b(t)a(t) \\
&\quad +(1+t)^{1/2}\int_{t/2}^{t} (t-s)^{-1/2}(1+t/2)^{-1}\,ds)\cdot b(t)^2 \\
&\le C_1\,(b(t)a(t)+b(t)^2),
\end{aligned}
$$

with a constant C_1 independent of t, and

$$
\begin{aligned}
&\left\|\int_0^t e^{(t-s)\partial_x^2}\partial_x(u^2(s))\,ds\right\|_{L^1} \le \int_0^t \|e^{(t-s)\partial_x^2}\partial_x\|_{L^1\to L^1}\|u^2\|_{L^1}\,ds \\
&\le \int_0^t (t-s)^{-1/2}(1+s)^{-1/2}\,ds\cdot b(t)a(t) \le C_2 b(t)a(t),
\end{aligned}
$$

with a constant C_2 independent of t such that finally

$$
\begin{aligned}
a(t) &\le a(0)+b(0)+C_2(b(t)a(t)+b(t)^{p-1}a(t)) \\
b(t) &\le a(0)+b(0)+C_1(b(t)a(t)+b(t)^2+b(t)^{p-1}a(t)).
\end{aligned}
$$

If $a(0)+b(0)<\delta$, with $\delta>0$ sufficiently small, we have the existence of a $C>0$ such that $a(t), b(t)\le C$ for all $t\ge 0$. □

Also here, whenever the L^1-L^∞ approach works, the renormalization approach works too. In [**BKL94**] the following diffusive stability result has been shown with the discrete renormalization approach from §14.1.6.

Theorem 14.3.4. *Let $b\in(0,1/2)$. Then there exist $C,\delta>0$ such that for all initial conditions u_0 of (14.45) with $\|u_0\|_{H_2^2}\le\delta$ there is a $z>-1$ such that the solution u of (14.45) with $u|_{t=1}=u_0$ satisfies*

$$\|\sqrt{t}u(x\sqrt{t},t)-f_z^*(x)\|_{H_2^2}\le Ct^{-1/2+b},$$

hence in particular

$$\|u(x,t)-t^{-1/2}f_z^*(t^{-1/2}x)\|_{L^\infty}\le Ct^{-1+b}.$$

Remark 14.3.5. An interesting application of the above theory is the inclined film problem, which is concerned with the flow of a viscous liquid down an inclined plane. It occurs in cooling processes, when wiping a blackboard, in coating processes, such as the painting of a wall, and it also appears on the windshield of a car on a rainy day. It is related to the other free surface problem which we considered in this book, namely the water wave problem which is studied in §12.2.1. However, for this problem viscous effects had been neglected completely.

In detail, we consider a film of viscous liquid flowing down an inclined plane with inclination angle θ and assume that for $|x| \to \infty$ the film is flat, i.e., $h(x,t) \to h_0$ for $|x| \to \infty$. See Figure 14.2 for a two-dimensional sketch. For such flat films there always exists the trivial so called Nusselt solution

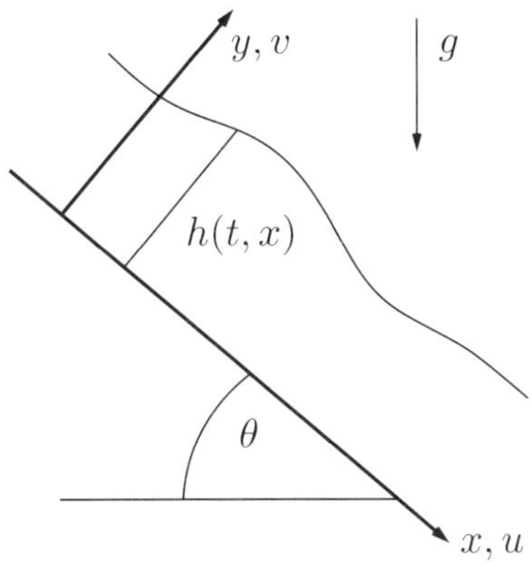

Figure 14.2. The inclined film problem; a fluid of height $y = h(x,t)$ runs down a plate with inclination angle θ subject to constant gravitational force g.

with a laminar flow and a flat surface. For sufficiently small Reynolds number this flow turns out to be stable with a self-similar decay as in Theorem 14.3.4. Using the discrete renormalization approach from §14.1.6, this has been proved for the so-called integrated boundary layer (IBL) approximation of the inclined film problem in [**Uec03**], for the free surface Navier-Stokes problem in [**Uec07**], and for the IBL over wavy bottoms in [**HSU12**], which adapts the Bloch wave analysis from §14.2.2 to quasilinear systems. ⌋

14.4. Phase diffusion equations

In the previous chapters we derived modulation equations, such as the GL equation, the KdV equation and the NLS equation for the description of the dynamics near a spatially homogeneous background state. If the background state is spatially periodic new modulation equations occur. In dissipative systems these are phase diffusion equations for the description of wave number modulations of a family of spatially periodic equilibria and depending on the scaling the Burgers equation or a conservation law for the description of wave number modulations of a family of traveling waves where the group velocity depends in a non-trivial way on the wave number. For conservative systems Whitham's equations for the description of wave number and amplitude modulations of a family of traveling waves appear, cf. [**Whi99**]. We explain the derivation of these equations for the real and complex GL equation. Moreover, we explain that due to the reconstruction of the solutions from the wave number, approximation results can only be expected locally, but not globally, in space. These equations give a feeling about the diffusive dynamics which can be expected in more complicated pattern forming systems.

14.4.1. The phase diffusion equation for the real GL equation. The real GL equation

$$\partial_T A = \partial_X^2 A + A - A|A|^2, \tag{14.47}$$

with $X \in \mathbb{R}$, $T \geq 0$, and $A(X,T) \in \mathbb{C}$ possesses the two-dimensional family of stationary solutions

$$A = A_{\text{per}}[q, \phi_0](X) = \sqrt{1-q^2} e^{iqX + i\phi_0}, \tag{14.48}$$

which is parameterized by $\phi_0, q \in \mathbb{R}$. In §8.3.6 we discussed the linearization around these equilibria and found diffusive modes and exponentially damped modes. In order to derive a nonlinear effective equation for the diffusive modes alone, namely the so called phase diffusion equation, we introduce polar coordinates

$$A(X,T) = r(X,T)e^{i\phi(X,T)}$$

in the real GL equation and obtain

$$\begin{aligned} \partial_T r &= \partial_X^2 r + r - (\partial_X \phi)^2 r - r^3, \\ \partial_T \phi &= \partial_X^2 \phi + \frac{2(\partial_X r)(\partial_X \phi)}{r}. \end{aligned} \tag{14.49}$$

We introduce the deviation (s, ϕ) from the equilibrium $(r, \phi) = (1, 0)$ by setting $r = 1 + s$ and find

$$\begin{aligned} \partial_T s &= \partial_X^2 s - 2s - (\partial_X \phi)^2 - (\partial_X \phi)^2 s - 3s^2 - s^3, \\ \partial_T \phi &= \partial_X^2 \phi + \frac{2(\partial_X s)(\partial_X \phi)}{1+s}. \end{aligned} \tag{14.50}$$

We replace the equation for the phase ϕ by an equation for the local wave number $\psi = \partial_X \phi$ and obtain

$$\begin{aligned} \partial_T s &= \partial_X^2 s - 2s - \psi^2 - \psi^2 s - 3s^2 - s^3, \\ \partial_T \psi &= \partial_X^2 \psi + 2\partial_X \left(\frac{(\partial_X s)\psi}{1+s} \right). \end{aligned} \tag{14.51}$$

To derive the phase-diffusion equation we make a long wave ansatz

$$\psi(X,T) = \check{\psi}(\delta X, \delta^2 T) \qquad \text{and} \qquad s(X,T) = \delta \check{s}(\delta X, \delta^2 T),$$

with $0 < \delta \ll 1$ a small perturbation parameter. This yields

$$(14.52) \qquad \begin{aligned} \delta^2 \partial_\tau \check{s} &= \delta^2 \partial_\xi^2 \check{s} - 2\check{s} - \check{\psi}^2 - \check{\psi}^2 \check{s} - 3\check{s}^2 - \check{s}^3, \\ \partial_\tau \check{\psi} &= \partial_\xi^2 \check{\psi} + 2\partial_\xi \left(\frac{\check{\psi} \partial_\xi \check{s}}{1 + \check{s}} \right), \end{aligned}$$

with $\tau = \delta^2 T$ and $\xi = \delta X$. After neglecting the terms of order $\mathcal{O}(\delta^2)$ for small $\check{\psi}$, the implicit function theorem yields that the first equation

$$(14.53) \qquad 0 = -2\check{s} - \check{\psi}^2 - \check{\psi}^2 \check{s} - 3\check{s}^2 - \check{s}^3$$

possesses a unique solution $\check{s} = s^*(\check{\psi})$ with $s^*(\check{\psi}) = -\check{\psi}^2/2 + \mathcal{O}(\check{\psi}^3)$. Due to the formula for the equilibria, we have in fact that $1 + s^* = \sqrt{1 - \check{\psi}^2}$. Inserting this into the equation for $\check{\psi}$ leads to the phase diffusion equation

$$(14.54) \qquad \partial_\tau \check{\psi} = \partial_\xi^2 \check{\psi} + 2\partial_\xi \left(\frac{\check{\psi} \partial_\xi (s^*(\check{\psi}))}{1 + s^*(\check{\psi})} \right) =: \partial_\xi^2 h(\check{\psi}),$$

where $h : \mathbb{R} \to \mathbb{R}$ is smooth. Indeed, h is the solution to the equation

$$h'(q) = 1 + \frac{2q}{1 + s^*(q)} s^{*\prime}(q) = \frac{1 - 3q^2}{1 - q^2}.$$

This can be verified as follows. For a basic equilibrium $r = r_q$, $\phi = qX + \phi_0$, $q \in (-1, 1)$ the corresponding phase diffusion equation is given by

$$(14.55) \qquad \partial_\tau \check{\psi} = \partial_\xi^2 \widetilde{h}(\check{\psi}, q)$$

where $\widetilde{h}(\check{\psi}, q) = h(\check{\psi} + q)$. For each q, let $k \mapsto \mu_{1,2}(k, q)$ denote the smooth curves of eigenvalues corresponding to the Fourier wave numbers k for the linearization of (14.49) around $(r, \phi) = (\sqrt{1 - q^2}, qX)$. In particular, let

$$\mu_1(k, q) = -k^2 - (1 - q^2) + \sqrt{(1 - q^2)^2 + 4q^2 k^2}$$

denote the critical curve for which $\mu_1(0, q) = 0$. Then the phase-diffusion equation (14.54) must give at lowest order a linear diffusion equation with diffusion coefficient

$$h'(q) = -\partial_k^2 \mu_1(0, q)/2 = (1 - 3q^2)/(1 - q^2).$$

An approximation result that the phase diffusion equation (14.54) makes correct predictions about the dynamics of the original real GL equation can be found in [**MS04b**, Theorem 3.1].

Theorem 14.4.1. *Let $2 \leq m \leq n-2$. Then there exists a $C_\psi^* > 0$ such that the following is true. Let ψ^* be a solution of the phase diffusion equation* (14.54) *with*

$$\sup_{\tau' \in [0,\tau_0]} \|\psi^*(\tau)\|_{H^n_{\mathrm{ul}}} \leq C_\psi^*,$$

and let s^* be defined by the solution $s^* = s^*(\psi^*)$ of (14.53)*. Then there exist $\delta_0 > 0$ and $C_2 > 0$ such that for all $\delta \in (0, \delta_0)$ we have solutions $(\check{s}, \check{\psi})$ of the $(\check{s}, \check{\psi})$-system* (14.52) *such that*

$$\sup_{\tau \in [0,\tau_0]} \|(\check{s}, \check{\psi})(\tau) - (s^*(\psi^*(\tau)), \psi^*(\tau))\|_{H^m_{\mathrm{ul}} \times H^m_{\mathrm{ul}}} \leq C_2 \delta^2.$$

Remark 14.4.2. Although the approximation result is uniformly in space for the (s, ψ)-variables, in the original A variable it is only locally in space due to the reconstruction of ϕ from ψ by spatial integration, see [**MS04b**, Theorem 4.1]. Since this is a serious restriction we recall the arguments. Our starting point is the relation

$$A(X,T) = (1 + \check{s}(\delta X, \delta^2 T)) \exp\Big(\mathrm{i} \int_0^X \check{\psi}(\delta X', \delta^2 T)\, dX'\Big)$$

which defines the solution A of the real GL equation (14.47) in terms of solutions $(\check{s}, \check{\psi})$ of (14.52). These solutions are approximated by

$$A_{\mathrm{app}}(X,T) = (1 + s^*(\delta X, \delta^2 T)) \exp\Big(\mathrm{i} \int_0^X \psi^*(\delta X', \delta^2 T)\, dX'\Big).$$

Then we obtain

$$\begin{aligned}
&|A(X,T) - A_{\mathrm{app}}(X,T)| \\
&\quad \leq \Big|(1 + \check{s}(\delta X, \delta^2 T)) \exp\Big(\mathrm{i} \int_0^X \check{\psi}(\delta X', \delta^2 T)\, dX'\Big) \\
&\qquad\qquad - (1 + s^*(\delta X, \delta^2 T)) \exp\Big(\mathrm{i} \int_0^X \psi^*(\delta X', \delta^2 T)\, dX'\Big)\Big| \\
&\quad \leq \Big|(1 + \check{s}(\delta X, \delta^2 T)) \exp\Big(\mathrm{i} \int_0^X \check{\psi}(\delta X', \delta^2 T)\, dX'\Big) \\
&\qquad\qquad - (1 + \check{s}(\delta X, \delta^2 T)) \exp\Big(\mathrm{i} \int_0^X \psi^*(\delta X', \delta^2 T)\, dX'\Big)\Big| \\
&\qquad + \Big|(1 + \check{s}(\delta X, \delta^2 T)) \exp\Big(\mathrm{i} \int_0^X \psi^*(\delta X', \delta^2 T)\, dX'\Big) \\
&\qquad\qquad - (1 + s^*(\delta X, \delta^2 T)) \exp\Big(\mathrm{i} \int_0^X \psi^*(\delta X', \delta^2 T)\, dX'\Big)\Big|
\end{aligned}$$

$$
\begin{aligned}
\leq & |1+\check{s}(\delta X,\delta^2 T))| \, \Big|\exp\Big(\mathrm{i}\int_0^X \check{\psi}(\delta X',\delta^2 T)\,dX'\Big) \\
& \qquad - \exp\Big(\mathrm{i}\int_0^X \psi^*(\delta X',\delta^2 T)\,dX'\Big)\Big| \\
& \quad + |s^*(\delta X,\delta^2 T)) - \check{s}(\delta X,\delta^2 T))| \\
\leq & C\Big|\int_0^X |\check{\psi}(\delta X',\delta^2 T) - \psi^*(\delta X',\delta^2 T)|\,dX'\Big| + C\delta^2 \\
\leq & \Big|\int_0^X C\delta^2 dX'\Big| + C\delta^2 \leq C\delta^2(1+|X|)
\end{aligned}
$$

using the approximation result of Theorem 14.4.1.

By adding higher order terms in (s^*, ψ^*), the approximation can be improved to hold uniformly on space intervals with length of order $\mathcal{O}(\delta^{-p})$ with $p \geq 0$ arbitrary. ⌋

14.4.2. The Burgers equation for the complex GL equation. We derive the Burgers equation for the complex GL equation

$$
\partial_T A = (1+\mathrm{i}\alpha)\partial_X^2 A + A - (1+\mathrm{i}\beta)A|A|^2, \tag{14.56}
$$

where $\alpha, \beta \in \mathbb{R}$, $X \in \mathbb{R}$, $T \geq 0$, and $A(X,T) \in \mathbb{C}$. It possesses wave trains

$$
A(X,T) = \sqrt{1-k^2}e^{\mathrm{i}(kX-\omega_{\mathrm{nl}}(k)T)}
$$

where $\omega_{\mathrm{nl}}(k) = \beta + (\alpha-\beta)k^2$. We introduce the amplitude and phase deviations (s,ϕ) of the wave train to $k=0$ via

$$
A(X,T) = (1+s(X,T))e^{\mathrm{i}(-\beta t+\phi(X,T))}
$$

which satisfy

$$
\begin{aligned}
\partial_T s =& \partial_X^2 s - 2s - (\partial_X\phi)^2 - (\partial_X\phi)^2 s - 2\alpha(\partial_X s)(\partial_X\phi) \\
& - \alpha\partial_X^2\phi - \alpha(\partial_X^2\phi)s - 3s^2 - s^3, \\
\partial_T\phi =& \partial_X^2\phi + \alpha\frac{\partial_X^2 s}{1+s} - \alpha(\partial_X\phi)^2 + \frac{2(\partial_X s)(\partial_X\phi)}{1+s} - 2\beta s - \beta s^2.
\end{aligned}
$$

Next, we replace the equation for the phase ϕ by an equation for the local wave number $\psi = \partial_X\phi$ and obtain

$$
\begin{aligned}
\partial_T s =& \partial_X^2 s - 2s - \psi^2 - \psi^2 s - 2\alpha(\partial_X s)\psi - \alpha\partial_X\psi - \alpha(\partial_X\psi)s - 3s^2 - s^3, \\
\partial_T\psi =& \partial_X^2\psi + \partial_X\left(\frac{\alpha\partial_X^2 s}{1+s} - \alpha\psi^2 + \frac{2(\partial_X s)\psi}{1+s} - 2\beta s - \beta s^2\right).
\end{aligned}
$$

To derive the Burgers equation, we assume that the wave number varies slowly in time and space and seek solutions of the form

$$
s(X,T) = \delta^2\check{s}(\delta X,\delta^2 T), \qquad \psi(X,T) = \delta\check{\psi}(\delta X,\delta^2 T;\delta).
$$

Substituting this ansatz into (14.56), we get

$$
\begin{aligned}
\delta^2\partial_\tau\check{s} =& \delta^2\partial_\xi^2\check{s} - 2\check{s} - \check{\psi}^2 - \delta^2\check{\psi}^2\check{s} - 2\delta^2\alpha(\partial_\xi\check{s})\check{\psi} \\
& - \alpha\partial_\xi\check{\psi} - \delta^2\alpha(\partial_\xi\check{\psi})\check{s} - 3\delta^2\check{s}^2 - \delta^4\check{s}^3, \\
\partial_\tau\check{\psi} =& \partial_\xi^2\check{\psi} - \partial_\xi(2\beta\check{s} + \alpha\check{\psi}^2) \\
& + \delta^2\partial_\xi\left(\frac{\alpha\partial_\xi^2\check{s}}{1+\delta^2\check{s}} + \frac{2(\partial_\xi\check{s})\check{\psi}}{1+\delta^2\check{s}} - \beta\check{s}^2\right),
\end{aligned}
\tag{14.57}
$$

where $\xi = \delta X$ and $\tau = \delta^2 T$. Neglecting terms of order $\mathcal{O}(\delta^2)$ and higher gives the equations

$$
\begin{aligned}
0 =& -2\check{s} - \check{\psi}^2 - \alpha\partial_\xi\check{\psi}, \\
\partial_\tau\check{\psi}_0 =& \partial_\xi^2\check{\psi} + \partial_\xi(-\alpha\check{\psi}^2 - 2\beta\check{s}).
\end{aligned}
\tag{14.58}
$$

Eliminating $\check{s}$ via the first equation shows that $\check{\psi}(\xi,\tau)$ has to satisfy in lowest order the viscous Burgers equation

$$
\partial_\tau\check{\psi} = (1+\alpha\beta)\partial_\xi^2\check{\psi} + (\beta-\alpha)\partial_\xi(\check{\psi}^2).
\tag{14.59}
$$

An approximation result that the Burgers equation (14.59) makes correct predictions about the dynamics of the original complex GL equation can be found in [**DSSS09**, Theorem 3.4]. As above the approximation result is uniformly in space for the (s,ψ)-variables, but in the original A variable it is only locally in space due to the reconstruction of ϕ from ψ by spatial integration.

Remark 14.4.3. For the description of long wave perturbations of amplitude $\mathcal{O}(1)$ a conservation law can be derived. In order to do so we make the long wave ansatz

$$
\psi = \check{\psi}(\delta X, \delta T) \qquad \text{and} \qquad s = \check{s}(\delta X, \delta T),
$$

and obtain

$$
\begin{aligned}
\delta\partial_\tau\check{s} =& \delta^2\partial_\xi^2\check{s} - 2\check{s} - \check{\psi}^2 - \check{\psi}^2\check{s} - 2\delta\alpha(\partial_\xi\check{s})\check{\psi} \\
& - \delta\alpha\partial_\xi\check{\psi} - \delta\alpha(\partial_\xi\check{\psi})\check{s} - 3\check{s}^2 - \check{s}^3, \\
\partial_\tau\check{\psi} =& \delta\partial_\xi^2\check{\psi} + \partial_\xi(-2\beta\check{s} - \alpha\check{\psi}^2 - \beta\check{s}^2) + \partial_\xi\left(\frac{\delta^2\alpha\partial_\xi^2\check{s}}{1+\check{s}} + \frac{2\delta(\partial_\xi\check{s})\check{\psi}}{1+\check{s}}\right),
\end{aligned}
$$

where $\tau = \delta T$, $\xi = \delta X$. Neglecting terms of order $O(\delta)$ and higher gives

$$
\begin{aligned}
0 =& -2\check{s} - \check{\psi}^2 - \check{\psi}^2\check{s} - 3\check{s}^2 - \check{s}^3, \\
\partial_\tau\check{\psi} =& \partial_\xi(-2\beta\check{s} - \alpha\check{\psi}^2 - \beta\check{s}^2).
\end{aligned}
$$

For small $\check{\psi}$ the first equation can be solved uniquely by the implicit function theorem, so that there exists a smooth even function $s^* : \mathbb{R} \to \mathbb{R}$ such that

$\check{s} = s^*(\check{\psi})$. Inserting this into the second equation gives the conservation law

$$\partial_\tau \check{\psi} = \partial_\xi(-2\beta s^*(\check{\psi}) - \alpha\check{\psi}^2 - \beta s^*(\check{\psi})^2) = \partial_\xi h(\check{\psi}) \tag{14.60}$$

where $h : \mathbb{R} \to \mathbb{R}$ is smooth. Estimates that (14.60) makes correct predictions about the dynamics of the complex GL equation can be found in [**MS04a**]. ⌋

14.5. Dispersive dynamics

As explained in the introduction of this chapter, the method of polynomial decay rates and asymptotically irrelevant nonlinearities plays an even more prominent role in dispersive systems, cf. [**Str89, Rac92, LP09**]. Here we explain similarities and differences between diffusive and dispersive dynamics. We restrict to one particular example, namely the nonlinear Klein-Gordon equation

$$\partial_t^2 u = \Delta u - u - u^p, \tag{14.61}$$

with $x \in \mathbb{R}^d$, $t \geq 0$, $p \in \mathbb{N}$, and $u(x,t) \in \mathbb{R}$, which we write as the first order system

$$\partial_t U = \mathrm{i}\Lambda U + N(U), \tag{14.62}$$

where

$$\widehat{\mathrm{i}\Lambda U} = \begin{pmatrix} 0 & \mathrm{i}\omega(k) \\ \mathrm{i}\omega(k) & 0 \end{pmatrix} \begin{pmatrix} \widehat{u} \\ \widehat{v} \end{pmatrix}, \qquad \widehat{N(U)} = -\frac{1}{\mathrm{i}\omega(k)} \begin{pmatrix} 0 \\ u^{*p} \end{pmatrix}.$$

We have the local existence and uniqueness of solutions in H^s, if $s > d/2$, since the skew symmetric Λ is the generator a unitary group $(\mathrm{e}^{\mathrm{i}t\Lambda})_{t\in\mathbb{R}}$ in every H^s, and since $N(U)$ is locally Lipschitz continuous from H^s to H^s, cf. §11.2. As before we are interested in the global existence and uniqueness of solutions. In contrast to the previously considered diffusive systems for general dispersive systems there is no L^∞-L^∞ estimate for the semigroup generated by the linearized system. Hence the scheme presented in §14.1.3 has to be modified. Instead of the two spaces L^1 and L^∞ we will work in three spaces, namely $H^{s_2} = W^{s_2,2}$, $W^{s_1,1}$, $W^{s_\infty,\infty}$, with s_1, s_2, s_∞ suitably chosen below. As before the $W^{s_\infty,\infty}$-norm will decay, the H^{s_2}-norm stays bounded, but for dispersive systems the $W^{s_1,1}$-norm will grow for $t \to \infty$. For (14.61) the modulus of the possible group velocities is bounded from above by 1 and so a L^∞-L^∞ estimate can be established for (14.61). However, we will not use this estimate in order to present a method which works for general dispersive systems, too.

Hence, for dispersive systems we consider

$$b(t) = \sup_{\tau\in(0,t)} (1+\tau)^{d/2} \|U(\tau)\|_{W^{s_\infty,\infty}} \quad \text{and} \quad c(t) = \sup_{\tau\in(0,t)} \|U(\tau)\|_{H^{s_2}}.$$

For the unitary group $e^{it\Lambda}$ we obtain

$$\|e^{it\Lambda}U_0\|_{H^{s_2}} \leq \|U_0\|_{H^{s_2}}$$

and

$$(1+t)^{d/2}\|e^{it\Lambda}U_0\|_{W^{s_\infty,\infty}} \leq \begin{cases} C(1+t)^{d/2}\|e^{it\Lambda}U_0\|_{H^{s_2}} \leq C\|U_0\|_{H^{s_2}}, & \text{for } t\in[0,1], \\ C(1+t)^{d/2}t^{-d/2}\|U_0\|_{W^{s_1,1}}, & \text{for } t\geq 1, \end{cases}$$

where for $t\in[0,1]$ we used Sobolev's embedding $H^{s_2}\subset W^{s_\infty,\infty}$ for $s_2\geq s_\infty+d/2$ and for $t\geq 1$ we used the L^1-L^∞ estimate from [**Sha85**] which holds for $s_1>s_\infty+d+2$.

As a consequence, we have to improve the previous estimates for the nonlinear terms. Using the product rule shows that $\partial_x^s(u^p)$ creates terms of the form

$$u^{p-1}\partial_x^s u,\, u^{p-2}(\partial_x u)\partial_x^{s-1}u, \ldots, u^{p-2}(\partial_x^{s/2}u)\partial_x^{s/2}u.$$

All these terms have in common that there is only one factor with more than $s/2$ derivatives. As a consequence we have the estimates

$$\begin{aligned} \|u^p\|_{H^s} &\leq C\|u\|^{p-1}_{W^{[s/2]+1,\infty}}\|u\|_{H^s}, \\ \|u^p\|_{H^s} &\leq C\|u\|^{p-1}_{H^{[s/2]+1}}\|u\|_{H^s}, \\ \|u^p\|_{W^{s,1}} &\leq C\|u\|^{p-2}_{W^{[s/2]+1,\infty}}\|u\|_{H^{[s/2]+1}}\|u\|_{H^s}. \end{aligned}$$

For $s_\infty\geq\lceil\frac{s_2}{2}\rceil+1$ we obtain for $G(t,\tau)=e^{i(t-\tau)\Lambda}N(u(\tau))$ that

$$\begin{aligned} \left\|\int_0^t G(t,\tau)\,d\tau\right\|_{H^{s_2}} &\leq \int_0^t \|G(t,\tau)\|_{H^{s_2}}\,d\tau \\ &\leq \int_0^t \|N(u(\tau))\|_{H^{s_2}}\,d\tau \leq \int_0^t C\|u(\tau)\|^{p-1}_{W^{s_\infty,\infty}}\|u(\tau)\|_{H^{s_2}}\,d\tau \\ &\leq \int_0^t C(1+\tau)^{-\frac{d}{2}(p-1)}\,d\tau\cdot b(t)^{p-1}c(t) \leq C_1 b(t)^{p-1}c(t), \end{aligned}$$

if $p>1+\frac{2}{d}$, and for $t\geq 1$ that

$$\begin{aligned} &(1+t)^{d/2}\left\|\int_0^t G(t,\tau)\,d\tau\right\|_{W^{s_\infty,\infty}} \\ \leq&(1+t)^{d/2}\int_0^{t/2}\|G(t,\tau)\|_{W^{s_\infty,\infty}}\,d\tau + (1+t)^{d/2}\int_{t/2}^t \|G(t,\tau)\|_{W^{s_\infty,\infty}}\,d\tau \\ \leq&(1+t)^{d/2}\int_0^{t/2}\|e^{i(t-\tau)\Lambda}\|_{W^{s_1,1}\to W^{s_\infty,\infty}}\|N(u(\tau))\|_{W^{s_1,1}}\,d\tau \\ &+(1+t)^{d/2}\int_{t/2}^t \|e^{i(t-\tau)\Lambda}N(u(\tau))\|_{H^{s_2}}\,d\tau, \end{aligned}$$

$$
\begin{aligned}
&\leq (1+t)^{d/2} \int_0^{t/2} C(t-\tau)^{-(d/2)} \|u(\tau)\|_{W^{s_\infty,\infty}}^{p-2} \|u(\tau)\|_{H^{s_2}}^2 \,\mathrm{d}\tau \\
&\quad + (1+\tau)^{d/2} \int_{t/2}^{t} C \|u(\tau)\|_{W^{s_\infty,\infty}}^{p-1} \|u(\tau)\|_{H^s} \,\mathrm{d}\tau \\
&\leq (1+t)^{d/2} \int_0^{t/2} C \left(\frac{t}{2}\right)^{-\frac{d}{2}} (1+\tau)^{-\frac{d}{2}(p-2)} \,\mathrm{d}\tau\, b(t)^{p-2} c(t)^2 \\
&\quad + (1+t)^{d/2} \int_{t/2}^{t} C \left(1+\frac{t}{2}\right)^{-\frac{d}{2}(p-1)} d\tau\, b(t)^{p-1} c(t) \\
&\leq C_2 \left(b(t)^{p-2} c(t)^2 + b(t)^{p-1} c(t) \right),
\end{aligned}
$$

where we used Sobolev's embedding theorem for $s_2 > s_\infty + \frac{d}{2}$ and for the existence of the integrals we assume $\frac{d}{2}(p-2) > 1$ and $\frac{d}{2}(p-1) \geq \frac{d}{2} + 1$. For $t \in (0, 1]$ we have

$$
\begin{aligned}
(1+t)^{d/2} \left\| \int_0^t \mathrm{e}^{\mathrm{i}(t-\tau)\Lambda} N(u(\tau)) \,\mathrm{d}\tau \right\|_{W^{s_\infty,\infty}} &\leq 2^{d/2} C \int_0^1 \|\mathrm{e}^{\mathrm{i}(t-\tau)\Lambda} N(u(\tau))\|_{H^{s_2}} \,\mathrm{d}\tau \\
&\leq 2^{d/2} C \int_0^1 \|N(u(\tau))\|_{H^{s_2}} \,\mathrm{d}\tau \leq C_3 b(t)^{p-1} c(t)
\end{aligned}
$$

Therefore we find

$$
\begin{aligned}
c(t) &\leq c(0) + C_1 b(t)^{p-1} c(t), \\
b(t) &\leq C(c(0) + b(0)) + (C_2 + C_3)(b(t)^{p-1} c(t) + b(t)^{p-2} c(t)).
\end{aligned}
$$

Hence $y(t) = b(t) + c(t)$ satisfies $y(t) \leq Cy(0) + Cy(t)^p$. It follows as above that $\sup_{t \geq 0} y(t) \leq 2Cy(0)$.

Theorem 14.5.1. *Consider* (14.61) *with* $p > 2 + \frac{2}{d}$, $s_2 > s_\infty + \frac{d}{2}$, $s_1 > s_\infty + d + 2$, *and* $s_\infty \geq [\frac{s_2}{2}] + 1$. *Then there exists* $\delta > 0$ *and* $C > 0$ *such that the following holds. If*

$$
\|U_0\|_{H^{s_2}} + \|U_0\|_{W^{s_\infty,\infty}} + \|U_0\|_{W^{s_1,1}} \leq \delta,
$$

then the associated solution U *with* $U|_{t=0} = U_0$ *exists globally and satisfies*

$$
\|u(t)\|_{H^{s_2}} \leq C \qquad \textit{and} \qquad \|u(t)\|_{W^{s_\infty,\infty}} \leq C(1+t)^{-\frac{d}{2}}.
$$

Remark 14.5.2. Although (14.61) can be solved forward and backward in time, Theorem 14.5.1 has a direction in time. The reason is the growth of the $W^{s_1,1}$-norm as time evolves. Since the $W^{s_1,1}$-norm appears in the assumptions of Theorem 14.5.1 the application of the Theorem cannot be iterated. ⌋

Remark 14.5.3. The condition $p > 1+2/d$ was used in the estimates for the H^{s_2}-norm. The stronger condition $p > 2+2/d$ was only used in the estimates for the $W^{s_\infty,\infty}$-norm. By a normal form transform [**Sha85**] the nonlinearity

u^p can be transformed into a nonlinearity of order $\mathcal{O}(u^{p+1})$. This transform leads to a loss of regularity and cannot be inverted. However, these problems can be avoided by considering (14.61) twice. The original equation is used to estimate the H^{s_2}-norm. In the second equation we perform the normal form transform and use it to estimate the $W^{s_\infty,\infty}$-norm. A loss of r derivatives is not a problem as long as $s_2 > s_\infty + r$. After the transform only the weaker condition $p > 1+2/d$ is necessary. The facts that the normal form transform has not be inverted and a loss of regularity can be accepted, also allows to handle quasilinear systems, where for the H^{s_2}-estimate an energy estimate instead of the variation of constant formula is used. The fact that the loss of regularity coming from the normal form transform is not a problem, is different from §11.5.4. ⌋

Further Reading. For the GL equation, the nonlinear diffusive stability of the spatially periodic equilibria has also been established in [**Kap94a**]. In [**EWW97**] the continuous renormalization approach has been used to give another proof of the stability of the bifurcating spatially periodic equilibria for the Swift-Hohenberg (SH) problem. The nonlinear stability of front solutions in the GL equation, i.e., of solutions which connect an Eckhaus-stable solution for $x \to -\infty$ with the trivial solution for $x \to \infty$ has been shown in [**EW94, Gal94, BK95**]. The associated results for the modulating fronts in the SH equation, respectively Couette-Taylor problem has been established in [**ES00, ES02**]. Diffusive mixing results for pattern with different phases, respectively wave numbers, for $x \to \pm\infty$ has been established in [**BKL94, GM98, SSSU12, IS16**]. A new suitable coordinate system for diffusive stability results of spatially periodic solutions has been constructed in [**GS11**]. The previous ideas have been transferred and extended to systems of conservation laws or systems with an additional conservation law in a number of papers. See [**JZ10, JZ11, JNRZ14**] for an overview. In all these papers about diffusive mixing of phases or wave numbers a more elaborate analysis is used to obtain more details about the effective nonlinear terms than we obtained in §14.2.

The method of polynomial decay rates and asymptotically irrelevant nonlinearities is also used in other situations, cf. [**JGK93, Kap94b**], especially, in dispersive systems it plays a very prominent role, cf. [**Str89, Rac92, LP09**]. The method of space-time resonances, cf. [**Sha10**], is a recent development which allows to control nonlinear terms by the linear evolution even in case when other advanced tools such as the normal forms method [**Sha85**] and the vector field method [**CK90**] fail.

Exercises

14.1. For $u(x,t) = ct^{-1/2}\mathrm{e}^{-x^2/(4t)}$ and $t \to \infty$, prove the estimates

$$\sup_{x\in\mathbb{R}} |\partial_t u| \sim t^{-3/2}, \qquad \sup_{x\in\mathbb{R}} |\partial_x^2 u| \sim t^{-3/2}, \qquad \text{and} \qquad \sup_{x\in\mathbb{R}} |u^p| \sim t^{-p/2}.$$

14.2. Compute the first four Hermite polynomials via the inverse Fourier transform of $\widehat{u}_j = (ik)^j \mathrm{e}^{-k^2}$ for $j = 0, 1, 2, 3$, cf. Exercise 7.19.

14.3. Prove the estimates (14.11) and (14.12).

14.4. Transfer the Lyapunov approach from §14.1.4 to 2D and 3D spatial domains.

14.5. Consider $\partial_t u = -\partial_x^4 u$, with $x \in \mathbb{R}$, $t \geq 0$, $u(x,t) \geq 0$. Make an ansatz $u(x,t) = \frac{1}{t^{1/4}} v(\frac{x}{t^{1/4}})$ for finding self-similar solutions. Show that v satifies

$$-v^{(4)} + \frac{1}{4}\xi v' + \frac{1}{4} v = 0.$$

Show that $\widehat{v}(k) = \mathrm{e}^{-k^4}$ satisfies the Fourier transformed equation. Discuss the smoothness and decay of the function $G(\xi) = c\int_{\mathbb{R}} \mathrm{e}^{-k^4} \mathrm{e}^{ik\xi} dk$ where $c \in \mathbb{R}$ is chosen such that $\int G(\xi) d\xi = 1$.

Bibliography

[AA88] H. Added and S. Added, *Equations of Langmuir turbulence and nonlinear Schrödinger equation: smoothness and approximation*, J. Funct. Anal. **79** (1988), no. 1, 183–210. MR950090 (89h:35273)

[AA11] A. Ambrosetti and D. Arcoya, *An introduction to nonlinear functional analysis and elliptic problems*, Progress in Nonlinear Differential Equations and their Applications, 82, Birkhäuser Boston Inc., Boston, MA, 2011.

[ABZ14] T. Alazard, N. Burq, and C. Zuily, *On the Cauchy problem for gravity water waves*, Invent. Math. **198** (2014), no. 1, 71–163.

[Ace00] A. B. Aceves, *Optical gap solitons: Past, present, and future; theory and experiments*, Chaos **10** (2000), no. 3, 584–589.

[ACF+99] M. S. Alber, R. Camassa, Y. N. Fedorov, D. D. Holm, and J. E. Marsden, *On billiard solutions of nonlinear PDEs*, Phys. Lett. A **264** (1999), no. 2-3, 171–178.

[ACF+01] ———, *The complex geometry of weak piecewise smooth solutions of integrable nonlinear PDE's of shallow water and Dym type*, Comm. Math. Phys. **221** (2001), no. 1, 197–227.

[ACHM94] M. S. Alber, R. Camassa, D. D. Holm, and J. E. Marsden, *The geometry of peaked solitons and billiard solutions of a class of integrable PDEs*, Lett. Math. Phys. **32** (1994), no. 2, 137–151.

[AD15] T. Alazard and J.-M. Delort, *Global solutions and asymptotic behavior for two dimensional gravity water waves*, Ann. Sci. Éc. Norm. Supér. (4) **48** (2015), no. 5, 1149–1238.

[ADN59] S. Agmon, A. Douglis, and L. Nirenberg, *Estimates near the boundary for solutions of elliptic partial differential equations satisfying general boundary conditions. I*, Comm. Pure Appl. Math. **12** (1959), 623–727.

[ADN64] ———, *Estimates near the boundary for solutions of elliptic partial differential equations satisfying general boundary conditions. II*, Comm. Pure Appl. Math. **17** (1964), 35–92.

[AGGP12] V. K. Andreev, Y. A. Gaponenko, O. N. Goncharova, and V. V. Pukhnachev, *Mathematical models of convection*, De Gruyter Studies in Mathematical Physics, vol. 5, De Gruyter, Berlin, 2012.

[Agr01] G. P. Agrawal, *Solitons and nonlinear wave equations*, Academic Press, 2001.

[Air45] G. B. Airy, *Tides and waves*, Enzyclopedia Metropolitana **5** (1845), 241–396.

[AJMR95] L. Arnold, C. Jones, K. Mischaikow, and G. Raugel, *Dynamical systems*, Lecture Notes in Mathematics, vol. 1609, Springer-Verlag, Berlin, 1995, Lectures given at the Second C.I.M.E. Session held in Montecatini Terme, June 13-22, 1994, Edited by R. Johnson.

[AK02] I. S. Aranson and L. Kramer, *The world of the complex Ginzburg-Landau equation*, Rev. Modern Phys. **74** (2002), no. 1, 99–143.

[AKN06] V. I. Arnold, V. V. Kozlov, and A. I. Neishtadt, *Mathematical aspects of classical and celestial mechanics*, third ed., Encyclopaedia of Mathematical Sciences, vol. 3, Springer-Verlag, Berlin, 2006, [Dynamical systems. III], Translated from the Russian original by E. Khukhro.

[AL08] B. Alvarez-Samaniego and D. Lannes, *Large time existence for 3D water waves and asymptotics*, Invent. Math. **171** (2008), no. 3, 485–541.

[ALB$^+$10] D. Avitabile, D. J. B. Lloyd, J. Burke, E. Knobloch, and B. Sandstede, *To snake or not to snake in the planar Swift-Hohenberg equation*, SIAM J. Appl. Dyn. Syst. **9** (2010), no. 3, 704–733.

[Alt16] H. W. Alt, *Linear functional analysis. An application-oriented introduction*, Springer, 2016.

[AM05] D. M. Ambrose and N. Masmoudi, *The zero surface tension limit of two-dimensional water waves*, Comm. Pure Appl. Math. **58** (2005), no. 10, 1287–1315.

[Ama95] H. Amann, *Linear and quasilinear parabolic problems. Vol. I. abstract linear theory*, Monographs in Mathematics, vol. 89, Birkhäuser Boston Inc., Boston, MA, 1995.

[Arn78] V. I. Arnold, *Mathematical methods of classical mechanics*, Graduate Texts in Mathematics. 60. Springer-Verlag, 1978.

[Arn88] V. I. Arnold, *Geometrical methods in the theory of ordinary differential equations*, second ed., Grundlehren der Mathematischen Wissenschaften [Fundamental Principles of Mathematical Sciences], vol. 250, Springer-Verlag, New York, 1988, Translated from the Russian by Joseph Szücs.

[AS81] M. J. Ablowitz and H. Segur, *Solitons and the inverse scattering transform*, SIAM, Philadelphia, Pa., 1981.

[Avi11] A. Avila, *Dynamics of renormalization operators*, Proceedings of the international congress of mathematicians (ICM 2010), Hyderabad, India, August 19–27, 2010. Vol. I: Plenary lectures and ceremonies, Hackensack, NJ: World Scientific; New Delhi: Hindustan Book Agency, 2011, pp. 154–175.

[Bar96] G. I. Barenblatt, *Scaling, self-similarity, and intermediate asymptotics*, Cambridge: Cambridge University Press, 1996.

[BB98] H. Brezis and F. Browder, *Partial differential equations in the 20th century*, Advances in Mathematics **135** (1998), 76–144.

[BBH94] F. Bethuel, H. Brezis, and F. Hélein, *Ginzburg-Landau vortices*, Progress in Nonlinear Differential Equations and their Applications, 13, Birkhäuser Boston Inc., Boston, MA, 1994.

[BBTW15] N. Bellomo, A. Bellouquid, Y. Tao, and M. Winkler, *Toward a mathematical theory of Keller-Segel models of pattern formation in biological tissues*, Math. Models Methods Appl. Sci. **25** (2015), no. 9, 1663–1763.

[BCBLS11] C. Blank, M. Chirilus-Bruckner, V. Lescarret, and G. Schneider, *Breather solutions in periodic media*, Comm. Math. Phys. **302** (2011), no. 3, 815–841.

[BCD+90] M. Bartuccelli, P. Constantin, C. R. Doering, J. D. Gibbon, and Magnus Gisselfält, *On the possibility of soft and hard turbulence in the complex Ginzburg-Landau equation*, Physica D **44** (1990), no. 3, 421–444.

[BDS16] R. Bauer, W.-P. Düll, and G. Schneider, *The KdV, Burgers, and Whitham limit for a spatially periodic Boussinesq model*, arXiv:1608.05632 (2016).

[Ben72] T. B. Benjamin, *The stability of solitary waves*, Proc. Roy. Soc. (London) Ser. A **328** (1972), 153–183.

[BEX14] W. A. Brock, G. Engström, and A. Xepapadeas, *Spatial climate-economic models in the design of optimal climate policies across locations*, European Economic Review **69** (2014), 78–103.

[BF06] A. Babin and A. Figotin, *Linear superposition in nonlinear wave dynamics*, Rev. Math. Phys. **18** (2006), no. 9, 971–1053.

[BFL+07] K. Busch, G. Von Freymann, S. Linden, S. F. Mingaleev, L. Tkeshelashvili, and M. Wegener, *Periodic nanostructures for photonics*, Physics reports **444** (2007), no. 3, 101–202.

[BGT96] B. Buffoni, M. D. Groves, and J. F. Toland, *A plethora of solitary gravity-capillary water waves with nearly critical Bond and Froude numbers*, Philos. Trans. R. Soc. Lond., Ser. A **354** (1996), no. 1707, 575–607.

[BHP05] D. Blömker, M. Hairer, and G. A. Pavliotis, *Modulation equations: Stochastic bifurcation in large domains*, Commun. Math. Phys. **258** (2005), no. 2, 479–512.

[BK92] J. Bricmont and A. Kupiainen, *Renormalization group and the Ginzburg-Landau equation*, Comm. Math. Phys. **150** (1992), 193–208.

[BK95] ______, *Renormalizing partial differential equations*, Constructive physics. Results in field theory, statistical mechanics and condensed matter physics. Proceedings of the conference, held at Ecole Polytechnique, Palaiseau, France, 25-27 July, 1994, Berlin: Springer-Verlag, 1995, pp. 83–115.

[BK00] J. Billingham and A. C. King, *Wave motion*, Cambridge University Press, Cambridge, 2000.

[BKL94] J. Bricmont, A. Kupiainen, and G. Lin, *Renormalization group and asymptotics of solutions of nonlinear parabolic equations*, Comm. Pure Appl. Math. **6** (1994), 893–922.

[BKL+09] M. Beck, J. Knobloch, D. J. B. Lloyd, B. Sandstede, and T. Wagenknecht, *Snakes, ladders, and isolas of localized patterns.*, SIAM J. Math. Anal. **41** (2009), no. 3, 936–972.

[BM13] D. Blömker and W. W. Mohammed, *Amplitude equations for SPDEs with cubic nonlinearities*, Stochastics **85** (2013), no. 2, 181–215.

[BMS01] D. Blömker, S. Maier-Paape, and G. Schneider, *The stochastic Landau equation as an amplitude equation*, Discrete Contin. Dyn. Syst., Ser. B **1** (2001), no. 4, 527–541.

[BMW94] B. Birnir, H. P. McKean, and A. Weinstein, *The rigidity of sine-Gordon breathers*, Commun. Pure Appl. Math. **47** (1994), no. 8, 1043–1051.

[BN85] N. Bekki and K. Nozaki, *Formations of spatial patterns and holes in the generalized Ginzburg-Landau equation*, Physics Letters A **110** (1985), no. 3, 133–135.

[BNSZ14] M. Beck, T. T. Nguyen, B. Sandstede, and K. Zumbrun, *Nonlinear stability of source defects in the complex Ginzburg-Landau equation*, Nonlinearity **27** (2014), no. 4, 739–786.

[Bou77] M. B. Boussinesq, *Essai sur la théorie des eaux courantes*, Inst. France (séries 2) **23** (1877), 1–680.

[Bou99] J. Bourgain, *Global solutions of nonlinear Schrödinger equations*, American Mathematical Society, Providence, RI, 1999.

[BR13] D. Blömker and M. Romito, *Local existence and uniqueness for a two-dimensional surface growth equation with space-time white noise*, Stochastic Anal. Appl. **31** (2013), no. 6, 1049–1076.

[Bra83] M. Bramson, *Convergence of solutions of the Kolmogorov equation to travelling waves*, Mem. Amer. Math. Soc. **44** (1983), no. 285.

[Bri13] Th. J. Bridges, *A universal form for the emergence of the Korteweg-de Vries equation*, Proc. R. Soc. Lond. Ser. A Math. Phys. Eng. Sci. **469** (2013), no. 2153, 20120707, 15.

[BS07] J. Bitzer and G. Schneider, *Approximation and attractivity properties of the degenerated Ginzburg-Landau equation*, J. Math. Anal. Appl. **331** (2007), no. 2, 743–778.

[BSTU06] K. Busch, G. Schneider, L. Tkeshelashvili, and H. Uecker, *Justification of the Nonlinear Schrödinger equation in spatially periodic media*, ZAMP **57** (2006), 1–35.

[BV90] A. V. Babin and M. I. Vishik, *Attractors of partial differential equations in an unbounded domain*, Proceedings of the Royal Society of Edinburgh **116A** (1990), 221–243.

[BV92] ______, *Attractors of evolution equations*, North Holland, 1992.

[BX08] W. Brock and A. Xepapadeas, *Diffusion-induced instability and pattern formation in infinite horizon recursive optimal control*, Journal of Economic Dynamics and Control **32** (2008), no. 9, 2745–2787.

[BX10] ______, *Pattern formation, spatial externalities and regulation in coupled economic–ecological systems*, Journal of Environmental Economics and Management **59** (2010), no. 2, 149–164.

[Car66] L. Carleson, *On convergence and growth of partial sumas of Fourier series*, Acta Math. **116** (1966), 135–157.

[Car81] J. Carr, *Applications of centre manifold theory*, Applied Mathematical Sciences, Vol. 35. New York Heidelberg Berlin: Springer Verlag. XII. 142 p. DM 32.00; $ 14.60 (1981)., 1981.

[CBCSU08] M. Chirilus-Bruckner, C. Chong, G. Schneider, and H. Uecker, *Separation of internal and interaction dynamics for NLS-described wave packets with different carrier waves*, J. Math. Anal. Appl. **347** (2008), no. 1, 304–314.

[CBS12] M. Chirilus-Bruckner and G. Schneider, *Detection of standing pulses in periodic media by pulse interaction*, J. Differential Equations **253** (2012), no. 7, 2161–2190.

[CBSU07] M. Chirilus-Bruckner, G. Schneider, and H. Uecker, *On the interaction of NLS-described modulating pulses with different carrier waves*, Math. Methods Appl. Sci. **30** (2007), no. 15, 1965–1978.

[CCPS12] M. Chirilus-Bruckner, C. Chong, O. Prill, and G. Schneider, *Rigorous description of macroscopic wave packets in infinite periodic chains of coupled oscillators by modulation equations*, Discrete Contin. Dyn. Syst., Ser. S **5** (2012), no. 5, 879–901.

[CdR15] G. M. Coclite and L. di Ruvo, *Well-posedness results for the short pulse equation*, Z. Angew. Math. Phys. **66** (2015), no. 4, 1529–1557.

[CDS15] M. Chirilus-Bruckner, W.-P. Düll, and G. Schneider, *NLS approximation of time oscillatory long waves for equations with quasilinear quadratic terms*, Math. Nachr. **288** (2015), no. 2-3, 158–166.

[CE80] P. Collet and J.-P. Eckmann, *Iterated maps on the interval as dynamical systems*, Progress in Physics, Birkhaeuser, 1980.

[CE86] P. Collet and J.-P Eckmann, *The existence of dendritic fronts*, Commun. Math. Phys. **107** (1986), no. 1, 39–92.

[CE90] P. Collet and J.-P. Eckmann, *The time dependent amplitude equation for the Swift-Hohenberg problem*, Comm. Math. Phys. **132** (1990), 139–152.

[CEE92] P. Collet, J.-P. Eckmann, and H. Epstein, *Diffusive repair for the Ginzburg-Landau equation*, Helv Phys Acta **65** (1992), 56–92.

[CG09] M. Cross and H. Greenside, *Pattern Formation and Dynamics in Nonequilibrium Systems*, Cambridge University Press, 2009.

[CH82] S. N. Chow and J. K. Hale, *Methods of bifurcation theory*, Springer-Verlag, New York, 1982.

[CH89] R. Courant and D. Hilbert, *Methods of mathematical physics. Vol. II*, Wiley Classics Library, John Wiley & Sons Inc., New York, 1989.

[CH93a] R. Camassa and D. D. Holm, *An integrable shallow water equation with peaked solitons*, Phys. Rev. Lett. **71** (1993), no. 11, 1661.

[CH93b] M. C. Cross and P. C. Hohenberg, *Pattern formation outside equilibrium*, Rev. Mod. Phys. **65** (1993), 854–1190.

[Chi06] C. Chicone, *Ordinary differential equations with applications*, second ed., Texts in Applied Mathematics, vol. 34, Springer, New York, 2006.

[CI94] P. Chossat and G. Iooss, *The Couette-Taylor problem*, New York: Springer-Verlag, 1994.

[CI75] N. Chafee and E. F. Infante, *A bifurcation problem for a nonlinear partial differential equation of parabolic type*, Applicable Anal. **4** (1974/75), 17–37.

[CK90] D. Christodoulou and S. Klainerman, *Asymptotic properties of linear field equations in Minkowski space*, Commun. Pure Appl. Math. **43** (1990), no. 2, 137–199.

[CK91] J. D. Crawford and E. Knobloch, *Symmetry and symmetry-breaking bifurcations in fluid dynamics*, Annual review of fluid mechanics, Vol. 23, Annual Reviews, Palo Alto, CA, 1991, pp. 341–387.

[CK99] T. K. Callahan and E. Knobloch, *Pattern formation in three-dimensional reaction-diffusion systems*, Physica D **132** (1999), 339–362.

[CK01] ———, *Long wavelength instabilities of three-dimensional patterns*, Phys. Rev. E **64** (2001), 036214.

[CK09] S. J. Chapman and G. Kozyreff, *Exponential asymptotics of localised patterns and snaking bifurcation diagrams*, Physica D **238** (2009), no. 3, 319–354.

[CL00] P. Chossat and R. Lauterbach, *Methods in equivariant bifurcations and dynamical systems*, Advanced Series in Nonlinear Dynamics, vol. 15, World Scientific Publishing Co., Inc., River Edge, NJ, 2000.

[CL09] A. Constantin and D. Lannes, *The hydrodynamical relevance of the Camassa-Holm and Degasperis-Procesi equations*, Arch. Ration. Mech. Anal. **192** (2009), no. 1, 165–186.

[Cod61] E. A. Coddington, *An introduction to ordinary differential equations*, Prentice-Hall Mathematics Series, Prentice-Hall Inc., Englewood Cliffs, N.J., 1961.

[Con01] P. Constantin, *Some open problems and research directions in the mathematical study of fluid dynamics*, Mathematics unlimited-2001 and beyond (Engquist et al., ed.), Springer, 2001, pp. 353–360.

[Con11] A. Constantin, *Nonlinear water waves with applications to wave-current interactions and tsunamis.*, Philadelphia, PA: Society for Industrial and Applied Mathematics (SIAM), 2011.

[CP89] J. Carr and R. Pego, *Metastable patterns in solutions of* $u_t = \varepsilon^2 u_{xx} - f(u)$, Commun. Pure Appl. Math. **42** (1989), no. 5, 523–576.

[CP90] ———, *Invariant manifolds for metastable patterns in* $u_t = \varepsilon^2 u_{xx} - f(u)$, Proc. R. Soc. Edinb., Sect. A **116** (1990), no. 1-2, 133–160.

[Cra85] W. Craig, *An existence theory for water waves and the Boussinesq and Korteweg-de Vries scaling limits*, Comm. Partial Differential Equations **10** (1985), no. 8, 787–1003.

[Cri95] D. G. Crighton, *Applications of KdV*, Acta Appl. Math. **39** (1995), no. 1-3, 39–67.

[CS98] C. Cercignani and D. H. Sattinger, *Scaling limits and models in physical processes*, DMV Seminar, vol. 28, Birkhäuser Verlag, Basel, 1998.

[CS13] C. Chong and G. Schneider, *Numerical evidence for the validity of the NLS approximation in systems with a quasilinear quadratic nonlinearity*, ZAMM Z. Angew. Math. Mech. **93** (2013), no. 9, 688–696.

[CSS92] W. Craig, C. Sulem, and P.-L. Sulem, *Nonlinear modulation of gravity waves: a rigorous approach*, Nonlinearity **5** (1992), no. 2, 497–522.

[CW16] P. Cummings and C. E. Wayne, *Modified energy functionals and the nls approximation*, arXiv:1606.00028v1, 2016.

[CZ15] V. Chepyzhov and S. Zelik, *Infinite energy solutions for dissipative Euler equations in* $\mathbb{R}^2$, J. Math. Fluid Mech. **17** (2015), no. 3, 513–532.

[DD13] T. Dohnal and W. Dörfler, *Coupled mode equation modeling for out-of-plane gap solitons in 2D photonic crystals*, Multiscale Model. Simul. **11** (2013), no. 1, 162–191.

[Dea16] E. J. Doedel et al, *AUTO: Continuation and bifurcation software for ordinary differential equations*, cmvl.cs.concordia.ca/auto/, 2016.

[Deb05] L. Debnath, *Nonlinear partial differential equations for scientists and engineers*, second ed., Birkhäuser Boston Inc., Boston, MA, 2005.

[Den93] J. Denzler, *Nonpersistence of breather families for the perturbed sine Gordon equation*, Commun. Math. Phys. **158** (1993), no. 2, 397–430.

[dES71] R. C. diPrima, W. Eckhaus, and L. A. Segel, *Non-linear wave-number interaction in near-critical two-dimensional flows*, Journal of Fluid Mechanics **49** (1971), 705–744.

[Dev89] R. L. Devaney, *An introduction to chaotic dynamical systems*, Addison-Wesley Publishing Company, 1989.

[DFKM96] G. Dangelmayr, B. Fiedler, K. Kirchgässner, and A. Mielke, *Dynamics of nonlinear waves in dissipative systems: reduction, bifurcation and stability*, Pitman Research Notes in Mathematics Series, vol. 352, Longman, Harlow, 1996, With a contribution by G. Raugel.

[DG95] Ch. R. Doering and J. D. Gibbon, *Applied analysis of the Navier-Stokes equations*, Cambridge Univ. Press, 1995.

[DGH03] H. R. Dullin, G. A. Gottwald, and D. D. Holm, *Camassa-Holm, Korteweg-de Vries-5 and other asymptotically equivalent equations for shallow water waves*, Fluid Dynam. Res. **33** (2003), no. 1-2, 73–95.

[DH16] W.-P. Düll and M. Heß, *Existence of long time solutions and validity of the Nonlinear Schrödinger approximation for a quasilinear dispersive equation*, arXiv:1605.08704, 2016.

[DHSZ16] W.-P. Düll, A. Hermann, G. Schneider, and D. Zimmermann, *Justification of the 2D NLS equation for a fourth order nonlinear wave equation - quadratic resonances do not matter much in case of analytic initial conditions*, J. Math. Anal. Appl. **436** (2016), no. 2, 847–867.

[DHV04] A. Doelman, G. Hek, and N. Valkhoff, *Stabilization by slow diffusion in a real Ginzburg-Landau system*, J. Nonlinear Sci. **14** (2004), no. 3, 237–278.

[DIM89] Y. Demay, G. Iooss, and A. Mielke, *Theory of steady Ginzburg-Landau equation in hydrodynamic stability problems*, Eur. J. Mech. **B 8** (1989), no. 3, 229–268.

[DJ89] P. G. Drazin and R. S. Johnson, *Solitons: an introduction*, Cambridge University Press, 1989.

[DK98] G. Dangelmayr and L. Kramer, *Mathematical tools for pattern formation*, Evolution of spontaneous structures in dissipative continuous systems, Lecture Notes in Phys. New Ser. m Monogr., vol. 55, Springer, Berlin, 1998, pp. 1–85.

[DKSZ16] W.-P. Düll, Kourosh Sanei Kashani, G. Schneider, and D. Zimmermann, *Attractivity of the Ginzburg-Landau mode distribution for a pattern forming system with marginally stable long modes*, J. Differ. Equations **261** (2016), no. 1, 319–339.

[DMCK11] A. D. Dean, P. C. Matthews, S. M. Cox, and J. R. King, *Exponential asymptotics of homoclinic snaking.*, Nonlinearity **24** (2011), no. 12, 3323–3351.

[Doe07] E. J. Doedel, *Lecture notes on numerical analysis of nonlinear equations*, Numerical continuation methods for dynamical systems, Springer, Dordrecht, 2007, pp. 1–49.

[DPS09] T. Dohnal, D. Pelinovsky, and G. Schneider, *Coupled-mode equations and gap solitons in a two-dimensional nonlinear elliptic problem with a separable periodic potential*, J. Nonlinear Sci. **19** (2009), no. 2, 95–131.

[DRUW14] T. Dohnal, J. Rademacher, H. Uecker, and D. Wetzel, *pde2path 2.0, in ENOC 14*, 2014.

[DS06] W.-P. Düll and G. Schneider, *Justification of the nonlinear Schrödinger equation for a resonant Boussinesq model*, Indiana Univ. Math. J. **55** (2006), no. 6, 1813–1834.

[DS09] ———, *Validity of Whitham's equations for the modulation of periodic traveling waves in the NLS equation*, J. Nonlinear Sci. **19** (2009), no. 5, 453–466.

[DSSS09] A. Doelman, B. Sandstede, A. Scheel, and G. Schneider, *The dynamics of modulated wave trains*, Mem. Amer. Math. Soc. **199** (2009), no. 934.

[DSW16] W.-P. Düll, G. Schneider, and C. E. Wayne, *Justification of the nonlinear Schrödinger equation for the evolution of gravity driven 2D surface water waves in a canal of finite depth*, Arch. Ration. Mech. Anal. **220** (2016), no. 2, 543–602.

[DU09] T. Dohnal and H. Uecker, *Coupled mode equations and gap solitons for the 2D Gross–Pitaevsky equation with a non-separable periodic potential*, Phys.D **238** (2009), 860–879.

[DU16] ______, *Bifurcation of Nonlinear Bloch waves from the spectrum in the nonlinear Gross-Pitaevskii equation*, JNS, to appear, 2016.

[Dül12] W.-P. Düll, *Validity of the Korteweg-de Vries approximation for the two-dimensional water wave problem in the arc length formulation*, Commun. Pure Appl. Math. **65** (2012), no. 3, 381–429.

[Dül16] ______, *Justification of the Nonlinear Schrödinger approximation for a quasilinear wave equation*, arXiv:1602.08016, 2016.

[Duo01] J. Duoandikoetxea, *Fourier analysis.*, Graduate Studies in Mathematics, vol. 29, American Mathematical Society, Providence, RI, 2001.

[DV15] A. Doelman and F. Veerman, *An explicit theory for pulses in two component, singularly perturbed, reaction-diffusion equations*, J. Dynam. Differential Equations **27** (2015), no. 3-4, 555–595.

[DvHX16] A. Doelman, P. van Heijster, and F. Xie, *A geometric approach to stationary defect solutions in one space dimension*, SIAM J. Appl. Dyn. Syst. **15** (2016), no. 2, 655–712.

[Eas73] M. S. P. Eastham, *The spectral theory of periodic differential equations*, Texts in Mathematics (Edinburgh), Scottish Academic Press, Edinburgh; Hafner Press, New York, 1973.

[Eck65] W. Eckhaus, *Studies in non-linear stability theory*, Springer Tracts in Nat. Phil. Vol.6, 1965.

[Eck93] ______, *The Ginzburg-Landau equation is an attractor*, J. Nonlinear Science **3** (1993), 329–348.

[Ein05] A. Einstein, *Über die von der molekularkinetischen Theorie der Wärme geforderte Bewegung von in ruhenden Flüssigkeiten suspendierten Teilchen*, Ann. Phys. **322** (1905), 549–560.

[EINP03] N. Ercolani, R. Indik, A. C. Newell, and T. Passot, *Global description of patterns far from onset: a case study*, Nonlinear dynamics. From lasers to butterflies. Selected lectures from the 15th Canberra international physics summer school, Canberra, Australia, 21 January – February 1, 2002, River Edge, NJ: World Scientific, 2003, pp. 411–435.

[ER99] J.-P. Eckmann and J. Rougemont, *Coarsening by Ginzburg-Landau dynamics*, Comm. Math. Phys. **199** (1999), 441–470.

[Erm02] B. Ermentrout, *Simulating, analyzing, and animating dynamical systems. a guide to xppaut for researchers and students*, Software, Environments, and Tools, vol. 14, SIAM, Philadelphia, PA, 2002.

[ES98] J. Escher and B. Scarpellini, *On the asymptotics of solutions of parabolic equations on unbounded domains*, Nonlinear Anal., Theory Methods Appl. **33** (1998), no. 5, 483–507.

[ES00] J.-P. Eckmann and G. Schneider, *Nonlinear stability of bifurcating front solutions for the Taylor-Couette problem*, ZAMM Z. Angew. Math. Mech. **80**

(2000), no. 11-12, 745–753, Special issue on the occasion of the 125th anniversary of the birth of Ludwig Prandtl.

[ES02] ———, *Non-linear stability of modulated fronts for the Swift-Hohenberg equation*, Comm. Math. Phys. **225** (2002), no. 2, 361–397.

[Eva98] L. C. Evans, *Partial differential equations*, Graduate Studies in Mathematics, vol. 19, American Mathematical Society, Providence, RI, 1998.

[EvH81] W. Eckhaus and A. van Harten, *The inverse scattering transformation and the theory of solitons. An introduction*, North-Holland Mathematics Studies, vol. 50, North-Holland Publishing Co., Amsterdam, 1981.

[EW91] J.-P. Eckmann and C. E. Wayne, *Propagating fronts and the center manifold theorem*, Comm. Math. Phys. **136** (1991), no. 2, 285–307.

[EW94] ———, *The nonlinear stability of front solutions for parabolic partial differential equations*, Comm. Math. Phys. **136** (1994), 285–307.

[EWW97] J.-P. Eckmann, C. E. Wayne, and P. Wittwer, *Geometric stability analysis of periodic solutions of the Swift-Hohenberg equation*, Comm. Math. Phys. **190** (1997), 173–211.

[Fef06] C. L. Fefferman, *Existence and smoothness of the Navier-Stokes equation*, The millennium prize problems, Clay Math. Inst., Cambridge, MA, 2006, pp. 57–67.

[Fen79] N. Fenichel, *Geometric singular perturbation theory for ordinary differential equations*, J. Differential Equations **31** (1979), no. 1, 53–98.

[FH54] K. O. Friedrichs and D. H. Hyers, *The existence of solitary waves*, Commun. Pure Appl. Math. **7** (1954), 517–550.

[Fib15] G. Fibich, *The nonlinear Schrödinger equation. singular solutions and optical collapse*, Applied Mathematical Sciences, vol. 192, Springer, Cham, 2015.

[Fis37] R. Fisher, *The advance of advantageous genes*, Ann. Eugenics **7** (1937), 355–369.

[Fit69] R. FitzHugh, *Mathematical models of excitation and propagation in nerve*, H.P. Schwan, ed., Biological Engineering, 1969, pp. 1–85.

[FL96] A. S. Fokas and Q. M. Liu, *Asymptotic integrability of water waves*, Phys. Rev. Lett. **77** (1996), no. 12, 2347–2351.

[FLS64] R. P. Feynman, R. B. Leighton, and M. Sands, *The Feynman lectures on physics. Vol. 2: Mainly electromagnetism and matter*, Addison-Wesley Publishing Co., Inc., Reading, Mass.-London, 1964.

[FMW92] W. E. Fitzgibbon, J. J. Morgan, and S. J. Waggoner, *Weakly coupled semilinear parabolic evolution systems*, Ann. Mat. Pura Appl. (4) **161** (1992), 213–229.

[Fow97] A. C. Fowler, *Mathematical models in the applied sciences*, Cambridge University Press, 1997.

[FPU55] E. Fermi, J. Pasta, and S. M. Ulam, *Studies in nonlinear problems*, Technical Report Los Alamos Sci. Lab.., 1955.

[FRMT01] C. Foias, R. Rosa, O. Manley, and R. Temam, *Navier-Stokes equations and turbulence*, Cambridge: Cambridge University Press, 2001.

[Gal94] Th. Gallay, *Local stability of critical fronts in nonlinear parabolic partial differential equations*, Nonlinearity **7** (1994), no. 3, 741–764.

[Gal11] G. P. Galdi, *An introduction to the mathematical theory of the Navier-Stokes equations. Steady-state problems*, second ed., Springer Monographs in Mathematics, Springer, New York, 2011.

[Gam01] T. W. Gamelin, *Complex analysis*, New York, NY: Springer, 2001.

[GCF+08] D. Grass, J. P. Caulkins, G. Feichtinger, G. Tragler, and D. A. Behrens, *Optimal control of nonlinear processes: With applications in drugs, corruption, and terror*, Springer Verlag, 2008.

[Geo15] S. G. Georgiev, *Theory of distributions*, Springer, Cham, 2015.

[GGKM74] C. S. Gardner, J. M. Greene, M. D. Kruskal, and R. M. Miura, *Korteweg-deVries equation and generalization. VI. Methods for exact solution*, Comm. Pure Appl. Math. **27** (1974), 97–133.

[GH83] J. Guckenheimer and Ph. Holmes, *Nonlinear oscillations, dynamical systems, and bifurcations of vector fields*, Applied Mathematical Sciences, 42, Springer, New York, 1983.

[GH03] M. D. Groves and M. Haragus, *A bifurcation theory for three-dimensional oblique travelling gravity-capillary water waves*, J. Nonlinear Sci. **13** (2003), no. 4, 397–447.

[GHS02] M. D. Groves, M. Haragus, and S. M. Sun, *A dimension-breaking phenomenon in the theory of steady gravity-capillary water waves*, R. Soc. Lond. Philos. Trans. Ser. A Math. Phys. Eng. Sci. **360** (2002), no. 1799, 2189–2243, Recent developments in the mathematical theory of water waves (Oberwolfach, 2001).

[GK14] B. Grébert and Th. Kappeler, *The defocusing NLS equation and its normal form*, Zürich: European Mathematical Society (EMS), 2014.

[GL50] V. L. Ginzburg and L. D. Landau, *On the theory of superconductivity*, Zhurnal Eksperimental'noi i Teoreticheskoi Fiziki **20** (1950), 1064.

[GL51] I. M. Gelńand and B. M. Levitan, *On the determination of a differential equation from its spectral function*, Izv. Akad. Nauk SSSR, Ser. Mat. **15** (1951), 309–360 (Russian).

[Gle88] J. Gleick, *Chaos - die Ordnung des Universums*, Knaur, 1988.

[GM72] A. Gierer and H. Meinhardt, *A theory of biological pattern formation*, Kybernetik **12** (1972), 30–39.

[GM98] Th. Gallay and A. Mielke, *Diffusive mixing of stable states in the Ginzburg-Landau equation*, Comm. Math. Phys. **199** (1998), 71–97.

[GMS12] P. Germain, N. Masmoudi, and J. Shatah, *Global solutions for the gravity water waves equation in dimension 3*, Ann. Math. (2) **175** (2012), no. 2, 691–754.

[GP14] Yan Guo and Xueke Pu, *KdV limit of the Euler-Poisson system*, Arch. Ration. Mech. Anal. **211** (2014), no. 2, 673–710.

[GPS17] S. Gilg, D. E. Pelinovsky, and G. Schneider, *Validity of the NLS approximation for periodic quantum graphs*, Nonl. Diff. Eq. and Appl. (2017), to appear.

[GR12] M. Ghergu and V. D. Radulescu, *Nonlinear PDEs, mathematical models in biology, chemistry and population genetics*, Springer Monographs in Mathematics, Springer, Heidelberg, 2012.

[Gri96] P. Grindrod, *The theory and applications of reaction-diffusion equations, patterns and waves*, second ed., The Clarendon Press, Oxford University Press, New York, 1996.

[GS83] P. Gray and S. K. Scott, *Autocatalytic reactions in the isothermal, continuous stirred tank reactor*, Chemical Engineering Science **38** (1983).

[GS85] M. Golubitsky and D. G. Schaeffer, *Singularities and groups in bifurcation theory. Vol. I*, Applied Mathematical Sciences, vol. 51, Springer-Verlag, New York, 1985.

[GS01a] Th. Gallay and G. Schneider, *KP description of unidirectional long waves. The model case*, Proc. R. Soc. Edinb., Sect. A, Math. **131** (2001), no. 4, 885–898.

[GS01b] M. D. Groves and G. Schneider, *Modulating pulse solutions for a class of nonlinear wave equations*, Commun. Math. Phys. **219** (2001), no. 3, 489–522.

[GS02] M. Golubitsky and I. Stewart, *The symmetry perspective. From equilibrium to chaos in phase space and physical space*, Progress in Mathematics, vol. 200, Birkhäuser Verlag, Basel, 2002.

[GS05] M. D. Groves and G. Schneider, *Modulating pulse solutions for quasilinear wave equations*, J. Differ. Equations **219** (2005), no. 1, 221–258.

[GS08] ———, *Modulating pulse solutions to quadratic quasilinear wave equations over exponentially long length scales*, Commun. Math. Phys. **278** (2008), no. 3, 567–625.

[GS11] Th. Gallay and A. Scheel, *Diffusive stability of oscillations in reaction-diffusion systems*, Trans. Am. Math. Soc. **363** (2011), no. 5, 2571–2598.

[GSS88] M. Golubitsky, I. Stewart, and D. G. Schaeffer, *Singularities and groups in bifurcation theory. Vol. II*, Applied Mathematical Sciences, vol. 69, Springer-Verlag, New York, 1988.

[GSU04] Th. Gallay, G. Schneider, and H. Uecker, *Stable transport of information near essentially unstable localized structures*, DCDS–B **4** (2004), no. 2, 349–390.

[GU17] D. Grass and H. Uecker, *Optimal management and spatial patterns in a distributed shallow lake model*, Electr. J. Differential Equations **2017** (2017), no. 1, 1–21.

[Gut94] A. Guthmann, *Einführung in die Himmelsmechanik und Ephemeridenrechnung*, Spektrum–Hochschultaschenbuch, 1994.

[GW05] Th. Gallay and C. E. Wayne, *Global stability of vortex solutions of the two-dimensional Navier-Stokes equation*, Comm. Math. Phys. **255** (2005), no. 1, 97–129.

[GW06a] ———, *Long-time asymptotics of the Navier-Stokes equation in $\mathbb{R}^2$ and $\mathbb{R}^3$ [Plenary lecture presented at the 76th Annual GAMM Conference, Luxembourg, 29 March–1 April 2005]*, ZAMM Z. Angew. Math. Mech. **86** (2006), no. 4, 256–267.

[GW06b] ———, *Three-dimensional stability of Burgers vortices: the low Reynolds number case*, Phys. D **213** (2006), no. 2, 164–180.

[Hal88] J. K. Hale, *Asymptotic behavior of dissipative systems*, Mathematical Surveys and Monographs, vol. 25, American Mathematical Society, Providence, RI, 1988.

[Ham99] M. Hamer, *Solitary killers*, New Scientist **163** (1999), 18–19.

[Han13] D. Han-Kwan, *From Vlasov-Poisson to Korteweg-de Vries and Zakharov-Kuznetsov*, Commun. Math. Phys. **324** (2013), no. 3, 961–993.

[HCS99] M. Haragus-Courcelle and G. Schneider, *Bifurcating fronts for the Taylor Couette problem in infinite cylinders*, ZAMP **50** (1999), no. 1, 120–151.

[Hen81] D. Henry, *Geometric Theory of Semilinear Parabolic Equations*, Springer Lecture Notes in Mathematics, Vol. 840, 1981.

[HI11] M. Haragus and G. Iooss, *Local bifurcations, center manifolds, and normal forms in infinite-dimensional dynamical systems*, Universitext. London: Springer; Les Ulis: EDP Sciences. xi, 329 p. , 2011.

[HK91] J. K. Hale and H. Kocak, *Dynamics and bifurcations*, Springer-Verlag, 1991.

[HN01] F. Hamel and N. Nadirashvili, *Travelling fronts and entire solutions of the Fisher-KPP equation in* $\mathbb{R}^N$, Arch. Ration. Mech. Anal. **157** (2001), no. 2, 91–163.

[HNM14] S. Hata, H. Nakao, and A. S. Mikhailov, *Sufficient conditions for wave instability in three-component reaction-diffusion systems*, Prog. Theor. Exp. Phys. (2014), 013A01.

[HO98] Y. Hayase and T. Ohta, *Sierpinski gasket in a reaction–diffusion system*, Phys. Rev. Let. **81** (1998), 1726.

[HO00] ______, *Self-replicating pulses and Sierpinski gaskets in excitable media*, Phys. Rev. E **62** (2000), 5998–6003.

[Hör83] L. Hörmander, *The analysis of linear partial differential operators. I. Distribution theory and Fourier analysis*, Grundlehren der Mathematischen Wissenschaften, vol. 256, Springer-Verlag, Berlin, 1983.

[Hor03] D. Horstmann, *From 1970 until present: the Keller-Segel model in chemotaxis and its consequences. I*, Jahresber. Deutsch. Math.-Verein. **105** (2003), no. 3, 103–165.

[Hor04] ______, *From 1970 until present: the Keller-Segel model in chemotaxis and its consequences. II*, Jahresber. Deutsch. Math.-Verein. **106** (2004), no. 2, 51–69.

[Hoy06] R. B. Hoyle, *Pattern formation. An introduction to methods*, Cambridge University Press, Cambridge, 2006.

[HPT00] J. K. Hale, L. A. Peletier, and W. C. Troy, *Exact homoclinic and heteroclinic solutions of the Gray-Scott model for autocatalysis*, SIAM J. Appl. Math. **61** (2000), no. 1, 102–130.

[HR90] J. Hale and G. Raugel, *Lower semicontinuity of the attractor for a singularly perturbed hyperbolic equation*, J. Dyn. Differ. Equations **2** (1990), no. 1, 19–67.

[HS12] M. Haragus and A. Scheel, *Dislocations in an anisotropic Swift-Hohenberg equation*, Comm. Math. Phys. **315** (2012), no. 2, 311–335.

[HSD04] M. W. Hirsch, S. Smale, and R. L. Devaney, *Differential equations, dynamical systems, and an introduction to chaos*, second ed., vol. 60, Elsevier/Academic Press, Amsterdam, 2004.

[HSU12] T. Häcker, G. Schneider, and H. Uecker, *Self-similar decay to the marginally stable ground state in a model for film flow over inclined wavy bottoms*, Electr. J. Diff. Eq. **61** (2012), 1–51.

[HSZ11] T. Häcker, G. Schneider, and D. Zimmermann, *Justification of the Ginzburg-Landau approximation in case of marginally stable long waves*, J. Nonlinear Sci. **21** (2011), no. 1, 93–113.

[HT00] K.-H. Hoffmann and Qi Tang, *Ginzburg-Landau Phase Transition Theory and Superconductivity*, International Series of Numerical Mathematics, Birkhäuser, 2000.

[HZ11] H. Hofer and E. Zehnder, *Symplectic invariants and Hamiltonian dynamics*, Modern Birkhäuser Classics, Birkhäuser Verlag, Basel, 2011, Reprint of the 1994 edition.

[IA98] G. Iooss and M. Adelmeyer, *Topics in bifurcation theory and applications. 2nd ed.*, Advanced Series in Nonlinear Dynamics. 3. Singapore: World Scientific. viii, 186 p., 1998.

[Igu01] T. Iguchi, *Well-posedness of the initial value problem for capillary-gravity waves*, Funkcial. Ekvac. **44** (2001), no. 2, 219–241.

[IK90] G. Iooss and K. Kirchgaessner, *Bifurcation d'ondes solitaires en présence d'une faible tension superficielle. (Bifurcation of solitary waves, for a low surface tension)*, C. R. Acad. Sci., Paris, Ser. I **311** (1990), 265–268 (French).

[IM91] G. Iooss and A. Mielke, *Bifurcating time-periodic solutions of Navier-Stokes equations in infinite cylinders*, J. Nonlinear Sci. **1** (1991), no. 1, 107–146.

[Ioo95] G. Iooss, *Capillary-gravity water-waves problem as a dynamical system*, A. Mielke (ed.) et al., Structure and dynamics of nonlinear waves in fluids. Proceedings of the IUTAM/ ISIMM symposium, Hannover, Germany, August 17-20, 1994. Singapore: World Scientific. Adv. Ser. Nonlinear Dyn. 7, 42-57 (1995), 1995.

[Ioo98] ______, *Travelling water-waves, as a paradigm for bifurcations in reversible infinite dimensional "dynamical" systems*, Doc. Math., J. DMV (1998), 611–622.

[IS16] S. Iyer and B. Sandstede, *Mixing in Reaction-Diffusion Systems: Large Phase Offsets*, arXiv:1610.06527, 2016.

[Jam03] G. James, *Centre manifold reduction for quasilinear discrete systems*, J. Nonlinear Sci. **13** (2003), no. 1, 27–63.

[JGK93] C. K. R. T. Jones, R. Gardner, and T. Kapitula, *Stability of travelling waves for non-convex scalar viscous conservation laws*, Commun. Pure Appl. Math. **46** (1993), no. 4, 505–526.

[JK84] F. John and S. Klainerman., *Almost global existence to nonlinear wave equations in three space dimensions*, Commun. Pure Appl. Math. **37** (1984), 443–455.

[JNRZ14] M. A. Johnson, P. Noble, L. M. Rodrigues, and K. Zumbrun, *Behavior of periodic solutions of viscous conservation laws under localized and nonlocalized perturbations*, Invent. Math. **197** (2014), no. 1, 115–213.

[Joh91] F. John, *Partial differential equations*, Springer, 1991.

[Joh02] R. S. Johnson, *Camassa-Holm, Korteweg-de Vries and related models for water waves*, J. Fluid Mech. **455** (2002), 63–82.

[Jon84] C. K. R. T. Jones, *Stability of the travelling wave solution of the FitzHugh-Nagumo system*, Trans. Amer. Math. Soc. **286** (1984), no. 2, 431–469.

[JR94] B. Jakobsen and F. Rosendahl, *The Sleipner Platform Accident*, Structural Engineering International **4** (1994), no. 3, 190–193.

[JZ10] M. A. Johnson and K. Zumbrun, *Nonlinear stability of periodic traveling wave solutions of systems of viscous conservation laws in the generic case*, J. Differ. Equations **249** (2010), no. 5, 1213–1240.

[JZ11] ______, *Nonlinear stability of periodic traveling-wave solutions of viscous conservation laws in dimensions one and two*, SIAM J. Appl. Dyn. Syst. **10** (2011), no. 1, 189–211.

[Kal88] L. A. Kalyakin, *Asymptotic decay of a one-dimensional wavepacket in a nonlinear dispersive medium*, Math. USSR Sbornik **60** (1988), no. 2, 457–483.

[Kal89] ______, *Long wave asymptotics. Integrable equations as asymptotic limits of non- linear systems*, Russ. Math. Surv. **44** (1989), no. 1, 3–42.

[Kap94a] T. Kapitula, *On the nonlinear stability of plane waves for the Ginzburg-Landau equation*, Commun. Pure Appl. Math. **47** (1994), no. 6, 831–841.

[Kap94b] ———, *On the stability of travelling waves in weighted L^∞ spaces*, J. Differ. Equations **112** (1994), no. 1, 179–215.

[Kat75] T. Kato, *Quasi-linear equations of evolution, with applications to partial differential equations*, Spectral theory and differential equations (Proc. Sympos., Dundee, 1974; dedicated to Konrad Jörgens), Springer, Berlin, 1975, pp. 25–70. Lecture Notes in Math., Vol. 448.

[Kat81] ———, *The Cauchy problem for the Korteweg-de Vries equation*, Nonlinear partial differential equations and their applications, Res. Notes in Math., vol. 53, Pitman, 1981, pp. 293–307.

[Kat95] ———, *Perturbation theory for linear operators. Reprint of the corr. print. of the 2nd ed. 1980*, reprint of the corr. print. of the 2nd ed. 1980 ed., Berlin: Springer-Verlag, 1995.

[Kaw72] T. Kawahara, *Oscillatory solitary waves in dispersive media*, J. Phys. Soc. Japan (1972), 260–264.

[KC13] G. Kozyreff and S. J. Chapman, *Analytical results for Front Pinning between an Hexagonal Pattern and a Uniform State in Pattern-Formation Systems*, Phys. Rev. Letters **111(5)** (2013), 054501.

[KDL90] E. Knobloch and J. De Luca, *Amplitude equations for travelling wave convection*, Nonlinearity **3** (1990), no. 4, 975–980.

[KdV95] D. J. Korteweg and G. de Vries, *On the change of form of long waves advancing in the rectangular canal and a new type of long stationary waves*, Phil.Magaz. **39** (1895), 422–443.

[Kee88] J. P. Keener, *Principles of applied mathematics*, Addison-Wesley, 1988.

[Kel77] H. B. Keller, *Numerical solution of bifurcation and nonlinear eigenvalue problems*, Application of bifurcation theory, Proc. adv. Semin., Madison/Wis. 1976, 359-384, 1977.

[Kel93] S. H. Kellert, *In the wake of chaos. Unpredictable order in dynamical systems*, Science and its Conceptual Foundations, University of Chicago Press, Chicago, IL, 1993.

[KH97] A. Katok and B. Hasselblatt, *Introduction to the modern theory of dynamical systems*, Cambridge University Press, 1997.

[Kir82] K. Kirchgaessner, *Wave-solutions of reversible systems and applications*, J. Differ. Equations **45** (1982), 113–127.

[KN79] T. Kano and T. Nishida, *Sur les ondes de surface de l'eau avec une justification mathématique des équations des ondes en eau peu profonde*, J. Math. Kyoto Univ. **19** (1979), no. 2, 335–370.

[KN86] ———, *A mathematical justification for Korteweg-de Vries equation and Boussinesq equation of water surface waves*, Osaka J. Math. **23** (1986), no. 2, 389–413.

[Kno92] E. Knobloch, *Nonlocal amplitude equations*, Pattern formation in complex dissipative systems (Kitakyushu, 1991), World Sci. Publ., River Edge, NJ, 1992, pp. 263–274.

[Koc15] H. Koch, *Nonlinear dispersive equations*, H. Koch, D. Tartaru, M. Visan, eds, Dispersive Equations and Nonlinear Waves, Birkhäuser, 2015.

[Kol27] A. Kolmogoroff, *Sur la convergence des séries de fonctions orthogonales*, Math. Z. **26** (1927), no. 1, 432–441.

[KP03] T. Kappeler and J. Pöschel, *KdV & KAM*, Springer., 2003.

[KP13] T. Kapitula and K. Promislow, *Spectral and dynamical stability of nonlinear waves*, Applied Mathematical Sciences, vol. 185, Springer, New York, 2013.

[KPP37] A. N. Kolmogorov, I. G. Petrovsky, and N. S. Piskunov, *Etude de l'équation de la diffusion avec croissance de la quantité de matière et son application à un problème boilogique*, Bulletin Université d'Etat a Moscou **A 1** (1937), 1–26.

[KPV91] C. E. Kenig, G. Ponce, and L. Vega, *Well-posedness of the initial value problem for the Korteweg-de Vries equation*, J. Amer. Math. Soc. **4** (1991), no. 2, 323–347.

[Kre85] U. Krengel, *Ergodic theorems*, De Gruyter Studies in Mathematics 6. Berlin-New York: Walter de Gruyter. VIII, 1985.

[KS05] M. Kunze and G. Schneider, *Estimates for the KdV-limit of the Camassa-Holm equation*, Lett. Math. Phys. **72** (2005), no. 1, 17–26.

[KSM92] P. Kirrmann, G. Schneider, and A. Mielke, *The validity of modulation equations for extended systems with cubic nonlinearities*, Proc. Roy. Soc. Edinburgh Sect. A **122** (1992), no. 1-2, 85–91.

[Kuc93] P. Kuchment, *Floquet theory for partial differential equations*, Operator Theory: Advances and Applications. 60. Basel: Birkhäuser Verlag, 1993.

[Kue15] C. Kuehn, *Multiple time scale dynamics*, Applied Mathematical Sciences, vol. 191, Springer, Cham, 2015.

[Kur80] Y. Kuramoto, *Instability and turbulence of wavefronts in reaction-diffusion systems*, Prog. Theor. Phys. **63** (1980), no. 6, 1885–1903.

[Kur84] ———, *Chemical oscillations, waves, and turbulence*, Springer Series in Synergetics, Vol. 19, Berlin etc.: Springer-Verlag. VIII, 156 p, 1984.

[Kuz04] Y. A. Kuznetsov, *Elements of applied bifurcation theory. 3rd ed.*, Springer, 2004.

[KY76] Y. Kuramoto and T. Yamada, *Pattern formation in oscillatory chemical reactions*, Prog. Theoret. Phys. **56** (1976), no. 3, 724–740.

[Lan05] D. Lannes, *Well-posedness of the water-waves equations*, J. Amer. Math. Soc. **18** (2005), no. 3, 605–654.

[Lan13] ———, *The water waves problem. Mathematical analysis and asymptotics*, Providence, RI: American Mathematical Society (AMS), 2013.

[Lap10] M. Lappa, *Thermal convection: patterns, evolution and stability*, John Wiley & Sons, Ltd., Chichester, 2010.

[Law90] I. D. Lawrie, *A unified grand tour of theoretical physics*, Adam Hilger Ltd., Bristol, 1990.

[Lax68] P. D. Lax, *Integrals of nonlinear equations of evolution and solitary waves*, Commun. Pure Appl. Math. **21** (1968), 467–490.

[LBCB$^+$09] V. Lescarret, C. Blank, M. Chirilus-Bruckner, Chr. Chong, and G. Schneider, *Standing generalized modulating pulse solutions for a nonlinear wave equation in periodic media*, Nonlinearity **22** (2009), no. 8, 1869–1898.

[LE91] I. Lengyel and I. R. Epstein, *Modeling of Turing structures in the chlorite-iodide-malonic acid-starch reaction system*, Science **251** (1991), 650–652.

[Lem16] P. G. Lemarié-Rieusset, *The Navier-Stokes problem in the 21st century*, Boca Raton, FL: CRC Press, 2016.

[Ler34] J. Leray, *Sur le mouvement d'un liquide visqueux emplissant l'espace*, Acta Math. **63** (1934), no. 1, 193–248.

[LeV92] R. J. LeVeque, *Numerical methods for conservation laws*, second ed., Birkhäuser Verlag, Basel, 1992.

[LL91] L. D. Landau and E. M. Lifschitz, *Lehrbuch der theoretischen Physik ("Landau-Lifschitz"). Band VI*, fifth ed., Akademie-Verlag, Berlin, 1991, Hydrodynamik. [Hydrodynamics], Translated from the Russian by Wolfgang Weller and Adolf Kühnel, Translation edited by Weller and with a foreword by Weller and P. Ziesche.

[LLS13] D. Lannes, F. Linares, and J.-C. Saut, *The Cauchy problem for the Euler-Poisson system and derivation of the Zakharov-Kuznetsov equation*, Studies in phase space analysis with applications to PDEs. In part selected papers based on the presentations at a meeting, Bertinoro, Italy, September 2011, New York, NY: Birkhäuser/Springer, 2013, pp. 181–213.

[Log15a] J. D. Logan, *Applied partial differential equations*, Undergraduate Texts in Mathematics, Springer, Cham, 2015, Third edition.

[Log15b] ———, *A first course in differential equations*, Undergraduate Texts in Mathematics, Springer, Cham, 2015, Third edition.

[Lor63] E. N. Lorenz, *Deterministic non-periodic flow*, J. Atmos. Sci. (1963).

[LP09] F. Linares and G. Ponce, *Introduction to nonlinear dispersive equations*, New York, NY: Springer, 2009.

[LSAC08] D. J. B. Lloyd, B. Sandstede, D. Avitabile, and A. R. Champneys, *Localized hexagon patterns of the planar Swift-Hohenberg equation*, SIAM J. Appl. Dyn. Syst. **7** (2008), no. 3, 1049–1100.

[Lun95] A. Lunardi, *Analytic semigroups and optimal regularity in parabolic problems*, Progress in Nonlinear Differential Equations and their Applications, 16, Birkhäuser Verlag, Basel, 1995.

[Lyn04] S. Lynch, *Dynamical systems with applications using MATLAB*, Birkhäuser, 2004.

[Man91] B. B. Mandelbrot, *Die fraktale Geometrie der Natur*, Birkhaeuser, 1991.

[Man92] P. Manneville, *Dissipative structures and weak turbulence*, Academic Press, 1992.

[Mar55] V. A. Marchenko, *On the reconstruction of the potential energy from phases of the scattered waves*, Dokl. Akad. Nauk SSSR **104** (1955), 695–698 (Russian).

[Mar07] P. A. Markowich, *Applied partial differential equations: a visual approach*, Springer, Berlin, 2007.

[Maz11] Vl. Maz'ya, *Sobolev spaces with applications to elliptic partial differential equations*, Grundlehren der Mathematischen Wissenschaften, vol. 342, Springer, Heidelberg, 2011.

[MB02] A. J. Majda and A. L. Bertozzi, *Vorticity and incompressible flow*, Cambridge Texts in Applied Mathematics, vol. 27, Cambridge University Press, Cambridge, 2002.

[MBK14] W. W. Mohammed, D. Blömker, and K. Klepel, *Multi-scale analysis of SPDEs with degenerate additive noise*, J. Evol. Equ. **14** (2014), no. 1, 273–298.

[Mel98] I. Melbourne, *Derivation of the time-dependent Ginzburg-Landau equation on the line*, J. Nonlinear Sci. **8** (1998), no. 1, 1–15.

[Mel00] ———, *Ginzburg-Landau theory and symmetry*, Nonlinear instability, chaos and turbulence, Vol. II, Adv. Fluid Mech., vol. 25, WIT Press, Southampton, 2000, pp. 79–109.

[Mer15] E. Meron, *Nonlinear physics of ecosystems*, CRC Press, 2015.

[MH92] K. R. Meyer and G. R. Hall, *Introduction to hamiltonian dynamical systems and the n-body problem*, Springer-Verlag, 1992.

[Mie88] A. Mielke, *Reduction of quasilinear elliptic equations in cylindrical domains with applications*, Math. Methods Appl. Sci. **10** (1988), no. 1, 51–66.

[Mie92] ———, *Reduction of PDEs on domains with several unbounded directions: a first step towards modulation equations*, Z. Angew. Math. Phys. **43** (1992), no. 3, 449–470.

[Mie97a] ———, *The complex Ginzburg-Landau equation on large and unbounded domains: sharper bounds and attractors*, Nonlinearity **10** (1997), no. 1, 199–222.

[Mie97b] ———, *Instability and stability of rolls in the Swift–Hohenberg equation*, Comm. Math. Phys. **189** (1997), 829–853.

[Mie97c] ———, *Mathematical analysis of sideband instabilities with application to Rayleigh–Bénard convection*, J. Nonlinear Science **7** (1997), 57–99.

[Mie98] ———, *Bounds for the solutions of the complex Ginzburg-Landau equation in terms of the dispersion parameters*, Physica D **117** (1998), no. 1-4, 106–116.

[Mie02] ———, *The Ginzburg-Landau equation in its role as a modulation equation*, Handbook of dynamical systems, Vol. 2, North-Holland, 2002, pp. 759–834.

[Mil06] P. D. Miller, *Applied asymptotic analysis*, Graduate Studies in Mathematics, vol. 75, American Mathematical Society, Providence, RI, 2006.

[Miu68] R. M. Miura, *Korteweg-de Vries equation and generalizations. I. A remarkable explicit nonlinear transformation*, J. Mathematical Phys. **9** (1968), 1202–1204.

[MK15] J. Müller and Chr. Kuttler, *Methods and models in mathematical biology. Deterministic and stochastic approaches*, Lecture Notes on Mathematical Modelling in the Life Sciences, Springer, Heidelberg, 2015.

[MM87] R. S. MacKay and J. D. Meiss (eds.), *Hamiltonian dynamical systems. A reprint selection*, IOP Publishing Ltd. Bristol, 1987.

[MN05] N. Masmoudi and K. Nakanishi, *From the Klein-Gordon-Zakharov system to the nonlinear Schrödinger equation*, J. Hyperbolic Differ. Equ. **2** (2005), no. 4, 975–1008.

[MN13] ———, *Multifrequency NLS scaling for a model equation of gravity-capillary waves*, Commun. Pure Appl. Math. **66** (2013), no. 8, 1202–1240.

[MS93] J. H. Maddocks and R. L. Sachs, *On the stability of KdV multi-solitons*, Commun. Pure Appl. Math. **46** (1993), no. 6, 867–901.

[MS95] A. Mielke and G. Schneider, *Attractors for modulation equations on unbounded domains - existence and comparison*, Nonlinearity **8** (1995), no. 5, 743–768.

[MS04a] I. Melbourne and G. Schneider, *Phase dynamics in the complex Ginzburg-Landau equation*, J. Differ. Equations **199** (2004), no. 1, 22–46.

[MS04b] ———, *Phase dynamics in the real Ginzburg-Landau equation*, Math. Nachr. **263-264** (2004), 171–180.

[MSU01] A. Mielke, G. Schneider, and H. Uecker, *Stability and diffusive dynamics on extended domains*, Ergodic theory, analysis, and efficient simulation of dynamical systems, Berlin: Springer, 2001, pp. 563–583.

[MU16] K. Matthies and H. Uecker, *Low regularity justification results for envelope approximations of nonlinear wave packets in periodic media*, Asymptotic Anal. **99** (2016), no. 1-2, 53–65.

[Mur89] J. D. Murray, *Mathematical Biology*, Springer, Berlin, 1989.

[MY60] L. Markus and H. Yamabe, *Global stability criteria for differential systems*, Osaka Math. J. **12** (1960), 305–317.

[Nal74] V. I. Nalimov, *The Cauchy-Poisson problem*, Dinamika Splošn. Sredy **254** (1974), no. Vyp. 18 Dinamika Zidkost. so Svobod. Granicami, 104–210.

[New85] A. C. Newell, *Solitons in mathematics and physics*, vol. 48, CBMS-NSF Reg. Conf. Ser. Appl. Math., 1985.

[New16] ______, *Short pulse evolution equation*, Laser filamentation, CRM Ser. Math. Phys., Springer, Cham, 2016, pp. 1–17.

[Nis02] Y. Nishiura, *Far-from-equilibrium dynamics*, Translations of Mathematical Monographs, vol. 209, American Mathematical Society, Providence, RI, 2002.

[Nti76] A. A. Ntinos, *Lengths of instability intervals of second order periodic differential equations*, Quart. J. Math. Oxford (2) **27** (1976), no. 107, 387–394.

[NTYU07] Y. Nishiura, T. Teramoto, X. Yuan, and Kei-Ichi Ueda, *Dynamics of traveling pulses in heterogeneous media*, Chaos **17** (2007), no. 3, 037104, 21.

[NW69] A. C. Newell and J. A. Whitehead, *Finite bandwidth, finite amplitude convection*, J. Fluid Mech. **38** (1969), 279–303.

[OD08] I. Oprea and G. Dangelmayr, *Dynamics and bifurcations in the weak electrolyte model for electroconvection of nematic liquid crystals: a Ginzburg-Landau approach*, Eur. J. Mech., B, Fluids **27** (2008), no. 6, 726–749.

[Olv14] P. J. Olver, *Introduction to partial differential equations*, Undergraduate Texts in Mathematics, Springer, Cham, 2014.

[Osb10] A. R. Osborne, *Nonlinear ocean waves and the inverse scattering transform*, International Geophysics Series 97. Amsterdam: Elsevier/Academic Press. xxvi, 2010.

[Paz83] A. Pazy, *Semigroups of linear operators and applications to partial differential equations*, Applied Mathematical Sciences, vol. 44, Springer-Verlag, New York, 1983.

[Pel11] D. E. Pelinovsky, *Localization in periodic potentials. From Schrödinger operators to the Gross-Pitaevskii equation*, London Mathematical Society Lecture Note Series 390. Cambridge: Cambridge University Press. x, 398 p., 2011.

[Pis06] L. M. Pismen, *Patterns and interfaces in dissipative dynamics*, Springer Series in Synergetics, Springer-Verlag, Berlin, 2006.

[PJS92] H.-O. Peitgen, H. Jürgens, and D. Saupe, *Fractals for the classroom. Part 2: Complex systems and the Mandelbrot set*, Springer, 1992.

[PL68] I. Prigogine and R. Lefever, *Symmetry breaking bifurcations in dissipative systems II*, J. Chem. Phys. **48** (1968), 1665–1700.

[Poi57] H. Poincaré, *Les methodes nouvelles de la mecanique celeste*, Dover Publications, 1957.

[Pom86] Y. Pomeau, *Front motion, metastability and subcritical bifurcations in hydrodynamics*, Physica D **23** (1986), 3–11.

[PS07] D. Pelinovsky and G. Schneider, *Justification of the coupled-mode approximation for a nonlinear elliptic problem with a periodic potential*, Appl. Anal. **86** (2007), no. 8, 1017–1036.

[PS13] D. Pelinovsky and G. Schneider, *Rigorous justification of the short-pulse equation*, NoDEA, Nonlinear Differ. Equ. Appl. **20** (2013), no. 3, 1277–1294.

[PW84] M. H. Protter and H. F. Weinberger, *Maximum principles in differential equations*, Springer-Verlag, New York, 1984.

[PW96] R. D. Pierce and C. E. Wayne, *On the validity of mean field amplitude equations for counterpropagating wavetrains*, Nonlinearity **8** (1996), 433–457.

[Rac92] R. Racke, *Lectures on nonlinear evolution equations. Initial value problems*, Braunschweig etc.: Vieweg, 1992.

[Rad13] T. Radko, *Double-diffusive convection*, Cambridge University Press, Cambridge, 2013.

[Rau91] J. Rauch, *Partial differential equations*, Graduate Texts in Mathematics. 128. New York etc.: Springer-Verlag. x, 263 p. with 42 ill. , 1991.

[Ray76] Lord Rayleigh, *On waves*, Phil.Magaz. **1** (1876).

[Ren92] M. Renardy, *A centre manifold theorem for hyperbolic PDEs*, Proc. Roy. Soc. Edinburgh Sect. A **122** (1992), no. 3-4, 363–377.

[Ren00] ______, *Mathematical analysis of viscoelastic flows*, CBMS-NSF Regional Conference Series in Applied Mathematics, vol. 73, Society for Industrial and Applied Mathematics (SIAM), Philadelphia, PA, 2000.

[Rob01] J. C. Robinson, *Infinite-dimensional dynamical systems. An introduction to dissipative parabolic PDEs and the theory of global attractors*, Cambridge Texts in Applied Mathematics, Cambridge University Press, Cambridge, 2001.

[Rob04a] ______, *An introduction to ordinary differential equations*, Cambridge University Press. , 2004.

[Rob04b] R. C. Robinson, *An introduction to dynamical systems. Continuous and discrete*, Prentice Hall, 2004.

[Rös76] O. Rössler, *An equation for continuous chaos*, Physics Letters A **57** (1976), 397–398.

[RR04] M. Renardy and R. C. Rogers, *An introduction to partial differential equations*, second ed., Texts in Applied Mathematics, vol. 13, Springer-Verlag, New York, 2004.

[RS75a] M. Reed and B. Simon, *Methods of modern mathematical physics. Four volumes*, Academic Press, 1975.

[RS75b] ______, *Methods of modern mathematical physics. II: Fourier analysis, self-adjointness*, Academic Press, 1975.

[RT71] D. Ruelle and F. Takens, *On the nature of turbulence*, Commun. Math. Phys. **20** (1971), 167–192.

[Rud73] W. Rudin, *Functional analysis*, McGraw-Hill, 1973.

[Rue89] D. Ruelle, *Chaotic evolution and strange attractors. the statistical analysis of time series for deterministic nonlinear systems, notes prepared and with a foreword by S. Isola*, Lezioni Lincee. [Lincei Lectures], Cambridge University Press, Cambridge, 1989.

[Rus44] J. S. Russell, *Report on waves*, Rep. 14th Meet. Brit. Assoc. Sci. (1844), 311–390.

[Sal08] S. Salsa, *Partial differential equations in action. From modelling to theory*, Universitext, Springer-Verlag Italia, Milan, 2008.

[San02] B. Sandstede, *Stability of travelling waves*, Handbook of dynamical systems. Volume 2, Amsterdam: Elsevier, 2002, pp. 983–1055.

[Sat76] D. H. Sattinger, *On the stability of waves of nonlinear parabolic systems*, Advances in Math. **22** (1976), no. 3, 312–355.

[Sca99] B. Scarpellini, *Stability, instability, and direct integrals*, Chapman & Hall, 1999.

[Sch79] J. Schnakenberg, *Simple chemical reaction systems with limit cycle behaviour*, J. Theoret. Biol. **81** (1979), no. 3, 389–400.

[Sch94a] G. Schneider, *Error estimates for the Ginzburg-Landau approximation*, ZAMP **45** (1994), 433–457.

[Sch94b] ———, *Global existence via Ginzburg-Landau formalism and pseudo-orbits of Ginzburg-Landau approximations*, Comm. Math. Phys. **164** (1994), 157–179.

[Sch94c] ———, *A new estimate for the Ginzburg-Landau approximation on the real axis*, J. Nonlinear Sci. **4** (1994), no. 1, 23–34.

[Sch95a] ———, *Analyticity of Ginzburg-Landau modes*, J. Diff. Eqns. **121** (1995), 233–257.

[Sch95b] ———, *Validity and Limitation of the Newell-Whitehead equation*, Math. Nachr. **176** (1995), 249–263.

[Sch96a] ———, *Diffusive stability of spatial periodic solutions of the Swift-Hohenberg equation*, Comm. Math. Phys. **178** (1996), 679–702.

[Sch96b] ———, *Limits for the Korteweg-de Vries-Approximation*, CIAM/GAMM 95 Applied Analysis, ZAMM - Journal of Applied Mathematics and Mechanics **76** (1996), no. S2, 341–344.

[Sch96c] ———, *The validity of generalized Ginzburg-Landau equations*, Math. Methods Appl. Sci. **19** (1996), no. 9, 717–736.

[Sch97] ———, *Justification of mean-field coupled modulation equations*, Proc. Royal Soc. Edinb. **127** (1997), no. A, 639–650.

[Sch98a] ———, *Approximation of the Korteweg-de Vries equation by the Nonlinear Schrödinger equation*, JDE **147** (1998), 333–354.

[Sch98b] ———, *Hopf bifurcation in spatially extended reaction-diffusion systems*, J. Nonlinear Sci. **8** (1998), no. 1, 17–41.

[Sch98c] ———, *Justification of modulation equations for hyperbolic systems via normal forms*, NoDEA **5** (1998), 69–82.

[Sch98d] ———, *Nonlinear stability of Taylor-vortices in infinite cylinders*, Arch. Rat. Mech. Anal. **144** (1998), no. 2, 121–200.

[Sch99a] ———, *Cahn-Hilliard description of secondary flows of a viscous incompressible fluid in an unbounded domain*, ZAMM Z. Angew. Math. Mech. **79** (1999), no. 9, 615–626.

[Sch99b] ———, *Global existence results in pattern forming systems - Applications to 3D Navier-Stokes problems -*, J. Math. Pures Appl., IX **78** (1999), 265–312.

[Sch99c] ———, *Some characterizations of the Taylor-Couette attractor*, Differential Integral Equations **12** (1999), no. 6, 913–926.

[Sch05] ———, *Justification and failure of the nonlinear Schrödinger equation in case of non-trivial quadratic resonances*, J. Diff. Eq. **216** (2005), no. 2, 354–386.

[Sch10] ———, *Bounds for the nonlinear Schrödinger approximation of the Fermi-Pasta-Ulam system*, Appl. Anal. **89** (2010), no. 9, 1523–1539.

[Sch11] ———, *Justification of the NLS approximation for the KdV equation using the Miura transformation*, Adv. Math. Phys. (2011), Art. ID 854719, 4.

[Sch15] R. Schnaubelt, *Center manifolds and attractivity for quasilinear parabolic problems with fully nonlinear dynamical boundary conditions*, Discrete Contin. Dyn. Syst. **35** (2015), no. 3, 1193–1230.

[SD98] R. Schielen and A. Doelman, *Modulation equations for spatially periodic systems: derivation and solutions*, SIAM J. Appl. Math. **58** (1998), no. 6, 1901–1930.

[Seg69] L. A. Segel, *Distant side-walls cause slow amplitude modulation of cellular convection*, Journal of Fluid Mechanics **38(1)** (1969), 203–224.

[Ser99] D. Serre, *Systems of conservation laws. I: Hyperbolicity, entropies, shock waves*, Cambridge: Cambridge University Press. xxii, 263 p., 1999.

[Ser00] ———, *Systems of conservation laws 2. Geometric structures, oscillations, and initial-boundary value problems*, Cambridge: Cambridge University Press. xi, 269 p., 2000.

[Ser15] S. Serfaty, *Coulomb gases and Ginzburg-Landau vortices*, Zürich: European Mathematical Society (EMS), 2015.

[Sey10] R. Seydel, *Practical bifurcation and stability analysis. 3rd ed.*, Springer, 2010.

[SGKM95] A. A. Samarskii, V. Galaktionov, S. Kurdyumov, and A. Mikhailov, *Blow-up in quasilinear parabolic equations*, De Gruyter, 1995.

[SH77] J. Swift and P. C. Hohenberg, *Hydrodynamic fluctuations at the convective instability*, Physical Review A **15** (1977), no. 1, 319–328.

[SH94a] P. J. Schmid and D. S. Henningson, *Optimal energy density growth in Hagen-Poiseuille flow*, J. Fluid Mech. **277** (1994), 197–225.

[SH94b] Y. Shi and J. E. Hearst, *The Kirchhoff elastic rod, the nonlinear Schrödinger equation, and DNA supercoiling*, J. Chem. Phys. **101** (1994), no. 6, 5186–5200.

[SH96] A. M. Stuart and A. R. Humphries, *Dynamical systems and numerical analysis*, Cambridge: Cambridge University Press, 1996.

[Sha85] J. Shatah, *Normal forms and quadratic nonlinear Klein-Gordon equations*, Comm. Pure Appl. Math. **38** (1985), no. 5, 685–696.

[Sha10] J. Shatah, *Space-time resonances*, Q. Appl. Math. **68** (2010), no. 1, 161–167.

[She97] A. Shepeleva, *On the validity of the degenerate Ginzburg-Landau equation*, Math. Methods Appl. Sci. **20** (1997), no. 14, 1239–1256.

[Shi76] M. Shinbrot, *The initial value problem for surface waves under gravity. I. The simplest case*, Indiana Univ. Math. J. **25** (1976), no. 3, 281–300.

[Sil65] L. P. Sil'nikov, *A case of the existence of a denumerable set of periodic motions*, Dokl. Akad. Nauk SSSR **160** (1965), 558–561.

[Siv77] G. Sivashinsky, *Nonlinear analysis of hydrodynamic instability in laminar flames. I - Derivation of basic equations*, Acta Astronautica **4** (1977), 1177–1206.

[Smo94] J. Smoller, *Shock waves and reaction-diffusion equations*, Springer, New York, 1994.

[Spa82] C. Sparrow, *The Lorenz equations: bifurcations, chaos, and strange attractors*, Applied Mathematical Sciences, 41. Springer-Verlag, 1982.

[SR89] X. Saint Raymond, *A simple Nash-Moser implicit function theorem*, Enseign. Math. (2) **35** (1989), no. 3-4, 217–226.

[SR10] P. L. Sachdev and Ch. Srinivasa Rao, *Large time asymptotics for solutions of nonlinear partial differential equations*, Springer Monographs in Mathematics, Springer, New York, 2010.

[SS99a] B. Sandstede and A. Scheel, *Essential instability of pulses and bifurcations to modulated travelling waves*, Proc. R. Soc. Edinb., Sect. A, Math. **129** (1999), no. 6, 1263–1290.

[SS99b] C. Sulem and P.-L. Sulem, *The nonlinear Schrödinger equation*, Springer-Verlag, New York, 1999.

[SS00] B. Sandstede and A. Scheel, *Absolute and convective instabilities of waves on unbounded and large bounded domains*, Phys. D **145** (2000), no. 3-4, 233–277.

[SS07] E. Sandier and S. Serfaty, *Vortices in the magnetic Ginzburg-Landau model*, Basel: Birkhäuser, 2007.

[SSSU12] B. Sandstede, A. Scheel, G. Schneider, and H. Uecker, *Diffusive mixing of periodic wave trains in reaction-diffusion systems*, J. Differ. Equations **252** (2012), no. 5, 3541–3574.

[SSW99] B. Sandstede, A. Scheel, and C. Wulff, *Bifurcations and dynamics of spiral waves*, J. Nonlinear Sci. **9** (1999), no. 4, 439–478.

[SSZ15] G. Schneider, Danish Ali Sunny, and D. Zimmermann, *The NLS approximation makes wrong predictions for the water wave problem in case of small surface tension and spatially periodic boundary conditions*, J. Dyn. Differ. Equations **27** (2015), no. 3-4, 1077–1099.

[Ste93] E. M. Stein, *Harmonic analysis: Real-variable methods, orthogonality, and oscillatory integrals. With the assistance of Timothy S. Murphy*, Princeton, NJ: Princeton University Press, 1993.

[STE$^+$10] I. Staude, M. Thiel, S. Essig, C. Wolff, K. Busch, G. Von Freymann, and M. Wegener, *Fabrication and characterization of silicon woodpile photonic crystals with a complete bandgap at telecom wavelengths*, Optics letters **35** (2010), no. 7, 1094–1096.

[Str26] D. J. Struik, *Détermination rigoureuse des ondes irrotationelles périodiques dans un canal à profondeur finie*, Math. Ann. **95** (1926), 595–634 (French).

[Str89] W. A. Strauss, *Nonlinear wave equations. Expository lectures from the CBMS regional conference held at George Mason University, January 16-22, 1989*, Providence, RI: American Mathematical Society, 1989.

[Str92] W. A. Strauss, *Partial differential equations - an introduction*, John Wiley & Sons Inc., New York, 1992.

[Str94] S. H. Strogatz, *Nonlinear dynamics and chaos*, Perseus, New York, 1994.

[Str04] B. Straughan, *The energy method, stability, and nonlinear convection*, second ed., Applied Mathematical Sciences, vol. 91, Springer-Verlag, New York, 2004.

[Str08] ______, *Stability and wave motion in porous media*, Applied Mathematical Sciences, vol. 165, Springer, New York, 2008.

[SU01] G. Schneider and H. Uecker, *Nonlinear coupled mode dynamics in hyperbolic and parabolic periodically structured spatially extended systems*, Asymptotic Anal. **28** (2001), no. 2, 163–180.

[SU03a] ______, *Almost global existence and transient self similar decay for Poiseuille flow at criticality over exponentially long times*, Physica D **185** (2003), no. 3-4, 209–226.

[SU03b] ______, *Existence and stability of exact pulse solutions for Maxwell's equations describing nonlinear optics*, ZAMP **54** (2003), 677–712.

[SU07] ———, *The amplitude equations for the first instability of electro-convection in nematic liquid crystals in the case of two unbounded space directions*, Nonlinearity **20** (2007), no. 6, 1361–1386.

[SUW11] G. Schneider, H. Uecker, and M. Wand, *Interaction of modulated pulses in nonlinear oscillator chains*, J. Difference Equ. Appl. **17** (2011), no. 3, 279–298.

[SVZZ13] S. Salsa, F. M. G. Vegni, A. Zaretti, and P. Zunino, *A primer on PDEs. Models, methods, simulations.* , Milano: Springer, 2013.

[SW04] T. Schäfer and C. E. Wayne, *Propagation of ultra-short optical pulses in cubic nonlinear media*, Phys. D **196** (2004), no. 1-2, 90–105.

[SW00a] G. Schneider and C. E. Wayne, *Counter-propagating waves on fluid surfaces and the continuum limit of the Fermi-Pasta-Ulam model*, International conference on differential equations. Proceedings of the conference, Equadiff '99, Berlin, Germany, August 1–7, 1999. Vol. 1, Singapore: World Scientific, 2000, pp. 390–404.

[SW00b] ———, *The long wave limit for the water wave problem I. The case of zero surface tension*, Comm. Pure. Appl. Math. **53** (2000), no. 12, 1475–1535.

[SW01] ———, *Kawahara dynamics in dispersive media*, Phys. D **152/153** (2001), 384–394, Advances in nonlinear mathematics and science.

[SW02] ———, *The rigorous approximation of long-wavelength capillary-gravity waves*, Arch. Rat. Mech. Anal. **162** (2002), 247–285.

[SW11] ———, *Justification of the NLS approximation for a quasilinear water wave model*, J. Differential Equations **251** (2011), no. 2, 238–269.

[SZ13] G. Schneider and D. Zimmermann, *Justification of the Ginzburg-Landau approximation for an instability as it appears for Marangoni convection*, Math. Methods Appl. Sci. **36** (2013), no. 9, 1003–1013.

[Tao06] T. Tao, *Nonlinear dispersive equations*, CBMS Regional Conference Series in Mathematics, vol. 106, CBMS, Washington, DC, 2006.

[Tao09] ———, *Why global regularity for Navier-Stokes is hard*, Structure and Randomness: Pages from Year One of a Mathematical Blog, AMS, 2009.

[Tay96] M. E. Taylor, *Partial differential equations. 3 volumes*, Applied Mathematical Sciences, Springer-Verlag, New York, 1996.

[TBD+96] P. Takáč, P. Bollerman, A. Doelman, A. van Harten, and E. S. Titi, *Analyticity of essentially bounded solutions to semilinear parabolic systems and validity of the Ginzburg-Landau equation*, SIAM J. Math. Anal. **27** (1996), no. 2, 424–448.

[Tem97] R. Temam, *Infinite-dimensional dynamical systems in mechanics and physics*, second ed., Applied Mathematical Sciences, vol. 68, Springer-Verlag, New York, 1997.

[Tem01] ———, *Navier-Stokes equations*, AMS, Providence, RI, 2001.

[Tes12] G. Teschl, *Ordinary differential equations and dynamical systems*, Graduate Studies in Mathematics, vol. 140, American Mathematical Society, Providence, RI, 2012.

[Thi88] W. Thirring, *Lehrbuch der mathematischen Physik, Band 1: Klassische dynamische Systeme*, Springer-Verlag, 1988.

[TM05] R. Temam and A. M. Miranville, *Mathematical modelling in continuum mechanics*, Cambridge University Press, 2005.

[Tot15] N. Totz, *A justification of the modulation approximation to the 3D full water wave problem*, Commun. Math. Phys. **335** (2015), no. 1, 369–443.

[Tuc02] W. Tucker, *A rigorous ODE solver and Smale's 14th problem*, Found. Comput. Math. **2** (2002), no. 1, 53–117.

[Tur52] A. Turing, *The chemical basis of morphogenesis*, Phil. Trans. Roy. Soc. B **237** (1952), 37–72.

[TW12] N. Totz and Sijue Wu, *A rigorous justification of the modulation approximation to the 2D full water wave problem*, Commun. Math. Phys. **310** (2012), no. 3, 817–883.

[TW15] Y. Tao and M. Winkler, *Boundedness vs. blow-up in a two-species chemotaxis system with two chemicals*, Discrete Contin. Dyn. Syst. Ser. B **20** (2015), no. 9, 3165–3183.

[TZ10] D. Turaev and S. Zelik, *Analytical proof of space-time chaos in Ginzburg-Landau equations*, Discrete Contin. Dyn. Syst. **28** (2010), no. 4, 1713–1751.

[Uec99] H. Uecker, *Diffusive stability of rolls in the two-dimensional real and complex Swift-Hohenberg equation*, Comm. Partial Differential Equations **24** (1999), no. 11-12, 2109–2146.

[Uec03] ______, *Approximation of the Integral Boundary Layer equation by the Kuramoto-Sivashinsky equation*, SIAM J. Appl. Math. **63** (2003), no. 4, 1359–1377.

[Uec07] ______, *Self-similar decay of spatially localized perturbations of the Nusselt solution for the inclined film problem*, Arch. Ration. Mech. Anal. **184** (2007), no. 3, 401–447.

[Uec09] ______, *A short ad hoc introduction to spectral methods for parabolic PDE and the Navier-Stokes equations*, Modern Computational Science 09 (R. Leidl and A. Hartmann, eds.), Universitätsverlag Oldenburg, Oldenburg, 2009, pp. 169–212.

[Uec16] ______, *Optimal harvesting and spatial patterns in a semi arid vegetation system*, Natural Resource Modelling **29** (2016), no. 2, 229–258.

[UGS+15] H. Uecker, D. Grieser, Z. Sobirov, D. Babajanov, and D. Matrasulov, *Soliton transport in tubular networks: transmission at vertices in the shrinking limit*, Phys. Rev. E **91** (2015), 023209.

[UW14] H. Uecker and D. Wetzel, *Numerical results for snaking of patterns over patterns in some 2D Selkov-Schnakenberg Reaction-Diffusion systems*, SIADS **13-1** (2014), 94–128.

[UWR14] H. Uecker, D. Wetzel, and J. Rademacher, *pde2path – a Matlab package for continuation and bifurcation in 2D elliptic systems*, NMTMA **7** (2014), 58–106, see also `www.staff.uni-oldenburg.de/hannes.uecker/pde2path`.

[Van89] A. Vanderbauwhede, *Centre manifolds, normal forms and elementary bifurcations*, Dynamics reported, Vol. 2, Dynam. Report. Ser. Dynam. Systems Appl., vol. 2, Wiley, Chichester, 1989, pp. 89–169.

[Vas15] A. Vasy, *Partial differential equations. An accessible route through theory and applications.*, Providence, RI: American Mathematical Society (AMS), 2015.

[VdP26] B. Van der Pol, *On relaxation-oscillations*, The London, Edinburgh and Dublin Phil. Mag. & J. of Sci. **2** (1926), no. 7, 978–992.

[Ver96] F. Verhulst, *Nonlinear differential equations and dynamical systems*, second ed., Springer-Verlag, Berlin, 1996.

[vH91] A. van Harten, *On the validity of Ginzburg-Landau's equation*, J. Nonlinear Science **1** (1991), 397–422.

[VI92] A. Vanderbauwhede and G. Iooss, *Center manifolds in infinite dimensions*, Springer, Dynamics reported **1** (1992), 125–163.

[vS95] W. van Saarloos, *The complex Ginzburg-Landau equation for beginners*, Spatio-temporal patterns in nonequilibrium complex systems (Santa Fe, NM, 1993), Santa Fe Inst. Stud. Sci. Complexity Proc., XXI, Addison-Wesley, Reading, MA, 1995, pp. 19–31.

[vS03] W. van Saarloos, *Front propagation into unstable states*, Physics Reports **386** (2003), 29–222.

[vSH92] W. van Saarloos and P. C. Hohenberg, *Fronts, pulses, sources and sinks in generalized complex Ginzburg–Landau equations*, Physica D **56** (1992), 303–367.

[VVV94] A. I. Volpert, V. A. Volpert, and V. A. Volpert, *Traveling wave solutions of parabolic systems*, Translations of Mathematical Monographs, vol. 140, American Mathematical Society, Providence, RI, 1994.

[vW85] W. von Wahl, *The equations of Navier-Stokes and abstract parabolic equations*, Vieweg, 1985.

[Wal82] P. Walters, *An introduction to ergodic theory*, Graduate Texts in Mathematics, Vol. 79. New York - Heidelberg -Berlin: Springer-Verlag. IX, 1982.

[Way96] C. E. Wayne, *An introduction to KAM theory*, Dynamical systems and probabilistic methods in partial differential equations. 1994 summer seminar on dynamical systems and probabilistic methods for nonlinear waves (Percy (ed.) et al. Deift, ed.), American Mathematical Society, 1996, pp. 3–29.

[Way97] ______, *Invariant manifolds for parabolic partial differential equations on unbounded domains*, Arch. Rat. Mech. Anal. **138** (1997), 279–306.

[Wei81] F. B. Weissler, *Existence and nonexistence of global solutions for a semilinear heat equation*, Israel Journal of Mathematics **38** (1981), 29–40.

[Wer00] D. Werner, *Funktionalanalysis*, extended ed., Springer-Verlag, Berlin, 2000.

[Whi99] G. B. Whitham, *Linear and nonlinear waves*, Pure and Applied Mathematics (New York), John Wiley & Sons, Inc., New York, 1999, Reprint of the 1974 original, A Wiley-Interscience Publication.

[Wie99] M. Wiegner, *The Navier-Stokes equations-a neverending challenge?*, Jahresber. Deutsch. Math.-Verein. **101** (1999), no. 1, 1–25.

[Wig88] S. Wiggins, *Global bifurcations and chaos. analytical methods*, Applied Mathematical Sciences, 73. New York etc.: Springer-Verlag., 1988.

[Wig03] ______, *Introduction to applied nonlinear dynamical systems and chaos*, second ed., Texts in Applied Mathematics, vol. 2, Springer-Verlag, New York, 2003.

[Wlo87] J. Wloka, *Partial differential equations*, Cambridge University Press, Cambridge, 1987.

[Wri06] J. D. Wright, *Corrections to the KdV approximation for water waves*, SIAM J. Math. Anal. **37** (2006), no. 4, 1161–1206.

[Wu97] Sijue Wu, *Well-posedness in Sobolev spaces of the full water wave problem in 2-D*, Invent. Math. **130** (1997), no. 1, 39–72.

[Wu99] ______, *Well-posedness in Sobolev spaces of the full water wave problem in 3-D*, J. Amer. Math. Soc. **12** (1999), no. 2, 445–495.

[Wu09] ______, *Almost global wellposedness of the 2-D full water wave problem*, Invent. Math. **177** (2009), no. 1, 45–135.

[Wu11] ______, *Global wellposedness of the 3-D full water wave problem*, Invent. Math. **184** (2011), no. 1, 125–220.

[WW02] C. E. Wayne and J. D. Wright, *Higher-order modulation equations for a Boussinesq equation*, SIAM J. Appl. Dyn. Syst. **1** (2002), no. 2, 271–302.

[WW15] C. E. Wayne and M. I. Weinstein, *Dynamics of partial differential equations*, Frontiers in Applied Dynamical Systems: Reviews and Tutorials, vol. 3, Springer, Cham, 2015.

[Yan10] J. Yang, *Nonlinear waves in integrable and nonintegrable systems*, SIAM, 2010.

[YLWM13] S. Yanchuk, L. Lücken, M. Wolfrum, and A. Mielke, *Spectrum and amplitude equations for scalar delay-differential equations with large delay*, Discrete Contin. Dyn. Syst. **35** (2015), no. 1, 537–553.

[Yos71] K. Yosida, *Functional analysis*, Springer, New York, 1971.

[Yos82] H. Yosihara, *Gravity waves on the free surface of an incompressible perfect fluid of finite despth*, Publ. Res. Inst. Math. Sci. **18** (1982), 49–96.

[Yos83] ______, *Capillary-gravity waves for an incompressible ideal fluid*, J. Math. Kyoto Univ. **23** (1983), 649–694.

[Zak68] V. E. Zakharov, *Stability of periodic waves of finite amplitude on the surface of a deep fluid*, Sov. Phys. J. Appl. Mech. Tech. Phys **4** (1968), 190–194.

[Zei71] E. Zeidler, *Existenzbeweis für cnoidal waves unter Berücksichtigung der Oberflächenspannung. (Existence proof for cnoidal waves under consideration of the surface tension.)*, Arch. Ration. Mech. Anal. **41** (1971), 81–107 (German).

[ZF71] V. E. Zakharov and L. D. Fadeev, *The Korteweg-de Vries equation is a fully integrable Hamiltonian system*, Functional Anal. Appl. **5** (1971), 280–287.

[Zim16] D. Zimmermann, *Justification of the Ginzburg-Landau approximation for quasilinear problems - including an application to Bénard-Marangoni convection*, Preprint Universität Stuttgart (2016), 89 pages.

[ZK65] N. J. Zabusky and M. D. Kruskal, *Interactions of solitons in a collisionless plasma and the recurrence of initial states*, Phys. Rev. Lett. **15** (1965), 240–243.

List of symbols

Acronyms
- BBM Benjamin-Bona-Mahony, 452
- GL Ginzburg-Landau, 9
- KdV Korteweg-de Vries, 9
- KPP Kolmogorov-Petrovsky-Piskounov, 12
- KS Kuramoto-Shivashinsky, 332
- NLS Nonlinear Schrödinger, 8
- SH Swift-Hohenberg, 10

Dynamicals systems
- (Z, Z_ρ)-attractor, attractor w.r.t. different norms, 286
- W_s, W_c, W_u, stable, center and unstable manifolds, 46
- $\mathcal{A} = \omega(B)$, attractor as ω-limit set of set B, 51
- $\mathcal{S}_t : X \to X$, general evolution semigroup, 49
- $\gamma_+(u_0), \gamma_-(u_0)$, forward, backward orbit through u_0, 38
- $\omega(u_0)$, ω-limit set of (the orbit through) u_0, 50
- $u(t, t_0, u_0)$ $[u(t, u_0)]$, solution of initial value problem [autonomous case], 15

Function spaces and norms
- $C^0(\overline{\Omega}, \mathbb{R})$, $\|u\|_{C_b^0}$, continuous functions, and their norm, 152
- $C_b^0(\overline{\Omega}, \mathbb{R})$, bounded continuous functions, 152
- $C_{b,unif}^0(\overline{\Omega}, \mathbb{R})$, bounded uniformly continous functions, 152
- $C_0^\infty(\Omega, \mathbb{R})$, infinitely differentiable functions with compact support, 153
- $C^{k,\alpha}(\overline{\Omega}, \mathbb{R})$, $\|u\|_{C^{k,\alpha}}$, Hölder spaces and norm, 153
- $C^m(\overline{\Omega}, \mathbb{R})$, continuously differentiable functions, 152
- $C_b^m(\overline{\Omega}, \mathbb{R})$, $\|u\|_{C_b^m}$, bounded continuously differentiable functions, and their norm, 152
- H_{per}^m, Sobolev space on torus, 158
- H_m^r, weighted Sobolev space of (differentiation) order r, 236
- $L^2((-1/2, 1/2], H^\theta([0, 2\pi), \mathbb{C}))$, basic Bloch space, 434
- L_m^2, weighted L^2 space, 235
- L_{loc}^2, local L^2 spaces, 277
- $L^p(\Omega, \mathbb{R})$, $\|u\|_{L^p}$, Lebesgue spaces and norms, 153
- $W^{m,p}(\Omega, \mathbb{R})$, $\|u\|_{W^{m,p}}$, $H^m(\Omega, \mathbb{R})$, $\|u\|_{H^m}$, Sobolev spaces and norms, 154
- $W^{m,p}(\mathbb{R}^d, \mathbb{R}), H^m(\mathbb{R}^d, \mathbb{R})$, Sobolev spaces over $\mathbb{R}^d$, 235
- $\mathcal{D}$, $\mathcal{D}'$, space of test functions, and its dual, i.e., space of distributions, 155
- $\mathcal{S}$, $\mathcal{S}'$, Schwartz space, and dual, i.e., the space of tempered distributions, 229
- $\widetilde{L}(\ell, \partial_x)$, Bloch operators, 435
- $\mathbb{R}^{\mathbb{N}}$, $\mathbb{R}^{\mathbb{Z}}$, $\mathbb{R}^\infty$, spaces of sequences, 135

$\ell_{p,\theta}(\mathbb{R}^{\mathbb{Z}})$, $\ell_{p,\theta}$, $\|u\|_{\ell_{p,\theta}(\mathbb{R}^{\mathbb{Z}})}$, weighted ℓ^p sequence spaces, and norm, 135
H^θ_{ul}, $\|\cdot\|_{H^\theta_{\mathrm{ul}}}$, uniformly local Sobolev spaces and norm, 278
L^2_{ul}, $\|u\|_{L^2_{\mathrm{ul}}}$, uniformly local L^2 spaces and norm, 277
$\|\cdot\|_\rho$, localized norm, 287
$\|u\|_{\ell_{\infty,\theta}(\mathbb{R}^{\mathbb{Z}})}$, weighted ℓ_∞ spaces, and norm, 135
$\mathrm{dist}_X(a, B)$, $\mathrm{dist}_X(A, B)$, distance in the Banach space X, 287
Miscellanous
$|n| = n_1 + \ldots + n_d$, for n a multi index, 151
$B_X(r, u_0)$ ball with radius r around u_0 in the Banach space X, 286
$\mathrm{Res}(\varepsilon\psi_{\mathrm{GL}})$, $\mathrm{Res}(\varepsilon\Psi_{\mathrm{GL}})$ residual of the modulation ansatz, 319
$\star$, convolution in Bloch space, 434
$\widehat{u}_k$, Fourier coefficients of u (Fourier series), 146
$u * v$, convolution of functions u, v, 231
$\mathrm{supp}(u)$, support of a function, 153
Erf, erf (error functions), 245
Operators
$\partial^n_x = \partial^{n_1}_{x_1} \ldots \partial^{n_d}_{x_d}$, for n a multi index, 151
E_c, E_s, E^h_c, E^h_s, mode filters, defined via cutoff in Fourier space, 338
$M = \mathcal{F}^{-1}\widehat{M}\mathcal{F}$, multiplier operator in physical space, 280
S_ε, scaling operators, 338
T_u regular distribution defined by function u, 155
T_y, translation operator, 277
$\mathcal{F}, \mathcal{F}^{-1}$, Fourier transform and inverse, 227
$\mathcal{T}$, $\mathcal{T}^{-1}$ Bloch transform and inverse, 434
$\Delta = \partial^2_{x_1} + \ldots + \partial^2_{x_d}$, Laplace operator, 1
e^{tA} exponential matrix, 24
$\mathrm{e}^{t\Lambda}$, semigroup generated by operator Λ, 338
$\widehat{M} = \mathcal{F}M\mathcal{F}^{-1}$, multiplier operator in Fourier space, 280

Index

absorbing set, 51
abstract pattern forming system, 339
action and angle variables, 114
Allen-Cahn equation, **222**
  bounded domain, 148, **167**
  front,kink,interface, 223
  local bifurcations, 478
attractor, 51
  (Z_u, Z_ρ), 287
  (Z, Z_ρ), 286
  abstract existence theorem, 51
  Chafee-Infante, 169
  for heat equation over $\mathbb{R}$, 285, 289
  for pattern forming systems, 378
  for PDEs on unbounded domains, 285
  for the SH equation over $\mathbb{R}$, 379
  global, 51
  lower semicontinuity, 54
  Navier-Stokes, 196
  shadowing, 52, 382
  upper semicontinuity, 54
autonomous, 15

Bénard's problem, 361
Bénard-Marangoni system, 396
basin of attraction, 51
Benjamin-Feir instability, 290, 294
Bessel's equality, 159
bifurcation, **76**
  Bogdanov-Takens, 95, 307
  genericity, 82
  global, 102
  Hopf, 78, **91**
  Newton-polygon, 81
  numerical analysis, 96
  pitchfork, 76
  reduced equation, 84
  saddle-node, 77
  scaling arguments, 80
  subcritical, 76
  supercritical, 76
  transcritical, 77
  Turing, 309
  Turing-Hopf, 309
  unfolding, 83
Biot-Savart law, 184
Bloch
  convolution, 434
  modes, 429
  operators, 435
  space, 434
  transform, 434, 518
Bose-Einstein condensates, 250
boundary conditions, 2
  Dirichlet, 2
  Neumann, 2
  periodic, **158**, 179, 292, 397
Boussinesq approximation, 356, 361
Boussinesq equation, 423, **451**, 472
breathers, 491
  generalized, 493
  sine-Gordon, 492
Brillouin zone, 430
Brownian motion, 207
Brusselator, 295
Burgers equation, 10, 12, **237**, 465, 526

bounded domain, 148, 167
integrated, 244
limit profile, 526
perturbed, 527

Cantor set, 61
carrier wave, 8, 401
celestial mechanics, 116
center manifold
cut off, 475
for SH equation, 480
center manifold, 47, **85**, 86
examples, 86–89
attractivity, 89
for maps, 89
in infinite dimensions, **473**
non-smoothness, 88
non-uniqueness, 87, 88
Chafee-Infante problem, 167
characteristics, 241
chemostat, 105
Cole-Hopf transformation, 244
compactness, **149**
competition, 68
conducting state, 361
conservation law, 262
conservation laws (Navier Stokes), 180
contracting set, 300
convection, **354**
Bénard, 361
cell, 355
porous media, 355
rolls, 361
convective stability, 222
convolution
in Bloch space, 434
in Fourier space, 231
of Fourier series, 146
Couette-Taylor problem, **370**
counter-example
for Ginzburg-Landau approximation, 365
for NLS approximation, 427
coupled mode equations, 440, 448
critical nonlinearity, 524, 526

Darcy's law, 355, 361
degree of irrelevance, 507
derivative
Gateaux and Fréchet, 140
distributional, 155
weak, 154

diffusion, 207
fundamental solution, 209
nonlinear, 498
on the real line, 208
polynomial decay, 497
selfsimilar decay, 498
semigroup, 211
smoothing properties, 210
diffusive mixing, 245
diffusive stability
in the GL equation, 507, 517
of Swift-Hohenberg rolls, 517
of Taylor vortices, 523
dispersion, 251, **401**
dispersion relation, 310, 356, 402, 404, 424, 432, 460
dispersive system
general, 412
distribution
δ, 234
Fourier transform of, 233
regular, 233
scaling of, 248
tempered, 234
dynamical system, 6, **38**
bistable, 305
chaotic, 59
completely integrable, 121
conservative, 109
continuous, 38
discrete, 38
dissipative, 51
equivariant, 371
excitable, 303
Hamiltonian, **109**
measure preserving, 112
oscillatory, 305
relaxational, 304
shift, 59

Eckhaus instability, 290
eigenmode, 4
electro convection, 355
electromagnetic pulse, 8
embedding
compact, 150
embedding, 281
Rellich, 150
Sobolev, 147, 150, 154, 156, 165, 200, 235
energy

kinetic, potential, total, for pendulum, 49
enstrophy, 203
equation
- Airy, 258, 259
- Allen-Cahn, *see also* Allen-Cahn equation
- Benjamin-Bona-Mahony, 466
- Boussinesq, *see also* Boussinesq equation
- Burgers, *see also* Burgers equation, 533
- Camassa-Holm, 470
- delay-differential, 397
- diffusion, heat, 3
- elliptic PDE, 2
- Euler, 182
- Fisher, *see also* KPP equation
- Ginzburg-Landau, *see also* Ginzburg-Landau equation
- Gross-Pitaevsky, 250, 439
- hyperbolic PDE, 4
- Kadomtsev-Petviashvili, 472
- Kawahara, 462, 472
- Klein-Gordon, 402, 535
- Kolmogorov-Petrovsky-Piskounov, *see also* KPP equation
- Korteweg-de Vries, *see also* KdV equation
- KS-KdV, 399
- Kuramoto-Sivashinsky, *see also* Kuramoto-Sivashinsky equation
- Laplace, 1
- lattice differential, 396
- Maxwell's, *see also* Maxwell's equations
- mKdV, 447
- Navier-Stokes, *see also* Navier-Stokes equations
- Newell-Whitehead, 364
- Nonlinear Schrödinger, *see also* NLS equation
- parabolic PDE, 3
- peakon, 471
- phase diffusion, 529
- plasma physics, 464
- reaction-diffusion, 212
- regularized long wave, 467, 471
- short pulse, 448
- sine-Gordon, 492
- Swift-Hohenberg, *see also* Swift-Hohenberg equation
- transport, 211

Euler-Poisson system, 464
exponential matrix, 23
exponentially long transients, 524

Feigenbaum constant, 101
finite wave length instability, 9
FitzHugh-Nagumo model, 296, **302**
fixed point, 27, 38
- asymptotically stable, 39
- hyperbolic, 43
- stability for iterations, 45
- stable, 26, 39
- unstable, 39

Floquet
- exponents, 30, 431
- multipliers, 30, 46, 431

flow
- conjugated, 59
- Couette, 185, 371
- inclined film, 396
- irrotational, 185
- Poiseuille, 185
- shear, 185
- strain, 185

flux function, 237
Fourier series, 146, 158
- C^m - $\ell_{1,m}$ estimate, 161
- H^m - $\ell_{2,m}$ isomorphy, 161
- coefficient, 146
- Gibbs phenomenon, 164
- main theorem, 160
- point-wise convergence, 164
- real, 163

Fourier transform, 225
- convolution, 231
- in $\mathcal{S}$ and L^p, 228
- in L^1, 226
- inverse, 228
- isomorphism property, 236
- of distributions, 233
- smoothness and decay, 228

Fourier's law, 208
FPU model, 264, 447, 451, **462**
- derivation of NLS and KdV, 464
- poly-atomic, 426

front, 216
- modulating, 322
- stability and instability, 216

fundamental matrix, 22

Galilei invariance, 293
gap solitons, 439, 447
Gelfand-Levitan equation, 271
Gierer-Meinhard model, 296
Ginzburg-Landau approximation, **315**
  abstract approximation theorem, 341
  attractivity, 383
  counter-example, 365
  for porous media flow, 361
  for Bénard convection, 363
  for Cahn-Hilliard problems, 396
  for degenerated problems, 395
  for delay-differential equations, 397
  for reaction diffusion systems, 347, 351
  for spatially periodic problems, 395
  for systems with neutral modes, 396
  for the Couette-Taylor problem, 373, 375
  for the Hopf bifurcation case, 354
  for the KS equation, 344
  for the Schnakenberg model, 350
  for the SH equation, 330
  for time periodic systems, 366
  Fourier mode distribution, 390
Ginzburg-Landau equation, 9, 12, **275**
  bounded domain, 148, 167
  derivation in general systems, 335
  derivation via solvability condition, 349
  front solutions, 322
  global existence, bounded domain, 283
  mean field coupling, 377
  roll solutions, 289
  stochastic version, 397
  universality of, 332
glass fibers, 440
global existence via GL approximation, 384
gradient system, 56, 168
Gray Scott model, 306
Greeks and Trojans, 121
Gronwall's inequality
  differential version, 37
  integral version, 21
group velocity, 253, 401, 404

Hölder spaces, 152
Hamiltonian system, 48, **109**
Heisenberg uncertainty principle, 247
Hermite polynomial, 499
hexagons (planar pattern), 363
high speed ferry, 9
Hodgkin-Huxley model, 302

inclined film problem, 529
inequality
  Gagliardo-Nirenberg, 201, 247
  Gronwall, *see also* Gronwall
  Hausdorff-Young, 163, 231
  Poincaré, 164
  Young's for convolutions, 146, 232
invariance (translational, $\mathcal{S}^1$, Galilei, scaling), 293
invariant manifolds
  for maps, 49
  for ODEs, 47
invariant region, 300
irrelevant nonlinearity, 500
  dispersive case, 535
  in Bloch space, 520
  in Fourier space, 511

Jordan normal form, 24

KdV approximation
  for bi-directional models, 466
  for the Boussinesq equation, 452
  for the Euler-Poisson system, 464
  for the FPU model, 464
  for water waves, 461
KdV equation, 9, 12, **259**
  2-soliton, 272
  as a Hamiltonian system, 261
  bounded domain, 148, 167
  hierarchy, 266
  inverse scattering, 271
  scattering data, 269
  systems of, 468
  universality, 456
Kepler's laws, 116
kink, *see also* front, 495
Kirkwood gaps, 127
Kolmogorov equations, 68, 296
Kolmogorov-Petrovsky-Piskounov equation, *see also* KPP equation
Korteweg-de Vries equation, *see also* KdV equation
KPP equation, 12, **206**, 299
  bounded domain, 148, 167
  traveling wave, 217
Kuramoto-Shivashinsky equation, 11, **332**

Lagrange multiplier, 274
Laplace's principle, 6
lattice differential equation, 447, **462**
law of mass action, 67, 297
Lax pair, 264
Lengyel-Epstein model, 295
Lennart-Jones potential, 130
lift (from GL to patterns), 383
linear planar systems, 26
  center, 28
  degenerate node, 29
  saddle, 28
  source, unstable node, 28
  vortex, 28
Lipschitz continuous, 34
liquid crystal, 365
localized pattern, 398
logarithmic time scale, 500
logistic map, 99
Lorenz attractor, 6, **101**
Lotka-Volterra system, 66
Lyapunov exponent, 60
Lyapunov function, 57, 502
Lyapunov-Schmidt reduction, **83**, 395

Mandelbrot set, 106
manifold
  center, *see also* center manifold
  stable, 46
  unstable, 46
Marangoni convection, 355
Marchenko equation, 271
mathematical pendulum, 47
maximal regularity, 157
maximum principle, 215, 391
Maxwell's equations, 5, **440**
Maxwell-Lorentz system, 444
mean flow, 188
Millennium problem, 5, 179, 196
Miura transformation, 263, 447
mode filters, 338
modulation equation, 7
momentum, 181
monodromy matrix, 30, 431
multi index, 151
multiplexing, 445
multiplier theory, 279

Navier-Stokes equations, 5, **179**, 370
Newton-polygon, 81
NLS approximation
  counter-example, 427
  for cubic nonlinearities, 407
  for multiplexing, 445
  for nonlinear wave equations, 410
  for photonic crystals, 439
  for quadratic systems, 416
  for quasilinear systems, 428
  for spatially periodic problems, 429
  for the Boussinesq model, 424
  for the FPU model, 464
  for the mKdV equation, 447
NLS equation, 12, **250**, 402
  as a Hamiltonian system, 255
  bounded domain, 148, 167
  derivation from general system, 412
  discrete, 293
  solitons, 251
  systems of, 445
  universality, 411
non-resonance conditions, 94, 406, 415
  weakening, 423
Nonlinear Schrödinger equation, *see also* NLS equation
norm, 18
  compatible, 19
  equivalent, 18, 134
  operator, 19
normal form transforms, **92**, 417
nullclines, 27
numerical path following, 96
Nusselt solution, 529

ODE
  n^{th} order linear, 23
  periodic solution, 43
  autonomous, 17
  blow up example, 36
  characteristic equation, 24
  direction field, 26
  example for non-unique solutions, 36
  first order linear system, 23
  flow, 27
  homogeneous, 17
  initial value problem, 15, 17
  integrated, 20
  singularly perturbed, 219
  vector field, 26, 34
omega-limit set, 50
operator
  elliptic, 2, 14
  hyperbolic, 14
  parabolic, 14
optical fiber, 7

optimal control, 313
orbit, 38
  heteroclinic, homoclinic, 46
orbital stability, 273
orthonormal system, 158
oscillator
  harmonic, 114, 130
  model for polarization, 442
  nonlinear, 47, 250
  van der Pol, 70
  weakly damped, 32

Parseval's equality
  for Fourier series, 159, 160
  for Fourier transform, 230
pattern forming systems, 9
period doubling, 98
phase locking, 398
phase portrait, 26
phase space, 38
phase velocity, 253, 401
photonic crystals, 429, 440, 447
  eigenvalue problem, 430
planar pattern (rolls, hexagons and squares), 363
Poincaré map, 45
Poiseuille flow, 524
polarization, 441
predator-prey system, 66, 68
pressure, 181, 187
pseudo orbits
  for the SH and the KS equation, 386
  for ODEs, 52

quadratic autocatalysis, 298

Rössler system, 99
random walk, 208
Rankine-Hugoniot, 248
rarefaction wave, 242
Rayleigh number, 355
Reaction-Diffusion systems, **295**
reduced flow, 86
relevant nonlinearity, 507
renormalization
  continuous, 499
  discrete, 504
resonance, *see also* non-resonance conditions
  non–trivial, 426
  trivial at $k = 0$, 423
restricted 3-body problem, 119
reversibility, 65
Reynolds number, 5, 183, 371
roll solution, 289
rolls (as planar patterns), 364

scattering transform
  AKNS scheme, 292
  ZS scheme, 292
Schnakenberg model, 295, **309**
Schrödinger equation (linear), 251
Schwartz space, 229
sectorial operator, 197
semigroup, 38
  diffusion, 211
  infinitesimal generator, 136
  linear wave equation, 409
  of bounded operators, 136
  operator-continuous, 137
  smoothing, 144
  strongly continuous, 137
  translation, 211
sensitive dependence, 60
shadowing, 53
shock, 241
SI and SIS deseases, 67
Silnikov chaos, 62
Sleipner crash, 6
Smale's horseshoe, 60, 128
small divisor problem, 125
snaking, 398
Sobolev spaces, 153, 235
  uniformly local, 276
  weighted, 236
solitary wave, 462
  KdV, 260
  stability, 273
  water wave problem, 484
soliton, 9, 264, *see also* solitary wave
solution
  mild, 143, 166, 213
  strong, 143, 166
spatial dynamics, **473**, 480
  (generalized) breathers, 491
  for elliptic problems in a strip, 482
  modulating front, 489
  water wave problem, 484
spatially extended domain, 7
spectral gaps, 433
squares (planar pattern), 363
stability
  asymptotic, 40
  diffusive, 290, **497**

exponential, 40
linearized, 40
stationary phase, **257**, 259
stream function, 356
superconductivity, 275
surface tension , 458
Swift-Hohenberg equation, 10, **319**, 480
attractor upper semicontinuity, 394
cubic-quintic, 397
global semiflow, absorbing ball, 380
modulating front, 489
spatially periodic equilibria, 480
symbiosis, 68
symplectic
structure, 122
transformation, 122

theorem
Beppo-Levi, 226
center manifold, 86, 476
contraction mapping, 20
dominated convergence, 226
Floquet, 30
Fubini, 226
Hartman-Grobman, 43
Hausdorff-Young, 231
Heine Borel, 150
KAM, 126
Liouville, 111, 123
Lyapunov's subcenter, 115
Nekhoroshov, 127
Paley-Wiener, 233
Picard-Lindelöf, 34
Poincaré-Bendixson, 55
Sarkovskii, 101
Smale-Birkhoff, 128
three wave interaction, 427, 448
traffic flow, 237
trapping region, 299
traveling wave
for reaction diffusion, 299
for the Burgers equation, 243
for the GL equation, 322
for the KdV equation, 261, 293
tsunami, 9
turbulence, 98
Turing instability, 307, 316, 347
Turing space, 310
Turing-Hopf, 375

ultra-short pulses, 446

van der Corput Lemma, 257
van der Pol oscillator, 70
variation of constant, 22
viscosity solution, 242
vorticity, 184

water wave problem, 448, **458**
linear, 460
solitary wave, 484
wave equation
linear, 3, 166, 206, 451
nonlinear, 402
nonlinear, 2D, 447
nonlinear, in Fourier space, 411
with periodic coefficients, 430, 494
Whitham system, 465

Selected Published Titles in This Series

182 **Guido Schneider and Hannes Uecker,** Nonlinear PDEs, 2017
180 **Joseph J. Rotman,** Advanced Modern Algebra: Third Edition, Part 2, 2017
179 **Henri Cohen and Fredrik Strömberg,** Modular Forms, 2017
178 **Jeanne N. Clelland,** From Frenet to Cartan: The Method of Moving Frames, 2017
177 **Jacques Sauloy,** Differential Galois Theory through Riemann-Hilbert Correspondence, 2016
176 **Adam Clay and Dale Rolfsen,** Ordered Groups and Topology, 2016
175 **Thomas A. Ivey and Joseph M. Landsberg,** Cartan for Beginners: Differential Geometry via Moving Frames and Exterior Differential Systems, Second Edition, 2016
174 **Alexander Kirillov Jr.,** Quiver Representations and Quiver Varieties, 2016
173 **Lan Wen,** Differentiable Dynamical Systems, 2016
172 **Jinho Baik, Percy Deift, and Toufic Suidan,** Combinatorics and Random Matrix Theory, 2016
171 **Qing Han,** Nonlinear Elliptic Equations of the Second Order, 2016
170 **Donald Yau,** Colored Operads, 2016
169 **András Vasy,** Partial Differential Equations, 2015
168 **Michael Aizenman and Simone Warzel,** Random Operators, 2015
167 **John C. Neu,** Singular Perturbation in the Physical Sciences, 2015
166 **Alberto Torchinsky,** Problems in Real and Functional Analysis, 2015
165 **Joseph J. Rotman,** Advanced Modern Algebra: Third Edition, Part 1, 2015
164 **Terence Tao,** Expansion in Finite Simple Groups of Lie Type, 2015
163 **Gérald Tenenbaum,** Introduction to Analytic and Probabilistic Number Theory, Third Edition, 2015
162 **Firas Rassoul-Agha and Timo Seppäläinen,** A Course on Large Deviations with an Introduction to Gibbs Measures, 2015
161 **Diane Maclagan and Bernd Sturmfels,** Introduction to Tropical Geometry, 2015
160 **Marius Overholt,** A Course in Analytic Number Theory, 2014
159 **John R. Faulkner,** The Role of Nonassociative Algebra in Projective Geometry, 2014
158 **Fritz Colonius and Wolfgang Kliemann,** Dynamical Systems and Linear Algebra, 2014
157 **Gerald Teschl,** Mathematical Methods in Quantum Mechanics: With Applications to Schrödinger Operators, Second Edition, 2014
156 **Markus Haase,** Functional Analysis, 2014
155 **Emmanuel Kowalski,** An Introduction to the Representation Theory of Groups, 2014
154 **Wilhelm Schlag,** A Course in Complex Analysis and Riemann Surfaces, 2014
153 **Terence Tao,** Hilbert's Fifth Problem and Related Topics, 2014
152 **Gábor Székelyhidi,** An Introduction to Extremal Kähler Metrics, 2014
151 **Jennifer Schultens,** Introduction to 3-Manifolds, 2014
150 **Joe Diestel and Angela Spalsbury,** The Joys of Haar Measure, 2013
149 **Daniel W. Stroock,** Mathematics of Probability, 2013
148 **Luis Barreira and Yakov Pesin,** Introduction to Smooth Ergodic Theory, 2013
147 **Xingzhi Zhan,** Matrix Theory, 2013
146 **Aaron N. Siegel,** Combinatorial Game Theory, 2013
145 **Charles A. Weibel,** The K-book, 2013
144 **Shun-Jen Cheng and Weiqiang Wang,** Dualities and Representations of Lie Superalgebras, 2012

For a complete list of titles in this series, visit the
AMS Bookstore at **www.ams.org/bookstore/gsmseries/**.